# Radiation Chemistry

## Principles and Applications

This book is dedicated to the memory of John Hodson Baxendale.

Future generations of radiation scientists will reap the benefits of Bax's energy, restlessness and incisive thought, as this generation has done.

# Radiation Chemistry

## Principles and Applications

Edited by

Farhataziz

and

Michael A.J. Rodgers

Farhataziz
2321 Georgetown Drive
Denton, Texas 76201

Michael A.J. Rodgers
Center for Fast Kinetics Research
University of Texas at Austin
Austin, Texas 78712

**Library of Congress Cataloging-in-Publication Data**

Radiation chemistry.

Includes bibliographies and index.
1. Radiation chemistry. I. Farhataziz, 1932-
II. Rodgers, M. A. J.
QD636.R3 1987 541.3'8 86-19061
ISBN 0-89573-127-4



Printed in the United States of America.

ISBN 0-89573-127-4 VCH Publishers
ISBN 3-527-26197-4 VCH Verlagsgesellschaft

Distributed in North America by:
VCH Publishers, Inc.
220 East 23rd Street, Suite 909
New York, New York 10010

Distributed Worldwide by:
VCH Verlagsgesellschaft mbH
P.O. Box 1260/1280
D-6940 Weinheim
Federal Republic of Germany

# Preface

Used in the widest sense, the phrase "radiation chemistry" implies the study of the chemical actions induced by the interaction of radiation with materials. Here "chemical action" is meant to suggest the breaking or making of chemical bonds and "radiation" would consist of all forms of radiant energy: photons from across the electromagnetic spectrum, particle emanations from nuclear processes, and particles from accelerating machines of one type or another. However, since the time that it became a sub-discipline of chemical kinetics, radiation chemistry has been taken as meaning chemical action resulting from the irradiation of materials with photons or particles having sufficient energy to induce ionization of the components of the material. More specifically, the overwhelming fraction of radiation chemical research has used photons and/or particles with energies of upwards of thousands of electron volts. Nevertheless, radiant energy of 10 electron volts, or less in some cases, is capable of causing molecular ionizations and therefore research employing entities with such low energy must be classified under the umbrella of radiation chemistry. Indeed, particles/photons at the low end of the energy scale would be more widely used if they were sufficiently penetrating in condensed phase and if sufficiently intense sources had been available. The preference for low energy radiation stems from the fact that understanding the initial events should be more straightforward. Deposition of energy from high energy particles causes complexities in event distribution which still evade a complete understanding.

In this book we have sought to bring together a balanced view of the subject as it is currently understood and to point out other areas of radiation science that depend on—to a greater or lesser extent—the principles of radiation chemistry. The first four chapters can be regarded as being the introduction to the field. These authors set the background and fundamentals, concerning themselves with how ionizing radiation interacts with matter, what at the initial consequences of that interaction, what kinetic laws govern the reactions of the early species produced and how particular initial spatial relationships become smeared out, and what instrumentation techniques are employed for quantitative studies in the area.

Subsequent chapters pick out specific fundamental areas and develop them in some depth thereby providing the reader with a thoroughgoing insight into the intricacies of the subject. The final six chapters can be loosely described as showing how the basic physicochemical and chemical principles are applicable in particular areas of radiation science. In this respect emphasis has been placed on the effects of radiation on biological systems and on polymer systems. The former has extremely important health-related facets such as tumor therapy; the latter is of technological significance.

As editors, we decided what areas should be covered in this book and where emphasis should be applied. Afterwards we made up a list of authors and asked them to provide a chapter under a given heading. We have been fortunate in that our authors are internationally recognized experts in their research areas and we know that they have approached their tasks with assiduousness. Thus any faults in the volume will stem from the editorial choice of material and emphasis. We hope that what we felt was the proper balance will be consonant with the feelings of our readership.

The volume is aimed at young scientists entering, or contemplating entering the field, or at professionals in other specialties who need information about radiation science or who are simply interested in other things. We have attempted to make it suitable for teaching at the graduate or upper-division undergraduate level. In all, it is meant to be a modern, serious exposition of a subject that started with the discovery of radioactivity at the turn of this century, developed only slowly until World War II, grew explosively in the early years of nuclear weapons development, and settled down to an academic discipline offering excitement and challenge in the sixth decade of this century.

The production of this book would not have been possible without the aid of several individuals. Prominent among these have been Joanne Duitsman, Marian Cross and Stephen J. Atherton. To these and those unnamed we offer our thanks.

Farhataziz
Denton, Texas

Michael A. J. Rodgers
Austin, Texas

# Contents

# 1

# Interaction of Ionizing Radiation with Matter

ALOKE CHATTERJEE

*Lawrence Berkeley Laboratory,*
*University of California at Berkeley*

## INTRODUCTION

Radiation chemistry is concerned with the interaction of energetic charged particles (electrons, protons, alpha and other heavy particles) and high-energy photons (x-rays and $\gamma$-rays) with matter. These interactions result in ionization (along with some excitation) of the medium; hence, charged particles, x-rays, and $\gamma$-rays are frequently called "ionizing radiations." In contrast, visible or near ultraviolet photons interact with matter by predominantly producing excited states (even though ionization can also be produced if there is enough energy in the photons). Thus, visible and ultraviolet photons are called "nonionizing radiations," and the chemical reactions induced by them are in the domain of photochemistry. In this chapter we concentrate on the interaction of ionizing radiation with matter. It is necessary to understand the various processes by which radiations interact with matter, because it is in these processes that energy is transferred to the medium; and this energy absorption leads to the chemical changes of radiation chemistry. We attempt to deal with the theory of the physical mechanisms of energy transfer. Some aspects of these mechanisms are well understood, and some are not. It is our hope that the material presented here will form a basis for understanding the many topics in this book.

X-rays and $\gamma$-rays interact with matter in three different ways, and the relative importance of the three processes depends on the energy of the


*Radiation Chemistry: Principles and Applications*

photons. The three modes of interaction are (a) the photoelectric effect, (b) the Compton effect, and (c) pair production. In the next section, entitled "Interaction of Photons with Matter," we discuss these three processes and consider the mechanism of energy transfer arising from them.

Energetic charged particles can deposit energy in a medium in several possible ways, but the one that concerns radiation chemistry the most is through the interaction with the electrons of the medium. Such electrons can be excited to higher energy levels, or they can be ionized; and these processes constitute the most important mechanism for the loss of energy by the charged particles. In the section, "The Atomic Stopping Power Theory," a discussion is presented in terms of these energy loss events, and we analyze the adequacy of stopping power theory as a basis for describing chemical effects of ionizing radiations.

It is well known that charged particles form tracks as they deposit energy in a medium. For example, particle tracks in photographic emulsions are well understood, and each track is characteristic of the charge and velocity of the particle that forms it. In high-energy particle studies, such tracks are used in particle identification. Similarly, many results of radiation chemistry can be understood in terms of "initial" track structure. By this term we mean the geometrical pattern of energy deposition at the instant the energy is transferred to the medium. In order to describe this pattern, we introduce the concept of spurs, blobs, short tracks, and branch tracks (commonly known as track entities). Based on these concepts, a general picture emerges for describing the track structures of all charged particles. A comprehensive discussion of the formation of tracks through interaction of charged particles with matter and related concepts is presented in the section entitled "Track Structure."

Starting with the phenomena of initial energy deposition, chemists, over the last several years, have increased their efforts to understand the difference between the interaction of charged particles with gas and liquid phases. For a basic understanding of the phenomenon of energy exchange, one needs to know the physical cross sections of various processes (excitation, ionization, etc). These cross sections are well known only for a few systems (water, ammonia, etc) in the gas phase. In order to obtain the corresponding information in the liquid phase, one generally modifies the gas phase data after making some simplifying assumptions. Such a procedure is theoretical in nature, and efforts to obtain more accurate values in the liquid phase are underway. A report on these efforts, including a Monte Carlo scheme of energy transport in the prethermal stage, will be presented in the section, "Spatial Distribution of Primary Species."

We intend to present a comprehensive account of the interaction of ionizing radiation with matter, with special emphasis on the physical mechanism of energy transfer and transport in the system. These are exceedingly early phenomena, and direct experimental information is unavailable and may never be available. Everything presented in this chapter is based primarily on theoretical considerations. Some of the material presented is based on the author's point of view, and, whenever possible, different points of view in the field will also be noted.

# INTERACTION OF PHOTONS WITH MATTER

When high-energy photons (x-rays or $\gamma$-rays) interact with matter, most of the energy is absorbed in the medium to eject electrons from the atoms of the material. This process is almost entirely dependent on the atomic composition and not on molecular structure. The absorption of energy from light waves (infrared, visible, and ultraviolet), including very soft x-rays, depends, in general, on the molecular structure of the medium and only indirectly on the atomic composition. Such a difference in the effects of atomic composition or molecular binding constitutes a fundamental difference between ionizing radiation and nonionizing radiation.

Sometimes no energy is absorbed from a photon, and a decrease in intensity can be due to scattering only. In this case the incident wave changes its direction with no effect in the medium. In this chapter we will not be dealing with this particular type of scattering because it induces no chemical change in the system. A photon cannot be partially stopped by the atoms of a medium. Some of the photons in a beam may be close enough to the location of an atom to interact and thus be removed from the rest, or they may not be affected at all. This all-or-none process results in an exponential law between the intensity of the beam before absorption, $I_0$, and the final intensity, $I$, after the absorption. This behavior can be expressed by the equation

$$I = I_0 \, e^{-\mu x} \tag{1-1}$$

where $\mu$ is the total absorption coefficient of the material, and $x$ is its thickness. This phenomenon is quite general, and if in a process (Compton effect) a scattered photon is produced the exponential law is still valid because it refers to the incident photon only. Various units have been used to measure the thickness $x$, and they are expressed as cm, g/cm$^2$, atoms/cm$^2$, electrons/cm$^2$, and so on. Because the exponential coefficient $\mu x$ must be dimensionless, the corresponding units for $\mu$ should be expressed as cm$^{-1}$, cm$^2$/g, cm$^2$/atom, cm$^2$/electron. For convenience, the following symbols have been conventionally used to indicate the unit being used:

$\mu_e$ for cm$^2$/electron

$\mu/\rho$ for cm$^2$/g (mass coefficient)

$\mu_a$ for cm$^2$/atom (atomic coefficient)

$\mu$ for cm$^{-1}$

All these coefficients can be converted from one to another if the atomic weight, $A$, and the atomic number, $Z$, are known. For example, to express the various symbols in terms of $\mu_e$, use the following relationships:

$$\mu_a = Z\mu_e \qquad \frac{\mu}{\rho} = N\left(\frac{Z}{A}\right)\mu_e \qquad \mu = \rho N\left(\frac{Z}{A}\right)\mu_e$$

where $N$ is Avogadro's number, and $\rho$ is the density expressed in $gm/cm^3$.

Note that the total absorption coefficient, $\mu$, is made up of contributions from three processes by which energy can be transferred from the electromagnetic radiations to the absorbing medium. These processes are called photoelectric effect, Compton effect, and pair production (electron–positron production). Regardless of which of these processes occurs, the exponential law given by Eq. (1-1) holds, provided one uses the corresponding absorption coefficient relative to a particular process. The different symbols of absorption coefficients corresponding to photoelectric, Compton effect, and pair-production processes are denoted by $\tau$, $\sigma$, and $\kappa$. We will now describe the three processes and their relative importance as it depends on the photon energy.

## *The Photoelectric Effect*

In this process a photon is absorbed completely with the ejection of an electron. The kinetic energy of this ejected electron is equal to the difference between the photon energy and the binding energy of the electron in the atom in accordance with the Einstein equation $E = h\nu - I_B$, where $E$ is the acquired kinetic energy of the ejected electron, $h\nu$ is the energy of a photon in the beam, and $I_B$ is the binding energy of the electron in the atom. In this process a negligibly small energy is imparted to the residual atom. Because various electrons in an atom have different binding energies (depending on the energy levels), the energy of the ejected electron, called the photoelectron, will vary.

In considerations of the photoelectric process, it is important to realize that for this effect to take place the electron to be ejected must be bound in a molecule or atom. A free electron cannot be given kinetic energy by this process because the energy and momentum cannot be conserved if the photon must be completely absorbed.

No rigorous theory exists to describe the photoelectric effect, but it is known from experiments that the *maximum* probability for this process occurs when the energy of the photon coincides with the binding energy of the electron with which it interacts. This phenomenon leads to sharp absorption edges known as $K$-edge, $L$-edge, $M$-edge, and so on, corresponding to absorption by the $K$-, $L$-, or $M$-shell electrons. In heavy elements such as lead, the binding energies for $K$- and $L$-shell electrons are quite high (between 10 keV and 100 keV), and these edges are quite noticeable in a graphic plot of the absorption coefficient against energy. In contrast, the binding energies of K-electrons in the atoms of a water molecule are relatively small (less than 500 eV). Hence, unless the photon energy is in this range, absorption edges do not occur (see Figure 1-1).

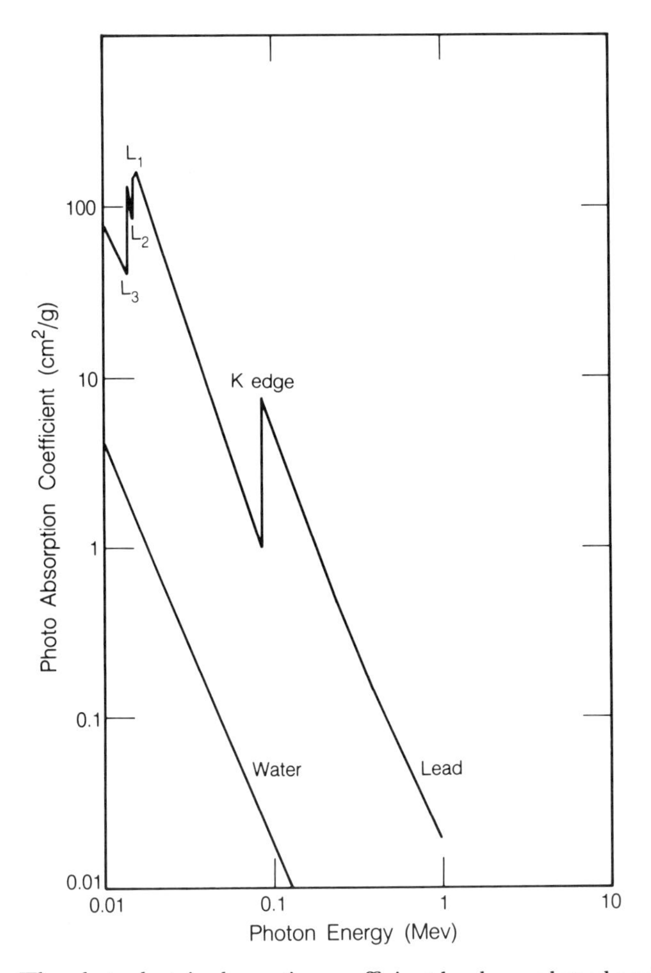


**Figure 1-1.** The photoelectric absorption coefficient has been plotted against photon energy for two media: lead and water. Lead being a heavy material the *K*- and *L*-edges are very pronounced. For water the *K*-edge is at 500 eV; hence it is outside the range of the scale used.

Although there is no rigorous theory for the photoelectric effect, it has been shown empirically that the photoelectric-absorption coefficient per electron is given by

$$\tau_e \sim \left(\frac{Z}{h\nu}\right)^3 (1 + 0.008Z) \qquad (1\text{-}2)$$

where $Z$ is the atomic number of the irradiated material, and $h\nu$ is the photon energy. Thus the probability for the photoelectric effect increases rapidly with $Z$ (power dependence of 3) and decreases with energy. For photons with energy greater than 1 MeV the contribution of the photoelectric effect in the total energy absorption process is small and can be neglected. Because Eq. (1-2) does not include edge effects, it is an approximation. For a molecular medium, $Z$ in Eq. (1-2) must be replaced by $Z_{\text{eff}}$. For water the most frequently used value of $Z_{\text{eff}}$ is 7.42. Other processes, such as Compton effect and pair production, have different values of $Z_{\text{eff}}$ because they have different $Z$ dependences.

### *The Compton Effect*

In contrast to the photoelectric effect, the characteristics of the Compton effect are such that only a part of the incident energy is absorbed to eject an electron, usually called the Compton electron. As a result of this interaction, the incident photon disappears, and a secondary photon is created with reduced energy and, in addition, propagating in a changed direction. The change of direction is required by the conservation of energy and momentum before and after the interaction.

In the Compton process the interaction takes place with the most loosely bound outer shell electrons, and for all practical purposes these electrons can be considered "free". For this approximation to be valid, the binding energy of these electrons should be much smaller than the incident photon energy. Because water is composed of atoms with relatively low atomic numbers and the threshold energy above which the Compton effect is important (for example, above 500 keV) is very small compared with most photon energies, all electrons in this medium are considered to be "free."

There is a simple consequence of this approximation. The contribution, $\sigma$, of the Compton process to the total absorption coefficient depends entirely on the number of electrons per gram. Except for hydrogen gas this value does not vary significantly for different elements and is nearly the same for water and most organic materials; $\sigma$ is independent of the atomic number of the material. The number of electrons varies with atomic number. Hence, if a radiation source of hard x-rays in which the Compton effect predominates has been calibrated by measuring energy dissipation per gram of air with a standard ionization gauge, this value can be converted by constant factors to determine the energy loss in other materials. Such a simplification does not exist for the photoelectric effect.

A Compton interaction can be considered to be an elastic collision between a photon and a loosely bound or unbound electron. In the process the photon energy is reduced, and the electron is set in motion. By applying the principle of conservation of energy and momentum, one can show that (i) the change in photon wavelength, $\Delta\lambda$, is given by

$$\Delta\lambda = \left(\frac{h}{m_0 c}\right)(1 - \cos\theta')$$

and (ii) the energy of the scattered photon is

$$h\nu' = \frac{h\nu}{1 + (h\nu/m_0c^2)(1 - \cos\theta')}$$

where $m_0$ is the rest mass of an electron, $\nu$ and $\nu'$ are, respectively, the frequencies of the incident and scattered photons, $c$ is the velocity of light, $h$ is Planck's constant, and $\theta'$ is the angle of the scattered photon measured from its original direction.

Let us now discuss the energy distribution of Compton electrons. This distribution can be calculated from a rigorous theory developed by Klein and Nishina,[1] a detailed description of which is given in many atomic physics textbooks.[2] According to this theory, Compton electrons have a fairly broad energy distribution that goes through a maximum at about three-fourths of the incident photon energy with an average energy of about one-half the photon energy. The average energy, of course, increases with photon energy; hence, the contribution of the Compton process to the total amount of absorbed energy increases with this increase in photon energy, although the total absorption coefficient gradually decreases. In Figure 1-2 the percentage contribution of the Compton effect for energy loss in water is plotted

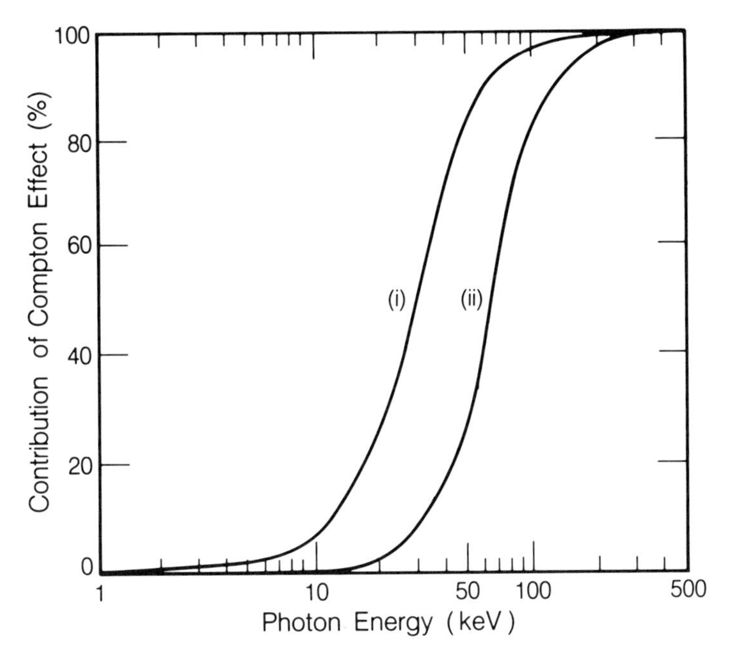


**Figure 1-2.** Relative importance of Compton and photoelectric interactions has been plotted in water for different photon energies. In (i) the plot shows the proportion of total number of electrons produced by the Compton effect; and in (ii) the proportion of total energy that appears in the recoil (Compton) electrons have been plotted. The difference from 100% at any photon energy gives the corresponding contribution for the photoelectric process.

for photon energies between 1 keV and 500 keV. This figure shows that at low energies the proportion of the total number of electrons produced in the Compton process does not contribute to the same proportion of the fraction of energy that appears in recoil electrons. For example, at 50 keV photon energy, 80% of the total electrons produced are from the Compton effect, but only 30% of the total energy transferred is contributed by these electrons. At high energies this is not the case. When the photon energy is 300 keV or more, the photoelectric effect essentially vanishes, and 100% of the electrons are produced by the Compton process. And of course the total energy appears in the Compton electrons.

In radiation chemistry, $^{60}Co$ $\gamma$-rays are used quite frequently as a source of ionizing radiation. The photons emitted from this source have an average energy of 1.2 MeV. At this energy nearly the entire interaction is through the production of Compton electrons. Because the scattered $\gamma$-rays have a high probability of penetrating the solution without further interaction, the radiation chemical effects are entirely due to the ejected Compton electrons. From this point of view it is quite similar to the photoelectric effect, because, in this process also, the chemical effects are caused entirely by the ejected electrons.

### *Pair Production*

In the cases of photoelectric and Compton effects, the interaction of the photons is always with electrons of the atoms. Pair production involves interaction of photons with the nucleus of the atom. Such an interaction results in the complete disappearance of the incident photon and the appearance of a positron and an electron pair. This is an example of a physical process in which energy is converted to mass.

The rest mass energy of an electron or a positron is 0.511 MeV. Hence, for pair production to occur, the minimum photon energy must be 1.02 MeV. This minimum energy requirement should not be confused with the cross section of the process. Even if the pair production is energetically possible for a photon of energy greater than or equal to 1.02 MeV, it does not necessarily mean that the probability of this process is significant. For example, a $^{60}Co$ $\gamma$-ray with an average photon energy of 1.2 MeV has a pair-production probability so small that virtually all the energy absorption process occurs through the Compton process. In water, pair production becomes significant for a photon energy of 3 MeV.

In pair production no net electronic charge is created because the positron and electron have opposite charges. For photons of energy larger than 1.02 MeV, the excess energy is shared by the positron and electron in the form of kinetic energy. Because the process must be considered as a collision between the photon and the nucleus, the nucleus also recoils with some momentum, but the energy involved is very small.

Following the creation of a positron and an electron pair, ionization and excitation are produced by the latter in the same manner as described in the section "Interaction of Charged Particles with Matter." Of course, chemical changes follow and are attributed generally to the electron produced in the pair production.

If the positron has kinetic energy, it slows down in a qualitatively similar way as an electron. During this time, however, the positron also has a finite probability of combining with one of the many electrons in the medium, and this annihilates the positron. The probability of annihilation is unity when the positron is at rest. In the annihilation process two photons, each with 0.511 MeV, are created. These photons are ejected in opposite directions from the point of their creation. At these photon energies, in water, the entire mode of interaction is through the Compton process.

Because pair production involves interaction with an atomic nucleus, the cross section for this process increases with atomic number. An increase in

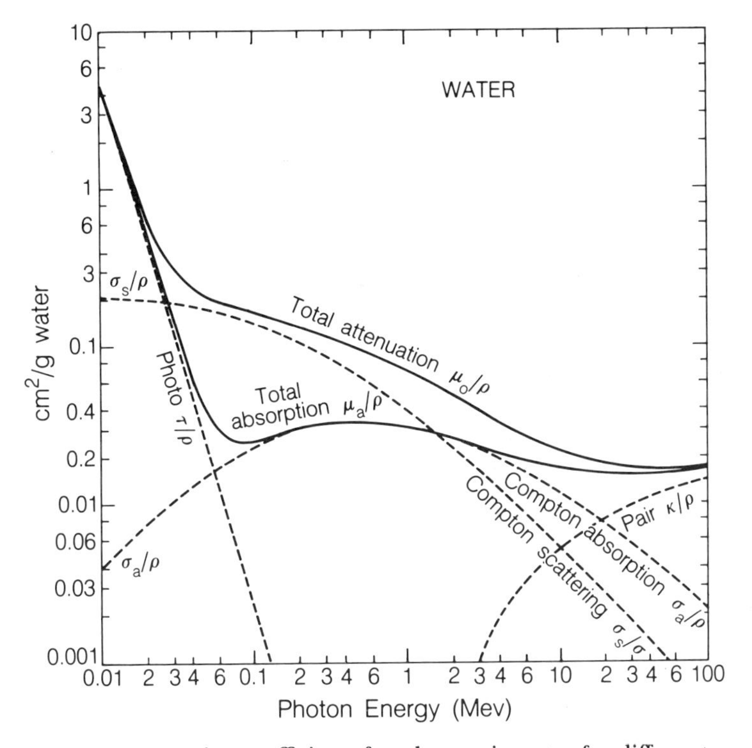


**Figure 1-3.** Mass attenuation coefficients for photons in water for different processes are shown. The curve labeled "total absorption" is $(\mu_a/\rho) = (\sigma_a/\rho) + (\tau/\rho) + (\kappa/\rho)$, where $\sigma_a$, $\tau$, and $\kappa$ are the corresponding linear coefficients for Compton absorption, photoelectric absorption, and pair production. When $\sigma_s$ (Compton scattering coefficient) is added to $\mu_a$, one gets the total attenuation coefficient. This plot has been obtained from ref. 2. (From "The Atomic Nucleus" by Robley D. Evans, Copyright (1955). Reprinted by permission of McGraw-Hill Book Company, Inc.)

atomic number implies an increase in the electric field around the nucleus and, hence a greater chance of an interaction. The absorption coefficient per atom in this process is proportional to the square of the atomic number ($Z^2$).

We can conclude that, regardless of the mechanism of interaction involving photons, generation of secondary electrons is an essential phenomenon. All the chemical changes in a medium following photon irradiation are attributed to these electrons. The cross section for production of secondary electrons in water is about $10^{-25}$ cm$^2$ per electron, which is significantly less than atomic cross sections ($\sim 10^{-16}$ cm$^2$) for ionization by the same electrons.

In Figure 1-3 the individual as well as the total mass absorption coefficient in water have been plotted against photon energies between 0.01 MeV and 100 MeV. At a photon energy of 500 keV there is a broad inflection, but, in general, above this energy this coefficient decreases with the increase in energy because the cross sections of both the photoelectric effect and the Compton effect decrease with increase in energy also.

## INTERACTION OF CHARGED PARTICLES WITH MATTER: THE ATOMIC STOPPING POWER THEORY

When a charged particle penetrates a medium, it interacts with the molecules, loses energy, and slows down. This slowing-down process has been investigated quite extensively, and the best-known physical quantity related to it is called the *stopping power* of the medium. Stopping power is defined as the rate of energy loss suffered by a charged particle in traversing a unit path length. This quantity is related to the charge and velocity of the incident particle and also to the physical properties of the medium. If the velocity is not too high, the energy lost by a charged particle is equal to the energy absorbed by the medium, in which case the stopping power is also called the linear energy transfer (LET).

A charged particle can lose its kinetic energy by interacting either with the electrons of the medium or with the nuclei of the medium. There are two types of nuclear interactions; one of them occurs at very high energies (bremsstrahlung production), and the other becomes important only at low energies (Rutherford type of collision). In this chapter only the electronic interactions will be discussed because they are of dominating importance for radiation chemistry. For irradiation with electrons in water, nuclear stopping dominates for energies above 100 MeV, and experimental studies in radiation chemistry have been restricted to energies less than 20 MeV. For heavier charged particles, the threshold in energy for nuclear interaction to dominate is greater than 600 MeV/$n$ (ie, energy per unit mass). In terms of velocity, the electronic stopping mechanism dominates over the entire velocity region from about 0.99 times the velocity of light ($c$) to about the first electronic excitation energy of the medium molecules (for electrons) or to a least-bound electron around the incident particle (for heavy charged particles). For even lower energies the stopping effects are complicated and quite dependent on the mass of the incident particles. For example, if the incident particle is a heavy ion, the phenomenon of charge exchange (capture and loss of electrons by the moving ion) dominates the stopping power mechanism. Eventually, this process is dominated by nuclear collisions of Rutherford type. Finally, the projectile is thermalized by elastic collisions of billiard ball type. For incident electrons at low velocities (subexcitation energy) the main stopping mechanism involves the excitation of molecular vibrations until the energy reduces to about 0.5 eV. Below this energy, and if the medium is a condensed phase, the incident electron may excite intramolecular vibrations on its way to becoming thermalized. Finally, either the electron reacts chemically or becomes attached to a molecule.

In this section we attempt to present the current status of the stopping power theory as it concerns radiation chemistry. Among the various mechanisms of energy losses involved, we have considered only interaction with the medium electrons (ionization and excitation). In radiation chemistry most studies are related to the formation of chemical species as a result of the energy deposition through this mechanism. This aspect of stopping theory is better understood from the view point of both theory and experiment.

### *The Classical Theory: Bohr Approach*

Let us consider a moving particle (velocity $v$) of electric charge $Z_1 e$ and mass $M_1$ penetrating a medium composed of atoms (or molecules) of atomic number $Z_2$ and atomic mass $M_2$. The moving charged particle generates an electromagnetic field and interacts with the electrons through this field. The projectile slows down through inelastic collisions—that is, by exciting and or ionizing the electrons and thereby changing the electronic state of the atoms. In the classical approach the theory is based on an impact parameter as the basic variable. We consider a collision with a single atom of the medium, and an impact parameter, $b$, is defined as the distance of closest approach between the charged particle projectile and the nucleus. It should be noted that the impact parameter is not an observable quantity. Early investigators[3,4] had considered the problem in a manner that involves collisions with free electrons (ie, they neglected the binding energy) and a maximum impact parameter chosen on a rather *ad hoc* basis. The imposition of a maximum impact parameter was necessary in order to prevent the energy loss from diverging. This divergence results from the integration of the Rutherford cross section, which yields infinity unless a cutoff distance is assigned. In his treatment, Bohr[5] realized that the consideration of binding effects is essential for a proper treatment of energy loss. He suggested that there exists an impact parameter $b_{\text{crit}}$ for which more distant collisions can be treated as electromagnetic excitations of charged harmonic oscillators in a spatially uniform field due to the incident particle. For collisions with the value of the impact parameter smaller than $b_{\text{crit}}$, the problem can be treated as the scattering of the projectile by free electrons. Bohr argued that for an effective momentum transfer, the collision must be sudden; that is, the transit time of approximately $2b/v$ must be smaller than the reaction time of the electron between the two stationary states; based on the uncertainty principle, the latter is $\hbar v/2E_1$, where $E_1$ is a typical atomic transition energy. Thus $b_{\max} \sim \hbar v/2E_1$.

We now derive a simplified version of the Bohr stopping power formula. According to Bohr theory, the differential cross section $d\sigma$ for the transfer of energy in the range between $\varepsilon$ and $\varepsilon + d\varepsilon$ to an electron from the incident charged particle is given by

$$d\sigma = \left(\frac{2\pi Z_1^2 e^4}{mv^2}\right)\frac{d\varepsilon}{\varepsilon^2} \tag{1-3}$$

where e and $m$ represent the charge and mass of an electron, respectively. For excitation and ionization of bound electrons, Bohr introduced the sum rule

$$\sum_n f_n \varepsilon_n = Z_2 \varepsilon \tag{1-4}$$

where $f_n$ is the "oscillator strength" and can be interpreted as the effective number of electrons in the atom that receives the energy gain $\varepsilon_n$. The stopping power can now be obtained by multiplying the cross section by the energy transfer and integrating it from the minimum energy loss, $\varepsilon_{\min}$, to the maximum energy loss, $\varepsilon_{\max}$. Such a procedure gives

$$-\frac{dE}{dx} = N \int d\sigma \sum_n f_n \varepsilon_n = \left(\frac{2\pi Z_1^2 e^4}{mv^2}\right) N Z_2 \ln\left(\frac{\varepsilon_{\max}}{\varepsilon_{\min}}\right) \tag{1-5}$$

where $N$ denotes the number density of the atoms of the stopping medium, and $\varepsilon_{\max}$ and $\varepsilon_{\min}$ are, respectively, the maximum and minimum energy transfers. Since $\varepsilon_{\max}$ corresponds to minimum impact parameters, one may neglect the atomic binding and consider the electrons as free. In a free collision, using the energy-momentum relation, one can obtain $\varepsilon_{\max} = 2mv^2$, since $M_1 \gg m$. For estimating $\varepsilon_{\min}$, which corresponds to the maximum impact parameter, $b_{\max}$, Bohr used the general relationship $\varepsilon = 2Z_1^2 e^4/mb^2v^2$ between energy transfer, $\varepsilon$, and the corresponding impact parameter, b. Based on the principle of sudden collision mentioned earlier, $b_{\max} = \hbar v/2E_1$. Thus, $\varepsilon_{\min} = 8Z_1^2 e^4 E_1^2/mv^2(\hbar^2 v^2)$. Substituting the values of $\varepsilon_{\max}$ and $\varepsilon_{\min}$ in Eq. (1-5), we see that the expression for stopping power reduces to

$$-\frac{dE}{dx} = \left(\frac{4\pi Z_1^2 e^4}{mv^2}\right) N Z_2 \ln\left(\frac{2mv^2}{E_1 \hbar v/4Z_1 e^2}\right) \tag{1-6}$$

Eq. (1-6) is the classical (or sometimes referred to as semiclassical) formula for stopping power as given by Bohr in the nonrelativistic approximation. For a more detailed and rigorous derivation of the stopping power in the classical approximation, the reader is referred to Jackson.[6]

As Mozumder[7] pointed out, it is sufficient to estimate the maximum and minimum impact parameters in the simplest of approximations because they appear in a logarithmic factor. It is quite conceivable that there may be situations where better estimates are required. According to the concepts developed by Bohr, the minimum impact parameter equals $Z_1e^2/mv^2$; however, according to the angular momentum consideration of quantum mechanics, it is given by $\hbar/mv$. In many situations the larger of these two should be taken. To calculate $b_{max}$, Bohr used the principle of sudden collision, sometimes referred to as the impulse approximation. According to this assumption, there is no interaction for larger values of $b$, and in reality the energy transfer falls off exponentially for impact parameters greater than $b_{max}$.[8] In spite of several limitations of Bohr's classical theory, the method yields results similar to those given by the more rigorous quantum mechanical theory for higher velocities and lower atomic numbers of the incident charged particles. At present the usual practice is to use Bethe's stopping power formula based on quantum mechanical considerations.

## *The Quantum Mechanical Approach: The Bethe Theory*

Beginning with the early 1920s, several attempts were made to use the principles of quantum mechanics to solve the problem of stopping power theory. In 1922 Henderson[9] applied the concept of discrete energy levels and restricted all the energy losses to the amount necessary for ionization. In this manner he deduced an expression for the stopping power that is roughly equal to one-half of the total stopping power. He essentially ignored the contribution due to distance collisions (electronic excitation of the atoms), which provides a mechanism for the other half of the energy loss. In 1930 Bethe,[10] using the first Born approximation, solved the problem within the framework of quantum mechanics. Essentially, the first Born approximation means that the scattering centers (the electrons in a stopping medium) act as small perturbations to the propagation of plane waves of the incident particle. This approximation is generally valid if the incident velocity is much higher than the velocity of an electron in the 1s-orbital around the incident particle. Since Bethe used the Born approximation, his solution is sometimes referred to as the Bethe–Born technique.

The important difference between Bethe's quantum mechanical approach and the classical approach of Bohr is the use of the momentum transfer rather than the impact parameter in characterizing collisions. The principles of wave mechanics forbid the consideration of a highly localized wave packet for a particle with well-defined momentum. In the classical treatment of Bohr, such highly localized wave packets have been assumed to exist in very close collisions, but they cannot be justified as a general rule. There can be special situations, as demonstrated by Mott,[11] where approximate justification for the use of classical formulas can be made. According to Mott, classical mechanics can be used if the incident heavy particle has a de Broglie wavelength much smaller than atomic dimensions. In a situation like this the incident particle can then be considered to be a slowly moving perturbing potential, for which the induced transitions are about the same as estimated by the Born approximation. It must be mentioned here that even within the first Born approximation, the Bethe formula consistently overestimates the stopping power because it includes energy losses to some states that are energetically inaccessible.

A complete derivation of the Bethe stopping power formula is outside the scope of this book. Readers interested in the derivation are referred to the excellent review by Fano.[12] A simplified version of the approach used by Bethe is given in Chapter 5 of this book. In the same chapter an approximate formula for the stopping power has been deduced in the nonrelativistic case. At relativistic speeds the maximum energy transfer is equal to $2mv^2/(1-\beta^2)$ instead of $2mv^2$, which is correct only for low velocities. A more rigorous expression using the relativistic limit is

$$-\frac{dE}{dx} = 2\kappa N Z_2\left[\ln\left(\frac{2mv^2}{I}\right) - \ln(1-\beta^2) - \beta^2\right] \tag{1-7}$$

where $\kappa$ is called the *stopping parameter* and is given by

$$\kappa = \frac{2\pi Z^2 e^4}{mv^2} \tag{1-8}$$

In Eq. (1-7), $I$ is defined as the mean excitation potential of the medium, and it determines the penetration of charged particles in a given material. It is an important parameter in Bethe's stopping power formula and is related to the distribution of oscillator strength of an atom or a molecule. It has to be either experimentally determined or calculated by using ground-state and excited-state atomic wave functions. Bethe was able to calculate a value of $I$ for atomic hydrogen by using exact wave functions. His results are only applicable to a gas of atomic hydrogen, a situation not encountered by radiation chemists. His attempt to extend his calculations for heavier atoms by using hydrogen-like wave functions was unsuccessful. In practice, $I$ is determined empirically from stopping power and/or range measurements by using the corrected stopping power formula [see Eq. (1-10)]. Table 1-1 represents a set of current estimates of $I$ for atomic and molecular media. Although we have developed stopping power theory for atoms only, some values of $I$ for molecular media are also provided. For molecular

**Table 1-1.** Mean Excitation Potentials

| Substance | $Z_2$ | $I$ (eV)[a] |
|---|---|---|
| Hydrogen | 1 | 15–18 |
| Helium | 2 | 39–45 |
| Carbon | 6 | 77–80 |
| Nitrogen | 7 | 79–102 |
| Oxygen | 8 | 91–101 |
| Air | — | 85 |
| Methane | — | 45 |
| Water | — | 70 |

[a] For atomic media ranges of values have been given.

effects and the subsequent modification of the atomic stopping power theory, please refer to the subsection on the Bragg rule.

We next discuss the corrections required for a more general application of the stopping power formula of Eq. (1-7).

## Low Velocities: Shell Correction

Many contributors in this field have emphasized that the derivation of the Bethe formula relies on the Born approximation and that this approximation is valid only when the speed of the incident particle is much greater than the orbital speed of electrons bound to an atom. The separation between low-momentum transfer and intermediate-momentum transfer is possible only in such a situation. The assumption of the Born approximation is difficult to justify for $K$-shell electrons for particles of low velocities except for light atomic media. In the case of penetration of high atomic number materials, even $L$-shell correction may be required. When great accuracy is not desired, the $K$-shell electrons may be omitted from the total value of $Z_2$. The simplest stopping power formula, including shell corrections, is

$$-\frac{dE}{dx} = 2\kappa N Z_2\left[\ln\left(\frac{2mv^2}{I}\right) - \ln(1-\beta^2) - \beta^2 - \frac{C}{Z_2}\right] \tag{1-7a}$$

where $C$ is the total shell correction factor.

The subject of shell correction has been extensively dealt with by many authors,[13,14,15] and fairly sophisticated results are available in graphs and tables. According to the view of this author, shell corrections do not have much significance in radiation chemistry unless a specific effect can be traced to $K$-shell excitation or ionization. In water the correction amounts to about 1% at low velocities.

## *The Condensed Phase: Density Correction*

So far we have made the implicit assumption that the medium through which a charged particle passes consists of a very dilute gas. Only for such a situation is it correct to add incoherently the energy lost to individual atoms to obtain the total stopping power. In a condensed medium the interspacing between atoms is much smaller; therefore it is no longer true that the projectile interacts with only one atom at a time. Also, the atoms cannot be considered to be independent of each other. Fermi pointed out that as $\beta \rightarrow 1$, the electronic stopping power would approach infinity except that the polarization screening of one atom from another by the medium electrons reduces the interaction. This effect is called the *density correction* because it is related to an aggregate of atoms. Generally, this correction is negligible at low velocities, but at high velocities Fano gives an expression for this correction factor, $\delta_2$, in terms of a complex dielectric constant that, in its simplified form, can be expressed as

$$\delta_2 = \ln\left[\frac{\hbar^2\omega_p^2}{I^2(1-\beta^2)}\right] - 1 \qquad (1\text{-}9)$$

where $\omega_p^2 = (4\pi N Z_2 e^2/m)^{1/2}$ is the plasma frequency and is a characteristic of the medium only. The values for density and plasma energy ($\hbar\omega_p$) are given in Table 1-2. At values of $\beta < 0.9$ this correction is negligible for water. Generally, the average error arising from the neglect of the density effect is claimed to be $\pm 2\%$ in the overall stopping power value. For most situations in radiation chemistry studies, this is not significant.

**Table 1-2:** Density and Plasma Energy for Various Substances

| Substance | Chemical Formulae | $\rho$ (gm/cm$^3$) | $h\omega_p$ (eV) |
|---|---|---|---|
| Hydrogen gas (STP) | $H_2$ | $8.96 \times 10^{-4}$ | 0.27 |
| Oxygen gas (STP) | $O_2$ | $1.42 \times 10^{-3}$ | 0.77 |
| Neon gas (STP) | Ne | $8.97 \times 10^{-4}$ | 0.61 |
| Polyethylene | $(CH_2)_n$ | 0.93 | 21.01 |
| Lucite, plexyglass | $(C_5H_8O_2)_n$ | 1.19 | 23.10 |
| Nuclear emulsion (G5) | — | 3.80 | 37.87 |
| Water | $H_2O$ | 1.00 | 21.46 |

If we now include all the corrections mentioned above, the total stopping power formula of a heavy charged particle can be written as

$$-\frac{dE}{dx} = 2\kappa N Z_2\left[\ln\left(\frac{2mv}{I}\right) - \ln(1-\beta^2) - \beta^2 - \frac{C}{Z_2} - \delta_2\right] \qquad (1\text{-}10)$$

## The Electron Stopping Power

Incident photons and incident heavy charged particles always create secondary electrons. In radiation chemistry many experimental studies involve energetic electrons as the source of radiation. Hence, a knowledge of the stopping power of a medium for incident electrons is essential.

Although they are similar, there are significant differences in the expressions for stopping power of electrons and heavy charged particles. The principal difference arises from the fact that primary electrons and secondary electrons are identical. A distinction can be made based on kinetic energy. Following a collision we say that the primary electron has the most energy. In an electron–electron collision, if we neglect the binding energy, the maximum energy that can be transfered is $\frac{1}{4}mv^2$ rather than $2mv^2$ as in the case of a heavy particle. Incorporating this distinction, for a nonrelativistic electron, the Bethe stopping power formula can be written as

$$-\left[\frac{dE}{dx}\right]_{\text{electron}} = 2\kappa NZ_2 \ln\left[\left(\frac{mv^2}{2I}\right)\left(\frac{e}{2}\right)^{1/2}\right] \tag{1-11}$$

where $e$ is the base of natural logarithms. The difference between stopping power of an incident electron and a heavy particle of the same charge and velocity is given by the quantity $\ln 4(2e)^{1/2}$ or 1.33, as compared to $\ln(mv^2)$. Hence, one can conclude that at comparable high velocities the stopping power due to an incident electron is always less than the stopping power of a heavy charged particle, but not significantly less. For example, at a velocity of $10^{10}$ cm s$^{-1}$, the electron stopping power in water is about 20% smaller than the stopping power of a heavy particle of the same charge.

The overall stopping power formula for an electron with corrections due to relativistic, shell, and density effects can be written as

$$-\left[\frac{dE}{dx}\right]_{\text{electron}} = \kappa NZ_2\left[\ln\left(\frac{mc^2\beta^2E}{2I^2(1-\beta^2)}\right) - (2(1-\beta^2)^{1/2} - 1 + \beta^2)\ln 2 + 1 - \beta^2 + \frac{1}{8}(1-(1-\beta^2)^{1/2}) - \frac{2C}{Z_2} - \delta_2\right] \tag{1-12}$$

where $E$ is the incident energy of an electron. The shell corrections for electrons and heavy particles are nearly the same.

## Molecular Stopping Power Theory: The Bragg Rule

Thus far our discussions have been relevant to the atomic stopping power theory. Most of the same discussions can be extended to molecular media by applying the Bragg rule. The Bragg rule provides a mechanism to determine the effective $I$ value of the molecules from a knowledge of the same values of its constituent atoms. The procedure involves the geometrical averaging of all the respective values of $I$ for the individual atoms in their free states over the electron numbers. The mathematical expression for this rule is

$$N' \ln I = N_1 \ln I_1 + N_2 \ln I_2 + \cdots$$

that is,

$$I^{N'} = I_1^{N_1} \cdot I_2^{N_2} \cdots \tag{1-13}$$

where $N'$ and $I$ are, respectively, the number of electrons and the mean excitation potential of the molecule, and $N_i$ and $I_i$ refer to the individual atom $i$. For many compounds and mixtures, Bragg's additivity rule is accurate within about 2%. It is not quite obvious why such a rule should at all be valid, because in a molecule the several atoms are held together by chemical bonds and it is well known that the energy levels of the valence electrons are altered significantly. On the other hand, for organic molecules and for water most of the oscillator strength involves higher energy states than the energy involved in the chemical binding, and, hence, specific details may not be important. There is evidence that the deviation from the Bragg rule is more apparent at low energies because the logarithmic term becomes more sensitive to the variation of $I$. Chan et al[16] designed experiments using helium ions (0.06 MeV/amu–0.5 MeV/amu) to measure the stopping powers and also to verify the Bragg rule in saturated alcohols and ethers in the gas phase. They found that the Bragg rule holds to within 1% in stopping power for single bonds at all energies. For double-bonded oxygen the deviation was found to be 6% more than that expected from the applications of the Bragg rule based on single-bond data at 0.5 MeV/amu. It is the view of this author that for applications in radiation chemistry, the Bragg rule provides a reasonable way of estimating the stopping power of all molecular media.

## *Capture and Loss: Charge Exchange*

In the stopping power expression Eq. (1-10), the stopping parameter $\kappa$ is a function of $Z_1$, where $Z_1$ is the charge on an incident heavy particle. At projectile velocities much greater than $Z_1 v_0$ ($v_0$ is the velocity of the orbital electron in the $K$ shell of a hydrogen atom) a heavy charged particle has no bound electrons; therefore, the full value of the nuclear charge should be substituted for $Z_1$. The situation gets complicated when the particle velocity is comparable to or less than $Z_1 v_0$. Positive ions have an appreciable probability of capturing an electron from the medium when slowed down to this velocity. While the projectile velocity is near $Z_1 v_0$, the captured electron is lost and recaptured in subsequent collisions. After an interplay of capture and loss events, an electron becomes permanently bound to the positive ion and thereby reduces its charge by 1. Then the whole sequence is repeated for charges $Z_1 - 1, Z_1 - 2, \ldots$, and eventually the ion is neutralized and subsequently thermalized. The cross sections for charge exchange processes are not very well known. Approximate solutions are available, but they are mostly suitable for very heavy charged particles. At a given velocity there are appreciable probabilities for a positive ion to exist in three or four different charge states. This charge then may be defined by $(Z_1^2)_{eff} = \langle Z_1^2 \rangle$ where the latter is the equilibrium average of the square of all the ionic charges weighted by their respective probabilities.

In the absence of any rigorous theory of charge changing collisions, several empirical or semiempirical expressions are available. According to Pierce and Blann,[17] the effective charge on a positive ion can be written as

$$(Z_1)_{eff} = Z_1 \left[ 1 - \exp\left( \frac{-130\beta}{Z_1^{2/3}} \right) \right] \tag{1-14}$$

A variant of this expression is used in Eq. (6-7).

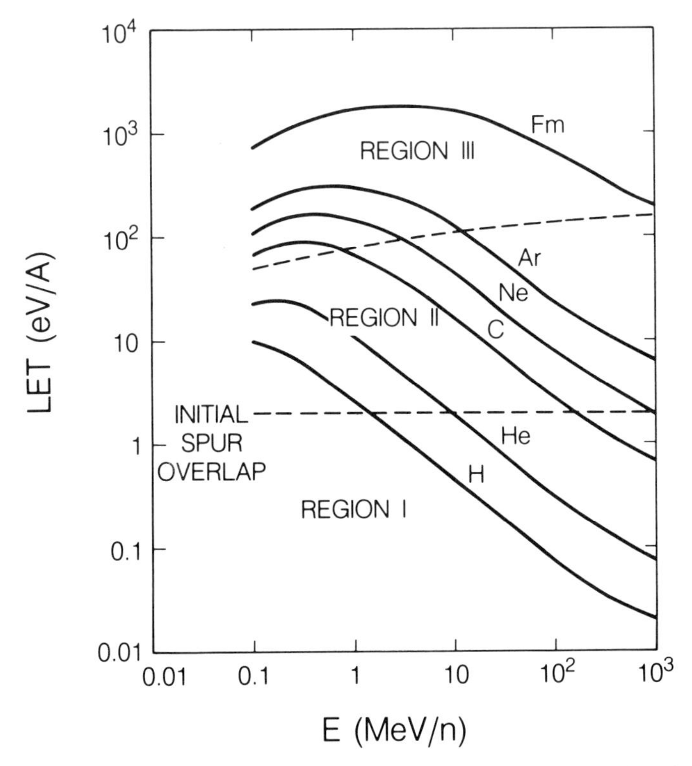


**Figure 1-4.** The stopping power (in eV/A) of water has been plotted for six different particles as a function of specific energy (in MeV per nucleon). In region I the LET is low enough that the spurs develop independently; in region II the spurs in the track core are merged right from the beginning, but the overlap is not excessive; ie, the diffusion of radicals competes with the process of radical recombination; in region III the spur overlap is excessive, and recombination of radicals dominate over diffusion. (Reprinted with permission from *Journal of Physical Chemistry 84*, 3531. Copyright 1980 American Chemical Society.)

It should be noted that most of these expressions are independent of the atomic number of the absorbing materials.

Figure 1-4 shows the calculated values of the stopping power or linear energy transfer (LET) of six selected heavy charged particles in water. All the discussions outlined in this section have been considered in estimating these values. As expected, the stopping power increases with a decrease in energy and finally reaches a maximum. With a further reduction in velocity the stopping power decreases also. When the charge exchange is not important, the stopping power at a given energy per nucleon is proportional to the square of the full charge on a moving ion. Figure 1-5 presents the stopping power of electrons in water.

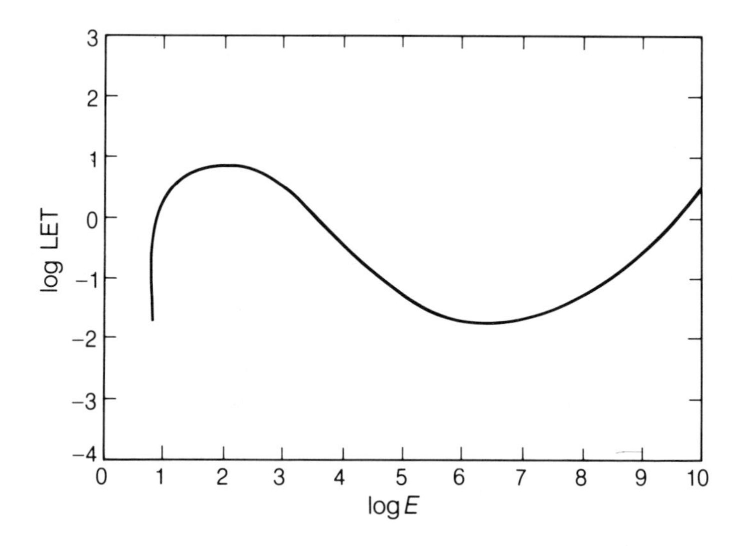


**Figure 1-5.** The stopping power of water against incident electron energy has been plotted on a log-log scale. The stopping power is best known in region $10^4$–$10^9$ eV, as given quite accurately by the Bethe stopping power formula when properly corrected. Below $10^4$ eV, values have been obtained by the method prescribed by Mozumder (ref. 7).

## TRACK STRUCTURE

Penetrating charged particles deposit energy along their paths in a medium, and the effect of this process leads to the formation of tracks. Such tracks can be recorded in cloud chamber measurements or photographed in emulsion plates. Beginning with a time scale of the order of $10^{-16}$ s, the structure of a track evolves and emulsion plate measurements generally refer to the steady state or long-term pattern of a track. In radiation chemistry and radiation biology we are interested in the structure of a track at a very early stage in the time scale (less than $10^{-14}$ s). This is commonly known as *latent* track structure, and it is essentially the geometrical pattern of the deposited energy. As a consequence of the energy gained by the medium, a sequence of nonhomogenous processes follows, which creates and transforms reactive intermediates until final radiation chemical products are formed. In this section we consider a particular theoretical model of track structure based on various considerations that have been extensively applied in the studies related to radiation chemical yields. Since the discussion relates to a time scale that is several orders of magnitude earlier than the experimentally accessible picosecond region, we have to depend upon theoretical developments only.

An ideal way to consider the problem of track structure is to begin with the excitation and ionization cross sections of the medium and relate them to the formation of tracks through a Monte Carlo procedure. Such a procedure has begun to emerge only recently, and significant improvements are expected in the near future.[18] One of the problems associated with the Monte Carlo technique is the nonavailability of proper cross sections in a condensed phase, such as liquid water. Another problem is that for a given particle of given energy in a given medium, each time a track is developed it will be different because of the statistical fluctuation in energy deposition, and one has to consider a large number of tracks before an average picture emerges.

The alternative to this procedure is the direct consideration of average tracks with structure obtained from the stopping power theory. This consideration forms the main discussion in this section. The usefulness of track structure for applications to radiation chemistry is demonstrated in Chapters 5 and 6.

### *Track Entities: Spurs, Blobs and Short Tracks*

Historically the development of the concept of track entities originated with energetic electrons. Hence, most of our discussions in this subsection refer to electrons; wherever possible, similarities with respect to tracks formed by heavy ions will be indicated. Heavy particle tracks have been separately discussed in the next subsection.

The stopping power formula as given by Eq. (1-10) or (1-12) is based on an implicit assumption that is called the *continuous slowing-down approximation* (c.s.d.a.). From this, one can calculate the average rate of energy loss, and for an understanding of the radiation chemical processes it can actually be inappropriate if one assumes that all track segments receive an average energy deposit. This would be justifiable if in reality the continuous slowing-down mechanism was an actual physical process. A more realistic way to consider the formation of a track is in terms of the energy loss spectrum considered implicitly in the Bohr theory. Distant collisions (or glancing collisions)—that is, collisions with large impact parameters—create excitations of the medium to energy levels that have oscillator strength for ordinary optical excitation. In materials with low atomic numbers, such as water, the oscillator strength distributions are essentially contained in the 100 eV region.[19] Nearly half of the energy loss of a charged particle occurs in loss events along the trajectory, which individually involve less than 100 eV [see Eq. (5-10)]. The average of all these loss events has been shown to vary between 30 and 40 eV,[20] depending on how the average is calculated. These low-energy-loss events along the trajectory of a moving particle create track entities called *spurs.* If the particle has a low stopping power or LET value, the spurs are widely spaced like a "string of beads," and they develop in time without any interference from the neighboring spurs. Intermediate- and high-energy electrons and also very high-energy, heavy charged particles create tracks of this nature. On the other hand, for a high-LET particle the spurs overlap and form a continuous column, giving rise to the consideration of another type of track entity called *short tracks.* It is clear from this discussion that a spur is a more fundamental concept than a short track. If 17 eV are required to create a radical pair in water, then on average a spur contains two or three radical pairs that can be considered isolated in a low-LET case. It is to be remembered that just like the spurs, the short tracks can also be created by both electrons and heavy particles of appropriate energies.

In the classical theory as well as in the quantum mechanical theory of stopping power, it has been recognized that the high-energy knock-on electrons are created as a result of close collisions, that is, collisions involving small impact parameters or large momentum transfers. The energy spectrum of these electrons vary between 100 eV and a maximum value of $2mv^2$ for heavy charged particles or $\frac{1}{4}mv^2$ for electrons. The cross section for the production of these electrons is governed by the Rutherford process and, hence, varies as $\varepsilon^{-2}$, where $\varepsilon$ is the energy of a knock-on electron. Such a dependence favors the production of low-energy electrons, although high-energy electrons are also produced. Above a critical energy (greater than 5 keV in water) these electrons produce their own tracks with the formation of isolated spurs. Hence, knock-on electrons with energy greater than 5 keV are considered as another track entity, called *branch tracks.* The problem of describing the radiation chemistry of these tracks is the same as that of describing the radiation chemistry of the primary electron track itself. Ordinarily, electrons with energies between 100 eV and 5 keV should have fallen in the category of short tracks, because the spurs created by these particles overlap with each other. However, Mozumder and Magee[21] considered these matters for electrons in water and decided that an energy loss greater than 500 eV should be classified as a short track. The track entities in the 100–500-eV range are called *blobs.* They argued that a blob is generated by a knock-on electron that has insufficient energy to escape the attraction of its sibling hole, although some screening may exist due to intermediate ionization processes. Their arguments led to the consideration of a blob as a separate entity.

In summary, the three different categories of track entities–spurs, blobs, and short tracks—can be considered to be based on the following energy criteria:

| | |
|---|---|
| spur: | energy deposited between 6–100 eV |
| blob: | energy deposited between 100–500 eV |
| short track: | energy deposited between 500–5000 eV |

Branch tracks are not considered separately because all electron tracks with energies greater than 5 keV can be broken down into similar patterns. It has to be noted that these demarcations cannot be rigid.

Following these concepts, Mozumder and Magee[22] calculated the partition of primary energy (when the particle is completely stopped in the medium) into spurs, blobs, and short tracks. Instead of reproducing his results, we give some numerical examples: (i) a 20-keV electron contributes 38% of its energy to produce isolated spurs, 12% for blobs and 50% for short tracks; (ii) a 1-MeV electron spends 65% of its energy to produce isolated spurs, 15% in blobs and 20% in short tracks; and (iii) a 10-MeV electron produces isolated spurs to the extent of 76% of its energy, 8% for blobs and 16% for short tracks.

It should be understood that the various track entities and their characteristics are clearly theoretical in nature. Experimentally it may be difficult to demonstrate their existence, although some measurements tend to give indirect support. These concepts have been applied quite extensively in calculating variation of radiation chemical yields with radiation quality. Some of the applications of these concepts are described in Chapters 5 and 6.

## *Heavy Charged Particle Tracks: Core and Penumbra*

The concept of the track entities for electron tracks introduced earlier also applies for heavy charged particles. However, a separate consideration of these tracks has proven to be helpful for two reasons: (i) These heavy ions scatter much less than the electrons, and therefore their tracks are more linear; (ii) for all practical purposes the fluctuation in energy loss by a heavy charged particle is not very significant, and therefore their tracks can be well approximated in terms of average spurs. For tracks generated by electrons, neither of these facts is true.

Even for moderate energies the LET of a heavy particle is quite high, and the spurs overlap with each other, creating a densely cylindrical region called the *core*. The energy loss events less than or equal to 100 eV are contained within this region, and those knock-on electrons that cannot penetrate very far because of their low energies deposit energy within the core. Hence, more than 50% of the total energy loss is enclosed within a small region and thereby creates a dense track core. Obviously, the radicals created in this region will also have a high density; hence, the subsequent chemical reactions can be understood in terms of track effects. The remaining fraction of the energy deposited by heavy charged particles is carried out by energetic secondary electrons (also called $\delta$-rays) into a region known as the *penumbra.* The LET of these electrons is at least an order of magnitude less than the LET restricted to energy loss within the core only. When restricted to the core, LET is sometimes referred to as the partial stopping power or $\mathrm{LET}_{r_c}$, where $r_c$ is the radius of the core. In the penumbra, electron tracks can be considered to be independent of each other, so each track develops its own chemistry. Further discussion on this aspect is presented in Chapter 6.

There are several ways to assign a radius to the core as a function of particle velocity and the properties of the medium. For example, if one recognizes that all the glancing collisions originate within a core, then one can apply Bohr's principle of sudden collision (as described in the section on stopping power) to estimate its radius. Based on this approach, the core radius can be written as

$$r_c = \frac{\hbar v}{2E_1} \tag{1-15}$$

where $E_1$ is the lowest electronic transition energy of the medium molecules. Another expression for $r_c$ can be written in terms of the plasma oscillation frequency of the medium as given by Eq. (6-4). Both methods yield similar results. Note that the size of the core is independent of the particle charge because it depends on the duration of interaction, which in turn depends on the velocity only.

In many situations one may need to know the variation in energy density around the trajectory of a heavy charged particle. One of the reasons for such a requirement comes from the inadequacy of LET as a unique physical parameter. Two particles having different charges may have the same LET but do not produce the same chemical yields. One way to understand these results is to consider the energy density parameter, $\rho(r)$, as a function of $r$, the radius of a cylinder around the particle trajectory. Chatterjee and Schaefer[23] and others demonstrated that $\rho(r)$ varies as $r^{-2}$ [see Eqs. (6-5) and (6-6)]. Just like the core, the penumbra also has a limiting radius, $r_p$, because all the high-energy electrons are contained within this radius due to their finite ranges.

Table 1-3 summarizes the track parameters that are characteristics only of the velocity of the charged particles. In the last column of this table, the symbol $\lambda$ gives the ratio of the initial energy density in the core to that in the penumbra near $r_c$, and it is significant that all concentrations of initial chemical species in the core are much larger than those in the penumbra.

The track picture that emerges from these discussions has been applied in the calculation of radiation chemical yields in aqueous solutions as described in Chapters 5 and 6.

**Table 1-3.** Track Parameters Independent of Particle Charge

| Energy (MeV/$n$)[a] | $\beta$ | $r_c$ (nm) | $r_p$ (nm) | $\varepsilon_{max}$ (eV)[a] | $\lambda$ |
|---|---|---|---|---|---|
| 1000 | 0.876 | 9.016 | $2.7 \times 10^5$ | $3.4 \times 10^6$ | 23.42 |
| 800 | 0.843 | 8.675 | $2.4 \times 10^5$ | $2.5 \times 10^6$ | 22.46 |
| 600 | 0.794 | 8.169 | $2.1 \times 10^5$ | $1.7 \times 10^6$ | 22.31 |
| 400 | 0.715 | 7.354 | $1.6 \times 10^5$ | $1.1 \times 10^6$ | 21.98 |
| 200 | 0.568 | 5.843 | $8.4 \times 10^4$ | $4.9 \times 10^5$ | 21.15 |
| 100 | 0.430 | 4.421 | $3.9 \times 10^4$ | $2.3 \times 10^5$ | 20.17 |
| 80 | 0.390 | 4.012 | $3.0 \times 10^4$ | $1.8 \times 10^5$ | 19.84 |
| 60 | 0.343 | 3.526 | $2.1 \times 10^4$ | $1.4 \times 10^5$ | 19.38 |
| 40 | 0.284 | 2.923 | $1.3 \times 10^4$ | $0.9 \times 10^5$ | 18.80 |
| 20 | 0.204 | 2.099 | $5.3 \times 10^3$ | $4.4 \times 10^4$ | 17.67 |
| 10 | 0.145 | 1.496 | $2.1 \times 10^3$ | $2.2 \times 10^4$ | 16.49 |
| 8 | 0.130 | 1.340 | $1.6 \times 10^3$ | $1.8 \times 10^4$ | 16.17 |
| 6 | 0.113 | 1.163 | $1.1 \times 10^3$ | $1.3 \times 10^4$ | 15.70 |
| 4 | 0.092 | 0.951 | $6.2 \times 10^2$ | $8.8 \times 10^3$ | 14.96 |
| 2 | 0.065 | 0.673 | $2.4 \times 10^2$ | $4.4 \times 10^3$ | 13.75 |
| 1 | 0.046 | 0.477 | $9.6 \times 10^1$ | $2.2 \times 10^3$ | 12.61 |
| 0.8 | 0.041 | 0.426 | $7.1 \times 10^1$ | $1.8 \times 10^3$ | 12.23 |
| 0.6 | 0.036 | 0.369 | $4.8 \times 10^1$ | $1.3 \times 10^3$ | 11.74 |
| 0.4 | 0.029 | 0.302 | $2.8 \times 10^1$ | $8.8 \times 10^2$ | 11.06 |
| 0.2 | 0.021 | 0.213 | $1.1 \times 10^1$ | $4.4 \times 10^2$ | 9.89 |
| 0.1 | 0.015 | 0.151 | 4.3 | $2.2 \times 10^2$ | 8.70 |

[a] 1 eV = 0.16 aJ.
Reprinted with permission from *Journal of Physical Chemistry 84,* 3530.


# SPATIAL DISTRIBUTION OF PRIMARY SPECIES

One of the most difficult questions remaining to be answered satisfactorily is the spatial distribution of "primary" species. The situation is made even worse by a lack of agreement on the chemical species to be called primary. Except for the hydrogen atom, Platzman considered this to be a nontrivial problem. The underlying reasons may be that for atomic hydrogen (or for noble gases) we know most of the possible excited states. According to Platzman, these states are very poorly known in the case of molecules in spite of the great progress made in molecular spectroscopy. To have a better idea for molecules, one must know the electronic structure as well as the excitation energy. However, for water it is customary to discuss the spatial distribution of primary species beginning with the ion $H_2O^+$, the excited neutral molecule $H_2O^*$, and subexcitation electrons.

Complementary to the question of the spatial distribution of primary species is the question of temporal distribution. This question generally refers to the time domain between $10^{-16}$ and $10^{-12}$ s, that is, prior to the thermalization of transient species. Unfortunately, this aspect of the problem is also quite poorly known, and one must rely on deductions based on theoretical concepts. Such a discussion has been presented in Chapter 5, and in this section we will simply concentrate on the spatial aspect of this distribution.

To evaluate the radiation chemical yields beginning with the track structure, one must make ad hoc assumptions regarding the spatial distribution of the initial species (some prefer to use "initial" instead of "primary"). It is usually assumed to be Gaussian about some center. In Chapters 5 and 6 these procedures are discussed. In this section we discuss a method developed by Turner et al[24] based on the Monte Carlo technique. The medium is water.

We mentioned earlier that in the Monte Carlo technique one needs to know the cross sections for ionization and excitations and that these elastic and inelastic collision processes are not very well known in liquid water. The procedure outlined by Turner et al used the low-energy photoabsorption cross-sectional data from experimental measurements and extrapolated them to high-energy values. Based on these cross sections, they developed an electron transport code for energetic incident electrons. Once the cross sections are available, the same procedure can be adopted for high-energy heavy charged particles.

For a primary or a secondary electron moving through a medium like water a flight distance is first selected through computer programming. This distance is based on the mean free path (inversely proportional to cross sections) calculated for elastic and inelastic collisions at the electron energy. If the collision turns out to be elastic, the scattering angle is chosen from a knowledge of scattering angle distribution with respect to cross section. The incident electron is then transported in the new direction, and the whole procedure is repeated. If the collision turns out to be inelastic, the code chooses the energy loss and decides whether the event is ionizing or nonionizing, based on the partition of the total cross sections between excitation and ionization. For an excitation event the position of the $H_2O^*$ is noted from the free path, and the electron is allowed to continue to travel in its original direction with its energy reduced by the excitation energy given up in the collision. For an ionization event the electron is assumed to be scattered by a free electron. The ejection of the so-called free electron is scattered as though initially unbound but with its energy reduced by the amount of the binding energy. All electrons are then followed or transported in a similar way until their energies fall below a subexcitation level of 7.4 eV in water. Thus, they obtain all the initial ($10^{-15}$ s) positions of $H_2O^*$, $H_2O^+$, and subexcitation electrons. The next step is to calculate the subsequent changes and the coordinates of the resulting chemical species present at about a picosecond.

If a water molecule is ionized, it reacts quickly with a neighboring neutral water molecule to form a hydronium ion and an OH radical:

$$H_2O^+ + H_2O \rightarrow H_3O^+ + OH$$

It is assumed that at $10^{-12}$ s the $H_3O^+$ ion is located at the site of the original $H_2O^+$ with the OH radical adjacent to it. The centers of the ion and radical are assumed to be separated by 2.9 Å, the diameter of a water molecule. The direction of the OH radical relative to the $H_3O^+$ is selected at random.

For an excited water molecule two channels for dissociation have been considered: (i) production of H and OH radicals, and (ii) production of $H_2$ and O. If H and OH radicals are produced, they are assumed to be separated by 5.8 Å, which is the size of two water molecules.

The subexcitation electrons are allowed to execute a random walk until they become hydrated at about $4 \times 10^{-13}$ s. At $10^{-11}$ s they are assumed to be present at a location selected at random from a Gaussian distribution about the original position with a diffusion length of 25 Å.

Following the above-mentioned procedures, Turner et al[24] have calculated the ferric yields in a Fricke dosimeter system for 5-keV and 1-keV electrons as well as energetic protons and $\alpha$-particles. The calculated yields agree with the data quite reasonably.

Thus, it is seen that except for the initial positions of $H_2O^*$, $H_2O^+$, and subexcitation electrons, no rigorous technique exists at present to localize the initial chemical species such as H, OH, and so forth. A large part of the present development is based on intuition. Nevertheless, the Monte Carlo technique provides us with some hope for future development in the knowledge of the initial distribution of the primary species.

# REFERENCES


1. Klein, O.; Nishina, Y. *Z. Phys.* **1929**, *52*, 853.
2. Evans, R. D. In "The Atomic Nucleus"; Schiff, L. I., Ed.; McGraw-Hill: New York, 1955.
3. Thomson, J. J. *Philos. Mag.* **1912**, *23*, 853.
4. Darwin, C. G. *Philos. Mag.* **1912**, *23*, 901.
5. Bohr, N. *Philos. Mag.* **1913**, *25*, 10.
6. Jackson, J. D. "Classical Electrodynamics", 2nd ed.; Wiley: New York, 1975.
7. Mozumder, A. In "Advances in Radiation Chemistry"; Burton, M.; and Magee J. L., Eds.; Wiley Interscience: New York, 1969.
8. Orear, J.; Rosenfeld, A. H.; Schluter, R. A. "Nuclear Physics", rev. ed.; University of Chicago Press: Chicago, 1956.
9. Henderson, G. H. *Philos. Mag.* **1922**, *44*, 680.
10. Bethe, H. *Ann. Phys.* **1930**, *5*, 325.
11. Mott, N. F., *Proc. Cambridge Philos. Soc.* **1931**, *27*, 553.
12. Fano, U. *Ann. Rev. Nucl. Sci.* **1963**, *13*, 1.
13. Walske, M. C. *Phys. Rev.* **1956**, *101*, 940.
14. Bonderup, E. *K. Dan. Vidensk. Selsk. Mat.-Fys. Medd.* **1967**, *35*, 17.
15. Bichsel, H. *Phys. Rev.* **1970**, *1*, 2854.
16. Chan, E. K. L.; Brown, R. B.; Lodhi, A. S.; Powers, D.; Matteson, S.; Eisenbarth, S. R. *Phys. Rev.* **1977**, *A16*, 1407.
17. Pierce, T. E.; Blann, M. *Phys. Rev.* **1968**, *173*, 990
18. Turner, J. E.; Magee, J. L.; Wright, H. A.; Chatterjee, A.; Hamm, R. N.; Ritchie, R. H. *Radiat. Res.* **1983**, *96*, 437.
19. Zeiss, G. D.; Meath, W. J.; MacDonald, J. C. F.; Dawson, D. J. *Radiat. Res.* **1977**, *70*, 284.
20. Zeiss, G. D.; Meath, W. J.; MacDonald, J. C. F.; Dawson, D. J. *Radiat. Res.* **1975**, *63*, 64.

21. Mozumder, A.; Magee, J. L. *Radiat. Res.* **1966**, *20*, 203.
22. Mozumder, A.; Magee, J. L. *J. Chem. Phys.* **1966**, *45*, 3332.
23. Chatterjee, A.; Schaefer, H. J. *Radiat Environ. Biophys.* **1976**, *13*, 215.
24. Turner, J. E.; Magee, J. L.; Hamm, R. N.; Chatterjee, A.; Wright, H. A.; Ritchie, R. H. In "Seventh Symposium on Microdosimetry", Oxford, U.K.; Booz, J.; Ebert, H. G.; Hartfield, H. D., Eds; Commission of European Communities: London, 1981, pp. 507–517.

# 2

# Primary Products in Radiation Chemistry

**NORMAN V. KLASSEN**

*Division of Physics,*
*National Research Council of Canada, Ottawa*

## WHAT ARE PRIMARY PRODUCTS?

In the preceding chapter we saw how energy is deposited in matter. In this chapter we shall examine the primary chemical products, the processes by which they are formed, and the reactions in which they participate. These primary products are electrons, ions, excited states, atoms, free radicals, and molecules. Any attempt to formulate an absolute definition of "primary product" quickly runs into problems because the meaning of the term for any given experiment is related to the objectives of the experimenter and the technical means at his disposal. Primary products are those that are present so early in the evolution of radiolysis that $G$(primary product)* is almost invariant over the range of conditions of the experiment. If it were possible to describe completely the evolution of an irradiated system, the "primary products" would be the first species created by energy absorption, that is, excited molecules and ions and electrons with greater than thermal energies. In the meantime, even though we know that the yield of the hydrated electron, $G(e_{aq}^-)$, is close to 4.6 several picoseconds (ps) after energy deposition, it will continue to be useful to describe its yield as 2.7 (its value at about 0.1 $\mu$s) for irradiated aqueous solutions in which dilute

* In radiation chemistry, $G$-values are defined as the number of molecules formed or lost per 100 electron volts of energy absorbed. See also Chapter 10.


*Radiation Chemistry: Principles and Applications*

solutes react with homogeneously distributed $e_{aq}^-$ with a yield of 2.7. For the same reason we shall continue to find it useful to call the hydroxyl radical a primary product in water with $G(OH) = 2.8$ even though most hydroxyl radicals are produced from another radiolytic species, the water cation, $H_2O^+$, in an almost inevitable reaction since it is believed to take place in the time required for a single vibration of a water molecule ($<10^{-13}$ s).

$$H_2O^+ + H_2O \rightarrow H_3O^+ + OH \tag{2-1}$$

Clearly, "primary product," "initial yield," or "primary process" must always be interpreted in the context in which they are found. Milton Burton[1] stated that "a process is secondary in a real and experimentally useful sense if it both follows a primary process and is subject to modification (in yield per unit time) by intrusion of a competitive process." He further stated that "if it is claimed that the secondary process is the sole and inevitable consequence of the first under any set of conditions, there would appear to be no pragmatic value in the definition."

When considering primary processes, it is useful to have an idea of the time frame within which radiolysis occurs. Radiation, be it electromagnetic like an x-ray, or a charged particle like a high-energy electron, traverses a molecular dimension of a few angstroms in $10^{-18}$–$10^{-17}$ s. Following energy deposition, the excited and charged species that are formed undergo the processes of deexcitation, thermalization, neutralization, and solvation to form well-known radiolytic species, such as the first excited singlet states and solvated electrons and cations. Some of these react within the spur; others diffuse out of the spur to become homogeneously distributed before reaction. A sampling of the time scale of representative physical and chemical events is given in Table 2-1.

**Table 2-1.** Chronology of Events during Radiolysis

| Event | Time (s) |
|---|---|
| Traversal of a molecular diameter by a high-energy electron | $10^{-18}$ |
| Ionization and excitation | $10^{-17}$–$10^{-16}$ |
| Relaxation of highly excited states | $10^{-15}$–$10^{-13}$ |
| Period of molecular vibration | $10^{-14}$ |
| Dissociation of excited molecules | $10^{-14}$ |
| Fastest ion–molecule reactions such as $H_2O^+ + H_2O \rightarrow H_3O^+ + OH$ | $10^{-14}$ |
| Thermalization of secondary electrons in water | $\leq 3 \times 10^{-13}$ |
| Solvation of electrons in water | $\leq 3 \times 10^{-13}$ |
| Thermal orientation of water molecules in liquid water | $10^{-12}$–$10^{-11}$ |
| Interval between molecular collisions in a gas at 1 atm | $10^{-10}$ |
| Fluorescence | $10^{-9}$ |
| Lifetime of the spur in water | $10^{-7}$ |
| Phosphorescence | $\sim 10^{-3}$ |

One of the earliest fascinations of radiation chemistry was the plethora of reactive species that could be created by ionizing radiation. On the one hand, this wealth of reactive species promised exciting research; on the other hand, the mixture seemed almost too rich to digest, and determining unique mechanisms an almost impossible task. Nowhere was this dilemma more apparent than in the radiolysis of hydrocarbons where the ever-increasing power of gas chromatography revealed an ever-increasing complexity of products necessitating increasingly complex reaction mechanisms. Methods other than the analysis of final products had to be developed to sort out the primary products and their reactions. The most widely used were scavengers and pulse radiolysis. *Scavenger* is the name given to a reagent that reacts rapidly with other species, in this case the primary products. For example, methyl chloride dissolved in water reacts with the hydrated electron, $e_{aq}^-$, to produce $Cl^-$. For 0.01 *M* methyl chloride $G(Cl^-) = 2.75$; that is, it is equal to the yield of $e_{aq}^-$ that escape spur recombination. For 0.1 *M* methyl chloride $G(Cl^-) = 3.1$, indicating significant scavenging of $e_{aq}^-$ that would have reacted inside the spur in pure water.[2] In this way scavengers can be used to intercept specific primary products, thereby creating a new set of products, and the primary products are then deduced from the new final products or from the final products that have been eliminated by the scavenger. In pulse radiolysis experiments a sample is irradiated with an intense pulse of radiation to produce a measurably large concentration of primary products. Ideally, the pulse is much shorter than the reaction time of the species of interest. The most widely used method of measuring species in pulse radiolysis studies is by their optical absorption, but electron spin resonance (ESR), conductivity, and luminescence are also used. An attractive feature of many pulse radiolysis experiments is the ability to measure the primary product itself instead of inferring its presence from other products. For example, the hydrated electron has a broad, intense, optical absorption band lying mostly in the visible and near-infrared.[3,4] Hundreds of rate constants of $e_{aq}^-$ reactions have been determined by measuring the increased decay rate of $e_{aq}^-$ caused by adding a reagent to water.

Over the past two decades, radiation chemistry has made great strides in elucidating primary products and measuring their yields at times closer and closer to the moment of energy deposition. The reaction rate of radiolytic species with one another in the same spur may be very rapid because of their high local concentrations. As the primary species in a spur diffuse away from one another, the probability that they will react with one another decreases, whereas the probability that they will react with homogeneously distributed solute molecules increases. If we wish to extend our examination of primary products into the inside of the spur, we must deal with reactions that do not obey homogeneous kinetics. They follow nonhomogeneous kinetics because the reactants are not randomly distributed.[5] Further details are presented in Chapter 4.

In the interests of keeping the number of references in this chapter under control, recent reviews, rather than original works are sometimes referenced. Included in the references are several published conference proceedings. These are often more difficult to obtain than regular journals but provide valuable surveys of the state-of-the-art.

## THE SOLVATED ELECTRON ($e_s^-$)

### *Liquids*

The electrons produced by the absorption of ionizing radiation are rapidly thermalized and then, in most media, localized and solvated as the dipoles of the surrounding molecules orient in response to the negative charge of the electron. The localized electron in rigid matrices is often long lived and referred to as a trapped electron. Trapped electrons may be only partially solvated if the viscosity of the matrix inhibits the movement of the solvating molecules. The solvated electron is the most studied of all primary products, and a discussion of its reactions presents a tour through much of radiation chemistry.

The solvated electron in water has a special name, the *hydrated* electron, denoted by $e_{aq}^-$ ("aq" stands for "aqueous" or "aquated"). It has a broad, intense absorption band, first discovered by Hart and Boag,[3] with a maximum at 720 nm,[4] which makes it readily measurable by pulse radiolysis. In picosecond studies using a mode-locked laser system in which electrons were produced by the photoionization at 265 nm of the ferrocyanide anion in aqueous solutions, the initially localized electron was observed at 1060 nm within 2 ps. The optical absorption evolved with time, shifting to lower wavelengths, reaching the normal absorption spectrum in 4 ps.[6] A more recent measurement with subpicosecond laser pulses indicated that the absorption of $e_{aq}^-$ at 615 nm is fully developed within the 0.3-ps time resolution of the experiment.[7] The authors argue that a solvation time of less than 1 ps is too fast for molecular orientation, so that electrons must be captured by preexisting deep traps in water. A recent report[8] puts an upper limit of $10^{-13}$ s on the appearance of $e_{aq}^-$ following photoionization of phenothiazine in a reverse micelle. The nature of $e_{aq}^-$ is still being actively debated. Early theories based on the polaron model treated $e_{aq}^-$ as an electron trapped in a potential well created by polarization of the continuous dielectric medium.[9] Fueki et al[10] proposed the semicontinuum model which treats the electron as an excess electron, in a cavity with a radius of about 0.1 nm, interacting strongly with the four to six water molecules closest to it by short-range charge-dipole attraction (both permanent and induced) and less strongly with the rest of the medium, which is treated as a continuum dielectric. The optical absorption spectrum of $e_{aq}^-$ has been interpreted as an electronic $1s \rightarrow 2p$ transition on the one hand and, on the other hand, as an electron transfer transition.[11] An examination of $e_{aq}^-$ spectra led Tuttle and Golden[12,13] to propose that the spectra at different temperatures are due to linear combinations of spectra belonging to two distinct forms of $e_{aq}^-$ and that not more than 25% of the absorption band can be ascribed to bound–bound transitions involving excited states with energies less than that of the photoejection threshold of the solvated electron. Theories of the solvated electron are examined in Chapter 8.

In the radiolysis of water the dominant species in the spur are $H_3O^+$, OH, and $e_{aq}^-$, produced by ionization and the rapid reaction of $H_2O^+$ [Reaction (2-1)]. The reactions of $e_{aq}^-$ with both $H_3O^+$ and OH have high rate constants.[14] Other species, reactive towards $e_{aq}^-$, and present in the spur, are $H_2O_2$[15] and $e_{aq}^-$ itself.

$$e_{aq}^- + H_3O^+ \rightarrow H_2O + H \qquad k = 2.3 \times 10^{10}\ M^{-1}\ s^{-1}$$

$$e_{aq}^- + OH \rightarrow OH^- \qquad k = 3 \times 10^{10}\ M^{-1}\ s^{-1}$$

$$OH + OH \rightarrow H_2O_2 \qquad k = 5 \times 10^{9}\ M^{-1}\ s^{-1}$$

$$e_{aq}^- + H_2O_2 \rightarrow OH + OH^- \qquad k = 1.2 \times 10^{10}\ M^{-1}\ s^{-1}$$

$$e_{aq}^- + e_{aq}^- \xrightarrow{H_2O} H_2 + 2\,OH^- \qquad k \simeq 5 \times 10^{9}\ M^{-1}\ s^{-1}$$

The combination of high rate constants and the close proximity of these species in the spur result in only about one-half of $e_{aq}^-$ escaping from the spur.

In 1970 Hunt and co-workers[16] created the ingenious stroboscopic method of pulse radiolysis, which allowed them to measure the yield of $e_{aq}^-$ and other solvated electrons, $e_s^-$, between 30 and 350 ps without a fast detection system. The time resolution was provided by using the Cerenkov light pulses (produced by the 10-ps-wide fine-structure pulses of high-energy electrons from the linac) as the analyzing light. They arrived at a value $G(e_{aq}^-) = 4.0$ at 30 ps. Later, Jonah et al[17] measured a somewhat higher value of $4.6 \pm 0.2$ at 100 ps using single fine-structure pulses to irradiate water. A *G*-value of $4.8 \pm 0.3$ at 30 ps for $e_{aq}^-$ has been reported.[18] A value of 4.6 is most commonly used. Little decay of $e_{aq}^-$ takes place between 30 and 100 ps.[16,19] The value of $G(e_{aq}^-)$ decreases to 2.7 at $10^{-7}$–$10^{-6}$ s. The decay of $e_{aq}^-$ before $10^{-7}$ s is relatively independent of absorbed dose except at a combination of high doses and dose rates or at high linear energy transfer (LET). For example, $G(e_{aq}^-) = 2.7$ at $10^{-7}$ s for doses from 54 to 7894 rad, but, beyond $10^{-7}$ s, the higher the dose the faster the decay.[20] An increased dose causes more decay of $e_{aq}^-$ only after $10^{-7}$ s, because at earlier times the electron is only reacting with species in the same spur. The spurs expand with time; that is, the species in the spur diffuse away from each other. Spur overlap will occur earlier at higher doses. Kinetically significant spur overlap for a pulse of 150 rad occurs after 500 ns.[20] There is a wide variation in the amount of energy, and hence the number of primary products, in a spur. Short et al[20] conclude from their computer simulation that about 30% of the energy from low-LET radiation is deposited in high-LET regions (blobs and short tracks). The more energy deposited in a spur, the more primary products in close proximity and the smaller the fraction of primary products that escape reaction in the spur. Only at very low dose rates is the reaction of $e_{aq}^-$ with water itself significant.

$$e_{aq}^- + H_2O \underset{2}{\overset{1}{\rightleftharpoons}} H + OH^- \qquad k_1 = 16 \pm 1\ M^{-1}\,s^{-1}$$ [14]

$$k_2 \simeq 1.5 \times 10^7\ M^{-1}\,s^{-1}$$ [21]

Boyd et al[22] developed a computer program to follow the nonspur reactions of all the important products in the radiolysis of water. Depending on the pH and the solute present, up to 41 reactions were used.

Fielden and Hart[23] measured $G(e_{aq}^-)$ in $D_2O$. They found $G(e_{aq}^-) = 2.90$ compared to 2.62 in $H_2O$ under the same conditions, in other words, an 11% greater yield in $D_2O$. The extinction coefficient of the absorption band at its maximum is also slightly greater in $D_2O$ than in $H_2O$. The solvation time of $e^-$ in $D_2O$ is <2 ps, just as in $H_2O$.

Alcohol molecules, being polar, solvate the electron strongly. This is reflected in the absorption spectra of $e_s^-$, with maxima in the visible for low-molecular-weight alcohols ($\lambda_{max} = 635$ nm in methanol[24]). Higher-molecular-weight alcohols do not solvate the electron as strongly ($\lambda_{max} = 1075$ nm in *t*-butyl alcohol[25]). $G(e_s^-)$ has been measured at very short times in several alcohols.[16,26] Electrons in alcohols provide a very interesting subject for picosecond pulse radiolysis because it is possible to observe the solvation process occurring. This process is evidenced by an absorption band that is strongest in the near-IR at 30 ps but that undergoes a subsequent blue shift attributed to increased solvation. The very broad absorption band at short times probably indicates a wide range of degrees of solvation. The solvation times for methanol, ethanol, l-propanol, and l-butanol at 293 K are 11, 23, 34, and 39 ps, respectively.[26] The values of $G(e_s^-)$ at 30 ps, 5 ns, and 1 $\mu$s are 3.3, 2.7, and 2.0, respectively, for methanol, and 3.2, $\simeq$3, and 1.7, respectively, for ethanol.[16]

The reactions occurring in the radiolysis of alcohols bear a striking resemblance to those in the radiolysis of water, but the reaction of $e_s^-$ with the solvent itself is much faster in alcohols.

$$e_s^- + CH_3OH \rightarrow CH_3O^- + H \qquad k \simeq 1.5 \times 10^4\ M^{-1}\,s^{-1}$$ [27]

$$e_s^- + C_2H_5OH \rightarrow C_2H_5O^- + H \qquad k \simeq 1 \times 10^4\ M^{-1}\,s^{-1}$$ [27]

In methanol, $e_s^-$ reacts primarily with $CH_3OH_2^+$ and the methoxy radical, $CH_3O$, in the spurs and primarily with the solvent after escape from the spur.[28]

$$e_s^- + CH_3OH_2^+ \rightarrow CH_3OH + H \qquad k = 4.8 \times 10^{10}\ M^{-1}\,s^{-1}$$

$$e_s^- + CH_3O \rightarrow CH_3O^-$$

The solvated electron in liquid ammonia has a secure place in the history books. The ammoniated electron, $e_{am}^-$, was discovered by Weyl in 1864 and identified by Kraus[29] in 1908 as an electron surrounded by ammonia molecules. These stable solvated electrons can be prepared by dissolving metallic sodium in liquid ammonia, producing a blue solution whose color is due to $e_{am}^-$.

$$Na \overset{NH_3}{\rightleftharpoons} Na^+ + e_{am}^-$$

The discovery of $e_{am}^-$ gave rise to a series of conferences called the Colloque Weyl (see Colloque Weyl V[30]), which provide a forum for the discussion of excess electrons. The stable $e_{am}^-$ in ammonia containing an alkali metal is at equilibrium. By contrast, $e_{am}^-$ produced in liquid ammonia by pulse radiolysis decays by reactions with other radiolysis products. Consistent with ammonia being less polar than water, $\lambda_{max}(e_{am}^-) \simeq 1800$ nm,[31,32] further into the IR than $e_{aq}^-$. $G(e_{am}^-) = 3.2$ at 5 ns.[33] A free ion yield of $3.5 \pm 1$ has been measured in $NH_3$ and $ND_3$ at 200–255 K.[32] The free ion yield of $e_{am}^-$ seems rather large for a solvent of dielectric constant 17. The reason suggested was the relatively low rate constant for the spur reaction between $e_{am}^-$ and $NH_4^+$ ($k = 1.2 \times 10^6\ M^{-1}\ s^{-1}$ at 238 K). The reaction, probably diffusion controlled, of $e_{am}^-$ with the $NH_2$ radical is the other important spur reaction of $e_{am}^-$.

The behavior of solvated electrons in liquid alkanes fits nicely into our general picture of what to expect in nonpolar, low-dielectric liquids. Weak solvation results in high mobilities and rapid and extensive geminate neutralization. The absorption maxima are well into the IR as befits weakly solvated electrons. The $\lambda_{max}$ of $e_s^-$ in methylcyclohexane at room temperature is approximately 3500 nm.[34] Sensitivity factors make conductivity more useful than optical absorption for $e_s^-$ measurements in alkanes.[35–38] Geminate recombination is rapid in liquid hydrocarbons. The geminate recombination times in cyclohexane and *n*-hexane are 2.5 ps and 15 ps, respectively.[35] The yield of free ions, that is, those that escape spur recombination, is small (less than 0.3 in *n*-alkanes from $C_4$ up) but becomes larger with branching, reaching 1.09 in neopentane.[39–41] The mobility of the electron also increases with increased hydrocarbon branching (eg, about 60 and 0.08 $cm^2\ V^{-1}\ s^{-1}$ in neopentane and *n*-hexane, respectively). The spur reactions of $e_s^-$ are dominated by recombination with positive ions. Reaction with free radicals occurs to a much lesser extent. Conductivity apparatus with a rise time of 60 ps has been used to measure excess electrons in dielectric liquids.[38]

Solvated electrons in polar, moderately polar, and nonpolar liquids follow a pattern. The initial yield of ion pairs is approximately 5. In polar liquids picosecond studies can detect most of the solvated electrons and follow their decay in the spur. In nonpolar liquids the spur reactions are faster and, in general, much lower yields of solvated electrons are detected even by picosecond techniques.

## *Solids*

The lifetime of the solvated electron can be greatly increased by increasing the viscosity of the medium. With certain liquids it is possible to form glasses. These liquids, when cooled rapidly in liquid nitrogen (77 K), do not crystallize. Instead, their viscosity rises rapidly to above $10^{12}$ poise (reached at $T_g$, the glass transition temperature), preventing the molecular movement necessary for crystallization to occur. The molecular structure of glasses is disordered, and akin to a cold liquid, rather than ordered like a crystal. In glasses the lifetime of solvated electrons is greatly increased, and a substantial yield may exist for hours. The solvation time is also greatly increased, and at 4.2 K (the temperature of liquid helium) partially solvated electrons can be studied hours after radiolysis. We shall use the term *trapped electron*, $e_t^-$, for both solvated and partially solvated electrons in solids. Hamill and co-workers,[42] who did pioneering work on trapped electrons in hydrocarbon glasses, recognized the importance of low analyzing light levels to avoid photobleaching. The long lifetime of $e_t^-$ in glasses permitted Kevan and co-workers[43] and others to study the structure of its solvation shell by ESR and related techniques.

Aqueous glasses can be made from certain concentrated solutions, 10 *M* NaOH, 9.5 *M* LiCl, 50/50 ethylene glycol/$H_2O$, to name but a few. A significant yield of $e_t^-$ is found in these glasses that may persist for days. At short times (100 ns) at 77 K the absorption band of $e_t^-$ is broader than that of $e_{aq}^-$ with more strength in the near-IR. After minutes a fairly stable spectrum evolves with a maximum in the visible.[44] The ESR spectrum of $e_t^-$ in 10 *M* NaOH has been interpreted in terms of a solvated electron in which six water molecules constitute the first solvation shell.[43] However, it must be borne in mind that these aqueous glasses are far from simply very cold water. They are very concentrated solutions, and one would expect the solutes, which are present to permit glass formation, to play some role in solvation. Another type of trapped electron exists in certain aqueous glasses[45–48] and in ice[49] with absorption maxima at 3600 and 2900 nm, respectively. This $e_t^-$ has been called $e_{ir}^-$, and the $e_t^-$ with the maximum in the visible has been called $e_{vis}^-$. In glasses it appears that $e_{vis}^-$ is solvated by water molecules strongly associated with the solute, and $e_{ir}^-$ with water molecules less affected by the solute.[46,48,50] Very substantial yields of trapped electrons can be detected in aqueous glasses by nanosecond pulse radiolysis. In ethylene glycol/water glasses at 6 K, 20 ns, a value of $G(e_{ir}^- + e_{vis}^-)$ as high as 4.2 was reported.[50] The decay of $e_t^-$ from about 100 ns onwards is approximately linear with the log of time. This has been described as a first-order decay with a time-dependent rate constant. The most popular[51,52] (but by no means the only[53,54]) interpretation of these kinetics is that $e_t^-$ decays by quantum mechanical tunneling to a reactant. In these glasses molecular diffusion is extremely slow because of the very high viscosities.

Glassy alcohols provide us with a chance to measure the solvation of excess electrons (accompanied by a narrowing and a blue shift of the optical absorption spectrum) on the nanosecond time scale.[55] The semicontinuum model has been used to correlate the spectral changes occurring in the $e_t^-$ absorption band in glassy ethanol with the progressive orientation of the polar solvent molecules surrounding the electron.[56] By using the same extinction coefficient for $e_t^-$ as for $e_s^-$ in liquid ethanol, we can estimate $G(e_t^-)$ in glassy ethanol at 76 K to be about 2.[55]

Although the reactive species in glasses may be the same as in the corresponding liquids, the immobility of the molecules and the activation energies involved may change drastically the relative importance of some reactions. A high activation energy may successfully inhibit a reaction at low temperatures. If the electron reacts by tunneling at low temperatures, its spatial relation to potential reactants is very important since the tunneling rate decreases by about an order of magnitude for each 0.2 nm increase in distance between the reactants.[57] Trapped electrons in very cold glasses are likely to undergo considerable reaction from the incompletely solvated state because the binding energy appears as a negative exponent in the tunneling expression.

In alkane glasses at early times (100 ns) $G$(trapped electrons) is often about 1.[58] In 3-methylpentane glass at 77 K, $G$(total ionization) was measured to be $5.4 \pm 0.5$ by the use of scavengers for both electrons and positive ions.[59] Because of the low dielectric constant of alkanes, the Onsager escape radius (the separation at which 37% of geminate ion pairs escape recombination with each other) is quite large—about 30 nm at room temperature and 100 nm at 77 K. At room temperature very few $e_s^-$ in liquid alkanes escape beyond the Onsager escape radius, as evidenced by the low values of $G$(free ion). At 77 K there are essentially none that escape. However, it is possible for both charged species to undergo other reactions before geminate neutralization occurs. In 3-methyloctane it has been possible to follow the decay of both the electron and the initial cation from 100 ns onwards over the temperature range from 6 to 293 K. In both glass and liquid the electron reacts to a significant extent with species other than the initial cation.[60] One possible reactant is a second cation, the product of the initial one.

The process of electron solvation is readily seen in hydrocarbon glasses. At 4 K, or at 100 ns at 76 K, $e_t^-$ has $\lambda_{max} \simeq 2000$ nm. At 76 K the spectrum of $e_t^-$ in 3-methylhexane and in 3-methylpentane glasses shifts with time to reach the stable value, $\lambda_{max} = 1600–1700$ nm, after several minutes.[61] A similar shift can also be observed on warming a glass to 77 K after irradiation at 4 K. Electron nuclear double resonance (ENDOR) experiments on the trapped electrons in 3-methylhexane glass showed that, after irradiation at 4 K, the electron–proton distance is 0.015 nm shorter than at 77 K.[62] Since H is the negative end of the C–H bond dipole, this change is in the expected direction for a higher degree of solvation at 77 K.

The measured yields of solvated electrons in crystals are often small due to the long distances that electrons in the conduction band may travel in the crystal lattice. In crystals $G(e_t^-)$ is sensitive to small amounts of impurity, to the accumulated effects of radiation damage, to crystal defects (eg, sample preparation), and to the temperature. At 6 K, 100 ns in $D_2O$ ice, $G(e_{ir}^-)$ is 0.27 for a previously unirradiated crystal but is 1.08 for a crystal previously irradiated to 300 krad.[63] The initial values of $G(e_{ir}^-)$ and $G(e_{vis}^-)$ and their decay kinetics are very dependent on the temperature of the ice. Near the melting point of ice, $G(e_{vis}^-) = 1.1$ at 269 K, 20 ns for a low dose per pulse.[64] Warman and Jonah[65] explained the results of a picosecond pulse radiolysis study of ice near its melting point by the production of two types of trapped electrons with similar absorption spectra. The absorption due to one species appears promptly (ie, 30 ps), whereas the absorption due to the other grows in over 2 ns. The latter species is responsible for the long-lived absorption observed at nanosecond and microsecond times in other studies.

Samskog, Lund, and Nilsson[66] found the same spectrum for $e_t^-$ in crystalline 1,8-octanediol at 294 K as in the liquid at 373 K ($\lambda_{max} = 625$ nm). Correcting for decay during the pulse, they calculated $G\varepsilon = 10^4$, which translates into a yield of 0.8 if we assume $\varepsilon_{max} = 1.3 \times 10^4$ as in liquid propanol.[67]

Gauthier[68] detected trapped electrons in polycrystalline 2,2-dimethylbutane, cyclohexane, methylcyclohexane, and 2,2,4-trimethylpentane but not in neopentane or tetramethylsilane. This agrees with conductivity results in liquid alkanes in which high conductivities are measured for excess electrons in neopentane and tetramethylsilane, indicating very little electron localization and, hence, rapid electron decay.

When ionic crystals, such as alkali halides, are subjected to ionizing radiation, trapped electrons, better known as F centers, and a myriad other color centers are produced.[69] A discussion of color centers in alkali halide crystals is presented in Chapter 14. Radiation workers are familiar with thermoluminescent detectors (TLDs). These are usually polycrystalline LiF pellets used as personal radiation monitors. Irradiated TLDs contain trapped electrons. When heated, recombination occurs with an accompanying emission of light, which is measured to determine the dose received by the pellet. The yield of F centers was measured with short laser pulses, at 46 ps following two-photon band-gap excitation of KCl from 12 to 880 K. The yield of F centers per ionizing event approaches unity near the melting point.[70] Illumination of F centers in their absorption band causes photoionization with a quantum efficiency of 0.2 to 0.5.[71]

Many theoretical investigations have considered the possibility that $e_{aq}^-$ might be $H_2O^-$. $H_2O^-$ has been produced in ionic crystals. KCl crystals, substitutionally doped with $OH^-$, were photolyzed at 77 K with ultraviolet (UV) light in the $OH^-$ band. Approximately 10% of the $OH^-$ decomposed to give H atoms and $O^-$. The interstitially trapped H atoms were mobilized by warming to 130 K and 25 to 50% reacted with $OH^-$ to form $H_2O^-$

centers.[72] Above 220 K the $H_2O^-$ center decomposed into an F center and a water molecule. The $H_2O^-$ center has $\lambda_{max} \simeq 550$ mm, and it has a narrower spectrum than $e_{aq}^-$.

### *Gases*

Electrons can also be solvated in the gas phase, aided by high pressures and the tendency for polar molecules in the vapor phase to cluster. Gaathon et al[73] measured the absorption spectrum of $e_{aq}^-$ in $D_2O$ vapor down to a density of 0.022 g $cm^{-3}$ at 523 K and found $\lambda_{max} \simeq 1400$ nm, similar to the values in supercritical $D_2O$.[74] A study of the mobilities of excess electrons in subcritical water vapor indicates that the transition from the quasi-free to the localized electron state occurs at about $10^{20}$ molecules per $cm^3$.[75] Recently, free negatively charged water clusters, $(H_2O)_n^-$ with $n \geq 8$, were found in irradiated water vapor.[76]

## QUASI-FREE ELECTRONS

A variety of names—quasi-free electrons, mobile electrons, dry electrons, damp electrons, and quasi-localized electrons—have been used to describe thermalized, but unsolvated, or incompletely solvated electrons. Ionizing radiation produces electrons with greater than thermal energies. Thermalization takes place by interaction with the medium and is followed in most condensed media by localization and solvation. However, there are some fluids in which electrons do not become localized and others in which both localized and nonlocalized states seem to play an important role. Hunt[16] wrote that there "are no clear demarcations between different classes of the electrons such as dry electrons, subexcitation electrons, hot or epithermal electrons, thermal electrons, mobile electrons and damp electrons." The classes sometimes overlap, and, in some cases, the specific terminology used relates to the type of experiment involved. Thermalization times depend greatly on the density and the nature of the medium; for example, energy loss by electrons to molecular compounds is much more efficient than to rare gas atoms. Sowada and Warman[77] have measured thermalization times as a function of density in argon in all three phases. Thermalization took 900 ns at $2.5 \times 10^{20}$ $cm^{-3}$ (molecules per cubic centimeter), decreasing to 10 ns at $10^{22}$ $cm^{-3}$. With a further increase in density, the thermalization time decreased even faster so that thermalization could no longer be explained by single atom scattering. These times should be compared to the less than 0.3 ps required for electron thermalization, localization, and solvation in water.

*Quasi-free electron*, $e_{qf}^-$, is a term widely used for thermalized, but not localized, electrons in fluids, notably rare gases and some alkanes in which the energy of the conduction band is below any localized state.[78] The $e_{qf}^-$ is in an extended, delocalized state with a plane wave function that extends over the entire fluid, whereas the wave function of the localized electron decreases exponentially away from one point in the fluid.[79,80] The most noticeable feature of quasi-free electrons is their extremely high mobilities, many orders of magnitude greater than those of $e_s^-$ in most liquids. Excess electrons in liquid xenon (161 K), krypton (115 K), argon (85 K), methane (111 K), and tetramethylsilane and neopentane (room temperature) have conductivities of 1900, 1800, 475, 400, 90, and 70 $cm^2\ V^{-1}\ s^{-1}$, respectively, compared to 0.09 in *n*-hexane and 0.0018 in water.[81] A popular, but not the only, model for electron transport in alkane liquids is that electrons are periodically ejected from the site of localization by thermal activation and move by band-type motion as $e_{qf}^-$ until they are retrapped. There are many mobility studies, for example, in ammonia,[82] that show the change from a very mobile excess electron to a localized one with increase in density.

The foregoing raises some questions for radiation chemists. Does $e_{qf}^-$ undergo the same sorts of reactions as $e_s^-$? Does it react faster because of its higher mobility? Does the fact that a fluid supports $e_{qf}^-$ rather than $e_s^-$ have an effect on *G*(free ion) and on the radiation products and their yields?

Huang and Freeman[83] and others[84] have determined the total ionization yield in liquid argon, krypton, and xenon by collecting charge in an electric field and extrapolating the yield to infinite field strength. The results, given in Table 2-2, show that *G*(total ion pairs) is as high as 6.6 in liquid xenon and that it is 20 to 50% higher in the liquids than in the corresponding low-density gases (a similar situation exists in water, in which *G*(total ion pairs) is 3.3 in the vapor[85] and 4.6 at 100 ps in the liquid[17]). Huang and Freeman[83] suggested that the reason for the higher yield in liquified rare gases is that the liquid has a smaller energy band gap between the top of the valence band and the bottom of the conduction band. They also suggested the reaction of higher excited states

$$M^* + M \rightarrow M_2 + e^-$$

as a possible additional source of ionization.

**Table 2-2.** Ionization Yields in Liquid and Gaseous Ar, Kr, and Xe[a]

| Medium | Temperature (K) | *G* (total ionization) |
|---|---|---|
| Liquid argon | 87 | 4.5 |
| Liquid krypton | 129 | 6.0 |
| Liquid xenon | 164 | 6.6 |
| Gaseous argon | room temp. | 3.8 |
| Gaseous krypton | room temp. | 4.1 |
| Gaseous xenon | room temp. | 4.5 |

[a] Taken from refs 83 and 84.

The high mobilities of $e_{qf}^-$ is translated into high diffusion coefficients by the Nernst–Einstein relation. This can lead to exceptionally high reaction rate constants for $e_{qf}^-$ as compared to $e_s^-$. The rate constant at 294 K for $e_{qf}^- + SF_6$ in tetramethylsilane is $2.1 \times 10^{14}\ M^{-1}\ s^{-1}$ [81] compared to $1.65 \times 10^{10}$ for $e_{aq}^- + SF_6$.[14] Clearly, quasi-free electrons are a chemically important primary product in systems in which localization is slow. An important exception is the low reaction rate of the precursor of $e_{aq}^-$ with $H_3O_{aq}^+$.[16] However, the reactions of $e_{qf}^-$ probably produce the same products as those of $e_s^-$ in most cases.

In a similar vein, incompletely solvated electrons will play a more important role in the overall reaction scheme if the solvation time is long, such as in some low-temperature glasses. In glassy ethanol at 77 K the reaction with weak electron acceptors such as acetone is preferentially with weakly solvated electrons.[86] These reactions proceed by tunneling rather than by diffusion. Since the tunneling rate depends strongly on the trap depth, incompletely solvated electrons will tunnel faster than completely solvated ones.

In 1969 Hamill[87] proposed a model for the radiolysis of water in which thermal electrons undergo many collisions with molecules before localization. Hamill concluded that there was strong evidence that these mobile "dry" electrons, $e^-$, react with concentrated electron scavengers, with the exception of hydrogen ions, $H_3O^+$. Results such as the lack of a pH effect on the yield of molecular $H_2$ in water argued against the reaction of $e^-$ with $H_3O^+$. These conclusions were verified by Wolff et al.[88] They found that electron scavengers such as acetone, $H_2O_2$, $Cd^{++}$, and $NO_3^-$ in high concentration not only react rapidly with $e_{aq}^-$ but also reduce the initial $G(e_{aq}^-)$ as measured at 24 ps. By contrast, $H_3O^+$, in the form of perchloric acid, reacts rapidly with $e_{aq}^-$ but does not reduce the initial value of $G(e_{aq}^-)$. They concluded that electron scavengers at high concentration, with the exception of $H_3O^+$, react with a precursor to $e_{aq}^-$, which seemed more likely to be the dry electron than excited water, $H_2O^*$.[16] Further work showed that the product of the reaction of the scavenger with an electron, for example $Cd^+$ from $Cd^{++}$, was observed immediately (ie, 24 ps) and that there existed a fairly close correlation between the "immediate" production of $Cd^+$ and the loss of "immediate" $e_{aq}^-$.[16] Jonah et al[17] estimated $G$(dry electron) = 5.4.

The presence of high concentrations of scavengers, S, reduces the picosecond value of $G(e_s^-)$ in water and alcohols.[16] Except for $H_3O^+$, it was found that in a given solvent $k(e_s^- + S) \times C_{37}$ = constant, the constants being $1.0 \times 10^{10}\ s^{-1}$ in water, $5 \times 10^9\ s^{-1}$ in ethylene glycol, and $1.3 \times 10^9\ s^{-1}$ in ethanol.[89] The value of $k(e_s^- + S)$ used was the rate constant measured at high concentration of S, and $C_{37}$ is the concentration of S required to reduce $G(e_s^-)$ to 37% (1/e) of its yield with no scavenger present. Hence, $k(e_s^- + S)$ is a quantity related to the reactivity of $e_s^-$, whereas $1/C_{37}$ relates to the reactivity of its precursor. The constancy of $k(e_s^- + S) \times C_{37}$ indicates that

$k(e^- + S)$ is proportional to the $k(e_s^- + S)$ measured in concentrated solutions. The picosecond results were discussed in terms of several models of fast electron processes, time-dependent rate constants, formation of encounter pairs between $e_s^-$ and S, quantum mechanical tunneling, and a reactive precursor to $e_s^-$. Interpreting their results in terms of a reactive precursor, Lam and Hunt[89] concluded that the precursor undergoes the same reactions as $e_s^-$, except with $H_3O^+$, but that it reacts at least 10 times faster, presumably due to rapid diffusion. In alcohols, but not in water, they found a much more rapid decay between 24 to 90 ps in the absorption spectrum of the electron at 1000 nm than at $\lambda_{max}$ (in the visible). This indicated the presence of partially solvated electrons (sometimes called "damp" electrons), which diffuse and react faster than $e_s^-$. Hence, it is necessary to consider all possibilities—"dry," "damp," and "wet" electrons in the picosecond time domain in alcohols.

The time scale of electron solvation in polar liquids by bulk polarization, induced by the localized electronic charge, is given by the parameter $\tau_r \varepsilon_\infty / \varepsilon_0$, where $\tau_r$ is the dielectric relaxation time of the medium, and $\varepsilon_\infty$ and $\varepsilon_0$ are the high- and low-frequency limits of the dielectric constant, respectively.[65] Solvation by bulk polarization is a possible explanation even in water, where the upper limit of the electron solvation time is 0.3 ps,[7] because this is close to the 0.5 ps given by $\tau_r = 10$ ps, $\varepsilon_\infty = 4$, and $\varepsilon_0 = 80$. Of course, this does not exclude the possibility that solvation actually occurs by orientation of the first-shell water molecules in a process involving proton jumps.[65] The half-time of localization of electrons in ice near its melting point has been estimated to be 80 ps.[90] Based on the trapped electron absorption that grows in over 2 ns in ice at 268 K, Warman and Jonah[65] estimate the half-time for electron solvation to be 400 ps. This is too rapid to be explained by bulk polarization. Therefore, electron solvation in ice is suggested to occur by a reorientation of the first-shell water molecules at a rate controlled by the frequency of proton jumps. Only a small decay of $e_s^-$ was observed between 2 ns and 4 ns.[65] This, coupled with $G(e_{vis}^-) = 1.1$ in ice at 269 K, 20 ns,[64] and a likely value of 4 to 5 for $G$(ionization), suggests that many excess electrons in ice react before solvation is complete. This is quite unlike water in which $G(e_{aq}^-) = 4.6$ at 100 ps, and high solute concentrations are required to bring reactions of the precursor of $e_{aq}^-$ into play.

## CATIONS AND ANIONS

Both primary cations and electrons are produced by ionizing radiation. Most primary cations are polyatomic and, as a result, their chemistry is more complicated than that of electrons. In the gas phase the slowness of recombination permits a wide variety of other cation reactions. Even in systems such as liquid alkanes, where cations and electrons undergo very rapid geminate recombination, other reactions of primary excited cations take place. Primary positive ions are, in general, ephemeral, often more difficult to observe by their optical absorption bands than the associated, solvated electrons. The complexity possible in ion–molecule reactions is indicated by the 76 reactions listed by Willis[91] to explain the ionic mechanism of the radiolysis of $CO_2$. Primary cations are usually quite reactive, and the effective primary cation in condensed media is often not simply the parent molecule minus an electron. Less commonly, electrons react very rapidly with the solvent to produce an anion that becomes the effective primary product.

A resurgence of interest in the role of ionic species in radiation chemistry took place in the 1950s. This resurgence was largely due to the rapid progress being made in understanding ion–molecule reactions through mass spectrometry studies. The development of high-pressure mass spectrometry and the studies of the equilibria in ion clustering were substantially motivated by the needs of radiation chemistry.[92] Since the 1950s, mass spectrometry has seen many innovations such as ion cyclotron resonance (ICR) and the selected ion flow tube (SIFT),[93] which permit the reactions of a specific ion to be clearly distinguished from those of other ions. The vacuum-UV sources needed for photoionization were developed. As well as elucidating the pathway of many ion–molecule reactions, mass spectrometry provides the thermochemical data often relied upon when postulating reaction mechanisms in radiation chemistry. An article by Wolkoff and Holmes[94] on the fragmentation of alkane molecular ions describes clearly the basis for the thermochemical assignments of species in mass spectrometry. Several recent reviews of ion chemistry are available.[95,96,97]

## *Excited Cations, Fragmentation, and Isomerization*

The cations initially produced by radiolysis are usually excited, electronically and vibrationally. Thermalization of vibrationally excited ions may occur in as few as 10 collisions but often requires 200 collisions or more, depending on the molecules involved.[98,99] Exothermic and thermoneutral ion–molecule reactions are usually fast processes with low activation energies. If the neutral molecule is polar, both the collision cross section and the likelihood of a reaction resulting from a collision are enhanced. Polarization attraction causes the colliding particles to gain considerable kinetic energy at small distances. Polarization attraction may also result in a fairly long-lived ionic complex in which exchange between the system's degrees of freedom is possible.[98]

The lifetimes of electronically excited cations can be quite appreciable in the gas phase. Many of the lifetimes of excited organic cations measured by Maier[100] by emission following electron impact excitation were in the 20–60-ns range. Fragmentation of excited molecular cations will occur to a larger extent in the gas phase than in the condensed phase because of the longer time between potentially deexciting collisions in the gas phase. On the other hand, bimolecular reactions of excited cations will be favored by the higher collision frequency in the liquids.

Bone et al[101] determined the fragmentation pattern for the parent cation in the radiolysis of propane gas. Koob and Kevan[102,103] measured the fragmentation in both the gas at 35°C, 1 atmosphere (atm), and in the liquid at 35°C. Wada et al[104] determined the ionic yields in liquid propane at 0°C and combined their results with those of Koob and Kevan and those of Bone, Sieck, and Futrell to produce the picture given in Table 2-3 of the

**Table 2-3.** The Fate of Primary Excited Species in the Radiolysis of Propane[a]

| | | | G | |
|---|---|---|---|---|
| | | | Liquid (35°C) | Gas (35°C) |
| $C_3H_8^{+*} \rightarrow$ | $C_2H_3^+ + CH_3 + H_2$ | | 0.13 | 0.27 |
| | $C_2H_4^+ + CH_4$ | | 0.15 | 0.62 |
| | $C_2H_5^+ + CH_3$ | | 0.3 | 1.30 |
| | $C_3H_5^+ + H_2 + H$ | | 0.13 | 0.25 |
| | $C_3H_7^+ + H$ | | 0.7 | 0.37 |
| | $C_3H_8^+$ | | 2.5 | 0.94 |
| | | Total positive ions | 3.9 | 3.8 |
| $C_3H_8^* \rightarrow$ | $C_2H_6 + CH_2$ | | | 0.1 |
| | $C_2H_4 + CH_4$ | | | 0.2 |
| | $C_2H_4 + CH_3 + H$ | | | 0.4 |
| | $C_3H_6 + H_2$ | | | 0.5 |

[a] Compiled from refs 101–104.

fate of excited propane ions in radiolysis. The fragmentation pattern differs in the two phases, and fragmentation is clearly less important in the liquid phase where deexcitation is more rapid. The breakdown diagram of excited propane cations has been studied by photoion–photoelectron coincidence (PIPECO)[105,106] and reviewed by Baer.[107] PIPECO shows that $C_3H_8^{+*}$ ions containing less than 1 eV excess internal energy fragment to give primarily $C_3H_7^+ + H$. At about 1 eV, $C_2H_4^+ + CH_4$ is the preferred process, and between 1 and 4 eV the dominant process is $C_2H_5^+ + CH_3$. At 3 eV the processes giving $C_3H_5^+ + H_2 + H$ and $C_2H_3^+ + CH_3 + H_2$ begin to show up. Another example of substantial radiolytic decomposition attributed to ion fragmentation was described by Wada et al.[104] From the final product analysis of irradiated liquid cyclohexane at room temperature, they estimated $G(c\text{-}C_6H_{12}^{+*} \rightarrow C_6H_{11}^+ + H)$ to be 0.9 and $G$(deexcitation to $c\text{-}C_6H_{12}^+$) to be 3.3.

If the parent cation does not represent the most stable structure, isomerization or rearrangement may occur. For example, the radical cations $C_3H_7^+$ and $C_4H_9^+$, given sufficient energy and time, rearrange to their most stable forms, the isopropyl cation, $(CH_3)_2CH^+$, and the $t$-butyl cation, $(CH_3)_3C^+$, respectively.[108,109] The higher the collision frequency the less isomerization that takes place, so that isomerization is usually less important in radiolysis than at the low pressures in most mass spectrometers. The fate of individual atoms in a molecule can be monitored by isotopic labeling. Although our examples have been drawn from hydrocarbon systems, fragmentation, and isomerization are quite general reactions for excited cations.

## *Charge Transfer*

Charge transfer is an obvious reaction of cations. If the parent molecule of the primary cation has a higher ionization potential than that of another molecule in the mixture, the charge transfer reaction is exothermic. For example,

$$Ar^+ + H_2 \rightarrow Ar + H_2^+$$ [110]

$$CO^+ + H_2O \rightarrow CO + H_2O^+$$ [93]

The net result is the transfer of an electron. Charge transfer may also take place between an excited ion, $A^{+*}$, and A,

$$A^{+*} + A \rightarrow A + A^+$$

where $A^{+*}$ can be electronically or vibrationally excited. Resonance charge transfer, a thermoneutral process, between A and ground-state unexcited $A^+$, has also been postulated. If the charge exchange reaction is more than 0.5 eV exothermic, the reaction in the gas phase usually proceeds with a collision efficiency of close to unity. It appears that in some cases a simple electron transfer suffices to transfer the positive charge, whereas in other cases a complex of the two reactants precedes the reaction. The dissociative charge transfer reaction,

$$He^+ + O_2 \rightarrow He + O^+ + O$$

is thought to occur through an excited state of $O_2^+$, which predissociates to $O^+ + O$ in a time comparable to, or less than, the vibration period.[110]

In most liquids the molecular cation, $M^+$, exhibits the mobility of a normal ion ($10^{-4}$–$10^{-3}$ cm$^2$ V$^{-1}$ s$^{-1}$). However, in cyclohexane and *trans*-decalin (*trans*-decahydronaphthalene) anomalously high mobilities of about $10^{-2}$ cm$^2$ V$^{-1}$ s$^{-1}$ were found.[35,111] In these solvents the rate constants for charge exchange to solutes such as pyrene are more than 10-fold greater than those expected for diffusion-controlled (the liquid equivalent of unit collision efficiency) reactions. In *trans*-decalin below 10°C the mobility of $M^+$ is greater than that of $e_s^-$. It was suggested that the rapid charge transfer is a resonant charge transfer process, facilitated by molecular cations that have very similar nuclear geometries to those of the neutral molecules. In ion cyclotron resonance studies the probability of a collision resulting in charge transfer from *cyclo*-$C_6D_{12}^+$ to *cyclo*-$C_6H_{12}$ is almost five times greater than the probability of charge transfer from $n$-$C_6D_{14}^+$ to $n$-$C_6H_{14}$, demonstrating that the high rate of charge transfer by the cyclohexane ion exists in the gas phase as well as in the liquid phase.[112,113] Alternative explanations to resonance charge transfer have been suggested to explain the unusually high mobility in cyclohexane. One possibility is that the lower mobilities measured in other alkanes are actually those of secondary cations.[60] Another possibility is that hydride transfer or proton transfer, involving $c$-$C_6H_{11}^+$ or $c$-$C_6H_{13}^+$, respectively, is responsible for the high mobility.[114]

Charge transfer from lower-molecular-weight alkanes, with higher ionization potential, to higher-molecular-weight alkanes, with lower ionization potential, takes place in irradiated mixed alkane glasses. Louwrier and Hamill[115] demonstrated this for $\gamma$-irradiated 3-methylpentane glasses at 77 K containing higher-molecular-weight alkanes and electron scavengers (to reduce the recombination of electrons with cations). The kinetics of charge transfer between the 3-methyloctane cation and squalane ($C_{30}H_{62}$) were examined by pulse radiolysis and found to occur efficiently even at 6 K.[116] In glasses it was found that the kinetics resembled those expected for tunneling. However, even though the net result of the charge transfer is the loss of an electron by a neutral molecule and its gain by the cation, the reaction kinetics conform to a tunneling barrier of only a few tenths of an electron volt, a far cry from the energy required to ionize the neutral molecule (~8 eV). For this reason it is not believed that the mechanism is electron tunneling. Rather, it seems that some form of hole tunneling, either direct or through a set of positive ion states of intervening molecules, is actually involved.

Any real understanding of the chemistry of the initial cations must include a knowledge of their structures and the location of the positive charge. Cations trapped in glasses have been examined by ESR to determine their structures. An important question is whether the positive charge is completely localized at a bond or whether it is delocalized over the entire molecule. The unpaired electron in linear alkane cations is delocalized over the in-plane C—H and C—C $\sigma$ bonds, whereas in highly branched alkane cations it is more confined to one of the C—C bonds.[117]

In the gas phase $H_2O^+$ can undergo a variety of reactions, proton transfer, H-atom transfer, and charge exchange.[118] In water the effective primary positive ion is usually $H_3O^+$, the result of an H-atom transfer from $H_2O$ to $H_2O^+$, which is believed to take place in less than $10^{-14}$ s. In aqueous solutions only a high concentration of another reactant can compete with $H_2O$ as a reaction partner for $H_2O^+$. Hunt and co-workers[16] investigated concentrated aqueous solutions of $Cl^-$, $Br^-$, and $CNS^-$ by picosecond pulse radiolysis to try to distinguish between the oxidizing reactions of $H_2O^+$ and OH but, partly because of the similarity of the optical absorption spectra of the products—for example, $Br_2^-$ and $BrOH^-$—they were unable to unmistakably determine the relative importance of $H_2O^+$ and OH.

$$H_2O^+ + Br^- \rightarrow H_2O + Br$$

$$Br + Br^- \rightarrow Br_2^-$$

$$OH + Br^- \rightarrow BrOH^-$$

Belevskii et al[119] investigated the identity of the primary oxidizing species in aqueous glasses containing concentrated scavengers by measuring, by ESR, the concentration of both the OH radical and the product of oxidation, both of which are stable at 77 K. A mechanism of rapid migration of $H_2O^+$, or the positive charge, through the matrix was suggested to explain the high rate constants for these scavengers with $H_2O^+$.

In $D_2O$ ice at 6 K, there is evidence for rapid and extensive charge transport and for reaction of $D_2O^+$ with radiolytically produced D atoms and OD radicals.[63] Warman et al[90] found evidence for a hole with a mobility of about $1 \times 10^{-6}$ $m^2$ $V^{-1}$ $s^{-1}$ and a lifetime of 50 to 100 ns in $H_2O$ ice above −40°C.

## *$H^+$, H, $H^-$, $H_2$, and $H_2^-$ Transfer*

Primary cations containing H atoms may undergo a proton transfer reaction in the presence of molecules with a sufficiently high proton affinity.[120] For example, an alcohol present in an alkane reacts with the primary alkane cation, $RH^+$, by proton transfer.

$$RH^+ + R'OH \rightarrow R + R'OH_2^+$$

Proton transfer reactions between primary cations and the parent molecule are common, especially in hydrogen-bonded systems. Proton and hydrogen atom transfers along H-bonds belong to the fastest chemical reactions because of the small moving mass and the short distance along the reaction coordinate. The primary cation in irradiated water is $H_2O^+$, but the transfer of a hydrogen atom from water to produce $H_3O^+$ and a hydroxyl radical [Reaction (2-1)] has been estimated to take place in $5 \times 10^{-15}$ s.[121] Solvation of $H_3O^+$ follows in less than $10^{-11}$ s.[16] The similar, rapid reactions in alcohols,[122] for example

$$CH_3OH^+ + CH_3OH \rightarrow CH_3OH_2^+ + CH_2OH$$

$$CH_3OH^+ + CH_3OH \rightarrow CH_3OH_2^+ + CH_3O$$

are often described as a proton transfer. The relative proportions of self-protonation and hydrogen atom transfer have not been determined in the liquid phase, but, in the gas phase, the value of proton transfer to hydrogen atom transfer in $H_2O$, $NH_3$, and $H_2S$ is 3.2, 3.6, and 1.1, respectively.[123] The best-known proton transfer in aqueous solutions is the acid–base equilibrium.

$$H_3O^+ + OH^- \overset{1}{\rightleftharpoons} 2\,H_2O$$

The forward rate constant of $1.4 \times 10^{11}\ M^{-1}\ s^{-1}$ is the highest known in aqueous solutions.

The transfer of protons, hydrogen atoms, hydride ions ($H^-$), hydrogen, and hydrogen molecule anions ($H_2^-$) in ion–molecule reactions in gaseous and condensed phases has been well studied in hydrocarbon systems, where it accounts for a significant fraction of the reactions of cations.

$$CH_4^+ + CH_4 \rightarrow CH_3 + CH_5^+ \qquad (H^+ \text{ transfer})$$

$$C_5H_{12}^+ + C_3H_6 \rightarrow C_5H_{11}^+ + C_3H_7 \qquad (H \text{ transfer})$$

$$C_5H_{12}^+ + C_3H_6 \rightarrow C_5H_{10}^+ + C_3H_8 \qquad (H_2 \text{ transfer})$$

$$C_2H_4^+ + C_2H_6 \rightarrow C_2H_5 + C_2H_5^+ \qquad (H^- \text{ transfer})$$

$$C_3D_6^+ + C_4H_{10} \rightarrow C_3D_6H_2 + C_4H_8^+ \qquad (H_2^- \text{ transfer})$$

A general scheme leading to a transfer reaction between $A^+$ and B as described by Meot-Ner[124] is the formation of a loose complex $(A^+ \cdot B)^*_{\text{loose}}$, held together by ion-induced dipole forces, when the relative translational energy of the collision is distributed into the internal degrees of freedom of the complex. Multiple collisions of $A^+$ and B occur in this loosely bound complex, and a collision at a favorable orientation may lead to a transfer reaction, or $(A^+ \cdot B)^*_{\text{loose}}$ may become a tightly bound (hydrogen or electronically bonded) complex $(A^+ \cdot B)^*_{\text{tight}}$ that can also lead to a transfer reaction. However, proton transfer reactions seem to occur preferentially from loose complexes, whereas $H^-$ transfers occur from tight complexes.

### *Dimer Cations and Condensation Reactions*

In some systems the ion–molecule complexes become so long lived as to be considered a viable chemical species. Primary cations in quite diverse systems are involved in such complexes. For example, argon ions combine with argon to form $Ar_2^+$. Olefins and aromatics also form dimer cations. Badger et al[125] measured absorption bands at 17,500 $cm^{-1}$ and 9800 $cm^{-1}$ due to the trapped naphthalene dimer cation produced in a glassy 1 : 1 mixture of isopentane and *n*-butylchloride containing $2 \times 10^{-2}\ M$ naphthalene irradiated at 77 K and warmed slightly to allow naphthalene cations and naphthalene molecules to diffuse together.

Condensation reactions of monoolefins lead to higher molecular weight monoolefins. Holroyd and Fessenden[126] irradiated liquid ethylene and ethylene-$d_4$ mixtures. The high yields of $C_4H_8$, $C_4H_4D_4$, and $C_4D_8$ indicated the formation of 1-butene by a condensation reaction. Likewise, the isotopic composition of other products, cyclobutane and *trans*-2-hexene, showed that they were formed by molecular dimerization and trimerization reactions, respectively.

## *Solvation and Clustering*

The concept of ion solvation in solutions is a common one in chemistry. We have already seen how the electron is stabilized by solvation in most condensed media and even in some gases. Cations, like other ions, are solvated in solution. The lifetime of the cation is increased by the shielding provided by its solvation shell unless the cation can react with the solvating molecules themselves. The diffusion of a cation is restricted by its solvation shell. The solvation of cations and anions also occurs in the gas phase. This phenomenon, generally referred to as clustering, has been extensively studied by Kebarle and co-workers[127] by measuring the equilibrium of reactions such as

$$H_3O^+(H_2O)_{n-1} + H_2O \rightleftharpoons H_3O^+(H_2O)_n$$

These measurements grew out of studies undertaken to understand ionic reaction mechanisms in the radiation chemistry of gases. The heat of reaction for the addition of the first, second, and third $H_2O$ molecules to $H_3O^+$ is $-36$, $-22.3$, and $-17$ kcal mole$^{-1}$, respectively;[128] that is, the exothermicity of the reaction decreases with the increase in the number of water molecules already present in the cluster. Conway[129] concluded that the most likely structure in liquid water is $H_9O_4{}^+$—that is, a proton hydrated by four water molecules or $H_3O^+$ hydrated by three water molecules. It is believed that rapid exchange of the proton can occur between $H_3O^+$ and the three surrounding $H_2O$ molecules, giving a kind of delocalization of $H^+$. A fourth water molecule is believed to be electrostatically coordinated to the central $H_3O^+$ but cannot undergo proton exchange with it except by reorientation.

## *Anions*

In some systems anions rather than electrons are considered to be the effective negatively charged primary species. This usually implies that the electron reacts very rapidly with the solvent, the reactions being electron attachment and dissociative electron attachment. Electron attachment produces a radical anion, that is, an anion containing an unpaired electron. Dissociative electron attachment usually produces a free radical plus a nonradical anion such as $Cl^-$. Obviously, compounds used as solutes to scavenge electrons will, when irradiated neat, be expected to produce primary anions.

Molecules containing halogens often capture electrons efficiently to produce the halide ion. Klots[130] noted that electrons react with $CCl_4$ in the gas phase to produce $Cl^-$,

$$e^- + CCl_4 \rightarrow CCl_3 + Cl^-$$

but if the $CCl_4$ is clustered with $CO_2$, sufficient collisional stabilization occurs to permit $CCl_4^-$ to be produced:

$$e^- + CCl_4(CO_2) \rightarrow CCl_4^- + CO_2$$

Shida and Hamill[131] found that pure acetone, $\gamma$-irradiated as a semicrystalline solid at 77 K, exhibited a $CH_3COCH_3^+$ band at 740 nm and a $CH_3COCH_3^-$ band at 460 nm. Aromatic molecules have a strong tendency to form both negative and positive ions. The dissociative attachment of electrons to benzyl chloride is an efficient process yielding the benzyl radical and the chloride ion. In glasses containing benzyl chloride, $\gamma$-irradiated at 77 K, the absorption band due to the benzyl radical is found. However, when $\gamma$-irradiated at 4 K, a somewhat different spectrum is found, which is probably due to a benzyl radical and a chloride ion trapped very close to one other.[132]

## EXCITED STATES

A clear indication that excited states are important primary products in radiolysis comes from the fact that $W$, the mean energy expended in a gas per ion pair formed, is much greater than the lowest ionization potential of the molecules concerned. The symbol $W$ was introduced in 1929 by Gray[133] in an important pioneering paper. The mean energy $W = E/N$, where $N$ is the mean number of ion pairs formed when the initial kinetic energy, $E$, of a charged particle is completely dissipated in the gas. The technique of measuring $W$ by charge collection is very well developed because of its extensive use in dosimetry.[85,134–136] If $W$ is known, it is a simple matter to calculate the primary yield of ion pairs in the gas phase.

$$G(\text{primary ion pairs}) = \frac{100}{W}$$

Values of $W$ for several pure gases and air at ambient temperature and pressure are shown in Table 2-4, as well as the lowest ionization potentials. Because the lowest ionization potentials are roughly half the values of $W$, about one-half of the absorbed energy results in translational, rotational, vibrational, and electronic excitations of neutral and charged species. Of these, electronically excited species are the most important precursors of radiation products.

Electronic excitation of a molecule involves the promotion of one or more electrons to higher energy orbitals.[137] Each electron in a molecule is characterized by a spin quantum number, $s$, that has the value 1/2. If any two electrons are compared, their spins may be parallel or antiparallel. According to the Pauli exclusion principle, each orbital can contain a maximum of two electrons—and, even then, only if their spins are antiparallel (called *paired* electrons). The total spin quantum number, $S$, of the molecule is the

**Table 2-4.** Ionization and Excitation in Some Common Gases

| Gas | IP[a] (eV) | W[b] (eV) | Distribution of 100 eV[c]: Ionization (eV) | Neutral excitation (eV) | Sub-excitation (eV) | Energy balance (%) |
|---|---|---|---|---|---|---|
| $O_2$ | 12.1 | 30.8 ± 0.4 | 61.3 | 20.2 | 7.8 | 89.3 |
| $N_2$ | 15.6 | 34.8 ± 0.2 | 56.6 | 39.2 | 6.9 | 105.5 |
| $N_2O$ | 12.7 | 32.6 ± 0.3 | 54.5 | 40.8 | 4.4 | 99.7 |
| CO | 14.0 | | 58.5 | 15.4 | 11.7 | 85.6 |
| $CO_2$ | 13.8 | 33.0 ± 0.7 | 53.4 | 41.4 | 6.6 | 101.4 |
| $H_2S$ | 10.5 | | 59.7 | 26.6 | 8.0 | 94.3 |
| $H_2O$ | 12.6 | 29.6 ± 0.3 | 56.8 | 35.8 | 6.6 | 99.2 |
| $NH_3$ | 10.2 | 26.6 ± 0.1 | 56.4 | 37.3 | 7.6 | 101.3 |
| $C_2H_6$ | 11.7 | 25.0 ± 0.6 | | | | |
| $C_2H_5OH$ | 10.5 | 24.8 ± 0.3 | | | | |
| Air | | $33.8_5$ ± 0.15 | | | | |

[a] The lowest ionization potential, ref. 135.
[b] Ref, 85.
[c] Ref. 136.

absolute value of the sum of the separate spins. For example, for two electrons $S = s_1 + s_2 = 1$ (parallel spins) or $S = s_1 - s_2 = 0$ (antiparallel spins). The multiplicity of a state is defined as $2S + 1$. If all the electrons in a molecule are paired, $S = 0$ and $2S + 1 = 1$, hence the name *singlet* state. The vast majority of molecules in their most stable state, called their *ground* state, are in the lowest singlet state, $S_0$. If an electron is promoted from the highest occupied orbital in the ground state to the lowest unoccupied orbital and retains a spin that is antiparallel to the electron remaining in the orbital, $2S + 1$ remains equal to 1 and the molecule is in the first excited singlet state, $S_1$. If the promoted electron attains a spin parallel to the remaining electron, $2S + 1 = 3$, hence the designation first excited *triplet* state, $T_1$. The best-known molecule with a triplet ground state, $T_0$, is oxygen. Because of its unpaired electrons, it is very reactive towards addition to free radicals that, by definition, contain an unpaired electron. Nitric oxide is the simplest stable molecule with an odd number of electrons. Like free radicals with a single unpaired electron, its ground state is a doublet, since $2S + 1 = 2$.

For molecules composed of atoms of low atomic number ($Z \leqslant 10$), the absorption of energy from a photon ($\gamma$-ray) or from the electric field of a fast-moving charged particle will most often produce transitions in which the total spin of the system does not change: $\Delta S = 0$.[138] Therefore, the promotion of an electron in an $S_0$ molecule by a $\gamma$-ray will usually produce an excited singlet state. However, when the energy is imparted by an electron, the total spin of an $S_0$ molecule plus an electron equals 1/2 so that both singlet and triplet excited states of the molecule can fulfill the condition $\Delta S = 0$ for the molecule and departing electron. For most molecular systems the cross section for triplet production is rather small for electron energies above 50 eV, so that excitation to singlet states is favored. In addition to a preference for transitions in which $\Delta S = 0$, there is a tendency for "vertical" transitions. According to the Franck–Condon principle, the relative positions and velocities of the nuclei will be the same before and after the absorption or emission of radiant energy.[137] If the nuclear coordinates of the lowest vibrational level of the excited state are not the same as those of the equilibrium ground state, a vertical transition from the ground state will result in a vibrationally as well as electronically excited state.

Deexcitation of an electronically excited state by a radiative transition (light emission) is called fluorescence for $\Delta S = 0$ and phosphorescence for $\Delta S \neq 0$. The lifetime for fluorescence, $S_1 \rightarrow S_0 + h\nu$, is typically $10^{-9}$ s, whereas the lifetime for phosphorescence, $T_1 \rightarrow S_0 + h\nu$, is typically $10^{-3}$ s but may be as long as seconds, especially in very viscous or solid media because the lifetime of triplet excited states is influenced by their environment. The longer lifetime of triplet excited states means that they have more time to take part in chemical reactions than do their excited singlet counterparts. The nonradiative relaxation of excited states in which $\Delta S = 0$ is called *internal conversion.* Internal conversion may occur by a transition to a vibrationally excited state, $S_1 \rightarrow S_0^v$. Internal conversion from higher singlet states to the $S_1$ state is usually sufficiently rapid ($\ll 10^{-12}$ s) that reactions of the higher states with solutes need not be considered. Often the transition $S_1 \rightarrow S_0$ requires about $10^{-9}$ s due to the substantial energy difference between $S_1$ and $S_0$. Under such circumstances, *intersystem crossing,* $S_1 \rightarrow T_1$, may occur at a comparable rate to internal conversion.[138] Electronically excited states with sufficient energy to break a bond may dissociate. This is especially likely for *repulsive* states, which have no stable nuclear configuration, and for states whose equilibrium geometry differs so greatly from the ground state that excitation to them leaves the nuclei with sufficient potential energy to dissociate.[138]

If a molecule absorbs sufficient energy to promote an electron to the free state, ionization may take place immediately or after a delay of $10^{-14}$ to $10^{-12}$ s, called autoionization.[138] If the delay in autoionization is sufficiently long, the molecule may dissociate into neutral fragments. If the promoted electron has sufficient energy to autoionize but does not, the molecule is said to be in a *superexcited* state. Given sufficient energy, ionization may produce a free electron with above-thermal kinetic energy and/or a vibrationally or electronically excited cation. We have already seen the considerable extent to which excited propane ions dissociate in both the gas and liquid phases. Multiply charged ions in the condensed phase rapidly become singly charged by charge transfer with neutral molecules because the second ionization potential is much higher than the first.

The ratio of excitations to ionizations formed in a gas by the action of a swiftly moving charged particle can be estimated from the optical approximation. According to this approximation, the ratio of excitations to ionizations equals the ratio of the optical oscillator strengths for absorptions of photons with energies less than the ionization potential to those with energies above the ionization potential. For saturated hydrocarbons in the gas phase, the ratio is commonly 0.2–0.3.[138] Willis and Boyd[136] examined the extent of excitation in a number of inorganic gases. They used fragmentation data from mass spectrometry and electron impact studies, *W*-values, rate data for ion–molecule and charge neutralization reactions, and quenching rate data for atoms, radicals, and excited molecules and combined these with product yields from single pulse, high-intensity irradiations (to reduce the extent of ion–molecule reactions between primary ions and radiolysis products). Table 2-4 summarizes their calculations of the distribution between ionization, neutral excitation, and subexcitation of 100 eV of energy deposited in several gases. An important technique in the radiolysis studies was the use of $SF_6$ and some fluorinated hydrocarbons, for example *cyclo*-$C_4F_8$, that can scavenge thermal electrons very rapidly to form $SF_6^-$, and so forth. Charge neutralization involving $SF_6^-$ normally does not lead to dissociation of the originally positively charged moiety. Often, by comparing product yields from the pure system with those from mixtures containing 0.01 to 0.1% of an electron scavenger such as $SF_6$, it is possible to deduce the product yields of neutralization reactions. Subexcitation electrons, by definition, cannot produce electronic excitation.

The radiolysis of $H_2O$ vapor is worth noting because of the great interest in liquid water, in which the role of excited states is still being debated. The important primary ionic processes and neutral excitation processes in gaseous $H_2O$ are shown in Table 2-5 along with the corresponding *G*-values calculated for each process by Willis and Boyd.[136] Rapid ion–molecule reactions convert all the primary positive ions to $H_3O^+$. In water vapor, primary neutral excitation processes are of major importance.

**Table 2-5.** Primary Processes in the Radiolysis of Water Vapor[a]

| Primary ionic processes | *G* (process) |
|---|---|
| $H_2O \rightarrow H_2O^+ + e^-$ | 1.99 |
| $H_2O \rightarrow OH^+ + H + e^-$ | 0.57 |
| $H_2O \rightarrow H^+ + OH + e^-$ | 0.67 |
| $H_2O \rightarrow H_2^+ + O + e^-$ | 0.01 |
| $H_2O \rightarrow H_2 + O^+(^2D) + e^-$ | 0.06 |
| Total primary ionization processes | 3.30 |
| **Primary neutral excitation processes** | |
| $H_2O \rightarrow H + OH$ | 3.58 |
| $H_2O \rightarrow H_2 + O$ | 0.45 |
| Total primary excitation processes | 4.03 |

[a] Ref. 136.

It has been stated that the optical approximation cannot be literally applied to the transfer of energy to molecules in the condensed phase. Klein and Voltz[139] suggested that the primary excitation in condensed media involves the excitation of collective plasma oscillations. These plasmons are very unstable and decay within about $10^{-15}$ s to produce highly excited states, distributed in a broad range centered at 20 eV, with a $G$ of about 4 to 5. Just as in the gas phase, these superexcited molecules decay by the competing processes of autoionization and internal conversion together with vibrational relaxation. The efficiency of autoionization is not notably influenced by the environment and is probably the principal monomolecular decay mode. However, there are bimolecular decay processes of upper excited states in condensed media that have no counterpart in the gases, such as the fission process in which a high singlet state is converted into a pair of lower-energy singlet or triplet states. Berg and Robinson[140] suggested that the energy loss peak around 20 eV in many condensed media is due to the ionization of single electrons from relatively deep valence levels. They described a process by which these excited ions can rapidly relax to give some of their energy to another electron, creating an excited molecule in a neighboring site. This process of "energy fission," $A^{+*}B \rightarrow A^{+}B^{*}$, should be extremely rapid ($10^{-14}$ s), involving as it does a purely electronic relaxation with, at most, low-order phonon processes for detailed energy conservation.

The role of excited states in the radiolysis of liquid water remains unclear. The lowest triplet state of water, and possibly a very weak singlet, are at about 4.5 eV,[141,142] an energy insufficient to break the H—OH bond in water (5.15 eV). All other electronically excited states lie above 6.6 eV. It is conceivable that higher excited states, even though very short lived, might react with the high concentrations of other primary products in the spur. Singh et al[143] have implicated excited water molecules in the radiolysis of liquid water to explain why $G(e_{aq}^-) = 4.6$ (at picoseconds) is much higher than the value of 3.3 found in the gas phase. They estimated a $G$ of 0.4 for the ionization of water from excited states above 6.5 eV.

$$H_2O^* \xrightarrow{H_2O} H_2O_{aq}^+ + e_{aq}^-$$

The ionization in liquid water is aided if the solvation energy of the resulting ions is available to lower the effective ionization potential.[144] In addition, Singh et al suggested that a significant yield of $e_{aq}^-$ could result from vibrational excitation of water by subexcitation electrons, $e_{se}^-$, followed by the ionization of $OH^-$ by $H_2O^*$ in the spur.[145]

$$H_2O + e_{se}^- \rightarrow H_2O^{*v} \xrightarrow{H_2O} H_{aq}^+ + OH_{aq}^-$$

$$H_2O^* + OH_{aq}^- \rightarrow H_2O + OH + e_{aq}^-$$

They found that 2 $M$ salicylate ions scavenged electrons on the picosecond time scale with $G$(dinegative salicylate ion) = 6.0 ± 0.4, well above the picosecond value of $G(e_{aq}^-) = 4.6$. They felt that the additional $e_{aq}^-$ might be produced by energy transfer from $H_2O^*$ to salicylate ions followed by the ionization of the excited salicylate ions. Bednar[146] concluded that, in the condensed state, vibrational, rotational, and collisional cooperation of the environment tends to shift the total energy of the Rydberg states of excited water to, or over, the first ionization limit, thus enhancing $G$(ionization). In addition, it was suggested that energy present in water as collective energy (plasmons) might be collected by a solute (eg, salicylate ion) to produce ionization.

Excited species in hydrocarbons have been extensively studied using many techniques. Direct observation of a species of interest is always desirable. In liquids a considerable fraction of the primary ions and excited species does not survive as long as a nanosecond, so that direct observation is technically difficult. A recent symposium[147] on fast processes in radiation chemistry contains many relevant studies. A widely used diagnostic procedure is to intercept the ionic species with scavengers: $SF_6$, $N_2O$, and so forth, scavenge electrons, while $NH_3$, cyclopropane, and so forth scavenge cations. This technique alters the products of ionic reactions. Unfortunately, the scavenger may also undergo excitation transfer as well as charge transfer; for example, both can produce $N_2$ from $N_2O$. The $S_1$ and $T_1$ states of many aromatics are well characterized, so aromatic hydrocarbons are widely used to scavenge excited states. Unfortunately, aromatics also scavenge electrons and undergo positive charge transfer, and neutralization of aromatic cations will also produce the excited aromatic so that it may be difficult to determine unambiguously the yield and nature of the precursors to the excited aromatic. Excited cations are studied in mass spectrometers. Photolysis of hydrocarbons has been carried out in the vacuum-UV. This has been done at energies just below and just above the ionization potential to distinguish between the products due to ions and those due to excited molecules. All of these methods, and others, have been used to untangle the mechanisms of hydrocarbon radiolysis. In spite of this it still remains difficult to obtain values for the yields of directly produced excited molecules that do not autoionize, but estimates of 15 to 30% seem reasonable when coupled with the fact that 100 eV leads to the primary decomposition of about 6.5 molecules as compared to a likely $G$(ionization) of about 5.[148] An excellent, encyclopedic review of hydrocarbon radiolysis up to 1977 is available.[39] Even for a relatively small molecule such as propane, the decay mechanism for the neutral and ionic excited species is quite complicated, as shown in Table 2-3.

The importance of a primary product as a reactive species depends, in part, on its yield and its lifetime. Katsumura et al[149] used picosecond pulse radiolysis to measure the fluorescence lifetimes ($S_1 \rightarrow S_0$) of a number of liquid $n$-alkanes. The lifetimes increased with increasing number of carbon atoms and ranged from 0.7 ns for $n$-hexane to 4.3 ns for $n$-pentadecane. The efficiency for emission is typically $10^{-3}$–$10^{-2}$, so that most $S_1$ states decay by nonradiative processes. Higher excited singlet states decayed very much faster. Because the ionization potential of molecules is greater than the energy of their $S_1$ states, recombination may lead to higher excited states. Walter and Lipsky[150] measured the yields of $S_1$ states produced in several liquid alkanes from both neutral and charged precursors. The values of $G(S_1)$ obtained for 2,3-dimethylbutane, cyclohexane, methylcyclohexane, dodecane, hexadecane, *cis*-decalin, and bicyclohexyl were 1.3, 1.4–1.7, 1.9–2.2, 3.3–3.9, 3.3–3.9, 3.4, and 3.5, respectively.

Walter et al[151] set out to determine (a) what fraction of the total yield of the lowest excited singlet state of a saturated hydrocarbon results from geminate ion pair recombination and (b) the probability that a geminate ion pair will generate an excited singlet state on recombination. To determine this, they observed the $S_1$ states by their fluorescence. The probability that $S_1$ will emit a photon was known from UV photolysis studies. Electrons were scavenged by cyclic perfluorocarbons that do not strongly quench the $S_1$ states. Bicyclohexyl was chosen as the solvent because it is known to convert all singlet states to $S_1$ with unit efficiency. The results indicated that the fraction of $S_1$ states resulting from recombination is $90 \pm 10\%$. The other $S_1$ states are formed from direct energy losses or, perhaps, partly by absorption of Cerenkov radiation. The answer to (b) was that $80 \pm 10\%$ or more of the cation–electron recombinations lead to $S_1$ formation. These results were calculated on the assumption that $G(S_1) = 3.5$ for bicyclohexyl and that $G$(geminate ion pairs) = 4. A comparison of these results with the theoretical study by Magee and Huang[152] suggested that most geminate ion pairs are initially singlet and that recombination takes place before spin relaxation occurs. Luthjens et al[153] measured the fluorescence in *cis*-decalin. They determined an efficiency of 0.85 for the formation of fluorescent excited states by recombination. They concluded that $G(S_1) = 3.5$ and that $G$(ion pairs) = 4.1. They speculated that the 15% of recombinations that do not lead to fluorescent states might involve positive ions that have fragmented.

Triplet production is very significant in aromatic hydrocarbons,[154] due, in large part, to efficient intersystem crossing. In the radiolysis of benzene $G(T_1) = 4.2$, whereas $G(S_1) = 1.6$, of which 1.2 arises from ion recombination and 0.4 from direct excitation.[155] The efficiency of intersystem crossing, $S_1 \rightarrow T_1$, in benzene is about 0.6. Electrons with insufficient energy to produce ionization may be a source of both triplet and singlet excited states. Brongersma and Oosterhoff[156] reported intense singlet–triplet transitions for gaseous saturated hydrocarbons excited by low-energy (5–15 eV) electrons based on trapping, and counting, the electrons that had lost most of their energy in a single collision. Sanche et al[157] measured the energy transfer from low-energy electrons (0–30 eV) to solid organic films. In some cases the energy transferred correlated with gas-phase energy levels, both singlet and triplet.

A single ion pair in a spur will normally be formed in a singlet state; that is to say, the spin of the free electron will be antiparallel to the spin of the unpaired electron in the geminate radical cation. Since an ion pair may retain its singlet character for tens of nanoseconds, fast geminate recombination will result in a singlet excited molecule. However, clusters of ions are common in spurs, and cross recombination will produce some triplets. Brocklehurst[158,159] has investigated the consequences of spin correlation in spurs. Recombination after complete spin randomization leads to a triplet–singlet ratio of 3:1.

In some systems an electronically excited molecule or atom and a ground-state molecule or atom can form an excited complex that is dissociative in the ground state.[138,160] This complex is called an *excimer* if both the ground state and excited molecules are the same species; it is called an *exciplex* if the two molecules are different.

## ATOMS, FREE RADICALS, AND MOLECULES

In most irradiated systems atoms, free radicals, and molecules are numbered among the primary products. They result from the decomposition of excited molecules and ions. However, the same free radicals and molecular products may well be produced by secondary reactions as well, complicating the determination of the primary yields. Common ways to determine the primary yields are to scavenge the precursors of the secondary yields and, especially in hydrocarbons, to use labeled compounds. The production of atoms, free radicals, and molecules by the decomposition of excited molecules arising from neutralization in the spur can usually be interfered with by ion scavengers. The radiolysis of suitably labeled compounds (eg, H atoms replaced by D atoms) and mixtures of labeled compounds permits one to deduce the origin of many products.

The stoichiometry of steady-state radiolysis (ie, spur reactions completed) of liquid water at room temperature may be written as[161]

$$4.14\ H_2O \overset{100\ eV}{=} 2.7\ e_{aq}^- + 2.7\ H^+ + 0.61\ H + 2.87\ OH + 0.43\ H_2 + 0.61\ H_2O_2 + 0.026\ HO_2$$

and at 1 ps (before completion of spur reactions) as

$$5.7\ H_2O \overset{100\ eV}{=} 4.78\ e_{aq}^- + 4.78\ H^+ + 0.62\ H + 0.15\ H_2 + 5.7\ OH$$

Detectable spur overlap and the changeover from spur kinetics to homogeneous kinetics occur at about $10^{-7}$ s for commonly used pulse radiolysis doses.[162] The steady-state *G*-values (above) have been successfully used as the primary yields for computer simulations of the radiolysis of dilute aqueous solutions.[22] However, studies of aqueous solutions saturated with $N_2O$ (which converts $e_{aq}^-$ into OH) by Schuler and co-workers[163,164] have shown that, above approximately $10^{-4}$ *M*, solutes reactive to OH can significantly scavenge OH from the track. This serves to emphasize the care that must be exercised when choosing the primary yields to be used in computer simulations.

In the vapor phase the *G*-values for ionization and excitation are similar, being 3.3 and 4.0, respectively (Table 2-5), and the stoichiometry of radiolysis at 160°C before ion–molecule reactions and neutralization may be written[161]

$$7.3\,H_2O \overset{100\,eV}{=} 3.3\,e^- + 0.6\,OH^+ + 2.7\,H_2O^+ + 0.45\,H_2 + 0.45\,O + 3.55\,OH + 4.15\,H$$

and after these processes,

$$7.3\,H_2O \overset{100\,eV}{=} 7.45\,H + 0.45\,H_2 + 1.05\,O + 6.25\,OH$$

The role of neutral excitation processes is a major one in water vapor and a minor one in liquid water at room temperature. The proportion of decomposition that begins as ionization and as excitation has a great influence on the yields of primary products.

The hydroxyl radical, OH, is a major primary product of water radiolysis. Jonah et al[17] measured $G(e_{aq}^-) = 4.6$ at 100 ps, and for the precursor to $e_{aq}^-$ they estimated *G*(dry electron) to be 5.4. A similar initial yield of $H_2O^+$ is expected. Most hydroxyl radicals are believed to be formed in about $10^{-14}$ s by Reaction (2-1). Hydroxyl radicals may also be formed by the decomposition of excited water molecules.

$$H_2O^* \rightarrow H + OH \tag{2-2}$$

Jonah and Miller[165] determined $G(OH) = 5.9 \pm 0.2$ at 200 ps by direct measurement of the optical absorption at 280 nm. The authors did not rule out the possibility of a systematic error in *G*(OH) greater than 0.2, due to the uncertainty in the extinction coefficient of OH. There is a satisfactory agreement between their value, $G(OH) = 5.9$, and the value of 5.7, relative to $G(e_{aq}^-) = 4.78$, calculated much earlier by Schwarz[166] in his application of the spur diffusion model to water radiolysis. Belevskii et al[119] reported that they were able to scavenge a large fraction of the precursor of OH in aqueous glasses at 77 K by the use of very high concentrations (~2 *M*) of scavengers such as $I^-$ and $Br^-$. They concluded, from a comparison of the reactivity of many scavengers, that $H_2O^+$ is the precursor to OH in aqueous glasses.

Hydrogen peroxide production in the spur is mainly by the dimerization of hydroxyl radicals. Hence, hydrogen peroxide is not a significant primary product of water radiolysis on the picosecond time scale.

If all primary hydroxyl radicals were produced by Reaction (2-1), the maximum $G(OH)$ should be no higher than $G(H_2O^+)$, which, in turn, should be no greater than $G$(dry electron), which was estimated as 5.4 by Jonah et al.[17] Since $G(OH) = 5.9$ at 200 ps,[165] we must consider the possibility that hydroxyl radicals are produced from excited water molecules [Reaction (2-2)], which also implies the production of a primary yield of hydrogen atoms. If we confine ourselves to $G(H)$ at picoseconds, we may neglect the reaction

$$e_{aq}^- + H_3O_{aq}^+ \rightarrow H + H_2O$$

Schwarz[166] concluded from spur diffusion calculations that the primary yields $G(H)$ and $G(H_2)$ equal 0.62 and 0.15, respectively. This did not include $H_2$ production by the reactions $e_{aq}^- + e_{aq}^-$, $e_{aq}^- + H + H_2O$, or $H + H$. In recent spur diffusion calculations Gopinathan and Girija[167] considered the primary yield of $H_2$ to be nil.

Brown and Hart[168] concluded that oxygen atoms are probably produced in pure water with $G(O) = 0.02$. This is such a small yield that it is normally ignored. However, in the vapor phase, O atoms are a significant primary product (see Table 2-5).

Much of the evidence regarding the primary yields in water has been reviewed by Draganic and Draganic,[169] by Schwarz,[166] and by Burns and Marsh.[161]

The extensive use of solutes to determine primary yields in crystalline ice is not possible because of the difficulty of incorporating solutes into the crystal lattice. In a pulse radiolysis experiment with $D_2O$ ice at 6 K, $G(OD)$ was found to be 3.8 at 6 $\mu$s.[63] It was concluded that both OD and D accumulated in the ice sample with successive pulses at 6 K and acted as effective scavengers of a highly mobile $D_2O^+$.

The radiolysis of liquid alkanes has been extensively studied. Obvious primary products are H atoms, alkyl radicals, $H_2$ molecules, and alkenes. Unfortunately, we do not yet have the advantage of direct picosecond determinations of these products as we have for $e_{aq}^-$ and OH in water. The brunt of proof for primary production rests with scavenger and isotopic labeling studies. Molecular hydrogen is a major product of alkane radiolysis. Therefore a resolution of how $H_2$ is produced is crucial to understanding the radiolysis mechanism. $H_2$ results from four processes:

$$A^* \rightarrow B + H_2 \tag{2-3}$$

$$H(\text{hot}) + A \rightarrow C + H_2 \tag{2-4}$$

$$H(\text{therm}) + A \rightarrow C + H_2 \tag{2-5}$$

$$H(\text{therm}) + H(\text{therm}) \rightarrow H_2 \tag{2-6}$$

$G(H_2)_{\text{primary}}$, or the so-called molecular hydrogen yield, is the yield that cannot be altered by adding small concentrations of H-atom scavengers.

This molecular yield results from Reactions (2-3) and (2-4). In Reaction (2-3) hydrogen is eliminated as $H_2$ from an excited molecule. Reaction (2-4) involves hot H atoms. Hot H atoms are those produced with sufficient kinetic energy ($\geq$0.25 eV[170]) to abstract an H atom from an alkane on the first collision. A very large concentration of scavenger would be needed to compete with Reactions (2-3) and (2-4). In Reactions (2-5) and (2-6) thermal H atoms diffuse until they react to form $H_2$. Thermal H atoms can be intercepted by modest concentrations of scavengers. The judicious replacement of H by D in the molecule is a good way to determine the origin of primary hydrogen. For example, in a mixture of completely deuterated and completely protiated molecules, Reaction (2-3) will produce only $H_2$ and $D_2$, whereas Reaction (2-4) will produce HD as well.

In the radiolysis of gaseous and liquid propane, $G(H_2) = 7.40$ and 4.80, respectively.[170] A good deal of the hydrogen is produced in secondary reactions. Fujisaki et al[170] irradiated gaseous and liquid propane in the presence of $SF_6$, to scavenge electrons, and $C_2H_4$, to scavenge thermal H atoms and cations, in order to determine $G(H_2)_{primary}$. They also irradiated $C_3H_8/C_3D_8$ mixtures to determine the unimolecular and bimolecular origins of the primary $H_2$. They concluded that, in the gas phase, $G(H_2)_{primary} = 2.3$, of which 1.4 is unimolecular and 0.9 is bimolecular. In the liquid phase $G(H_2)_{primary} = 1.6$, of which 0.4 is unimolecular and 1.2 is bimolecular. The values found for $G$ of primary thermal H atoms in gaseous and liquid propane were 2.5 and 0.9, respectively. A large fraction of the thermal H atoms produces $H_2$. The remaining $H_2$ production probably stems largely from processes following ion recombination.

Wojnarovits and Foldiak[171] compared the photolysis of liquid alkanes excited to the $S_1$ state with the radiolysis of the same alkanes. They concluded that unimolecular $H_2$ production is the characteristic decomposition path for the $S_1$ excited state of *n*-alkanes and $C_5$–$C_{10}$ cycloalkanes. In *cyclo*-$C_6H_{12}$ the $S_1$ state produces $H_2$ + *cyclo*-$C_6H_{10}$ with an efficiency of 86%. Using this information, they calculated the total (not just primary) production of $S_1$ for cyclohexane to be $G(S_1) = 1.5 \pm 0.4$ and the total for other *n*-alkanes and cycloalkanes to be in the range $G(S_1) = 1.5$–$2.5$. In the case of alkanes with tertiary or quaternary carbon atoms, the $S_1$ states usually decompose by unimolecular fragment alkane elimination. Upon investigating some 30 alkanes, Wojnarovits and Foldiak concluded that $G$ for primary C–H or C–C scission has an approximately constant value of $6.5 \pm 0.1$. The relative importance of C–C scission increases with increased branching of the alkane.

H atoms react with alkanes mostly by H abstraction. Except for methane and ethane, [172] this reaction takes place even in cold matrices. The loss of H or $H_2$ by alkanes or alkane cations generates free radicals or molecular products (see Table 2-3).

Isildar and Schuler[173] determined the primary yields of fragment radicals, that is, those resulting from C–C scission, in the radiolysis of liquid *n*-hexane by using tritium iodide as a radical scavenger. The reaction between the alkyl radicals (R) and TI, incorporating the radioactive tracer into the final products, was considered to be quantitative.

$$R + TI \rightarrow RT + I$$

The products were separated by gas chromatography, and the radioactivity of each product measured. The results were $G(CH_3T) \sim 0.7$, $G(C_2H_5T) \sim 0.4$, $G(C_3H_7T) \sim 0.3$, $G(C_5H_9T) \sim 0.3$, and $G(C_5H_{11}T) \sim 0.06$. It was concluded that C—C bond rupture results predominantly either from high-energy processes that do not directly involve ionic precursors or, more likely, from the dissociation of the initial ions at very early times ($\leq 10^{-11}$ s) before a substantial fraction of the geminate ions undergo neutralization.

# REFERENCES


1. Burton, M. *Faraday Discuss Chem. Soc.* **1963**, *7*, 1.
2. Balkas, T. I.; Fendler, J. H.; Schuler, R. H. *J. Phys. Chem.* **1970**, *74*, 4497.
3. Hart, E. J.; Boag, J. W. *J. Am. Chem. Soc.* **1962**, *84*, 4090.
4. Michael, B. D.; Hart, E. J.; Schmidt, K. H. *J. Phys. Chem.* **1971**, *75*, 2798.
5. Freeman, G. R. *Chemistry in Canada*, **1982**, *34*, 18.
6. Rentzepis, P. M.; Jones, R. P.; Jortner, J. *J. Chem. Phys.* **1973**, *59*, 766.
7. Wiesenfeld, J. M.; Ippen, E. P. *Chem. Phys. Lett.* **1980**, *73*, 47.
8. Gauduel, Y.; Migus, A.; Martin, J. L.; Antonetti, A. *Chem. Phys. Lett.* **1984**, *108*, 319.
9. Pikaev, A. K. "The Solvated Electron in Radiation Chemistry"; Keter Press: Jerusalem, 1971, Chapter 7.
10. Fueki, K.; Feng, D.; Kevan, L. *J. Phys. Chem.* **1970**, *74*, 1976.
11. Funabashi, K. *J. Phys. Chem.* **1981**, *85*, 2734.
12. Tuttle, T. R.; Golden, S. *J. Phys. Chem.* **1980**, *84*, 2457.
13. Golden, S.; Tuttle, T. R. *J. Chem. Soc., Faraday Trans. 2* **1979**, *75*, 474.
14. Anbar, M.; Bambenek, M.; Ross, A. B. In "Selected Specific Rates of Transients from Water in Aqueous Solution. 1. Hydrated Electron"; U.S. Dept. of Commerce: Washington, D.C., May 1973, Report NSRDS-NBS 43.
15. Dorfman, L. M.; Adams, G. E. In "Reactivity of the Hydroxyl Radical in Aqueous Solutions"; U.S. Dept. of Commerce: Washington, D.C., June 1973, Report NSRDS-NBS 46.
16. Hunt, J. W. In "Advances in Radiation Chemistry"; Burton, M.; and Magee, J. L., Eds.; John Wiley: 1976, Vol. 5, pp. 185–315.
17. Jonah, C. D.; Matheson, M. S.; Miller, J. R.; Hart, E. J. *J. Phys. Chem.* **1976**, *80*, 1267.
18. Sumiyoshi, T.; Katayama, M. *Chem. Lett.* **1982,** 1887.
19. Burns, W. G.; Sims, H. E.; Goodall, J. A. B. *Radiat. Phys. Chem.* **1984**, *23*, 143.
20. Short, D. R.; Trumbore, C. N.; Olson, J. H. *J. Phys. Chem.* **1981**, *85*, 2328.
21. Anbar, M.; Farhataziz; Ross, A. B. In "Selected Specific Rates of Reactions of Transients from Water in Aqueous Solution. 2. Hydrogen Atom"; U.S. Dept. of Commerce: Washington, D.C., May 1975, Report NSRDS-NBS 51.
22. Boyd, A. W.; Carver, M. B.; Dixon, R. S. *Radiat. Phys. Chem.* **1980**, *15*, 177.
23. Fielden, E. M.; Hart, E. J. *Radiat. Res.* **1968**, *33*, 426.
24. Jha, K. N.; Bolton, G. L.; Freeman, G. R. *J. Phys. Chem.* **1972**, *76*, 3876.
25. Teather, G. G.; Klassen, N. V. *Int. J. Radiat. Chem.* **1975**, *7*, 475.
26. Chase, W. J.; Hunt, J. W. *J. Phys. Chem.* **1975**, *79*, 2835.
27. Watson, Jr., E.; Roy, S. In "Selected Specific Rates of Reactions of the Solvated Electron in Alcohols"; U.S. Dept. of Commerce: Washington, D.C., August 1972, Report NSRDS-NBS 42.
28. Johnson, D. W.; Salmon, G. A. *Radiat. Phys. Chem.* **1977**, *10*, 294.
29. Kraus, C. A. *J. Am. Chem. Soc.* **1908**, *30*, 1323.

30. Colloque Weyl V, *J. Phys. Chem.* **1980**, *84*, No. 10.
31. Farhataziz; Perkey, L. M.; Hentz, R. R. *J. Chem. Phys.* **1974,** *60*, 4383.
32. Jou, F.; Freeman, G. R. *J. Phys. Chem.* **1981**, *85*, 629.
33. Farhataziz; Perkey, L. M.; Hentz, R. R. *J. Chem. Phys.* **1974**, *60*, 717.
34. Ahmad, M. S.; Atherton, S. J.; Baxendale, J. H. In "Proceedings of the Sixth International Conference of Radiation Research", Tokyo, 1979; Okada, S.; Imamura, M.; Terishima, T.; and Yamaguchi, H., Eds.; Toppan Printing: Tokyo, 1979, pp. 220–227.
35. Warman, J. M. In "The Study of Fast Processes and Transient Species by Electron Pulse Radiolysis"; Proceedings of the NATO Advanced Study Institute, Capri, Italy, Sept. 1981; Baxendale, J. H.; and Busi, F., Eds.; Reidel: Dordrecht, Holland, 1982, pp. 433–533.
36. Schmidt, W. F.; Allen, A. O. *J. Phys. Chem.* **1968**, *72*, 3730.
37. Dodelet, J. P.; Freeman, G. R. *Can. J. Chem.* **1972**, *50*, 2667.
38. Beck, G. *Radiat. Phys. Chem.* **1983**, *21*, 7.
39. "Radiation Chemistry of Hydrocarbons" Foldiak, G. Ed.; Elsevier: Amsterdam, 1981.
40. Gyorgy, I. In ref. 39, pp. 63, 65.
41. Namba, H.; Chiba, M.; Nakamura, Y.; Tezuka, T.; Shinsaka, K.; Hatano, Y. In "Proceedings of the Seventh International Conference on Conduction and Breakdown Dielectric Liquids"; Schmidt, W. F., Ed.; West Berlin, 1981, pp. 41–45.
42. Hamill, W. H. In "Radical Ions"; Kaiser, E. T.; and Kevan, L., Eds.; Wiley Interscience: New York, 1978, Chapter 9.
43. Lin, D.; Kevan, L. *Radiat. Phys. Chem.* **1981**, *17*, 71.
44. Klassen, N. V.; Adams, R. J.; Teather, G. G.; Ross, C. K. *J. Phys. Chem.* **1980**, *84*, 3609.
45. Buxton, G. V.; Gillis, H. A.; Klassen, N. V. *Can. J. Chem.* **1976**, *54*, 367.
46. Gillis, H. A.; Teather, G. G.; Buxton, G. V. *Can. J. Chem.* **1978**, *56*, 1889.
47. Trudel, G. J.; Gillis, H. A.; Klassen, N. V.; Teather, G. G. *Can. J. Chem.* **1981**, *59*, 1235.
48. Cygler, J.; Klassen, N. V.; Teather, G. G. *Radiat. Phys. Chem.* **1986**, *27*, 47.
49. Buxton, G. V.; Gillis, H. A.; Klassen, N. V. *Can. J. Chem.* **1977**, *55*, 2385.
50. Wu, Z.; Klassen, N. V.; Gillis, H. A.; Teather, G. G. *Can. J. Chem.* **1983**, *61*, 189.
51. Beitz, J. V.; Miller, J. R. *J. Chem. Phys.* **1979**, *71*, 4579.
52. Rice, S. A. *J. Phys. Chem.* **1980**, *84*, 1280.
53. Hamill, W. H.; Funabashi, K. *Phys. Rev.* **1977**, *B16*, 5523.
54. Shlesinger, M. S. *J. Chem. Phys.* **1979**, *70*, 4813.
55. Klassen, N. V.; Gillis, H. A.; Teather, G. G.; Kevan, L. *J. Chem. Phys.* **1975**, *62*, 2474.
56. Fueki, K.; Feng, D. F.; Kevan, L. *J. Chem. Phys.* **1972**, *56*, 5351.
57. Miller, J. R.; Beitz, J. V. In "Proceedings of the Sixth International Conference of Radiation Research"; Okada, S.; Imamura, M.; Terishima, T.; and Yamaguchi, H., Eds.; Toppan Printing: Tokyo, 1979, pp. 301–308.
58. Klassen, N. V.; Teather, G. G. *J. Phys. Chem.* **1979**, *83*, 326.
59. Battacharya, D.; Willard, J. E. *J. Phys. Chem.* **1980**, *84*, 146.
60. Klassen, N. V.; Teather, G. G. *J. Phys. Chem.* **1985**, *89*, 2048.
61. Klassen, N. V.; Gillis, H. A.; Teather, G. G. *J. Phys. Chem.* **1972**, *76*, 3847.
62. Pin, D.; Kevan, L. *Chem. Phys. Lett.* **1976**, *40*, 517.
63. Wu, Z.; Gillis, H. A.; Klassen, N. V.; Teather, G. G. *J. Chem. Phys.* **1983**, *78*, 2449.
64. Gillis, H. A.; Teather, G. G.; Ross, C. K. *J. Phys. Chem.*, **1980**, *84*, 1248.
65. Warman, J. M.; Jonah, C. D. *Chem. Phys. Lett.* **1981**, *79*, 43.
66. Samskog, P. O.; Lund, A.; Nilsson, G. *Chem. Phys. Lett.* **1981**, *79*, 447.
67. Dorfman, L. M. In "Solvated Electron"; Gould, R. F., Ed.; American Chemical Society: Washington, D.C., 1965, Chapter 4.
68. Gauthier, M. *Radiat. Phys. Chem.* **1979**, *13*, 139.
69. "Point Defects in Solids"; Crawford, J. H.; and Slifkin, L. M., Eds.; Plenum Press: New York, 1972, Vol. 1.
70. Williams, R. T.; Bradford, J. N.; Faust, W. L. *Phys. Rev. B.* **1978**, *18*, 7038.
71. Gudat, A. E.; Scott, A. B.; Wagner, M. *J. Chem. Phys.* **1974**, *60*, 4396.
72. Rush, W.; Siedel, H. *Phys. Status Solidi B* **1974**, *63*, 183.
73. Gaathon, A.; Czapski, G.; Jortner, J. *J. Chem. Phys.* **1973**, *58*, 2648.
74. Michael, B. D.; Hart, E. J.; Schmidt, K. H. *J. Phys. Chem.* **1971**, *75*, 2798.
75. Giraud, V.; Krebs, P. *Chem. Phys. Lett.* **1982**, *86*, 85.
76. Armbruster, M.; Haberland, H.; Schindler, H-G. *Phys. Rev. Lett.* **1981**, *47*, 323.
77. Sowada, U.; Warman, J. M. In "Proceedings of the Seventh International Conference on Conduction and Breakdown in Dielectric Liquids"; Schmidt, W. F., Ed.; West Berlin, 1981, pp. 14–18.

78. Holroyd, R. A.; Russel, R. L. *J. Phys. Chem.* **1974**, *78,* 2128.
79. "Electron-Solvent and Anion-Solvent Interactions"; Kevan, L.; and Webster, B. C., Eds.; Elsevier: Amsterdam, 1976.
80. Kestner, N. R. In ref. 79, Chapter 1.
81. Schmidt, W. F. In ref. 79, Chapter 7.
82. Krebs, P.; Wantschick, M. *J. Phys. Chem.* **1980**, *84,* 1155.
83. Huang, S. S.-S.; Freeman, G. R. *Can. J. Chem.* **1977**, *55,* 1838.
84. Takahashi, T.; Konno, S.; Hitachi, A.; Hamada, T.; Nakamoto, A.; Miyajima, M.; Shibamura, E.; Hoshi, Y.; Masuda, K.; Doke, T. *Sci. Papers Inst. Phys. Chem. Res.* **1980**, *74,* 65.
85. "Average Energy Required to Produce an Ion Pair"; International Commission on Radiation Units and Measurements: Washington, D.C., May 1979, Report 31.
86. Miller, J. R. *J. Phys. Chem.* **1978**, *82, 767.*
87. Hamill, W. H. *J. Phys. Chem.* **1969**, *73,* 1341.
88. Wolff, R. K.; Bronskill, M. J.; Hunt, J. W. *J. Chem. Phys.* **1970**, *53,* 4211.
89. Lam, K. Y.; Hunt, J. W. *Int. J. Radiat. Phys. Chem.* **1975**, *7,* 317.
90. Warman, J. M.; de Haas, M. P.; Verbene, J. B. *J. Phys. Chem.* **1980**, *84,* 1240.
91. Willis, C. *J. Chem. Ed.* **1981**, *58,* 88.
92. Hiraokao, K.; Kebarle, P. *Radiat. Phys. Chem.* **1982**, *20,* 41.
93. Adams, N. G.; Smith, D.; Grief, D. *Int. J. Mass Spectrom. Ion Phys.* **1978**, *26,* 405.
94. Wolkoff, P.; Holmes, J. L. *J. Am. Chem. Soc.* **1978**, *100,* 7346.
95. "Mass Spectrometry and Radiation Chemistry", Meisels, G. G., Ed.; *Radiat. Phys. Chem.* **1982**, *20(1).*
96. "Gas Phase Ion Chemistry"; Bowers, M. T., Ed.; Academic Press: New York, 1979, Vols. 1, 2.
97. Harrison, A. G. "Chemical Ionization Mass Spectrometry", CRC Press; Boca Raton, Fl., 1983.
98. Talrose, V. L.; Vinogradov, P. S.; Larin, I. K. In ref. 96, Vol. 1, p. 305.
99. Meot-Ner, M. In ref. 96, Vol. 1, Chapter 6.
100. Maier, J. P. *Chimia,* **1980**, *34,* 219.
101. Bone, L. I.; Sieck, L. W.; Futrell, J. H. In "The Chemistry of Ionization and Excitation"; Johnson, G. R. A.; Scholes, G., Eds.; Taylor and Francis: London, 1967, pp. 223–235.
102. Koob, R. D.; Kevan, L. *Trans. Faraday Soc.* **1968**, *64,* 706.
103. Koob, R. D.; Kevan, L. *Trans. Faraday Soc.* **1968**, *64,* 422.
104. Wada, T.; Shida, S.; Hatano, H. *J. Phys. Chem.* **1975**, *79,* 561.
105. Brehm, B.; Eland, J. H. D.; Frey, R; Schulte, H. *Int. J. Mass Spectrom. Ion Phys.* **1976**, *21,* 373.
106. Stockbauer, R.; Inghram, M. G. *J. Chem. Phys.* **1976**, *65,* 4081.
107. Baer, T. In ref. 96, Vol. 1, Chapter 5.
108. Herod, A. A.; Harrison, A. G.; O'Malley, R. M.; Ferrer-Correia, A. J.; Jennings, K. R. *J. Phys. Chem.* **1970**, *74,* 2720.
109. Cacace, F. *Radiat. Phys. Chem.* **1982**, *20,* 99.
110. Gentry, W. R. In ref. 96, Vol. 2, Chapter 15.
111. Zador, E.; Warman, J. M.; Hummel, A. *Chem. Phys. Lett.* **1973,** *23,* 363.
112. Lias, S. G.; Ausloos, P.; Horvath, Z. *Int. J. Chem. Kinet.* **1976**, *8,* 725.
113. Ausloos, P. *Radiat. Phys. Chem.* **1982**, *20,* 87.
114. Trifunac, A. D.; Sauer, M. C.; Jonah, C. D. *Chem. Phys. Lett.* **1985**, *113,* 316.
115. Louwrier, P. W. F.; Hamill, W. H. *J. Phys. Chem.* **1968**, *72,* 3878; *ibid.* **1970**, *74,* 1418.
116. Cygler, J.; Teather, G. G.; Klassen, N. V. *J. Phys. Chem.* **1983**, *87,* 455.
117. Toriyama, K.; Nunome, K.; Iwasaki, M. *J. Chem. Phys.* **1982**, *77,* 5891.
118. Dotan, I.; Lindinger, W.; Rowe, B.; Fahey, D. W.; Fensenfeld, F. C.; Albritton, D. L. *Chem. Phys. Lett.* **1980**, *72,* 67.
119. Belevskii, V. N.; Belopushkin, S. I.; Bugaenko, L. T. *High Energy Chem.* **1981**, *14,* 231.
120. Aue, D. H.; Bowers, M. T. In ref. 96, Vol. 2, Chapter 9.
121. Ogura, H.; Hamill, W. H. *J. Phys. Chem.* **1973**, *77,* 2952.
122. Dainton, F. S.; Salmon, G. A.; Wardman, P. *Proc. Roy. Soc. London Ser. A* **1969**, *313,* 1.
123. Huntress, W. T.; Pinizzotto, R. F. *J. Chem. Phys.* **1973**, *59,* 4742.
124. Meot-Ner, M. In ref. 96, Vol. 1, pp. 264–268.
125. Badger, B.; Brocklehurst, B.; Russell, R. D. *Chem. Phys. Lett.* **1967**, *1,* 122.
126. Holroyd, R. A.; Fessenden, R. W. *J. Chem. Phys.* **1963**, *67,* 2743.
127. Kebarle, P. *Ann. Rev. Phys. Chem.* **1977**, *28,* 445.

128. Good, A.; Durden, D. A.; Kebarle, P. *J. Chem. Phys.* **1970**, *52,* 212.
129. Conway, B. E. "Ionic Hydration in Chemistry and Biophysics"; Elsevier: Amsterdam, 1981.
130. Klots, C. E. *Radiat. Phys. Chem.* **1982**, *20,* 51.
131. Shida, T.; Hamill, W. H. *J. Am. Chem. Soc.* **1966**, *88,* 3683.
132. Higashimura, T.; Namiki, A.; Noda, M.; Hase, H. *J. Phys. Chem.* **1972**, *76,* 3744.
133. Gray, L. H. *Proc. Roy. Soc. London Ser. A,* **1929**, *122.* 647.
134. Knoll, G. F. "Radiation Detection and Measurement"; John Wiley: New York, 1979, Chapter 5.
135. Lind, S. C. "Radiation Chemistry of Gases"; Reinhold: New York, 1961.
136. Willis, C.; Boyd, A. W. *Int. J. Radiat. Phys. Chem.* **1976**, *8,* 71.
137. Calvert, J. G.; Pitts, J. N. "Photochemistry"; John Wiley: New York, 1966.
138. Lipsky, S. *J. Chem. Ed.* **1981**, *58,* 93.
139. Klein, G.; Voltz, R. *Int. J. Radiat. Phys. Chem.* **1975,** *7,* 155.
140. Berg, J. O.; Robinson, G. W. *Chem. Phys. Lett.* **1975**, *34,* 211.
141. Knoop, F. W. E.; Brongersma, H. H.; Oosterhoff, L. J. *Chem. Phys. Lett.* **1972**, *13,* 20.
142. Chutjian, A.; Hall, R. I.; Trajmar, S. *J. Chem. Phys.* **1975**, *63,* 892.
143. Singh, A.; Chase, W. J.; Hunt, J. W. *Faraday Discuss. Chem. Soc.* **1977**(63), 28–37; see also the discussion on pp. 70–79.
144. Boyle, J. W.; Ghormley, J. A.; Hochanadel, C. J.; Riley, J. F. *J. Phys. Chem.* **1969**, *73,* 2886.
145. Natzle, W. C.; Moore, C. B.; Goodall, D. M.; Frisch, W.; Holzwarth, J. F. *J. Phys. Chem.* **1981**, *85,* 2882.
146. Bednar, J. *Radiat. Phys. Chem.* **1978**, *12,* 79.
147. "International Symposium on the Fast Processes in Radiation Chemistry" *Radiat. Phys. Chem.* **1983**, *21,* Nos. 1, 2.
148. Wojnarovits, L. In ref. 39, p. 182.
149. Katsumura, Y.; Yoshida, Y.; Tagawa, S.; Tabata, Y. *Radiat. Phys. Chem.* **1983,** *21,* 103.
150. Walter, L.; Lipsky, S. *Int. J. Radiat. Phys. Chem.* **1975**, *7,* 175.
151. Walter, L.; Hirayama, F.; Lipsky, S. *Int. J. Radiat. Phys. Chem.*, **1976**, *8,* 237.
152. Magee, J. L.; Huang, J-T. J. *J. Phys. Chem.* **1972**, *76,* 3801.
153. Luthjens, L. H.; Codee, H. D. K.; De Leng, H. C.; Hummel, A.; Beck, G. *Radiat. Phys. Chem.* **1983**, *21,* 21.
154. Thomas, J. K. *Int. J. Radiat. Phys. Chem.* **1976**, *8, 1.*
155. Roder, M. In ref. 39, pp. 401, 405.
156. Brongersma, H. H.; Oosterhoff, L. J. *Chem. Phys. Lett.* **1969**, *3,* 437.
157. Sanche, L.; Bader, G.; Caron, L. *J. Chem. Phys.* **1982**, *76,* 4016.
158. Brocklehurst, B. *Nature* **1977**, *265,* 613.
159. Brocklehurst, B. *Radiat. Phys. Chem.* **1983**, *21,* 57.
160. Oka, T.; Rama Rao, K. V. S.; Redpath, J. L.; Firestone, R. F. *J. Chem. Phys.* **1974**, *61,* 4740.
161. Burns, W. G.; Marsh, W. R. *J. Chem. Soc., Faraday Trans. 1* **1981**, *77,* 197.
162. Trumbore, C. N.; Short, D. R.; Fanning, J. E.; Olson, J. H. *J. Phys. Chem.* **1978**, *82,* 2762.
163. Schuler, R. H.; Patterson, L. K.; Janata, E. *J. Phys. Chem.* **1980**, *84,* 2088.
164. Schuler, R. H.; Hartzell, A. L.; Behar, B. *J. Phys. Chem.* **1981**, *85,* 192.
165. Jonah, C. D.; Miller, J. R. *J. Phys. Chem.* **1977**, *81,* 1974.
166. Schwarz, H. A. *J. Phys. Chem.* **1969**, *73,* 1928.
167. Gopinathan, C.; Girija, G. *Radiat. Phys. Chem.* **1983**, *21,* 209.
168. Brown, W. G.; Hart, E. J. *Radiat. Res.* **1972**, *51,* 249.
169. Draganic, I. G.; Draganic, Z. D. "The Radiation Chemistry of Water"; Academic Press: New York, 1971.
170. Fujisaki, N.; Shida, S.; Hatano, Y. *J. Chem. Phys.* **1970**, *52,* 556.
171. Wojnarovits, L.; Foldiak, G. *Radiat. Res.* **1982**, *91,* 638.
172. Iwasaki, M.; Toriyama, K. *J. Phys. Chem.* **1978**, *82,* 2056.
173. Isildar, M.; Schuler, R. H. *Radiat. Phys. Chem.* **1978**, *11,* 11.

# 3

# Instrumentation for Measurement of Transient Behavior in Radiation Chemistry

L. K. PATTERSON

*Radiation Laboratory, University of Notre Dame*

## INTRODUCTION

Radiation-induced chemical processes involve intermediates that are reactive in most environments and, hence, can persist only for short periods. In fact, the course of reaction from the moment of energy absorption by the medium to the appearance of stable products may involve several transient species in sequence or in parallel. Although some information relating to reaction intermediates may be inferred from the results of steady-state experiments, detailed identification and characterization of these unstable species require the ability to monitor transient behavior with a high degree of precision over extremely short intervals. Because the concentration of such intermediates may influence the mechanistic pathway as well as product yields, detection techniques sensitive to very low concentrations of radical are often required for characterization of transient behavior.

The development of sophisticated techniques for time-resolving radiation-induced chemical events has undergone spectacular growth over the past 25 years. Starting with the observation of benzyl radical in organic solution


*Radiation Chemistry: Principles and Applications*

by pulse radiolysis[1a], as well as the illustration of transient behavior in aqueous solution,[1,b,c,d] this field has seen the appearance of pulsed irradiation sources in the nanosecond and picosecond time range as well as a range of monitoring techniques that can elucidate many different facets of transient behavior. Because so much of our understanding of radiation chemistry has come to depend on these pulse radiolysis techniques, the bulk of this discussion will be devoted to outlining the principles involved and the kinds of information that may be obtained.

Here, some sources of pulsed irradiation and monitoring techniques (though not every monitoring technique) will be discussed. Because of space limitations, the figures will focus mainly on techniques rather than resultant spectra and kinetic traces. The types of spectra produced resemble those obtained by application of the various modes of detection to steady-state measurement; in most cases, inclusion of kinetic traces adds little without a discussion of the particular mechanism involved. The description of the physical principles underlying each detection technique has been condensed to a qualitative outline. More rigorous descriptions are given in the literature cited. Most of the material will focus on experiments in liquids—mainly aqueous systems—although a significant number of time-resolved studies in gases and solids may be found in the literature. A few relevant steady-state procedures are also discussed.

## PULSE RADIOLYSIS

As with the perturbation source in other relaxation techniques, the delivery of ionizing radiation to a chemical system in the form of a short pulse allows one to achieve a nonequilibrium system in which significant concentrations of transient species may be produced and then monitored in subsequent time by the technique of choice. Given that the monitoring technique is capable of following the transient behavior generated, a limit to the time resolution of the measurement is set by the shape of the radiolytic pulse.

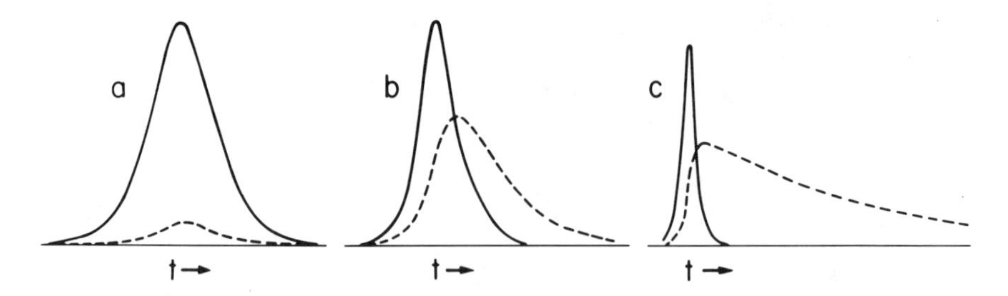


**Figure 3-1.** Dependence of radical behavior [– – –] on radiolytic pulse shape [——]: (a) half-life for first-order decay ($t_{1/2}$) of intermediate is one-tenth of pulse width at half-height ($t_{hw}$); (b) $t_{1/2} = t_{hw}$; (c) $t_{1/2} = 10 \cdot t_{hw}$.

Should the pulse be long with respect to the relaxation time of the system, insufficient transient population will be produced for measurement. This condition is shown in Figure 3-1(a). For the intermediate case in which the relaxation time is comparable to that for the pulse, processes of generation and disappearence will overlap, making difficult the analysis of transient behavior, especially for complex kinetics [Figure 3-1(b)]. In the best case the pulse is very short compared to relaxation; the transient population is immediately produced, and kinetic analysis will be dependent only on subsequent chemical events [Figure 3-1(c)].

## *Accelerators*

The function of an accelerator is to impart high kinetic energy to an ion; in radiation chemistry such particles, most generally electrons, are then directed through the experimental medium, where they lose some fraction of their energy in ionization events. Each of the devices considered here is capable of delivering these particles over very short intervals.

Aside from pulse length, there are several accelerator parameters important to pulse radiolysis measurements. First, the pattern of energy deposition will depend on particle velocity, the density of the medium, and the length of passage through the chemical system. A few examples of energy deposition for high-energy electrons as a function of passage through the cell are shown in Figure 3-2.[2] For ions of greater mass with similiar energy, the energy will be deposited over a shorter pathway. The ion energy must be sufficient to assure penetration of the irradiation cell and ideally, provide homogeneous distribution throughout the medium used. Parenthetically, any irradiation chamber design must take this into account. Uncharacterized, inhomogeneous radical concentrations will lead to errors, especially in analysis of second-order kinetics.[3] On the other hand, pulses of particles having excessively high energy are capable of inducing hazardous radioactivity in targets.

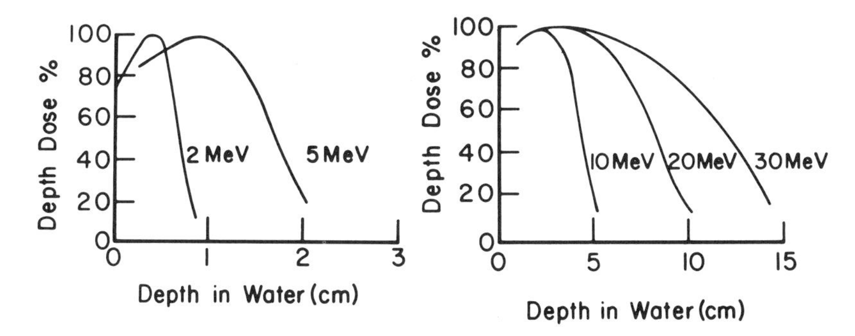


**Figure 3-2.** Dose depth curves for electrons of various energies (ref. 2).

Another characteristic of high-energy electrons is their emission of Cherenkov light when they enter a medium in which they are moving faster than the speed of light. This emission exhibits the same time-intensity behavior as the electron pulse, and the light emitted is a continuum with intensity increasing toward the violet end of the spectrum. This emission can perturb transient measurements at short times, if it overloads a light-sensitive detector. Conversely, it provides a short, well-defined light pulse that is useful in some measurements.

The quantity of radiation absorbed by a unit of mass per pulse will determine the *concentration* of transient produced. This quantity is termed "dose." The SI unit of absorbed dose now coming into use is the Gray (Gy), defined as 1 Gy = 1 J/kg. However, the traditional unit employed is the rad or, in pulse work, the kilorad (1 Krad = 10 Gy). Although it is customary to speak of peak current output from an accelerator, it is more useful from a chemical point of view to discuss dose generated in the chamber used. With appropriate calculations one may show that a dose of 1 krad generating a radical with a $G$-value (number of radicals generated per 100 ev absorbed) of, say, 6 will produce a radical concentration of $6 \times 10^{-6}$ M. The variation in sensitivity of detection methods makes it desirable to have available pulses in the range 0.1 to 10 krad per pulse. Although one must assure adequate dose for the method of detection being employed, it is often desirable to limit dose to the minimum level required instrumentally in order to avoid radical interactions involving products of the radiation chemistry. In systems where first- and second-order processes are in competition, low doses will slow the second-order components and may provide better resolution of first-order processes.

Additionally, because much of the detection equipment now in use is automated, it becomes advantageous to be able to average many experiments together. For efficient averaging it becomes desirable to use an accelerator with high repetition rates.

Although cyclotrons and pulsed x-ray sources have been employed in specialized pulse radiolysis measurements, three types of electron accelerators are used most often in time-resolved studies: the linear accelerator (linac), the Van de Graaff generator, and the Febetron. Some of the operating characteristics of these machines are given in Table 3-1.

The linac operates by using an arrangement of microwave cavities in series to accelerate the electrons along a path defined by the oscillating polarity of the field. The principles of this equipment are outlined in Figure 3-3. Streams of electrons, whose lengths are determined by the switching of the electron gun, are introduced into the accelerator tube. In the prebuncher section each stream is chopped and compressed into electron bunches by the oscillating microwave field in each of the cavity elements. The stream is now a train of electron bunches whose overall length is the length of the original stream and whose internal spacing in time is the reciprocal of the microwave frequency. In this figure the individual clusters will be 770 ps

**Table 3-1.** Operating Parameters for Some Representative Accelerator Facilities[a]

| | Linac | | Febetron (706) | Van de Graaff | | | X-ray |
|---|---|---|---|---|---|---|---|
| Particle | Electron | | Electron | Electron | | Proton | Photon |
| Length of pulse (ns) | 0.03 | 10 | 50 | 1 | 100 | 0.2 | $10^6$ |
| Rep rate/s | 800 | 360 | $10^{-2}$ | $10^5$ | $10^5$ | $10^6$ | 10 |
| Energy (MeV) | 20 | 10 | 1.8 | 3 | 3 | 2 | 0.12 |
| Typical dose (krad) | 2 | 10 | $2 \times 10^3$ | 1 | 100 | $10^{-4}$ | $10^{-4}$ |
| Peak current (A) | 200 | 20 | $5 \times 10^3$ | 5 | 5 | $10^{-5}$ | — |

[a] Much of this data is taken from compilation in ref. 4(b) and references therein.

apart. The width of each bunch is governed by the electrical parameters of the cavity arrangement but is in the picosecond region. Cavity tuning in the buncher section is arranged in such a way as to accelerate the electrons; however, in the last, or "linear," section the velocity is increased very little, and the increased electron energy is exhibited by relativistic changes in mass. Subsequent magnetic focusing must take this into account.

Techniques have been developed for producing single picosecond pulses for radiolysis;[4] alternatively, the entire train may be used as a longer, more intense electron pulse on the nanosecond or even microsecond scale. If the whole train is used, one must be sure that the intensity fluxuations associated with the structure of the train do not perturb the chemistry under study. The energy spectrum in such a pulse of electrons may be rather broad [eg, 10 MeV (±0.5)]. A special bending magnet at the end of the accelerator may be used for focusing electrons from the entire energy spectrum.

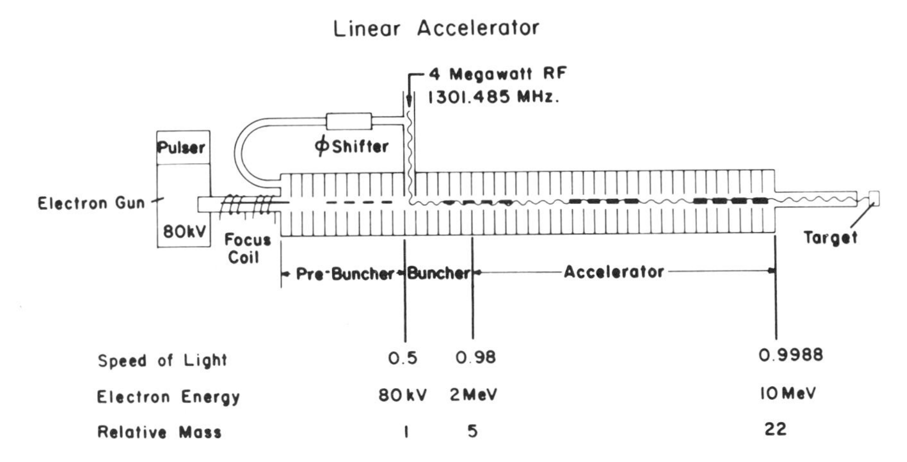


**Figure 3-3.** Principles of operation for the linear accelerator.

The particular advantages of the linac come from the ease with which very high-energy electrons may be produced. Electron energies in the range of 10 to 20 MeV may be readily obtained. Such energies provide the desired deep penetration of the electrons and uniform distributions of ionization events throughout the irradiation chamber. Integrated dose for a pulse train will depend on its length; the table gives characteristic doses available. The repetition rates of the pulse trains can be very high, say 360 Hz for 5–50-ns pulses. This often exceeds considerably the rate at which chemical systems can be renewed after irradiation. The linac has been used in the high-rep-rate mode as a "steady-state" source but, as such, more nearly approximates a rotating sector source and is subject to similar restrictions. One difficulty with this accelerator arises from the electronic noise produced by the emerging electrons as a consequence of their picosecond rise times. Such noise may seriously interfere with signals generated in monitoring devices, and it is often necessary to go to considerable lengths to electrically isolate the monitoring electronics.

The Van de Graaff, whose operation is outlined in Figure 3-4, is essentially designed around an electrostatic generator. A high potential is built up on a metal hemisphere by conveying to it electrons that have been sprayed onto a moving belt (by the principle of charged concentric spheres, the electrons will always flow from the belt to the hemisphere; see, for example, ref. 5(a) or any elementary book on electricity and magnetism). This potential is connected to ground through a voltage divider whose components govern acceleration stages in an evacuated tube. Electrons are deposited in the tube and accelerated toward the target. Electronically, it is much simpler

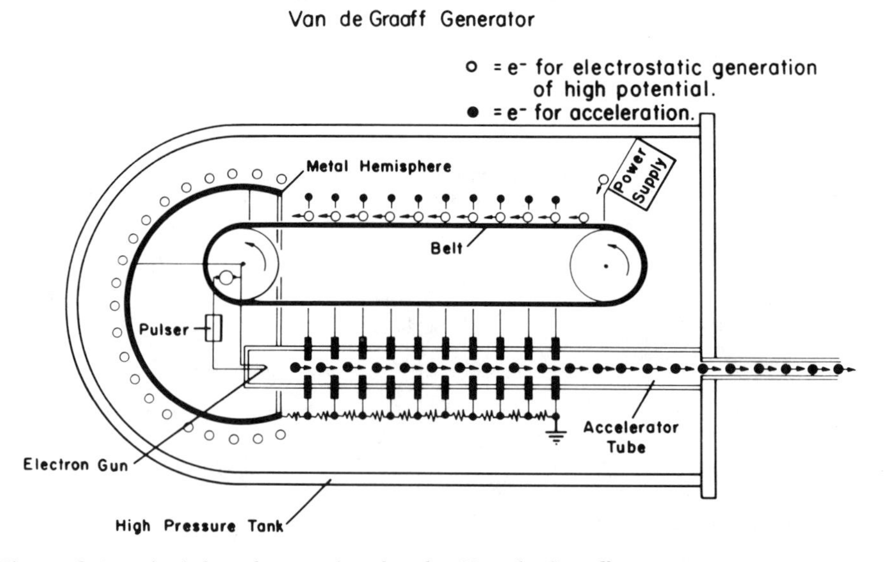


**Figure 3-4.** Principles of operation for the Van de Graaff generator.

than the linac and generates a great deal less electronic noise which may interfere with monitoring devices. For most research in physics the Van de Graaff is operated in the dc mode and is used as a steady-state source. With proper isolation, potentials up to 9 MeV may be produced. In the discussion of steady-state in situ ESR measurements, it will be seen that the Van de Graaff is the accelerator of choice. By modulation of the grid voltage, however, pulses of the electrons may be readily produced on the nanosecond scale and longer. For pulse radiolysis, electrons are generated with energies characteristically in the range 1.8 to 3.0 MeV. In special cases, however, pulses of 100 ps have been produced.[5b] Currents of approximately 5 A are readily obtained on the nanosecond scale. From the table this may be seen to be rather less than the peak current generated by a linac. The Van de Graaff is also readily applicable to acceleration of various other particles. Some positive-ion pulse radiolysis studies have been carried out using such a source in a somewhat different configuration than shown here, for example ref. 6.

Another accelerator used for pulse radiolysis is the Febetron (brand name for the impulse generator produced by Field Emission Corp.), which operates on the principle of the Marx-bank circuit. A number of connected high-voltage capacitors are charged in parallel and then switched, through spark gaps, to a series arrangement, producing a high potential drop from one end of the circuit to the other. This generates electron emission from a cathode and acceleration toward a window which functions as the anode. Voltages of 2.0 MeV may be obtained, and extremely high currents (see Table 3-1) generated. These characteristics make the machine especially attractive for gas phase studies. The Febetron also suffers from the marked electromagnetic noise in the switching of the spark gaps and the subsequent high-current pulse. Further, the time required to recharge the capacitors is quite long. Pulses may be generated with this machine on the order of one per minute. This is particularly problematic if one wishes to average the results of many experiments.

## *Dose Monitors*

Because the magnitude of the radiolytic pulse is not absolutely reproducible in most machines, it is desirable to measure, at least on a relative basis, the magnitude of each individual pulse. The signal generated by the monitor may then be calibrated to some chemical dosimeter and used as an indirect indicator of individual dose. Several devices may be used. Simply by passing the pulse through a toroid, one may induce a current that is proportional to the radiolytic pulse (ref. 7 and references therein). Alternatively, one may position a metal target in the beam to gather electrons and directly produce a measurable charge. However, the secondary emission monitor placed at the end of the accelerator tube is probably the most widely used device.[8] It has the advantage of measuring each pulse used, and it does not exhibit a great dependence on focusing of the electron beam. A thin foil (characteristically 25 $\mu$m Ni) is placed perpendicular to the electron beam so that the beam must pass through it before emerging at the accelerator exit. A small fraction (2–3%) of the beam is intercepted as it passes through, generating low-velocity secondary electrons from the foil. The secondary electrons emitted from this foil are collected by applying a suitable voltage to adjacent foils; a current proportional to the electron beam is produced. With proper amplification a signal is produced that is a measurable representation of dose per pulse.

# TRANSIENT MONITORING TECHNIQUES

## *Optical Methods*

### Absorption Spectroscopy

The most widely used technique for monitoring transient behavior in pulse radiolysis is absorption spectroscopy; a present day configuration incorporating computer control is given in Figure 3-5. One may see in the top portion that the light source, optical train, monochromator, detector, and recorder comprise an absorption spectrophotometer. The irradiation cell is subjected to a pulse of high-energy particles (electrons), and a transient recorder is simultaneously triggered to store the detector signal as a function of time. If light of the monitoring wavelength is absorbed by the transient species produced, a deflection of the detector signal (from either a photomultiplier or photodiode) will track transmittance changes, $\Delta I$, in the cell. These changes will parallel the formation and subsequent kinetic behavior of the transient population by changes in monitoring light transmittance. From a record of these values and measurements of incident light intensity, $I_0$, absorbance as a function of time can be obtained. Additionally, with measurement of the dose per pulse, these absorbance values can be normalized to the intensity of the electron beam. From subsequent samples of kinetic behavior at other wavelengths, one may plot the absorbance at a selected time after the pulse as a function of wavelength to build a time-resolved transient spectrum. A representative example of time-resolved spectra produced in this way is given in Figure 3-6.[9] More rarely now, a spectrographic technique is employed in which a short pulse of broadband monitoring light interrogates the chemical system at a preselected time after the radiolytic pulse. When passed through a spectrograph instead of a monochromator, this single flash produces a spectrum of intermediates over a time interval equal to the length of the monitoring flash.[2] Sample spectra at given delays after irradiation can now also be obtained by means of an optical multichannel analyzer (OMA). It follows, of course, that data on

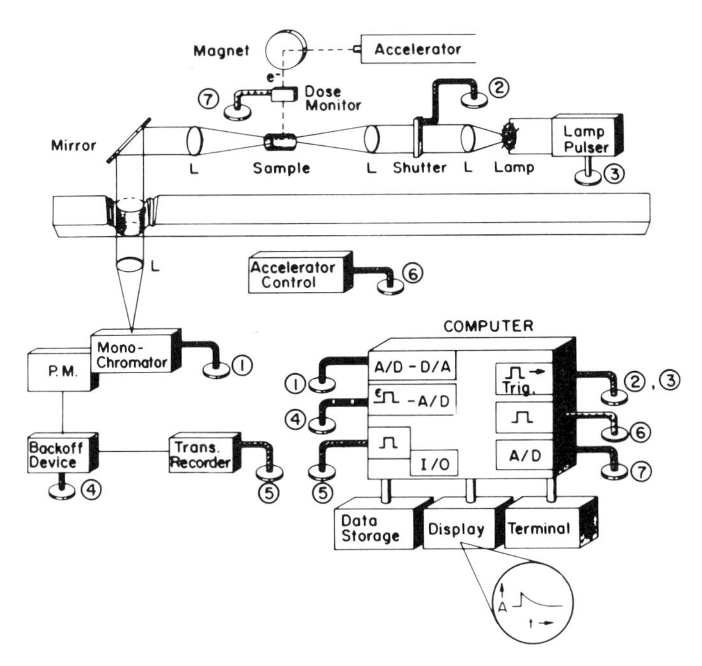


**Figure 3-5.** Experimental arrangement for pulse radiolysis using optical absorption spectroscopy (upper portion) in computerized mode (lower portion). Numbers indicate device activated in various operations in the experiment whose time course is shown in Fig. 3-8 below. Numbers also indicate type of computer operation utilized in experiment control: A/D = analog-to-digital conversion, D/A = digital-to-analog conversion, ⊓ = logic-signal.

kinetic behavior may be extracted from a series of experiments in which spectra are taken at different times after the radiolytic pulse. In a few laboratories both kinetics and spectra are obtained simultaneously by using a streak camera. See Figure 3-7 and below.

Sampling procedures, such as those discussed above, build up a record of representative behavior by fitting together small portions of data, each taken from a sequence of measurements on the assumption that the kinetic and spectroscopic behaviors of the system are reproducible from experiment to experiment. Whether spectral behavior is sampled as a function of time or kinetic behavior is sampled at a fixed spectroscopic parameter (wavelength here), the conditions of reproducibility determine reliability of the data. Small fluctuations in some parameters such as dose or monitoring source can be removed by normalization of data. Others, such as chemical interference from buildup of reaction products must be avoided.

In principle, the absorption technique is applicable to all transients or combinations of transients, although in practice the wavelengths of absorption and/or the extinction coefficients of the species under study may be less than optimal. Both time and spectral response will depend on several

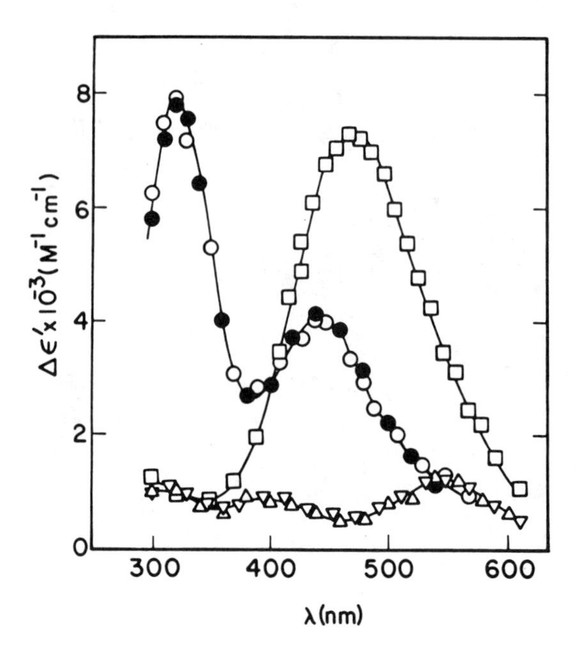


**Figure 3-6.** Typical time-resolved spectra taken from a series of kinetic traces. Reaction of $(CNS)_2^-$ (□) with a Ni(II) (CR + 4H) macrocyclic complex to give Ni(III) and addition of $CNS^-$;[9] (○) represents data 15 μs after the pulse. A subsequent configuration of the Ni(III) complex is indicated by (Δ).

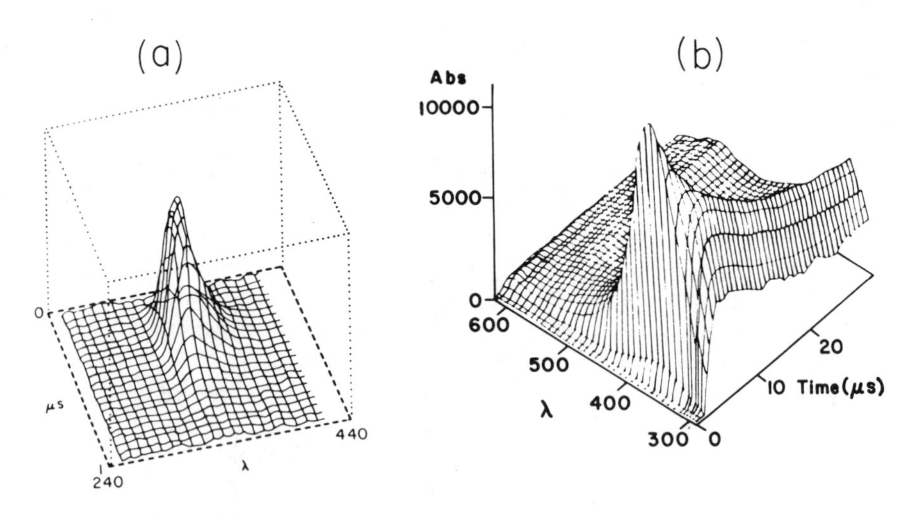


**Figure 3-7.** (a) Spectrum from streak camera: irradiation of naphthalene in benzene;[16b] (b) spectrum from computer system: $Br_2$-reaction with Ni(CR + 4H) complex in aqueous solution.

design parameters. The choice and arrangement of optical components—lamp, lens (or, better, mirrors), irradiation cell, and monochromator—are all critical for determining the range of chemical systems to which this technique may be applied. Discussions of optical design may be found in the literature.[10,11]

There are, of course, some cases in which measurement of emission from transient species is desired. Intensity of emitted light often can be tracked by eliminating the monitoring light and, if necessary, changing the optics to intercept photons from the irradiation chamber over a greater solid angle. Application to kinetics of an irradiated bioluminescent enzyme is a case in point.[12] On a few occasions the photoexcited fluorescence properties of transient species have been studied where the parent molecule itself is not fluorescent.[13] This latter measurement requires some rearrangement of the monitoring apparatus.

Clearly, no theoretical limits are imposed on measurements by the time range of the absorption phenomenon ($10^{-15}$ s). However, behavior of the detector and/or recorder can impose very real limits on the time response of the instrument; obviously, the detection system must have a fast response time with respect to the kinetic processes under study. For most applications in the nanosecond region and longer, a properly chosen photomultiplier and circuit will exhibit a sufficient response time. In the picosecond range, diodes may be used, although they must be carefully evaluated for time response.[11] Historically, it is worth noting that in some of the initial picosecond measurements, Bronskill and Hunt devised an ingenious technique by which picosecond pulses of Cherenkov light associated with a linac radiolytic pulse train were sent through a system of mirrors to interrogate the transient population at prechosen times after the radiolysis. The speed of light and the light path through the mirror assembly determined the synchronization of interrogating light with transient behavior. This approach obviated the need for fast electronics in the study of certain chemical systems.[14]

Finally, it is often desirable to be able to monitor very low absorbances, on the order of 1% or less. In some cases the system under study may exhibit very small extinction coefficients. In others it may be desirable to produce low radical concentrations to isolate competing first- and second-order processes. This will require techniques for producing acceptable signal-to-noise ratios in very small signals. For very long-lived transients one may remove a high-frequency noise by appropriate *RC* filtering (as long as the response time of the detector remains about 10 times less than the lifetime of the process being observed).

The signal-to-noise ratio for a photomultiplier will be related to the intensity of the light striking the cathode and will increase as the square root of that intensity. It is often useful, for transient measurements up through the microsecond region, to include a device that generates very high monitoring light output during transient measurement. Several types of light pulsers exist that can deliver to the monitoring lamp, for the duration of measurement, a high-current pulse with a relatively square output profile.[15] Further, pulsing raises the lamp temperature, and the relative gains in light intensity will be greatest toward the violet edge of the lamp output, where greater intensity is often needed most.

Averaging provides another approach to achieving high signal-to-noise ratios in very weak transient absorption buried in random detector noise. Such noise will average out when a number of experiments are added together. Again, the signal-to-noise ratio will improve by the square root of the number of experiments included. This approach is, of course, applicable to all types of "noisy" data, and its utility will be best illustrated below.

### *Transient Recorders in Pulse Radiolysis*

Over the past 15 years sophisticated transient recorders have been developed that now largely replace the conventional oscilloscope–Polaroid photograph technique traditionally used for recording transient signals. As will be seen later, these have found application throughout the area of fast kinetics and not just in the measurements described above. The true digital recorder, such as in the Biomation series, digitizes the input signal in real time and can resolve the signal into 2000 separate channels. Characteristically, for an 8-bit recorder such as the Biomation 8100, the interpoint time can be as short as 10 ns, but for the 6500, 6-bit and 2-ns resolution may be achieved. Although the Biomation is the most popular of its type in this country, similar devices are offered by other vendors. The Tektronix 7912 series recorders, on the other hand, emulate an oscilloscope storage by using a two-dimensional diode array that replaces the phosphorescent screen. The charged diodes interact with the oscilloscopic electron beam, recording its passage in a pattern of discharged elements within the array. Afterward, the "on–off" condition of each element is interrogated, and a digital record of the "trace" is created in a temporary memory. Response times of 0.5 ns can be achieved with this device. Both types of recorders may be interfaced for computer control of operational parameters: time base, sensitivity, delays.

To use an electronic recorder in an absorption measurement, it is undesirable to input the full photomultiplier signal corresponding to $I_0$. To do so in a voltage range that properly amplifies $\Delta I$ would saturate the recorder. A "backoff device" is then needed to compensate for the $I_0$ signal by an equal signal of opposite polarity that remains constant over the period of the recording.[11,17a] In that way the change in transmission may be selectively amplified and stored. Additionally, the backoff signal gives a measure of $I_0$ that may be digitized and stored. The function of the backoff is illustrated in Figure 3-8.

For experiments in which one wishes to record both kinetics and spectra simultaneously, one may employ a streak camera.[16] Here, the polychromatic

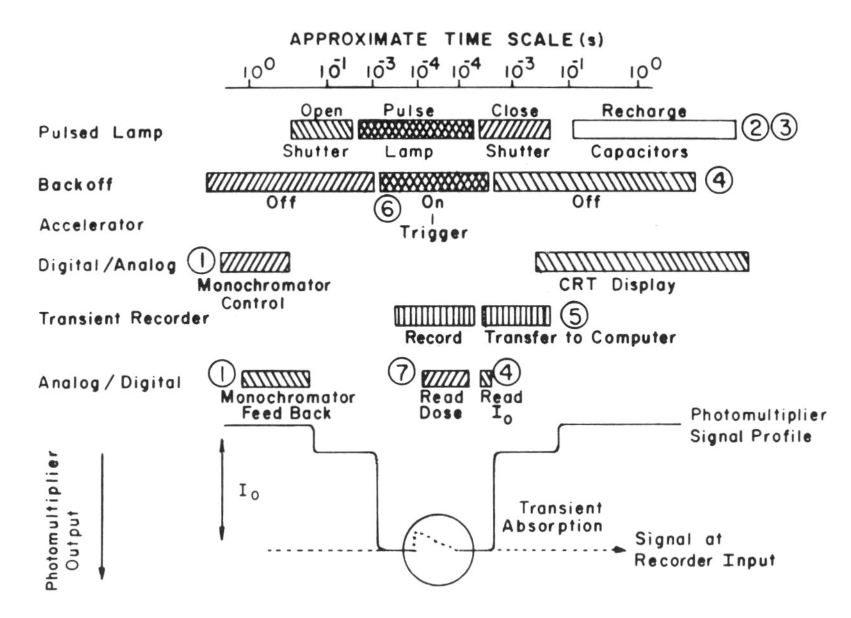


**Figure 3-8.** Sequence of computer procedures in pulse radiolysis measurement; numbers correspond to operations and devices shown in Fig. 3-5.

analyzing light emerging from the irradiation cell is first passed through a spectrograph to wavelength-resolve it into a band spectrum. The band of dispersed light—changing heterogeneously in intensity over time as a function of transient behavior—is focused on a cathode surface that emits electrons in a continuous pattern carrying both the wavelength and time-dependent intensity information. This pattern is swept across a phosphor surface perpendicular to its path by a deflection system. A glowing streak is produced on the two-dimensional phosphor with wavelength plotted in one direction and time plotted in the other. The intensity at any point within the streak is proportional to the intensity of the monitoring light at a given wavelength and time after the radiolytic pulse. The streak pattern is then a record of the transient spectrum over the interval of the streak. This information is recorded by scanning with a TV camera and storing on a video disk. With data from a blank experiment in which no radiation was given, one can calculate absorbance values. A transient spectrum generated in this fashion is given in Figure 3-7(a). Such an approach is most useful for systems in which one has a limited amount of material and cannot expose it to a large dose of radiation. The streak camera is mainly limited by the degree of intensity resolution that may be achieved in the streak; it must be remembered that measurement of $\Delta I$ involves resolving small deviations in large signals. However, the time resolution that may be achieved here is extremely high, and picosecond measurements are possible.

***Computerization in Pulse Radiolysis***

The application of computer technology to pulse radiolysis measurements over the past decade has significantly altered and simplified the experimentalist's role in gathering and processing data.[17] This development, of course, has been linked to the appearance of transient recorders that present data to the computer in digital form. Such systems, if well designed, reduce the complexity of data gathering to the level of a simple laboratory spectrometer. Whereas it was formerly necessary for the person making measurements to exert manual control over various experimental parameters and to record transient data by oscilloscopes and photography, an appropriately designed computerized apparatus can provide almost all necessary interaction through a keyboard. Such a system can provide timing signals for synchronization of lamp pulsing, radiolysis, and data gathering. The functions that may be brought under computer control are outlined in Figure 3-5, and the sequence of operations for generation of kinetic data by absorption spectroscopy is given in Figure 3-8. Further, one may build into the system certain monitoring functions by which the computer may recognize an invalid set of experimental parameters generated in a particular experiment and automatically correct instrument settings. Such a system also organizes the data in a single format from which kinetic and spectral calculations of choice may be made either during or after a series of experiments. One may combine both types of information into a single data display that presents the experimentalist with the overall time course of the reaction studied. An example of such displays is given in Figure 3-7(b).

In initial installations the computers used were stand-alone systems totally dedicated to the experiment underway. The investment in such systems was relatively large, considering that use was restricted to one operation at a time. Today, with the advent of microcomputers, it is possible to design a satallite data-gathering station whose operating program is loaded from a larger host system and to which data is passed for storage and analysis. Such a system minimizes the time that must be dedicated by the host computer to the experiment and, in institutions with need for computerization of several types of experiments, provides the most efficient and inexpensive course to pursue. Most recently, a new generation of personal computers is appearing with powerful data storage, graphics and communications capabilities which are stimulating a re-evaluation of computer strategies for automation.

***Applications***

Of the approximately 4600 pulse radiolysis papers published to date,[18] the vast majority report absorption spectroscopic measurements. Probably the largest application of this particular technique has been to aqueous systems in which reactions subsequent to production of primary water radicals ($e_{aq}^-$, ·OH, ·H) have been investigated. This, of course, includes most work relevant to radiobiology. The titles of several recent reviews are given in the references.[19] Some literature on nonaqueous systems is also cited.[20]

A significant amount of work has also been carried out in the gas phase, making it possible to study the behavior of excited states, recombination of charged species, electron capture, ion–molecule reactions, and excitation transfer.[21] In some cases it is possible to study many of these processes on a long time scale, removing the requirement for having fast time detection.

**Resonance Raman Spectroscopy**

Rather than being absorbed in an electronic transition, a photon interacting with a molecule or radical may emerge from the interaction at the incident wavelength and scattered at various angles. However, the incident photon may stimulate transitions determined by the vibrational-rotational character of the molecule. Such transitions can originate either from ground-state species or from species already excited. If such transitions occur, the scattered photon will be altered in wavelength by an amount corresponding to the energy involved in the transition. This, qualitatively, is the basis for Raman scattering; a spectrum of such "altered" photons can be resolved in terms of molecular vibrational and rotational parameters. Although signals associated with such spectra are extremely weak, significantly stronger signals may be produced by using the phenomenon of *resonance enhancement.* If the incident light coincides with the absorption band of the molecule, Raman transitions involving only vibrational modes coupled to the electronic excitation will be selectively enhanced; this provides signals that are orders of magnitude more intense than those produced under nonresonant conditions. Indeed, to a first approximation, the sensitivity of detection for resonance Raman is a function of the square of the extinction coefficient–bandwidth ratio.[22] Detailed mathematical treatment of Raman data may be found in texts on spectroscopy (eg, ref. 23). In addition to carrying information about the details of molecular structure, the Raman spectrum with its wealth of fine structure approaches much more nearly the concept of a "molecular fingerprint" than does the broad, featureless electron absorption spectrum observed in solution studies. It may be seen, then, that the principal advantages of this approach are in spectroscopic rather than kinetic studies.

Resonance Raman spectroscopy suffers from low sensitivity because the signals produced are considerably weaker than those obtained from absorption. Because the scattered photons are sampled over a very short interval in a single time-resolved measurement, the integrated signal is small. An optimal experimental arrangement requires, then, an intense, tunable, narrow band-pass light source to produce a significant signal-to-noise ratio. Additionally, as per the discussion above, it is useful to average many measurements. To gather spectral information at a particular time after the pulse, one uses a technique analogous to the absorption spectrographic method described above. Here an OMA may be employed to advantage. This device provides, typically, an array of 500 diodes capable of recording intensity as a function of wavelength when linked to a spectrograph.

In an instrument designed at the Notre Dame Radiation Laboratory, a Xe–F laser coupled to a dye laser is used for excitation. With appropriate dyes, wavelengths covering the range from 250 to 800 nm are accessible. This provides the intense pulsed excitation source required. The Raman signal is captured on an OMA whose sensitivity covers the region from 300–900 nm (tubes with response further into the UV are available); the device can be gated in synchrony with the laser at any point following the radiolytic pulse to capture the Raman spectrum at that time. Additionally, to facilitate averaging, measurements may be carried out at a frequency of 1 to 60 Hz. This apparatus provides a high signal-to-noise ratio, and data obtained is of the quality needed for meaningful structural calculations. An outline of the system is given in Figure 3-9, and a well-defined spectrum obtained by summing many experiments together is presented. It may be seen that the high resolution obtained by averaging allows determination of two very close peaks.

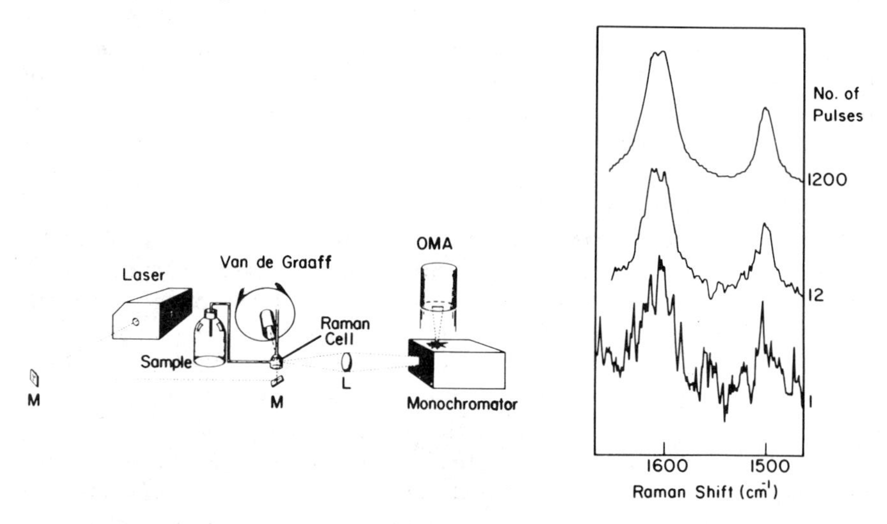


**Figure 3-9.** Apparatus for time-resolved resonance Raman spectroscopy. Spectra from ·OH attack on alkaline methoxylphenol. Spectral resolution reflects number of experiments averaged to distinguish two peaks around 1600 $cm^{-1}$. (Courtesy of G.N.R. Tripathi)

The number of laboratories involved in this type of measurement is not large. Representative papers (including some instrument design) are listed in the references. These include studies of $(CNS_2)^-$,[24a] $\beta$-carotene,[24b] semiquinones and phenoxyl radicals,[24c] and viologen radical.[24d] For phenoxyl radicals vibrational assignments in some detail have been made.[24c] One might anticipate that a potentially important application would exist for cases in which transient absorption bands may represent more than one species but that cannot be resolved by kinetics. It would be possible to determine whether the absorption bands have a single origin by comparing Raman scattering excitation from various regions of the spectrum.

### Rayleigh Scattering

Some classes of macromolecules exhibit changes of shape and size upon being subjected to ionizing radiation; time resolution of such behavior is possible by using Rayleigh scattering. Here, photons are scattered in all directions by the interacting molecules without change of wavelength and, when integrated over all angles, without loss of total incident light. However, the intensity observed at an angle to the incident beam will be a function of the number, size, and shape of scattering particles present. In pulse radiolysis, techniques have been developed to exploit this phenomenon.[25a] The pulsed laser is replaced by a CW system, typically an Ar-ion laser (514 nm), and the scattering intensity is monitored by a photomultiplier because there is only interest in kinetic data. A light guide whose entrance surface is placed at right angles to the irradiation cell gathers the scattered light. This technique has been applied to studies of polymer degradation[25a,b] and then later to macromolecules of biological interest.[25c,d]

## *Electrochemical Techniques*

### Conductivity

Ionizing radiation can lead to the generation of charged species in all media, as may be ascertained from discussions of processes in gases, liquids, and solids to be found in other sections of this book. In water, for example, $H^+$ and $e_{aq}^-$ (as well as a little $OH^-$) are produced immediately after the pulse. Subsequent reactions involving oxidation by ·OH, reduction by $e_{aq}^-$ or protonation by $H^+$ can alter the charge of secondary species. Changes in populations and mobilities of ions following the pulse may be monitored with considerable precision by conductivity methods. Such measurements may either help to clarify the charge behavior of species identified by other methods or may provide insight into the characteristics of intermediates not readily detected by optical means. There are two principle difficulties that must be overcome in the application of this technique. First, the electrical circuit for transient measurement must include the conductivity cell in which the high-energy elections are deposited and, therefore, is subject to considerable noise at the time of irradiation. Second, the measurement must differentiate between the conductivity of the system before and after irradiation; hence, as the conductivity of the sample becomes more significant, the sensitivity of the method falls off.

The change in conductivity of a solution, $\Delta L$, following the radiolytic pulse will be determined by the relationship

$$\Delta L = \frac{1}{(k_c \times 10^3)} (\Delta C_1 Z_1 \lambda_1 + \Delta C_2 Z_2 \lambda_2 + \cdots) \tag{3-1}$$

where $\Delta C$ is the change in concentration of each ion in the system, $Z$ is charge, $\lambda$ is the equivalent conductance, and $k_c$ is an experimental constant in $cm^{-1}$ related to the configuration of the cell. Circuits used for pulse conductivity measurements are outlined in Figure 3-10. When a voltage,

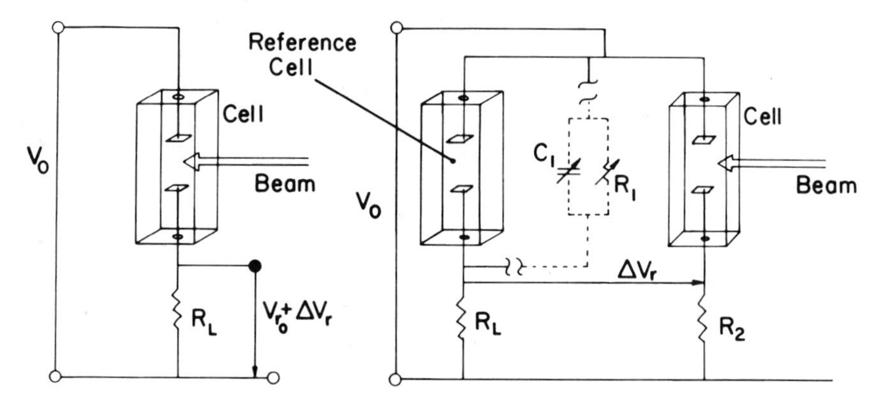


**Figure 3-10.** Circuits used in pulse conductivity: (left) voltage divider arrangement; (right) bridge circuit: reference cell or balancing circuit ($C_1$, $R_1$ shown in dots) can be used.

$V_0$, is applied to the circuit, the corresponding voltage, $V_r$, measured across the load resistor will contain contributions from both the preirradiation system, $V_{r0}$, and the change due to irradiation, $\Delta V_r(t)$,

$$V_r = V_{r0} + \Delta V_r(t) = \frac{V_0(L_c + \Delta L_c(t))}{L_R + L_c + \Delta L_c(t)}$$

$$= \frac{V_0 L_c}{L_R + L_c + \Delta L_c(t)} + \frac{V_0 \Delta L_c(t)}{L_R + L_c + \Delta L_c(t)} \tag{3-2}$$

where $L_R$ is the reciprocal of the load resistor, $L_c$ is the prepulse conductivity of the cell, and $\Delta L_c$ is the transient conductance. Normal experimental conditions dictate that $\Delta L_c(t) \ll L_R$, $L_c \ll L_R$, and $\Delta L_c(t) \ll L_c$. If one compensates for $V_{r0}$ by the bridge technique—correspondingly removing $L_c$ from the numerator—then one may measure $\Delta V_r(t)$ alone, and

$$\Delta V_r(t) = \frac{V_0 R_R}{k_c \times 10^3} (\Delta C_1 Z_1 \lambda_1 + \Delta C_2 Z_2 \lambda_2 \cdots) \tag{3-3}$$

provides a linear relationship between ion concentration change and $\Delta V$

measured. In practice, the term $k_c$ is not evaluated independently; the system may be calibrated using a known reaction that produces a given ion in established yield. The reactions

$$(CH_3)_2HCOH \xrightarrow{OH} (CH_3)_2\dot{C}OH$$

$$\xrightarrow{C(NO_2)_4} \cdot C(NO_2)_3^- + H^+ + NO_2 + (CH_3)_2CO \qquad (3\text{-}4)$$

and

$$CH_3Cl \xrightarrow{H^+, e_{qq}^-} H^+Cl^- + CH_3 \qquad (3\text{-}5)$$

have been widely used for such a purpose. If desired, $k_c$ may be obtained from the conductivity associated with a known dose.[26] In the former case the $C(NO_2)_3^-$ ion absorbs light ($\varepsilon_{350} = 1.5 \times 104\ M^{-1}\ cm^{-1}$); simultaneous detection of absorbance and $\Delta V_r$ can be used to determine both $k_c$ and dose.

It may be readily seen that $\lambda$ determines the types of ions most readily measured by this technique. For example, $H^+(314\ \Omega^{-1}\ cm^2)$ will generate five times more signal than $NH_4^+$ ($64.5\ \Omega^{-1}\ cm^2$), and either $e_{aq}^-$ ($190\ \Omega^{-1}\ cm^2$) or $OH^-(172\ \Omega^{-1}\ cm^2)$ will produce about that much more signal than will $Cl^-(65\ \Omega^{-1}\ cm^2)$. Hence, exchange of one ion for another of the same charge can provide a marked change in solution conductivity and a means for monitoring many reactions. Especially with ions of high specific conductance, it may be shown that good signals can be obtained with $\Delta C$ of $10^{-6}$–$10^{-7}$ $M$. Also, it may be noted from Eq. (3-2) that as the ionic strength of the sample solution increases, $L_c$ becomes prominent and $\Delta V$ will fall with a corresponding loss in sensitivity. This behavior presents a limit to the solution conditions under which transient conductivity measurements can be carried out with the circuitry described. If one increases $V_0$ to compensate, problems of electrolysis will become important.

There are several variations on the basic apparatus described above that are employed for special experimental conditions. The time resolution for pulse conductivity systems is, in theory, an $RC$ product of the cell capacitance multiplied by the parallel combination of the resistive load and the cell resistance. However, for work in short times one must also contend with spurious signals arising at the time of the pulse due to deposit of electrons in the cell and air ionization outside the cell. One system has been developed that applies a large voltage pulse of positive or negative polarity for the duration of the measurement; a backoff system is used to compensate for this large signal and provide measurement of the baseline.[27] Data from two measurements with pulses of opposite sign are subtracted to eliminate the spurious signals. The short duration of the voltage pulse avoids the problems of electrolysis. Experiments in the nanosecond to millisecond range have been shown possible with this arrangement. See Figure 3-11. Elimination of artifacts associated with the radiolytic pulse have been approached with use of transformer circuitry and limiting amplifiers.[28,29] The latter is capable of picosecond resolution.

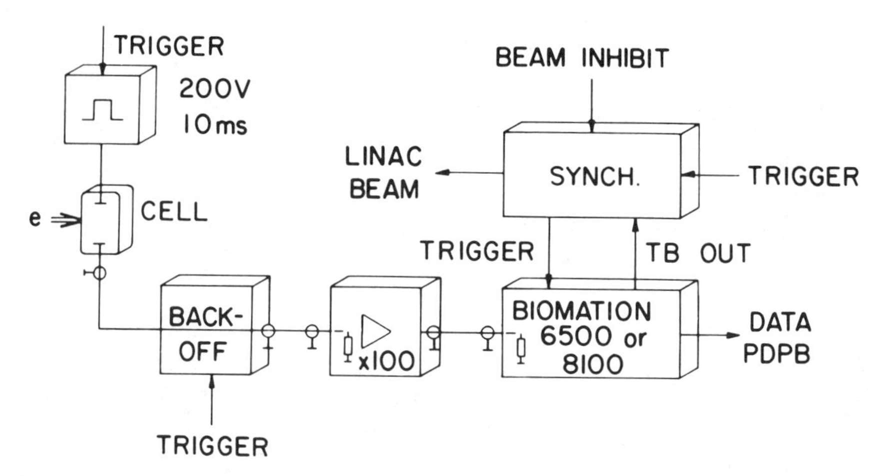


**Figure 3-11.** Arrangement for application of monitoring pulse in conductivity measurement (ref. 29).

Using some of the circuits described above in extended times, that is, in the millisecond range and longer, can lead to problems with polarization and electrolysis. Some laboratories have employed alternating-current (ac) circuitry to avoid these problems.[30] In this technique the voltage applied to the radiation and reference cells is high-frequency ac (eg, 10 MHz), which also provides the reference for a synchronous detector rectifying the signal from the bridge circuit. Such systems are best applied in the range from microseconds to seconds and have response times limited by the ac frequency, $\tau = 1/f$.

Conductivity measurements are also carried out in nonpolar media by using variations on the instrumentation described. Picosecond processes have been measured by using a coaxial cell with parallel plate detectors.[31] Conductivity may also be monitored at short times by following microwave power loss in a cavity containing a very low dielectric gas or medium.[32] Such a technique has been applied to the study of geminate processes and behavior of electrons in these types of media; instruments with nanosecond and subnanosecond response times have been constructed. Parenthetically, it might be noted that the microwave approach has also been applied to studies in ice.[33]

Conductivity measurements have been used to characterize a wide variety of phenomena in aqueous systems: neutralization, deprotonation,[34] dissociation of radicals, electron reactions,[26,30] and ligand exchange in inorganic systems.[35] Studies of ion and electron behavior have also been carried out in a wide variety of nonpolar solvents with both conventional and microwave techniques.[32,36] Various types of processes have been studied in gases by microwave techniques.[37]

**Polarography**

The polarographic technique, used to determine the redox properties of stable molecules, may also be applied to transient species in systems where significant populations of intermediates can be produced near the electrode. This is accomplished by passing the high-energy electron beam through the polarographic cell with the proper geometry.[38,39] The experimental arrangement for such a device is shown in Figure 3-12. Instruments incorporating dropping mercury electrodes and hanging drop electrodes have been used.

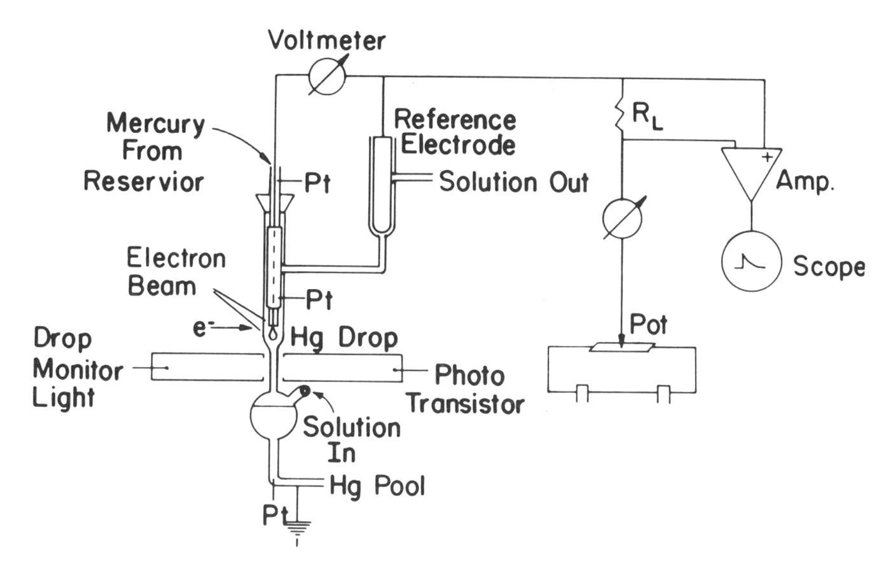


**Figure 3-12.** Irradiation chamber and circuit outline for pulse polarography (refs. 38, 39).

The former has the advantage of renewing the surface but exhibits the disadvantage of capacitive current flow due to growth of the drop at the time of the measurement. As in the case of conductivity, the time resolution of the instrument will depend on the *RC* product of the system, where *R* is the circuit resistance and *C* is the capacitance of the double layer associated with mercury drop electrode.

In normal operation a given electrode potential is set at a selected value, and the polarographic current due to transient oxidation or reduction is then measured as a function of time after the radiolytic pulse. However, the relationship between measured current as a function of time and radical concentration is more complicated than in either conductivity or optical measurements. Here, the signal will depend on concentration of species at the electrode and, of course, the rate of the redox reaction. The radical concentration at the electrode is altered not only by chemical reaction and redox processes but also by diffusion, which replenishes radicals at the electrode surface. Further, the reversibility of the radical redox reaction at the electrode will also determine the form of the relationship.

An example is selected in which it is assumed that the radicals are energetically stable for the period of measurement and that the reaction is irreversible.[38,40] Many carbon-centered radicals, indeed, do exhibit irreversible redox behavior. In this case radiolytic generation of radicals around an electrode, whose potential is appropriate for redox processes in the radical, becomes equivalent to a standard polarographic experiment in which the electrode potential is suddenly shifted to a level at which electron transfer can occur. The mathematics of the latter experiment can be applied to the pulse radiolysis data. In Figure 3-13(a) one may see the effect that

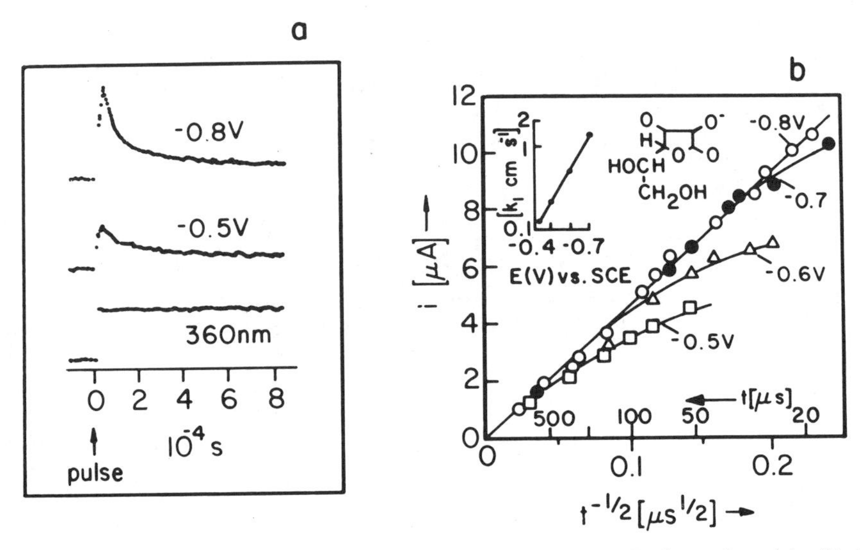


**Figure 3-13.** Data from pulse polarography of ascorbic radical produced by $Br_2^-$ (a) cathode current and absorbance vs. time; (b) current vs. $t^{1/2}$ for various potentials (ref. 38).

applied potential has on the decay of ascorbic acid radicals. At various potentials the current measured as a function of time after the pulse is plotted against $t^{-1/2}$. The straight line represents current, $i_d$, due to diffusion alone and allows calculation of the radical diffusion constant, $D$. From values of $i/i_d$ taken from this graph and $D$, one may extract the transfer rate constants, $k_f$, as a function of potential; the slope of a resultant plot of $k_f$ vs. potential gives the transfer coefficient, $\alpha$, of the cathodic process.[38] For systems in which either the radical is short lived or in which the assumption of irreversibility does not hold, the data treatment changes. Space does not permit the arguments required to be appropriately developed; the reader is encouraged to consult the literature cited for the treatment of such cases.[38,41]

This technique has been used to extract information concerning diffusion and redox behavior from a variety of transient systems. Some of the most prominent work has involved alcohol and other aliphatic radicals. The reviews cited provide a survey of the types of systems to which the technique has been applied and quantitative information on the parameters obtained.[38,41,42]

## *Magnetic Resonance Methods*

### Electron Spin Resonance

The methods described above may be used in myriad systems and exhibit sensitivities that may be applied to very small populations of transient species. However, for none of those methods, except to some extent with Raman spectroscopy, is there data that may be directly interepreted to give information on the structure of transients.

By contrast, magnetic resonance techniques display more limited sensitivity in most systems when compared to optical spectroscopy and are subject to very real time limits imposed by physical events at the molecular level. However, the patterns of data obtained from electron spin resonance (ESR), and to some extent specialized nuclear magnetic resonance (NMR) measurements may be used to describe the molecular structure of radicals. Further, this data can be used to characterize radical events that, in perturbing the magnetic character of the system, provide chemical information inaccessible by other means.

The techniques of ESR have been widely applied to studies of free radicals which, by definition, exhibit unpaired electrons. The main components of the apparatus are shown in Figure 3-14 along with the appropriately oriented coordinate frame for the vector model; this model, although not a rigorous treatment of magnetic behavior, is useful for simple descriptions of ESR phenomena, such as those given below. In the magnetic field, $H_0$ (characteristically 1–10 kG), applied to a radical population, the spin magnetic moment of an unpaired electron can take two principal orientations to the field; the associated energies may then be described by

$$E = 2\beta mH \qquad \text{with } m = \pm 1/2 \tag{3-6}$$

corresponding to the two orientations. (In fact, though the integer is written here as 2, the true value is 2.0023.) Here $\beta$ is the Bohr magnetron. Relative populations of the two states in an ensemble of radicals will be determined by the Boltzmann distribution. The magnetization, $M$, for the overall radical ensemble may be visualized in the vector precessing about the direction of the field $H_0$ at a frequency proportional to the strength of $H_0$, and its magnitude (hence component along the $z$-axis) is determined by the difference in populations of the two states. The transitions that may occur,

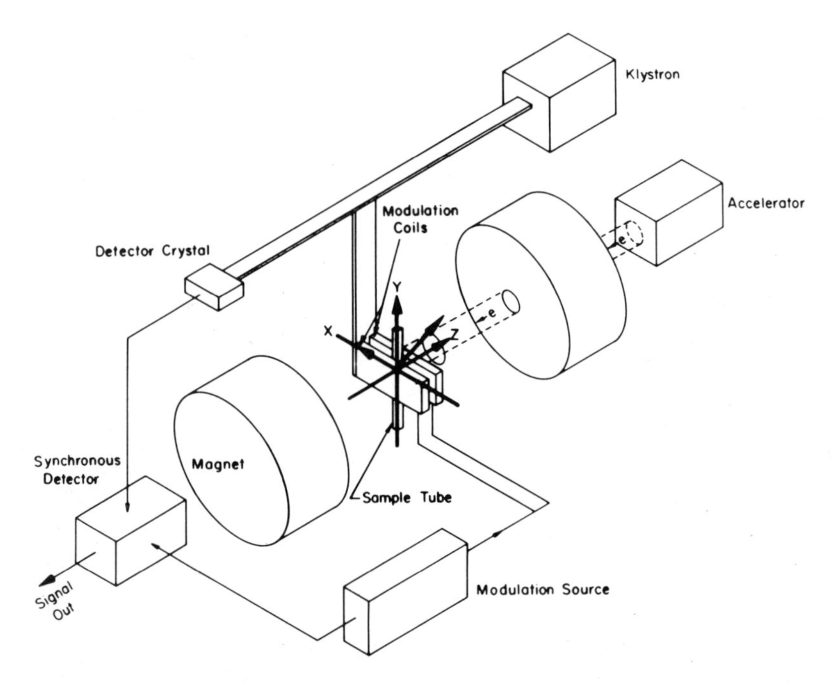


**Figure 3-14.** Physical arrangement for irradiation in ESR cavity. Vector model is superimposed on cavity.

either by absorption or emission, are defined by

$$\Delta E = E_{+1/2} - E_{-1/2} = 2\beta H \tag{3-7}$$

and will take place if, when bathed in radiation of frequency $\nu$, $H_0$ is adjusted to the resonance condition

$$h\nu = \Delta E = 2\beta H_0 \tag{3-8}$$

Such radiation is provided by a microwave cavity normally operating in the range of 10 GHz. In the vector model the sinusoidal magnetic component of the microwave radiation may be visualized as oscillating in the *x-y* plane. When this oscillation frequency equals the precession frequency of $M$, the condition of resonance is reached. Absorbance or emission (depending on the relative populations of the upper and lower states) "tips" the vector $M$ toward the *x-y* plane. A condition of equal populations would be represented by $M$ lying in that plane. Experimentally, absorption or emission may be readily monitored by changes in transmitted or reflected microwave power (depending on cavity design.) These data can be plotted as a function of applied magnetic field to give the ESR spectrum.

In reality, the factor in Eqs. (3-7) & (3-8) must be replaced by a constant, $g$, since there will be deviations from the free electron value of 2.0023 arising from interactions between the electron and the molecular environment in which it is found. Most importantly, the field local to the electron can experience discrete perturbations due to nuclear magnetic moments of atoms with which the electron interacts. These perturbations give rise to patterns of lines within the spectrum that may readily be interpreted in terms of molecular structure associated with the unpaired electron. This structural information, which may be derived from ESR data, then, constitutes the real power of the technique. The reader is encouraged to seek out some of the many excellent rigorous treatments of magnetic phenomena and their interpretation in terms of molecular structure.[43]

At the most elementary level it is possible to prepare stable populations of radicals at low temperatures in which diffusion is minimized and, consequently, in which the lifetimes can be long enough for measurement. This technique, however, exhibits severe limitations as to the type of system that can be investigated and the measurement of secondary radicals that must be created by diffusional processes.

Although not providing time-resolved measurements, techniques have been developed to emulate by flow-mixing systems the generation of ·OH. These may be used in some systems to provide ESR data on unstable radicals in solution. The first utilizes reactions of the form[44]

$$M^{+n} + H_2O_2 \rightarrow M^{+(n+1)} + OH^- + \cdot OH \quad (3\text{-}9)$$

Alternatively, hydrogen peroxide can be photolytically cleaved to provide ·OH radicals for ESR measurements.

Of more general application is the formation of radicals in the ESR cavity by actual in situ radiolysis.[45] For proper focusing of the electron beam in the cavity, a coaxial passage for the beam through the magnet is provided. Solutions containing the parent molecule flow through the cavity and are irradiated by a dc beam from the Van de Graaff generator. The experimental approach used here is shown in Figure 3-14. If a steady-state population of about $10^{11}$ spins can be maintained (given, say, a single line spectrum 1 G wide), spectra can be gathered. Some kinetic parameters can also be obtained from this method. For example, second-order rate constants may be derived from signal dependence on electron beam intensity. As a further example, many relative rate constants for H-atom reaction have been measured by using competition techniques.[46]

In true time-resolved measurements there are several further considerations bearing on the origins of line intensity and width in a given ensemble of radicals. Because the applied microwave field induces both absorption and emission by the spin states of the electron, net microwave power change will depend on populations of the $m = \pm 1/2$ states. Because of the Boltzmann distribution of spins, the lower energy state is the more populated, and absorption may be expected. In normal, steady-state measurements, the relative population of states will remain essentially constant as transitions from the lower to higher states are balanced by relaxation processes returning the system to equilibrium.

In the period immediately after pulsed irradiation, however, the time-dependent behavior of the ESR spectrum may reflect several simultaneous processes that perturb the population of spin states. There will, of course, be an overall loss resulting from radical reactions resulting in products with no unpaired spins. The rate of line intensity change due to this effect may be described in simple kinetic terms. Second, where the radical production process results in a nonequilibrium population of the two spins, there will be a relaxation toward the equilibrium condition; this rate term may be expressed simply by $k(r)(M - M_0)$, the rate constant for relaxation multiplied by the deviation of the number of spins from its equilibrium population, $M_0$, at any moment, $t$. Finally, since the energies of the two spin states differ slightly, their reactivities may be expected to differ also. If, for example, a second-order reaction is involved, the rate of the difference in disappearance for each spin may be shown to be $d(r_2 - r_1)/dt = (k_1 - k_2)R^2$, depending on the sign of $(k_2 - k_1)$; here, $R$ is the total radical concentration, and $k_2$ and $k_1$ are the rate constants for radicals in the upper and lower spin states.[47] The terms $r_2$ and $r_1$ represent the populations of upper and lower states. This will result in an excess population of one spin state and an overall shift in the magnetization. This latter effect is termed chemically induced dynamic electron polarization (CIDEP) and must be taken into account in time-resolved ESR measurements. (There are also other factors involved in the definition of CIDEP.) The contributions from these various factors will be reflected in the populations of spin states and govern both the sign and intensity of the microwave interaction with the spin system. A rigorous mathematical treatment of these contributions to time-resolved behavior and examples of their application to experimental data may be found in the literature.[47] Obviously the quantitative interpretation of such complications in real cases can yield additional insights into chemical dynamics. For example, the CIDEP phenomenon provides information on the relative reactivity of the two spin states. Information can also be gathered on nonreactive encounter of radicals. Such events will enhance spin relaxation through Heisenberg exchange; hence, the radical concentration dependence of $k(r)$—the reciprocal of the relaxation time—provides a measure of these encounters.[48]

There are two principal experimental techniques for applying pulse radiolysis to ESR measurements. These differ by the way in which the radical population is interrogated after the radiolytic pulse. The first involves essentially the normal experimental arrangement described above, using CW microwave radiation.[49] However, this approach dispenses with modulation of the magnetic field. If this microwave power is too high, saturation of the spin system (roughly equal populations of states) can occur, leading to emission by the system, and oscillations can follow. In general, the time resolution of this approach is not the highest of the two ($\mu$s) but the line resolution is generally better. In fact, submicrosecond measurements of ESR spectra are subject to uncertainty broadening that limits the available resolution.[49] Additionally, this approach allows one to sit at a single field strength and scan the whole time course for *every* measurement in a manner analogous, in some respects, to that employed for optical absorption.

Time resolution on the nanosecond scale may be achieved with the second technique, which takes advantage of the spin echo phenomenon to make a "snapshot" of the radical population with high time resolution and to "develop" it later.[50] Events outlined in Figure 3-15 using the vector illustrate qualitatively the principle of the echo technique.[51] A radical ensemble

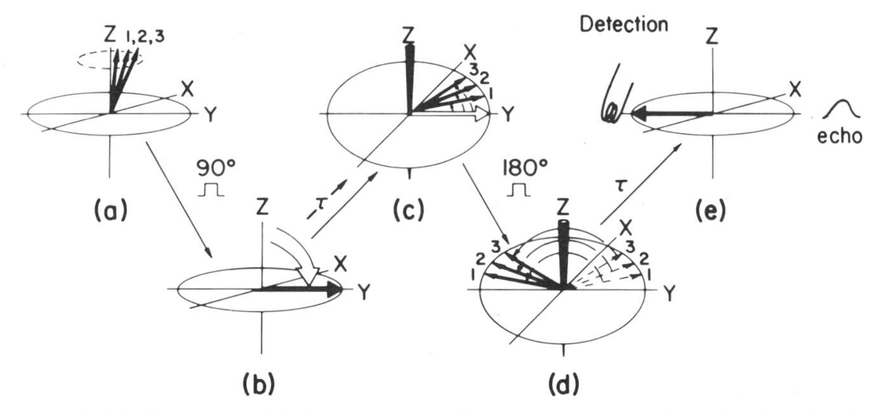


**Figure 3-15.** Vector model for sequence of events in generation of spin echo for nanosecond ESR measurements.

produced by pulse radiolysis in the cavity precesses about the $z$-axis, the direction of the external magnetic field (a). At the time for which one wishes a measurement of the radicals present, the system is subjected to a microwave pulse (termed the 90° pulse) 30 to 100 ns in duration. Microwave absorption isolates a population of radicals, which, in the model, is described by magnetic moments bent 90° into the $x$-$y$ plane (b). This population is now collectively in phase with respect to the interaction with $H_0$ and the direction of the microwave input. The population in the $x$-$y$ plane represents a "snapshot" of the radical population (as reflected through its magnetization) at the time of the 90° pulse. Due to slightly differing precession frequencies, this population subsequently undergoes an orderly dephasing as its components orbit the applied field, $H_0$ (c). After a time interval, $\tau$ (0.2–1 $\mu$s), the system is subjected to a second, more powerful pulse (termed the 180° pulse) that throws this population into a mirror image relationship but still in the $x$-$y$ plane. Since the direction and frequencies of precession have not been changed, this population will rephase and generate an echoed signal, again after time $\tau$. Detection is symbolized in the figure by the presence of a coil. The echo represents a valid picture of the radical system at the time of the 90° pulse. Other radicals generated and magnetization changes due to CIDEP after the 90° pulse will not be synchronized to the pulse sequence and therefore will not alter the relative composition of this "picture" population. The absolute strength of the echo is, however, subject to losses arising from spin–spin interaction and spin–lattice relaxation. In most cases such losses are minimal for the intervals, $\tau$, employed. This technique cannot be applied when the sample population disappears over time $2\tau$. A time profile is built by the sampling technique; the experiment is reproduced at a high repetition rate and a representative picture of the time profile from behavior of the echo as a function of varying the delay between the radiolytic event and the 90° pulse. The particular strengths of the method lie in the available time resolution and isolation of the actual detection events until initial electronic perturbations associated with the radiolytic pulse have settled.

**Other Techniques**

There are several other magnetic resonance techniques relevant to special situations. One such technique is fluorescence detected magnetic resonance (FDMR).[52] Radiolysis in nonpolar solvents generates solvent ions that may interact with dissolved solutes to give secondary ions. Charge transfer recombination between solute and solvent ion can generate charge-neutral excited species that subsequently fluoresce. If radiolysis of the system in a variable magnetic field is followed by a microwave pulse, coupling between singlet and triplet states of the ions can occur at the resonance field strength during the period of the microwave pulse. The products of recombination will reflect these perturbations, and the fluorescence will be decreased. By monitoring the fluorescence intensity as a function of applied field, one may map the ESR spectrum of the geminate ions. For these special applications, where radical ions produce excited states, high sensitivity and time resolution can be achieved. The apparatus is similar to that given in Figure 3-14, except that the microwave cavity is pulsed and a system for gathering fluorescence is used.

***Nuclear Magnetic Resonance***

Although normally used to study diamagnetic systems, NMR has been applied indirectly to the characterization of radical species.[53] Even as one observes one form of CIDEP in radicals, one may similarly observe a nonequilibrium population of nuclear spins in products of radical reactions occurring in a magnetic field (CIDNP). Because the relaxation times for nuclear spins are so much longer than those for electrons, the induced "memory" of the radical reaction persists over longer times. In this type of pulse radiolysis experiment, two sets of magnets may be employed: one in which the radiolysis takes place, and the second associated with the NMR spectrometer. It is necessary to isolate the radiolysis from the spectrometer because the homogenity requirements of the magnetic field preclude introducing a coaxial pathway through the system. Because of the slow relaxation times (1–3 s), it is possible to flow material from the irradiation chamber to the NMR spectrometer in a time that allows the polarized products to be observed. The CIDNP exhibits a strong field dependence that also carries information about the history of the radical mechanism. Time resolution may be introduced into this approach at intervals after the radiolysis by means of a microwave pulse in the irradiation cavity.[54]

### *Applications*

The types of spectral patterns generated by the species under study limit application of magnetic resonance to pulse radiolysis measurements. The method is not as generally applicable as, say, optical absorption (though, when feasible, the data gathered often provides more insight into the nature of the intermediates). Nevertheless, data has been gathered from a variety of radicals in aqueous systems and hydrocarbon solvents. The literature cited, principally reviews, gives a reasonable sampling of the systems treated.[45,47,52,53]

## SUMMARY

Over the past two decades pulse radiolysis studies have provided insights into the unique modes of behavior exhibited by radiolytically generated transients. These insights have formed the basis for our understanding of the mechanisms by which ionizing radiation can transform materials. Experimental techniques have extended the accessible time resolution even into the picosecond domain. Additionally, these techniques have evolved from the initial pioneering measurements of transient optical absorption to include highly sophisticated, if sometimes less generally applicable, characterization of transient electron and nuclear resonance phenomena. No small fraction of these advances have been made possible by the electronic revolution of these 20 years, which has provided components with wide functional scope, high sensitivity, small size, and remarkable time resolution. Additionally, the development of laboratory computers and the associated peripherals incorporating digital logic—transient recorders, digital-to-analogue converters, timing devices—have made possible many of the procedures necessary for more sophisticated measurements. Without them, sampling techniques, averaging, and many instrument control functions would not be possible. Of course, they have enhanced beyond measure our ability to deal with complex data sets.

It should be clear from the limited discussion that to press beyond his present understanding of transient behavior the experimentalist will be well served by a knowledge of analogue and digital electronics even if he is not principally involved with instrument design. Only with an understanding of the phenomena involved and their electronic manifestations may one hope to add to the array of tools at hand or extend present techniques into new areas.

## ACKNOWLEDGMENTS

Helpful conversations with many members of the laboratory as well as assistance with figure design from M. Blank are gratefully acknowledged. Support for manuscript preparation was provided by the Office of Basic Energy Sciences, Department of Energy. This is Document No. NDRL-2526 from the Notre Dame Radiation Laboratory.

## REFERENCES

1. (a) McCarthy, R. L.; MacLachlan, A. *Trans. Faraday Soc.* **1960**, *56*, 1187; (b) Keene, J. P. *Nature* **1960**, *188*, 843; (c) Matheson, M. S.; Dorfman, L. *J. Chem. Phys.* **1960**, *32*, 1870; (d) Boag, J. W.; Steel, R. E. *Brit. Emp. Cancer Campaign Rept., Part II* **1960**, *38*, 251.
2. Boag, J. W. In "Actions Chimiques et Biologiques des Radiations"; Haissinsky, M., Ed.; Masson: Paris, 1963.
3. Boag, J. W. *Trans. Faraday Soc.* **1968**, *64*, 677.
4. (a) Ramler, W.; Mavrogenes, G.; Johnson, K. In "Fast Processes in Radiation Chemistry and Biology"; Adams, G. E.; Fielden, E. M.; and Michael, B. D., Eds,; John Wiley: Bristol, 1975, p. 25; (b) Sauer, M. C. In "The Study of Fast Processes and Transient Species by Electron Pulse Radiolysis"; Baxendale, J. H.; and Busi, F., Eds; Reidel: Dordrecht, 1981, p. 35 and references therein.
5. (a) Halliday D.; Resnick, R. "Physics, Part II"; John Wiley; New York, 1960, pp. 733–736; (b) Deal, T. R., private communication.
6. Christensen, H. C.; Nilsson, G.; and Thomas, K.-A. *Chem. Phys. Lett.* **1973**, *22*, 533.
7. Zimek, Z. *Radiat. Phys. Chem.* **1978**, *11*, 179 and references therein.
8. Karzmark, C. J. *Rev. Sci. Instr.* **1964**, *35*, 1645.
9. Morliere, P.; Patterson, L. K. *Inorg. Chem.* **1982**, *21*, 1837.
10. Hunt, J. W.; Greenstock, C. L.; Bronskill, M. J. *Int. J. Radiat. Phys. Chem.* **1972**, *4*, 87.
11. Roffi, G. In "The Study of Fast Processes and Transient Species by Electron Pulse Radiolysis"; Baxendale, J. H.; and Busi, F., Eds.; Reidel: Dordrecht, 1982, p. 63.
12. Bell, D. H.; Gould, J. M.; Patterson, L. K. *Radiat. Res.* **1982**, *90*, 518.
13. Hodgson, B. W.; Keene, J. P.; Land, E. J.; Swallow, A. J. *J. Chem. Phys.* **1975**, *63*, 3671.
14. Bronskill, M. J.; Hunt, J. W. *J. Phys. Chem.* **1968**, *72*, 3762.
15. (a) Hodgson, B. W.; Keene, J. P. *Rev. Sci. Inst.* **1972**, *43*, 493; (b) Beck, G. *Rev. Sci. Instr.* **1974**, *45*, 318; (c) Luthjens, L. H. *Rev. Sci. Instr.* **1973**, *44*, 1661; (d) Fenger, J. *Rev. Sci. Instr.* **1981**, *52*, 1847.
16. (a) Gordon, S.; Schmidt, K. H.; Martin, J. E. *Rev. Sci. Instr.* **1974**, *45*, 552; (b) Schmidt, K. H.; Gordon, S.; Mulac, W. A. *Rev. Sci. Instr.* **1976**, *47*, 356.
17. (a) Patterson, L. K.; Lillie, J. *Int. J. Radiat. Phys. Chem.* **1974**, *6*, 129; (b) Eriksen, T. E.; Lind. J.; Reitberger, T. *Chem. Scr.* **1976**, *10*, 5; (c) Thornton, A. T.; Laurence, G. S. *Radiat. Phys. Chem.* **1978**, *11*, 311; (d) Ross, C. K.; Lokan, K. H.; Teather, G. G. *Comput. Chem.* **1979**, *3*, 89; (e) Foyt, D. C. In "The Study of Fast Processes and Transient Species by Electron Pulse Radiolysis"; Baxendale, J. H.; and Busi, F., Eds; Reidel: Dordrecht, 1982, p. 213.
18. Compilation by the Notre Dame Radiation Laboratory Data Center.
19. (a) Land, E. J. *Current Topics Radiat. Res. Quart.* **1972**, *7*, 105; (b) Baxendale, J. H.; Rodgers, M. A. J. *Chem. Soc. Rev.* **1978**, *7*, 235; (c) Buxton, G. V. *Inorg. React. Mech.* **1981**, *7*, 106; (d) Benssason, R. V.; Land, E. J.; Truscott, T. G. "Flash Photolysis and Pulse Radiolysis, Contributions to the Chemistry of Biology and Medicine", Pergamon Press: New York, 1983.
20. Busi, F. In "The Study of Fast Processes and Transient Species by Electron Pulse Radiolysis"; Baxendale, J. H.; and Busi, F., Eds; Reidel: Dordrecht, 1982, p. 417; Freeman, G. R. *ibid.*, p. 399.

21. (a) Sauer, M., Jr. In "Advances in Radiation Chemistry"; Burton, M.; and Magee, J. L., Eds; John Wiley: New York, 1976, Vol. 5, p. 97; (b) Willis, C.; Boyd, A. W. *Int. J. Radiat. Phys. Chem.* **1976**, *8*, 71; (c) Sauer, M. C. In "The Study of Fast Processes and Transient Species by Electron Pulse Radiolysis"; Baxendale, J. H.; and Busi, F., Eds; Reidel; Dordrecht, 1982, p. 601.
22. (a) Albrecht, A. C. *J. Chem. Phys.* **1961**, *34*, 1476; (b) Albrecht, A. C.; Hutley, M. C. *ibid.* **1971**, *55*, 4438.
23. Herzberg, G. "Molecular Spectra and Molecular Structure. 2. Infrared and Raman Spectra of Polyatomic Molecules" Van Nostrand: Princeton, 1951.
24. (a) Pagsberg, P.; Wilbrandt, R.; Hansen, K. B.; Weisberg, K. V. *Chem. Phys. Lett.* **1976**, *39*, 538; (b) Dallinger, R. F.; Guanci, J. J.; Woodruff, W. H.; Rodgers, M. A. J. *J. Am. Chem. Soc.* **1979**, *101*, 1355 (c) Tripathi, G. N. R.; Schuler, R. H. In "Time Resolved Vibrational Spectroscopy"; Atkinson, G. H., Ed; Academic Press: New York, 1983 (and references therein); (d) Lee, P. C.; Schmidt, K.; Gordon, S.; Meisel, D. *Chem. Phys. Lett.* **1981**, *80*, 242.
25. (a) Beck, G.; Kiwi, J.; Lindenau, D.; Schnabel, W. Z. *Eur. Poly. J.* **1974**, *10*, 1069; (b) Schnabel, W. In "Developments in Polymer Degradation"; Grassie, N. Ed; Applied Science Publ.: London, 1979 (and references therein); (c) Lindenau, D.; Hagen, U.; and Schnabel, W. Z. *Radiat. Environ. Biophys.* **1976**, *13*, 287 (and subsequent papers by Schnabel); (d) Hashimoto, S.; Seki, H.; Masuda, T.; Imamura, M.; Kondo, M. *Int. J. Radiat. Biol. Relat. Stud. Phys., Chem. Med.* **1981**, *40*, 31.
26. (a) Asmus, K.-D. *J. Radiat. Phys. Chem.* **1972**, *4*, 417 (and references therein); (b) Balkas, T. I.; Fendler, J. H.; Schuler, R. H. *J. Phys.* **1970**, *74*, 4497.
27. Janata, E. *Radiat. Phys. Chem.* **1982**, *19*, 17.
28. Maughan, R. L.; Michael, B. D.; Anderson, R. F. *Radiat. Phys. Chem.* **1978**, *11*, 229.
29. Janata, E. *Radiat. Phys. Chem.* **1980**, *16*, 37.
30. Asmus, K.-D.; Janata, E. In "The Study of Fast Processes and Transient Species by Electron Pulse Radiolysis"; Baxendale, J. H.; and Busi, F., Eds; Reidel; Dordrecht, 1982, p. 91 (and references therein).
31. Beck, G. *Rev. Sci. Instr.* **1979**, *50*, 1147.
32. Warman, J. M. In "The Study of Fast Processes and Transient Species by Electron Pulse Radiolysis"; Baxendale, J. H.; and Busi, F., Eds.; Reidel: Dordrecht, 1982, p. 129 (and references therein).
33. Verberne, J. B.; Loman, H.; Warman, J. M.; de Hass, M. P.; Hummel, A.; Prinsen, L. *Nature* **1978**, *272*, 343.
34. Janata, E.; Schuler, R. H. *J. Phys. Chem.* **1980**, *84*, 3351.
35. Simic, M.; Lilie, J. *J. Am. Chem. Soc.* **1974**, *96*, 291.
36. (a) Allen, A. O.; Holroyd, R. A. *J. Phys. Chem.* **1974**, *78*, 796 (and references therein); (b) Dodelet, J.; Freeman, G. R. *Can. J. Chem.* **1977**, *55*, 2264.
37. Shimamori, H.; Fessenden, R. W. *J. Phys. Chem.* **1978**, *69*, 4732; *ibid.* **1979**, *70*, 1137; *ibid.* **1981**, *74*, 453; (b) Warman, J. M.; Sennhauser, E. S.; Armstrong, D. A. *ibid.* **1979**, *70*, 995.
38. Henglein, A. In "Electroanalytical Chemistry—A Series of Advances"; Bard, A. J., Ed.; Marcel Dekker: New York, 1976, Vol. 9, p. 163 and references therein.
39. Lilie, J. *J. Phys. Chem.* **1972**, *76*, 1487.
40. Gratzel, M.; Henglein, A. *Ber. Bunsenges. Phys. Chem.* **1973**, *77*, 2.
41. Asmus, K.-D.; Janata, E. In "The Study of Fast Processes and Transient Species by Electron Pulse Radiolysis"; Baxendale, J. H.; and Busi, F., Eds.; Reidel: Dordrecht, 1982, p. 11 and references therein.
42. Pleticha-Lansky, R. *Radiat. Res. Rev.* **1974**, *5*, 301.
43. Grody, W. In "Techniques of Chemistry", West, W. and Weissberger, A. Eds.; Wiley Interscience: New York, 1980, Vol. 15.
44. Czapski, G. *J. Phys. Chem.* **1971**, *75*, 2957 (and references therein).
45. Fessenden, R. W.; Schuler, R. H. In "Advances in Radiation Chemistry Vol. 2"; Burton, M.; and Magee, J. L., Eds; Wiley R. W. *J. Chem. Phys.* **1963**, *39*, 2147.
46. Neta, P. *Chem. Rev.* **1972**, *72*, 533.
47. Fessenden, R. W. In "Chemically Induced Magnetic Polarization"; Reidel: Dordrecht, 1977, p. 11 and references therein.
48. Verma, N. C.; Fessenden, R. W. *J. Chem. Phys.* **1976**, *65*, 2139; (b) Syage, J. A.; Lawler, R. G.; Trifunac, A. D. *ibid.* **1982**, *77*, 4774; (c) Fessenden, R. W.; Hornak, J. P.; Vankataraman, B. *ibid.* **1981**, *74, 3694.*

49. (a) Fessenden, R. W. *J. Chem. Phys.* **1973**, *58*, 2489; (b) Verma, N. C.; Fessenden, R. W. *ibid.* **1973**, *58*, 2501.
50. Trifunac, A. D.; Norris, J. R.; Lawler, R. G. *J. Chem. Phys.* **1979**, *71*, 4380.
51. Macomber, J. D. "The Dynamics of Spectroscopic Transitions"; Wiley Interscience: New York, 1976, p. 226.
52. Trifunac, A. D.; Smith, J. P. In "The Study of Fast Processes and Transient Species by Electron Pulse Radiolysis"; Baxendale, J. H.; and Busi, F., Eds; Reidel: Dordrecht, 1981, p. 179 and references therein.
53. Trifunac, A. D. *ibid.*, p. 163 and references therein.
54. Trifunac, A. D. *ibid.*, p. 347 and references therein.

# 4

# Kinetics in Radiation Chemistry

**A. HUMMEL**

*Interuniversitair Reactor Instituut*
*Delft, The Netherlands*

## INTRODUCTION

In radiation chemistry we deal with reactive species that are not formed at random positions in space. Excitation and ionization occur along the track of the primary particle in single events and groups of events of various size, the spacing along the track depending on the velocity of the primary particles and the density of the medium. The electrons of low energy ejected in the ionization processes mostly get thermalized in the neighborhood of the positive ion, except in very low-density media, and groups of correlated pairs of charged species are formed. These pairs may either recombine or escape from the "initial" recombination, and eventually become homogeneously distributed in space, when recombination with "ions" from other groups can take place. In the meantime, reaction of the ions also may occur, either unimolecularly, as happens with excited positive ions, or in a reaction with molecules of the medium.

Excited states, formed either directly or as a result of charge neutralization, frequently dissociate and give rise to spatially correlated radicals. Single pairs of radicals and groups of various numbers of radicals are formed in this way along the track. Also in this case the initial distribution of the reactive species is nonhomogeneous, with regions of high local concentration along the track, where mutual reaction of the species is favored. The time


*Radiation Chemistry: Principles and Applications*

scale at which the nonhomogeneity persists is, of course, dependent on the diffusion coefficients of the reactive species and, therefore, may differ appreciably for different species. (When comparing, for example, the kinetics of the excess electron in liquid cyclohexane and that of the cyclohexyl radical, this differs by three orders of magnitude.) The kinetics of the formation and decay of the various reactive species in a given medium in space and time (charged species, excited states, radicals) is partially interlinked and may be rather complex.

In the observable effects of radiation the effects of the chemical properties of the species formed and the effects of the spatially inhomogeneous formation are often intricately mixed. When studying the chemical properties of the transient species by observation of changes in concentration of reactants and products with time during or after irradiation, the effects of the nonhomogeneous formation of species may present a complication in the analysis of the kinetics. Vice versa, information about the spatial inhomogeneity may be obtained from the kinetics if sufficient knowledge exists about the properties of the species involved.

In this chapter we shall first briefly review the kinetics of first- and second-order processes for continuous and pulsed irradiation, without taking the effects of nonhomogeneous formation of the species into consideration. We also discuss diffusion controlled reactions under conditions where interactions of more than two particles can be neglected, first the kinetics of the diffusion-controlled reaction of randomly generated species ("homogeneous" reaction) and then that of isolated pairs of reactants. The latter is often called *geminate kinetics* when dealing with pairs of oppositely charged species; we shall use this term for the kinetics of isolated pairs in general. In the last section we discuss briefly the kinetics of groups of more than two reactants.

## FIRST- AND SECOND-ORDER PROCESSES

### *First-Order Processes*

We consider a collection of species A that disappear by reaction in some way while no new species A are being formed. The reaction of the species A is first order when the rate of reaction, defined as the decrease in the concentration of A due to reaction per unit time, $-dC_A/dt$, is proportional to the concentration of A:

$$-\frac{dC_A}{dt} = kC_A \tag{4-1}$$

The symbol $k$ is the rate constant or the specific rate of reaction and has a dimension 1/time. It is equal to the fractional decrease of the concentration of species A due to reaction per unit time, $-(1/C_A)\,(dC_A/dt)$, or the probability of reaction per unit time for the species present (which is independent of the concentration of species and of time).

Integrating Eq. (4-1) gives

$$\ln\left(\frac{C_A(t)}{C_A(0)}\right) = -kt \tag{4-2}$$

which shows that the fraction of species surviving after a time $t$ is independent of the initial concentration. If the time dependence of $C_A$ is known, the rate constant $k$ can be determined from a plot of $\ln C_A$ against time, which is linear [Eq. (4-2)]; the slope is equal to $k$. Observe that to determine $k$ one need know only the relative decay with time, not the absolute value of $C_A$. The average lifetime of the species present at a given time is independent of that time and is given by

$$\tau = \frac{-1}{C_A(0)} \int_0^\infty t \frac{dC_A}{dt}\, dt = \frac{1}{k} \tag{4-3}$$

The fraction of survival after a time equal to the average lifetime $\tau$ is $C_A(\tau)/C_A(0) = e^{-1} = 0.37$ (and after $t = n\tau$ the fraction is $e^{-n}$). The time at which $C_A(t)/C_A(0)$ has decreased to 0.5 is given by $t_{1/2} = -(1/k)\ln 0.5 = \tau \ln 2 = 0.69\tau$.

When the species A are being formed homogeneously in space (ie, at random positions) with a rate of formation $q_A$ per unit time and per unit volume while they decay in a first-order process, we can write

$$\frac{dC_A}{dt} = q_A - kC_A \tag{4-4}$$

If we, assume a constant rate of formation $q_A$ to start at $t = 0$ ($q_A = 0$ for $t < 0$, $q_A = \text{const}$ for $t > 0$) and $C_A = 0$ at $t = 0$, we obtain the solution of Eq. (4-4) as

$$C_A = \frac{q_A}{k}(1 - e^{-kt}) \tag{4-5}$$

For values of $t$ where $t \ll 1/k$, $e^{-kt} \simeq 1 - kt$, and it follows that we have $C_A \simeq q_A t$. For very large values of $t$ a stationary concentration will be reached, where $dC_A/dt = 0$, with $C_A = q_A/k$.

The average lifetime of a species is, of course, also here equal to $1/k$, independent of time and concentration, since formation of new species A does not affect the chance process of decay of a particular species A. In Figure 4-1 the growth of the concentration of A towards a stationary concentration as a result of a constant rate of production, according to Eq. (4-5), is illustrated. Also the decay of $C_A$ after discontinuation of the production at $t = t_1$ is shown, where $C_A(t) = C_A(t_1)\, e^{-k(t-t_1)}$.

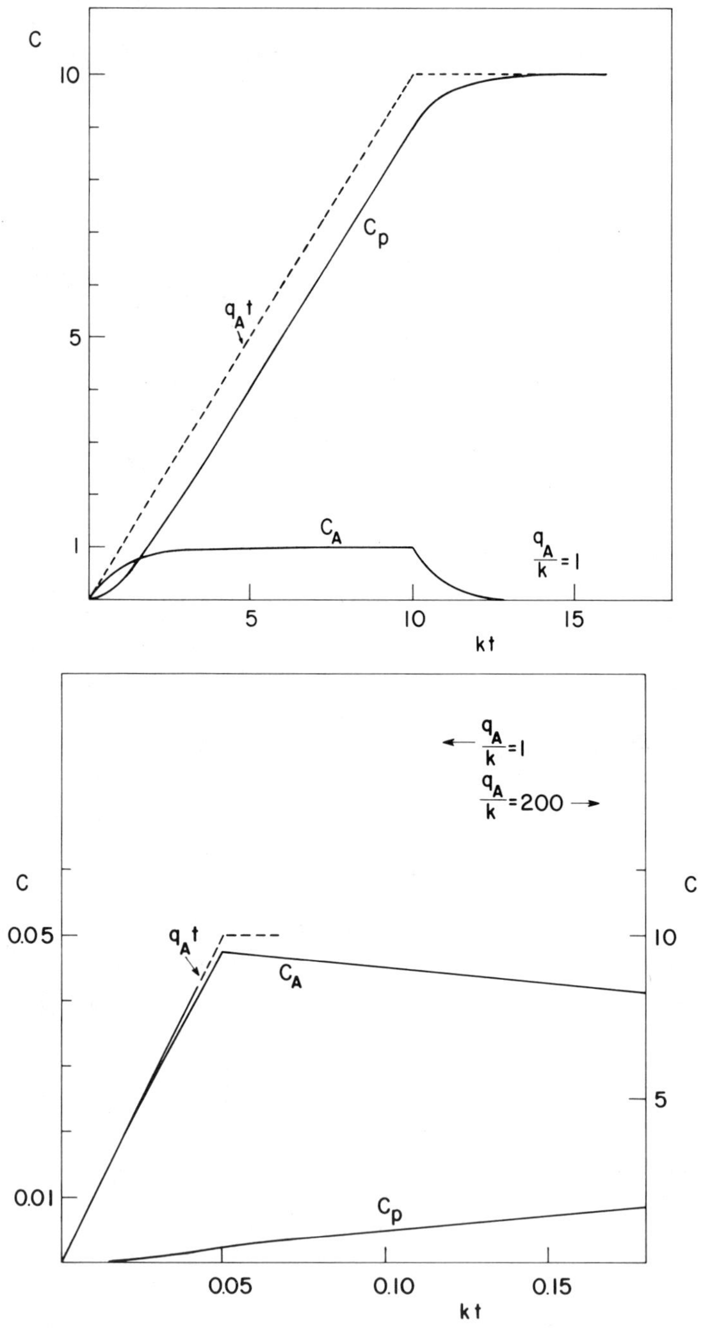


**Figure 4-1.** The concentration of A, $C_A$, and P, $C_p$ resulting from the production of A with a rate $q_A$ during $0 \leq t \leq t_1$ and a decay of A in a first-order process with a specific rate $k$, giving rise to a product P, plotted as a function of $kt$ for two values of $t_1$.

If the decay of the species A results in formation of a product P, the rate of formation of P is

$$\frac{dC_P}{dt} = kC_A \tag{4-6}$$

While A is being formed, $C_A$ is given by Eq. (4-6); therefore

$$\frac{dC_P}{dt} = q_A(1 - e^{-kt}) \tag{4-7}$$

which gives for the concentration of product, $C_P$,

$$C_P = \frac{q_A}{k}[kt - (1 - e^{-kt})] \tag{4-8}$$

The growth of $C_P$ with time, as given by Eq. (4-8), is also shown in Figure 4-1. The linear increase represents the stationary rate of formation of P, which is, of course, equal to $q_A$. After discontinuation of the production of A at $t = t_1$, $C_P$ keeps increasing, while A decreases according to $C_A(t) = C_A(t_1)\, e^{-k(t-t_1)}$, and $C_P$ is given by

$$C_P = C_P(t_1) + [1 - e^{-k(t-t_1)}] \cdot C_A(t_1) \tag{4-9}$$

At $t = \infty$ the concentration of P must be equal to the total number of species formed during $t_1$, per unit volume, multiplied by $f$: $C_P(\infty) = fq_A t_1$.

In Figure 4-1 we have pictured a situation where the duration of the production period ($t_1$) is considerably larger than the average lifetime of transient species A ($1/k$), which results in a stationary concentration of A. Also in Figure 4-1 we show what is observed when the production period, with the same lifetime, is shortened. With this pulse length very little decay takes place during the pulse of the species A, and hence also very little production of P. It should be remarked that the relative form of the curves does not change with the production rate $q_A$, since both $C_A(t)$ and $C_P(t)$ scale with $q_A$, which results from the fact that the species A do not interact.

## *Pseudo-First-Order Reactions; Competition*

For a first-order process the probability of reaction per unit time of the species involved is independent of time and the concentration of that species. This will be the case for truly unimolecular reactions, where the reaction of the species does not involve other molecules, such as the dissociation of an isolated excited molecule. A bimolecular reaction manifests itself as a first-order process, when one of the reacting species is present in large excess, so that the concentration of that species can be considered constant. The rate equation for such a pseudo-first-order reaction of A with S, where $C_S \gg C_A$, is

$$\frac{dC_A}{dt} = -k_S C_S C_A = -k' C_A \tag{4-10}$$

and $k'$ is the pseudo-first-order specific rate (rate constant) with dimension $(\text{time})^{-1}$, and the second-order specific rate $k_s$ has a dimension $(\text{concentration} \times \text{time})^{-1}$. The probability of reaction per unit time for a species A (which is equal to $k' = k_S C_S$) is proportional to the probability of meeting a species S and to the chance of reaction taking place during encounter. When the efficiency of reaction on encounter is very high, the rate of reaction will be determined by the encounter frequency only, and the reaction is said to be diffusion controlled. This gives rise to some peculiarities in the kinetics, which manifest themselves at extremely short times and at very high concentrations of solute. In the following sections we deal with this in some detail. For the moment we disregard these complications.

**Instantaneous Formation of A**

We now consider the simultaneous decay of a species A due to a first-order process with specific rate $k_0$ and a pseudo-first-order reaction with species S with a second-order specific rate $k_S$, after instantaneous formation of $A$ $(C_S \gg C_A)$

$$\text{A} \rightarrow \text{P} \cdots k_0 \tag{4-11}$$

$$\text{A} + \text{S} \rightarrow \text{Q} \cdots k_S \tag{4-12}$$

The rate equation for $C_A$ is

$$\frac{dC_A}{dt} = -k_0 C_A - k_S C_S C_A \tag{4-13}$$

with solution

$$C_A(t) = C_A(0)\, e^{-(k_0 + k_S C_S)t} \tag{4-14}$$

It follows that $\ln C_A(t) = \ln C_A(0) - (k_0 + k_S C_S)t$. A plot of the logarithm of $C_A(t)$ vs. $t$ is therefore linear, and the slope is equal to $k_0 + k_S C_S$; by plotting the slope of the semilogarithmic plots of the decay obtained at different solute concentrations, $C_S$, against $C_S$, values of $k_0$ and $k_S$ can be obtained.

For the rate of formation of the product Q of the reaction of A with S, we have

$$\frac{dC_Q}{dt} = k_S C_S C_A \tag{4-15}$$

or, with Eq. (4-13),

$$\frac{dC_Q}{dt} = -\frac{k_S C_S}{k_0 + k_S C_S} \frac{dC_A}{dt} \tag{4-16}$$

where $k_S C_S/(k_0 + k_S C_S)$ is the fraction of the disappearing A's that react with S to give Q. Likewise, for the rate of formation of P we have

$$\frac{dC_P}{dt} = k_0 C_A = -\frac{k_0}{k_0 + k_S C_S} \frac{dC_A}{dt} \tag{4-17}$$

After substituting Eq. (4-14) into Eq. (4-15) and integrating, we get

$$C_Q(t) = \frac{k_S C_S}{k_0 + k_S C_S} C_A(t = 0)[1 - e^{-(k_0 + k_S C_S)t}] \tag{4-18}$$

It is shown that the growth kinetics of Q is characterized by the same exponential term that describes the decay of A [Eq. (4-14)]. For the final concentration of $C_Q$, at $t = \infty$, we now find that

$$C_Q(t = \infty) = \frac{k_S C_S}{k_0 + k_S C_S} C_A(t = 0) \tag{4-19}$$

The fraction of A that has reacted with S to give the product Q is

$$\frac{C_Q(t = \infty)}{C_A(t = 0)} = f_Q(C_S) = \frac{k_S C_S}{k_0 + k_S C_S} \tag{4-20}$$

Since $f_P + f_Q = 1$, for the probability of formation of P in the presence of S we have

$$f_P(C_S) = \frac{C_P(t = \infty)}{C_A(t = 0)} = \frac{k_0}{k_0 + k_S C_S} \tag{4-21}$$

**Continuous Formation of A**

The fraction of A that has reacted to give the product Q, $f_Q$, is the probability for a species A to give Q. This probability is independent of the presence of other species A. If now species A are formed with a steady rate $q_A$, we shall expect to find a steady production of $Q$, $f_Q q_A$. As we have shown in the previous section, for the case of one single decay process, after a period of time a stationary state is reached where $dC_A/dt = 0$ and $C_A = q_A/k$. Now we find for the stationary concentration that

$$C_A = \frac{q_A}{k_0 + k_S C_S} \tag{4-22}$$

The rate of formation of Q resulting from the reaction of A with S is given by Eq. (4-15). After substituting Eq. (4-22) in (4-15), we obtain the rate of formation of Q under steady-state conditions:

$$\frac{dC_Q}{dt} = \frac{k_S C_S}{k_0 + k_S C_S} q_A \tag{4-23}$$

and the growth of the concentration of Q with time:

$$C_Q(t) = \frac{k_S C_S}{k_0 + k_S C_S} q_A t \tag{4-24}$$

From Eqs. (4-23) and (4-24) we see that the fraction of the species A reacting with S to give Q is

$$f_Q(C_S) = \frac{k_S C_S}{k_0 + k_S C_S} = \left(1 + \frac{k_0}{k_S C_S}\right)^{-1} \tag{4-25}$$

as we have found before [Eq. (4-20)]. In a similar way, for the rate of formation of P, we get

$$\frac{dC_P}{dt} = \frac{k_0}{k_0 + k_S C_S} q_A \tag{4-26}$$

and for the fraction of A's giving P's in the presence of S, we get

$$f_P(C_S) = \frac{k_0}{k_0 + k_S C_S} = \left(1 + \frac{k_S C_S}{k_0}\right)^{-1} \tag{4-27}$$

From Eqs. (4-23) through (4-27) we see that the ratio of rate constants $k_0/k_S$ can be determined from the rate of formation of either P or Q. If the rate of formation of P or Q as well as of A ($q_A$) can be measured, $k_0/k_S$ can be obtained from the results of a measurement at one concentration of S. If only relative values of the rates of formation are available $k_0/k_S$ can be determined from results obtained at different $C_S$, as is shown in the following. Since from Eq. (4-23) it follows that if we measure $\alpha_1(dC_Q/dt)$ and $\alpha_2 q_A$, where $\alpha_1$ and $\alpha_2$ are unknown constants,

$$\alpha_1 \frac{dC_Q}{dt} = \left(1 + \frac{k_0}{k_S C_S}\right)^{-1} \frac{\alpha_1}{\alpha_2} (\alpha_2 q_A) \tag{4-28}$$

a plot of $(\alpha_1\, dC_Q/dt)^{-1}\ \alpha_2 q_A$ against $1/C_S$ will show a slope of $(k_0/k_S)(\alpha_2/\alpha_1)$ and an intercept of $\alpha_2/\alpha_1$, which enables us to determine $k_0/k_S$ without knowing either $\alpha_1$ or $\alpha_2$. If the rate of formation of A, $q_A$, is kept constant, a plot of $(\alpha_1 dC_Q/dt)^{-1}$ vs. $1/C_S$ is also satisfactory. If a relative value of $dC_p/dt$ can be obtained $(\alpha_3\, dC_P/dt)^{-1}$, the ratio of this quantity in the presence and absence of S gives us directly the value of $k_0/(k_0 + k_S C_S)$, as can be seen from Eq. (4-26).

Let us consider the case when A is an excited species that fluoresces ($A \rightarrow P + h\nu$). The fluorescence intensity, $I$, is proportional to the rate of decay of A due to fluorescence, $k_0 C_A$, and also to the rate of formation of any product from that decay, $dC_P/dt$ (the light quanta are themselves a product!), or

$$I = \alpha_3 \frac{dC_P}{dt} = \alpha_3 \frac{k_0}{k_0 + k_S C_S} q_A$$

The ratio of the fluorescence intensity in the absence and presence of a quencher, $I_0/I$, is now given by

$$\frac{I_0}{I} = \frac{k_0 + k_S C_S}{k_0}$$

which is equal to $1/f_p(C_S)$ [Eq. (4-27)]. In Figure 4-2 this ratio is plotted against the quencher concentration $C_S$ [curve (a)]. From Eq. (4-27) we see that the slope is equal to $k_S/k_0$. (This is sometimes called a Stern–Volmer plot.)

In the case where two solutes are present that react with A in a pseudo-first-order fashion, we find that for the stationary rate of reaction of the species

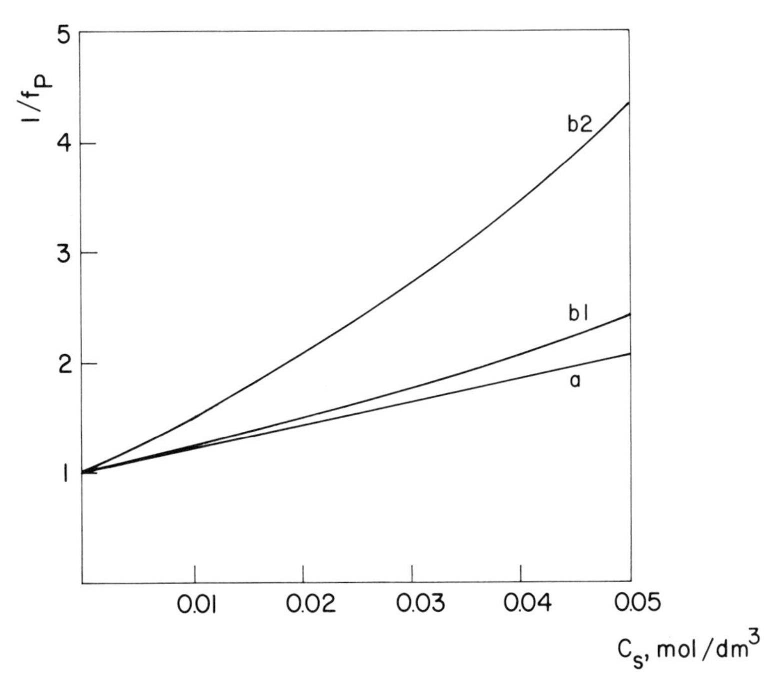


**Figure 4-2.** The inverse of the probability of not reacting with a solute, $1/f_p$, in a competition between a first-order decay and a pseudo-first-order reaction with a solute, as a function of solute concentration: calculated by using Eq. (4-27) with $k_0 = 3.5 \times 10^8\ s^{-1}$ and $k_S = 7.53 \times 10^9\ M^{-1}\ s^{-1}$ (a); by Eq. (4-91) with $k_0 = 3.5 \times 10^8\ s^{-1}$, $R = 5 \times 10^{-10}$ m, $D = 2 \times 10^{-9}\ m^2/s$, $(4\pi RD) = 7.53 \times 10^9\ M^{-1}\ s^{-1}$ (b1), and with $R = 15 \times 10^{-10}$ m, $D = 6.7 \times 10^{-10}\ m^2/s$, $(4\pi RD) = 7.53 \times 10^9\ M^{-1}\ s^{-1}$ (b2).

A with S, giving rise to formation of Q, in the presence of both S and S′

$$\frac{dC_Q}{dt} = \frac{k_S C_S}{k_0 + k_S C_S + k_{S'} C_{S'}} q_A \tag{4-29}$$

By determining the rate of formation of product in the presence of various concentrations of the solutes, the ratio of the specific rates can be determined.

## *Second-Order Processes*

We consider two reactants A and B distributed homogeneously in space and reacting in a second-order fashion:

$$A + B \rightarrow R \tag{4-30}$$

When formation of new species A and B does not take place, we can write

$$\frac{dC_A}{dt} = \frac{dC_B}{dt} = -\frac{dC_R}{dt} = -kC_AC_B \tag{4-31}$$

We see that the probability of A reacting per unit time, $-(1/C_A)(dC_A/dt)$, is proportional to $C_B$.

A special case arises when $C_A(0) = C_B(0)$. In this case

$$C_A(t) = C_B(t) \tag{4-32}$$

and Eq. (4-31) reads

$$\frac{dC_A}{dt} = -kC_A{}^2 \tag{4-33}$$

which has the solution

$$\frac{1}{C_A(t)} - \frac{1}{C_A(0)} = kt \tag{4-34}$$

or

$$\frac{C_A(t)}{C_A(0)} = \frac{1}{1 + kC_A(0)t} \tag{4-35}$$

which shows that the fraction surviving after a time $t$ is dependent on $C_A(0)$, in contrast with what has been observed for first-order decay. The time at which the concentration of A (and B) has decreased to one-half of the initial value, $t_{1/2}$, is given by

$$t_{1/2} = \frac{1}{kC_A(0)} \tag{4-36}$$

The next half-life is twice as long, and so on. From Eq. (4-34) we see that $k$ is equal to the slope of a plot of $1/C_A$ against time.

When we deal with a reaction

$$A + A \rightarrow R \tag{4-37}$$

it is customary to write

$$-\frac{dC_A}{dt} = 2\frac{dC_R}{dt} = 2kC_A{}^2 \tag{4-38}$$

When A and B are being formed homogeneously with a rate $q$, while they react with each other with a specific rate $k$, the rate equation is

$$\frac{dC_A}{dt} = \frac{dC_B}{dt} = q - kC_AC_B \tag{4-39}$$

Because $C_A = C_B$, Eq. (4-39) can be rewritten as

$$\frac{dC_A}{dt} = q - kC_A{}^2 \tag{4-40}$$

The solution is

$$C_A(t) = \left(\frac{q}{k}\right)^{1/2} \frac{e^{2t\sqrt{qk}} - 1}{e^{2t\sqrt{qk}} + 1} \tag{4-41}$$

For very small values of $t$, $e^{2t\sqrt{qk}} \simeq 1 + 2t\sqrt{qk}$, and Eq. (4-41) reduces to

$$C_A(t) \simeq \left(\frac{q}{k}\right)^{1/2} \frac{t\sqrt{qk}}{1 + t\sqrt{qk}} \simeq qt,$$

as expected for very small times where the recombination is negligible. For $t = \infty$ it follows from Eq. (4-41) that $C_A = (q/k)^{1/2}$, the stationary-state concentration, which can also be obtained from Eq. (4-40) with $dC_A/dt = 0$. We see that the stationary-state concentration is found to be proportional to the square root of the rate of formation $q$, in contrast to what we have found earlier for a first-order decay, where the stationary-state concentration is equal to $q/k$.

## *Competition of a Second-Order Reaction and a Pseudo-First-Order Reaction*

We now consider a second-order reaction

$$A + B \rightarrow R \tag{4-42}$$

where A and B are formed instantaneously with equal concentrations, $C_A(0) = C_B(0)$, in competition with the reaction

$$A + S \rightarrow Q \tag{4-43}$$

where $C_S$ is present in excess ($C_S \gg C_A(0)$), and

$$Q + B \rightarrow T \tag{4-44}$$

takes place with the same rate constant as Reaction (4-42), $k_{QB} = k_{AB} = k$. Because $C_A + C_Q = C_B$,

$$\frac{dC_B}{dt} = -k_{AB}C_AC_B - k_{QB}C_QC_B = -kC_B{}^2 \tag{4-45}$$

and therefore

$$C_B = \frac{C_B(0)}{1 + ktC_B(0)} \tag{4-46}$$

For the rate of disappearance of A we now obtain

$$\frac{dC_A}{dt} = -\, k_{AB}C_AC_B - k_{AS}C_SC_A$$

$$= -\,(k\frac{C_B(0)}{1 + ktC_B(0)} + k_{AS}C_S)C_A \tag{4-47}$$

which on integration gives

$$\frac{C_A}{C_A(0)} = e^{-k_{AS}C_S t} \cdot \frac{1}{1 + ktC_B(0)} \tag{4-48}$$

which shows that the decay of A is a product of the decay due to the pseudo-first-order reaction and the second-order reaction. If the initial concentration of B, $C_B(0)$ (and A) is sufficiently low, the term $1/(1 + ktC_B(0)) \approx 1$, and the decay is effectively first order, due to the fact that the second-order decay contributes negligibly to the overall decay of A.

If A and B are formed continuously with a rate $q_A = q_B = q$, we have, in the steady state,

$$\frac{dC_B}{dt} = q - (k_{AB}C_A + k_{QB}C_Q)C_B = q - kC_B{}^2 = 0 \tag{4-49}$$

and $C_B = (q/k)^{1/2}$. Since

$$\frac{dC_A}{dt} = q - (k_{AB}C_B + K_{AS}C_S)C_A = 0 \tag{4-50}$$

we have

$$C_A \frac{q}{kC_B + k_{AS}C_S} = \frac{q}{k(q/k)^{1/2} + k_{AS}C_S} \tag{4-51}$$

and for the stationary rate of formation of Q, we get

$$\frac{dC_Q}{dt} = k_{AS}C_SC_A = \frac{k_{AS}C_S}{k(q/k)^{1/2} + k_{AS}C_S} \tag{4-52}$$

We see that now the fraction of species A reacting with S depends on the rate of production of A, in contrast to what was observed with first-order decay only [Eq. (4-23)]. Analogously, with two solutes S and S′ reacting with A to give Q and Q′, respectively, for reaction with S, we obtain

$$\frac{dC_Q}{dt} = \frac{k_{AS}C_S}{k(q/k)^{1/2} + k_{AS}C_S + k_{AS'}C_{S'}} \cdot q \tag{4-53}$$

provided $k_{AS'} = k_{AS} = k_{AB} = k$. The ratio of the rates of formation of Q and Q′ is now equal to $k_{AS}C_S/k_{AS'}C_{S'}$, which provides a means to determine the ratio $k_{AS}/k_{AS'}$. In practice, the equality $k_{AS} = k_{AS'} = k_{AB}$ often does not exist, and the kinetics is more complicated. To circumvent this problem, chemists often carry out competition experiments under such conditions that the rate of the second-order reaction $A + B \rightarrow R$ is negligible compared to the pseudo-first-order decay ($k(q/k)^{1/2} \ll k_{AS}C_S$, $k_{AS'}C_{S'}$). This involves working with a sufficiently low rate of formation $q$ and with a sufficiently large concentration of scavenger S.

### *Nonhomogeneous Formation of Reactive Species*

In radiation chemistry the initial formation of the reactive species takes place inhomogeneously along the track; pairs of spatially correlated species are formed, and regions of relatively high concentrations of species are created. After a period of time in which mutual reaction of these spatially correlated species dominates, due to diffusion the inhomogeneity gradually disappears and eventually the distribution may be considered homogeneous. As mentioned in the introduction, this time varies appreciably for different species. The fraction of species that escapes the mutual reactions in the nonhomogeneous phase depends, of course, on the initial spatial distribution and, therefore, on the linear energy transfer and will be different for different species (eg, charged species vs. radicals). Our knowledge about the "initial" yields of the various species is mostly very incomplete; however, the "homogeneous" yields are often well known.

As we have seen above, the relative decay in a (pseudo-) first-order process during a given period of time is independent of the concentration. This implies that with a nonhomogeneous distribution the relative decay will be the same in regions of different concentration and equal to the overall (observed) decay.

For second-order decay with equal concentrations of both reactants the relative decay rate depends on the concentration of the reactants ($C^{-1}\, dC/dt) = kC$); therefore, when local concentration differences exist, the decay rate will also differ locally. The decay of the average concentration, in general, will not even obey second-order kinetics. In radiation chemistry the local second-order decay rate may be orders of magnitude larger than expected on the basis of the decay of the average concentration. Fortunately, the time domains in which the nonhomogeneous and the homogeneous decays take place can often be separated (by having a sufficiently low dose, or dose rate), so that these processes can be studied separately. With scavenging and competition studies complications may arise when high concentrations are used, so that competition in the nonhomogeneous regime takes place, and the homogeneous competition kinetics presented earlier no longer apply. We shall return to this when discussing scavenging of geminately recombining species.

## HOMOGENEOUS DIFFUSION-CONTROLLED REACTIONS

### *Introduction*

The species formed following the primary interaction of high-energy radiation with matter are mostly unstable and often react with molecules in the medium or with each other on first encounter. As a result, the diffusion towards each other of the reacting species will determine the rate at which the reaction proceeds. In this section we discuss the kinetics of such reactions when the reactants are initially homogeneously distributed in space. The first treatment of this problem dates back to the beginning of this century when Smoluchowski, considering the reaction of solute molecules with a colloid particle, realized that the rate of reaction was determined by the diffusion flow of solute towards the particle.[1] Debye, in 1942, extended the treatment to the interaction of species with an arbitrary spherical potential.[2] Although Smoluchowski and Debye conjectured that at the moment the reacting species would encounter each other they would react infinitely fast, it was Collins and Kimball[3] in 1949 who pointed out that this conjecture was physically unrealistic and would lead to an anomaly at $t = 0$. They introduced a finite rate of reaction of the encountering species and expressed this in the so-called radiation boundary condition. With this boundary condition an expression for the specific rate is obtained that is also valid for reactions that are not diffusion controlled.

In the following treatment the continuous diffusion equation will be used. For times on the order of the diffusive jump time this treatment breaks down. For a discussion of the validity of this equation for diffusion in general, the reader is referred to the paper of Chandrasekhar,[4] and for the application to reaction kinetics problems at very short time to the article of Noyes.[5] More recently, a general mathematical description of diffusion-controlled reactions has been given by Wilemski and Fixman, who also critically evaluated the existing theories.[6]

For a general introduction on the effect of diffusion on the rate of reactions see refs. 7 and 8.

## *Neutral Species; Stationary Rate*

We consider the reaction of two neutral species A and S in a liquid. We assume the species S to be present in large excess over A, so that reaction does not markedly affect the overall concentration of S. Also, we consider such a low concentration of A that the particles A are so far away from each other that they can be considered independent of each other. The liquid may be thought to be partitioned up in volumes (cells) with each one species A (and a large number of S). We choose the species A at the center of a spherically symmetrical coordinate system and investigate the distribution of species S in the space around A. We consider the probability density of species S, or the local concentration of S, at a position $P$ with respect to A, which is the number of species S found in a small volume $v$ around $P$ divided by that volume $v$, averaged over the ensemble, in the limit as $v \to 0$. The local concentration will be denoted by $c(\bar{r}, t)$, and $C(t)$ will be used for the case where the probability density is independent of position (homogeneous, or random, distribution). We return to the ensemble of cells with A at the origin and with a random distribution of S in the space around A at $t = 0$. The local concentration of S at a distance $r$ from A at $t = 0$, $c(r, t = 0)$, will be independent of $r$ and equal to the bulk concentration $C_S$ down to a small distance $R$, the distance of closest approach (or the encounter radius). This is shown in Figure 4-3, curve a. A finite local concentration of A at $r = R$ means that some A's of the ensemble have an S adjacent to them. Now if reaction takes place among the A's that have an S adjacent to them, the contribution of cells with an S at a distance $r = R$ from the ensemble will decrease and therefore cause the (average) local concentration $c_S(R)$ to decrease, and on the average a depletion layer will result around A. Due to the random diffusive moment, however, some species A and S that initially were at some distance apart from each other will move towards each other, and more reaction becomes possible. Eventually the flow of S into A due to diffusion becomes equal to the rate of reaction at the encounter radius, and a stationary state arises with a stationary concentration distribution $c_S(r)$, as shown in Figure 4-3, curve b. The average amount of S moving towards A through a sphere at $r$ due to diffusion during $\Delta t$ is $4\pi r^2 D$ $(dc_S/dr)\,\Delta t$, where $D$ is the sum of the diffusion coefficients of A and S. In the stationary state this amount is equal to the fraction of A reacting during

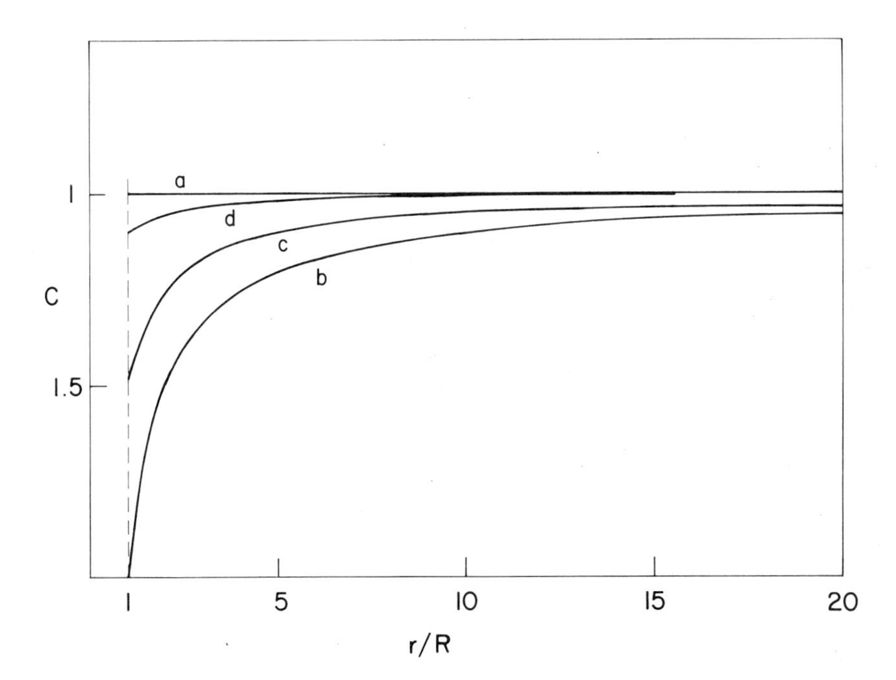


**Figure 4-3.** The probability density of S as a function of distance from A for a diffusion-controlled reaction of neutral particles, with a random distribution at $t = 0$ (a) and at time $t \gg R^2/\pi D$, when the stationary state is established, for different values of $k$ [Eq. (4-59)]; (b) $k/4\pi RD = 1$; (c) $k/4\pi RD = 0.5$; (d) $k/4\pi RD = 0.1$.

$\Delta t$, which in turn is equal to the fractional decrease of the overall concentration of A during $\Delta t$; that is, $(dC_A/dt)(1/C_A)\,\Delta t$. This can be written as

$$4\pi r^2 D \frac{dc_S}{dr} = -\frac{dC_A}{dt}\frac{1}{C_A} \tag{4-54}$$

The rate of reaction of A with S is defined as the decrease of the overall concentration of A per unit time and can be written as a product of the overall concentrations $C_A$ and $C_S$ and the specific rate $k$:

$$-\frac{dC_A}{dt} = kC_A C_S \tag{4-55}$$

Combining Eqs. (4-54) and (4-55) gives

$$4\pi r^2 D \frac{dc_S}{dr} = kC_S \tag{4-56}$$

Because we have assumed $C_S \gg C_A$, the local concentration $c_S$ at the boundary of our cells can be considered constant, and we take $c_S(r = \infty) = C_S$. The solution of Eq. (4-56) is now

$$c_S(r) = C_S - \frac{kC_S}{4\pi Dr} \tag{4-57}$$

and for $r = R$ we have

$$c_S(R) = C_S\left(1 - \frac{k}{4\pi DR}\right) \tag{4-58}$$

which shows a relation between the local concentration of S at the reaction radius and the overall specific rate. This can also be expressed as

$$k = 4\pi RD\left\{1 - \frac{c_S(R)}{C_S}\right\} \tag{4-59}$$

In Figure 4-3 we show concentration profiles $c_S(r)$ for different values of $k/4\pi RD$. If we assume that $c_S(R = 0)$ (the Smoluchowski boundary condition), we find for the stationary diffusion-controlled specific rate that

$$k = 4\pi RD \tag{4-60}$$

where $D$ is the sum of the diffusion coefficient of A and S: $D = D_A + D_S$.

This boundary condition effectively means that the rate of reaction on encounter is infinitely fast. A better boundary condition, which accounts for the finite reaction rate on encounter is the so-called radiation boundary condition

$$4\pi R^2 D\left(\frac{dc_S}{dr}\right)_R = k_R c_S(R) \tag{4-61}$$

where the probability of reaction per unit time is set equal to the product of the concentration of S at $R$ and a specific rate $k_R$. With Eq. (4-56) we now have

$$kC_S = k_R c_S(R) \tag{4-62}$$

This shows that the concentration of S at $R$ relative to $C_S$ is equal to the ratio $k/k_R$. Substituting Eq. (4-61) in Eq. (4-62) yields, after rearrangement,

$$k = \frac{4\pi DR}{1 + 4\pi DR/k_R} \tag{4-63}$$

or

$$\frac{1}{k} = \frac{1}{4\pi RD} + \frac{1}{k_R} \tag{4-64}$$

We see that for $k_R \gg 4\pi RD$ Eq. (4-63) reduces to $k = 4\pi RD$ again. When the reaction on encounter is slow and $k_R \ll 4\pi RD$, then $k = k_R$. In this case no depletion layer is formed, as can be seen from Eq. (4-62). The reaction is no longer controlled by diffusion.*

In the above derivation we have assumed that S is present in excess over A, so that $c_S$ ($r = \infty$) can be assumed constant and equal to the bulk concentration $C_S$. In most practical situations, however, this requirement is unnecessary, and the above equations for the specific rate are even applicable when $C_A = C_S$. This is probably because the concentration gradient is established on a time scale that is usually orders of magnitude smaller than the time necessary for appreciable changes in the bulk concentration. For a discussion of the complications as a result of higher concentrations see ref. 9.

## *Neutral Species; Transient Effects*

Thus far we have considered the rate of reaction in the stationary state after the profile of the local concentration of S around A has become constant in time. We now consider the rate of reaction at shorter times. The radiation boundary condition is now written as

$$4\pi R^2 D\left(\frac{\partial c_S}{\partial r}\right)_R = k_R c_S(R, t) \tag{4-65}$$

* Frequently one finds the stationary diffusion-controlled specific rate expressed in terms of viscosity, $\eta$. Using the approximate expression $D = k_B T/6\pi\eta r$, for the diffusion coefficients of the reactants, where $r$ is the Stokes radius, we obtain

$$k = 4\pi R(D_A + D_S) = \frac{2Rk_B T}{3\eta}\left(\frac{1}{r_A} + \frac{1}{r_S}\right)$$

When $r_A = r_S = R/2$, this reduces to $k = 8k_B T/3\eta$. (If $k_B T$ and $\eta$ are expressed in CGS units, $k$ is in $cm^3\,s^{-1}$. One also finds $k = 8RT/3000\eta$, where $k$ is in $M^{-1}\,s^{-1}$ if $RT$ and $\eta$ are again expressed in CGS units.)

The probability of reaction of A per unit time, $k_R c_S(R, t)$, where

$$k_R c_S(R, t) = -\frac{dC_A}{dt}\frac{1}{C_A} \tag{4-66}$$

is now time dependent, and if we write

$$-\frac{dC_A}{dt} = kC_A C_S \tag{4-67}$$

we see from Eqs. (4-66) and (4-67) that

$$k_R c_S(R, t) = kC_S \tag{4-68}$$

where $k$ is time dependent.

Because $c_S(R) = C_S$ at $t = 0$, we see that $k = k_R$ at $t = 0$. This shows that $k_R$ is equal to the overall specific rate of reaction when the concentration at the encounter radius is equal to the overall average concentration. We see that for diffusion-controlled reactions the specific rate $k$ is time dependent, starting at a high value of $k_R$ at $t = 0$ and decreasing to a steady-state value given by Eq. (4-63).

If $k_R \ll 4\pi RD$, no depletion layer will be formed, and it follows that $k$ will not be time dependent.

The specific rate as a function of time, $k(t)$, is related to the diffusive flow at $r = R$,

$$k(t)C_S = 4\pi R^2 D\left(\frac{\partial c_S}{\partial r}\right)_R \tag{4-69}$$

which follows from Eqs. (4-65) and (4-68). An expression for $k(t)$ can be obtained by solving the diffusion equation

$$\frac{\partial c_S}{\partial t} = D\left(\frac{\partial^2 c_S}{\partial r^2} + \frac{2}{r}\frac{\partial c_S}{\partial r}\right) \tag{4-70}$$

with $c_S(r, t = 0) = C_S$ and $c_S(r = \infty, t) = C_S$, and by using the radiation boundary condition, Eq. (4-65). An analytical solution for $c(r, t)$ can be obtained, and $k(t)$ is found to be[3]

$$k(t) = \frac{k_D}{1 + k_D/k_R}\left[1 + \frac{k_R}{k_D}\exp\left\{\frac{Dt}{R^2}\left(1 + \frac{k_R}{k_D}\right)^2\right\}\operatorname{erfc}\left\{\frac{\sqrt{Dt}}{R}\left(1 + \frac{k_R}{k_D}\right)\right\}\right] \tag{4-71}$$

where $k_D = 4\pi RD$, and

$$\operatorname{erfc} x = (2/\sqrt{\pi})\int_x^\infty e^{-x^2}\,dx$$

(for $x = 0$, erfc $x = 1$; for $x = \infty$, erfc $x = 0$). A rather good approximation for $e^{x^2}\operatorname{erfc} x$ is

$$e^{x^2}\operatorname{erfc} x = \frac{2}{\sqrt{\pi}}\frac{1}{x + \sqrt{x^2 + 4/\pi}}$$

Since, for large $x$, $e^{x^2}$ erfc $x \simeq (1/\sqrt{\pi}) \cdot (1/x)$, we have, for large $t$, approximately

$$k(t) = \frac{k_D}{1 + k_D/k_R}\left(1 + \frac{k_R}{k_D + k_R} \cdot \frac{R}{\sqrt{\pi D t}}\right) \qquad (4\text{-}72)$$

which reduces to

$$k(t) = 4\pi R D\left(1 + \frac{R}{\sqrt{\pi D t}}\right) \qquad (4\text{-}73)$$

for $k_R \gg k_D$. For $t = 0$ in Eq. (4-71) we obtain $k = k_R$, as expected. To get an estimate of the time scale where $k(t)$ differs significantly from $k_D$, we consider Eq. (4-73) and calculate the value of $t$ for which $k(t) = 2k_D$. With an encounter radius $R$ of $6 \times 10^{-8}$ cm and a sum of diffusion coefficients $D = 1 \times 10^{-5}\ \text{cm}^2\ \text{s}^{-1}$, which are common values for molecules in normal liquids, we find that $t = R^2/\pi D = 1.1 \times 10^{-10}$ s. At $t = 10^{-11}$ s we obtain $k/k_D = 3.5$; and at $t = 10^{-9}$ s, $k/k_D = 1.3$. We see that for "normal" molecules the effects of time dependence play a role below times of the order of magnitude of a nanosecond.

## *A Force Between the Species*

In the foregoing we have discussed the reaction of species that do not interact with each other when they are not in contact. When the species exert a force on each other, the force causes an additional flow of species S towards A, and we now have, in the steady state,

$$kC_S = 4\pi r^2 D\left(\frac{dc_S}{dr} + \frac{c_S}{k_B T}\frac{dV}{dR}\right) \qquad (4\text{-}74)$$

where $V$ is the potential energy of the interaction. Using the substitution $c(r) = c'(r)\, e^{-V(r)/kT}$ with boundary conditions $c_S(r = \infty) = C_S$ and $kC_S = k_R c_S(R)$, we get

$$k = \frac{4\pi D\beta}{1 + 4\pi D\beta/k_R\, e^{-V(R)/k_B T}} \qquad (4\text{-}75)$$

or

$$\frac{1}{k} = \frac{1}{4\pi D\beta} + \frac{1}{k_R\, e^{-V(R)/k_B T}} \qquad (4\text{-}76)$$

where

$$\beta = \left\{\int_R^\infty \mathrm{e}^{V(r)/k_B T} r^{-2}\, dr\right\}^{-1} \qquad (4\text{-}77)$$

For $V = 0$ it follows that $\beta = R$; in this case Eq. (4-75) reduces to Eq. (4-63). For $k_R\, e^{-V(R)/k_B T} \gg 4\pi D\beta$, Eq. (4-75) reduces to

$$k = 4\pi D\beta \tag{4-78}$$

which is equal to the solution found with the Smoluchowski boundary condition.

When A and S are charged species with charge $e$, the potential energy is $V(r) = s\, e^2/4\pi\varepsilon_0\varepsilon_r r$, where $s = -1$ for oppositely charged species and $s = +1$ for species with like charge. Using Eq. (4-77), we find $\beta = -sr_c(1 - e^{sr_c/R})^{-1}$, where $r_c$ is the distance at which $V(r) = kT$, $r_c = e^2/4\pi\varepsilon_0\varepsilon_r k_B T$. If we now consider oppositely charged ions and assume that $k_r\, e^{r_c/R} \gg 4\pi D\beta$, we find that using (4-75), for the diffusion-controlled stationary specific rate of reaction,

$$k = \frac{4\pi D r_c}{1 - e^{-r_c/R}} \tag{4-79}$$

For a dielectric liquid, where $r_c/R \gg 1$ (eg, with $\varepsilon_r = 2$ and $T = 300$ K, $r_c = 3 \times 10^{-6}$ cm, and $R$ is of the order of $10^{-8}$ cm), Eq. (4-79) reduces to

$$k = 4\pi D r_c \tag{4-80}$$

Realizing that the sum of the diffusion coefficients of the ions, $D$, is related to the sum of the mobilities, $u$, by $D/u = k_B T/e$, and using $r_c = e^2/4\pi\varepsilon_0\varepsilon_r k_B T$, we can write Eq. (4-80) as

$$k = \frac{eu}{\varepsilon_0 \varepsilon_r} \tag{4-81}$$

It is interesting to see that in the case of oppositely charged ions the reaction radius $R$ does not appear in the expression for the stationary diffusion-controlled rate (provided $k_R$ is large). The theoretical expression for this case has been verified experimentally for reactions of oppositely charged molecular ions as well as for reactions of positive ions with excess electrons in various nonpolar liquids.[10,11]

For nonpolar liquids the term $e^{-r_c/R}$ in Eq. (4-79) vanishes because $r_c/R \gg 1$, but in a highly polar liquid like water this is no longer true (with $\varepsilon_r = 80$, $r_c = 7 \times 10^{-8}$ cm). For a reaction between species with like charges, the specific rate is $k = 4\pi D r_c/(e^{r_c/R} - 1)$. This shows that for nonpolar liquids $k$ becomes extremely small, due to the repulsion of the Coulomb forces; for water, however, the Coulomb interaction lowers the rate, but only by an order of magnitude or so.

An interesting effect is observed when considering the rate of reaction of ions in aqueous solution while an "inert" electrolyte is present. Due to the screening of the Coulomb field by the ions of the electrolyte, we find that for increasing ionic strength the rate increases for reactions between like charges and decreases for reactions between opposite charges. The observed ionic strength effect agrees with our expectation on the basis of Eq. (4-75), using the potential obtained from Debye–Hückel theory.[7,8]

For neutral particles with a random distribution as initial condition an exact solution for the time dependence of the specific rate has been found for the whole time domain by using the radiation boundary condition. For particles with a potential such a solution has not been found. An approximate lower limit has been found for $k(t)$ for longer times ($t > r_c^2/\pi D$) for the case of a Coulomb potential with the random initial condition by using the radiation boundary condition[12]

$$k(t) \geqslant 4\pi A D\left[1 + \frac{A}{(\pi D t)^{1/2}}\, e^{s r_c/R}\left(1 + \frac{4\pi s r_c D}{k_R}\right)\right] \tag{4-82}$$

where

$$A = s r_c\left[\left(1 + \frac{4\pi s r_c D}{k_R}\right) e^{s r_c/R} - 1\right]^{-1}$$

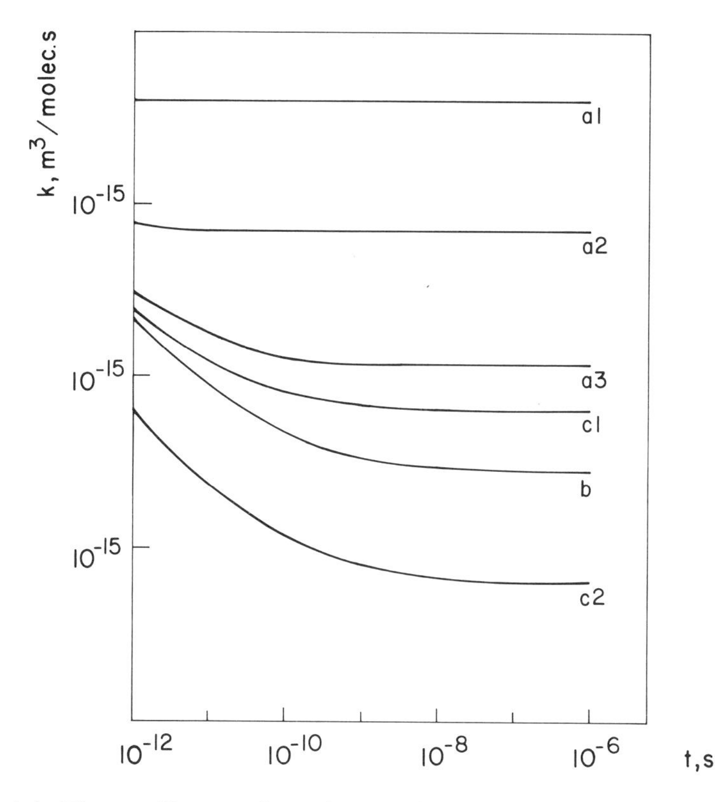


**Figure 4-4.** The specific rate of reaction as a function of time for a random initial distribution for oppositely charged particles for $r_c = 30$ nm (a1), 5.6 nm (a2), 0.7 nm (a3) and for particles with the same charge for $r_c = 0.7$ nm (b) calculated numerically with the Smoluchowski boundary condition for $R = 5 \times 10^{-10}$ m and $D = 10^{-8}$ m²/s,[12] and for neutral particles with $R = 5 \times 10^{-10}$ m, $D = 10^{-8}$ m²/s (c1), and $R = 5 \times 10^{-10}$ m, $D = 10^{-9}$ m²/s (c2), according to Eq. (4-73).

(with $s = -1$ for oppositely charged species and $s = +1$ for like charges). It can be shown that in the limit as $t \rightarrow \infty$, Eq. (4-82) reduces to the same expression obtained from Eq. (4-75) for a Coulomb potential. For $k_R \rightarrow \infty$, which corresponds to the Smoluchowski boundary condition, Eq. (4-82) becomes

$$k(t) \geqslant 4\pi A' D\left[1 + \frac{A' \, e^{sr_c/R}}{(\pi D t)^{1/2}}\right] \tag{4-83}$$

with $A' = sr\{e^{sr_c/R} - 1\}$.

To illustrate the time dependence of $k$, we show in Figure 4-4 results for like and unlike charges with different values of $r_c$, obtained by numerical solution of the time-dependent diffusion equation,[12] and also for neutral particles, obtained by using Eq. (4-73).

An exact expression has been obtained for the Laplace transform of the reaction rate for the case of a Coulomb potential with the Boltzmann distribution as initial condition by using the radiation boundary condition.[13]

Also for an arbitrary spherical potential with a Boltzmann distribution as initial distribution, an approximate solution has been found by using the Smoluchowski boundary condition.

## *Reaction at a Distance*

When the emission spectrum of an excited molecule and the absorption spectrum of an acceptor molecule overlap, long-range energy transfer may take place. In this case the concept of a reaction radius is unsatisfactory. The same is true when the reaction involves electron transfer by means of a tunneling process. In the absence of long-range interaction, in the stationary state the diffusion flow of S across a sphere at $r$ around the origin is independent of $r$, but now the flow across $r - \Delta r$ will be smaller than that across $r$, the difference being due to reaction of S with A while S is at distance $r$ from A. If the long-range transfer is very efficient, very few (unreacted) solute molecules will reach the encounter radius and a depletion of solute molecules takes place around the particle at the origin. This effect can clearly be seen from the results of Pilling and Rice for electron tunneling.[14] The stationary rate of reaction takes place with an effective reaction radius larger than the sum of the radii of the reactants. It is clear that in this case the picture used above for the derivation of the time dependence of the specific rate of reaction [Eq. (4-71)] is incorrect. At short times deviations of Eq. (4-71) must be expected.

For a further discussion of long-range effects on the stationary rate of reaction, see ref. 15; on the time-dependent rate and competition kinetics, see refs. 16 and 17.

For the extreme case, where diffusion of the reactants does not take place, which happens with tunneling reactions in the solid, and when no new reactants are being formed, the depletion extends in space when time progresses. For a discussion of this see Refs. 15 and 18.

## *Competition with Diffusion-Controlled Reactions*

We have already discussed the competition of first-order processes without taking the time dependence of the specific rate into consideration. Now considering this time dependence, we obtain the total rate of reaction of A at time $t$ after formation due to the reactions $A \rightarrow P$ and $A + S \rightarrow Q$:

$$-\frac{dC_A}{dt} = \{k_0 + k_S(t)C_S\}C_A \tag{4-84}$$

and for the rate of reaction $A \rightarrow P$,

$$\frac{dC_P}{dt} = k_0 C_A = \frac{k_0}{k_0 + k_S(t)C_S} \frac{dC_A}{dt} \tag{4-85}$$

which shows that the fraction of A's reacting to give P is no longer independent of time and, furthermore, that the relative effect of the time dependence of $k$ depends on $C_S$. By integrating Eq. (4-84) we find the time dependence of $C_A$:

$$C_A(t) = C_A(0) \exp\left[-\left\{k_0 t + \int_0^t k_S(t) C_S \, dt\right\}\right] \tag{4-86}$$

which shows that the decay is no longer exponential. Substituting the approximate expressions $k(t) = 4\pi RD(1 + R/\sqrt{\pi Dt})$ [Eq. (4-73)] into Eq. (4-86) and integrating over $t$ yield

$$C_A(t) = C_A(0) \exp\left\{-k_0 t - 4\pi RDC_S t\left(1 + \frac{2R}{\sqrt{\pi Dt}}\right)\right\} \tag{4-87}$$

For $t$ large, $2R/\sqrt{\pi Dt}$ becomes much less than 1, and the decay reduces to a normal first-order exponential decay

$$C_A(t) = C_A(0) \exp\{-(k_0 + 4\pi RDC_S)t\}.$$

The deviation from first-order decay of the reaction of reactive species with solutes, due to the time dependence of the specific rate, has been observed experimentally in studies of the decay of fluorescing excited species,[17,19] as well as in studies of the reaction of excess electrons in a glass, after formation during a short pulse.[20]

Substituting (4-86) into (4-85) gives the rate of formation of P:

$$\frac{dC_P}{dt} = k_0 C_A(0) \exp\left[-\left\{k_0 t + \int_0^t k_S(t) C_S \, dt\right\}\right] \tag{4-88}$$

The final concentration of P ($C_P(t = \infty)$) is found by integrating this expression, and the fraction of A's giving $\dot{P}$'s is now

$$f_P(C_S) = \frac{C_P(t=\infty)}{C_A(t=0)} = \int_0^\infty k_0 \exp\left[-\left\{k_0 t + \int_0^t k_S(t) C_S \, dt\right\}\right] dt \quad (4\text{-}89)$$

When we have a continuous production of A with a rate $q_A$, the stationary concentration of A is a sum of contributions of species A produced at different times before. During $dt$ an amount $q_A \, dt$ is formed, and a period of time $t$ later the probability of survival is $\exp[-\{k_0 t + \int_0^t k_S(t) C_S \, dt\}]$ [Eq. (4-86)]. Integrating the product over time gives the stationary concentration

$$C_A = \int_0^\infty q_A \exp\left[-\left\{k_0 t + \int_0^t k_S(t) C_S \, dt\right\}\right] dt \quad (4\text{-}90)$$

The rate of production of P is $k_0 C_A$, and inspection of Eqs. (4-90) and (4-89) shows that this rate of production in the stationary state is just equal to $f_P(C_S) \cdot q_A$, as expected.

Using Eq. (4-73) in Eq. (4-89), we obtain the probability that A will react to give P, in the presence of S, or the probability that A will not react with S:

$$f_P(C_S) = \frac{k_0}{k_0 + 4\pi RDC_S}(1 - p\sqrt{\pi}\, e^{p^2} \operatorname{erfc} p) \quad (4\text{-}91)$$

with

$$p = 4R^2 C_S \sqrt{\pi D}(k_0 + 4\pi RDC_S)^{-1/2}$$

The term within parentheses in Eq. (4-91) represents the effect of the time dependence of $k$. For $p \ll 1$, $(e^{p^2} \operatorname{erfc} p)$ approaches 1, and the term within parentheses also approaches 1. Therefore for small $C_S$ the relative effect of the time dependence of $k$ decreases, because with smaller $C_S$ the reaction $A + S \rightarrow Q$ takes place predominantly at longer times (and with a smaller probability because more A has decayed via $A \rightarrow P$). In Figure 4-2 the effect of the time dependence of the specific rate on the competition is illustrated. The inverse of the probability of not reacting with the scavenger S is plotted against $C_S$. Curves b1 and b2 have been calculated with Eq. (4-91) for different values of $R$ and $D$ but for the same value of $4\pi RD$. Curve a was calculated with Eq. (4-27), with $k_S$ equal to the value of $4\pi RD$ used above, and therefore corresponds to Eq. (4-91) for the case of a very large value of $D$ (and small $R$). The limiting slope at $C_S = 0$ is the same for all three curves; the effect of the time dependence increases with larger $R$ and smaller $D$.

It is clear that at higher concentrations Eq. (4-91) is not correct, because the contribution of reactions taking place at very early times increases, and Eq. (4-73) is no longer applicable; Eq. (4-71) should be used instead. When the reactants react at a distance exceeding the encounter distance, at very short times rates will be observed that are larger than predicted by Eq.

(4-71) for diffusion-controlled reactions. This results in an additional curving upwards of the competition curves, as presented in Figure 4-2. In connection with fluorescence quenching this effect is sometimes called "static" quenching in contrast to the "dynamic" quenching, which is described by Eq. (4-71). For further discussion of this effect see refs 15 through 17.

# ISOLATED PAIRS OF REACTANTS

## *Introduction*

In this section we discuss the kinetics of the reaction of a pair of reactants initially separated by a distance $r_0$. With fast electrons the primary energy deposition in the condensed phase results predominantly in the formation of small groups of pairs of oppositely charged species that can be considered independently of each other. Approximately 60% of these groups contains only one pair of "ions," initially separated by a few to several nanometers, and the single ion pairs represent approximately one-third of the total energy deposition. The kinetics of the reaction of isolated pairs is therefore of great importance for radiation chemistry. The application of the theory has been especially successful for hydrocarbon liquids, where due to the long range of the Coulomb interaction only a relatively small fraction of the initially formed pairs of oppositely charged species escape from the other's field.

The first treatment of geminate ion kinetics was given in 1938 by Onsager,[21] who calculated the probability of escape from geminate recombination in the presence of an external field using the Smoluchowski boundary condition. In 1956, Monchick[22] made an important contribution to the study of geminate reaction kinetics by considering an arbitrary potential with the radiation boundary condition and including scavenging by a solute. He obtained an exact solution for the scavenging probability for neutral particles and for the escape probability for an arbitrary potential. Although a rather simple analytical solution for the time-dependent problem for neutral particles was found,[23] insight into the time dependence of the kinetics of ion pairs was only obtained after applying numerical techniques.[24] Recently, exact but rather complicated solutions have also become available.[25]

## *Reaction and Survival, without a Scavenger*

We consider the diffusion in each other's field of two particles initially separated at a distance $r_0$. We take the origin of the coordinate system at one of the particles. When we are interested not in the angular dependence of the relative motion but only in the relative distance of the particles, we consider the ensemble of the cases where the second particle is initially at all possible positions on a sphere at $r = r_0$, or $c(r, t = 0) = \delta(r - r_0)/4\pi r^2$. We may now write the expression for the flow through a sphere at $r$ due to diffusion and field:

$$J = -4\pi r^2 \left( D\frac{\partial c}{\partial r} + \frac{D}{k_B T}\frac{dV}{dr} c \right) \tag{4-92}$$

where $D$ is again the sum of the diffusion coefficients and $V$ is the potential energy. Since the difference between the flow through the sphere at $r$ and $r + dr$ during $dt$ is equal to the change in the number of species in the shell with volume $4\pi r^2\,dr$,

$$-\frac{\partial J}{\partial r}\,dr\,dt = 4\pi r^2\,dr\frac{\partial c}{\partial t}\,dt \tag{4-93}$$

it follows from Eq. (4-93) and (4-92) that

$$\frac{\partial c}{\partial t} = \frac{D}{r^2}\frac{\partial}{\partial r}\left\{ r^2\left(\frac{\partial c}{\partial r} + \frac{c}{k_B T}\frac{dV}{dr}\right)\right\} \tag{4-94}$$

which is the spherically symmetrical representation of what is often referred to as the Smoluchowski or Debye–Smoluchowski equation. More generally, this is written as

$$\frac{\partial c}{\partial t} = \operatorname{div}\left\{ D\left(\operatorname{grad} c + \frac{c}{k_B T}\operatorname{grad} V\right)\right\} \tag{4-95}$$

We now have a sink at $r = \infty$, representing the particles that altogether escape reaction, and a sink close to the origin, representing the reaction between the particles. The first sink is described by

$$c(\infty, t) = 0 \tag{4-96}$$

The second boundary condition describing the reaction may take different forms. With the radiation boundary condition, for the rate of reaction of the two particles, $R(t)$, we can write

$$R(t) = 4\pi R^2 D\left(\frac{\partial c}{\partial r} + \frac{c}{k_B T}\frac{dV}{dr}\right)_{r=R} = k_R c(R, t) \tag{4-97}$$

With the Smoluchowski boundary condition we have

$$c(R, t) = 0 \tag{4-98}$$

For neutral particles ($V = 0$) Infelta has obtained an analytical solution in a closed form for $c(r, t)$, using the radiation boundary condition;[23c] for the Smoluchowski boundary condition the expression is very simple:[23a,b]

$$c(r, t) = \frac{1}{8\pi r r_0\sqrt{\pi D t}}\left[\exp\left\{-\frac{(r - r_0)^2}{4Dt}\right\} - \exp\left\{-\frac{(r + r_0 - 2R)^2}{4Dt}\right\}\right] \tag{4-99}$$

Also, for a Coulomb potential with the radiation boundary condition Eq.

(4-94) has been solved analytically.[25] Expressions have been given for $c(r, t)$ that can be evaluated numerically. In Figure 4-5 we plot the development of the probability of finding the species in a shell at $r$ in space and time for oppositely charged species, expressed in reduced units.[26] In Figure 4-5 (c and d) we show the effect of a relatively small $k_R$, which results in a temporary accumulation of species at the origin. The particles "stick together" due to the Coulomb field, but have not yet reacted. Such reactions occur in the radiolysis of hydrazine and amines, as shown by Delaire et al.[27]

The probability that the particles have not yet reacted together at a given time, or the survival probability, $W(t)$, is equal to the probability density $c(r, t)$ integrated over space:

$$W(t) = \int_R^\infty c(r, t) 4\pi r^2 \, dr \tag{4-100}$$

The probability that reaction has taken place at time $t$, $P(t)$, which is of course equal to $1 - W(t)$, is also equal to the total flow of probability across the reaction radius during the time $t$:

$$P(t) = \int_0^t R(t) \, dt \tag{4-101}$$

which, on substitution of Eq. (4-97) for $R(t)$, gives

$$P(t) = \int_0^t k_R c(R, t) \, dt \tag{4-102}$$

The total probability for reaction, $P(t = \infty)$, is now

$$P_R = \int_0^\infty k_R c(R, t) \, dt \tag{4-103}$$

where $P_R = P(t = \infty)$. The probability that the particles escape reaction, $W_{esc} = W(t = \infty)$, is $W_{esc} = 1 - P_R$.

Note that since $L\{\int_0^t F(t) \, dt\} = f(s)/s$, Laplace transformation of Eq. (4-101) yields $\tilde{P}(s) = \tilde{R}(s)/s$, and it follows from Eq. (4-97) that $\tilde{R}(s) = k_R \tilde{c}(R, s)$. If $\tilde{c}(r, s)$ can now be obtained by solving the Laplace-transformed Eq. (4-94), $\tilde{P}(s)$ can be obtained, and therefore $P(t)$ can be found by inverse Laplace transformation of $\tilde{P}(s)$.

It is clear that we do not need to know the complete time dependence of the probability density distribution in space in order to get the reaction probability at infinite time. Since it follows from Eq. (4-103) that we only need to know $\int_0^\infty c(R, t) \, dt$, we could integrate both sides of the diffusion equation (4-94) over time and solve for $c'(r) = \int_0^\infty c(r, t) \, dt$. Since the Laplace transform of $c(r, t)$ is defined as

$$\tilde{c}(r, s) = \int_0^\infty c(r, t) \, e^{-st} \, dt \tag{4-104}$$

we see that for $s = 0$, $\tilde{c}(r, s = 0) = c'(r)$, and therefore $P_R = k_R \tilde{c}(R, s = 0)$.

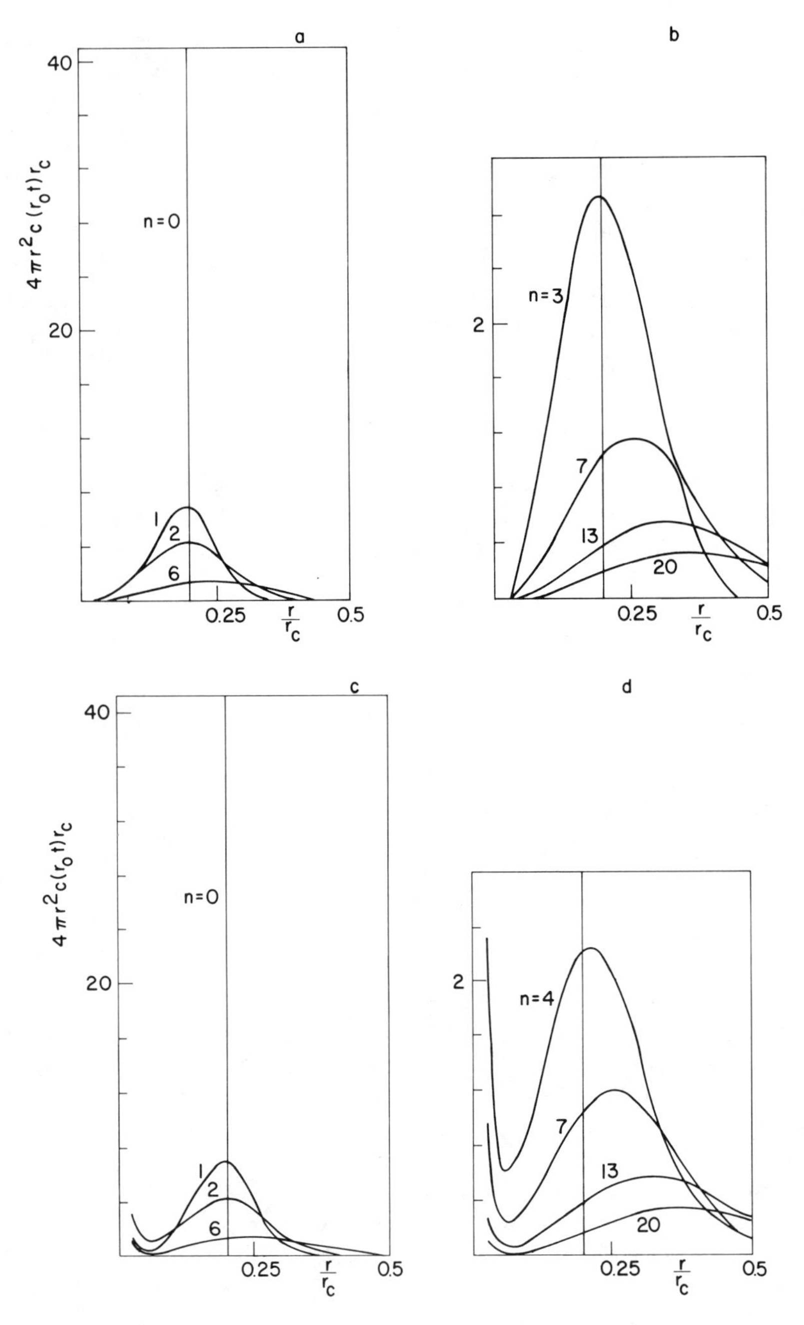


**Figure 4-5.** The probability that an ion pair with initial separation $r_0$ has a separation $r$ at different times $t$, for two values of $k_R$, expressed in reduced variables, calculated numerically;[26] $4\pi r^2 c(r, t) r_c$, which is the probability per unit distance $r_c$, is plotted as a function of $r/r_c$ for different $Dt/r_c$, with $r_0/r_c = 0.2$, $R/r_c = 0.01$; $k_R/Dr_c = 1.03 \times 10^2$ (a), (b), and $k_R/Dr_c = 1.03$ (c), (d); $Dt/r_c = 0.001$ n.

Monchick[22] has obtained $P_R$ for an arbitrary spherical potential $V(r)$ (using the radiation boundary condition) as a limiting case of a more general treatment by substituting $c(r, t) = c_1(r, t)\, e^{-V(r)/kT}$ in Eq. (4-94) and by solving the Laplace-transformed equation. The result is

$$P_R = \frac{S(r_0)}{S(R) + (4\pi D/k_R)\, e^{V(R)/k_B T}} \tag{4-105}$$

with

$$S(r) = \int_r^\infty \frac{1}{z^2}\, e^{V(z)/k_B T}\, dz.$$

The reaction probability $P_R$ can also be obtained by considering a continuous production of charged species at the sphere $r = r_0$ which results in a steady flow towards the sink at the origin and a flow towards infinity. The fraction of the production going inward and outward, equal to the probability of reaction and escape, respectively, can be calculated.[28] Recently, Tachiya and Sano have used another mathematical treatment to calculate the escape probability, which makes it possible to deal with nonspherical problems.[29–31]

For neutral particles, $V(r) = 0$, researchers have found analytical expressions for $P(t)$, using the different boundary conditions. For the radiation boundary condition we have[23]

$$P(t) = \frac{R}{r_0(1 + k_D/k_R)}\left[\operatorname{erfc}\frac{r_0 - R}{\sqrt{4Dt}} - \exp\left\{\left(1 + \frac{k_R}{k_D}\right)\left(\frac{r_0}{R} - 1\right) + \left(1 + \frac{k_R}{k_D}\right)^2 \frac{Dt}{R^2}\right\}\right.$$
$$\left. \times \operatorname{erfc}\left\{\left(1 + \frac{k_R}{k_D}\right)\frac{\sqrt{Dt}}{R} + \frac{r_0 - R}{\sqrt{4Dt}}\right\}\right] \tag{4-106}$$

For $t = \infty$ we obtain the reaction probability

$$P_R = \frac{R}{r_0}\frac{1}{1 + 4\pi RD/k_R} \tag{4-107}$$

Using the Smoluchowski boundary condition, we get

$$P(t) = \frac{R}{r_0}\operatorname{erfc}\frac{r_0 - R}{\sqrt{4Dt}} \tag{4-108}$$

which for $t = \infty$ gives

$$P_R = \frac{R}{r_0} \tag{4-109}$$

for the reaction probability.

The difference between the solution for $k_R$ finite [Eq. (4-107)] and $k_R = \infty$ [Eq. (4-109)] is the factor $(1 + 4\pi DR/k_R)^{-1}$. Note that the stationary diffusion-controlled rate for finite $k_R$, $k = 4\pi DR/(1 + 4\pi DR/k_R)$ [Eq. (4-63)], differs from the rate for $k_R = \infty$, $k = 4\pi DR$, by the same factor. A general relation between $P_R$ for isolated pairs of reactants and $k$ for bulk reaction of pairs has been shown to exist.[31–33]

At long times a fraction of the pairs of reactants escapes geminate recombination, and these species ultimately react with escaped members of other geminate pairs. It is clear that the distribution in space of these escaped species that react "homogeneously" will never be random. The transient behavior of the homogeneous specific rate, expressed by Eq. (4-71), is therefore not expected to hold. In this case the probability density rather grows in from $r = \infty$ towards smaller $r$. To my knowledge, this problem has not been treated theoretically.

For oppositely charged ions with charge $e$,

$$\frac{V(r)}{k_B T} = -\frac{e^2}{k_B T 4\pi\varepsilon_0\varepsilon_r r} = -\frac{r_c}{r},$$

and it follows from Eq. (4-105) that

$$P_R = \frac{1 - e^{-r_c/r_0}}{1 - e^{-r_c/R} + (4\pi D r_c/k_R)\, e^{-r_c/R}} \tag{4-110}$$

which for $k_R \gg 4\pi D r_c$ and $r_c \gg R$ reduces to $P_R = 1 - e^{-r_c/r_0}$, which corresponds to Onsager's result for the escape probability[11]

$$W_{esc} = e^{-r_c/r_0} \tag{4-111}$$

Since $r_c = 30$ nm at room temperature, in $n$-hexane, where the primary ionization leads to pairs with separations of about 5 nm, $W_{esc} = 2.5 \times 10^{-3}$, whereas in water, taking $r_0 = 2$ nm, $r_c = 0.75$ nm, $W_{esc} = 0.69$.

The expressions for the survival probability as a function of time for charged particles are rather cumbersome;[25] $W(t)$ has also been calculated numerically.[24] Plots calculated by Noolandi et al[25] are given for $r_0 = r_c/2$ and $r_0 = r_c/4$ (with $R = 0$) in Figure 4-6. For long times the survival probability can be written as

$$\lim_{R\to 0} W(t) = e^{-r_c/r_0}\left\{1 + \frac{r_c}{(\pi D t)^{1/2}}\right\} \tag{4-112}$$

which shows that for large $t$ the survival probability at a given time divided by the escape probability is independent of the initial separation. Note, however, that this equation holds only for values of $W(t)$ slightly above $W(t = \infty)$, as shown in Figure 4-6. Also the approximate analytical solution, obtained with the so-called prescribed diffusion approximation by Mozumder,[34]

$$W(t) \simeq \exp\left[-\frac{r_c}{r_0}\left\{1 - \operatorname{erf}\frac{r_0}{(4Dt)^{1/2}}\right\}\right] \tag{4-113}$$

has been plotted in Figure 4-6.

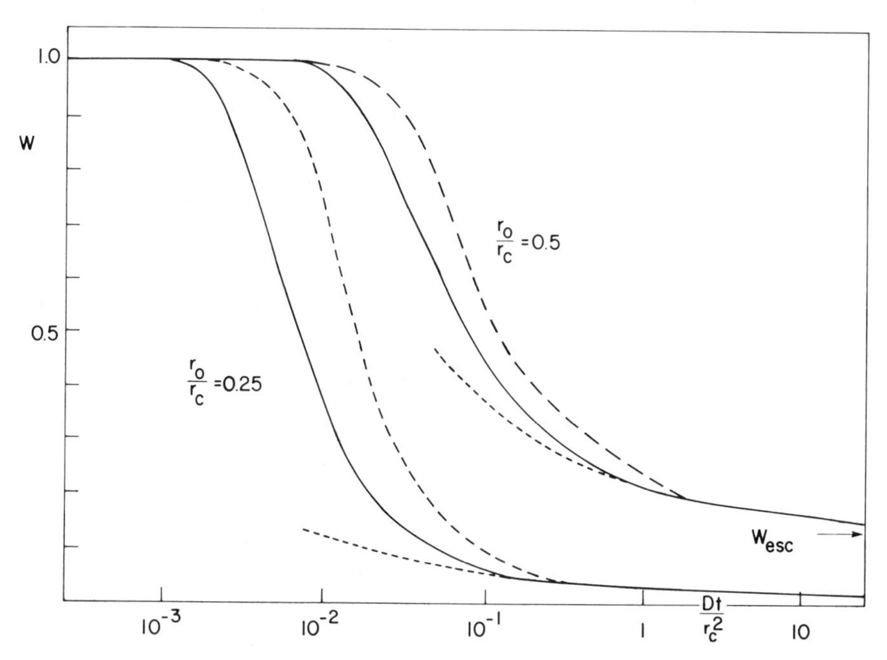


**Figure 4-6.** The probability of survival against recombination of a pair of oppositely charged ions with initial separation $r_0$ as a function of time, expressed as $Dt/r_c^2$, for $r_0/r_c = 0.5$ and $r_0/r_c = 0.25$.[25b] The long-time approximation [Eq. (4-112)] and the results obtained with the prescribed diffusion approximation [Eq. (4-113)] are also indicated.

## *Geminate Reaction with a Scavenger Present*

We now consider the case where a solute (S) is present that may react with the particle not at the origin. This can be accounted for in the diffusion equation (4-94) with an additional sink term, $-k_S C_S c$, where $k_S$ is the specific rate of reaction with the scavenger and $C_S$ is the scavenger concentration:

$$\frac{\partial c}{\partial t} = \frac{D}{r^2}\frac{\partial}{\partial r}\left(r^2 \frac{\partial c}{\partial r} + \frac{c}{kT}\frac{dV}{dr}\right) - k_S C_S c \tag{4-114}$$

We assume $k_S$ may be a constant. For diffusion-controlled reactions at comparatively large concentrations, this is obviously incorrect. At this point, however, we shall not deal with this complication.

The solution of this equation, $c(r, t)$, is now simply related to the case where no solute is present, $c_0(r, t)$, by

$$c(r, t) = c_0(r, t)\, e^{-k_S C_S t} \tag{4-115}$$

When we substitute Eq. (4-115) into Eq. (4-114) we find the original diffusion equation in $c_0$, without the scavenging term.

The probability for reaction with the scavenger per unit time is equal to the local rate of reaction, $k_S C_S c(r, t)$, per unit volume and time integrated over space:

$$S(t) = \int_R^\infty k_S C_S c(r, t) 4\pi r^2 \, dr \tag{4-116}$$

Substituting Eq. (4-114) into Eq. (4-116) gives

$$S(t) = \int_R^\infty k_S C_S c_0(r, t) \, e^{-k_S C_S t} 4\pi r^2 \, dr \tag{4-117}$$

This can be written as

$$S(t) = k_S C_S W_0(t) \, e^{-k_S C_S t} \tag{4-118}$$

where $W_0(t)$ is the probability of survival in the absence of $S$. We obtain the probability for the particle to have reacted with the solute at time $t = \infty$, $P_S(k_S C_S)$, by integrating the rate of reaction $S(t)$ over time. Thus, using Eq. (4-33) for $S(t)$ gives

$$P_S(k_S C_S) = \int_0^\infty k_S C_S W_0(t) \, e^{-k_S C_S t} \, dt = k_S C_S \tilde{W}_0(k_S C_S) \tag{4-119}$$

This shows that the scavenging probability is related to the survival probability function by the Laplace transformation (provided $k_S$ is constant).

Because the fraction that does not react with the solute reacts with the particle A at the origin, $P_S(k_S C_S)$ can also be obtained by considering the flow of unreacted particles toward the origin. Since the rate of reaction with A at time $t$ in the presence of a scavenger, $R(t)$, is equal to the rate in the absence of a scavenger, $R_0(t)$, multiplied by the probability that during time $t$ scavenging has not taken place, $R(t) = R_0(t) \, e^{-k_S C_S}$, and since the total probability for reaction with A is $P_R(k_S C_S) = \int_0^\infty R(t) \, dt$, we have

$$P_R(k_S C_S) = \int_0^\infty R_0(t) \, e^{-k_S C_S t} \, dt = \tilde{R}_0(k_S C_S) \tag{4-120}$$

With the radiation boundary condition, we have $R_0(t) = k_R c_0(R, t)$ [Eq. (4-97)]; therefore, $P_R(k_S C_S) = k_R \tilde{c}_0(R, k_S C_S)$.* If we substitute $R_0(t) = -dW_0/dt$ in Eq. (4-120), integrate by parts, and with Eq. (4-110), we see that $P_R + P_S = 1$, as expected; the sum of the probability of reaction with the scavenger and with the particle at the origin (escape of reaction with the

* It has been shown that the Laplace-transformed rate of recombination, $\tilde{R}(s)$, as obtained with the radiation boundary condition, can be related in a simple way to the Laplace-transformed rate obtained with the Smoluchowski boundary condition, $\tilde{R}_S(s)$:

$$\tilde{R}(s) = \frac{\tilde{R}_S(s)}{1 - (k_D R/k_R)(\partial \tilde{R}_S/\partial r)_{r=R}}$$

where $k_D = 4\pi RD$.[35]

scavenger) must be equal to 1. Note that when no scavenger is present, this equality does not hold because $p_S(0) = 0$ and $P_R(0) = 1 - W_{esc}$.

### Neutral Particles

For two neutral particles, $V(r) = 0$, initially at a distance $r_0$ the probability for geminate reaction in the presence of a scavenger is[22]

$$P_R(k_S C_S) = \frac{R}{r_0} \cdot \frac{\exp\{-(r_0 - R)\sqrt{k_S C_S/D}\}}{1 + k_D/k_R + (k_D R/k_R)\sqrt{k_S C_S/D}} \tag{4-121}$$

For $k_S C_S = 0$, Eq. (4-121) reduces to $P_R(0) = (P/r_0)(1 + k_D/k_R)^{-1}$, which we found in Eq. (4-107).

For extremely efficient geminate reactions $k_R \gg k_D$,

$$P_R(k_S C_S) = \frac{R}{r_0} \exp\left\{-(r_0 - R)\sqrt{\frac{k_S C_S}{D}}\right\},$$

and the scavenging probability becomes

$$P_S(k_S C_S) = 1 - \frac{R}{r_0} \exp\left\{-(r_0 - r)\sqrt{\frac{k_S C_S}{D}}\right\} \tag{4-122}$$

At low concentration, or rather if $(r_0 - R)\sqrt{k_S C_S/D} \ll 1$, Eq. (4-122) can be written as

$$P_S(k_S C_S) = 1 - \frac{R}{r_0} + \frac{R}{r_0}(r_0 - R)\sqrt{\frac{k_S C_S}{D}} \tag{4-123}$$

or, with $W_{esc} = 1 - R/r_0$,

$$P_S(k_S C_S) = W_{esc}\left(1 + R\sqrt{\frac{k_S C_S}{D}}\right) \tag{4-124}$$

This shows that for low scavenger concentration the scavenging probability is proportional to the square root of $k_S C_S/D$ and, furthermore, that the scavenging probability relative to the escape probability is independent of the initial separation.

### Ion Pairs

For $V \neq 0$ no closed solution has been found for the scavenging probability for the entire range of concentrations. For a Coulomb potential, however, the scavenging probability has been calculated in various ways.[24,26,36] Since, for a pair of oppositely charged ions $V(r) = -e^2/4\pi\varepsilon_0\varepsilon_r r$, and therefore $(1/k_B T)(dV/dr) = r_c/r$, Eq. (4-114) can be written as a function of the reduced variables $r/r_c$, $Dt/r_c^2$, $cr_c^3$, and $k_S C_S r_c^2/D$. The scavenging probability can therefore be expressed as a function of $k_S C_S r_c^2/D$. We do this in Figure 4-7

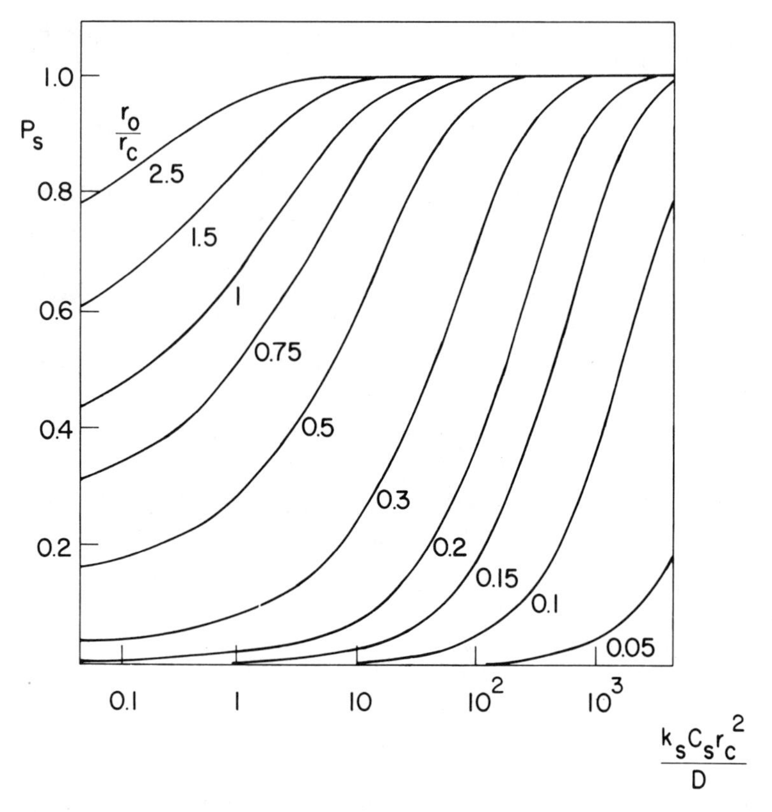


**Figure 4-7.** The probability for reaction with a solute of one ion of a pair of oppositely charged ions, for different initial separations $r_0$, as a function of the solute concentration,[36d] expressed as $k_S C_S r_c^2/D$.

for different initial separations. For small concentrations

$$P_S(k_S C_S) = e^{-r_c/r_0} \left\{ 1 + \left( \frac{k_S C_S r_c^2}{D} \right)^{1/2} \right\} \tag{4-125}$$

and because $e^{-r_c/r_0} = W_{esc}$, the ratio of the scavenging probability and the escape probability is again independent of the initial separation $r_c$.

### Two Scavengers

When two scavengers, S1 and S2, are present that react with the same species, the probability for reaction of S1 is

$$P_S(k_{S1} C_{S1}) = \frac{k_{S1} C_{S1}}{k_{S1} C_{S1} + k_{S2} C_{S2}} P_S(k_{S1} C_{S1} + k_{S2} C_{S2}) \tag{4-126}$$

We have seen that the scavenging probability is determined by $k_S C_S r_c^2/D$. The probability of scavenging of one of the charged particles is determined by the sum of the diffusion coefficients of both. As a result the probability of scavenging of one particle will, in general, depend on the scavenging of the other particle. This effect is very pronounced when the mobilities of the charged particles before and after scavenging differ considerably, as is often the case with scavenging of excess electrons and electron holes. In this way the scavenging of one particle may be increased by scavenging of the other, due to a decrease of the mobility and a resulting increase in lifetime. If the second scavenger reacts with both particles, a decrease due to straightforward competition occurs, in addition to the enhancement due to increased lifetime. For a further discussion of the competition kinetics of the scavenging of geminately recombining pairs of charged species, the reader is referred to refs. 26 and 37.

## GROUPS OF MORE THAN TWO SPECIES

### *Introduction*

As we mentioned earlier, with fast electrons in the condensed phase about one-third of the energy loss leads to isolated ion pairs. The frequency of occurrence of groups with more pairs decreases rapidly with the number of pairs, which is illustrated by the fact that the formation of groups of more than six pairs accounts for only about 15% of the energy absorption (and that the average number in the groups of up to six pairs is approximately 1.6). For one to understand radiation chemistry with fast electrons, analysis of the kinetics of groups of a few pairs is therefore crucial. With slower-moving charged particles the average distance between the groups of ionizations along the track is smaller, and merging of the groups takes place. The degree to which the groups can be considered independent of each other depends, of course, on the range of the Coulomb interaction and, therefore, on the dielectric constant. With high linear energy transfer (LET) radiation the groups due to small energy losses along the track are no longer independent, and the track has to be considered as one group.

In this section we briefly discuss the kinetics of groups of more than two species. We direct ourselves to outlining some of the problems involved. Applying the methods to the radiation chemistry of particular systems is dealt with in other chapters.

### *Independent Particle Approximation*

We consider a collection of particles in one another's neighborhood that carry out a diffusion motion while they also may react with each other. The development of the probability density distribution in time of each of the particles depends on that of all other particles, the degree of correlation depending on the initial distribution. An exact solution for the kinetics in this case has not been given, not even when there is no field between the particles.

To make the problem tractable, we often approximate the probability density distribution for all the ($n$) particles of the group as a product of the probability densities of the $n$ individual particles (which, of course, is true only when the particles do not interact and are independent). Now if $c_i(\bar{r}_i, t)$ is the probability density of particle $i$ at location $\bar{r}_i$ at time $t$ and we assume that $i$ reacts with $j$, when a particle $j$ exists at a distance $|\bar{r}_i - \bar{r}_j| = R_{ij}$ from $i$, we may write, following Monchick et al,[38]

$$\frac{\partial}{\partial t} c_i(\bar{r}_i, t) = \nabla \cdot D \nabla c_i(\bar{r}_i, t) - c_i(\bar{r}_i, t) \sum_{j \neq i} \frac{k_{ij}}{4\pi R_{ij}^2} \int c_j(\bar{r}_j, t) \delta(|\bar{r}_i - \bar{r}_j| - R_{ij})\, d\bar{r}_j \qquad (4\text{-}127)$$

The first term on the right side represents the free diffusion of $i$, the second term the reaction with a specific rate $k_{ij}$ when a particle $j$ is present at a distance $R_{ij}$ around $i$. The specific rate $k_{ij}$ is defined in the same way as $k_R$ in the two-particle problem (note that this is different in ref. 38). The probability density of finding $i$ at $\bar{r}_i$ and $j$ at $\bar{r}_j$ is here written as the product $c_i(\bar{r}_i, t)c_j(\bar{r}_j, t)$. This is in contrast to our earlier treatment of the two-particle problem, where the density distribution as a function of the distance between the particles was found by considering the flow due to reaction and diffusion. As a result the physical significance of the rate parameter $k_{ij}$ is vague (if we are dealing with efficient reactions). If we further assume that we can write the integral as $4\pi R_{ij}^2 c_j(\bar{r}_i, t)$, which means that the average value of $c_j(\bar{r}_j, t)$ at the sphere is equal to the value in the center, Eq. (4-127) becomes

$$\frac{\partial}{\partial t} c_i(\bar{r}_i, t) = \nabla D \cdot \nabla c_i(\bar{r}_i, t) - c_i(\bar{r}_i, t) \sum_{j \neq i} k_{ij} c_j(\bar{r}_i, t) \qquad (4\text{-}128)$$

which shows that the effect of finite size of the particles has disappeared. After integrating Eq. (4-128) over space, we obtain the rate of change with time of the probability, $w_i$, that the species $i$ has not reacted:

$$\frac{d}{dt} w_i = -\sum_{j \neq i} k_{ij} \int c_i(\bar{r}_i, t) c_j(\bar{r}_i, t)\, d\bar{r}_i \qquad (4\text{-}129)$$

For $N_0$ identical particles with identical probability density distributions at $t = 0$, Eqs. (4-128) become identical for all $i$ and $c_i(\bar{r}_i = \bar{r}, t) = c_j(\bar{r}_j = \bar{r}, t)$. The probability of finding any of the particles in a unit volume, which is equal to the concentration, is

$$c(\bar{r}, t) = \sum_{i=1}^{N_0} c_i(\bar{r}_i = \bar{r}, t) = N_0 c_i(\bar{r}_i = \bar{r}, t).$$

It now follows from Eq. (4-128) that

$$\frac{\partial c}{\partial t} = D\nabla^2 c - \frac{N_0 - 1}{N_0} kc^2 \tag{4-130}$$

where we have written $c$ for $c(\bar{r}, t)$ and have assumed $D$ constant in space. For large $N_0$, $(N_0 - 1)/N_0 \rightarrow 1$, and Eq. (4-130) reduces to

$$\frac{\partial c}{\partial t} = D\nabla^2 c - kc^2 \tag{4-131}$$

For two kinds of particles, A and B, with initial numbers $N_{A0}$ and $N_{B0}$, we get

$$\frac{\partial c_A}{\partial t} = D\nabla^2 c_A - \frac{N_{A0} - 1}{N_{A0}} k_{AA} c_A{}^2 - k_{AB} c_A c_B \tag{4-132}$$

If a scavenger is present at a concentration $C_S$, an additional term, $-k_{AS} c_A C_S$, is introduced. The number of particles present at any time is obtained by integrating over space, $N_A = \int c_A \, dv$, and from Eq. (4-132) it follows that

$$\frac{dN_A}{dt} = -\frac{N_{A0} - 1}{N_{A0}} k_{AA} \int c_A{}^2 \, dv - k_{AB} \int c_A c_B \, dv \tag{4-133}$$

Clearly, the number of product molecules AB formed from the reaction A + B is

$$\frac{dN_{AB}}{dt} = k_{AB} \int c_A c_B \, dv \tag{4-134}$$

## *Prescribed Diffusion*

To deal with equations like (4-134) we often make the prescribed diffusion or Jaffé approximation.[38,39] We assume that the concentration distribution of the reactive species in space and time is

$$c_A(\bar{r}, t) = N_A(t) c_A{}^0(\bar{r}, t) \tag{4-135}$$

where $c_A{}^0(\bar{r}, t)$ is the solution for the case of free diffusion

$$\frac{\partial}{\partial t} c_A{}^0 = D_A \nabla^2 c_A{}^0 \tag{4-136}$$

The integral $\int c_A{}^2 \, dv$ in Eq. (4-133) now reduces to $\{N_A(t)\}^2 \int \{c_A{}^0(\bar{r}, t)\}^2 \, d\bar{r}$. For the initial distributions $c_A{}^0\,(\bar{r}, t = 0)$ we often take Gaussian distributions with spherical or cylindrical symmetry for which the solutions are known. Equation (4-133) can now be written as

$$\frac{dN_A}{dt} = -f_{AA} k_{AA} N_A{}^2 - f_{AB} k_{AB} N_A N_B \tag{4.137}$$

where $f_{AA}$ and $f_{BB}$ are known (time-dependent) parameters. Equations of this form have been used extensively to analyze nonhomogeneous kinetics in the radiation chemistry of water, where long-range Coulomb interactions of the species can be neglected.[40,41]

## *Stochastic Treatments*

In the prescribed diffusion treatment the spatial distribution of the species is determined by the diffusion equation (4-136), without reaction. This is obviously incorrect, because the rate of reaction between species is higher in regions where the local concentration is higher than in regions where the concentration is lower, which must result in an additional flattening of the spatial distribution that is not accounted for by Eq. (4-136).

Another shortcoming is the use of an average specific rate of reaction. As we saw when discussing the diffusion-controlled reaction of species, the specific rate of reaction is time dependent because the species that happen to be created next to each other can react very fast, without diffusion, whereas species that are formed away from each other first have to find each other by diffusion. Taking this effect into consideration resulted in the development of concentration gradients. Schwarz has used the time-dependent specific rate for initially homogeneous distributions of species [Eq. (4-73)] in the prescribed diffusion treatment.[40] For a small number of particles initially present this represents an assumption of which the physical significance is rather unclear, because the expression for the time-dependent rate is obtained for a constant concentration of species at infinite distance.

Because the independent particle approximation neglects correlation between the species, this treatment will be increasingly unsatisfactory for smaller numbers of particles. Therefore, Monchick et al[38] proposed an approximation in which correlation between pairs of species is taken into account. More recently, Clifford et al[42] have introduced a new approach. In this treatment the rate of the reaction of each pair is assumed to be independent of the other particles in the group. Clifford et al have used this "independent pair approximation" in two different treatments. In the first they used the stochastic theory of chemical kinetics, employing the method of generating functions, as outlined in the review article by McQuarrie,[43] where the probability for the occurrence of a discrete number of particles as a function of time is calculated. In the second they used a Monte Carlo method to choose the moment of reaction for each of the pairs (first passage time) from the known probability of survival as a function of time for each pair. The results obtained with the two methods for small numbers of particles initially agree well. The same authors have also performed a Monte Carlo calculation of the diffusion in space together with reaction on encounter. The results agree quite satisfactorily with the independent pair treatment.

Comparing these results with those obtained from treatments based on the independent particle approximation is difficult because in the latter treatments the probability distributions of the different particles overlap; in other words, two particles have a finite probability of being found at the same place, whereas in the independent pair treatment such pairs are not considered. This has especially important consequences for the choice of the initial spatial distributions, which is discussed in detail in Green et al.[44]

Recently, Bartczak and Hummel[45] have simulated the diffusion and drift in the mutual field of small groups of ions by a Monte Carlo method. In this way the probabilities of survival of the ions as a function of time have been calculated for one, two, and three ion pairs initially. The remarkable result has emerged that for various initial spatial distributions the probability of survival as a function of time for two and three pairs is not very different from that obtained for the pairs when treated as independent.

## *The Track*

In radiation chemistry the primary reactive species are formed non-homogeneously along the track of the primary particle. The initial spatial distribution is always complex. With fast electrons there is a relatively large contribution of pairs and small groups of species that can be considered independently; with slower particles the contribution of larger groups increases, and with high-LET particles the track consists of a cylindrical core of reactive species with high density with occasional sidetracks due to high-energy secondary electrons. The kinetics in radiation chemistry reflects the initial spatial distribution (and is a primary source of information about it). Studies of early processes in radiation chemistry by investigating the kinetics, therefore, must always involve consideration of the track structure. This will be dealt with in the next chapter.

# REFERENCES


1. Smoluchowski, M. *Z. Phys. Chem.* **1917**, *92*, 129.
2. Debye, P. *Trans. Electrochem. Soc.* **1942**, *82*, 265.
3. Collins, F. C.; Kimball, G. E. *J. Colloid Sci.* **1949**, *4*, 425.
4. Chandrasekhar, S. *Rev. Mod. Phys.* **1943**, *15*, 1.
5. Noyes, R. M. *Prog. React. Kinet.* **1961**, *1*, 129.
6. Wilemski, G.; Fixman, M. *J. Chem. Phys.* **1973**, *58*, 4009.
7. Lin, S. H.; Li, K. P.; Eyring, H. In "Physical Chemistry. An Advanced Treatise"; Eyring, H., Ed.; Academic Press: 1975, Vol. 7, Chapter 1.
8. Weston, R. Jr.; Schwarz, H. A. "Chemical Kinetics"; Prentice-Hall: Englewood Cliffs, N.J., 1972.

9a. Felderhof, B. U.; Deutch, J. M. *J. Chem. Phys.* **1976**, *64*, 4551.
9b. Deutch, J. M.; Felderhof, B. U.; Saxton, M. J. *J. Chem. Phys.* **1976**, *64*, 4559.
9c. Peak, D.; Pearlman, K.; Wantuck, P. J. *J. Chem. Phys.* **1976**, *65*, 5538.

10. Hummel, A.; Allen, A. O. *J. Chem. Phys.* **1966**, *44*, 3426.

11a Wda, T.; Shinsaka, K.; Namba, H.; Hatano, Y. *Can. J. Chem.* **1977**, *55*, 2144.
11b. Nakamura, Y.; Namba, H.; Shinsaka, K.; Hatano, Y. *Chem. Phys. Lett.* **1980**, *76*, 311.
12. Rice, S. A.; Butler, P. R.; Pilling, M. J.; Baird, J. K. *J. Chem. Phys.* **1979**, *70*, 4001.
13a. Friauf, R. J.; Noolandi, J.; Hong, K. M. *J. Chem. Phys.* **1978**, *68*, 5169.
13b. Hong, K. M.; Noolandi, J. *J. Chem. Phys.* **1978**, *68*, 5172.
14. Pilling, M. J.; Rice, S. A. *J. Chem. Soc., Faraday Trans. 2* **1975**, *71*, 1563.
15. Pilling, M. J.; Rice, S. A. *Prog. React. Kinet.* **1978**, *9*, 93.
16a. Andre, J. C.; Niclause, M.; Ware, W. R. *Chem. Phys.* **1978**, *28*, 371.
16b. Andre, J. C.; Niclause, M.; Ware, W. R. *Chem. Phys.* **1979**, *37*, 103.
16c. Sipp, B. Ph.D. Dissertations, University Louis Pasteur, Strasbourg, France, 1981.
17. Czapski, G.; Peled, E. *J. Phys. Chem.* **1973**, *77*, 893.
18. Bartczak, W. M.; Kroh, J.; Romanowska, E.; Stradowski, G. *Current Topics Radiat. Res. Quart.* **1977**, *11*, 307.
19. Nemzek, T. L.; Ware, W. R. *J. Chem. Phys.* **1975**, *62*, 477.
20. Buxton, G. V.; Cattell, F. C. R.; Dainton, F. S. *J. Chem. Soc., Faraday Trans. 1* **1975**, *71*, 115.
21. Onsager, L. *Phys. Rev.* **1938**, *54*, 554.
22. Monchick, L. *J. Chem. Phys.* **1956**, *24*, 381.
23a. Pitts, E. *Trans. Faraday Soc.* **1969**, *65*, 2372.
23b. Yguerabide, J. *J. Chem. Phys.* **1967**, *47*, 3049.
23c. Infelta, P. P. Unpublished.
24a. Abell, G. C.; Mozumder, A.; Magee, J. L. *J. Chem. Phys.* **1972**, *56*, 5422.
24b. Hummel, A.; Infelta, P. P. *Chem. Phys. Lett.* **1974**, *24*, 559.
25a. Hong, K. M.; Noolandi, J. *J. Chem. Phys.* **1978**, *69*, 5026.
25b. Hong, K. M.; Noolandi, J. *J. Chem. Phys.* **1978**, *68*, 5163.
26. Hummel, A. In "Advances in Radiation Chemistry"; Burton, M.; and Magee, J. L., Eds.; John Wiley: New York, 1974, Vol. 4, Chapter 1.
27. Delaire, J. A.; Croc, E.; Cordier, P. *J. Phys. Chem.* **1981**, *85*, 1549.
28. Magee, J. L.; Mozumder, A. In "Advances in Radiation Chemistry"; Burton, M.; Magee, J. L., Eds.; Wiley Interscience, 1969, p. 83.
29. Tachiya, M. *J. Chem. Phys.* **1978**, *69*, 2375.
30. Sano, H. *J. Chem. Phys.* **1981**, *74*, 1394.
31. Sano, H.; Tachiya, M. *J. Chem. Phys.* **1979**, *71*, 1276.
32. Rice, S. A.; Baird, J. K. *J. Chem. Phys.* **1978**, *69*, 1989.
33. Berlin, Y. A.; Cordier, P.; Delaire, J. A. *J. Chem. Phys.* **1980**, *73*, 4619.
34. Mozumder, A. *J. Chem. Phys.* **1968**, *48*, 1959.
35a. Pedersen, J. B. *J. Chem. Phys.* **1980**, *72*, 771.
35b. Pedersen, J. B. *J. Chem. Phys.* **1980**, *72*, 3904.
36a. Magee, J. L.; Tayler, A. B. *J. Chem. Phys.* **1972**, *56*, 3061.
36b. Infelta, P. P. *J. Chem. Phys.* **1972**, *69*, 1526.
36c. Tachiya, M. *J. Chem. Phys.* **1979**, *70*, 238.
36d. Friauf, R. J.; Noolandi, J.; Hong, K. M. *J. Chem. Phys.* **1979**, *71*, 143.
37a. Rzad, S. J.; Schuler, R. K.; Hummel, A. *J. Chem. Phys.* **1969**, *51*, 1369.
37b. Davids, E. L.; Warman, J. M.; Hummel, A. *J. Chem. Soc., Faraday Trans. 1* **1975**, *71*, 1252.
38. Monchick, L.; Magee, J. L.; Samuel, A. H. *J. Chem. Phys.* **1957**, *26*, 935.
39. Jaffe, G. *Ann. Phys.* **1913**, 303.
40. Schwarz, H. A. *J. Phys. Chem.* **1969**, *73*, 1928.
41a. Kuppermann, A. In "The Chemical and Biological Action of Radiations"; Haissinsky, M., Ed.; Academic Press: New York, 1961, Vol. 5, Chapter 3.
41b. Kuppermann, A. "Radiation Research", Proc. of the Third International Congress of Radiation Research, Amsterdam, 1968.
42a. Clifford, P.; Green, N. J. B.; Pilling, M. J. *J. Phys. Chem.* **1982**, *86*, 1318.
42b. Clifford, P.; Green, N. J. B.; Pilling, M. J. *J. Phys. Chem.* **1982**, *86*, 1322.
43. McQuarrie, D. A. *J. Appl. Prob.* **1967**, *4*, 413.
44. Green, N.; Pilling, M. J.; Clifford, P.; Burns, W. G. *J. Chem. Soc., Faraday Trans. 1* **1984**, *8*, 1313.
45. Bartczak, W. M.; Hummel, A. *Radiat. Phys. Chem.* **1986**, *27*, 71.

# 5

# Theoretical Aspects of Radiation Chemistry

J. L. MAGEE AND A. CHATTERJEE

*Lawrence Berkeley Laboratory,*
*University of California at Berkeley*

## INTRODUCTION AND OVERVIEW

Essentially all effects of high-energy radiation on matter occur through the action of fast charged particles (if the primary radiation is electromagnetic, that is, x- or $\gamma$-rays, the energy is first used to produce fast electrons by the photoelectric or the Compton effects). The charged particles affect the matter by creation of electronic excitation, positive ions and recoil electrons. The latter species are transformed in sequences of processes leading to the ultimate radiation chemical products. Radiation chemistry is concerned with all of the species and processes in such chains of events.

Many separate topics can be identified for special consideration, and some of them are covered in other chapters of this book. In this chapter the focus is on a theoretical description of the entire sequence of events in irradiated systems, and we are particularly interested in the relationships of the events as they depend on the qualities and energies of the radiations. The point of view is that of the authors, and there has been more effort expended in trying to make this view clear than in trying to present a coherent treatment of radiation chemical theory that integrates work of other authors. Throughout this chapter and the next, water has been used extensively in examples. This is done because so much of radiation chemical theory deals with water, perhaps in large part because of this substance's importance in radiation biology.


*Radiation Chemistry: Principles and Applications*

High-energy radiations form tracks as their energy is transmitted to the matter, and the structure of particle tracks (or the geometrical distribution of energy) is greatly important in determining radiation chemical yields. All transient reactive species are formed in the tracks, and their initial energies are higher than the thermal energy distributions of the environment. The natural development of a track involves three processes occurring simultaneously with the transient species: (a) thermalization, (b) transformation and reaction, (c) diffusion away from locus of formation (see Figures 5-1 and 5-2). Thermalization is complete in a picosecond or so at any locality, but the chemical transformation and diffusion continue for relatively long times—microseconds or even milliseconds. The early phenomena, particularly the processes occurring before thermalization, are known mostly from theory and conjecture. This situation is discussed in the section entitled "Early Phenomena." The chemical reactions of thermalized transient species (radicals) in tracks have been the centerpiece of radiation chemical theory. This topic is discussed in the section called "Track Reactions." Much of radiation chemistry is concerned with effects that are determined by single tracks, but there are phenomena that depend on more general conditions of an irradiation. This situation is considered in a section entitled "Theoretical Treatment of Background Reactions in Irradiated Systems." In the last section an attempt is made to summarize the status of radiation chemical theory.

The objectives of theoretical radiation chemistry are to understand all chemical phenomena initiated by the absorption of high-energy radiations and their dependencies on the chemical composition of the medium and the geometrical pattern of energy deposit (ie, track structure). Most experimental investigations in radiation chemistry have involved irradiation by electrons. The track structure of this type of radiation is typical of "low-LET" (linear energy transfer) radiations. To a rough approximation electron tracks can be taken as made up of isolated ionizations. Why such an approximation works at all, and its relationship to a more realistic model are questions addressed in Chapter 6.

The first theoretical considerations of radiation chemistry were involved with the construction of "radical diffusion models" to explain the radiation chemical yields in the radiolysis of water. It was thought at first that the only important radiation-produced species were the H and OH radicals. Later, other species were observed, and now the equation for the radiation action on water is usually written

$$H_2O \rightsquigarrow H,\ e_{aq}^-,\ OH,\ H_3O^+,\ H_2O,\ H_2,\ H_2O_2,\ OH^-,\ HO_2,\ O_2,\ O_2^-,\ HO_2^- \tag{5-1}$$

In this equation (which is not chemically balanced) the observed species are essentially listed on the right, and it is understood that they have arisen in different ways following the initial radiation action. Under the conditions of observation the species on the right are normal chemical species in thermal equilibrium with their environments (ie, they are "thermalized"). Chemical reactions among them are clearly important in radiation chemistry, and the most important reactions they undergo are listed in Table 5-1.

**Table 5-1.** Reactions in Irradiated Neutral Water

| Reactions | Reaction Rate Constant[a] ($1\ M^{-1}\ s^{-1}$) |
|---|---|
| A. *Recombination of primary radicals* | |
| *1. $H + H \rightarrow H_2$ | $1 \times 10^{10}$ |
| 2. $e_{aq}^{-} + H \rightarrow H_2 + OH^{-}$ | $2.5 \times 10^{10}$ |
| 3. $e_{aq}^{-} + e_{aq}^{-} \rightarrow H_2 + 2\,OH^{-}$ | $6 \times 10^{9}$ |
| 4. $e_{aq}^{-} + OH \rightarrow OH^{-} + H_2O$ | $3 \times 10^{10}$ |
| *5. $H + OH \rightarrow H_2O$ | $2.4 \times 10^{10}$ |
| *6. $OH + OH \rightarrow H_2O_2$ | $4 \times 10^{9}$ |
| 7. $H_3O^{+} + e_{aq}^{-} \rightarrow H + H_2O$ | $2.3 \times 10^{10}$ |
| 8. $H_3O^{+} + OH^{-} \rightarrow H_2O$ | $3 \times 10^{10}$ |
| B. *Other reactions of radicals* | |
| *9. $H + H_2O_2 \rightarrow H_2O + OH$ | $1 \times 10^{10}$ |
| 10. $e_{aq}^{-} + H_2O_2 \rightarrow OH + OH^{-}$ | $1.2 \times 10^{10}$ |
| *11. $OH + H_2O_2 \rightarrow H_2O + HO_2$ | $5 \times 10^{7}$ |
| *12. $OH + H_2 \rightarrow H_2O + H$ | $6 \times 10^{7}$ |
| *13. $HO_2 + H \rightarrow H_2O_2$ | $1 \times 10^{10}$ |
| 14. $e_{aq}^{-} + O_2 \rightarrow O_2^{-} + H_2O$ | $1.9 \times 10^{10}$ |
| *15. $HO_2 + OH \rightarrow H_2O + O_2$ | $1 \times 10^{10}$ |
| *16. $HO_2 + HO_2 \rightarrow H_2O_2 + O_2$ | $2 \times 10^{6}$ |
| *17. $H + O_2 \rightarrow HO_2$ | $1 \times 10^{10}$ |
| 18. $O_2^{-} + H_3O^{+} \rightarrow HO_2$ | $3 \times 10^{10}$ |
| 19. $H_3O^{+} + HO_2^{-} \rightarrow H_2O_2$ | $3 \times 10^{10}$ |
| 20. $HO_2 + O_2^{-} \rightarrow HO_2^{-} + O_2$ | $5 \times 10^{7}$ |
| 21. $H_2O_2 + HO_2 \rightarrow H_2O + O_2 + OH$ | 530 |
| 22. $H_2O_2 + O_2^{-} \rightarrow OH + OH^{-} + O_2$ | 16 |
| C. *Dissociation reactions*[b] | |
| 23. $H_2O \rightarrow H_3O^{+} + OH^{-}$ | $5.5 \times 10^{-6}$ |
| 24. $HO_2 \rightarrow H_3O^{+} + O_2$ | $1 \times 10^{6}$ |
| 25. $H_2O_2 \rightarrow H_3O^{+} + HO_2^{-}$ | $3 \times 10^{-2}$ |

[a] The rate constants were selected from the collection published by Ross et al.[14–17]
[b] Rate constants for Eqs. 23–25 are $s^{-1}$.
* Equations occurring in acid water.

## *The Picosecond Barrier*

It was soon realized that the transient species observed in track reactions were generally not the species initially created by the radiation, and the search for "precursors" of the observed radicals was initiated. The invention of nano- and picosecond pulse radiolysis made this search practical, and much information was obtained on early radiation chemical processes. There are, however, limitations on how early a direct observation can be made on a process occurring in a macroscopic sample that has received a radiation pulse. We can understand this limitation by a simple consideration.[1]

Take as a minimal reaction cell in a pulse radiolysis experiment a cube with 3-mm sides. To make the conditions optimal for time resolution, assume the energy deposit to be made by a $\delta$-pulse of electrons traveling with the speed of light (this means that the excitation is created by a plane of electrons that has zero thickness). The time to deposit energy in the cell by the electron beam is 10 ps ($0.3/3 \times 10^{10}$). The analyzing light passes through the cell parallel to and in the same direction as the electron beam; its speed, however, is smaller than the speed of light in a vacuum at which the electron beam travels. Depending on the frequency of the analyzing light, its speed is about two-thirds as great. Thus, events in the cell are inevitably recorded as simultaneous events that actually differ by a good fraction of $10^{-11}$ s. A limitation, therefore, exists for the resolution of early events in pulse radiolysis. We call this limitation the *picosecond barrier.*[1]

In the section entitled "Early Phenomenona" we consider the use of theory and conjecture to obtain information on events that occur earlier than a picosecond.

The stopping power of matter for charged particles is very well known for gaseous media. There are problems associated with the energy loss in condensed media that are beginning to be understood. The response of the matter that absorbs the energy is a much more difficult problem and one that has had less attention than the stopping power problem. On the other hand, this is a problem of broader interest than radiation chemistry, and developments can be expected in the future with the help of theorists from diverse fields.

The theoretical treatment of irradiated systems by Monte Carlo techniques is perhaps the most logical method of attack, and some beginnings have been made in this direction. The technique is highly intuitive and involves marching the transient species through space and time. One sees the evolution in track structure with time and the reactions as they occur. The great attraction of this technique is that new information on processes (cross sections, etc) and properties of transient species can be incorporated as they become available. Preliminary work has been reported and seems promising.[2]

At some time, perhaps in the near future, a presentation of the theory of radiation chemical effects in terms of a Monte Carlo framework will be possible. The emphasis throughout such a treatment will be on cross sections, reaction probabilities, and so forth. The structure of individual tracks will be examined, and many varieties of tracks described; there will be no model tracks in this type of theory because each track developed by the Monte Carlo technique will have its own characteristics, and there will be no need to develop an average picture of the initial energy deposit. At the present time, however, the understanding of radiation chemistry using track models is the only practical way to attain a reasonable perspective. All heavy particles have similar track structure, and we can say that a general track model exists. Radical diffusion models can be constructed to explain observed chemical yields in irradiated systems using this track model. We close this introduction with a brief presentation of the basis for the track model and its relationship to stopping power theory. In the next chapter we discuss the use of the model in radiation chemistry.

### *Basis of Track Model: Core and Penumbra*

An approximate form of the equation for the energy loss per unit path of a particle of charge $ze$ in a gas of atoms of atomic number $Z$ is

$$-\frac{dE}{dx} = 2\kappa ZN \ln\left(\frac{2mv^2}{I}\right) \tag{5-2}$$

where $\kappa$ is the stopping parameter

$$\kappa = \frac{2\pi z^2 e^4}{mv^2} \tag{5-3}$$

e, $m$ are the electronic charge and mass, respectively, $v$ is the particle velocity, $N$ is the number of atoms per unit volume, and $I$ is the "mean excitation potential." A more rigorous form of Eq. (5-2) is given in the treatment of stopping power theory in Chapter 1.

Equation (5-2) is valid only for nonrelativistic particles with velocity $v$ greater than the classical velocities of the bound electrons. The quantity $I$ determines the penetration of such particles in a given material. A theoretical expression relates $I$ to the distribution of oscillator strength of an atom,

$$Z \ln I = \sum f_j \ln \varepsilon_j + \int f'(\varepsilon) \ln \varepsilon \, d\varepsilon \tag{5-4}$$

where $f_j$ is the oscillator strength for the $j$th discrete transition and $f'(\varepsilon)$ is the oscillator strength per unit energy interval in the continuum. The sum extends over all discrete transitions, and the integral extends over the continuum.

To a good first approximation the stopping power of any molecular material is determined by the overall atomic composition and, therefore, is almost independent of specific chemical binding of the atoms. This fact is frequently called the "Bragg rule." The approximate independence of stopping power on chemical binding has been interpreted to mean that the most important primary loss processes occur for energies that are high compared with energies of chemical binding, and, at such excitation energies, it is reasonable to assume that the density of oscillator strength depends only on atomic composition.

The gross structure of the track of a high-energy particle depends on the primary loss spectrum. To an approximation we shall consider below, this spectrum is determined by the distribution of oscillator strength in the stopping material.

It is well known that approximately equal contributions to the stopping power of a material for high-energy particles come from glancing collisions, which have low energy loss per event, and head-on collisions, which have high energy loss per event. Bethe[3] gives

$$\sigma_j = \kappa\left(\frac{f_j}{\varepsilon_j}\right)\ln\left(\frac{2mv^2}{\varepsilon_j}\right) \tag{5-5}$$

as the approximate cross section for excitation of a discrete level with frequency $\nu_j = \varepsilon_j/h$ and oscillator strength $f_j$ by a particle of velocity $v$. In our designation this excitation resulted from a "glancing" collision. We can also say that the differential cross section of an atom for glancing collisions in the continuum at energy $\varepsilon$ is given by

$$d\sigma_g(\varepsilon) = \kappa\left[\frac{f'(\varepsilon)}{\varepsilon}\right]\ln\left(\frac{2mv^2}{\varepsilon}\right)d\varepsilon \tag{5-6}$$

Although Eqs. (5-5) and (5-6) for "glancing" collisions are only approximately valid, they allow the cross section estimate to be made from the oscillator strength distribution.

The differential cross section for the energy loss $\varepsilon$ in the interval $d\varepsilon$ by a free electron in collision with a heavy particle is given by classical considerations as

$$d\sigma = \kappa\left(\frac{d\varepsilon}{\varepsilon^2}\right) \tag{5-7}$$

In accordance with the customary assumptions of dispersion theory, we take the number of electrons that receive such losses to be equal to the oscillator strength of all transitions with energies less than $\varepsilon$.[4] Thus, for head-on collisions we get

$$d\sigma_h = n(\varepsilon)\kappa\left(\frac{d\varepsilon}{\varepsilon^2}\right) \tag{5-8}$$

where

$$n(\varepsilon) = \sum f_i + \int^{\varepsilon} f'(\varepsilon)\, d\varepsilon \tag{5-9}$$

It is clear that the total contribution of each of these processes to the stopping power, $-dE/dx$, is the same; that is,

$$-\left(\frac{dE}{dx}\right)_{\mathrm{g}} = -\left(\frac{dE}{dx}\right)_{\mathrm{h}} = N\int \varepsilon\, d\sigma_{\mathrm{g}}(\varepsilon) = N\int \varepsilon\, d\sigma_{\mathrm{h}}(\varepsilon)$$
$$= \kappa ZN \ln\left(\frac{2mv^2}{I}\right) \tag{5-10}$$

and the sum gives the result of Eq. (5-1). On the other hand, the average energy loss in a single event is widely different for the two processes. The glancing collisions have a large cross section for low-energy processes, and many such losses occur; the head-on collisions have relatively small cross section at all energies up to a maximum, but the few primary events that occur account for a large energy loss.

We are interested in materials with low atomic number, such as water or simple organic molecules. The oscillator strength distribution is confined effectively to excitation energies below 100 eV or so; say the average excitation energy for glancing collisions in water is approximately 40 eV. Consider a fast particle with an average energy loss of 0.04 eV/Å in liquid water as an example of a "low-LET" particle. According to the "continuous slowing down approximation," one might think that in every angstrom unit of path the particle would lose 0.04 eV. Actually, however, losses that occur must create excited states of the water molecule, and the average loss for a glancing collision is approximately equal to 40 eV. Half the energy is lost in glancing collisions, so the average distance between such loss events is about $40/0.02 \simeq 2000$ Å. The other half of the energy is lost in knock-on collisions with a much higher average loss, say 800 eV; the average distance between such loss events is about $800/0.02 \simeq 40{,}000$ Å. The track model that seems indicated from this brief consideration is a string of widely spaced track entities (~2000 Å apart), which have been called "spurs," and interspersed among these with a spacing about 20 times wider is a second type of track entity created by recoil electrons. The recoil electrons move away from the particle trajectory and form tracks of their own.

In the track model used throughout this chapter and the next, we call the loss pattern of the glancing collisions (ie, the string of spurs) the track *core*; and we call the loss pattern of the recoil electrons of the knock-on collisions the *penumbra*. The track of this model is the *average track*. We assume that the essentials of radiation chemistry are to be obtained by using an average track picture. If fluctuations in energy deposit, original radical concentrations, and so on, are important, they can be obtained by special considerations. In contrast, the Monte Carlo method deals directly with tracks that contain fluctuating energy deposits.

All heavy particle tracks are similar. Track cores are made up of strings of spurs with average spacings that depend on the quantity $-dE/dx$ of the particles. Track penumbras are made up of electron tracks. This fact tells us that the electron track has a special role in radiation chemistry. In Chapter 1 it is shown how any electron track is composed of track entities called "spurs," "blobs," and "short tracks." The variation of the fraction of energy in these three types of entities as a function of electron energy is shown in that chapter. We also know that track segments of high-energy electrons (say 10 MeV) are similar to track segments of high-energy protons with the same value of $-dE/dx$. Based on the equipartition principle of stopping power (ie, equal amounts of energy are lost in glancing and knock-on collisions), the number of spurs per unit track length is approximately the same, and the amount of energy in knock-on electrons per unit track length is also approximately the same. There are differences arising from differences in primary particle scattering and the maximum energy in the knock-on electron spectrum. In any case we can relate the effects of heavy particle tracks very precisely with the effects of high-energy electrons, and experimental results on aqueous solutions agree with this analysis.

Finally, we should note that the simplest model of the track of high-energy electrons, or in fact the track of any low-LET radiation, is a collection of isolated spurs. Such a track model is actually a good approximation for low-LET radiations for situations in which the blobs and short tracks introduce no qualitatively different effects.

## EARLY PHENOMENA

All events involved in the utilization of the energy deposited in a spur can be described on a single time scale; this is the simplest example of what we mean by "local time." The earliest significant event is the energy deposit by the primary particle, and the uncertainty relationship

$$\Delta t\, \Delta E \simeq \hbar \simeq 6.6 \times 10^{-16}\ \mathrm{eV\ s} \qquad (5\text{-}11)$$

applies to it. The *most probable* energy of a spur in water is about 20 eV, so we have

$$\Delta t \simeq 0.33 \times 10^{-16}\ \mathrm{s}$$

and we often say that the earliest time is $10^{-16}$ s.

Another aspect of the formation of a spur is the uncertainty in position $x$ of the energy deposit arising from the uncertainty relationship, where $p$ is the momentum of the particle

$$\Delta p\, \Delta x \simeq \hbar \qquad (5\text{-}12)$$

which applies to the primary particle. Calling $u$ the speed of the particle, we have, from elementary relativistic mechanics,

$$\Delta p = \frac{\Delta E}{u} \qquad (5\text{-}13)$$

so that

$$\Delta x \simeq \frac{\hbar u}{\Delta E} \qquad (5\text{-}14)$$

and for a relativistic particle that loses 20 eV,

$$\Delta x \simeq 10^{-6}\ \mathrm{cm}$$

Equations (5-11) and (5-13) are clearly consistent: they just give different aspects of the uncertainty. We can see this as follows:

$$\Delta t \simeq \frac{\Delta x}{u} \simeq \frac{\hbar}{\Delta E} \tag{5-15}$$

The large value of $\Delta x$ given by Eq. (5-14) for a fast particle (ie, $\Delta x \simeq 100$ Å) has sometimes been taken to mean that the state formed in the matter is *delocalized.* Actually, this $\Delta x$ is just the uncertainty in the position of the loss, and the correlation of events occurring after the loss makes the process compatible with a spur of considerably smaller size.

In the subpicosecond region we can distinguish three time domains.[1] In the earliest (starting at $10^{-16}$ s) only electronic processes are possible. The time is too short for molecular motions of any kind. In the second (starting at $10^{-14}$ s or so), molecular vibrations become important, and energy flows into them. In the third (starting shortly after $10^{-13}$ s), the lower-grade molecular motions become excited, and a local temperature becomes meaningful. (Note that $\Delta t \simeq \hbar/kT \simeq 0.3 \times 10^{-13}$ *s* at $T = 300$ K.) Diffusion processes start in the picosecond time scale when all degrees of freedom of the system become activated.

Figure 5-1 gives a schematic description of the energy flow through the various degrees of freedom of the molecular material (ie, $H_2O$) accepting the energy.[5] It is generally believed that at the time of thermalization most

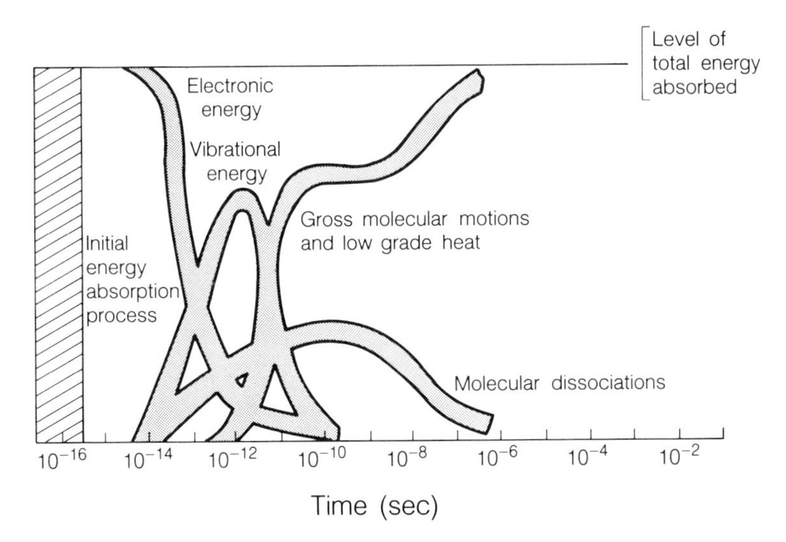


**Figure 5-1.** Schematic description of the partition of energy in a spur as a function of time $t$ (in s). The vertical dimension of a given region is proportional to the fraction of the deposited energy that is in a particular form. This figure is intended to describe the situation for a spur having average energy. Note that, in exothermal reactions, the line showing low-grade heat can terminate above the level of total energy absorbed. (Reproduced with permission.[5])

of the deposited energy has already been transformed into heat. The early reactions of the radicals are exothermic, so as the radicals disappear the radiation energy is all found in the form of heat.

On the time scale of Figure 5-1, chemical processes are also occurring. Let us consider water as an important example. On the time scale of $10^{-16}$ s, excited ($H_2O^*$), superexcited ($H_2O^{\ddagger}$), and ionized ($H_2O^+$, $H_2O^{+*}$) states of the water molecule are produced together with the $\delta$-rays and other (relatively) slow electrons as a result of energy absorption. According to Platzman[6] the positive ion of water has, on an average, 8 eV of excitation energy ($H_2O^{+*}$). The various excited and ionized states are not clearly discernible, because on this time scale the energy uncertainty is approximately 10 eV. On the other hand, a statistical correlation problem is already evident and important at this time scale in relation to the spur theory, that is, the theory of the chemical effects following an isolated energy loss in the range 6 to 100 eV. It may be stated as follows: Given an ionization or excitation at a place, what is the probability of having another event (excitation or ionization) produced by secondary processes usually involving low-energy electrons in the immediate vicinity *consistent* with a given average energy loss? Another way of looking at the same problem is to find the distribution in species and space for spurs of given average energy. This problem can be handled with Monte Carlo techniques, and exploratory work is actually underway. As a practical matter we have assumed that the average number of species, $n_x(\varepsilon)$, of a given type $x$ in a spur of average energy $\varepsilon$ is given by the relationship

$$n_x(\varepsilon) \propto G(x)\varepsilon \tag{5-16}$$

where $G(x)$ is the $G$-value for production of the species. Very little is actually known of the spatial distributions, but it has usually been assumed that they are Gaussian about some center and that the maximum *concentration* of any species is the same in all track entities. All of these assumptions require further investigation.

In the time interval of $10^{-15}$ s nothing of great chemical significance happens. This time interval may be considered as an incubation period. According to Kalarickal and Magee,[7] the dry hole ($H_2O^+$) can move by exact resonance on this time scale. Excited states can, in principle, also migrate if the resonant transfer time is shorter than the vibrational time.

In the time interval of $10^{-14}$ s the radicals produced from the dissociation of $H_2O^*$ and $H_2O^{\ddagger}$ initially have the same distribution as those of the excited states themselves. The latter are, in principle, derivable from the distribution of energy deposition of the primary particle and the secondary electrons. In practice, however, this is largely unknown. The $H_2O^+$ ions react with $H_2O$ to form $H_3O^+$ and OH in an ion–molecule reaction that is perhaps the fastest chemical process known. The $H_3O^+$ ion produced by this ion–molecule reaction is initially unhydrated. The distribution of OH radicals from this channel is the same as the distribution of $H_2O^+$.

Consideration of thermalization ($\Delta E \simeq 10^{-2}$ eV) is not meaningful for times less than $10^{-13}$ s, although it is probable that thermalization, trapping, and hydration can follow in quick succession. It is also plausible that the electron is never fully thermalized but is simply trapped and hydrated. On this time scale radicals diffuse a little [root mean square (rms) displacement ~0.5 Å]; however, they may react in a concentrated solution in the fashion of static quenching.[8] On the other hand, the dry (prehydrated) electron can move significantly (rms displacement ~30 Å) and therefore has a significant probability of reacting in a concentrated solution in competition with hydration and neutralization.[8a]

During the time interval of $10^{-12}$ s (ps) direct observability of radiation-produced species becomes possible (actually $\simeq$ 10 ps in current experiments). On this time scale diffusion-kinetic reactions take place with time-dependent rate coefficients. The time dependence, sometimes referred to as the Smoluchowski transient, is a result of the changeover, during the diffusion process, of the discontinuous concentration gradient at the reaction radius (at $t = 0$) to a smoothly varying concentration gradient given as a solution of Smoluchowski's equation ($t > 0$).

The most important transients in irradiated water are $H_2O^+$, $e^-$, $H_3O^+$, OH, $e_{aq}^-$, H, $H_3O_{aq}^+$. Figure 5-2 gives current ideas of the development of a spur, including the transformation of the intermediate species and expansion by diffusion. Many considerations of the radiolysis of water involve only the thermal reaction period, and the initial action of the radiation is indicated by Eq. (5-1). It is clear that the production of the indicated species involves processes requiring nanoseconds to complete; for example, most of the molecular products $H_2$ and $H_2O_2$ are created by radical combination. In acid water the hydrated electron reacts in $10^{-10}$ s or less to form H atoms, and the initiation reaction has sometimes been given as

$$H_2O \rightsquigarrow H, OH, H_2, H_2O_2 \tag{5-17}$$

From a radiation chemical point of view, the distribution in space of $e_{aq}^-$, $H_3O_{aq}^+$ and OH at $10^{-11}$ s is what is meant by "initial track structure." The radical diffusion model uses such distributions to calculate the reactions during track expansion and the radical reactions with scavengers; all of these processes are complete in a few microseconds, and the product yields are determined. The events occurring before thermalization (a few picoseconds) have not, of course, been observed directly. However, the ion–molecule reaction, which forms $H_3O^+$ and OH at $\simeq 2 \times 10^{-14}$ s (it is the fastest chemical process of radiation chemistry), and the hydration of the electron at the longitudinal relaxation time for water ($\simeq 4 \times 10^{-13}$ s) are widely accepted.[9] In any case the reactive intermediates are formed in the prethermal period. At thermalization, all of the deposited energy, except that tied up in dissociation of water to form the intermediates, is distributed as a local temperature rise in the track.

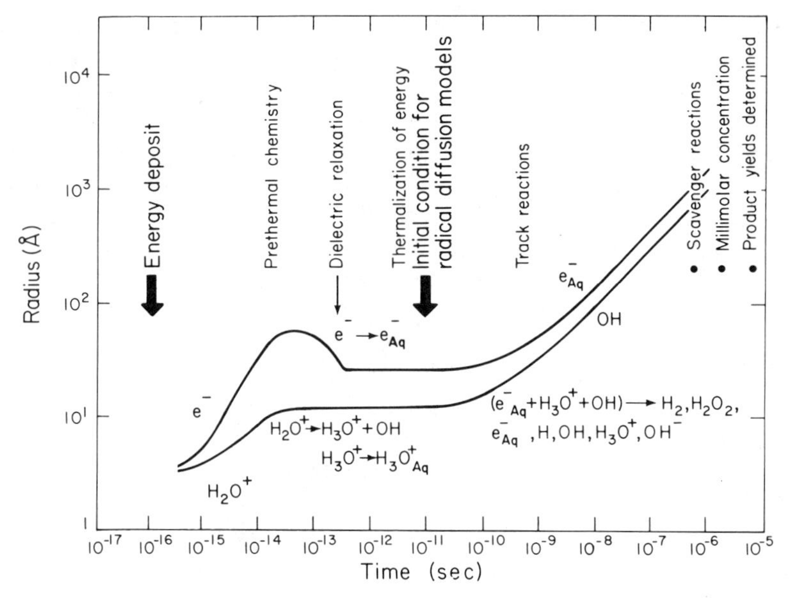


**Figure 5-2.** Space-time development of small isolated energy deposit (spur) in a dilute aqueous system. In the prethermal time period ($t < 3 \times 10^{-12}$ s), phenomena involve electronic motion for the most part; the ion–molecule reaction creating $H_3O^+$ and the longitudinal relaxation of $H_2O$ require only proton motion. In the postthermal period the species already created undergo diffusion-controlled reactions with each other (indicated by the parentheses on the left of the equation). (From "Radiation Research Proceedings of Sixth International Congress of Radiation Research"; Okada, S.; Imamura, M.; Terashima, T.; and Yamaguchi, H., Eds.; Japanese Association for Radiation Research, 1979. Reproduced by permission.)

## *Sources of Information on Early Times*

There are two general methods for obtaining information on early times. One is the real-time observation made possible by the pulse techniques of radiolysis; the other is the indirect method using presumed relationships between concentration and time to obtain information about early time phenomena from measurements at much longer times. We have seen that the picosecond barrier limits the early observations in real time. The most common indirect methods make use of data taken essentially at infinite times after energy deposit to obtain information at early times. These methods depend upon the conjugate relationship between concentration and time and, therefore, have limitations of a different type. Systems with various compositions must be used in the same study of early events to complete the analysis, so it may be difficult to ascertain the pattern of events for the system of interest (ie, a pure component). The time scale measured is a "local" time, and resolution usually can be made finer than that obtained in real time.

All indirect methods use chemical competitions of one sort or another. Laplace-transform methods use the competition of electron scavengers with electron–ion recombination, and the method is virtually limited to this case. The recombination times are all relatively long in the thermalized electron time domain. Other electron reactions that probably involve an earlier period have been used by Hunt[8a] and Hamill.[10]

Radical diffusion models for track reactions in water start with thermalized radicals at $10^{-11}$ s, and thus, in a way, these calculations also give information on the early thermal period.

## *The Thermal Time Scale*

The Laplace-transform method for information at early times has been used principally for obtaining the lifetime distribution for free electrons in liquid hydrocarbons. In such systems electrons are formed by ionization, become thermalized before they escape from the fields of their geminate ions, and recombine on the picosecond time scale. Various scavengers are used that can react with the electrons; the probability per unit time that an electron will react with a scavenger is

$$-\frac{dF(t)}{dt} = k_s'[\mathrm{S}]F(t) \tag{5-18}$$

where $F(t)$ is the probability that an electron thermalized at $t = 0$ is still uncombined at time $t$, $k_s'$ is the reaction rate coefficient, and [S] is the scavenger concentration. The function $F(t)$ is the desired information to be obtained by the scavenger studies. The total probability that the electron reacts with the scavenger before it is neutralized is equal to

$$\omega(\lambda) = \lambda \int_0^\infty e^{-\lambda t} F(t)\, dt \tag{5-19}$$

where $\lambda = k_s'[\mathrm{S}]$. Clearly, if $\omega(\lambda)$ is measured experimentally for all scavenger concentrations from zero to $\infty$, $F(t)$ can be obtained by an inverse Laplace transform. This method has been used quite extensively to study electrons in organic liquids, and much is known concerning their lifetime distributions.

There is a fundamental difficulty in the application of the Laplace-transformation method in systems, such as water, that have second-order reactions. We illustrate this difficulty by considering the diffusion-controlled reactions of a species with itself in competition with a homogeneously distributed scavenger. The relevant equations are

$$\frac{\partial c_0}{\partial t} = D\nabla^2 c_0 - k' {c_0}^2 \tag{5-20}$$

and

$$\frac{\partial c}{\partial t} = D\nabla^2 c - k' c^2 - k_s' c_s c \tag{5-21}$$

where $c_0$ and $c$ are, respectively, the concentrations of the reactant species in the absence of scavenger and in the presence of a scavenger at concentration $c_s$, $D$ is the diffusion coefficient, and $k'$ and $k'_s$ are, respectively, the specific rate constants for reaction with itself and with the scavenger. The total amount of scavenger reaction in a volume $V$ is

$$N(\lambda) = \lambda \int_0^\infty dt \int_0^\infty c\, dV = \lambda[\text{Laplace transform of } f_1(t)] \qquad (5\text{-}22)$$

where

$$f_1(t) = \int_0^\infty c_1\, dV \qquad (5\text{-}23)$$

and $c = c_1\, e^{-\lambda t}$. However, $c_1$ is the solution *not* of Eq. (5-20) but of the equation

$$\frac{\partial c_1}{\partial t} = D\nabla^2 c_1 - k'\, e^{-\lambda t} c_1^{\,2}, \qquad (5\text{-}24)$$

which, in reality, does not describe the chemical reaction in any system. It is thus clear that the Laplace-transform procedure used for aqueous solutions can at best be a crude approximation valid in the limit that nonlinear (second-order) reactions are negligible, and in any case rather large errors are to be expected if such a procedure is used in a straightforward manner.

The use of scavenger measurements at very long times as a function of scavenger concentration to predict from indirect calculations the real-time observations cannot in any case yield valid results for times shorter than a picosecond because thermal reactions do not take place at shorter times. This is a fundamental difficulty with such attempts to obtain information about early times; one cannot obtain anything about the precursors of the radicals one is observing in real time. There are, in addition, other serious problems with both the direct and indirect methods. One is time dependence of reaction rate coefficients. This is an intrinsic difficulty.

There are two sources of time variation of a reaction rate coefficient. One is the relaxation time for the establishment of a steady-state concentration gradient about the reacting species. Although we know that such an effect must exist, the case for an inhomogeneous distribution, such as found in a track, has not been solved theoretically. It is presumed that the general magnitude of the effect is the same as that for the initially homogeneous system; that is,

$$k'(t) = k'_\infty \left[1 + \frac{R}{(\pi D t)^{1/2}}\right] \qquad (5\text{-}25)$$

where $R$ is the reaction radius, $D$ is the diffusion constant, and $k'_\infty$ is the reaction rate coefficient at long times. The second source of time variation is the relaxation of the ion atmospheres and applies to reactions between charged species, particularly in water. This problem has not been considered explicitly for tracks.

Another intrinsic difficulty is the possibility of a "static" reaction. Attention has been called to this problem by Czapski and Peled.[8] A transient intermediate such as a hydrated electron may be formed within the reactive radius of a scavenger. Clearly, the reaction of such a pair will not be described by a second-order rate equation.

Before the Laplace-transform method can be used in any particular case, it must be demonstrated that these various difficulties do not invalidate the application.

## *The Prethermal Time Scale*

In the Laplace-transform method the scavenger reaction is competing with the recombination of thermalized electrons and geminate ions. If earlier times are of interest another method must be used. We have noted that in this earlier time scale scavenger reactions are not thermal but involve species at higher energy. They are "hot" or "epithermal" processes.

Hamill[10] has suggested that in the earliest time scale the epithermal electrons (he calls them "dry" electrons) react with scavengers in competition with the hydration process according to the following mechanism:

$$e^- + S \xrightarrow{k_s'} S^- \tag{5-26}$$

$$e^- \xrightarrow{k'} e_t^- \rightarrow e_{aq}^- \tag{5-27}$$

where $e_t^-$ indicates a state that irreversibly goes to hydration. The reaction rate coefficients are to be taken as the appropriate averages $\langle \sigma v \rangle$, where $\sigma$ is the cross section and $v$ is the electron velocity; a trap concentration also appears in $k'$.

In this simple mechanism the fraction of reaction in the scavenger channel is

$$f_s = \frac{k_s'[S]}{k_s'[S] + k'} \tag{5-28}$$

The ordinary form of this expression used in the analysis is

$$\frac{1}{f_s} = 1 + \frac{k'}{k_s'[S]} \tag{5-29}$$

and one expects to obtain a linear plot for $f_s^{-1}$ vs. $[S]^{-1}$. In Eq. (5-29) $f_s$ is proportional to $G(s)$, the measured yield at long times. The competition described by Eqs. (5-26) and (5-27) is usually called a Stern–Volmer mechanism.

Sawai and Hamill[11] and Ogura and Hamill[12] have shown that in many concentrated solutions in water and alcohols, the experimental results for the yields of hydrated electrons seem to be in agreement with this mechanism. If one assumes that the epithermal electron hops about from a molecule to neighboring molecules, that the trapping process [Eq. (5-27)] can occur at any site, and that reaction with S occurs on every encounter, the time scale can be established. The analysis sets the trapping time at about $10^{-13}$ s in both water and alcohol.

A similar analysis can be made for the motion of the epithermal $H_2O^+$ ion. According to Ogura and Hamill, in concentrated chloride solutions the reactions

$$H_2O^+ + Cl^- \xrightarrow{k'_{Cl}} H_2O + Cl \tag{5-30}$$

$$H_2O^+ + H_2O \xrightarrow{k'} H_3O^+ + OH \tag{5-31}$$

are in competition, followed by reaction of Cl atoms with chloride ions:

$$Cl + Cl^- \rightarrow Cl_2^- \tag{5-32}$$

Measuring the yield of $Cl_2^-$ gives the effectiveness of the scavenger reaction (5-30). Here it is reasonable to assume that the $H_2O^+$ hole jumps between neighboring molecules randomly and reacts on each encounter with $Cl^-$. The result for the ratio $k'/k'_{Cl}$ was found to be 14. This means that the $H_2O^+$ jumps on the average 14 times before Reaction (5-31) occurs. Estimates of the jump time are $10^{-15}$ s, and the time scale of the ion–molecule reaction (5-30) is $10^{-14}$ s.

Another unresolved problem exists with respect to the prethermal period. The Stern–Volmer mechanism of Hamill and his associates requires an electron reaction proportional to the scavenger concentration because of the competitive nature of the depletion [Eqs. (5-26) and (5-27)]. His experimental results seem to be compatible with this mechanism. On the other hand, Hunt et al[8a] have reported an exponential depletion of electrons,

$$G_e[S] = G_e^{\,0}\, e^{-[S]/S_{37}} \tag{5-33}$$

where $G_e^{\,0}$ is the observed yield in the absence of a scavenger, and $[S_{37}]$ is the concentration of scavenger for which the initial yield is reduced to $e^{-1}$ (ie, 37%). Hunt et al[8a] have not proposed a mechanism that would lead to this result, but Czapski and Peled[8] have argued that it agrees with the reaction of hydrated electrons with ions in contact at the times of formation. They derive an exponential function of the type in Eq. (5-33), using Poisson statistics for the probability of formation of contact pairs. Although one can criticize the use of Poisson statistics, the basic result remains valid: In concentration solutions the formation of contact pairs is to be expected.

## TRACK REACTIONS

The special feature of the kinetics of radiation chemistry is that the initial reactions take place in *tracks*. The radiation chemical system considered in most detail is that in which each track is completely independent, which means that the intermediates formed in one track react completely with each other and constituents of the medium before they can diffuse far enough to encounter intermediates from another track. This situation is required for the yield to be independent of the dose rate, and this is found experimentally to be the most common case. (It is called the "low background" case (Magee[13]), which means that concentrations of intermediates do not build up outside tracks.)

The complete set of chemical reactions occurring in the irradiated system must be obtained. This information can come from radiation chemistry or from general knowledge of the chemical nature of the system. Table 5-1 gives the reactions occurring in irradiated water. The reaction rate constants listed are obtained from ref. 14 through 17. The reactions included are presumably the ones that occur between thermalized species at times greater than $10^{-11}$ s after energy deposit. For neutral water there are 12 such species: H, $e_{aq}^-$, OH, $H_3O^+$, $H_2O$, $H_2$, $H_2O_2$, $OH^-$, $HO_2$, $O_2$, $O_2^-$, $HO_2^-$. In 0.8 *N* acid hydrated electrons react in less than $10^{-10}$ s to form H atoms, and only seven species are present at longer times; H, OH, $H_2O$, $H_2$, $H_2O_2$, $HO_2$, $O_2$. Actually, one can never be sure that all reactions have been identified, and truncated systems are inevitably used. An extreme case of truncation is the "one-radical model," which has only a single radical intermediate (say R) and a single radiation chemical product (say $R_2$) arising from combination of the radicals.

The set of $P$ species that react in an irradiated system undergoes $Q$ reactions. In the case of water $P = 12$ and $Q = 25$ (see Table 5-1). Let us write a system of differential equations for the $P$ species based on the $Q$ reactions as if the system were homogeneous. Table 5-2 gives this system of equations for water; equations that apply to the acid case are indicated by asterisks. In the actual irradiated system the transient species are formed in tracks, and it has generally been assumed that we have a set of partial differential equations

$$\frac{\partial c_i}{\partial t} = D_i \nabla^2 c_i + \frac{dc_i}{dt} \qquad i = 1, 2, 3, \ldots, P \tag{5-34}$$

to determine the track reactions; here $dc_i/dt$ is to be obtained from Table 5-2.

In radiation chemistry it has usually been assumed that the initial concentrations of species (ie, at $10^{-11}$ s) can be given in terms of continuous functions in space. Gaussian functions have always been used in the past. Perhaps with the increasing application of Monte Carlo techniques the initial conditions will be better known in the near future.

The set of equations (5-34) is nonlinear, and no general method of solution exists. The straightforward numerical procedure is to convert them to *difference* equations and to solve them by the method of the march of steps. In this process it must be remembered that the resulting *parabolic partial*

**Table 5-2.** Differential Equations for Transient Species and Radiation Products in Irradiated Neutral Water

$$
\frac{{}^*d(\mathrm{H})}{dt} = -2k_1(\mathrm{H})^2 - k_2(\mathrm{e_{aq}}^-)(\mathrm{H}) - k_3(\mathrm{H})(\mathrm{OH}) + k_7(\mathrm{H_3\overset{+}{O}})(\mathrm{e_{aq}}^-) - k_9(\mathrm{H})(\mathrm{H_2O_2}) + k_{12}(\mathrm{OH})(\mathrm{H_2}) - k_{13}(\mathrm{HO_2})(\mathrm{H}) - k_{17}(\mathrm{H})(\mathrm{O_2})
$$

$$
\frac{d}{dt}(\mathrm{e_{aq}}^-) = -k_2(\mathrm{e_{aq}}^-)(\mathrm{H}) - 2k_3(\mathrm{e_{aq}}^-)^2 - k_4(\mathrm{e_{aq}}^-)(\mathrm{OH}) - k_7(\mathrm{H_3\overset{+}{O}})(\mathrm{e_{aq}}^-) - k_{10}(\mathrm{e_{aq}}^-)(\mathrm{H_2O_2}) - k_{14}(\mathrm{e_{aq}}^-)(\mathrm{O_2})
$$

$$
\frac{{}^*d}{dt}(\mathrm{OH}) = -k_4(\mathrm{e_{aq}}^-)(\mathrm{OH}) - k_5(\mathrm{H})(\mathrm{OH}) - 2k_6(\mathrm{OH})^2 + k_9(\mathrm{H})(\mathrm{H_2O_2}) + k_{10}(\mathrm{e_{aq}}^-)(\mathrm{H_2O_2}) - k_{11}(\mathrm{OH})(\mathrm{H_2O_2}) - k_{12}(\mathrm{OH})(\mathrm{H_2}) - k_{15}(\mathrm{HO_2})(\mathrm{OH}) + k_{21}(\mathrm{H_2O_2})(\mathrm{HO_2}^-) + k_{22}(\mathrm{H_2O_2})(\mathrm{O_2}^-)
$$

$$
\frac{d}{dt}(\mathrm{H_3\overset{+}{O}}) = -k_7(\mathrm{H_3\overset{+}{O}})(\mathrm{e_{aq}}^-) - k_8(\mathrm{H_3\overset{+}{O}}(\mathrm{OH}^-) - k_{18}(\mathrm{H_3\overset{+}{O}})(\mathrm{O_2}^-) + k_{23}(\mathrm{H_2O}) + k_{24}(\mathrm{HO_2}) - k_{19}(\mathrm{H_3\overset{+}{O}})(\mathrm{HO_2}^-) + k_{25}(\mathrm{H_2O_2})
$$

$$
\frac{{}^*d}{dt}(\mathrm{H_2O}) = k_4(\mathrm{e_{aq}}^-)(\mathrm{OH}) + k_5(\mathrm{H})(\mathrm{OH}) + k_7(\mathrm{e_{aq}}^-)(\mathrm{H_3\overset{+}{O}}) + k_8(\mathrm{H_3\overset{+}{O}})(\mathrm{OH}^-) + k_9(\mathrm{H})(\mathrm{H_2O_2}) + k_{11}(\mathrm{OH})(\mathrm{H_2O_2}) + k_{12}(\mathrm{OH})(\mathrm{H_2}) + k_{14}(\mathrm{e_{aq}}^-)(\mathrm{O_2}) + k_{15}(\mathrm{HO_2})(\mathrm{OH}) - k_{23}(\mathrm{H_2O}) + k_{21}(\mathrm{H_2O_2})(\mathrm{HO_2})
$$

$$
\frac{{}^*d}{dt}(\mathrm{H_2}) = k_1(\mathrm{H})^2 + k_2(\mathrm{e_{aq}}^-)(\mathrm{H}) + k_3(\mathrm{e_{aq}}^-)^2 - k_{12}(\mathrm{OH})(\mathrm{H_2})
$$

$$
\frac{{}^*d}{dt}(\mathrm{H_2O_2}) = k_6(\mathrm{OH})^2 - k_9(\mathrm{H})(\mathrm{H_2O_2}) - k_{10}(\mathrm{e_{aq}}^-)(\mathrm{H_2O_2}) - k_{11}(\mathrm{OH})(\mathrm{H_2O_2}) + k_{13}(\mathrm{HO_2})(\mathrm{H}) + k_{16}(\mathrm{HO_2})^2 + k_{19}(\mathrm{H_3\overset{+}{O}})(\mathrm{HO_2}^-) - k_{21}(\mathrm{H_2O_2})(\mathrm{HO_2}) - k_{22}(\mathrm{H_2O_2})(\mathrm{O_2}^-) - k_{25}(\mathrm{H_2O_2})
$$

$$
\frac{d}{dt}(\mathrm{OH}^-) = k_2(\mathrm{e_{aq}}^-)(\mathrm{H}) + 2k_3(\mathrm{e_{aq}}^-)^2 + k_4(\mathrm{e_{aq}}^-)(\mathrm{OH}) - k_8(\mathrm{H_3\overset{+}{O}})(\mathrm{OH}^-) + k_{10}(\mathrm{e_{aq}}^-)(\mathrm{H_2O_2}) + k_{23}(\mathrm{H_2O}) + k_{22}(\mathrm{H_2O_2})(\mathrm{O_2}^-)
$$

$$
\frac{{}^*d}{dt}(\mathrm{HO_2}) = k_{11}(\mathrm{OH})(\mathrm{H_2O_2}) - k_{13}(\mathrm{HO_2})(\mathrm{H}) - k_{15}(\mathrm{HO_2})(\mathrm{OH}) - 2k_{16}(\mathrm{HO_2})^2 + k_{17}(\mathrm{H})(\mathrm{O_2}) + k_{18}(\mathrm{H_3\overset{+}{O}})(\mathrm{O_2}^-) - k_{24}(\mathrm{HO_2}) - k_{20}(\mathrm{HO_2})(\mathrm{O_2}^-) - k_{21}(\mathrm{HO_2})(\mathrm{H_2O_2})
$$

$$
\frac{{}^*d}{dt}(\mathrm{O_2}) = -k_{14}(\mathrm{e_{aq}}^-)(\mathrm{O_2}) + k_{15}(\mathrm{HO_2})(\mathrm{OH}) + k_{16}(\mathrm{HO_2})^2 - k_{17}(\mathrm{H})(\mathrm{O_2}) + k_{20}(\mathrm{HO_2})(\mathrm{O_2}^-) + k_{21}(\mathrm{H_2O_2})(\mathrm{HO_2}) + k_{22}(\mathrm{H_3O})(\mathrm{O_2}^-)
$$

$$
\frac{d}{dt}(\mathrm{O_2}^-) = k_{14}(\mathrm{e_{aq}}^-)(\mathrm{O_2}) - k_{18}(\mathrm{H_3\overset{+}{O}})(\mathrm{O_2}) + k_{24}(\mathrm{HO_2}) - k_{20}(\mathrm{HO_2})(\mathrm{O_2}^-) - k_{22}(\mathrm{H_2O_2})(\mathrm{O_2}^-)
$$

$$
\frac{d}{dt}(\mathrm{HO_2}^-) = -k_{19}(\mathrm{H_3\overset{+}{O}})(\mathrm{HO_2}^-) + k_{20}(\mathrm{HO_2})(\mathrm{O_2}^-) + k_{25}(\mathrm{H_2O_2})
$$

*difference* equations may generate convergent solutions that are not the correct solutions of the corresponding differential equations. To safeguard against this difficulty, one must maintain a certain relationship between increments in the spatial and temporal directions. In some sense Eqs. (5-34) are like the self-consistent field equations of quantum chemistry; in other words, the mechanism, rate constants, and so forth, and initial conditions can be varied until the results agree with experiment. No one has thought about the general problem in such a disciplined way that a satisfactory analysis can be made at this time. However, it is possible to get an indication of the validity of the entire scheme, and this has been done in the case of water, as we shall see later.

## *Reactions in the Spur*

A key concept in the understanding of track reactions in low-LET irradiations is the insensitivity of the yields of scavenger reactions to scavenger concentration when this concentration is low. Consider a model system in which a single kind of radical R is formed that can react either with another radical or with a scavenger S:

$$\mathrm{R} + \mathrm{R} \rightarrow \mathrm{R}_2 \tag{5-35}$$

$$\mathrm{R} + \mathrm{S} \rightarrow \mathrm{RS} \tag{5-36}$$

The track is taken to be a "string of beads;" that is, it is composed only of widely spaced spurs. It seems reasonable to take the initial radical distribution in a spur as Gaussian. The set of Eqs. (5-34) is now reduced to four equations. We consider in some detail only the equation for the radical R. To make any progress, we must introduce a simplification, and this is done in the form of an approximation called *prescribed diffusion.* Prescribed diffusion has been used widely in studies of the track reactions of radiation chemistry. The approximation is largely intuitive and justified only in that it gives reasonable results. We assume that the radical concentration is determined by diffusion alone except for a depletion by the recombination reaction. Thus, we say that the radical concentration in each spur is given by

$$c(r, t) = \frac{N(t)e^{-r^2/4Dt}}{(4\pi Dt)^{3/2}} \qquad t_0 \leqslant t \tag{5-37}$$

where $r$ is the distance from the center of the spur, $N(t)$ is the number of radicals that exist in the distribution at time $t$, and $t_0$ is the time that gives the correct initial concentration:

$$r_0^2 = 4Dt_0 \tag{5-38}$$

The radicals were created by processes that did not involve diffusion, so the equation has no meaning for $t < t_0$.

In this model the track equation for the radical concentration becomes

$$\frac{\partial c}{\partial t} = D\nabla^2 c - kc^2 - k_s c_s c \tag{5-39}$$

Another approximation, neglect of scavenger depletion, is introduced here.

This means that the scavenger is assumed to remain at the same constant concentration at all times. These approximations allow us to solve Eq. (5-39). Integration is first carried out over space, yielding the ordinary differential equation

$$\frac{dN}{dt} = -\frac{kN^2}{(8\pi Dt)^{3/2}} - k_s c_s N \tag{5-40}$$

If the variable $u = N^{-1}$ is introduced, Eq. (5-40) is transformed into a linear equation that can be solved analytically to give

$$N(x) = N_0 \, e^{-q(x-1)} \left[ 1 + \left\{ \frac{kN_0 t_0}{(8\pi D t_0)^{3/2}} \right\} J(x, 1) \right]^{-1} \tag{5-41}$$

where $q = k_s c_s t_0$, $x = t/t_0$, and $J(x, 1) = \int_1^x e^{-qx}(dx/x^{3/2})$. The fraction of radicals reacting with the scavenger is given by

$$f_s = k_s c_s \int_1^\infty \frac{N(x)}{N_0} \, dx \tag{5-42}$$

Although the equation for $f_s$ is a closed-form expression, it is not simple. The limit of zero scavenger concentration, $c_s = 0$, is an interesting special case in which

$$N(x) = N_0 \left\{ 1 + \left[ \frac{2kN_0 t_0}{(8\pi D t_0)^{3/2}} \right] [1 - x^{-1/2}] \right\}^{-1} \tag{5-43}$$

and, at infinite time,

$$N(\infty) = N_0 \left\{ 1 + \frac{2kN_0 t_0}{(8\pi D t_0)^{3/2}} \right\}^{-1} \tag{5-44}$$

which shows that only partial recombination of radicals occurs in a spur expanding into a three-dimensional scavenger-free region. If a small scavenger concentration exists, the uncombined radicals will react with the scavenger by default, having nothing else to do. At very small scavenger concentrations, therefore, the fraction reacting with the scavenger is

$$f_s = \left\{ 1 + \frac{2kN_0 t_0}{(8\pi D t_0)^{3/2}} \right\}^{-1} \tag{5-45}$$

The near constancy of the fraction of intermediates that react with the scavenger is essentially explained by this example. Experimentally, there may be several decades of scavenger concentrations, say from $10^{-6}$ to $10^{-3}$ *M*, in which $f_s$ is essentially the same for low-LET radiations.

To get a simple estimate of the variation of $f_s$ with $c_s$, let us note that the lifetime of a radical with respect to the scavenger reaction is

$$\tau = \frac{1}{k_s c_s} \tag{5-46}$$

We can argue that all the recombination that takes place must be over at this time, and using Eq. (5-43) we get, for the fraction uncombined at the time $\tau$,

$$f_s = f_s^0[1 + (1 - f_s^0)q^{1/2} + \cdots] \tag{5-47}$$

and this is the fraction that reacts with the scavenger.

The variation of $f_s$ with the square root of the scavenger concentration is also obtained in more rigorous treatments of the model. Balkas et al[18] showed that their scavenging function introduced for hydrocarbons

$$f_s = f_s^0 + (\alpha c_s)^{1/2}[1 + (\alpha c_s)^{1/2}]^{-1} \tag{5-48}$$

can also be applied in water. We have noted that the Laplace-transform method, which yields a $c_s^{1/2}$ dependence in hydrocarbons, does not apply to water; the $c_s^{1/2}$ dependence is obtained in water for a different reason.

Actually, much of the early work on scavengers in water used the equation

$$f_s = f_s^0 + \beta c_s^{1/3} \tag{5-49}$$

where $\beta$ is an arbitrary constant. This equation never had a theoretical interpretation, and it is now quite clear that the $c_s^{1/2}$ variation is in better agreement with experiment.

The results of radiolysis using low-LET radiations are in apparent agreement with the string-of-beads model for the track. More sophisticated models take into account the knock-on electrons and penumbra effects. There are very significant differences between the string-of-beads track model and the more correct model discussed in Chapter 6.

## *Reactions in a Cylindrical Track*

It is clear from the above considerations that information can be obtained from scavenger studies on the initial sizes of spurs. Changing the LET in the string-of-beads model for the track means changing the spur separations. As the spacing decreases, the radicals from neighboring spurs along the track can intermingle and react, increasing Reaction (5-35) over Reaction (5-36) in our one-radical model. Tracks with variable interspur spacing are discussed in Chapter 6. It is clear, however, that in the limit of small spur separation, a cylindrical geometry is obtained. Such a cylindrical track was considered in the early days of radiation chemical track theory as the model for the high-LET track. It can be considered in prescribed diffusion in the same way as the spur. Suppose the radical concentration is given by

$$c(r, t) = \frac{N(t)e^{-r^2/4Dt}}{4\pi Dt} \qquad t_0 \leq t \tag{5-50}$$

where $N(t)$ is now the number of radicals per unit distance along the track.

We get an equation similar to (5-40) for $N(t)$:

$$\frac{dN}{dt} = \frac{kN^2}{8\pi Dt} - k_s c_s N \tag{5-51}$$

If we make the same transformations as before, we can integrate this equation to get

$$N = N_0\, e^{-q(x-1)} \left[1 + \frac{kN_0}{8\pi D} K(x, 1)\right]^{-1} \tag{5-52}$$

where $K(x, 1) = \int_1^x e^{-q(y-1)}(dy/y)$ is an exponential integral. The fraction of radicals that react with the scavenger is given by Eq. (5-42) as before.

In the limit of zero scavenger we get

$$N = N_0\left[1 + \frac{kN_0}{(8\pi D)\ln x}\right]^{-1} \tag{5-53}$$

and we see that in the limit of infinite time (ie, $x \to \infty$) all radicals recombine. The simplest approximation for the fraction of radicals reacting with a scavenger is obtained by using the value $x = (k_s c_s t_0)^{-1}$ in Eq. (5-52). The essential difference between a cylindrical track and a spherical track (spur) is that radical recombination goes to completion in the former, whereas it is always partial in the latter.

## *Studies of the Track Partial Differential Equations*

Kuppermann and Belford[19–21] solved the set of Eqs. (5-34) numerically for water both in one-radical and multiradical approximations, starting with Gaussian initial distributions of intermediates. They explored many aspects of the problem, such as comparison of numerical solutions of the partial differential equation with results obtained by using prescribed diffusion; the latter was found to be a reasonably good approximation for the conditions expected in tracks.

Figures 5-3 and 5-4 show the calculated time development of a spur and a cylindrical track, respectively. The two cases are qualitatively similar. At first the radical recombination is dominant, and at a later stage the scavenger reaction occurs. In these figures $N(t)$ is the number of radicals surviving at time $t$; $N_{R_2}(t)$ is the number of $R_2$ molecules at time $t$; $N_R(t)$ is the number of RS molecules, also at time $t$.

There are two ways to probe the concept that the core is made up of spurs spaced at random separations (with a well-defined average value) along the track axis. One way is to change the average separation of spurs or the LET. Figure 5-5 shows the results of a multiradical calculation[21] for the variation of LET on primary water decomposition products with an experimental comparison. The other way is to vary the scavenger concentration for the same radiation, and low-LET radiation gives the best example.

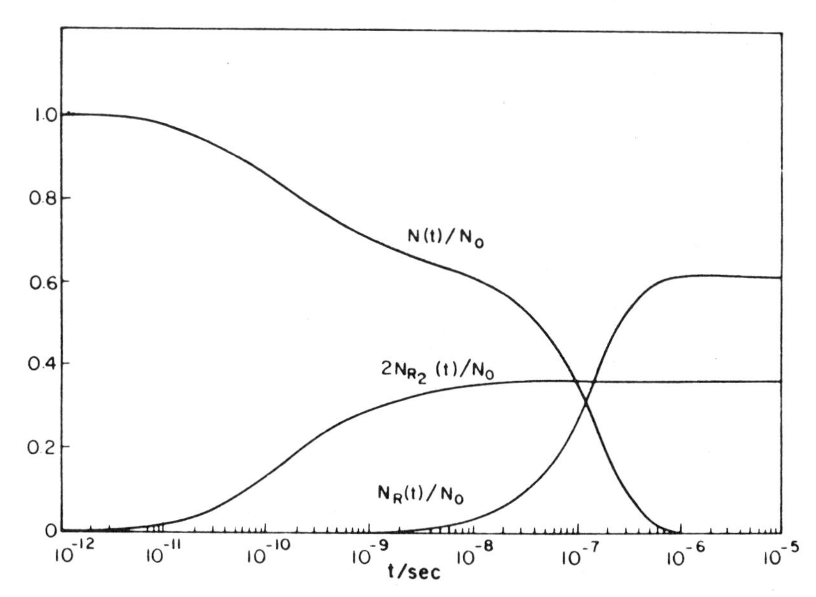


**Figure 5-3.** Variation of $N(t)/N_0$, $2N_{R_2}(t)/N_0$, and $N_R(t)/N_0$ with time for a spherical spur. Gaussian initial distribution. $N_0 = 12$ radicals, $r_0 = 10$ Å, $D_R = 4 \times 10^{-5}$ cm$^2$/s, $D_s = 4 \times 10^{-6}$ cm$^2$/s, $k_{RR} = k_{RS} = 6 \times 10^9$ M$^{-1}$ s$^{-1}$, $c_{s_0} = 10^{-3}$ M. (From "Actions Chimiques et Biologiques des Radiations"; Haissinsky, M., Ed.; Masson, 1961. Reproduced with permission.)

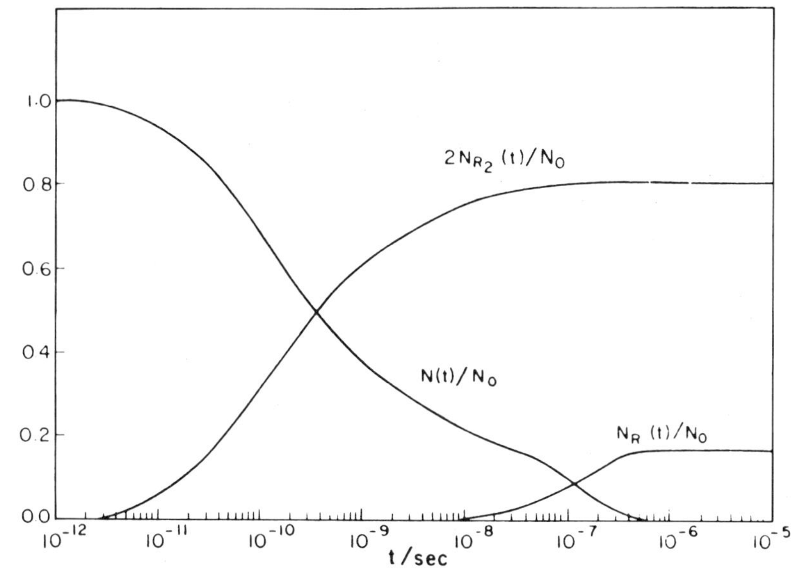


**Figure 5-4.** Variation of $N(t)/N_0$, $2N_{R_2}(t)/N_0$, and $N_R(t)/N_0$ with time for an axially homogeneous cylindrical track. Gaussian initial distribution. $N_0 = 8.5 \times 10^7$ radicals/cm. Other parameters the same as for Fig 3. (From "Actions Chimiques et Biologiques des Radiations"; Haissinsky, M., Ed.; Masson, 1961. Reproduced with permission.)

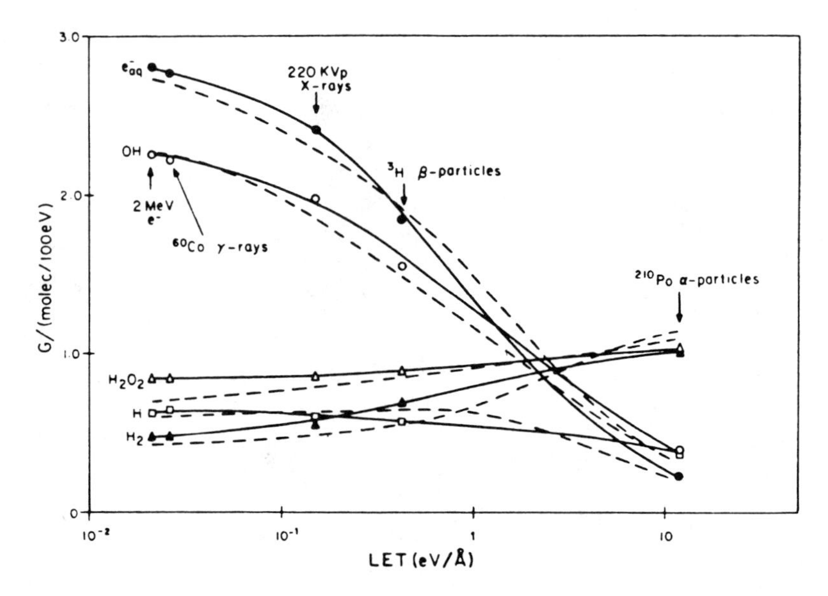


**Figure 5-5.** Variation of $G_{e_{aq^-}}$, $G_{OH}$, $G_{H_2O_2}$, $G_H$, and $G_{H_2}$ with average LET. The points and full lines represent theoretical calculations corresponding to the radiations indicated. The dashed lines are experimental.[23]

Figure 5-6 shows a calculation of the expected effect of scavenger on the molecular yield for γ-radiation in the solid curve and gives an experimental comparison with points plotted according to the method suggested by Schwarz.[22,23] In a "Schwarz plot" it is not the scavenger concentration plotted on the abscissa but rather the scavenger concentration times the relative reactivity. Allen[24] and Draganic and Draganic[25] have reconsidered the data of this figure.

The decrease in molecular yields with scavenger concentration varies as the square root of the latter at small concentrations, as one would expect from Eq. (5-48). Figure 5-6 covers six decades of scavenger concentrations, and the quantitative nature of the curve at large concentrations depends in a more complicated manner on the track expansion.

Calculations using the oversimplified model with a track taken as a string of spurs give results in rough agreement with experiment. Of course, there are always parameters to be fitted; the situation is complicated, and no one has analyzed the parameter-fitting problem in a very sophisticated manner.

A more elaborate calculation of the radiolysis of water has been presented by Schwarz,[26] who considered the tracks created by γ-rays to be composed of spurs, blobs, and short tracks, as suggested by Mozumder and Magee,[27] and he used a sophisticated version of prescribed diffusion in calculating the radiation chemical yields of the various track entities. This calculation is generally considered to demonstrate that the radiolysis of water is actually understood.

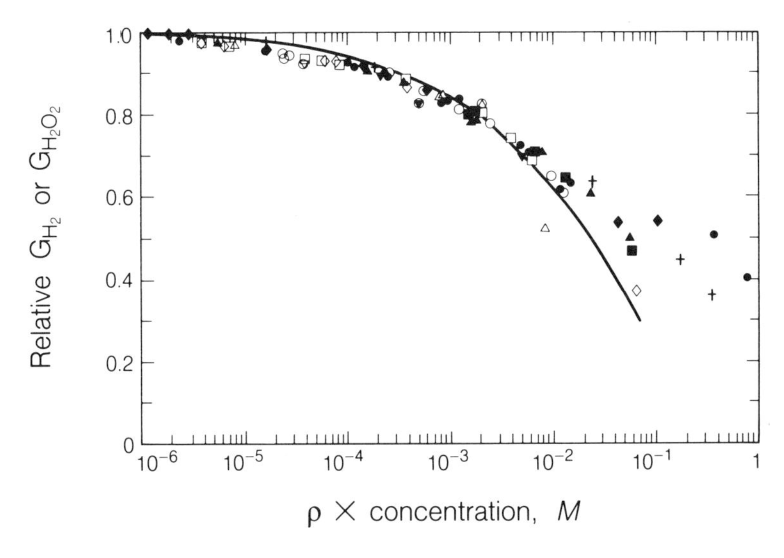


**Figure 5-6.** Schwartz plot showing fractional lowering of the molecular yields, plotted against (concentration) $\times \rho$ (on a logarithmic scale), where $\rho$ is an arbitrary constant chosen to bring values for all substances onto the same curve. References to the literature for the experimental points are given by Allen.[24] (Reproduced with permission.[24])

Figures 5-3 and 5-4 show that the radical recombination and scavenger reactions are separated in time. In calculations of radiation chemical yields it is usually assumed that such a separation is possible, and the radical processes are considered without scavenger competition until a "scavenging time" is reached, after which only scavenger reactions occur. As an example, consider the reactions in the Fricke dosimeter where the scavenger dominates.

$$H + O_2 \rightarrow HO_2 \tag{5-54}$$

$$HO_2 + Fe^{2+} \rightarrow HO_2^- + Fe^{3+} \tag{5-55}$$

$$HO_2^- + H^+ \rightarrow H_2O_2 \tag{5-56}$$

$$OH + Fe^{2+} \rightarrow OH^- + Fe^{3+} \tag{5-57}$$

$$H_2O_2 + Fe^{2+} \rightarrow OH^- + OH + Fe^{3+} \tag{5-58}$$

The yield of oxidation of $Fe^{2+}$ is given by

$$G(Fe^{3+}) = 2G_{H_2O_2} + 3G_H + G_{OH} \tag{5-59}$$

which, for $\gamma$-irradiation, adds up to 15.6. As the LET changes, the $G$'s of the molecular products and the radical products change in a way to make the model consistent with experiment.

# THEORETICAL TREATMENT OF BACKGROUND REACTIONS IN IRRADIATED SYSTEMS

The track reactions of electrons and heavy particles in water are understood well enough to furnish a basis for the explanation of the radiation chemistry of a large class of dilute aqueous solutions. In the latter the water radicals are scavenged effectively, and their concentrations outside of tracks are negligible. In such systems all radiation chemical phenomena occur in tracks, and radiation chemical yields are proportional to the number of tracks. Chemical dosimeters (Fricke, etc; see refs. 24, 25, and 28) belong to this class of systems.

More general considerations of irradiated systems require explicit treatment of background reactions. For example, if pure $H_2O$ is irradiated, initially there are no scavengers to react with the radicals as the tracks expand; radicals from different tracks intermingle, and the concentrations build up until radical–radical reactions occur homogeneously. The molecular products ($H_2$, $H_2O_2$) also react with the radicals in homogeneous reactions. All of these phenomena must be considered in adequate treatments of such systems.

Let us consider the irradiation of an aqueous system at a constant dose rate. We assume that tracks fall randomly in space and time with a well-defined average rate of energy deposition. A background radical concentration is established, and it is expected to fluctuate around some average value. A track that falls in such a system at first has concentrations of product molecules and radicals much larger than the background; thus, it can be considered as an isolated track as far as its initial development is concerned. At a later stage its concentrations are more comparable to the background, and, finally, it must merge with the background; at some time (the track lifetime) we say that the track no longer exists, and the molecular and radical products are considered as additions to the background. Within this framework of theoretical concepts, several hierarchies of approximation are possible. The lowest involves the interaction of individual tracks with the background treated as spatially homogeneous; higher approximations involve interactions of the tracks in pairs, triplets, and so forth, and explicit treatment of fluctuations in background concentrations. Here we consider only the lowest approximations; that is, we take the yields of the track reactions as source terms for a set of reactions for the system assumed to occur homogeneously in space.

Consider a system irradiated by a particle beam that delivers $\nu$ tracks per unit area per unit time. Each track expands at a rate determined by the diffusion of its radicals, and we can say that the cross-sectional area of the expanding cylindrical track is $4\pi Dt$, where $D$ is the diffusion coefficient and $t$ is the time of expansion. There is a correction for the *initial* track area, but, under conditions of interest here, this is negligible. At the time equal to the *track lifetime* $\tau$ the track cross-sectional area is $4\pi D\tau$. We take $\tau$ as the time in which the average number of tracks falling in the irradiated volume, each taken at maximum expansions, just fills the volume. Thus, the condition is $\nu\tau 4\pi D\tau = 1$ and $\tau = (4\pi D\nu)^{-1/2}$. This definition of track lifetime is made so that, in the track lifetime, radicals can diffuse all over the volume, and the assumption that the products created in tracks are homogeneously distributed is reasonable.

We are interested in "pure" water, that is, water in which the only radical scavengers are the "molecular" decomposition products $H_2$, $H_2O_2$, and $O_2$. Experimental studies on this system have been reported by several authors.[29,30]

Two situations have been found in the decomposition of water by radiations. If the tracks produce more radicals (H, OH) than molecules ($H_2$, $H_2O_2$), as for low-LET radiations (electrons, x-, and $\gamma$-rays), a chain "back-reaction" destroys the molecules ($H_2$, $H_2O_2$) and limits their buildup. A steady state in which no net decomposition occurs is reached quickly. If the molecular yields are larger than the radical yields, as in the case of heavy particle tracks, the back-reaction is not effective at first and a net decomposition of water occurs.[24,30–32]

On general grounds we believe that any closed system has radiation stationary states. In the case of water, low-LET radiations have such states with very little water decomposition. In the early phase of a heavy particle irradiation it would appear that the particles are decomposing the water because the stationary state is so far away.

Two types of theoretical studies are possible for systems under steady irradiation; they are involved with transients and steady states, respectively. Initial-value problems are concerned with transient behavior; a system under constant irradiation is followed in time for some initial condition until a steady state is reached. All of the calculations reported here are of this type.

## *Equations for Steadily Irradiated Water*

We are interested in water of all possible pHs, and we actually consider two special cases: acid water at pH $< 1$, and neutral water at pH $= 7$. These two cases illustrate all principles involved, and using the methods presented we can make treatments at any desired pH.

In acid water under low-LET irradiation, track reactions create molecular ($H_2$, $H_2O_2$) and radical (H, OH) products that can be treated as if formed homogeneously in the system. The actual track mechanisms are well understood,[33,34] and the assumption made here leads to results consistent with them. The radiation creation process in acid water is usually indicated by the equation

$$H_2O \rightsquigarrow H, OH, H_2, H_2O_2 \tag{5-17}$$

The amount of decomposition produced by 1-MeV $\gamma$-rays is known, and for the absorption of 100 eV of energy we can write

$$4.51\ H_2O \rightsquigarrow 3.71\ H + 0.40\ H_2 + 2.95\ OH + 0.78\ H_2O_2 \qquad (5\text{-}60)$$

This equation gives the products released into the dilute scavenger system by the tracks.

The treatment of neutral water is somewhat more complicated because it involves more species. This situation can be understood most simply by noting that the $H^+$ ion concentration is so low that the hydrated electrons are not converted into H atoms before track reactions occur, and charged species must be treated explicitly. Whereas in acid water we need seven species, in neutral water many more are involved. Bielski and Gebicki[35] have considered the reactions occurring in irradiated water containing oxygen, and they introduce 27 species that are involved in 99 reactions. Of course, it is not known that this system of reactants and reactions is complete, and so any actual calculation we make must involve a truncated system. At first we consider an approximation that has 12 reactants, the 12 species with differential equations in Table 5-2. Seven of them (including water) are involved in Eq. (5-61), and the other five ($O_2$, $HO_2$, $O_2^-$, $OH^-$, $HO_2^-$) are formed in secondary reactions.

The equation for the initial radiation decomposition can be written

$$H_2O \rightsquigarrow H_3O^+,\ e_{aq}^-,\ H,\ OH,\ H_2,\ H_2O_2 \qquad (5\text{-}61)$$

and we have the quantitative relationship for the decomposition produced by the absorption of 100 eV of energy from 1-MeV $\gamma$-rays:

$$4.10\ H_2O \rightsquigarrow 2.65\ H_3O^+ + 2.65\ e_{aq}^- + 0.55\ H + 2.70\ OH + 0.45\ H_2 + 0.70\ H_2O_2 \qquad (5.62)$$

Table 5-1 lists the thermal reactions in which the radiation products participate; the neutral case includes all of the reactions and the acid case only the ones with an asterisk. Table 5-2 gives the differential equations that describe the concentration changes; again, as in Table 5-1, the equations that apply to the acid cases are marked with an asterisk. In the equations of Table 5-2 the direct radiation creation (those involving the rate of irradiation, $I$) must give the concentrations of created species in moles per liter per second; thus, $I$ is the rate of radiation absorption in units of $6 \times 10^{25}$ eV per liter per second.

We are interested in the solution of these equations for several special conditions, which can be classified in the categories:

1. Continuous irradiation at constant rate that starts with the molecular concentrations of $H_2$, $H_2O_2$, and $O_2$ equal zero.
2. Continuous irradiation at constant rate that starts with various initial concentrations of $H_2$, $H_2O_2$, and $O_2$.

Radiation chemists have treated the problem intuitively using stationary states for all transient species.[29,30] However, such techniques are not adequate; at best they are ad hoc with limited usefulness. We introduce in the next section mathematical methods powerful enough to solve all problems that are likely to arise.

For irradiated pure water (zero initial values of $H_2$, $H_2O_2$, and $O_2$) the radical concentrations (H, OH) approach "stationary values" very quickly, and $H_2$, $H_2O_2$ build up linearly at first and then approach stationary values on a longer time scale. We know from the chemical nature of the system that the latter will occur because $H_2$ and $H_2O_2$ are destroyed in a chain reaction and thus must approach limiting concentrations. The chain reaction decomposition is

$$\begin{aligned} H_2 + OH &\rightarrow H_2O + H \\ H_2O_2 + H &\rightarrow H_2O + OH \end{aligned} \tag{5-63}$$

This decomposition is, of course, embedded in the equations of Table 5-2.

We include as "pure" water that which contains sulfuric acid, even in molar concentration. The latter is believed not to interfere with the water chemistry as described here.

We expect that the general behavior of the two systems (acid and neutral water) will be similar. Under $\gamma$-irradiation both will approach stationary states with concentrations of decomposition products limited by the back-reactions. Heavy particles will appear to produce decomposition in both systems. The positions of the steady states and their relaxation times will, of course, be different for acid and neutral water.

## *Numerical Treatment of Equations*

We can write the two sets of equations (for acid and neutral water, respectively) given in Table 5-2 in the same form:

$$\frac{dx_j}{dt} = f_j(x_1, x_2, \ldots, x_p) + IG_j \qquad j = 1, 2, \ldots, p \tag{5-64}$$

where $x_j$ is the concentration of the $j$th species, $IG_j$ its source term as obtained from Eqs. (5-60) or (5-62) and $p$ is the number of species (7 for acid water; 12 for neutral water). This set of ordinary differential equations (ODE) is unusually difficult to solve. The equations are nonlinear and involve variables and parameters (concentrations, rate constants) that differ by many orders of magnitude. Such equations are called *stiff*, a concept introduced by Curtiss and Hirschfelder.[36] The term "stiff" applies to systems of equations in which there is inherently a large negative feedback, such as that arising from the balancing of opposing chemical reactions. See ref. 37 for recent references on chemical systems.

The equations considered here also have a sensitivity to initial conditions that makes them also "ill conditioned." Here we discuss briefly the numerical methods[38,39] we have used to obtain solutions for several special cases and that we believe will be adequate to treat most of the problems of interest to radiation chemists.

A more compact form for Eq. (5-64) is

$$\frac{dx_j}{dt} = F_j \tag{5-65}$$

where $F_j = f_j(x_1, x_2, \ldots, x_p) + IG_j$. We are interested in obtaining values of the $x_j$'s over a time interval $0 < t < T$. We divide $T$ into $N$ equal small intervals and so obtain $N + 1$ points along the time axis: $t_0, t_1, t_2, \ldots, t_N$. We deal only with initial-value problems, so we know the $x_j$'s at $t_0$. We calculate the set of $x_j$'s at time $t_1$ using the relationship

$$\frac{dx_j}{dt} = \frac{x_j(t_1) - x_j(t_0)}{t_1 - t_0} = F_j(t_1) \tag{5-66}$$

where $F_j(t_1) = F_j(x_1(t_1), x_2(t_1), \ldots, x_p(t_1))$.

If we express Eq. (5-66) as

$$x_j(t_1) = x_1(t_0) + F_j(x_1(t_1), x_2(t_1), \ldots, x_p(t_1))(t_1 - t_0) \qquad j = 1, 2, \ldots, p \tag{5-67}$$

we obtain a set of $p$ nonlinear equations that must be solved for $x_j(t_1)$ $(j = 1, 2, \ldots, p)$, where the $x_j(t_0)$'s are known.

If $\Delta t_m = t_{m+1} - t_m$, Eq. (5-66) is expressed as

$$x_j^{m+1} = x_j^m + \Delta t_m F_j(x_1^{m+1}, x_2^{m+1}, \ldots, x_p^{m+1}) \tag{5-68}$$

where superscripts are used to indicate the time step; that is, $x_j^m = X_j(t_m)$, $j = 1, 2, \ldots, p$. This is known as the backward Euler approximation of Eq. (5-64), and this difference scheme has a better stability property than the forward Euler approximation of Eq. (5-64). Actually, the backward Euler approximation is just the simplest of all approximations that can be made. The same methods can be employed with any other approximation scheme (trapezoidal, Simpson's rule, etc).

Equation (5-67) is an implicit relationship and must be solved by an iterative scheme for each time step $m = 1, 2, 3, \ldots$. A perturbed functional iterative scheme developed and discussed in detail in ref. 38 has been applied to solve Eq. (5-68). The fundamental concept is this: If at some $t_{m+1}$ our actual solution is $x_j^{m+1,*}$, then at each iteration (designated by superscript $k = 1, 2, \ldots$) $x_j^{m+1,k}$ is given a small perturbation $w_j^{m+1,k}$ so that $x_j^{m+1,k}$ can be brought closer and closer to the true solution $x_j^{m+1,*}$. This mechanism is discussed in a rather simple format in ref. 39. Some discussion of the applicability of this method to solve chemical kinetic problems may be found in ref. 40. The method is attractive because it can be used for chemical systems of essentially arbitrary complexity. There are theoretical complexities, however, that remain to be clarified. We are looking for unique solutions, but from a mathematical points of view it is not known that such uniqueness exists; the system of equations is nonlinear and stiff, and there must be at least several solutions. We think we can rule out all other solutions on the grounds that they do not correspond to reality.

## *Results of Calculations*

Figure 5-7 gives the result of a numerical solution of the equations for $\gamma$-irradiation of acid water at a rate of 40 krad/min. Clearly, a steady state is reached in about 5 s, so no net decomposition of water occurs after that time. Figure 5-8 gives the result of a similar calculation for the irradiation of acid water by Ne particles at 10 MeV per nucleon and also at 40 krad/min. We see that after about 0.2 s the situation in both systems (Figures 5-7 and 5-8) is about the same, but for the Ne irradiation, instead of approaching a steady state, the molecular products $H_2O_2$, $H_2$, and $O_2$ continue to be formed and the radicals continue to decrease. We expect that the radicals H, OH, and $HO_2$ will reach steady values on a shorter time scale than the molecular products. This is a qualitatively different solution than for the $\gamma$-ray case. The calculation has not been continued long enough so that we can estimate the approximate length of the transient time.

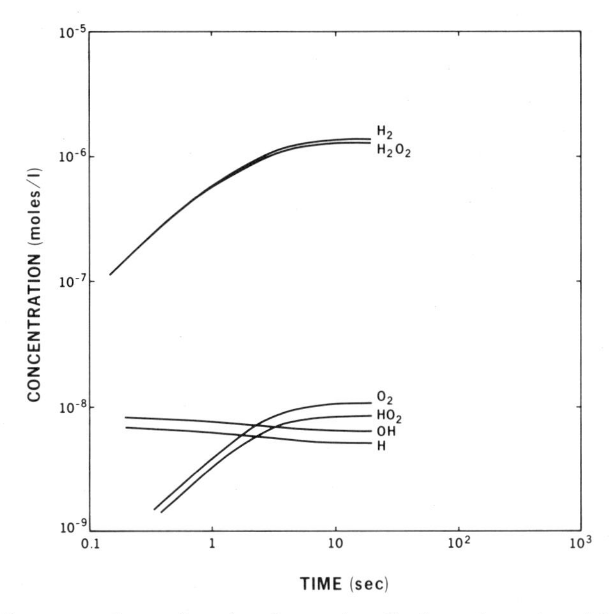


**Figure 5-7.** Concentrations of molecular and radical products in acid water under $\gamma$-irradiation at 40 krad/min. (Reproduced with permission.[40])

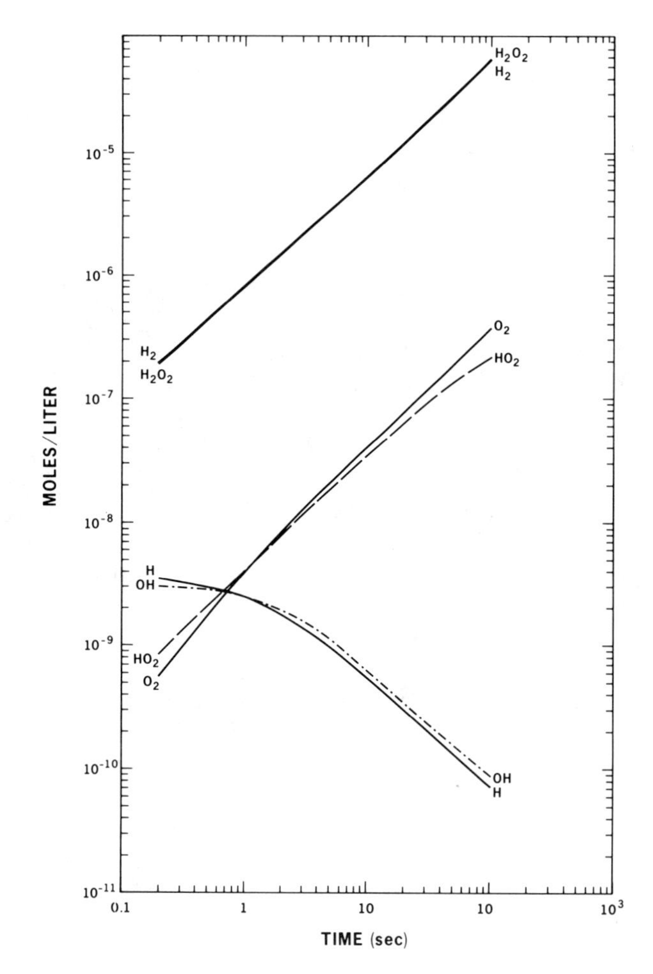


**Figure 5-8.** Concentrations of molecular and radical products in acid water under Ne irradiation (at 10 MeV per nucleon, 45 eV per Å) at 40 krad/min. (Reprinted with permission.[40])

A complete treatment of irradiated systems requires a self-consistent approach with separate calculations of the track expansions and of the homogeneous reactions of the background. Methods for treating the track reactions are presented in refs. 33 and 34, and the companion problem—the background reaction—is discussed in ref. 40. We have not yet made a complete treatment in which the track calculations are made consistent (by an iterative process) with the system of background reactions; such a study can now be made profitably because the stiff equations of the background can be solved in a reliable manner by the methods introduced here. These methods give us a powerful tool for theoretical radiation chemistry; the difficulties experienced here with obtaining agreement between calculation and experiment arose from inadequacies in the chemical description of the system, not in the computational aspects.

This is a powerful technique for investigating irradiated systems. It is particularly attractive because the stiff equations for the homogeneous reactions can be solved with good accuracy, and we can concentrate on the physical and chemical problems. However, we still do not know the most effective way to define a minimal system that can actually be used to approximate a real system nor how to obtain a self-consistent set of parameters for such a system by using experimental data. These questions require further consideration.

## STATUS OF RADIATION CHEMICAL THEORY

The description of all processes occurring between absorption of energy from high-energy radiation and the formation of radiation chemical products is a formidable task involving many disciplines and specialities. After 35 years of theoretical effort we find that a framework for the understanding of the phenomena of radiation chemistry is pretty well known in terms of a track theory using prescribed diffusion, but even a rigorous statement of most of the problems has not yet been made. Table 5-3 contains a listing of events occurring in irradiated water in terms of local time; some of the most important problems for theory are shown in the third column.

Since the interpretation of the radiation chemical yields in water in terms of track theory became established, only slow progress has been made in refinements of this theory. Now in a rough way we understand LET and radiation quality effects.

A large part of the research in radiation chemistry has been involved in pulse radiolysis studies and, particularly, in studies of the hydrated and solvated electrons. This work, of course, has been important in its own right, but it has also been invaluable in the elucidation of the early phenomena and in the establishment of initial conditions for track model calculations.

With regard to track structure concepts themselves, the good state of stopping power theory has been the most important ingredient. It is true that further development is still required, but the fact that tracks are composed of cores and penumbras is clear. The reconciliation of track models with Monte Carlo calculations that use the most reliable data on cross sections is an important matter for the future.

Many of the problems listed in Table 5-3 challenge the radiation chemical theorist. The most important is the general formulation of the track problem. How does one go about calculating the chemical effects of an energy deposit in matter that is widely distributed in space with many different magnitudes of energy density? It is an interesting fact that we think we actually know a great deal about this matter, but we have not formulated the problem rigorously.

**Table 5-3.** Problems of Theoretical Radiation Chemistry

| Time (s) | Events in Water | Problems of Theoretical Radiation Chemistry |
|---|---|---|
| $10^{-16}$ | Ionization, excitation | Stopping power—relationship of gas and liquid state |
| $10^{-15}$ | Dry hole ($H_2O^+$) migrates | Response of matter to energy absorption |
| | Spur electrons fall to subexcitation range | Physical track structure<br>Ranges, penetration of low-energy electron |
| $10^{-14}$ | Dissociations, ion–molecule reactions $H_2O^+ + H_2O \rightarrow H_3O^+ + OH$ | Energy mechanisms of subexcitation electron |
| $10^{-13}$ | Electron thermalized and hydrated $e^- \rightarrow e_{aq}^-$ | Mechanism of thermalization and hydration |
| $10^{-12}$ | Diffusion kinetics with time-dependent rate coefficients | Shape and sizes of track entities<br>Chemical track structure |
| $10^{-11}$ | First real-time observations<br>Track reactions underway | Theoretical basis for track reactions |
| $10^{-10}$ | | Relationships of partial differential equations to track models in prescribed diffusion and Monte Carlo techniques |
| $10^{-9}$ | Spur reaction complete | |
| $10^{-8}$ | Track end (blob, short track) reactions complete | Treatment of diffusive intratrack overlap |
| $10^{-7}$ | Radical reactions with scavenger at millimolar concentration | |
| $10^{-6}$ | Heavy particle track reactions complete | Treatment of diffusive intertrack overlap |
| $10^{-5}$ | | |
| $10^{-4}$ | Radical reactions with scavenger at micromolar concentration | Treatment of stiff equations of the homogeneous reaction system |
| $10^{-3}$ | | |

# REFERENCES


1. Mozumder, A.; Magee, J. L. *Int. J. Radiat. Phys. Chem.* **1975**, *7*, 83.
2. Turner, J. E.; Magee, J. L.; Wright, H. A.; Chatterjee, A.; Hamm, R. N.; Ritchie, R. H. *Radiat. Res.* **1983**, *96*, 437.
3. Bethe, H. A. *Ann. Phys. (Leipzig)* **1930**, *5*, 325.
4. Magee, J. L. *Ann. Rev. Phys. Chem.* **1961**, *12*, 389.
5. Burton, M.; Funabashi, K.; Hentz, R. R.; Ludwig, P. K.; Magee, J. L.; Mozumder, A. "Transfer and Storage of Energy by Molecules"; Burnett, G. M.; and North, A. M., Eds.; Wiley Interscience: New York, 1969, Vol. 1, pp. 161ff.
6. Platzman, R. L. "Radiation Research"; Silini, G., Ed.; North-Holland: Amsterdam, 1967, pp. 20–42.
7. Kalarickal, S. Ph.D. Dissertation, University of Notre Dame, 1959.
8. Czapski, G.; Peled, E. *J. Phys. Chem.* **1973**, *77*, 893.

8a. Hunt, J. W. In "Advances in Radiation Chemistry"; Burton, M.; and Magee, J. L., Eds.; Wiley Interscience: New York, 1976, Vol. 5, pp. 185–315.
9. Mozumder, A. *J. Chem. Phys.* **1971**, *55*, 3020, 3026.
10. Hamill, W. H. *J. Phys. Chem.* **1969**, *73*, 1341.
11. Sawai, T.; Hamill, W. H. *J. Phys. Chem.* **1970**, *74*, 3914.
12. Ogura, H.; Hamill, W. H. *J. Phys. Chem.* **1974**, *78*, 504.
13. Magee, J. L. *J. Am. Chem. Soc.* **1951**, *73*, 3270.
14. Anbar, M.; Bambenek, M.; Ross, A. B. In "Selected Specific Rates of Reactions of Transients from Water in Aqueous Solution. 1. Hydrated Electron"; U.S. Dept. of Commerce/National Bureau of Standards: Washington, D.C., 1973, Report NSRDS-NBS 43.
15. Ross, A. B. In "Selected Specific Rates of Reactions of Transients from Water in Aqueous Solution. 1. Hydrated Electron"; U.S. Dept. of Commerce/National Bureau of Standards: Washington, D.C., 1975, Supplemental Data, Report NSRDS-NBS 43.
16. Anbar, M.; Farhataziz; Ross, A. B. In "Selected Specific Rates of Reaction of Transients from Water. 2. Hydrogen Atom"; U.S. Dept. of Commerce/National Bureau of Standards: Washington, D.C., 1975, Report NSRDS-NBS 51.
17. Farhataziz; Ross, A. B. In "Selected Specific Rates of Reactions of Transients from Water in Aqueous Solutions. Hydroxyl Radical and Perhydroxyl Radical and Their Radical Ions"; U.S. Dept. of Commerce/National Bureau of Standards: Washington, D.C., 1977, Report NSRDS-NBS 59.
18. Balkas, T. I.; Fendler, J. H.; Schuler, R. H. *J. Phys. Chem.* **1970**, *74*, 4497.
19. Kuppermann, A.; Belford, G. G. *J. Chem. Phys.* **1962**, *36*, 1412, 1427.
20. Kuppermann, A. *Act. Chim. Biol. Radiat.* Haissinsky, M., Ed.; Masson, **1961**, *5*, 85.
21. Kuppermann, A. In "Radiation Research"; Silini, G., Ed.; North-Holland: Amsterdam, pp. 212–234.
22. Schwarz, H. A. *J. Am. Chem. Soc.* **1955**, *77*, 4960.
23. Schwarz, H. A.; Caffrey, J. M.; Scholes, G. *J. Am. Chem. Soc.* **1959**, *81*, 1801.
24. Allen, A. O. "The Radiation Chemistry of Water and Aqueous Solutions"; Van Nostrand: Princeton, 1961.
25. Draganic, I. G.; Draganic, Z. D. "The Radiation Chemistry of Water"; Academic Press: New York, 1971.
26. Schwarz, H. A. *J. Phys. Chem.* **1969**, *73*, 1928.
27. Mozumder, A.; Magee, J. L. *Radiat. Res.* **1966**, *20*, 203.
28. Fricke, H.; Hart, E. J. In "Radiation Dosimetry"; Attix, F. H.; and Roesch, W. C., Eds.; Academic Press: New York, 1966, Vol. 2, Chapter 12, pp. 167–239.
29. Allen, A. O.; Hochanadel, C. J.; Ghormley, J. A.; Davis, T. W. *J. Phys. Chem.* **1952**, *56*, 575.
30. Hochanadel, C. J. *J. Phys. Chem.* **1952**, *56*, 587.
31. Allen, A. O. *J. Phys. Colloid Chem.* **1947**, *52*, 479.
32. Allen, A. O. *Disc. Faraday Soc.* **1952**, *12*, 79.
33. Magee, J. L.; Chatterjee, A. *J. Phys. Chem.* **1978**, *82*, 2219.
34. Magee, J. L.; Chatterjee, A. *J. Phys. Chem.* **1980**, *84*, 3529.
35. Bielski, B. H. J.; Gebicki, J. M. In "Advances in Radiation Chemistry"; Burton, M.; and Magee, J. L., Eds.; Wiley Interscience: New York, 1970, Vol. 2, pp. 177–279.
36. Curtiss, C. F.; Hirschfelder, J. O. *Proc. Natl. Acad. Sci. U.S.A.* **1952**, *38*, 235.
37. Edelson, D. *Science* **1981**, *214*, 981.
38. Dey, S. K. *J. Comp. Appl. Math.* **1977**, *3*, 17.
39. Dey, S. K. *J. Franklin Inst.* **1979**, *307*, 21.
40. Chatterjee, A.; Magee, J. L.; Dey, S. K. *Radiat. Res.* **1983**, *96*, 1.

# 6

# Track Models and Radiation Chemical Yields

**A. CHATTERJEE AND J. L. MAGEE**

*Lawrence Berkeley Laboratory, University of California at Berkeley*

## INTRODUCTION

In Chapter 5 it was shown that all heavy particle tracks are similar in structure: They all consist of a core composed of the excitations produced by "glancing" collisions (also called "resonant" losses) and a penumbra consisting of the tracks of knock-on electrons. Another important principle is that low-$z$ particles at high energy have tracks that are very similar to high-energy electron tracks. It should be possible to discuss track models in a unified manner, and the objective of this chapter is to review attempts of the authors to do this.

The initial radiation chemical track models were essentially of two types. One was constructed to apply to low-energy heavy particle tracks and treated radical diffusion and reaction in cylindrical symmetry. The other was constructed for low-LET radiations and treated the isolated spur as the track model. Secondary electrons were not treated explicitly in these initial track models. In this chapter we develop a more sophisticated model that takes into account explicitly the core and penumbra of the general track and attempts to assign roles to them separately in the creation of chemical yields. Each track is composed of regions that vary significantly in properties, and a successful track model must bring out this fact.


*Radiation Chemistry: Principles and Applications*

The radiation chemical yield of a radiation-produced product is given in terms of $G$-values, that is, the number of molecules created per 100 eV absorbed. We speak in terms of two kinds of $G$-values: the integral yield, $G$, and the differential (or "local") yield, $G'$, also sometimes called the "track segment" yield. These quantities are functions of the same variables, such as the type of particle, the particle energy, the observed product, and so forth; the notation we employ is flexible, and any number of the variables may appear as arguments; for example, the differential ferric yield for a particle of charge $z$ and energy $E$ per nucleon in a solution with scavenger concentration $c_s$ may be given as $G'(Fe^{3+}; E)$ when the other conditions are understood, or as $G'(Fe^{3+}; E, z, c_s)$ when the particle and the concentration of solute ($Fe^{2+}$) need to be indicated explicitly. We are usually interested in the variation of a particular yield with heavy particle energy and merely write $G'(E)$ or $G(E)$. The two kinds of $G$-values are related as follows:

$$G'(E) = \left(\frac{d}{dE}\right) EG(E) \tag{6-1}$$

From a theoretical point of view $G'(E)$ is the more fundamental quantity.

Throughout the chapter the symbol $E$ for energy of a heavy particle means specific energy, that is, energy per nucleon.

We are concerned only with systems in which single track effects dominate and radiation chemical yields are sums of yields for individual tracks. We know from Chapter 5 that the energy deposits of heavy particle tracks are composed of spurs along the particle trajectory (about one-half of the energy) and a more diffuse pattern composed of the tracks of knock-on electrons, called the penumbra (about one-half of the energy). The simplest way to introduce the concept of a unified track model for heavy particles[1,2] is to consider the special case of the track of a heavy particle with an LET below 0.2–0.3 eV/Å, which in practice limits us to protons, deuterons, or particles with energy above 100 MeV per nucleon. At these LET values, to a good approximation, spurs formed by the main particle track can be considered to remain isolated throughout the radiation chemical reactions. For such a case the $G'$-value of any radiation chemical product can be explicitly given if the $G_e(\varepsilon)$-value of the same product is known for electron tracks. It is understood that we are always talking about the same radiation chemical product (not indicated explicitly), and we can write

$$G'(E) = (1-f)G_{sp} + f\frac{\int_{\varepsilon_0}^{\varepsilon_{max}} G_e(\varepsilon)w(E, \varepsilon)\,d\varepsilon}{\int_{\varepsilon_0}^{\varepsilon_{max}} w(E, \varepsilon)\,d\varepsilon} \tag{6-2}$$

where $G'(E)$ is the differential yield per 100 eV of the particle with energy $E$; $G_{sp}$ is the isolated spur yield; $G_e(\varepsilon)$ is the $G$-value of an electron track that starts with energy $\varepsilon$; $w(E, \varepsilon)$ is the fraction of energy lost by the particle of energy $E$ in creation of knock-on electrons per unit energy interval at $\varepsilon$; $\varepsilon_0$ is 100 eV, the low-energy limit of knock-on electrons; $\varepsilon_{max} =$

$2mc^2\beta^2/(1-\beta^2)$, the maximum knock-on electron energy; $f$ is the fraction of energy expended in the creation of knock-on electrons.

We should perhaps note that all spur properties, such as energy distribution, are expected to be independent of the particle that creates them. The first term on the right of Eq. (6-2) is the core contribution, and the second term is the penumbra contribution. The validity of Eq. (6-2) is seen from elementary considerations: The total yield consists of contributions from a finite number of track entities (spurs and knock-on electron tracks) that develop and produce their yields independently of one another.

Equation (6-2) is important for heavy particle track theory. It gives an explicit relationship between the differential yield of a heavy particle track in the low-LET region in terms of the yields of electron tracks. Furthermore, a similar equation applies to electron tracks themselves[3] (see Eq. (6-25) and discussion thereof). Thus, Eq. (6-2) is an anchor point in track theory.

Most heavy particle tracks have LETs too large for Eq. (6-2) to apply directly; the spurs in their cores are close enough together that they interact as they expand, and there is also an interaction between the core and penumbra. We show later that a more general equation can be written for $G'(E)$:

$$G'(E) = F_{\text{chc}}G'_{\text{chc}}(E) + [1 - F_{\text{chc}}(E)]G'_{\text{pen}}(E) \tag{6-3}$$

where $F_{\text{chc}}(E)$ is the fraction of energy in the track assignable to the track core on the basis of a chemical criterion (discussed in the next section). This energy includes the energy of the physical core and part of the penumbra energy with which the physical core interacts in a chemical way as the expansion occurs. We call this core the *chemical* core in contrast to the *physical* core, which essentially consists of the spurs of the initial deposit. We show later that a radius $r_1$ (larger than the size of the physical core) can be defined in such a manner that it encloses the chemical core.

$G'_{\text{chc}}(E)$ is the differential yield attributable to the chemical core; it is discussed at more length later. $G'_{\text{pen}}(E)$ is the penumbra yield attributable to the portions of the electron tracks of the penumbra that do not interact with the core; it is also discussed in the later sections.

In order to understand the interaction between core and penumbra in a heavy particle track, we must know the initial energy distribution explicitly. Chatterjee and Schaefer[4] have proposed that the core is initially contained within a radius $r_c$ given by

$$r_c = \frac{\beta c}{\Omega_p} \tag{6-4}$$

where $\beta$ is the particle velocity in terms of the velocity of light, $c$, and $\Omega_p$ is the plasma frequency of the stopping medium. Strictly speaking, $\Omega_p$ only applies to an electron gas and is given by $\Omega_p = (4\pi ne^2/m)^{1/2}$, where $n$ is the number density of the electrons, e is the electron charge, and $m$ is the electron mass. In current condensed phase theory, however, even molecular solids have plasma frequencies that are significant in considerations of energy loss phenomena. In this chapter the only stopping medium considered is water.

The penumbra consists of the tracks of knock-on electrons, but for our purposes it is convenient to think of an energy distribution that is continuous in space. Chatterjee et al[5] have used stopping power theory to make an estimate of such an energy distribution applicable to the heavy particle track. Based on this estimate, Chatterjee and Schaefer[4] have proposed the following average (initial) energy densities in the track:

$$\rho_{\text{core}} = \left(\frac{\text{LET}}{2}\right)[\pi r_c^2]^{-1} + \left(\frac{\text{LET}}{2}\right)\left[2\pi r_c^2 \ln\left(\frac{e^{1/2} r_p}{r_c}\right)\right]^{-1} \qquad r \leq r_c \qquad (6\text{-}5)$$

$$\rho_{\text{pen}}(r) = \left(\frac{\text{LET}}{2}\right)\left[2\pi r^2 \ln\left(\frac{e^{1/2} r_p}{r_c}\right)\right]^{-1} \qquad r_c < r \leq r_p \qquad (6\text{-}6)$$

Here, LET is the average energy loss per unit path length, and $r$ is the radial distance from the particle trajectory. Values of $r_c$ and $r_p$ as functions of the specific energy (energy per nucleon) of heavy particles are given in Table 1-3. The first term in Eq. (6-5) is the average energy density due to the glancing collisions; the second term gives a small contribution from the energy lost by the knock-on electrons that do not penetrate the core. The energy density of Eq. (6-6) arises from knock-on electrons that are stopped between $r_c$ and $r_p$.

It is clear that the track dimensions $r_c$ and $r_p$ depend only on particle velocity $\beta$, not on charge or mass. The energy density, however, depends on the LET and thus on the particle charge. The general particle track, therefore, depends on two parameters, which can be taken as the energy per unit mass ($E$) and the atomic number of the particle ($z$). The actual effective charge $z_{\text{eff}}$ is determined by the interaction of the particle with the medium; it depends on the velocity and cannot be set arbitrarily. Values of some of the track quantities that depend on velocity only are given in Table 1-3.

Initial concentrations of reactive intermediates (radicals) in the core are expected to be much larger than those in the penumbra. The actual location of the chemical intermediates formed by the resonant electronic excitations is not necessarily within the core of radius $r_c$; the early processes involve subexcitation electron transport to distances of 20 to 50 Å or so; although these electrons return toward their loci of formation before they become hydrated, the initial core radius for electrons is not likely to be much less than 30 Å. In any case, however, the core has a much larger radical concentration than the adjacent penumbra.

The densities given by Eqs. (6-5) and (6-6) are *average* values, and the actual initial values may fluctuate greatly around them. For example, the core is composed of spurs containing, on the average, 40 eV; if the LET is small enough, the core consists of a string of spurs rather than a continuous distribution of intermediates; this phenomenon has been considered in Chapter 5. The penumbra is composed of tracks of knock-on electrons, and it is always more inhomogeneous than the core. Here the individual tracks have their recombination reactions, which take place early, and then the escaping radicals intermingle. The extent and importance of the intermingling of radicals from neighboring electron tracks of the penumbra depend on the rate of the scavenger reaction in a manner considered below.

The linear energy transfer (LET) for six selected particles in water as a function of energy per unit mass ($E$) is shown in Figure 1-4. These curves were obtained by using a recipe proposed by Blann.[6–8] The effective charge is given by Eq. (6-7)

$$z_{\text{eff}} = z(1 - e^{-0.95 v_r}) \tag{6-7}$$

where $z$ is the atomic number of the particle and $v_r = v/(v_0 z^{2/3})$; $v$ is the particle velocity, and $v_0$ is the average velocity of the electron in the H atom ($v_0 = \mathrm{e}^2/h$). The stopping power (LET) of a particle with energy per nucleon $E$ is given in terms of the stopping power of the proton at the same energy [Eq. (6-8)]:

$$\left(\frac{dE}{dx}\right)_{\text{ion}(z_{\text{eff}})} = \left(\frac{dE}{dx}\right)_{\text{proton}} z_{\text{eff}}^2 \tag{6-8}$$

The proton stopping power was calculated by using the Bragg rule with data for the atomic targets H and O obtained from ref. 9. The experimental data of ref. 9 scatter very widely, and a considerable amount of smoothng was made; the low-energy stopping power of heavy particles in water cannot be considered as well known.

## THE CHEMICAL CORE OF A HEAVY PARTICLE TRACK

The equations with asterisks in Tables 5-1 and 5-2 give the chemical reactions of the seven species that must be considered in acid water. At this point, however, we consider a much simpler system, which has radicals of one type only. At any position in a track, radicals recombine with a rate given by $2kc^2$ (because two radicals disappear) and react with the scavenger with a rate $k_s c_s c$, where $c$ is the local radical concentration, $c_s$ is the scavenger concentration, and $k$ and $k_s$ are rate constants. A radical concentration that we use as a reference is given by Eq. (6-9):

$$c_1 = \frac{k_s c_s}{2k} \tag{6-9}$$

For this particular radical concentration the radical recombination and the scavenger reaction have the same importance. If the radical concentration is larger than $c_1$, recombination is more important than scavenging; if it is less than $c_1$, the reverse is true. The mean time for the scavenger reaction is given by Eq. (6-10), and we use $t_1$ as a reference time:

$$t_1 = (k_s c_s)^{-1} \tag{6-10}$$

The track reactions cannot last for times appreciably larger than $t_1$. The distance that radicals can diffuse in time $t_1$, $l_1$, is given by Eq. (6-11):

$$l_1^2 = 4Dt_1 \tag{6-11}$$

## *The Effects of Increasing LET*

Let us consider the changing phenomena in tracks as LET increases. At the lowest LET (much less than 2 eV/Å), particle tracks are closely related to high-energy electron tracks. In this range of LET, tracks are clearly separated into entities that develop independently; as LET varies, $G'$-values change because the spectrum of knock-on electrons changes. This is the LET region in which Eq. (6-2) is valid.

As LET increases, spurs (though initially separate in the track core) overlap (due to diffusion of radicals) before their "forward" reactions (ie, radical combination reactions) are complete. As LET continues to increase, the spurs are merged initially to form a cylindrical track. If we say that the average spur contains 40 eV and has a radius of 20 Å, the initial merging occurs when the energy deposit in the core is about 1 eV/Å, or when the total LET is about 2 eV/Å. We call the region in which LET is less than 2 eV/Å "region I" (see Figure 1-4). In region I the cores of tracks are made up of spurs that are initially separate.

Other important "high-LET effects" are expected to arise from the "engulfing" by the core (due to diffusion of radicals) of the penumbra. Of course, the penumbra is actually composed of a statistical distribution of electron tracks; in the low-LET track they develop independently, but at higher LET (where they are deposited closer together) they overlap with each other and with the core before radical recombination is terminated by scavenger reaction. A measure of this kind of interaction is given by the average energy density in the track, and we can use the energy densities given in Eqs. (6-5) and (6-6). From Table 1-3, column 4, note that the values of $r_p$ turn out to be related to the maximum electron energies in column 5 approximately as 1 Å/eV.

Let us consider the radical recombination in the penumbra of a track, using the one-radical model and the average energy densities given by Eq. (6-6). The partial differential equation that applies is Eq. (6-12), where the $c$'s are concentrations already defined and $D$ is the diffusion coefficient:

$$\frac{\partial c}{\partial t} = D\nabla^2 c - 2kc^2 - k_s c_s c \tag{6-12}$$

The first term on the right describes the diffusion, and the second and the third describe the radical recombination and scavenger reactions, respectively. In our exploratory consideration we take the scavenger reaction as a small perturbation and look at the equation obtained by neglecting the last term of Eq. (6-12):

$$\frac{\partial c}{\partial t} = D\nabla^2 c - 2kc^2 \tag{6-12a}$$

The average initial radical concentration compatible with Eq. (6-6) is

$$c(r) = c_0\left(\frac{r_0}{r}\right)^2 \tag{6-13}$$

where $r_0$ is an arbitrary reference radius, and $c_0$, the initial radical concentration at $r_0$, is to be obtained by using Eq. (6-6); $c_0$ is proportional to LET. Using Eq. (6-13) in Eq. (6-12a) yields

$$\frac{\partial c}{\partial t} = \left[\frac{4D}{c_0 r_0^2} - 2k\right]c^2 \qquad \text{at } t = 0 \tag{6-14}$$

This equation is valid only initially, because at later times Eq. (6-13) does not generally hold. The diffusion and recombination terms in Eq. (6-12a) have the same initial dependence on $r$, and this dependence is the same as $c^2$, that is, $r^{-4}$. This means that, on the average, the rates of change in concentration because of diffusion and recombination have the same relative importance throughout the penumbra. There is a value of LET for each particle energy $E$ for which the bracket is zero. The condition for zero bracket can be written as

$$2k(4D)^{-1}c_0 r_0^2 = 1 \tag{6-15}$$

For the Fricke system a reasonable value for $2k/(4D)$ to be used in a one-radical treatment of water is 15 Å. The value of $c_0 r_0^2$ is somewhat more difficult to estimate. The energy density of Eq. (6-6) and the radical concentration of Eq. (6-13) correspond to average values, that is, values existing after the separate electron tracks of the penumbra intermingle. Thus, the "initial" condition of Eq. (6-14) is a fictitious one in which the radicals escaping from the electron tracks of the penumbra form the average distribution of Eq. (6-13). In track studies in water we have used 17 eV as the energy requirement for the formation of a radical pair. The different significance of the initial condition in the penumbra, which actually occurs after the forward reaction of the electron tracks is more or less complete, means that effectively a very different magnitude of energy per radical pair is required. We take 34 eV as the energy per radical pair. This choice of parameters allows a calculation of the position of zero bracket in Eq. (6-14), and it is indicated in Figure 1-4 by the line labeled $\alpha = 1$. The region below this line (and above region I) is called "region II"; the region above is called "region III".

Equation (6-14) gives the initial $\partial c/\partial t$ in the penumbra for the partial differential Eq. (6-12a). Consider the early development in time of this equation in the penumbra for $r$-values sufficiently distant from $r_c$ or $r_p$. If the bracket is positive, diffusion dominates recombination initially; if the bracket is negative, recombination dominates diffusion initially. An examination of $\partial c/\partial t$ for small values of the time shows that, as time increases, $\partial c/\partial t$ becomes more positive if it is initially positive and more negative if it is initially negative. We call region II *a region of diffusion domination*, and region III *a region of recombination domination.*

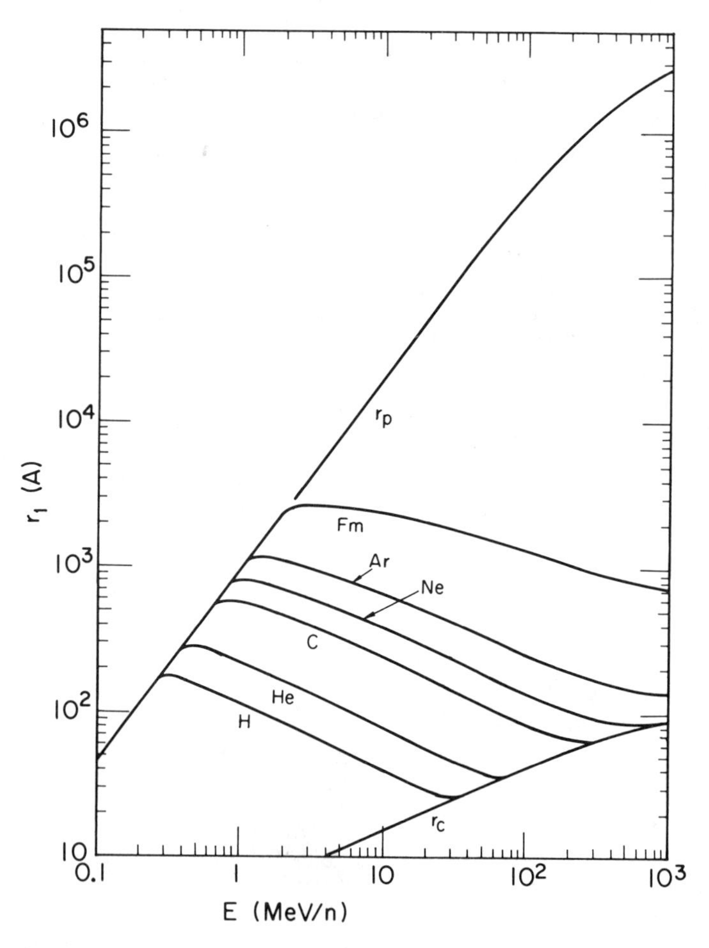


**Figure 6-1.** Chemical core radii $r_1(E)$ (in Å) for the six selected particles vs. specific energy $E$ (in MeV/n). At low energies all values of $r_1(E)$ are equal to the penumbra radius, $r_p$; at high energies all values of $r_1(E)$ become equal to the physical core radius, $r_c$. Of course, $r_c$ and $r_p$ are common for all particles. (Reproduced with permission.[1])

The importance of the scavenging reaction depends on the radical concentration as compared with $c_1$ [Eq. (6-9)]. The values of the radii for the six representative particles for which $c = c_1$ are given in Figure 6-1; here $c_1$ was chosen to be compatible with the parameters of the Fricke dosimeter for a scavenging time of $3 \times 10^{-7}$ s. We call $r_1$ the radius of the "chemical core" because it is based on a chemical-reaction criterion, whereas the core radius $r_c$ was based on a physical criterion. Figure 6-1 shows that $r_1$ is equal to $r_c$ at high particle energies; as particle energy decreases, $r_1$ increases and $r_p$ decreases; at some energy, $r_1$ becomes equal to $r_p$; for smaller values of $E$, $r_1 = r_p$.

The fraction of the total energy of the tracks enclosed within the radius $r_1$ for this particular set of parameters is shown in Figure 6-2 for each of the six representative particles. At values of $E$ for which $r_1$ is equal to or less than the core radius $r_c$, the fraction of energy in the chemical core ($F_{chc}$) is approximately 0.52; at values of $E$ for which $r_1$ is equal to $r_p$ or $k_s c_s/(2k)$ is less than the radical concentration at $r = r_p$, $F_{chc} = 1$.

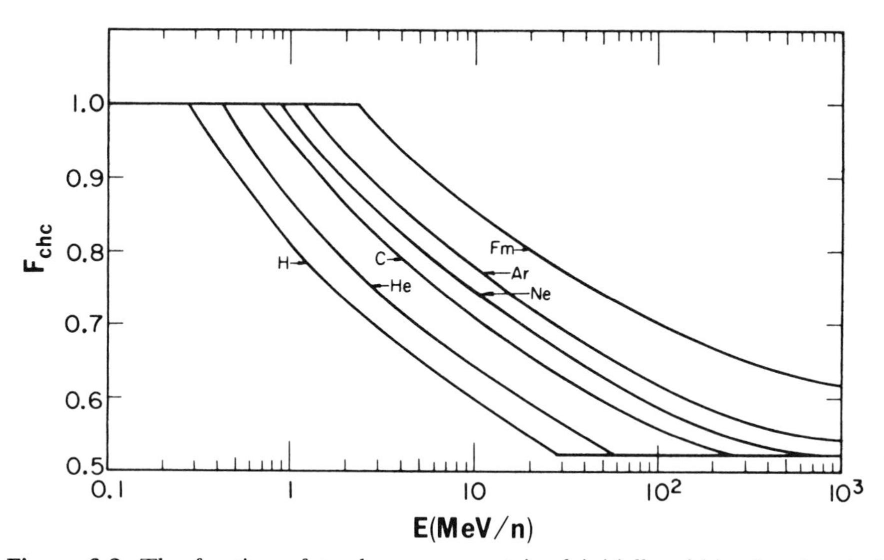


**Figure 6-2.** The fraction of track energy contained initially within the chemical core radius $r_1(E)$ for the six particles vs. the specific energy $E$. (Reproduced with permission.[1])

Using Eqs. (6-9)–(6-11) and (6-13) allows us to write

$$r_1^2 = \left(\frac{2kc_0 r_0^2}{4D}\right) l_1^2 \tag{6-16}$$

This relationship shows that $r_1$ and $l_1$ vary in the same way with scavenger concentration. We also see that for the condition of zero bracket in Eq. (6-14),

$$r_1^2 = l_1^2 \tag{6-16a}$$

When applying Eq. (6-12) to the portion of the track with $r > r_1$ (penumbra portion), to a good approximation we can neglect the recombination term $2kc^2$. The consideration leading to Eq. (6-14) shows that the diffusion and recombination terms are of comparable importance, so the diffusion term can also be neglected; thus, Eq. (6-17) appears to be a reasonable approximation.

$$\frac{\partial c}{\partial t} = -k_s c_s c \tag{6-17}$$

This means that the scavenger reacts completely with the radicals, and thus the yields are just the same as for the separate tracks forming the penumbra.

This treatment suggests that the "chemical core" (ie, the portion of the track contained within $r = r_1$) reacts in a typical kind of track process, and the "chemical penumbra" (the portion beyond $r = r_1$) reacts as if the electron tracks were isolated. Thus, the approximate separation of the heavy particle track into two noninteracting parts appears to be reasonable.

### *Unified Treatment of the Chemical Core*

As we have seen, the chemical core is composed of the spurs formed in the physical core plus that part of the neighboring penumbra with which they interact before the scavenger reactions become dominant. A large fraction of the chemical core energy and a correspondingly large fraction of the radicals are contained in the spurs.

In low-LET tracks, spurs are well separated from each other, and they never have a chance to intermingle before the scavenger reaction becomes important. For the intermediate-LET case, initially the spurs are separated from each other, but they do overlap before the reactions with the scavenger begin. The mean separation distance between the spurs is so small for high-LET tracks that from the time of their creation they overlap with each other, forming a cylindrical track for all practical purposes.

In spite of the substantial differences in the properties of tracks that have extreme values of LET, a rather generalized treatment is feasible. The method is highly intuitive and very similar to the one developed by Ganguly and Magee.[10] We continue using the one-radical model for simplicity, but extension to a multiradical treatment is straightforward. Such a treatment is used in this chapter.

In one-radical prescribed diffusion the concentration of radicals in a track of $N$ spurs is given by

$$c(r, z, t) = \sum_{i=1}^{N} \nu_i \frac{\exp[(-r^2 - (z - z_i)^2)/(r_i^2 + 4Dt)]}{[\pi(r_i^2 + 4Dt)]^{3/2}} \tag{6-18}$$

The $N$ spurs have centers located at positions on the $z$-axis at $z_1, z_2, \ldots, z_N$; the $i$th spur has $\nu_i$ radicals and a radius parameter $r_i$. If we use this concentration in the partial differential equation (6-12a) and first integrate over all space, we get Eq. (6-19), which is an equation summed over all the $N$ spurs.

$$\frac{d}{dt}\sum_i \nu_i = -2k \sum_{i=1}^{N} \frac{\nu_i^2}{[2\pi(r_i^2 + 4Dt)]^{3/2}} + \sum_{\substack{i=1 \\ j\neq i}}^{N} \frac{\nu_i \nu_j \exp[-(z_i - z_j)^2/(r_i^2 + r_j^2 + 8Dt)]}{[\pi(r_i^2 + r_j^2 + 8Dt)]^{3/2}} \tag{6-19}$$

In prescribed diffusion the $\nu_i$ values are considered as functions of time to be determined. We need a set of differential equations to determine the $\nu_i$, and we propose the following:

$$\frac{d}{dt}\nu_i = \frac{-2k\nu_i^2}{[2\pi(r_i^2 + 4Dt)]^{3/2}} + \nu_i \sum_{j\neq i} \frac{\nu_j \exp -(z_j - z_i)^2/(r_i^2 + r_j + 8Dt)}{[\pi(r_i^2 + r_j^2 + 8Dt)]^{3/2}} \qquad i = 1, 2, 3, \ldots, N \tag{6-20}$$

In justifying the choice of these equations, we note that all possible spur–spur interaction terms are included, and summation of both sides of Eq. (6-20) over $i$ leads to Eq. (6-19).

A rigorous treatment of this set of equations has not been made, but the following approximation leads to a result that has intuitive appeal. Consider the equation for an interior spur, the $i$th, and approximate the summation in Eq. (6-20) by an integration. We are interested in the average value of $d\nu_i/dt$, and we take the probability that the $j$th spur occurs in the interval $dz_j$ as $(dz_j/S_1)$, where $S_1$ (average distance between adjacent spurs) is a parameter depending on the LET. Substituting the appropriate integral gives Eq. (6-21).

$$\frac{d\nu_i}{dt} = -2k\left[\frac{\nu_i^2}{[2\pi(r_i^2 + 4Dt)]^{3/2}} + \nu_i\left\langle\frac{\nu_j}{[\pi(r_i^2 + r_j^2 + 8Dt)]S_1}\right\rangle\right] \tag{6-21}$$

It is clear that the first term of Eq. (6-21) gives the ordinary prescribed diffusion result for a spur, and the second term gives its interaction with the other spurs; the bracket ⟨ ⟩ on the second term indicates an average over the spur distribution function. Equation (6-21) applies to a spur of any size. Spurs are created by resonant energy losses in the range 6 to 100 eV, and, with 17 eV required to form a radical pair, $\nu_i$ varies from 2 to 12; to get the track yield, we must average this expression over the spur distribution function. Although such a distribution function has been proposed,[11] we do not consider the average explicitly at this time but, for simplicity, assume that the following equation applies to the low-LET average spur that has $\nu$ radicals:

$$\frac{d\nu}{dt} = -\frac{2k\nu^2}{[2\pi(r^2+4Dt)]^{3/2}}\left[1+\frac{[2\pi(r^2+4Dt)]^{1/2}}{S_1}\right] \tag{6-22}$$

where we choose $\nu$ as the average number of radicals in a spur and the average $r^2$ parameter so that reasonable spur yields[3] are obtained (eg, 18.4 for the $Fe^{3+}$ yield of the Fricke system); the parameter $S_1$ varies inversely with LET, and the proportionality constant can be adjusted to agree with experiment.

When $S$ gets so small with the increase of LET that the second term in the bracket is much larger than unity, we get

$$\frac{d\nu}{dt} \simeq \frac{-2k\nu^2}{[2\pi(r^2+4Dt)]S_1} \tag{6-23}$$

and we can write the equation in terms of the variable $\nu/S_1 = N$, the number of radicals per unit distance

$$\frac{dN}{dt} \simeq \frac{-2kN^2}{2\pi(r^2+4Dt)} \tag{6-24}$$

which is the ordinary prescribed diffusion equation for cylindrical symmetry (without considering scavenger reaction or interaction with penumbra). Equation (6-22), therefore, has a convenient form, which allows the low-LET core equation to remain valid as LET increases from intermediate to high. This equation contains the essence of the Ganguly–Magee track treatment;[10] here, however, we recognize that only the part of the track that does not include high-energy $\delta$-rays is involved. This treatment can also be extended to apply to a multiradical case; in a later section this is done, and the $G'$-value for the Fricke dosimeter is calculated. We note that the yield of the core depends only on $LET_{core}$ (because the penumbra is not included).

# RADIATION CHEMISTRY OF HIGH-ENERGY ELECTRONS

Consider an electron with energy $E$ that is completely absorbed in a system; it has a 100-eV yield for some particular radiation product, which we designate $G_e(E)$. The only example we shall use is the Fricke system; so the $G_e(E)$ value will always be that for $Fe^{+3}$, although the reslts of this section are much more general than this. The subscript e refers to the electron. We are interested in the most general relationship between the $G_e$ values for electrons that have different energies. The concept of an integral equation relating the two yields is best introduced through a consideration of Eq. (6-2). This equation applies to a low-LET heavy particle, but it seems reasonable to expect that a similar equation should apply to a high-energy electron (which also has low LET).

The elementary arguments used in the derivation of Eq. (6-2) depend on the smallness of the individual energy losses with respect to the total energy of the particle under consideration. For all heavy particles the maximum energy loss, $(2m/M)E$, where $m/M$ is the mass ratio of the electron to the particle, is a very small fraction of the particle energy. Thus, considering the right side of Eq. (6-2) as a differential effect is well justified.

The next step is to derive a similar equation for the electron as primary particle. In this case the maximum energy loss is one-half the incident energy, and the possibility for using such an equation is not at all clear. It can, however, be shown that the similar equation[3]

$$\begin{aligned} G_e'(E) &= \frac{d}{dE} EG_e(E) \\ &= (1-f)G_{sp} + f\frac{\int_{E_0}^{E/2} G_e(\varepsilon)w_e(E,\varepsilon)\,d\varepsilon}{\int_{E_0}^{E/2} w_e(E,\varepsilon)\,d\varepsilon} \end{aligned} \tag{6-25}$$

is valid for energies $E$ that are sufficiently high; for example, $E \geqslant E_1$, where $E_1$ is an energy large enough so that the spurs formed by the electron are widely spaced and essentially finish recombination before they overlap. At $E = 20$ keV the LET is 0.135 eV/Å, and spurs are expected to be, on the average, about 500 Å apart. Recombination of H atoms, with an initial radius of 15 Å or so, is expected to be 90% complete before overlapping. This value of $E_1$ would seem to be reasonable, at least in the radiolysis of water. Equation (6-25) is an asymptotic equation that becomes more valid at high energies.

Equation (6-25) can be used to obtain $G_e(E)$. Suppose that $G_e(E)$ is known below $E = E_1$. Integrate Eq. (6-25) to obtain

$$EG_e(E) = E_1G_e(E_1) + (1-f)(E-E_1)G_{sp} + f\int_{E_1}^{E} \Gamma(E')\,dE' \tag{6-26}$$

where

$$\Gamma(E') = \frac{\int_{E_0}^{E'/2} G_e(\varepsilon)w_e(E',\varepsilon)\,d\varepsilon}{\int_{E_0}^{E'/2} w_e(E',\varepsilon)\,d\varepsilon} \tag{6-27}$$

Equations (6-26) and (6-27) give the relationship between the function $G_e(E)$ at pairs of energies above $E = E_1$.

The function $w_e(E, \varepsilon)$ for the electron or positron is complicated, and full numerical procedures should be used in consideration of Eqs. (6-26) and (6-27). However, it is instructive to use a simple approximation that arises from the Rutherford cross section for knock-on collisions and leads to an analytical treatment. This approximation is

$$w_e(E, \varepsilon) \sim \frac{1}{\varepsilon} \tag{6-28}$$

Inserting Eq. (6-28) into Eq. (6-25) and transforming to the variable $\eta = \ln(E/2E_0)$ leads to the differential equation

$$\frac{d}{d\eta}\eta\, e^{-\eta}\frac{d}{d\eta}e^{\eta}[G_e(\eta)] = (1-f)G_{sp} + fG_e(\eta - b) \tag{6-29}$$

where $b = \ln 2$. This is an equation valid for $E > E_1$ or $\eta > \ln(E_1/2E_0)$.

Equation (6-29) is an unusual equation in that the arguments of the function $G_e(\eta)$ are different on the left and right sides. It has a solution in the form of an infinite series. However, this solution is not convenient to use, and the approximate solution

$$G_e(\eta) = A + \left(\frac{1}{\eta^{\alpha}}\right)\left(a_0 + \frac{a_1}{\eta} + \frac{a_2}{\eta^2} + \frac{a_3}{\eta^3}\right) \tag{6-30}$$

has been found. Here $A$, $\alpha$, $a_0$, $a_1$, $a_2$, and $a_3$ are constants to be determined. This solution can be shown to be correct to about $\eta^{-4}$. It cannot be improved by the addition of more terms because it diverges as the number of terms increases without limit.

If a Taylor series is used for $G_e(\eta - b)$, the following values are found for the constants in Eq. (6-30):

$$A = G_{sp} \qquad \alpha = 1 - f \qquad a_1 = \alpha(\alpha - bf)a_0$$

$$a_2 = \frac{\alpha(1+\alpha)}{2}\left\{(1+\alpha-bf)(\alpha-bf) - f\left(\frac{b^2}{2}\right)\right\}a_0$$

$$a_3 = \frac{\alpha(1+\alpha)(2+\alpha)}{6}\left[(\alpha+2-bf)\left\{(\alpha+1-bf)(\alpha-bf) - f\left(\frac{b^2}{2}\right)\right\} - (\alpha-bf)fb^2 - f\left(\frac{b^3}{3}\right)\right]a_0$$

Substituting $f = 0.4$ (ca. 40% of the energy is spent in producing another generation of electrons) leads to the explicit expression

$$\begin{aligned} G_e(\eta) &= G_{sp} + \left(\frac{a_0}{\eta^{0.6}}\right)\left\{1 + \frac{0.1936}{\eta} + \frac{0.1587}{\eta^2} + \frac{0.2752}{\eta^3}\right\} \\ &= G_{sp} + \left(\frac{a_0}{\eta^{0.6}}\right)F(\eta) \end{aligned} \tag{6-31}$$

where $F(\eta)$ has been substituted for the expression in braces.

## *Model Calculation for Electron Tracks*

Ideally, we would like to find a self-consistent solution for $G_e(E)$ that satisfies Eqs. (6-25) and (6-26) for all energies. In principle, a scheme can be devised for using experimental data to find the best solution $G_e(\varepsilon)$ that satisfies both the data and Eqs. (6-25) and (6-26). However, experimental data are not known precisely enough for this method to be used effectively. The most straightforward procedure is to use a model calculation. Ferrous oxidation in acid solution (the Fricke dosimeter) has been chosen as the example to investigate.

From general considerations we know that tracks of electrons starting with small energies (less than 5 keV according to Mozumder and Magee[12]) are single entities. Those having less than 100 eV are spurs; those with energies of 100 to 5000 eV are blobs and short tracks. The model calculation we have used separates low-energy tracks into three energy regions: (a) 100–1600 eV; (b) 1600–5000 eV; (c) above 5000 eV. It is a two-radical model that uses prescribed diffusion. The initial water decomposition corresponds to $G_{-H_2O} = 5.88$ or the requirement for 17 eV to create a radical pair (H, OH). Hydrated electrons are not considered explicitly because in acid solution they are rapidly transformed into H atoms. Presentation of the details of this model is beyond the scope of this treatment, and readers are referred to ref. 3 for details.

Figure 6-3 shows the calculated $G_e(Fe^{3+}; E)$ curves for the Fricke solutions containing oxygen or nitrogen. The model calculation was extended from low energies to the high-energy region by using Eqs. (6-26) and (6-27) and the most accurate inelastic cross section (Møller formula) in the $w_e(E, \varepsilon)$ function. We have found, however, that the analytical expression of Eq. (6-31) is in excellent agreement if proper constants are chosen. This agreement with the approximate analytical expression obtained by using the Rutherford cross section for the electron knock-on spectrum is not entirely understood. Perhaps it is largely a result of normalization (on the left side of Eq. (6-25) there is a ratio of integrals, both of which involve the cross section). Table 6-1 summarizes the constants to be used in the high-energy regions of the curves in Figure 6-3.

The emphasis in this treatment has been on the general nature of the energy dependence of the yields rather than on the quantitative values, and the latter are still subject to a certain amount of revision. On the other hand, the authors believe that the $G_e(E)$ curves of Figure 6-3 are in good agreement with the experimental data of radiation chemistry. It has actually not been customary to report experimental yields in terms of $G_e(E)$; values obtained with monoenergetic electron beams from accelerators are such yields, but values obtained with x-rays or $\gamma$-radiations are not. These radiations produce their own characteristic spectra of recoil electrons that are actually not always well known. Table 6-2 contains some yields that were calculated by using Figure 6-3 and spectra assumed for several radiation

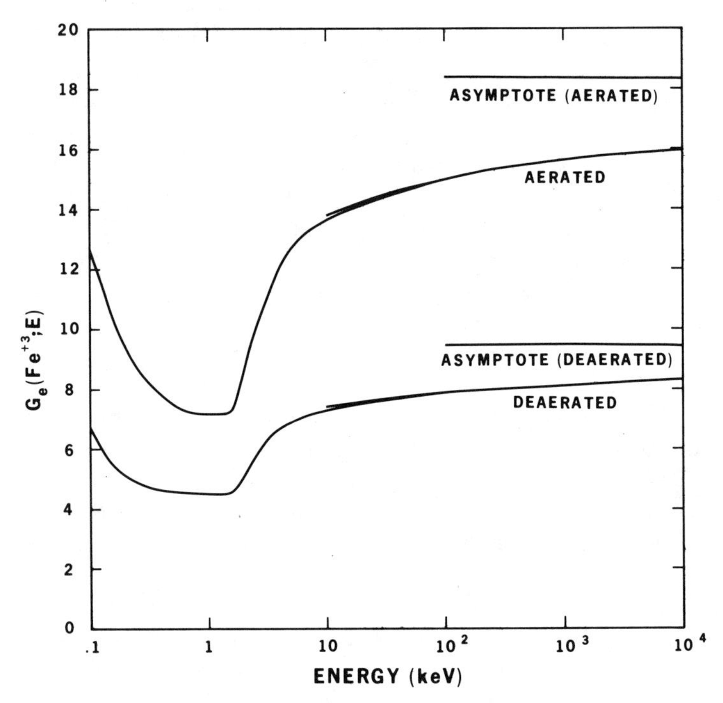


**Figure 6-3.** $G_e(Fe^{2+}; E)$ curves calculated from the electron track model. Above 10 keV the curves calculated from the asymptotic equations are shown: for the aerobic case,

$$G_e(Fe^{3+};\ E) = 18.40 - \frac{9.7337}{\eta^{0.6}} F(\eta)$$

for the deaerated case,

$$G_e(Fe^{3+};\ E) = 9.44 - \frac{4.3026}{\eta^{0.6}} F(\eta)$$

The asymptotes are indicated. (Reproduced with permission.[3])

**Table 6-1.** Constants for $G_e(E)$ Curves and Values at $10^6$ eV[a]

| $G$(Product) | $G_{sp}$ | $a_0$ | $G_e(10^6$ eV) |
|---|---|---|---|
| $G_{-H_2O}$ | 4.96 | −1.5870 | 4.51 |
| $G_{OH}$ | 3.92 | −3.4209 | 2.95 |
| $G_H$ | 4.48 | −2.7156 | 3.71 |
| $G_{H_2O_2}$ | 0.52 | 0.9169 | 0.78 |
| $G_{H_2}$ | 0.24 | 0.5643 | 0.40 |
| $G(Fe^{3+}, O_2)$ | 18.40 | −9.7337 | 15.64 |
| $G(Fe^{3+}, N_2)$ | 9.44 | −4.3026 | 8.22 |

[a] Reproduced with permission.[3]

**Table 6-2.** Calculated $G(Fe^{3+})$ Compared with Experiment[a]

| Radiation | $G(Fe^{3+})$ Model Calculated | $G(Fe^{3+})$ Experiment | |
|---|---|---|---|
| $^{60}$Co $\gamma$-ray | 15.61 | $15.5 \pm 0.3$[15] | $15.6 \pm 0.3$[27] |
| $^{32}$P $\beta$-ray | 15.34 | $15.35 \pm 0.5$[15] | $15.4 \pm 0.8$[28] |
| 220-kV x-rays | 14.70 | $14.6 \pm 0.3$[15] | $15.0 \pm 0.5$[29] |
| Tritium $\beta$-rays | 12.94 | $12.8 \pm 0.4$[15] | $12.9 \pm 0.2$[30] |

[a] Reproduced with permission.[3]

sources, and a comparison is made with experiment. The spectra for $^{60}$Co $\gamma$-radiation and 220-kV x-radiation were obtained from Cormack and Johns;[13] the spectra for the $\beta$-radiation were obtained from Nelms.[14]

The Fricke dosimeter has been investigated extensively. Errors are large enough in the reported $G$-values so that it is impossible to improve our $G_e(E)$ curves of Figure 6-3 by requiring them to conform more closely to experiment. Fregene[15] carried out an investigation spanning a larger range of energies than any other single investigation; $\beta$-, x-, and $\gamma$-radiations were used. In Figure 6-3, 11 values measured by Fregene[15] are plotted along with the curve $G_e(E) = 18.40 - (9.7337/\eta^{0.6})F(\eta)$, which we have chosen as the asymptotic yield curve for the aerated Fricke dosimeter. The points are plotted at the mean energies as listed by Fregene (Table I, ref. 15). The error bars of the points (not shown in Figure 6-3) are large enough so that the solid curve agrees with them.

The expression for the $G$-value for the Fricke dosimeter is $G(Fe^{+3}) = 2G_{H_2O_2} + G_{OH} + 3G_H$, and we see that the larger the radical yields the larger the $G$-value. Low-LET radiations have larger $G$-values than high-LET radiations. Traditionally, we think of electrons with energies in the megaelectronvolt range to be typical low-LET radiations, but in fact most low-LET radiations are high-energy electrons. We notice that the $G$-value for megaelectronvolt electrons is 15.6, and the $G$-value for the spur is 18.4. If electron tracks were completely composed of spurs, the Fricke $G$-values for all would be 18.4. It is an important fact that there are higher-LET portions of electron tracks that have much lower Fricke $G$-values. We see from Figure 6-3 that electrons in the energy range 0.5 to 1.5 keV have $G$-values below 12, and the $G$-value 15.6 of the megaelectronvolt electron is an average of 18.4 for spurs and about 10 to 12 for track ends.

It is an important fact that all radiations have substantial variations of LET in their tracks.

# FRICKE DOSIMETER YIELDS FOR HEAVY PARTICLES

## *Outline of the Model*

Considerations presented above based on a one-radical treatment of the core expansion led to a way to separate a chemical core from the penumbra of a heavy particle track. Figure 6-1 gives the values of the radius, $r_1$, that makes the separation, and Figure 6-2 gives the fraction of energy in the chemical core for the set of representative particles we consider. We can now use Eq. (6-3) to obtain the $G'$-values for our representative heavy particles. First, however, we need $G'_{\text{chc}}(E)$ and $G'_{\text{pen}}(E)$.

In the calculation of $G'_{\text{chc}}(E)$ we use the complete chemical mechanism given in Table 5-1, and we use a multiradical scheme based on Eq. (6-22). To get the ferric yield of the Fricke dosimeter, we need $G'_{\text{H}}$, $G'_{\text{OH}}$, and $G'_{\text{H}_2\text{O}_2}$. The differential equations for the seven species of acid water are shown in Table 6-3. In our treatment we use all of the acid water reactions.

In Table 6-3 the reactions involving primary radical recombination, usually considered to be the most important track reactions, are enclosed in a box. Table 6-4 shows the multispecies track reactions resulting from treatment of the reactions in the box according to the prescribed diffusion core model (see Eq. (6-22) and discussion thereof). In our treatment we actually used all of the equations of Table 6-3; the system of differential equations shown in Table 6-4 can be completed by inspection to produce the set used in the calculations.

**Table 6-3.** Differential Equations for Transient Species and Radiation Products in Track[a]

| | (boxed terms) | |
|---|---|---|
| $\frac{d}{dt}(\text{H})$ | $= -2k_1(\text{H})^2 - k_5(\text{H})(\text{OH})$ | $-k_9(\text{H})(\text{H}_2\text{O}_2) + k_{12}(\text{OH})(\text{H}_2) - k_{13}(\text{HO}_2)(\text{H}) - k_{17}(\text{H})(\text{O}_2)$ |
| $\frac{d}{dt}(\text{OH})$ | $= -k_5(\text{H})(\text{OH}) - 2k_6(\text{OH})^2$ | $+ k_9(\text{H})(\text{H}_2\text{O}_2) - k_{11}(\text{OH})(\text{H}_2\text{O}_2) - k_{12}(\text{OH})(\text{H}_2) - k_{15}(\text{HO}_2)(\text{OH})$ |
| $\frac{d}{dt}(\text{H}_2\text{O})$ | $= k_5(\text{H})(\text{OH})$ | $+ k_9(\text{H})(\text{H}_2\text{O}_2) + k_{11}(\text{OH})(\text{H}_2\text{O}_2) + k_{12}(\text{OH})(\text{H}_2) + k_{15}(\text{HO}_2)(\text{OH})$ |
| $\frac{d}{dt}(\text{H}_2)$ | $= k_1(\text{H})^2$ | $- k_{12}(\text{OH})(\text{H}_2)$ |
| $\frac{d}{dt}(\text{H}_2\text{O}_2)$ | $= k_6(\text{OH})^2$ | $- k_9(\text{H})(\text{H}_2\text{O}_2) - k_{11}(\text{OH})(\text{H}_2\text{O}_2) + k_{13}(\text{HO}_2)(\text{H}) + k_{16}(\text{HO}_2)^2$ |
| $\frac{d}{dt}(\text{HO}_2)$ | $= k_{11}(\text{OH})(\text{H}_2\text{O}_2) - k_{13}(\text{HO}_2)(\text{H})$ | $- k_{15}(\text{HO}_2)(\text{OH}) - 2k_{16}(\text{HO}_2)^2 + k_{17}(\text{H})(\text{O}_2)$ |
| $\frac{d}{dt}(\text{O}_2)$ | $= k_{15}(\text{HO}_2)(\text{OH}) + k_{16}(\text{HO}_2)^2$ | $- k_{17}(\text{H})(\text{O}_2)$ |

(The first five rows of the middle column are enclosed in a box in the original.)

[a] Reproduced with permission.[2]

**Table 6-4.** Track Equations in Prescribed Diffusion[a]

$$\frac{d\nu_1}{dt} = -\frac{2k_1\nu_1^2}{[2\pi(r_1^2+4D_1t)]^{3/2}}\left[1+\frac{[2\pi(r_1^2+4D_1t)]^{1/2}}{S_1}\right]$$
$$-\frac{k_5\nu_1\nu_2}{[\pi(r_1^2+r_2^2+4D_1t+4D_2t)]^{3/2}}\left[1+\frac{[\pi(r_1^2+r_2^2+4D_1t+4D_2t)]^{1/2}}{S_1}\right]$$

$$\frac{d\nu_2}{dt} = -\frac{k_5\nu_1\nu_2}{[\pi(r_1^2+r_2^2+4D_1t+4D_2t)]^{3/2}}\left[1+\frac{[\pi(r_1^2+r_2^2+4D_1t+4D_2t)]^{1/2}}{S_1}\right]$$
$$-\frac{2k_6\nu_2^2}{[2\pi(r_2^2+4D_2t)]^{3/2}}\left[1+\frac{[2\pi(r_2^2+4D_2t)]^{1/2}}{S_1}\right]$$

$$\frac{d\nu_3}{dt} = \frac{k_5\nu_1\nu_2}{[\pi(r_1^2+r_2^2+4D_1t+4D_2t)]^{3/2}}\left[1+\frac{[2\pi(r_2^2+4D_2t)]^{1/2}}{S_1}\right]$$

$$\frac{d\nu_4}{dt} = \frac{k_1\nu_1^2}{[2\pi(r_1^2+4D_1t)]^{1/2}}\left[1+\frac{[2\pi(r_1^2+4D_1t)]^{1/2}}{S_1}\right]$$

$$\frac{d\nu_5}{dt} = \frac{k_6\nu_2^2}{[2\pi(r_2^2+4D_1t)]^{3/2}}\left[1+\frac{[2\pi(r_2^2+4D_2t)]^{1/2}}{S_1}\right]$$

[a] Reproduced with permission.[2]

The separation of core and penumbra is a valid approximation largely because the track reactions occur early. We use the concept that radical–radical reactions dominate at first and scavenger reactions can be neglected. After a characteristic time given in the one-radical approximation by Eq. (6-10), the scavenger reaction dominates. In a multiradical system such a time is actually difficult to define precisely, and we choose $t_1 = 3 \times 10^{-7}$ s for use in the calculations. One can think of $t_1$ as an adjustable parameter. The value we have chosen is compatible with the ferrous concentration found in the Fricke dosimeter, with $10^{-3}$ molar concentration of $Fe^{2+}$ ions.

The scavenger reactions of $Fe^{3+}$ are not explicitly considered. It is, of course, assumed that they are dominant for times larger than $t_1$. (They are given in Eqs. (5-51) to (5-56).)

A string of spurs is produced within the core, and each spur contains, on the average, 40 eV of energy. Thus, the average number of a given radical (H or OH) in a spur is equal to 40/17 at the onset of radical reactions. At low LET, spurs are formed separate from one another, and, even when they expand because of the diffusion of their radicals, they remain isolated while the recombination reactions continue essentially to completion. Perhaps we should note that "completion" of the recombination reaction in a spur means that the maximum recombination has occurred, not that all radicals have recombined. We follow the expansion of the spur until $t = t_1$, the scavenger reaction time. At intermediate LET, spurs are initially isolated from one another, but as time progresses they grow large enough so that radicals intermingle between neighboring spurs. Such a track core goes over to cylindrical expansion before the time $t = t_1$ is reached. At high LET, spurs overlap initially, and the typical high-LET effects of radiation chemistry occur.

The interspur-separation parameter, $S_1$, has been taken as $50/\mathrm{LET}_{\mathrm{core}}$ on the basis of the agreement of calculation with experimental results. The average energy of a spur is 40 eV, so one might expect that $S_1$ should be $40/\mathrm{LET}_{\mathrm{core}}$. Actually, the consideration leading to the introduction of $S_1$ involves several averages over distribution functions, and the situation is not so simple that the parameter can be chosen completely a priori. In this consideration $\mathrm{LET}_{\mathrm{core}}$ is not the energy deposit per unit distance in the chemical core but rather the initial energy deposit per unit distance in the physical core.

The various $r_j$ values appearing in the differential equations in Table 6-4 are parameters in our calculation associated with the initial distribution of radicals. In Table 6-5 these initial radii for the various radical and molecular projects are given along with the diffusion constants used. The radii for hydrogen and hydroxyl radicals are adjusted for very energetic protons in such a manner that the ferric yield in an isolated spur is 18.4.

**Table 6-5.** Values of Initial Radii and Diffusion Constants

| Number of Species in Table 6-4[a] | Species | Initial Radius (Å) | Diffusion Constant ($cm^2/s$) |
|---|---|---|---|
| 1 | H | 26 | $8 \times 10^{-5}$ |
| 2 | OH | 13 | $2 \times 10^{-5}$ |
| 4 | $H_2$ | 26 | $8 \times 10^{-5}$ |
| 5 | $H_2O_2$ | 13 | $2 \times 10^{-5}$ |
| 6 | $HO_2$ | 26 | $2 \times 10^{-5}$ |
| 7 | $O_2$ | 26 | $2 \times 10^{-5}$ |

[a] Numbers apply to Tables 6-3 and 6-4. Of the seven species in acid water, water itself is number 3 and does not require radius and diffusion constants.

The system of differential equations of Table 6-4 (expanded to include all reactions of Table 6-3) is solved numerically from $t = 10^{-12}$ to $3 \times 10^{-7}$ s. These equations apply to a string of spurs, but it is shown that, as the LET increases, they go over into a form that is correct for a cylindrical distribution of radicals. We actually use these equations for all core calculations. The radicals of the penumbra that belong to the chemical core (ie, in the region $r < r_1$) are included in the core as they become engulfed. At the cutoff time, $t = 3 \times 10^{-7}$ s, the remaining H, OH, $HO_2$, and $H_2O_2$ react with $Fe^{2+}$ to give the "calculated" $Fe^{3+}$ yields of the chemical core.

The above discussion of the core processes would tend to justify this procedure for regions I and II (see Figure 1-4). Region III, however, is one in which recombination dominates, and it should be calculated for nondiffusing radicals. Actually, recombination is so fast and so nearly complete that both calculations give essentially the same results. For convenience, the model used in regions I and II goes into region III without modifications.

$G'_{pen}(E)$ is obtained by a modification of the integral expression of Eq. (6-2). The basic electron yield curve as a function of energy and the electron spectrum of the electrons absorbed in the region $r_1 \leq r \leq r_p$ must be used. Let us consider the following expression for the penumbra yield:

$$G'_{pen}(E) = \frac{\int_{\varepsilon_1}^{\varepsilon_{max}} G_e(\varepsilon')w(E, \varepsilon)\, d\varepsilon}{\int_{\varepsilon_1}^{\varepsilon_{max}} w(E, \varepsilon)\, d\varepsilon} \tag{6-32}$$

All of the knock-on electrons originate on the track axis and go into the chemical core region $r \leq r_1$; $\varepsilon_1$ is the minimum energy that allows them to reach $r_1$ and must be taken as the lower limit of the integral. The penetration of electrons in the penumbra is a statistical problem that has been considered by Chatterjee et al.[5] These authors found that the value of $r_p$ is essentially proportional to $\varepsilon_{max}$. This result means that, on the average, the penetration of an electron is linear in energy. Thus, we take $\varepsilon' = \varepsilon - \varepsilon_1$ as the appropriate argument in the electron yield expression, that is, $G_e(\varepsilon')$, because the electrons that go beyond $r_1$ leave this amount of energy in the penumbra.

## *Results*

The results of the basic calculations of $G'_{chc}(E)$ are presented in Figure 6-4. The calculational procedures strongly suggest that $G'_{chc}$ should be mainly a function of $LET_{chc}$, and a plot of $G'_{chc}$ vs. $LET_{chc}$ (Figure 6-5) shows that a unique function is in fact obtained. The range of $LET_{chc}$ values is very large (from $10^{-2}$ to $10^3$ eV/Å), and the regions of the curve that apply to the individual particles are indicated. At low particle energies, where $F_{chc} = 1$, the $G'_{chc}$ values of Figure 6-5 are expected to be the same as the total $G'$. At high particle energies, on the other hand, there is always a relatively large contribution from the penumbra, and $G'_{chc}$ is significantly different from the total $G'$. The low-LET limit for $G'_{chc}$ is the same as $G_{spur}$, that is, 18.4, whereas the total $G'$ for the proton is about 15; under these conditions the penumbra yield is lower than $G'_{chc}$ because of the track end (short tracks) contributions to the latter.

In Figure 6-1 at all values of the abscissa the particles have the same velocity and, thus, the same values of $r_c$ and $r_p$. The penumbra contributions to $G'(E)$, however, are not precisely the same because $r_1$ and $F_{chc}$ are different for particles with different atomic numbers. Figure 6-6 shows the $G'_{pen}$ values for the six particles. The vertical arrows mark the energies for the various particles below which there is no penumbra (ie, $F_{chc} = 1$). $G'_{ko}(E)$ is the $G$-value for an ideal penumbra, which contains all knock-on electrons from 100 eV to $\varepsilon_{max}$.

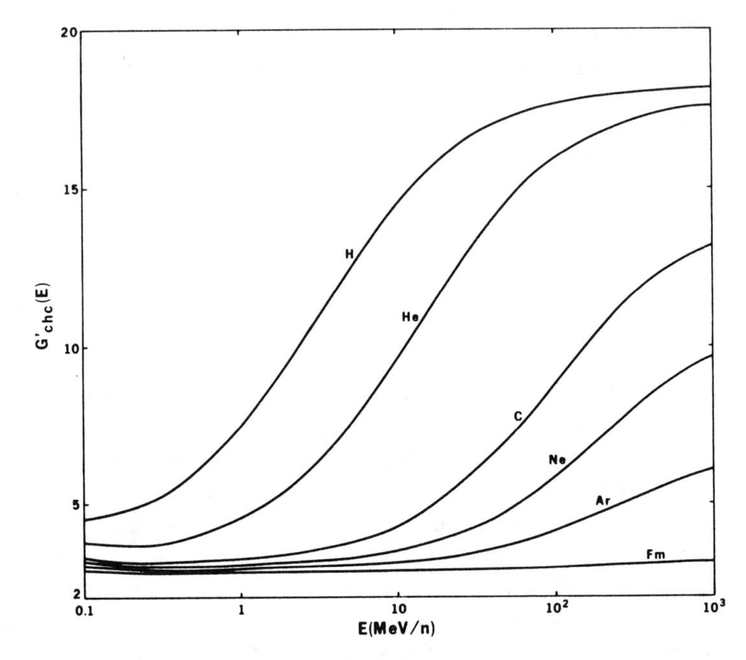


**Figure 6-4.** Calculated differential chemical core yields, $G'_{chc}(E)$, for the six representative particles vs. the energy per nucleon, $E$. (Reproduced with permission.[2])

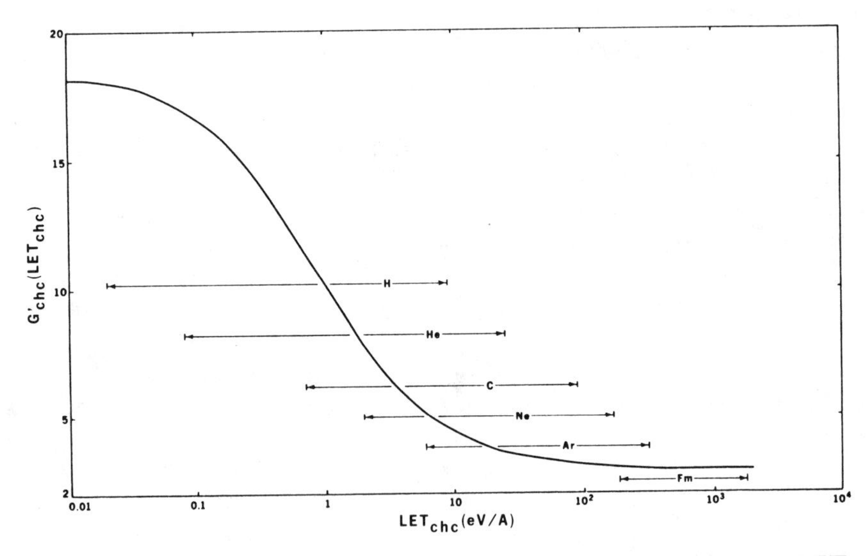


**Figure 6-5.** Calculated differential chemical core yield, $G'_{chc}(\mathrm{LET}_{chc})$ vs. $\mathrm{LET}_{chc}$. (Reproduced with permission.[2])

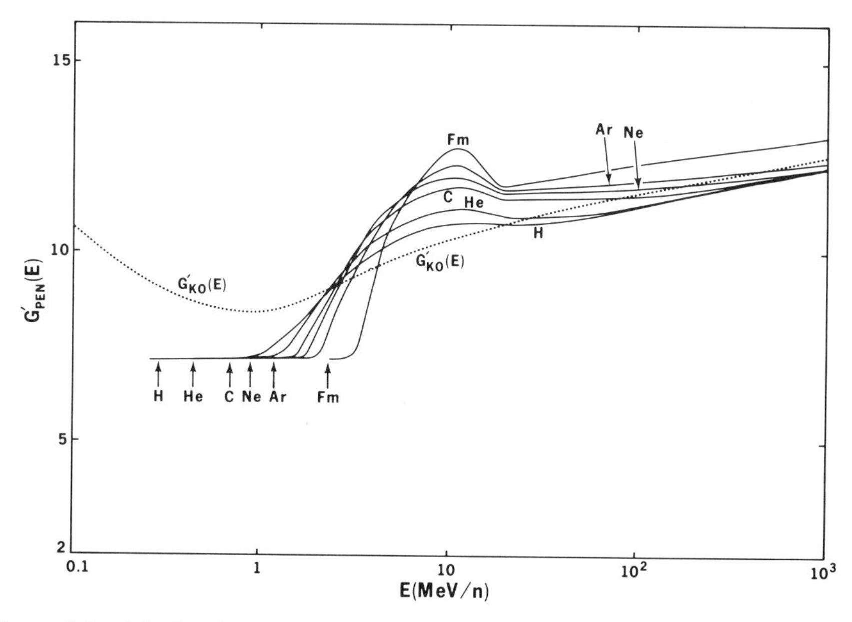


**Figure 6-6.** Calculated penumbra yields, $G'_{pen}(E)$, for the six representative particles vs. the energy per nucleon, $E$. The vertical arrows mark the energy limits below which no penumbra exists. The curve marked $G'_{ko}(E)$ gives the yield of the complete spectrum of knock-on electrons from 100 eV to $\varepsilon_{max}$ corresponding to the energy $E$. (Reproduced with permission.[2])

Figure 6-7 gives the primary results of the calculations $G'(E)$ vs. $E$ for the six particles. Data from Figures 6-1, 6-4, and 6-6 have been substituted into Eq. (6-3) to obtain the curves of this figure. It gives a broad view of the differential ferric yields for heavy particles. Some experimental data points are given for comparison with the calculations. At high energies $G'(E)$ can be measured directly, but at lower energies the data points are estimated from measured integral yields [ie, $G(E)$]. The directly measured values of Jayko et al[16,17] for C, Ne, and Ar are indicated by the dashed lines a, b, and c, respectively. The experimental values are higher than the calculated curves by more than the estimated experimental error. It is known that nuclear fragmentation (fragmentation of the beam into lower-atomic-number particles) of the beam creates an error in the direction of increasing the yield, and part of the discrepancy lies here. A study of the effects of beam fragmentation by Christman et al is underway, and the data will be reevaluated.

Four sets of experimental data furnish estimates of the locations of the curve for proton in the low-energy region. The dashed curve d is an estimate by Schuler and Allen[18] of $G'(E)$ from their experimental data on $G(E)$ for the deuteron; the open and closed circles are data points calculated from experimental $G(E)$ values for the deuteron and proton, respectively, by

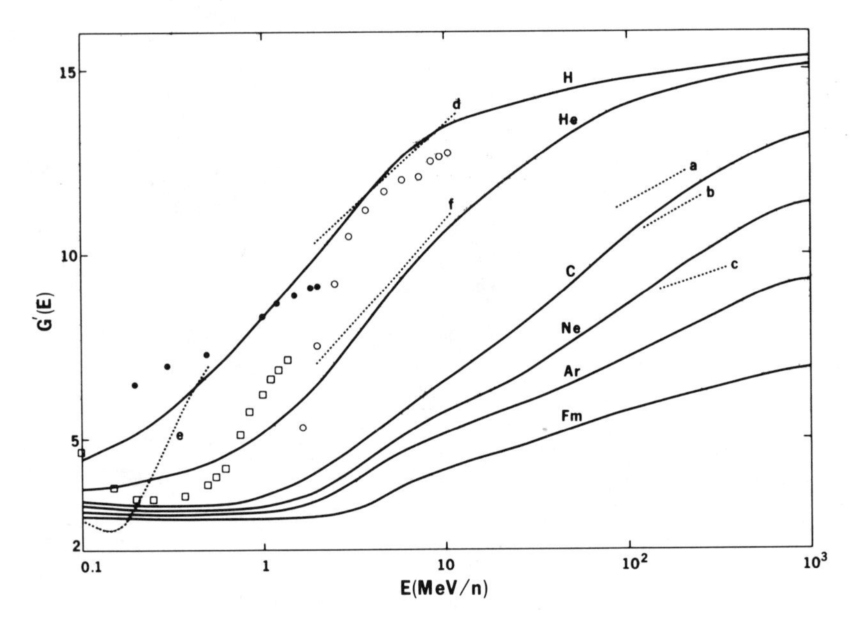


**Figure 6-7.** Calculated differential yields, $G'(E)$, for the six representative particles vs. energy per nucleon, $E$, are given by the solid curves. Estimates of values for C, Ne, and Ar from direct experimental measurements by Appleby et al are indicated by the dashed lines a, b, and c, respectively. In the low-energy region (below 10 MeV/n) estimates have been made from analysis of integral $G$-values by various authors. For H: (●) Hart et al[19]; (dashed curve e) Pucheault and Julien. For D: (○) Hart et al[19]; (dashed curve d) Schuler and Allen. For He: (□) Gordon and Hart; (dashed curve f) Schuler and Allen. (Reproduced with permission.[2])

Hart et al,[19] the dashed curve e is an estimate of Pucheault and Julien[20] from experimental $G(E)$ values for accelerated protons.

Two sets of experimental data furnish information on the location of the curve for He in the low-energy region. The dashed curve f is the estimate by Schuler and Allen[18] of $G'(E)$ based on an analysis of experimental $G(E)$ for accelerated helium ions; the open squares are data points calculated from experimental $G(E)$ values for $\alpha$ particles by Gordon and Hart.[21]

The experimental data points in the low-energy region ($E < 10$ MeV/n) scatter so much that they do not furnish a real check on the model calculations. Although the model is rough and the qualitative aspects of track structure are not well known at low energies, calculated values are perhaps preferable to any particular set of experimental values because of an overall consistency in the former.

The solid curves of Figure 6-8 give $G(E)$-values calculated from the data of Figure 6-7 by using Eq. (6-1). All of the experimental data points plotted for comparison are from direct measurements of $G(E)$-values; data from which the estimates of $G'$-values shown in Figure 6-4 were made are shown; in addition, $G(E)$-values for accelerated protons by Kochanny et al,[22] He

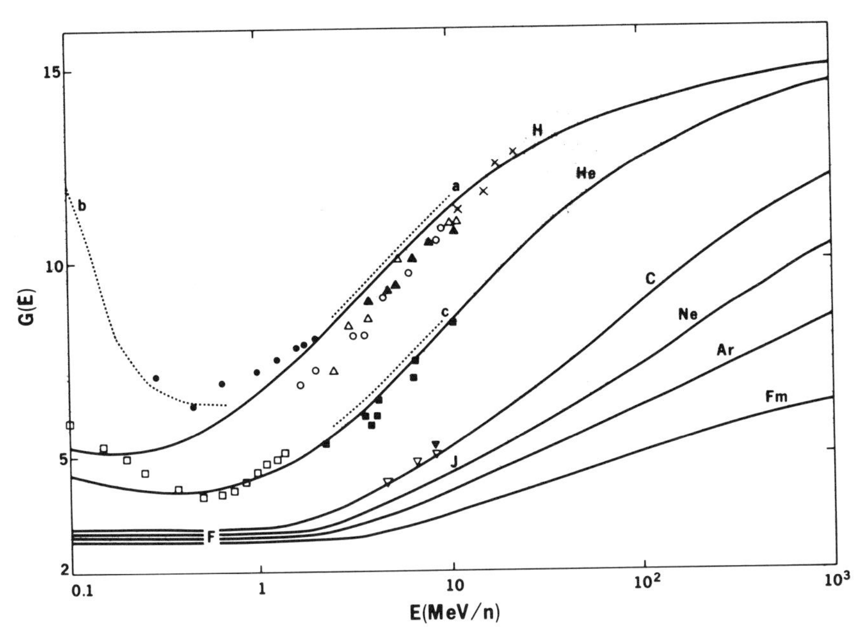


**Figure 6-8.** Calculated integral yields, $G(E)$, for the six representative particles vs. the energy per nucleon, $E$, are given by the solid curves. Experimental measurements of various authors are indicated by the plotted points; dashed curves give some of the authors' estimates of best curves drawn through their data points. For H: (●) Hart et al[19]; (▲) Anderson and Hart; (×) Kochanny et al[22]; (dashed curve b) Pucheault and Julien. For D: (○) Hart et al[22]; (▲) Anderson and Hart; (dashed curve a) Schuler and Allen. For He: (■) Anderson and Hart; (□) Gordon and Hart; (dashed curve c) Schuler and Allen. For C: (▽) Schuler; (▼) Jayko et al[16]. For Ne: (J) Jayko et al.[16] For fission fragments: (F) Bibler. (Reproduced with permission.[2])

nuclei by Anderson and Hart,[23] carbon by Schuler,[24] carbon and neon by Jayko et al,[16] and data on fission fragments by Bibler[25] are given. Data obtained by Matsui et al[26] for H and He are not indicated in the figures; they agree well with the H data of Hart et al[19] and the He data of Anderson and Hart.[23]

The scatter of the experimental data points for $G(E)$ is as large as that for $G'(E)$. There is a suggestion, however, that $G(E)$ rises more at low energies than our calculations show. If it is further verified, such a result could point the way for modifying the track structure model presented here.

## SUMMARY OF TRACK MODELS AND LET EFFECTS IN RADIATION CHEMISTRY

Tracks are composed of cores and penumbras, and the results of radiation chemistry can be explained on this basis. We do not expect that many effects will be simple functions of the LET of particles; all particle tracks have both "low-LET" and "high-LET" portions.

Figure 6-5 shows the differential yields of the Fricke dosimeter for the chemical cores of the various particles used in the calculations of this chapter. There is a common curve when the results are plotted vs. LET *of the chemical core.* Clearly, all of these tracks are essentially the same. Their penumbras, on the other hand, are different, and the plot of the total differential yields vs. the total LET given in Figure 6-9 shows that the same results are not obtained for the various particles.

From Figure 6-9 note that the chemical core for an LET below about 1 eV/Å is effectively a lower-LET track than the penumbra. At these low-energy deposits the core is a string of spurs that do not overlap effectively before they react with scavengers, whereas the penumbras have track ends (blobs or short tracks) that are regions with LET higher than spurs. In this same connection we have noted several times that the spur has a $G$-value of 18.4 for the Fricke dosimeter and when the track end effects are added for megaelectronvolt electrons the yield is reduced to 15.6.

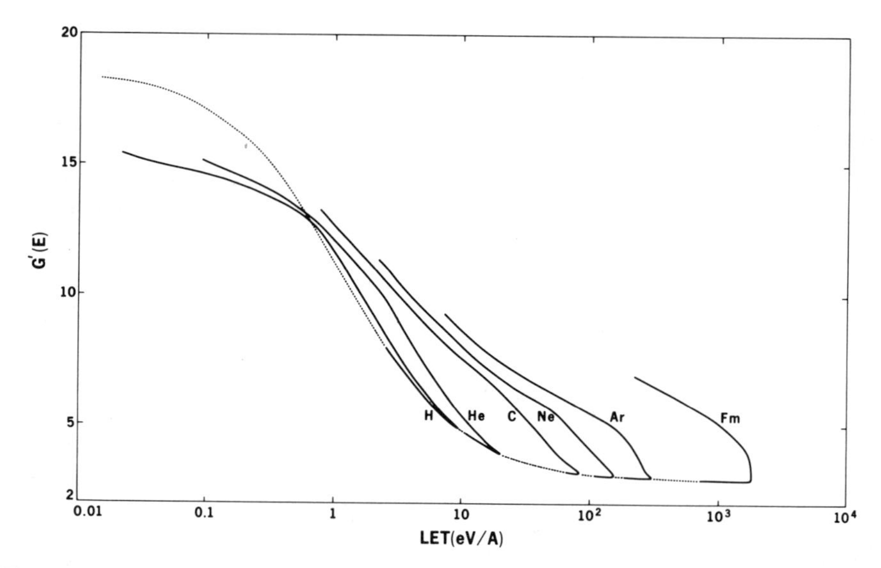


**Figure 6-9.** Calculated differential yields, $G'(E)$, for the six representative particles vs. total LET are given by the solid curves. The calculations for C, Ne, Ar, and Fm are in the range $0.1 \leqslant E \leqslant 100$; the calculation for H and He are in the range $0.001 \leqslant E \leqslant 100$. The dashed curve indicates the common curve obtained under conditions in which the penumbra vanishes. (Reproduced with permission.[2])

In this chapter we have examined track effects in a unified manner and in a way that should allow treatment for any particle. Effects attributed to LET and to radiation quality are obtained from track models without introduction of additional assumptions. The track models themselves can be improved, but the message of this chapter is that the radiation chemistry of particles should be considered as a unified subject.

# REFERENCES


1. Magee, J. L.; Chatterjee, A. *J. Phys. Chem.* **1980**, *84*, 3529.
2. Chatterjee, A.; Magee, J. L. *J. Phys. Chem.* **1980**, *84*, 3537.
3. Magee, J. L.; Chatterjee, A. *J. Phys. Chem.* **1978**, *82*, 2219.
4. Chatterjee, A.; Schaefer, H. J. *Radiat. Environ. Biophys.* **1976**, *13*, 215.
5. Chatterjee, A.; Maccabee, H. D.; Tobias, C. A. *Radiat. Res.* **1973**, *54*, 479.
6. We thank Professor Blann for this personal communication.
7. Pierce, T. E.; Blann, M. *Phys. Rev.* **1968**, *173*, 390.
8. Equation 7 differs from the equation given in the abstract of ref. 7, reflecting revision by the authors of ref. 7.
9. (a) Anderson, H. H.; Ziegler, J. F. In "The Stopping Power and Ranges of Ions in Matter"; Ziegler, J. F., Ed.; Pergamon Press: New York, 1977, Vol. 3. (b) Ziegler, J. F., *ibid.*
10. Ganguly, A. K.; Magee, J. L. *J. Chem. Phys.* **1956**, *25*, 129.
11. Magee, J. L.; Chatterjee, A. *Radiat. Phys. Chem.* **1980**, *15*, 125.
12. Mozumder, A.; Magee, J. L. *Radiat. Res.* **1966**, *20*, 203.
13. Cormack, D. V.; Johns, H. E. *Brit. J. Radiol.* **1953**, *25*, 369.
14. Nelms, A. T. *Natl. Bur. Stand. Circ.* **1953**, No. 524.
15. Fregene, A. O. *Radiat. Res.* **1967**, *31*, 256.
16. Jayko, M.; Appleby, A.; Christman, E; Chatterjee, A; Magee, J. L. Lawrence Berkeley Laboratory, University of California, Berkeley, CA, 1978, Report LBL-7432.
17. Christman, E. A.; Appleby, A.; Jayko, M. *Radiat. Res.* **1981**, *85*, 443.
18. Schuler, R. H.; Allen, A. O. *J. Am. Chem. Soc.* **1957**, *79*, 1565.
19. Hart, E. J.; Ramler, W. J.; Rocklin, S. R. *Radiat. Res.* **1956**, *4*, 378.
20. Pucheault, J.; Julien, R. *Tihany Symp. Radiat. Chem.* **1972**, *2, 1191.*
21. Gordon, S.; Hart, E. J. *Radiat. Res.* **1961**, *15*, 440.
22. Kochanny, G. L., Jr.; Timnick, A.; Hochanadel, C. J.; Goodman, C. D. *Radiat Res.* **1963**, *19*, 462.
23. Anderson, A. R.; Hart, E. J. *Radiat. Res.* **1961**, *14*, 689.
24. Schuler, R. H. *J. Phys. Chem.* **1967**, *71*, 3712.
25. Bibler, N. E. *J. Phys. Chem.* **1975**, *79*, 1991.
26. Matsui, M.; Seki, H.; Karasawa, T; Imamura, M. *J. Nucl. Sci. Technol.* **1970**, *7*, 97.
27. Fricke, H.; Hart, E. J. "Radiation Dosimetry"; Attix, F. H.; and Roesch, W. C., Eds.; Academic Press: New York, 1966, Vol. 2, Chapter 12, pp. 167–239.
28. Donaldson, D. M.; Miller, N. *J. Chim. Phys.* **1955**, *52*, 578.
29. Haybittle, J. L.; Saunders, R. D.; Swallow, A. J. *J. Chem. Phys.* **1956**, *25*, 1213.
30. Sutton, H. C. *Phys. Med. Biol.* **1956**, *1*, 153.

# 7

# The Electron: Its Properties and Reactions

R. A. HOLROYD

*Department of Chemistry, Brookhaven National Laboratory*

## INTRODUCTION

The electron is universally produced as an intermediate in the radiolysis of all gases, liquids, and solids. It is therefore of prime importance to learn about the chemistry of the electron; the subsequent reactions of this species will determine a significant fraction of the overall radiation chemical changes. It is also important to understand that while in some cases electrons may live for only a few picoseconds before geminate recombination occurs, in other cases free electrons may live for milliseconds or longer in nonpolar liquids.

This chapter begins with a brief account of the processes occurring subsequent to ionization, involving thermalization of the electron, geminate recombination of the electron with ions, and escape to form free electrons. The times involved, as established by experiment, as well as the reactions that can occur in each of these stages are discussed. To understand the unusual chemical behavior of the electron it is important to first understand its physical properties: drift mobility and energy levels. How these properties are measured is described in each case. The discussion shows how temperature, density, phase, and molecular structure affect each property, and


*Radiation Chemistry: Principles and Applications*

the existing data are interpreted in terms of a modern theoretical framework. Finally, the recombination and attachment reactions of electrons are described and related to the physical properties. The story is incomplete in several aspects. One is that much remains undiscovered about the reactions and behavior of electrons. Also, space allows only brief generalizations and summaries; for more detailed information the student is referred to several excellent reviews on the subject.[1–6]

# DYNAMICS OF IONIZATION

The role of electrons in the radiation chemistry of liquids can be better understood if we have a mental picture of the overall ionization process. The early events leading to positive ions and electrons have been described (see Chapters 1 and 2).

## *Thermalization*

In general, the electron will have some small kinetic energy left over from the ionization event, and as a consequence of this energy the electron can separate from the positive ion. During separation and while still hot, the electron may undergo chemical reactions as well as energy loss processes that cause it to become thermalized. Energy sufficient for electronic excitation will be quickly lost to nearby molecules. Thus the electron has only a few tenths of an electronvolt of energy during most of the separation process and is often dubbed a *subexcitation* electron. Eventually, after numerous elastic and inelastic collisions, the electron will thermalize.

The time required for thermalization (defined as the time when the electron's energy is 10% above thermal) has been measured and is several nanoseconds for the rare gas liquids (see Table 7-1). The thermalization

**Table 7-1.** Electron Thermalization

| Liquid | Thermalization Time (ps) | | Thermalization Distance (Å)[c] |
|---|---|---|---|
| | Calculated[a] | Measured[b] | |
| Xenon | 5500 | 6500 | 700 |
| Krypton | 2600 | 4400 | 900 |
| Argon | 1000 | 900 | 1600 |
| Neopentane | 1.8 | — | 212 |
| 2,2,4-Trimethylpentane | <0.8 | — | 90 |

[a] Ref. 1.
[b] From time-resolved experiment, ref. 6.
[c] *b*-value determined from free-ion yields in Table 2.

time is much shorter (about a picosecond) in the molecular hydrocarbon liquids according to theoretical estimates because the electron can also lose energy to vibrational modes. The longer times in Xe, Kr, and Ar come about because only elastic collisions are available for energy loss. In liquids composed of unsymmetrical or polar molecules the thermalization times are expected to be much less than a picosecond.

On thermalization the electron will have separated some distance from the positive ion. This distance is about 20 Å for polar liquids, between 50 and 200 Å for hydrocarbons, and longer for rare gas fluids (Table 7-1, last column). Of course, these are representative distances; in reality, a distribution of distances is expected. Thus, in summary, the hot electron released in an ionization event is transported some distance from the cation, and a significant time is required for thermalization.

An important question regarding thermalization is, Can chemical reactions occur while the electron is hot? Evidence supporting such reactions comes from free-ion yields (see below). The addition of electron-attaching solutes like $SF_6$, $CCl_4$, and $N_2O$ in small concentrations (0.01 $M$) reduces free-ion yields for 2-MeV x-ray radiolysis substantially.[7,8] The free-ion yield is a sensitive measure of the separation distance, $r$, of the electron from the positive ion. Thus, this reduction indicates that subexcitation electrons react before thermalization with such solutes and are thereby trapped close to the positive ion, which reduces the escape probability. An additional source of information is the observation that perfluoroalkanes reduce the yield of photoionization, at photon energies above the threshold energy, from tetramethyl-*p*-phenylenediamine in nonpolar liquids.[9] Thus, in a manner similar to the behavior of the above solutes, perfluoroalkanes attach subexcitation (or epithermal) electrons, thereby decreasing the ion-pair separation distance ($r$) at thermalization and hence the escape probability.

## Free-Ion Yields

Upon thermalization the electron is separated from its geminate positive ion by a distance, $r$. At this distance the electron and ion are coulombically bound together. A free electron is defined as one that has escaped its counterion. The probability of escape ($P_{esc}$) depends both on the dielectric constant of the liquid, $\varepsilon$, and on $r$ according to Eq. (7-1):

$$P_{esc} = e^{-r_c/r} \qquad \text{where } r_c = \frac{e^2}{\varepsilon k T} \tag{7-1}$$

The effects of these parameters for two typical cases are illustrated in Figure 7-1. For a hydrocarbon of low dielectric constant, $P_{esc}$ is negligible for $r \leqslant 40$ Å and rises to only 0.2 at 200 Å, whereas for a polar liquid like water $P_{esc}$ is already 0.84 at 40 Å.

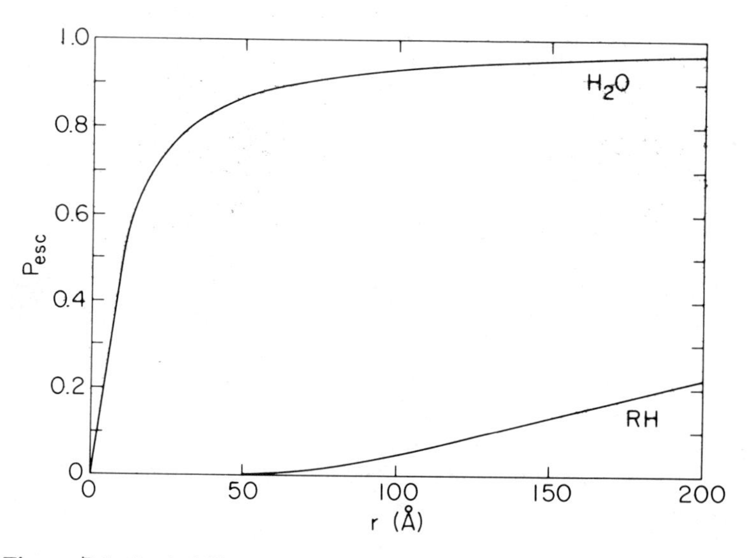


**Figure 7-1.** Probability of escape ($P_{esc}$) vs. ion separation distance ($r$).

Accurate determination of free-ion yields usually are made by conductivity experiments but can also be obtained by pulse radiolysis or product analysis.[10] Free-ion yields for representative liquids are shown in Table 7-2. The yields for nonpolar liquids tend to increase with the sphericity of the molecules, being highest for neopentane. The fact that the free-ion yield for *cis*-butene-2 is greater than that for the isomeric 1-butene is attributed to the sphericity of the former.[11]

**Table 7-2.** Zero-Field Free-Ion Yields at 296 K[a]

| Liquid | $G_{fi}$ |
|---|---|
| Neopentane | 1.1 |
| 2,2,4,4-Tetramethylpentane | 0.83 |
| 2,2-Dimethylbutane | 0.30 |
| 2,2,4-Trimethylpentane | 0.33 |
| Cyclohexane | 0.15 |
| *n*-Pentane | 0.15 |
| *n*-Hexane | 0.10 |
| *n*-Decane | 0.12 |
| 1-Butene | 0.09 |
| *cis*-Butene-2 | 0.23 |
| Hexyne-1 | 0.10 |
| Benzene | 0.05 |
| $CS_2$ | 0.31 |
| Perfluoro-*n*-pentane | 0.035 |
| Ammonia | 3.2 |
| Ethyl alcohol | 1.8 |

[a] Ref. 10.

The free-ion yield may be used to derive separation distances with Eq. (7-1). Some distribution of distances $F(r)$ is assumed; recent studies of photoionization indicate that an exponential distribution is appropriate.[12] Then the free-ion yield is given by

$$G_{\text{fi}} = G_{\text{total}} \int_0^{\infty} F(r) P_{\text{esc}} \, dr \tag{7-2}$$

## Geminate Recombination

The probability that an electron will escape geminate recombination in a dielectric liquid is given by Eq. (7-1). This is a steady-state yield and gives no information about the yield at short times nor about the time dependence of the geminate population. Information about the time dependence is available from theory, product yields, and pulse radiolysis.

A theoretical description of the time dependence requires a solution of the differential equation that describes the diffusional motion of two oppositely charged species in a dielectric liquid due to their mutual coulombic field. Early solutions to this problem involved the method of prescribed diffusion[13] or numerical integration;[14,15] more recently, an exact analytical solution has been obtained.[16] This latter solution gives the probability of survival, as a function of time, of an ion-pair separated by a distance, $r$. In practice, a distribution of distances needs to be considered. The theory predicts that the time at which the survival probability is 50% is between 0.4 and 28 ps for typical dielectric liquids (see Table 7-3). This theory also shows that geminate recombination is characterized by a fast initial decay followed by a long tail of recombining pairs.

**Table 7-3.** Geminate Recombination Times

| | Recombination Time (ps)[a] | |
|---|---|---|
| Liquid | Experimental[b] | Theory[c] |
| | *Electrons* | |
| Neopentane | — | 0.4 |
| 2,2,4-Trimethylpentane | 0.7 | — |
| Cyclohexane | 2.8 | 9.0 |
| $n$-Hexane | 15.0 | 28.0 |
| | 9.0 | |
| | *Ions* | |
| $CCl_4$[d] | — | 680.0 |

[a] Defined as time required for 50% to recombine.
[b] Data from refs. 1 and 17.
[c] Refs. 15 and 16.
[d] Ref. 18.

Another source of information on geminate recombination is from measurements of yields of products formed in reactions of the geminate ions with solutes: for example, from yields of methyl radicals from added methylbromide. Experimentally it is found for many solutes that the yield of product $G_i(R)$ is given by a square root dependence:[1,19]

$$G_i(R) = G_{fi} + G_{gi}\frac{(\alpha_s[S])^{1/2}}{1 + (\alpha_s[S])^{1/2}} \tag{7-3}$$

where $G_{fi}$ is the free-ion yield, $G_{gi}$ the geminate yield, and [S] the concentration of solute. The parameter $\alpha_s$ is the product of the geminate lifetime with the rate constant for reaction of the ion with S. Thus, where such rates have been measured, the lifetime may be calculated from values of the reactivity parameter $\alpha_s$. The results of such a calculation (Table 7-3) roughly agree with theoretical calculations.

The third approach is to experimentally time-resolve the geminate decay by following the absorption of the ions in a pulse radiolysis experiment. This has been done for *n*-hexane and cyclohexane, and for the former a "geminate spike" of fast decaying ions is observed immediately following a 30-ps pulse of electrons.[20] The data for longer times (>120 ps) fit a $t^{-0.5}$ dependence in accord with theoretical expectations. So far, pulse experiments have only time-resolved the tail of the geminate decay. Much shorter pulses and better time resolution are needed to observe the decay of the majority of the geminate ions. A recent study[17] of geminate recombination following photoionization of anthracene in *n*-hexane showed that 50% of the cation-electron pairs recombine in 9 ps or less. Furthermore, the temperature dependence showed that the lifetime depends inversely on electron mobility, as expected from theory.

More success has been obtained in time-resolving the decay of geminate solute *ions*. The first observation was for solutions of biphenyl (0.1 *M*) in cyclohexane.[21] The biphenyl anion is formed by reaction with the electron, and the yield of anions after a 12-ns pulse was found by pulse radiolysis to be $G = 1.1$. The biphenyl anion decayed rapidly over 50 ns. A more recent study, utilizing microwave conductivity to detect the geminate ions in $CCl_4$,[18] found that the results at 20° fit the equation

$$G(\tau) = 0.095(1 + 0.6\tau^{-0.6}) \tag{7-4}$$

where $\tau = Dt/r_c^2$ and $r_c = 2.56 \times 10^{-6}$ cm. From this equation the time required for 50% of an initial yield of $G = 5.0$ ion pairs per 100 electronvolts to recombine is 680 ps; the yield at 12 ns would be $G = 0.5$. Although Eq. (7-4) is only truly correct for longer times, it does give an idea of the time dependence of the geminate ion population in $CCl_4$.

# ELECTRON MOBILITY

Many of the unusual chemical properties of the electron can be better understood through a knowledge of its physical properties. These include the drift mobility and the energy level structure of the electron in the liquid. In 1969 it was discovered that excess electrons in nonpolar liquids are from 100 to 100,000 times more mobile than ions in the same liquids. Thus, under the influence of an applied electric field, the drift velocities of electrons are correspondingly greater than the drift velocities of ions. Mobility is defined simply as the ratio of the drift velocity to the applied electric field,

$$\mu = \frac{v_d}{E} \tag{7-5}$$

and is related to the more familiar unit of equivalent ionic conductivity ($\lambda$) encountered with electrolyte solutions. The units of mobility are $cm^2/V\text{-}s$ whereas the units of $\lambda$ are $cm^2$/ohm-gram equivalent. The ratio of the two

**Table 7-4.** Electron Mobilities and Band Energies at 295 K

| Nonpolar Liquids | $\mu$ ($cm^2\ V^{-1}\ s^{-1}$) | $E_a$ (eV) | Ref. | $V_0$ (eV) | Ref. |
|---|---|---|---|---|---|
| Tetramethylsilane | 100 | — | 22 | −0.55 | 23 |
| Neopentane | 70 | — | 24 | −0.43 | 23 |
| 2,2,4,4-Tetramethylpentane | 24 | 0.06 | 25 | −0.33 | 23 |
| 2,2-Dimethylbutane | 12 | 0.07 | 26 | −0.20 | 23 |
| 2,2,4-Trimethylpentane | 5.6 | 0.05 | 23,27 | −0.24 | 28 |
| Cyclopentane | 1.1 | — | 24 | −0.19 | 23 |
| Cyclohexane | 0.24 | — | 27 | +0.01 | 23 |
| Ethane | 28.0 | — | 29 | +0.12[a] | 23 |
| *n*-Pentane | 0.15 | 0.20 | 27 | 0 | 23 |
| *n*-Hexane | 0.093 | 0.16 | 27 | +0.1 | 23 |
| *n*-Decane | 0.038 | 0.22 | 27 | +0.18 | 23 |
| *n*-Pentadecane | 0.03 | 0.16 | 30 | — | — |
| 1-Butene | 0.064 | — | 23 | — | — |
| *cis*-Butene-2 | 2.2 | 0.16 | 23 | −0.16[b] | 23 |
| *trans*-Butene-2 | 0.029 | 0.23 | 23 | — | — |
| 2,3-Dimethylbutene-2 | 5.8 | — | 23 | −0.25 | 23 |

| Polar Liquids | $\mu$ | Ref. | Band[c] Energy | Ref. |
|---|---|---|---|---|
| $H_2O$ | $1.8 \times 10^{-3}$ | 31 | − 1.14 | 32–35 |
| $NH_3$ | $1.0 \times 10^{-2}$ | 36 | − 1.13 | 37,38 |
| $CH_3OH$ | $6.0 \times 10^{-4}$ | — | − 1.02 | 33 |
| $(C_2H_s)O$ | $5.1 \times 10^{-3}$ | 39 | — | — |

[a] $V_0$ measured at 271 K.
[b] At 243 K.
[c] See text for definition.

is thus the Faraday:

$$\frac{\lambda}{\mu} = 96{,}500 \text{ coulombs/gram equivalent} \tag{7-6}$$

It is instructive to compare the mobility of electrons in familiar conductors, making use of Eq. (7-6), with excess electron mobility. Thus, as an example the equivalent conductivity ($\lambda$) of copper is $4.1 \times 10^6$ cm$^2$/ohm-gram equivalent. From Eq. (7-6) the mobility of electrons in copper is 42 cm$^2$/V-s, a value similar to the mobility of excess electrons in some hydrocarbons.

Data on the mobility of excess electrons are now available for more than 50 liquids. These include alkanes, alkenes, cycloalkanes, benzene, methyl substituted benzenes, various tetraalkylsilanes, as well as liquid rare gases. A selected list is shown in Table 7-4 giving measurements at 295 K.

## *Measurement of Mobility*

Most methods of determining the electron mobility are pulse techniques in which the concentration of electrons is followed during or after the pulse. Either a short pulse of x-rays may be used to create excess charges or, alternatively, electrons can be produced by an ultraviolet pulse that either ejects photoelectrons from a metal surface or photoionizes a solute like tetramethyl-*p*-phenylenediamine (TMPD).[27] In the latter case it is important that the solute not react too rapidly with the electron. The concentration of electrons is determined either by measuring the current due to the electron drifting in the applied electric field between two electrodes or by microwave absorption.[40]

The mobility of electrons can be derived in several ways. The most direct is the time-of-flight method in which a pulse of radiation produces ionization and the space between the electrodes is uniformly filled with electrons. The electrons drift toward the positive electrode, producing a current in the external circuit. A "break" in the current-vs.-time curve is observed when all of the electrons have reached the anode. The time at which this break occurs after the pulse is the drift time, $t_d$, and the mobility is related to it by

$$\mu = \frac{d^2}{t_d V} \tag{7-7}$$

where $d$ is the electrode spacing and $V$ is the applied voltage.

Another method is to measure the change in conductivity in a sample induced by a pulse of radiation. The conductivity is proportional to the product of the mobility and the concentration of electrons. To use this method requires that the total number of electrons created be independently determined; this involves a knowledge of both the radiation dose and the yield ($G$) of ion pairs per 100 electronvolts.

### Solvent Effects

A wide range of electron mobilities is observed, depending on the solvent. An obvious dependence of electron mobility on the structure of the molecules may be noted. For example, the mobility in *n*-alkanes decreases rapidly with increasing chain length up to *n*-hexane, and then there is very little decrease beyond *n*-decane. The mobility invariably increases with branching, being larger for the more spherical molecules; the mobility in neopentane is 400 times that in *n*-pentane. The mobility in the isomeric butenes ranges from 0.029 $cm^2/V$-s in *trans*-butene-2 to 5.8 $cm^2/V$-s in 2,3-dimethylbutene-2. Again the mobility is highest for the more branched and spherelike of the olefins.

A rough correlation of the room temperature mobility with the mean separation distance between geminate ions (see above) has been noted.[25] These two effects are correlated because the separation of geminate ions involves the transport of epithermal electrons and the mobility involves the scattering of thermal electrons. However, the nature of the liquid has the biggest influence on mobilities that differ by a factor of over 2000 (from 0.03 to 70) for hydrocarbons; whereas, the separation distances differ by only a factor of 5 (from 40 to 200 Å) for these compounds.

### Polar Liquids

Information on the mobility of solvated electrons ($e_{sol}^-$) is available for a few polar liquids. The mobilities are generally slightly greater than ionic mobilities, which are about $7 \times 10^{-4}$ $cm^2/V$-s.[1] For water the observed mobility of $e_{aq}^-$ is $1.8 \times 10^{-3}$ (see Table 7-4). This corresponds to an equivalent ionic conductivity of 174 $cm^2$/ohm, which is higher than the conductivity of many ions in water but comparable to that of $OH^-$. For ammonia the mobility of $e_s^-$ is even higher. Viscosity has a bearing on the mobility of solvated electrons because mobility increases with temperature, as does the mobility of ions. However, the transport is not totally ionic; the results suggest the electron can jump or tunnel from its cavity to a neighboring site. In ethers, which are only slightly polar, the electron mobility is larger, and migration via the conduction band is indicated.[39]

### Temperature Effects

For most of the hydrocarbon liquids the electron mobility increases with temperature, and the temperature dependence is described by an Arrhenius-like equation

$$\mu = \mu_0 \, e^{-E_a/kT} \tag{7-8}$$

where $\mu_0$ has values around 100 $cm^2/V$-s, and typically the activation energy is not more than a few tenths of an electronvolt (Table 7-4). Over a larger temperature range Eq. (7-8) is not strictly valid, however. An Arrhenius plot of the mobility for ethane is concave downward, whereas for many other liquids, such as $n$-pentane, cyclopentane, and neohexane, such plots are concave upward when high-temperature data are included.[4,41]

## Density Effects

As the temperature increases, the density of a liquid decreases, and therefore the increase in mobility with increasing temperature may be due in part to the density decrease. Very complete density studies have been made for the inert gases and several hydrocarbons,[41–45] where the mobility has been measured as a function of density from the dilute gas to liquid density. In the case of symmetrical molecules like methane[45] and neopentane, the mobility goes through a pronounced maximum at an intermediate density, usually somewhat above the critical density (Figure 7-2). The positions of these maxima correlate with the position of conduction-band minima, an important fact in the understanding of electron transport (see below). Other hydrocarbons that are less symmetrical usually show only small maxima or inflections in the $\mu$-vs.-density plots.[44]

Similar maxima in $\mu$ are observed for Ar, Kr, and Xe. In these fluids the mobilities at the maxima are an astonishing 1700, 2300, and 3900 $cm^2/V$-s, respectively. Here, again, the positions of the maxima correlate with minima in the conduction-band energy.[46] Figure 7-3 shows this correlation for argon. The evidence indicates the electron is quasi-free in these fluids, a conclusion supported by the observation that the conduction band is below any trapped state that may exist.

The situation is quite different in liquid He and Ne.[52] Figure 7-4 shows the density dependence of the mobility in liquid He at temperatures near 4 K. Like most gases at low density, below $10^{21}$ $cc^{-1}$, the mobility is proportional to $1/N$ and the electron is quasi-free. However, at an intermediate density the mobility drops by four orders of magnitude, and in the liquid $\mu = 0.02$ $cm^2/V$-s. Neon is similar with a mobility of 0.0016 $cm^2/V$-s at 25 K.[53] The explanation for these low mobilities in He and Ne is that the electron is trapped in these liquids in *bubbles* (see the section "Trapped Electrons" below). The trapped or bubble state is well below the conduction band, and quasi-free transport is almost impossible.

## Electric Field Effects

As indicated in Eq. (7-5) the drift velocity of an electron is directly proportional to the electric field. However, this proportionality breaks down at high fields, generally when the drift velocity is greater than $10^5$ cm/s. For high-mobility liquids like argon and neopentane the drift velocity at high

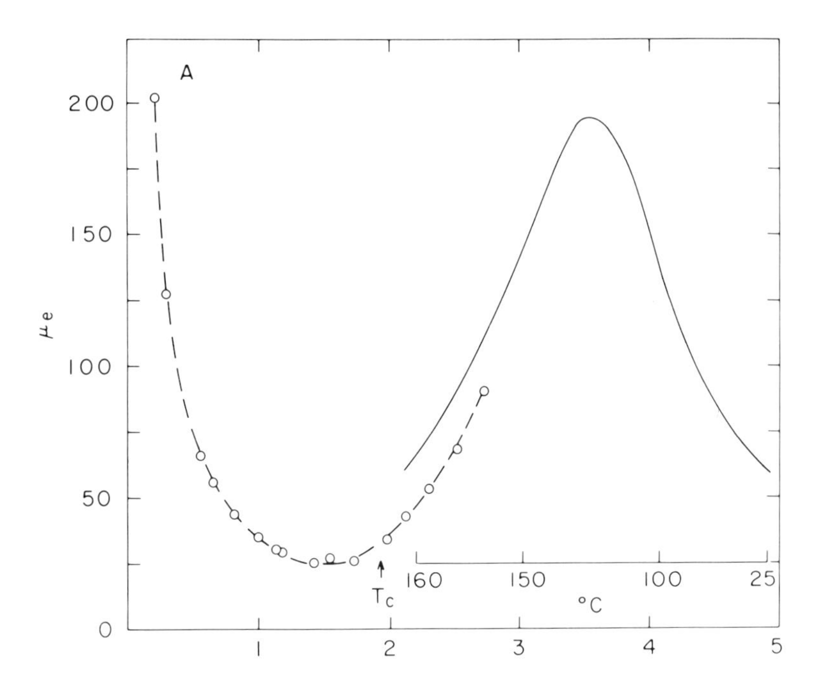


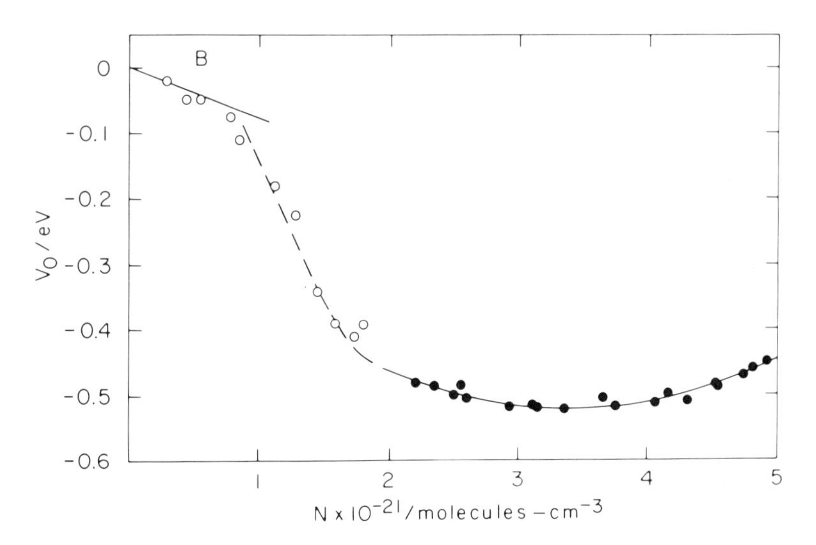


**Figure 7-2.** (A) Electron mobility in $cm^2\ V^{-1}\ s^{-1}$ vs. number density in neopentane; ○ stands for gas phase data at 161°C (±1°C); solid line — ref. 41; (B) Work function shifts ($V_0$) vs. number density in neopentane; ○ stands for gas phase at 166°C (±1°C); ● stands for liquid phase data. (Reproduced with permission.[43b])

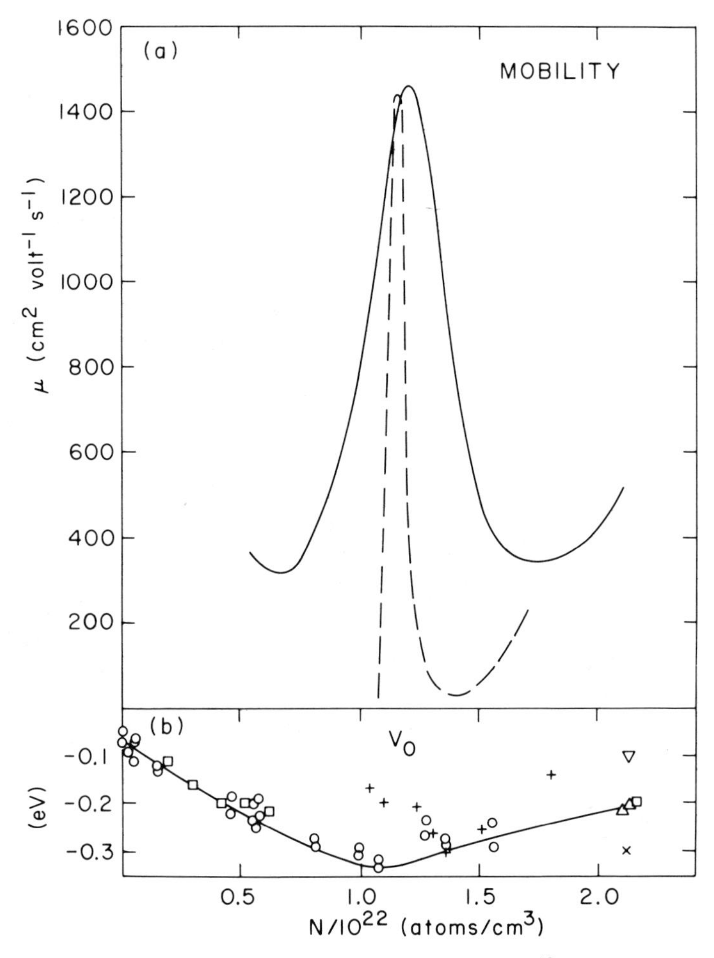


**Figure 7-3.** (a) Electron mobility vs. atom density in argon;[42] (b) $V_0$ vs. atom density in argon. ○;[46b] ×;[47] △,▽;[48,49] □;[50] +.[51]

fields is sublinear and increases as $E^{1/2}$. The explanation of this phenomena is that electrons pick up energy from the field during the time between collisions, and thus their mean energy is greater than thermal.[36] The opposite field effect is observed for low-mobility liquids like pentane and cyclopentane; at high fields the drift velocity increases faster than that given by Eq. (7-5); that is, a supralinear dependence is observed. This has been attributed to the fact that since electrons are trapped in low-mobility liquids the high electric field can increase the tendency of the electron to hop or jump from trap to trap.[36]

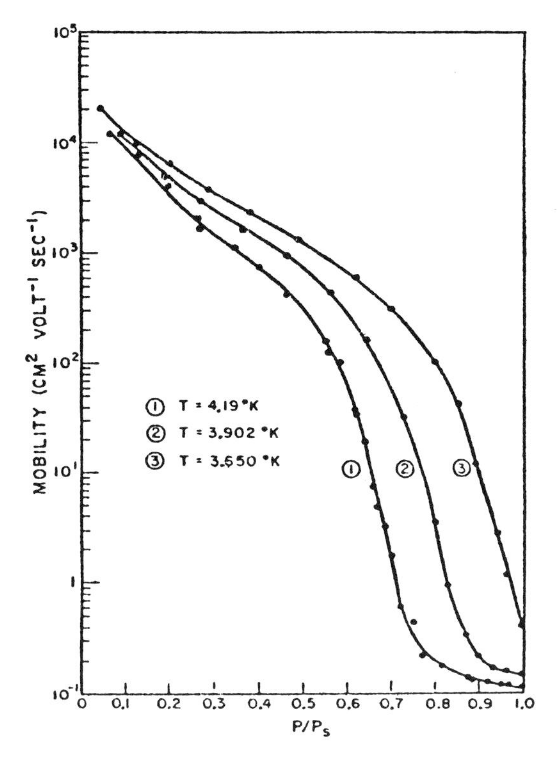


**Figure 7-4.** Mobility vs. pressure of He at constant temperature. The solid curves have no theoretical significance. (Reproduced with permission.[52])

## *Solids and Glasses*

Only limited data exist on the mobility of electrons in solids, and it is therefore risky to generalize. The results available (Table 7-5) indicate that for liquids consisting of spherical molecules the mobility increases on freezing. For neopentane the increase is a factor of 3, and for Ar, Kr, and Xe, increases by similar factors are observed. In contrast, for low-mobility liquids like cyclohexane and 3-methylpentane the mobility is less in the solid than in the liquid.

At very low temperatures, below 35 K, the mobility is constant, independent of temperature; in both 3-methylpentane and 2-methyltetrahydrofuran $\mu$ is about 0.02 $cm^2/V$-s below 35 K. Above this temperature the mobility increases in these compounds. In alkaline ice the low field mobility is 1.8 $cm^2/V$-s at 80 K and decreases as the temperature is increased to 120 K.[64] In pure ice the electron mobility is reported to be much higher (see Table 7-5).

**Table 7-5.** Low-Temperature Data on $\mu$ and $V_o$

| | Mobility | | | Conduction Band | | | |
|---|---|---|---|---|---|---|---|
| | $\mu$ (cm$^2$ V$^{-1}$ s$^{-1}$) | Temp. | Ref. | $V_0$ | Temp. | Method[a] | Ref. |
| *Liquids* | | | | | | | |
| Methane | 400 | 111 | 23 | −0.25 | 100 | $\phi$ | 49 |
| Helium | 0.02 | 4.2 | 23 | +1.05 | 4.2 | $\phi$ | 23 |
| Neon | 0.002 | 25 | 23 | +0.5 | 22 | — | 54 |
| Argon | 475 | 85 | 23 | −0.2 | 84 | $\phi$ | 49 |
| | | | | −0.33 | 84 | $\phi$ | 23 |
| Krypton | 1300 | 120 | 23 | −0.45 | 123 | $\phi$ | 49 |
| | | | | −0.52 | 118 | $\phi$ | 48 |
| Xenon | 2200 | 163 | 23 | −0.66 | 162 | $\phi$ | 48 |
| *Solids* | | | | | | | |
| Argon | 1000 | 82 | 23 | +0.3 | 6 | E | 55 |
| | | | | 0 | 82 | $\phi$ | 48 |
| Krypton | 3700 | 113 | 23 | −0.25 | 20 | E | 55 |
| | | | | −0.4 | 114 | $\phi$ | 48 |
| Xenon | 4500 | 157 | 23 | −0.46 | 40 | E | 55 |
| | | | | −0.57 | 160 | $\phi$ | 48 |
| 3-Methylpentane | 0.02 | 35 | 56 | — | — | — | |
| Cyclohexane | 0.07 | 275 | 57 | +0.48 | 279 | PI | 58 |
| Neopentane | 170 | 253 | 59 | +0.34 | 77 | PI | 60 |
| 2,2,4-Trimethylpentane | — | — | — | +0.46 | 166 | PI | 58 |
| | — | — | — | 0.40 | 80 | ET | 61 |
| *n*-Hexane | — | — | — | +0.98 | 77 | PI | 60 |
| Ethanol | — | — | — | +0.24 | 77 | PI | 62 |
| Methanol | — | — | — | −0.01 | 77 | PI | 62 |
| Water | 25 | 208 | 63 | — | — | — | — |
| 10 *M* NaOH | 1.8 | 80 | 64 | −0.1 | 77 | PI | 65 |
| | | | | −0.36 | 77 | PI | 62 |

[a] $\phi$=work-function shift; E=exciton spectra; PI=photoionization; ET=electron transmission.

## *Hall Mobility*

Measurements of the Hall mobility of excess electrons in hydrocarbon liquids are few, reflecting the extreme experimental difficulties. What measurements are available provide considerable insight into electron transport in these liquids. The Hall mobility, $\mu_H$, which is determined in crossed electric and magnetic fields, measures directly the scattering of the quasi-free electron moving in the conduction band. This is in contrast to the drift mobility, $\mu$, discussed above, which reflects both scattering of the electron as well as time spent in traps. The Hall mobilities of the electron in neopentane[66] and tetramethylsilane[67] were recently found to be 90 and 124 cm$^2$/V-s, respectively. These values are 25 to 70% higher than the drift mobilities; they demonstrate that traps are only of small importance in these liquids and support the notion that excess electrons move in conduction bands. The electron Hall mobility has also been measured in alkaline ice and is $4.7\ cm^2/V\text{-}s$ at 80 K, where $\mu$ is $1.8\ cm^2/V\text{-}s$. The temperature and field dependence indicate that electron transport is also characteristic of a band model in this solid.[64]

# THE CONDUCTION BAND

Excess electrons exist in liquids in a variety of states. These may be classified broadly into either localized or extended states, depending on the size of the electron wave function. The lowest extended state is at the bottom of the conduction band, and the energy of this state, defined relative to vacuum, is designated by $V_0$. This state is empty in a pure liquid and is only occupied when excess electrons are present. When $V_0$ is negative, the injection of an electron into the liquid from vacuum is exothermic; that is, the liquid can be considered to have an electron affinity. An example is liquid argon for which $V_0$ is $-0.3$ eV. If $V_0$ is positive, injection is endothermic, which is the case for liquid He, where $V_0 = +1.05$ eV.

Generally, values of the conduction-band energies for nonpolar liquids are between $-0.6$ and $+0.2$ eV at room temperature (see Table 7-4). Straight-chain $n$-alkanes have high band energies, and the value of $V_0$ tends to increase as the length of the chain increases. Branching lowers $V_0$, and the lowest values are observed for the most symmetrical molecules like neopentane and $Si(CH_3)_4$.

## *Methods of Measurement*

Several methods have been employed to measure the conduction-band energy in fluids. The most common is the photoelectric method in which the work function, $\phi$, of a metal is measured first in vacuum and then in the liquid of interest. The difference is the band energy, $V_0$:

$$V_0 = \phi_{liq} - \phi_{vac} \tag{7-9}$$

According to photoelectric theory, at photon energies above the work function electrons emerge from the metal with excess kinetic energy. In a liquid these electrons lose energy as a result of scattering events and thermalize at a distance from the metal surface. The electrons are subject to the image potential in the metal but can escape the field by diffusion and be measured as a photocurrent. The situation is very similar to the geminate recombination and escape described earlier. This method has been applied to numerous hydrocarbons[68] as well as to polar fluids.[69] Most of the $V_0$ values in Table 7-4 were obtained by this method.

Another method used to measure $V_0$ is photoionization of a solute. The energetics of ionization change in solution because the electron needs only enough energy to reach the conduction band ($V_0$) and because the positive ion formed polarizes the solvent and reduces the energy requirement by $P_i^+$. In general, the ionization threshold, $E_i$, in solution is given by

$$E_i = \text{I.P.} + P_i^+ + V_0 \tag{7-10}$$

where I.P. is the gas-phase ionization potential. To calculate $V_0$ from threshold measurements requires that $P_i^+$ be known or calculated. The same general concept applies to photoionization of a pure liquid. The energy required to remove an electron from a molecule of the liquid, frequently called the band-gap energy, $E_g$, is given by

$$E_g = \text{I.P.} + P^+ + V_0 \tag{7-11}$$

A method that allows determination of $V_0$ directly is the combined detection of exciton spectra and photoemission threshold. This method has been applied primarily to rare gas solids.[55] In photoemission the electron is detected external to the solid, and the positive ion is left in the solid. Thus, for a solute, i, the threshold for ionization $E_i^{th}$ is given by

$$E_i^{th} = \text{I.P.} + P_i^+ \tag{7-12}$$

If on the same sample, the energy gap $E_i$ is also measured, for example by detection of the limit of excitation levels, then combining Eq. (7-10) and (7-12) gives $V_0$ directly:

$$V_0 = E_i - E_i^{th} \tag{7-13}$$

Results of this method are shown in Table 7-5 for the rare gas solids.

## *Correlation of $V_0$ with Mobility*

There is a general correlation of the value of $V_0$ at 295 K with the mobility ($\mu$) of the electron in nonpolar liquids at that temperature. This correlation can be seen by inspecting Table 7-4. Those liquids for which the electron mobility is high have low values of $V_0$. As $V_0$ increases, the mobility *generally* decreases. This relation of $\mu$ and $V_0$ has been attributed to the role of traps: the trap energy is assumed to be linearly related to $V_0$.[70,71] That is, the higher the conduction band, the deeper the traps, the greater the activation energy, and therefore the lower the mobility. This dependence of $\mu$ on $V_0$ for hydrocarbons is roughly given by the empirical equation[71]

$$\mu = \frac{125}{1 + 360e^{15V_0}} \tag{7-14}$$

which is approximately correct for most hydrocarbons at room temperature.

## Temperature Effects

According to theory,[72] $V_0$ consists of two terms: $V_0 = \underline{T} + \underline{U}$, where $\underline{U}$ is the average polarization energy, which can be calculated from the molecular polarizability and the density, and $\underline{T}$ is the kinetic energy of the electron in the potential well. This latter term is like the energy level of an electron in a box, where the walls of the box are in this case molecules of the liquid. In the theory the molecules are considered as hard spheres, and if the molecular separation increases, on average, the "box" in which the electron moves is larger and $V_0$ would be expected to decrease because $\underline{T}$ decreases. Experimentally, $V_0$ has been observed to decrease with increasing temperature for several liquids, as shown in Figure 7-5. This decrease is related to the decrease in density, which increases the average separation between molecules. The solid lines in the figure are theoretical and agree well with the experimental data.

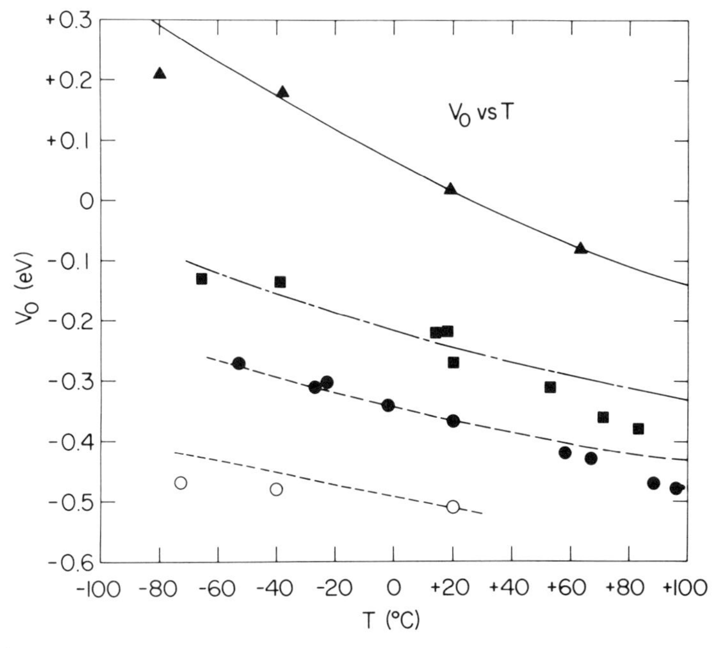


**Figure 7-5.** Plot of $V_0$ vs. temperature. Points are experimental; lines are calculated from Wigner–Seitz method: *n*-hexane (▲, —); 2,2,4-trimethylpentane (■, – – –); 2,2,4,4-tetramethylpentane (●, — — —); tetramethylsilane (○, ——). Reproduced with permission.[28]

## Density Effects

Because $V_0$ decreases as the density decreases and because $V_0$ goes to zero as the density approaches zero for liquids, there must be a minimum in $V_0$

at some intermediate density. These minima have been experimentally observed in several liquids including ethane,[73] propane,[74] neopentane,[43] $Si(CH_3)_4$,[43] Kr,[46a] and Xe.[46a] The minima occur at a density between 60 and 80% of the normal liquid density. The results for neopentane are shown in Figure 7-2. At the density at which the conduction band goes through a minimum, the mobility in neopentane goes through a maximum. Similar results have been obtained for $Si(CH_3)_4$, Ar, Kr, and Xe. The maximum in the mobility is attributed to the fact that potential fluctuations are minimal and scattering of the electrons is less at this density (see deformation potential model below).

Because theory predicts that the conduction level will decrease with decreasing density, it conversely predicts that $V_0$ will increase with increasing density. This is verified by low-temperature studies of photoionization in hydrocarbons[58] and by studies of solid rare gases.[48,55] As the temperature is lowered, the ionization threshold of TMPD in hydrocarbon solvents increases[58] as does the work function of a metal immersed in a hydrocarbon.[28] Both of these phenomena indicate that $V_0$ is higher in the more dense, low-temperature fluid. Also, on solidification there is a further density increase, and $V_0$ increases further. By comparing Table 7-4 and 7-5 for the hydrocarbons we see that $V_0$ is approximately 0.8 eV higher at 77 K than at 295 K. These results have been confirmed by transmission spectra of low-energy electrons through thin films at 80 K.[61] The $V_0$-values determined by transmission agree with the values determined by photoionization. Similarly, for the rare gas solids $V_0$ is higher than for the rare gas liquids (see Table 7-5). For example, for argon $V_0$ is −0.20 eV at 85 K (liquid) and increases to 0.0 eV in the (solid) at 82 K and to +0.3 eV at 6 K (solid).

## *Polar Fluids*

The work-function method has also been applied to polar liquids, gases, and solids. Very clearly defined thresholds are observed here, too, but states other than the extended states of the conduction band may be involved, as will be discussed later. First, let us consider the results.

A technique used in solids at low temperature is photoionization. The results obtained by Bernas et al[75] for TMPD in several glassy matrices at 77 K have been evaluated[62] by Eq. (7-10) with estimated values of $P^+$. The results for a few solids are shown in Table 7-5. The $V_0$-value derived this way decreases as the polarity of the matrix increases. Because these values are for 77 K, one would expect much lower values at 295 K, as is the case for nonpolar compounds.

Photoelectron emission from metal electrodes into polar fluids has been studied for several liquids. In this case the electrode potential, $\underline{\phi}$, must be included, and the photocurrent, $i$, according to the theory proposed by Brodski et al,[32] is given by

$$i = A(h\nu - h\nu_0 - e\underline{\phi})^{5/2} \tag{7-15}$$

where $h\nu$ is the photon energy and $h\nu_0$ is the threshold frequency. This equation has been experimentally verified for photoemission from Hg and amalgams.[35] In this way the lowest level of the "band" of states into which electrons may be injected has been measured. Values reported for the energy of this band for injection into $H_2O$ from a metal electrode range from $-1.0$ to $-1.26$ eV.[32–35] The value for the threshold energy for injection into $NH_3$ is $-1.13$ eV relative to vacuum.[37,38]

The photoemission method has also been applied to polar gases that have been studied as a function of density. In this case the Fowler function is used to determine thresholds; that is, a square root dependence of the photocurrent on the parameter $h\nu/kT$ is observed. For ammonia vapor at 400 K two thresholds are observed; at the highest density studied of 0.1 g/cm$^3$ the thresholds are $-0.9$ and $-0.33$ eV.

The above ammonia vapor study as well as one on liquid ammonia[37] suggests that electrons can be injected into polar fluids not only into the conduction band but also into a lower "band" of preexisting trapped states.[76] The results place the bottom of the conduction band at $-0.4$ eV in liquid ammonia, and the lower threshold, at $-1.13$ eV, is the preexisting trapped state level. These results indicate that the lower thresholds observed in other polar liquids, including water, represent a band of trapped states that is only a few tenths of an eV above the solvated electron level.

## TRAPPED ELECTRONS

As indicated at the end of the previous section, not all excess electrons occupy the conduction band where the wave function is extended, but in many cases the electrons are trapped. In the trapped state the electron wave function is localized in space. There are two general types of trapped electrons: presolvated and solvated. The former imply localization of the electron in a preexisting site or cavity without any relaxation of solvent molecules; the latter imply solvent molecules have obtained their equilibrium configuration, which is a lower energy state. In the following we present the evidence for trapped electrons in fluids, examine the theoretical models for trapped electrons in nonpolar liquids (for polar liquids, see Chapter 8), and discuss the dynamics of the trapping process in hydrocarbons.

### *Evidence for Trapping*

Pulse radiolysis experiments have provided clear evidence for solvated electrons in alkanes. Absorption spectra with a maxima in the infrared (near $2\mu$) have been observed at room temperature for *n*-hexane,[77,78] cyclohexane,[77] methylcyclohexane,[77,79] isopentane,[77] and 3-methyloctane,[80] and at low temperatures in liquid propane.[81] The spectrum detected in methyl-

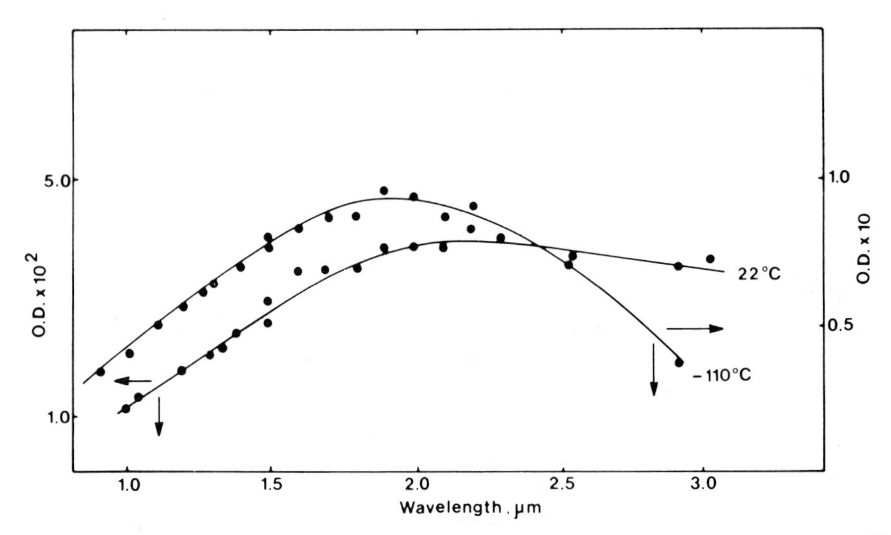


**Figure 7-6.** Spectrum of $e_s^-$ in methylcyclohexane. Reproduced with permission.[79]

cyclohexane is shown in Figure 7-6. Some alkanes like 3-methylpentane and 3-methylhexane form glasses at low temperature and the absorption spectra of the solvated electron are observed in these glasses.[82,83] The absorption maximum in 3-methylhexane is at 1.78 $\mu$ at 77 K. Also of interest is the observation that the solvated electron absorption is *not* observed at 77 K in glassy mixtures of 2,2,4-trimethylpentane and 2,2-dimethylbutane.[83] The significance of the apparent lack of trapping in this mixture is probably related to the high electron mobility in these liquids and to the low conduction-band energy. Alkanes for which absorption spectra have been detected have high conduction-band energies.

Electron spin resonance (ESR) experiments provide further evidence for trapped electrons and give specific information about the structure of the trap and changes in structure occurring on solvation. A significant fraction of the electrons produced can be trapped in matrices at low temperatures; for example, $G(e_t^-) = 0.6/100$ eV in 3-methylpentane at 77 K.[84] The ESR spectrum of the trapped electron is a singlet occurring near the free-spin value; there is no hyperfine interaction. But the line shape can be interpreted, and the average electron–proton distance evaluated. For 3-methylhexane this distance is 3.5 Å at 77 K.[85] From electron spin–echo data it has been determined that for alkanes and ethers the number of nearest-neighbor protons is 20,[86] corresponding to approximately six solvent molecules in the first solvation shell with one methyl group from each molecule directed toward the electron. These studies also show that small structural changes occur on solvation, and the electron–proton distances increase slightly.

Mobility measurements as a function of density provide another source of evidence for trapping of electrons. As discussed earlier and shown in Figure 7-4, the electron mobility drops off precipitously in He gas above a density of $10^{21}$ atoms/cc. Above this density localized and quasi-free states are both stable, and the electron spends more and more time in the localized state as the density increases.[87] A very similar behavior is observed in some hydrocarbons, except that localization sets in at a higher density, above $6 \times 10^{21}$ $cc^{-1}$ in ethane.[44]

## *Theoretical Models*

### Bubbles

The mobility results for He, Ne, and $H_2$ suggest that the electron is trapped in these fluids at high densities and in the liquids in *bubbles.* The bubble radii are 6.4 Å in Ne, 11 Å in $H_2$, and 16 Å in He at one atmosphere pressure. This trapped state occurs because the conduction-band level is quite high: +1.05 eV in He and 0.5 eV in Ne (Table 7-5). Consequently, the most stable configuration for the electron is in a cavity where the repulsive electron–helium potential is balanced by pressure–volume and surface tension energies.[72]

The bubble model has also been proposed for hydrocarbons.[88] A cavity radius of 4.5 Å was shown to be consistent with the room-temperature mobility data in a series of hydrocarbons. However, detailed calculations of a cavity of this dimension show that the bubble state is not stable unless the dielectric constant just outside the cavity is larger than the macroscopic value.[89] But this is expected because the electron in the cavity interacts with uncompensated C—H dipoles in the first solvation shell and raises the dielectric constant locally. The mobility data can be accounted for by a dielectric constant of 3 and a radius of 3.5 Å.

### Semicontinuum Model

For hydrocarbons one of the most successful models for explaining the trapped state is the semicontinuum model.[86] In this model the interactions of the electron with the molecules in the first solvation shell are taken into account specifically in terms of the electron-polarizability interaction. The rest of the solvent is treated as a continuous dielectric medium. The total energy of the trapped electron also includes the kinetic energy, medium-rearrangement energy, repulsive interaction with medium electrons, the energy required to form a cavity, and the proton–proton repulsive energy. With this model the ground state energy, $E_t$, of the trapped electron has been calculated for several alkanes. The results, shown in Table 7-6, are consistent with localized electrons in liquid ethane, propane, and *n*-pentane; that is, $E_t < V_0$ for these liquids, and the electron mobility is correspondingly low. Quasi-free electrons are expected in neopentane and tetramethylsilane because $E_t > V_0$, and these are high-mobility liquids. We can generalize for other liquids and postulate a *localization criteria*: Localization occurs when $E_t < V_0$. Further, one should also note that for *n*-alkanes the difference

**Table 7-6.** Semicontinuum Calculations[a]

| Liquid | $T$ | $V_0$ | $E_t$ | State |
|---|---|---|---|---|
| Ethane | 182 | 0.02 | −0.25 | L |
| Propane | 179 | 0.11 | −0.18 | L |
| $n$-Pentane | 296 | 0.04 | −0.16 | L |
| Neopentane | 296 | −0.45 | −0.40 | qf |
| Tetramethylsilane | 293 | −0.55 | −0.54 | qf |

[a] Refs. 86 and 90.
[b] L = localized; qf = quasi-free.

$V_0 - E_t$ is comparable to the observed activation energy of the mobilities (Table 7-4).

## Dynamics of Solvation

A general picture of the trapping process is now available from pulse radiolysis and ESR studies. Trapping in an organic liquid occurs very fast (in ps) at room temperature[78] but is slowed down enough at low temperatures to observe the process. For example, in the pulse radiolysis of 3-methylhexane at 76 K, the maximum in the absorption spectrum of $e_t^-$ is near 2000 nm at 10 $\mu$s after the pulse and gradually shifts to shorter wavelengths (1700 nm) at longer times.[84] In more-polar organic glasses like ethanol, similar but larger shifts are observed. Another way of observing the changes in trapping is to irradiate at an even lower temperature (4 K) and warm to 77 K. In this case changes have been observed in the ESR line width and in the optical absorption maximum, which shifts irreversibly to the blue after warming.

Traps in these liquids occur as a result of fluctuations in both density and solvent molecule orientation. Thus, initially there will be a range of potential-well depths; some electrons will be in shallow traps and some in deeper traps. With time, or with an increase in temperature, the results show that a shift to deeper traps occurs. This shift can be visualized to occur in either of two ways. The first would be reorientation of the molecules in the solvation shell of the electron by electrostatic forces. The second would be a redistribution of the electrons among the traps to favor the deeper traps. The mechanism for the second process may involve thermal activation from the traps or tunneling from trap to trap. In general, it is possible that the shift to deeper traps may occur in both ways, depending on conditions. However, pulse radiolysis results favor the first mechanism.

Polarity of the matrix has a profound effect on the time required to cause a shift in the absorption maximum of $e_t^-$. At 77 K the times range from 3 $\mu$s in a methanol–water glass to more than 1 s in alkane glasses. Similar results were obtained in liquid alcohols at low temperatures, where the shift is fastest in methanol and slowest in $n$-propyl alcohol.[91] The greater the polarity the faster the shifts occur—a result that demonstrates the importance of molecular rearrangement of dipoles in the solvation shell of the electron.

Effects of solutes like biphenyl on the spectral changes in $e_t^-$ also lead to a similar conclusion. Electrons react to form the biphenyl anion, which can be detected optically. In the presence of such solutes the shift in absorption maxima of $e_t^-$ is still observed, but the biphenyl anion absorption does not increase during this time period. This provides further evidence that electrons do not move from trap to trap but stay in the same trap. Thus, at least at 77 K, the electron solvation process involves primarily the electrostatic reorientation mechanism.

## ELECTRON TRANSPORT THEORY

The mobility of an electron drifting in an electric field is given by

$$\mu = \frac{e}{m}\tau \tag{7-16}$$

where e and $m$ are the charge and mass of the electron, respectively, and $\tau$ is the average time between collisions. This equation is valid where the scattering events are isolated, as in a dilute gas. However, the situation is analogous in nonpolar liquids that consist of atoms (Ar, Kr, Xe) or symmetrical molecules ($CH_4$, $C(CH_3)_4$, $Si(CH_3)_4$). In such liquids the electron is quasi-free and scattered only by potential fluctuations caused by fluctuations in density.[92]

### *Deformation Potential Theory*

For nonpolar liquids composed of atoms or symmetrical molecules, the mobility goes through a maximum at a density somewhat above the critical density. Another observation relevant to our understanding of transport in these liquids is that the minimum in the energy of the electron ($V_0$) occurs at the same density as the maximum in the mobility.[43] In the deformation potential model the electron is scattered elastically by fluctuations in potential occurring on the scale of the electron wavelength. The magnitudes of these fluctuations, which are due to density variations, are given by changes in $V_0$ and depend on the derivatives of $V_0$:

$$\Delta V_0 = \frac{\partial V_0}{\partial N}\Delta N + \frac{1}{2}\frac{\partial^2 V_0}{\partial N^2}\Delta N + \cdots \tag{7-17}$$

At the minimum in $V_0$ the leading term goes to zero, and, since the mean free path, $\Lambda$, is inversely proportional to $\Delta V_0^2$, the mobility will be a maximum at this density. The situation may be compared to a gas: In a low-density gas the electron mobility is given by the Lorentz equation

$$\mu = \frac{2e}{3}\left(\frac{2}{\pi mkT}\right)^{1/2}\Lambda \tag{7-18}$$

where $\Lambda = 1/4\pi Na^2$ and the electron is scattered by individual molecules of dimension $a$. In the dense fluid a similar equation is invoked, but $\Lambda$ is given by[92]

$$\Lambda = \frac{\pi h^4}{m^2N^2kT\chi_T l} \tag{7-19}$$

where

$$l = V_0'^2 + V_0''^2\frac{\beta}{kT\chi_T} + \cdots, \tag{7-20}$$

$\chi_T$ is the isothermal compressibility, and $\beta$ is a parameter involving the structure factor and momentum transfer changes in the scattering events. The ellipsis in Eq. 7-20 refers to additional cross-derivative terms. Berlin et al showed that if only the first term involving $V_0'$ is included, this type of theory predicts the room-temperature quasi-free mobility in hydrocarbon liquids reasonably well.[93]

Because the first derivative of $V_0$ goes to zero near where the mobility is a maximum, Eq. (7-20) would lead to an infinite mobility at this density unless additional terms are included. With higher terms included this model accounts very well for the maximum in the mobility of the electron in liquid argon.[92] Subsequent to the publication of this model, $V_0$ measurements as a function of density became available for Ar, Kr, and Xe (see Figure 7-3). This provided a better test of the model because experimentally determined values of the derivatives could be used in Eq. (7-20). Mobilities calculated this way are in rough agreement with experiment for all three rare gas fluids (ref. 46(b) and Figure 7-3).

Thus potential fluctuations arising from density fluctuations seem to be the sole scattering centers of quasi-free electrons. This theory is analogous to the deformation potential model for electron transport in solids.[94]

## *Trap Model*

For nonpolar liquids composed of unsymmetrical molecules the mobility is generally low and temperature dependent. Further, studies of the density dependence indicate that at an intermediate density the mobility drops off steeply. At and above this density the trapped state is below the conduction band energetically. Trapping has been confirmed in many of these liquids, including propane, *n*-hexane, cyclohexane, 2-methylbutane, methylcyclohexane, and 3-methyloctane, by the observation of the absorption spectrum of the solvated electron (see the section entitled "Trapped Electrons").

The trapped or solvated electron is relatively immobile, like an ion, and transport is assumed to occur when the electron is thermally activated from the trap to the conduction band. Then the mobility is given by

$$\mu = \mu_0 \, e^{-E_a/kT} \tag{7-21}$$

The difference in energy between the trapped state and conduction band is comparable to the observed activation energy for ethane, propane, and *n*-pentane.[86]

### *Hopping Model*

For low-mobility liquids another transfer mechanism that does not involve the electron in the quasi-free state has been proposed. In this "hopping" model the electron jumps from one localized state to another nearby. This mechanism is especially applicable to solids and somewhat-polar liquids where the electron mobility is low and thermally activated.[2]

The equation for transport in the hopping model can be derived from the diffusion coefficient $D$, which is equal to $\lambda^2\nu$, where $\lambda$ is the mean jump distance and $\nu$ is the jump frequency. Because mobility is related to $D$ by the Nernst–Einstein relation, $\mu = eD/kT$, one obtains

$$\mu = \frac{e\lambda^2\nu}{kT} \tag{7-22}$$

The process of new trap formation and hopping of the electron involves an activation energy; thus the equation becomes

$$\mu = \frac{e\lambda^2\nu_0}{kT} e^{-E_a/kT} \tag{7-23}$$

This model also provides an explanation of the effect of high fields, which increase the mobility in low-mobility liquids like ethane and propane. The increase in $\mu$ by the electric field is attributed to an enhanced hopping probability. Comparing theory and experiment shows[4] rather curiously that $\lambda$, the jump distance, varies between 10 and 40 Å, depending on temperature and solvent; $\nu_0$ is of the order of $10^{13}$ to $10^{14}$ s$^{-1}$.

Both the trap model and the hopping model account adequately for the available low-field-mobility data. More mobility data over a wider range of experimental variables as well as more Hall mobility data would be useful for fully understanding transport in these liquids.

## ELECTRON REACTIONS

The excess electron is an uncommon reactant and its chemistry in nonpolar solvents has only recently been investigated. Electrons react readily with many substances, but they do not react with such compounds as alkanes, alkenes, inert gases, benzene, and tetraalkylsilanes. These can be used as solvents in electron reactivity studies. Such solvents must be very highly purified, however, to remove trace reactive impurities (like $O_2$); otherwise the electron will disappear rapidly by reacting with the impurities.

Various reactions of electrons are discussed below. One of the most important is recombination with positive ions, which necessarily are produced in equal numbers in radiation chemistry. Another reaction is electron attachment; some of these reactions occur at diffusion-controlled rates, whereas others are slower and energetics determine the rate. Entropy effects are also very significant, particularly in equilibrium reactions of electrons.

## *Measurement of Rates*

Determining the second-order rate constant $k$ for the reaction

$$e^- + S \xrightarrow{k} S^- \tag{7-24}$$

where S is a solute, requires a method of measuring the concentration of electrons as a function of time after a pulse of radiation. The electron decay is actually first order because the concentration of the solute changes little during the experiment; thus the first-order rate constant is $k'$, which is equal to $k$[S].

Several methods of measuring the electron concentration are currently used. One method measures electrical conductivity, as described earlier for mobility measurements. The conductivity signal in a pulse experiment is proportional to the product of the concentration of electrons and the drift velocity, $v_d$, in the electric field. The conductivity signal decays with time $t$:

$$i = i_0 e^{-k't}\left(1 - \frac{t}{t_d}\right) \tag{7-25}$$

where $t_d$ is the time required for the electron to drift the distance $d$ in the electric field, $t_d = d/v_d$. Generally, $t \ll t_d$, so that the decay is nearly first order due to reaction with the solute.

Another method specific to charged species is microwave absorption. This technique is also used to determine mobilities of ions and electrons. This is a very sensitive technique and can detect ions or electrons at the 50-picomolar level.[40]

Pulse radiolysis is used to measure electron reaction rates in solvents where the electron is trapped and has an optical absorption spectrum. 3-Methyloctane is an example of such a solvent, and the absorption of the electron in this solvent[80] is similar to that in Figure 7-6. The rate of decay of the electron absorption in the presence of solutes provides data on reaction rates.[94,95] Much higher concentrations (micromolar) of electrons are required in this method, and, consequently, volume recombination of electrons occurs. Significant correction for recombination is necessary unless detection is in the nanosecond time regime.

### Homogeneous Recombination

In nonpolar liquids the single most important reaction of electrons in the absence of a reactive solute is recombination with positive ions,

$$e^- + RH^+ \xrightarrow{k_R} RH^* \tag{7-26}$$

The rate constant is large—for example, $5 \times 10^{13}\ M^{-1}\ s^{-1}$ in $n$-hexane.[96] This large rate is attributed to a large (~300 Å) capture radius of the positive ion. That is, the coulombic field extends a long way from the positive ion; the coulombic potential drops to $kT$ at a distance $r_c$; see Eq. (7-1). A diffusion-controlled reaction with a reaction radius $r_c$ would have a rate constant

$$k_R = \frac{4\pi D e^2}{\varepsilon k T} \tag{7-27}$$

This becomes

$$k_R = \frac{4\pi e \mu}{\varepsilon} \tag{7-28}$$

because $D = \mu kT/e$. Thus, the recombination rate constant is proportional to the mobility. In molar units Eq. (7-28) becomes

$$k_R = \frac{1.09 \times 10^{15}}{\varepsilon} \mu\ M^{-1}\ s^{-1}$$

This is consistent with experiment; that is, $k_R$ is proportional to $\mu$ at least for mobilities up to $100\ cm^2/V\text{-}s$,[71,96,97] but deviations have been noted for liquids of higher electron mobility, such as argon[98] and methane.[99]

Homogeneous or volume recombination of electrons with ions leads to products as in the geminate case. If $RH^+$ remains as the molecular ion, then recombination produces an excited state that generally decomposes into radicals or fragments (or fluoresces). If the molecular ion reacts or fragments, then recombination with the product ion leads to fragment radicals directly. Overall, electron–ion recombination is one of the most significant radiation chemical processes in these liquids.

### Diffusion-Controlled Attachment

Attachment reactions of electrons with some solutes may occur at diffusion-controlled rates; that is, the rate constant is given by

$$k_a = 4\pi R D \tag{7-29}$$

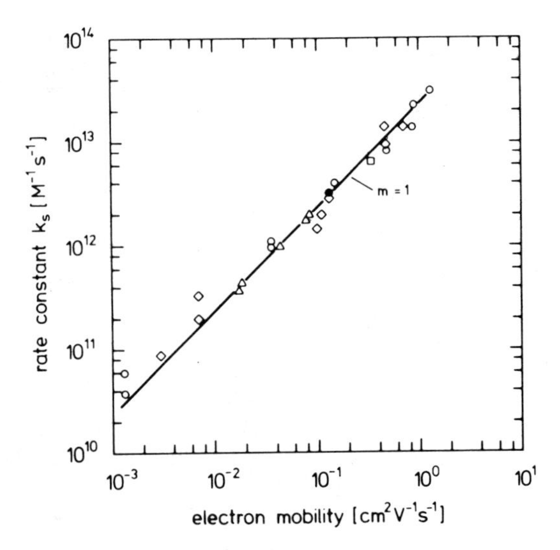


**Figure 7-7.** Electron scavenging rate constant as a function of electron mobility: ethane (○), propane (◇), *n*-pentane (●), *n*-butane (□), and *n*-hexane (△). Reproduced with permission.[100]

The reactions of electrons with nitroaromatic compounds and $SF_6$ are diffusion limited in the low-mobility liquids. Again, since $D = \mu kT/e$, such reaction rates should be proportional to $\mu$. Rate data for the reaction of $e^-$ with $SF_6$ are shown in Figure 7-7, where $k_a$ is seen to be proportional to $\mu$ at least up to values of $\mu = 1.0$ $cm^2$/V-s. From the slope of the line in Figure 7-7 the radius, $R$ is found to be 14.5 Å for $SF_6$.[101]

Additional support for the role of diffusion is obtained from studies of the effect of temperature on electron rates. In *n*-hexane the activation energies for attachment to pyrene, biphenyl, perylene, and $CCl_4$ are all approximately 3.5 kcal/mole.[95] This value is very similar to the energy of activation of the electron mobility in *n*-hexane, which is 3.8 kcal/mole.[27] This correspondence indicates that the rate determining step is diffusion. A study of electron attachment to nitroaromatic compounds[102] in cyclohexane showed that the rate depended on the position of nitration, the rate for *o*-dinitrobenzene being nearly twice the rate for *p*-dinitrobenzene. This was shown to be an effect of dipole moment. A theoretical calculation showed that the reaction radius, $R$, increased as the dipole moment increased, and diffusion-controlled rates are proportional to the reaction radius, as shown by Eq. (7-29).

## *Role of Energetics*

The rate constants for reaction of the electron with many other solutes are large but not diffusion controlled. The rates are very solvent dependent (see Table 7-7), and, in general, the chemistry of the electron is quite

**Table 7-7.** Electron Reaction Rates

| Solute | Solvent | $k$ ($M^{-1}$ $s^{-1}$) | Ref. |
|---|---|---|---|
| Biphenyl | $n$-Hexane | $1.0 \times 10^{12}$ | 95 |
| | 2,2,4-TMP | $1.0 \times 10^{13}$ | 103 |
| | $Si(CH_3)_4$ | $7.0 \times 10^{12}$ | 103 |
| Pyrene | $n$-Hexane | $1.1 \times 10^{12}$ | 95 |
| Naphthalene | 2,2,4-TMP | $6.4 \times 10^{12}$ | 104 |
| | $Si(CH_3)_4$ | $1.0 \times 10^{13}$ | 104 |
| $O_2$ | $n$-Hexane | $9.1 \times 10^{10}$ | 95 |
| | 2,2,4-TMP | $1.4 \times 10^{11}$ | 105 |
| | | $2.5 \times 10^{11}$ | 95 |
| | $Si(CH_3)_4$ | $6.0 \times 10^{11}$ | 105 |
| $CO_2$ | 2,2,4-TMP | $6.9 \times 10^{12}$ | 106 |
| | $Si(CH_3)_4$ | $1.3 \times 10^{11}$ | 106 |
| $Cr(CO)_6$ | 2,2,4-TMP | $7.8 \times 10^{13}$ | 107 |
| $SF_6$ | $CH_4$ | $8.0 \times 10^{14}$ | 100 |

complex. This complexity is attributed to a dominant role of energetics. Figure 7-8 is a plot of the rate constant for reaction with several solutes in different solvents at room temperature. The data are plotted as a function of the conduction-band energy of the solvent, $V_0$. In some cases the change in rate can be attributed to changes in $V_0$. The reason the conduction band is so important is that these are fast vertical attachment reactions.[109,110] To understand this, consider the data for ethyl bromide; the rate goes through a maximum in Figure 7-8. Figure 7-9 shows schematically the potential energy curves for ethyl bromide and ethyl bromide anion. For vertical attachment an energy of 0.76 eV is required (even though overall the reaction is exothermic). Indeed, it has been observed that for the gas phase this reaction goes through a maximum when the kinetic energy of the electron is 0.76 eV. In the liquid phase the electron is thermal, and the 0.76-eV energy is not available as kinetic energy. However, the potential energy of the electron changes from its energy in the conduction band, $V_0$, to its energy on the anion; that is, the anion polarizes the solvent and this polarization energy, $P^-$, is about $-1$ eV in nonpolar solvents. The rate of attachment will be a maximum when $V_0$ is equal to $\varepsilon_l^{max}$ (see Figure 7-9). In $Si(CH_3)_4$, $V_0$ is too low ($-0.55$ eV), and the rate of attachment to ethyl bromide is small. As the conduction-band energy increases, the rate increases and in 2,2,4-trimethylpentane (TMP) $V_0$ is $-0.24$ eV and the rate is a maximum as in the gas. For $n$-hexane $V_0$ is too high, and the rate of attachment again decreases. Similar maxima are observed in the rates of attachment to $C_2HCl_3$, $N_2O$, and *cyclo*-$C_4F_8$ and attributed to the same effect.[105,111]

The kinetic energy (K.E.) of the electron should be included and added to the potential terms; thus the rate really depends on $V_0 - P^- + \text{K.E.}$ The electron can gain energy at high fields in Ar, Kr, and Xe and at very high

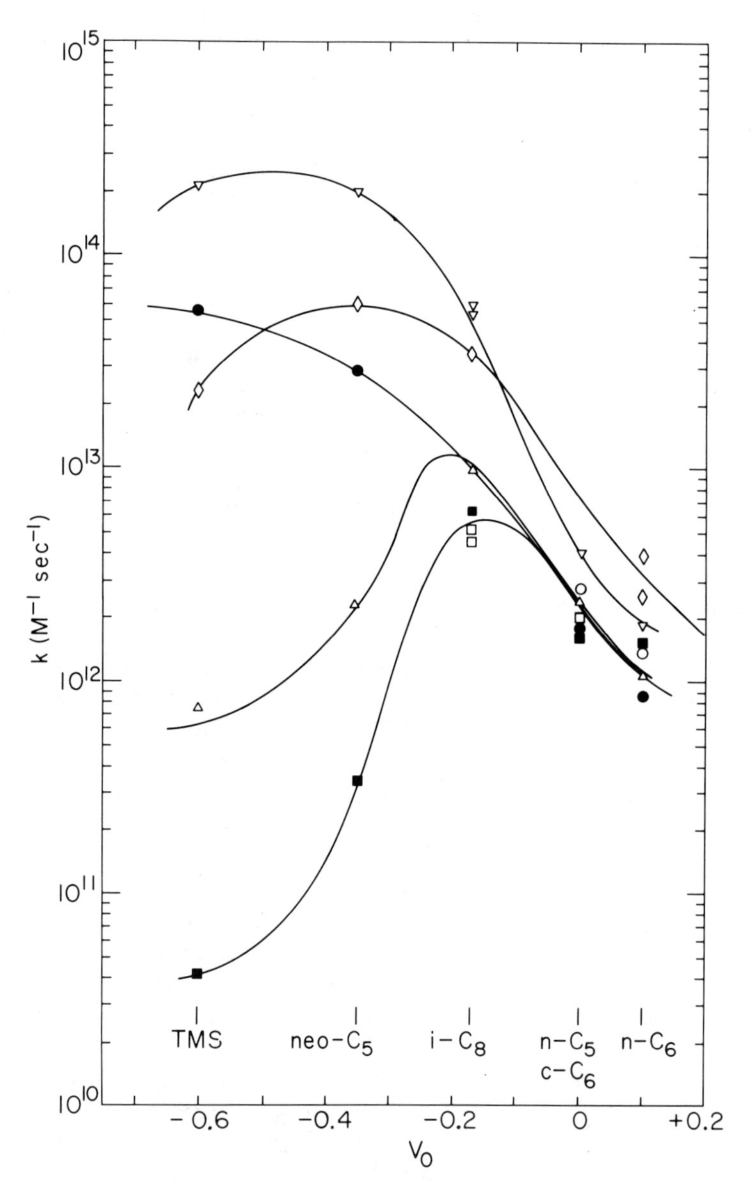


**Figure 7-8.** Electron rate constants at room temperature as a function of $V_0$ (in eV) for reaction with (▽) $SF_6$, (◇) $C_2HCl_3$, (○) $CCl_4$, (△) $N_2O$, and (□) $C_2H_5Br$: open symbols ref. 108, filled symbols ref. 96. Reproduced with permission.[108]

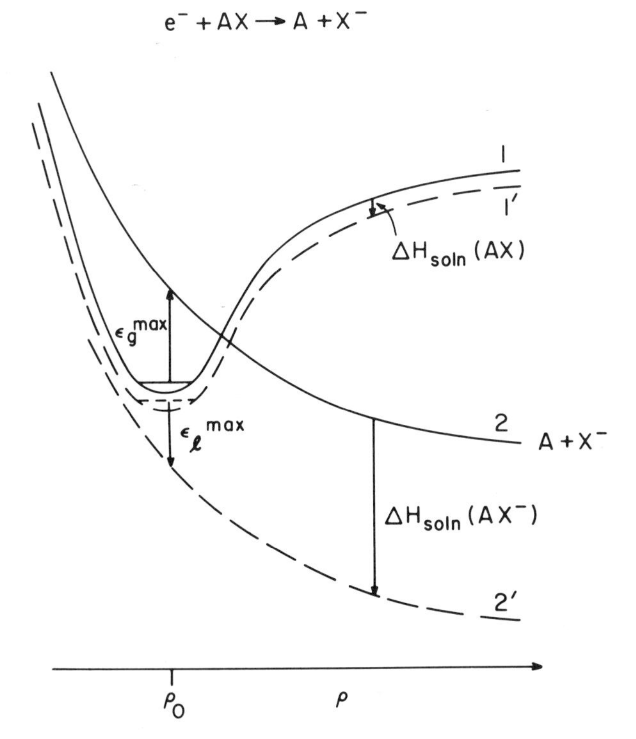


**Figure 7-9.** Schematic potential energy diagram for AX (curves 1 and 1′) and $AX^-$ (curves 2 and 2′); solid lines for gas phase; dotted lines for liquid.

fields in $Si(CH_3)_4$. This kinetic energy can compensate for low values of $V_0 - P^-$. For example, the rate of attachment to ethylbromide is low in $Si(CH_3)_4$ but increases at high fields as the kinetic energy of the electron increases.[112] Similarly, the rate of reaction with $N_2O$ increases at high fields in liquid argon and xenon due to the increase in kinetic energy of the electron.[113]

A very dramatic solvent effect is observed for certain nondissociative attachment reactions which are reversible:

$$e^- + A \rightleftharpoons A^- \tag{7-30}$$

That is, for some solutes the anion formed, $A^-$, thermally autodetaches an electron. The lifetimes of such anions depend on solvent and temperature and range from $10^{-3}$ to $10^{-9}$ s.

The equilibrium constants, $K$, are shown in Figure 7-10 as a function of $1/T$ for a series of aromatic hydrocarbons in tetramethylsilane as solvent. The naphthalene and biphenyl data are in the middle of the figure, where $K$ is approximately $10^6$ at room temperature. If the temperature is increased, the equilibrium favors the free electron, indicating the attachment reaction is exothermic. The triphenylene and phenanthrene data are on the left,

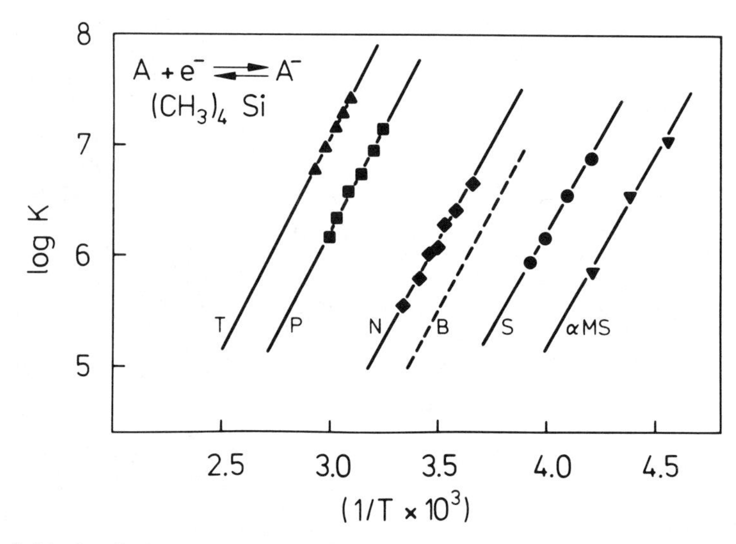


**Figure 7-10.** $\log(k_1/k_2)$ vs. $1/T$ for equilibria $A + e^- \rightleftharpoons A^-$ in $(CH_3)_4Si$ as solvent; ▲ A is triphenylene (1.0 $\mu$mol · $L^{-1}$), ■ phenanthrene (1.3 $\mu$mol · $L^{-1}$), ◆ naphthalene (2.3 $\mu$mol · $L^{-1}$), ● styrene (2.0 $\mu$mol · $L^{-1}$), ▼ $\alpha$-methylstyrene (0.8 $\mu$mol · $L^{-1}$) Dotted line is biphenyl data. Reproduced with permission.[104]

and the equilibrium constants are greater than $10^6$ at room temperature. These compounds have a higher electron affinity than naphthalene. The styrenes with lower electron affinity exhibit a smaller equilibrium constant at room temperature. Thus the value of the equilibrium constant increases with electron affinity or conversely, compounds of the highest electron affinity have the most negative free energy, $\Delta G^0$.

The effect of solvent on the equilibrium constant is very large. For a given reactant $K$ is approximately $10^4$ times greater in TMP than in $Si(CH_3)_4$. That is, the anion is more stable or $\Delta G^0$ is more negative in TMP. This effect is observed for naphthalene and styrene, as shown by the data in Table 7-8. These shifts with solvent are attributed primarily to shifts of the ground state of the free electron. Thus, the ground state of the electron in TMP is approximately 0.3 eV higher than in $Si(CH_3)_4$. In *n*-hexane and cyclohexane the ground state is even higher.[104,115] Note that these equilibria data reflect the ground-state energy of the electron, which is not necessarily the same as $V_0$. In $Si(CH_3)_4$ the ground state *is* the conduction band, but in *n*-hexane and cyclohexane the ground state is the trapped state that lies below the conduction-band energy ($V_0$).

Entropy plays a major role in these equilibria, as shown by the data in Table 7-8. The enthalpies are large and negative, indicating that the negative ions are very stable in solution. These large enthalpies are compensated by large negative entropies that provide the driving force for detachment of the electron and make the equilibria observable. The entropy change is associated primarily with polarization of solvent molecules around the anion.

**Table 7-8.** Thermodynamic Parameters for $e^- + A \rightleftharpoons A^-$[a]

| Solvent | Solute | $E_a$ (eV) | $\Delta G^0_{298}$ (kcal/mole) | $\Delta H^0$ | $\Delta S^0$ (cal/mole-K) |
|---|---|---|---|---|---|
| $Si(CH_3)_4$ | Triphenylene | 0.28 | −11.5 | −17.7 | −20.3 |
| | Phenanthrene | 0.31 | −10.1 | −16.8 | −22.2 |
| | Naphthalene | 0.15 | −7.6 | −16.1 | −28.4 |
| | Biphenyl | 0.08 | −6.9 | −17.5 | −35.6 |
| | Styrene | −0.01 | −5.5 | −15.4 | −33.5 |
| | $CO_2$ | −0.44 | −7.3 | −21.1 | −46.0 |
| 2,2,4-TMP | Naphthalene | 0.15 | −13.0 | −19.7 | −22.6 |
| | Styrene | −0.01 | −11.1 | −22.7 | −39.0 |
| | $CO_2$ | −0.44 | −12.7 | −24.9 | −40.9 |
| Cyclohexane | $C_6F_6$ | −0.35 | −7.8 | −21.3 | −45.5 |
| *n*-Hexane | Styrene | −0.01 | −12.8 | −26.0 | −44.4 |
| | $C_6F_6$ | −0.35 | −7.6 | −16.8 | −30.9 |

[a] Refs. 114, 104, 106.

Entropy is also a rate-determining factor in many of the reactions of solvated electrons.[3] The reactions of solvated electrons in polar solvents and of trapped electrons in glasses are not considered here. The reader is referred to Chapters 2, 11, and 13 and to several excellent reviews[3,116] for more details on this topic.

# REFERENCES


1. Warman, J. M. In "The Study of Fast Processes and Transient Species by Electron Pulse Radiolysis"; Baxendale, J. H.; and Busi, F. Eds.; Reidel: Dordrecht, 1982, ASI, Vol. 86, pp. 437–527.
2. Yakovlev, B. S. *Russ. Chem. Rev.* **1979**, *48*, 615.
3. Schindewolf, U. *Angew. Chem. Int.* (Engl. Edition) **1978**, *17*, 887.
4. Schmidt, W. F. *Can. J. Chem.* **1977**, *55*, 2197.
5. Davis, H. T.; Brown, R. G. *Adv. Chem. Phys.* **1975**, *31*, pp. 329–464.
6. Sowada, U.; Warman, J. M.; DeHaas, M. P. *Phys. Rev. B* **1982**, *25*, 3434.
7. Schmidt, W.; Allen, A. O. *J. Chem. Phys.* **1970**, *52*, 2345.
8. Schmidt, W. *Radiat. Res.* **1970**, *42*, 73.
9. Lee, K.; Lipsky, S. *J. Phys. Chem.* **1982**, *86*, 1985.
10. Allen, A. O. "Yields of Free Ions Formed in Liquids by Radiation"; NSRDS-NBS-57, 1976.
11. Dodelet, J.-P.; Shinsaka, K.; Kortsch, U.; Freeman, G. R. *J. Chem. Phys.* **1973**, *59*, 2376.
12. Choi, H. T.; Sethi, D. S.; Braun, C. L. *J. Chem. Phys.* **1982**, *77*, 6027.
13. Mozumder, A. *J. Chem. Phys.* **1971**, *55*, 3026.
14. Abell, G. C.; Funabashi, K. *J. Chem. Phys.* **1973**, *58*, 1079.
15. Hummel, A.; Infelta, P. P. *Chem. Phys. Lett.* **1974**, *24*, 559.
16. Hong, K. M.; Noolandi, J. *J. Chem. Phys.* **1978**, *68*, 5163.
17. Braun, C. L.; Scott, T. W. *J. Phys. Chem.* **1983**, *87*, 4776.
18. van den Ende, C. A. M.; Luthjens, L. H.; Warman, J. M.; Hummel, A. *Radiat. Phys. Chem.* **1982**, *19*, 455.
19. Warman, J. M.; Asmus, K.-D.; Schuler, R. H. *J. Phys. Chem.* **1969**, *73*, 931.
20. Jonah, C. D. *Radiat. Phys. Chem.* **1983**, *21*, 53.
21. Thomas, J. K.; Johnson, K.; Klippert, T.; Lowers, R. *J. Chem. Phys.* **1968**, *48*, 1608.

22. Cipollini, N. E.; Allen, A. O. *J. Chem. Phys.* **1977**, *67*, 131.
23. Allen, A. O. "Drift Mobilities and Conduction Band Energies of Excess Electrons in Dielectric Liquids"; NSRDS-NBS-58, 1976.
24. Schmidt, W. F.; Allen A. O. *J. Chem. Phys.* **1970**, *52*, 4788.
25. Dodelet, J.-P.; Freeman, G. R. *Can. J. Chem.* **1972**, *50*, 2667.
26. Dodelet, J.-P.; Shinaska, K.; Freeman, G. R. *Can. J. Chem.* **1976**, *54*, 774.
27. Nyikos, L.; Zádor, E.; Schiller, R. "Proceedings of 4th International Symposium of Radiation Chemistry"; Keszthely, Hungary, 1976.
28. Holroyd, R. A.; Tames, S.; Kennedy, A. *J. Phys. Chem.* **1975**, *79*, 2857.
29. Döldissen, W.; Schmidt, W. F.; Bakale, G. *J. Phys. Chem.* **1980**, *84*, 1179.
30. Boriev, I. A.; Yakovlev, B. S. *Int. J. Radiat. Phys. Chem.* **1976**, *8*, 511.
31. Schmidt, K. H.; Buck, W. L. *Science* **1966**, *151*, 70.
32. Brodsky, A. M.; Gurevich, Y. Y. *Soviet Phys. JETP* **1968**, *27*, 114.
33. Yamashita, K.; Imai, H. *Bull. Chem. Soc. Japan* **1977**, *50*, 1066.
34. Schiffrin, D. J. *J. Electroanal. Chem.* **1975**, *63*, 283.
35. Pleskov, Y. V.; Rotenberg, Z. A. *J. Electroanal. Chem.* **1969**, *20*, 1.
36. Schmidt, W. F. In "Electron-Solvent and Anion-Solvent Interactions"; Kevan, L.; Webster, B. C.; Eds.; Elsevier: Amsterdam, 1976, p. 213.
37. Bard, A. J.; Itaya, K.; Malpas, R. E.; Teherani, T. *J. Phys. Chem.* **1980**, *84*, 1262.
38. Itaya, K.; Malpas, R. E.; Bard, A. J. *Chem. Phys. Lett.* **1979**, *63*, 411.
39. Dodelet, J.-P.; Freeman, G. R. *Can. J. Chem.* **1975**, *53*, 1263.
40. Warman, J. M.; Infelta, P. P.; DeHaas, M. P.; Hummel, A. *Can. J. Chem.* **1977**, *55*, 2249.
41. Dodelet, J.-P.; Freeman, G. R. *Can. J. Chem.* **1977**, *55*, 2264.
42. Jahnke, J. A.; Meyer, L.; Rice, S. A. *Phys. Rev. A* **1971**, *3*, 734.
43. Cipollini, N. E.; Holroyd, R. A. a) *J. Chem. Phys.* **1977**, *67*, 4636; b) Proceedings of the Sixth International Congress on Radiation Research, Tokyo, Japan, May 1979, p. 228.
44. Nishikawa, M.; Holroyd, R. A.; Sowada, U. *J. Chem. Phys.* **1980**, *72*, 3081.
45. Cipollini, N. E.; Holroyd, R. A.; Nishikawa, M. *J. Chem. Phys.* **1977**, *67*, 4638.
46. Reininger, R.; Asaf, U.; Steinberger, I. T. (a) *Chem. Phys. Lett.* **1982**, *90*, 287; (b) Reininger, R.; Asaf, U.; Steinberger, I. T.; Basak, S. *Phys. Rev. B* **1983**, *28*, 4426.
47. Gomer, R.; Lekner, J. *Phys. Rev.* **1967**, *156*, 351.
48. von Zdrojewski, W.; Rabe, J. G.; Schmidt, W. F. *Z. Naturforsch. A.* **1980**, *35a*, 672.
49. Tauchert, W. K.; Jungblut, H.; Schmidt, W. F. *Can. J. Chem.* **1977**, *55*, 1860.
50. Messing, I.; Jortner, J. *Chem. Phys.* **1977**, *24*, 183.
51. Allen, A. O.; Schmidt, W. F. *Z. Naturforsch. A.* **1982**, *37a*, 316.
52. Levine, J. L.; Sanders, T. M., Jr. *Phys. Rev.* **1967**, *154*, 138.
53. Loveland, R. J.; LeComber, P. G.; Spear, W. E. *Phys. Lett.* **1972**, *39A*, 225.
54. Raz, B.; Jortner, J. In "Electrons in Fluids"; Jortner, J.; and Kestner, N. R.; Eds.; Springer-Verlag: Heidelberg, 1973, p. 413.
55. Jortner, J.; Gaathon, A. *Can. J. Chem.* **1977**, *55*, 1801.
56. Maruyama, Y.; Funabashi, K. *J. Chem. Phys.* **1972**, *56*, 2342.
57. Namba, H.; Chiba, M.; Nakamura, Y.; Tezuka, T.; Shinsaka, K.; Hatano, Y. *J. Electrostatics* **1982**, *12*, 79.
58. Bullot, J.; Gauthier, M. *Can. J. Chem.* **1977**, *55*, 1821.
59. Namba, H.; Shinsaka, K.; Hatano, Y. *J. Chem. Phys.* **1979**, *70*, 533.
60. Grand, D.; Bernas, A. *J. Phys. Chem.* **1977**, *81*, 1209.
61. Hiraoka, K. *J. Phys. Chem.* **1981**, *85*, 4008.
62. Noda, S.; Kevan, L.; Fueki, K. *J. Phys. Chem.* **1975**, *79*, 2866.
63. Warman, J. M.; Kunst, M.; DeHaas, M. P. *J. Electrostatics* **1982**, *12*, 115.
64. Huang, T.; Eisele, I.; Kevan, L. *J. Chem. Phys.* **1973**, *59*, 6334.
65. Grand, D.; Bernas, A. *Chem. Phys. Lett.* **1983**, *97*, 119.
66. Muñoz, R. C.; Ascarelli, G. *Phys. Rev. Lett.* **1983**, *51*, 215.
67. Muñoz, R. C.; Ascarelli, G. *Chem. Phys. Lett.* **1983**, *94*, 235.
68. Holroyd, R. A.; Allen, M. *J. Chem. Phys.* **1971**, *54*, 5014.
69. Krebs, P.; Bukowski, K.; Girand, V.; Heintze, M. *Ber Bunsenges. Phys. Chem.* **1982**, *86*, 879.
70. Dodelet, J.-P.; Shinsaka, K.; Freeman, G. R. *J. Chem. Phys.* **1975**, *63*, 2765.
71. Wada, T.; Shinsaka, K.; Namba, H.; Hatano, Y. *Can. J. Chem.* **1977**, *55*, 2144.
72. Springett, B. E.; Jortner, J.; Cohen, M. H. *J. Chem. Phys.* **1968**, *48*, 2720.
73. Yamaguchi, Y.; Nakajima, T.; Nishikawa, M. *J. Chem. Phys.* **1979**, *71*, 550.
74. Nakagawa, K.; Ohtake, K.; Nishikawa, M. *J. Electrostatics* **1982**, *12*, 157.

75. Bernas, A.; Gauthier, M.; Grand, D. *J. Phys. Chem.* **1972**, *76*, 2236.
76. Krohn, C. E.; Thompson, J. C. *Phys. Rev. B* **1979**, *20*, 4365.
77. Baxendale, J. H.; Bell, C.; Wardman, P. a) *J. Chem. Soc. Faraday Trans. 1* **1973**, *69*, 776; b) *Chem. Phys. Lett.* **1971**, *12*, 347.
78. Kenney Wallace, G. A.; Jonah, C. D. *J. Phys. Chem.* **1982**, *86*, 2572.
79. Busi, F. In "The Study of Fast Processes and Transient Species by Electron Pulse Radiolysis"; Baxendale, J. H.; and Busi, F.; Eds.; Reidel, Dordrecht: 1982, Vol. 86, pp. 417.
80. Gillis, H.; Klassen, N. V.; Woods, R. J. *Can. J. Chem.* **1977**, *55*, 2022.
81. Gillis, H. A.; Klassen, N. V.; Teather, G. G.; Lokan, K. H. *Chem. Phys. Lett.* **1971**, *10*, 481.
82. Hase, H.; Higashimura, T.; Ogasawara, M. *Chem. Phys. Lett.* **1972**, *16*, 214.
83. Kimura, T.; Ogawa, N.; Fueki, K. *Bull. Chem. Soc. Japan* **1981**, *54*, 3854.
84. Kevan, L. In "Advances in Radiation Chemistry"; Burton, M.; and Magee, J. L.; Eds.; Wiley: New York, 1974, Vol. 4, p. 181.
85. Lin, D. P.; Kevan, L. *Chem. Phys. Lett.* **1976**, *40*, 517.
86. Kevan, L. *J. Phys. Chem.* **1978**, *82*, 1144.
87. Schwarz, K. W.; Prasad, B. *Phys. Rev. Lett.* **1976**, *36*, 878.
88. Schiller, R. *J. Chem. Phys.* **1972**, *57*, 2222.
89. Hammer, H.; Schoepe, W.; Weber, D. *J. Chem. Phys.* **1976**, *64*, 1253.
90. Kimura, T.; Fueki, K.; Kevan, L. *J. Chem. Phys.* **1978**, *68*, 3945.
91. Baxendale, J. H.; Wardman, P. *J. Chem. Soc., Faraday Trans 1* **1973**, *3*, 584.
92. Basak, S.; Cohen, M. H. *Phys. Rev. B* **1979**, *20*, 3404.
93. Berlin, Y. A.; Nyikos, L.; Schiller, R. *J. Chem. Phys.* **1978**, *69*, 2401.
94. Schockley, W. "Electrons and Holes in Semiconductors"; Van Nostrand, Princeton, 1950.
95. Baxendale, J. T.; Geelen, B. P. H. M.; Sharpe, P. H. G. *Int. J. Radiat. Phys. Chem.* **1976**, *8*, 371.
96. Allen A. O.; Holroyd, R. A. *J. Phys. Chem.* **1974**, *78*, 796.
97. Nakamura, Y.; Shinsaka, K.; Hatano, Y. *J. Chem. Phys.* **1983**, *78*, 5820.
98. Fuochi, P. G.; Freeman, G. R. *J. Chem. Phys.* **1972**, *56*, 2333.
99. Yakovlev, B. S.; Boriev, I. A.; Novikova, L. I.; Frankevich, E. L.; *Int. J. Radiat. Phys. Chem.* **1972**, *4*, 395.
100. Bakale, G.; Sowada, U.; Schmidt, W. F. *J. Phys. Chem.* **1975**, *79*, 3041.
101. Bakale, G.; Schmidt, W. F. *Naturforsch. A.* **1981**, *36a*, 802.
102. Bakale, G.; Gregg, E. C.; McCreary, R. D. *J. Chem. Phys.* **1977**, *67*, 5788.
103. Zador, E.; Warman, J. M.; Hummel, A. *Chem. Phys. Lett.* **1973**, *23*, 363.
104. Holroyd, R. A. *Ber. Bunsenges. Phys. Chem.* **1977**, *81*, 298.
105. Gangwer, T. E.; Holroyd, R. A. *J. Chem. Phys.* **1979**, *70*, 3586.
106. Holroyd, R. A.; Gangwer, T. E.; Allen, A. O. *Chem. Phys. Lett.* **1975**, *31*, 520.
107. Kang, Y. S.; Holroyd, R. A. *Radiat. Phys. Chem.* **1982**, *20*, 237.
108. Allen, A. O.; Gangwer, T. E.; Holroyd, R. A. *J. Phys. Chem.* **1975**, *79*, 25.
109. Henglein, A. *Ber. Bunsenges. Phys. Chem.* **1975**, *79*, 129.
110. Funabashi, K.; Magee, J. L. *J. Chem. Phys.* **1975**, *62*, 4428.
111. Holroyd, R. A. In "Proceedings Fourth International Symposium of Radiation Chemistry"; Keszthely, Hungary, 1976.
112. Beck, G.; Bakale, G. "Abstracts of Papers", 31st National Meeting of the Radiation Research Society, San Antonio, Texas; 1983, p. 74.
113. Bakale, G.; Sowada, U.; Schmidt, W. F. *J. Phys. Chem.* **1976**, *80*, 2556.
114. Warman, J. M.; DeHass, M. P.; Zador, E.; Hummel, A. *Chem. Phys. Lett.* **1975**, *35*, 383.
115. Holroyd, R. A.; McCreary, R. D.; Bakale, G. *J. Phys. Chem.* **1979**, *83*, 435.
116. Anbar, M. In "Solvated Electron" R. F. Gould, Ed.; American Chemical Society: Washington, D.C., 1965, p. 55.

# 8

# Theories of the Solvated Electron

**N. R. KESTNER**

*Department of Chemistry,*
*Louisiana State University*

## INTRODUCTION

In previous chapters we have already had some discussions of trapped electrons in liquids. We have seen examples of their spectra and even of their kinetics. The media involved can be of various types, and the electron can be introduced into the liquid by any of a number of schemes. In the process of introducing the electron, numerous changes occur in the medium and with the electronic species itself as a function of time. In this chapter we will address only the final state of the electron, that is, the "solvated state," which, if no chemical reaction would occur, is a stable entity with well-defined characteristics. Except for some metal–ammonia solutions, and possibly a few other cases, such "stable" species, in reality, exist but a short time (often as short as microseconds). Nevertheless, this chapter will only deal with this final "time-independent," "completely solvated," "equilibrium" species. The last statement is added to indicate that the solvent around the electron has also come to thermal equilibrium with the field of the charge.

We have read in previous chapters that an electron can exist in either a localized state or a free state, that is, in a conduction band. The bottom of this conduction band, the lowest energy possible for this state, will be designated as $V_0$. This quantity can be measured but is somewhat more difficult to define theoretically in a real finite-temperature macroscopic medium.


*Radiation Chemistry: Principles and Applications*

An electron is trapped or solvated if the system can have a free energy less than that which characterizes the bottom of the conduction band. In most simple models the role of temperature and especially entropy effects are neglected, and thus this stability is usually expressed in terms of energy or enthalpy alone, despite the violations of thermodynamic consistency. Thus far, no clear-cut examples have been provided to indicate that this is a very poor approximation, but such situations probably exist. As a working definition for a localized state, we will use

$$E_T < V_0 \tag{8-1}$$

where $E_T$ is the total energy of stabilization of the electron, that is, heat of solution. Quantum statistical methods are now being developed to consider the proper free-energy expressions, but such work is still in its infancy.

To understand the nature of the trapped electron in different solvents, we must know more about the effective electron–molecule interaction in the liquid as well as the correct molecule–molecule interactions. To do the analysis in detail, we would need to perform a quantum statistical mechanical calculation of the system of an electron and the liquid. Except for a few rather specialized, rather different situations this has not been done; it probably cannot be done at present for the most interesting problems. In the cases studied the interactions were approximated rather drastically. Even if the machinery for doing the complete calculation were available, we would have some major difficulties, raised by the very nature of quantum phenomena, in any simple analysis of the results. The most serious issue is that it is impossible to write down the interaction of an electron and a molecule in a simple form, depending on, say, the distance of the electron from the molecule because the electron cannot be localized and all electrons are equivalent, those in the molecule and the extra electron. To obtain a simple relationship we need to make approximations. Accurate calculations on large systems appear to be very difficult, and thus these approximations are one's only hope to systematically study trapped electrons. Most studies of trapped electrons approximate some parts of the total problem. In many cases there is an extensive literature on each approximation, but much is only of historical interest.[1,2,3] We will try to present in this brief review the general results in a more consistent manner. There are many good reviews that treat the problem in a more historical manner. Many significant papers are not mentioned here despite their important role in the development of the subject because we want to present a short logical treatment of the major ideas as they are now perceived.

## BACKGROUND: FUNDAMENTAL FEATURES

Basically, the major problem in presenting a simple qualitative picture is the definition of the effective electron–molecule interaction. This issue can be avoided if all electrons of the entire medium or large cluster are treated simultaneously. This has been done in a few cases and confirms the general ideas we are about to present; however, it does not lend itself to making general predictions. For that purpose we must introduce a "pseudopotential." This word is often used in a very qualitative sense, but it can be defined quite precisely. The pseudopotential introduces the quantum nature, which resides in the wave function, into the potential. That is, a term is added to the classical potential to introduce those quantum effects. For example, if one tries to add an electron to a helium atom, that extra electron is required to be in an orbital orthogonal to those occupied by the electrons in the usual ground-state helium atom. Such orthogonality was introduced first by solid-state physicists as orthogonalized plane waves[4], (OPW), but later work indicated that one could put those effects directly in the potential, leaving the electron state functions almost unconstrained (ie, almost plane waves).

There are, in fact, an infinite number of possible ways to define a pseudopotential. Cohen and Heine[5] have shown that the simplest and most useful is to define the effective potential as

$$V\Psi + V_R\Psi = V\Psi - \sum \langle \Psi | V | \phi \rangle \phi \qquad (8\text{-}2)$$

where $\Psi$ includes all orbitals occupied by electrons in the atom or molecule. Those authors have shown that the resulting wave function is then the smoothest possible. From now on when we speak of effective potentials, we will mean something like $V + V_R$; this potential will include all quantum effects, such as the Pauli exclusion principle, antisymmetry, and others.

With this reference potential in mind let us consider the factors that cause an electron to become localized in a neutral fluid. Depending on the sign, size, and magnitude of $V + V_R$, there can be two different situations. The first case obtains if the effective potential is large and positive. Helium is the ideal example. The classical interaction between the electron and a helium atom is very weak (even the polarizability is very small), but the helium wave function is tightly bound with a large ionization potential. It strongly excludes, via antisymmetry, all electrons from penetrating into the core. In energy terms this means that the term $V_R$ in the potential is large and positive. Substituting realistic wave functions, polarizations, and the like, one finds that the effective potential is almost totally repulsive and rises to almost 8.2 eV at 0.5 Å as the separation decreases.[6,7]

The other case is more common; namely, the interaction is negative. In such cases the electron can be localized or not, but the decision is based on the structure of the medium as well. The medium must have some asymmetry, even if only at the microscopic level, in order to localize the electron. These cases are more complicated and will be considered after briefly discussing simple nonpolar fluids.

Consider now the series of nonpolar fluids, helium through xenon. The first two or three members are the most interesting. However, since they have been discussed in other chapters we will be very brief. We have already pointed out that the electron–helium interaction is effectively repulsive and quite strong. In the case of neon the polarizability of the atom is larger, and therefore the electron–neon interaction includes some attractive terms. However, the situation for the total potential is not clear cut, and the repulsive forces appear to win out, but only slightly. If we look to argon, we now have a clear example of an effective attraction at reasonable electron–argon separations. Following the above arguments it should not be surprising that the electron in liquid argon is not localized. There are no factors that strongly attract an electron to one region of the fluid. This is in contrast with the situation in helium: If you put an electron in liquid helium, it will try to force other atoms away; "it will dig a hole." As has been pointed out previously, that hole in helium is quite large (because the interactions are quite strong).[8] The electron in neon appears to be more weakly trapped, consistent with our discussion. In other rare gases the electron is delocalized.[9]

When the effective interaction between the electron and the molecule is negative, the electron will not become localized unless it can have a strong interaction with an asymmetric molecule. The interaction of an electron and argon is not capable of causing any major change in the liquid structure that would trap an electron in any one region. However, if the molecule has a more complicated structure, such that the electron can exert a torque on the molecule and therefore upset the local liquid structure, hopefully generating a small void, the electron can find a low free-energy state. That initial trap can then deepen and lead to the stable solvated electron. For very large molecules that cannot pack well in the liquid state, small void regions exist even in the normal liquid; often bond dipoles may point toward these void areas, and thus this is a weak trap even without any cooperative action from the electron. In such cases the electron becomes weakly trapped at such a site, and then, as its wave function becomes more localized, it is able to exert a more dominant force. It can then reorganize the medium to stabilize the electron further. This is a cooperative phenomena. It is quite likely that the initial stages of localization in all liquids involve these preexisting traps—traps that exist because of the complicated molecular structure of the molecule, or traps that might exist because of transitory solvent fluctuations. In all of these cases the electron interacts not with one molecule at a time but with many. Thus, we have a large statistical mechanical problem if we want to treat it in full detail.

There is another situation, probably very rare, in which the picture is simpler. We refer to the case where the electron is so tightly bound to one molecule that the species is more properly called a solvated anion and not a solvated electron. To arrive at such a situation the electron affinity of the molecule must be extremely large. This will then force the extra electron to be highly localized. If the electron were not highly localized, its wave function would overlap large numbers of adjacent molecules and we would thus have an electrostrictive region at most, but more likely a very delocalized state. In some amine and ether solutions there are well-known and well-documented examples of metal anions. The best current review of such species is by Edwards.[10] In these cases the negative ion is not the same species as the solvent; hence, delocalization is less likely for symmetry reasons. However, even when species like $S^-$ or $M^-$ (solvent or metal anion) exist, their properties are usually strongly influenced by the medium, and therefore the negative species is also highly influenced by a variety of factors, including solvent structure, temperature, and pressure.

There is currently a great deal of controversy concerning the solvated anion model.[11] A goodly amount of the confusion concerns semantics. In this chapter we insist that the solvated anion term is appropriate only if the trapped electron has properties similar to the free anion (modified slightly by solvent effects), there is strong evidence of electron spin interaction with but one molecule, and, furthermore, that the local structure of the solvent is not drastically perturbed except as needed by electrostriction. With that definition the solvated anion can be dropped from our discussion because no such evidence exists. Some authors have generalized the term so much that it is not easy to distinguish it from the more conventional trapped electron model we will now consider. Their major points appear to be that there may be no void and that the excited state may be a continuum state. They have not proven the last point for any detailed molecular model, but the first point can be approximately true not only for the solvated anion but for a great many models. It is true in some cases considered later, when the localized semimolecular models are used.

# CALCULATIONS

## *Models*

Let us now consider a model of a trapped electron in a polar fluid containing anisotropic molecules. We will first discuss a very simple model and then look at the more detailed calculations that have been performed for electrons in water, ammonia, and especially alcohols.

The following factors must be considered:

a. Electron–molecule interactions for molecules in the first coordination layer.
b. Electron–molecule interactions for electrons further removed from the trapping site. In many cases this will be treated as a continuum.
c. Molecule–molecule interactions, a complicated function of the orientation of the molecules relative to the radius vector drawn from the center of the trapping site.
d. Any effects related to solvent reorganization not included in item c.

This is a listing of what one needs to consider for a model semiclassical calculation. For quantum or cluster-type calculations some of these effects are combined. In any case the most serious problems relate to electron solvent interactions, especially those at large distances. Furthermore, the exact statistical mechanics of such regions and the proper inclusion of temperature are difficult to treat properly. This is an area in which research is now beginning, but so far all of that work has also made serious approximations.

For simple model calculations we can write the total energy of the system as

$$E_{\mathrm{T}} = E_{\mathrm{E}} + E_{\mathrm{MM}} + E_{\mathrm{S}} \tag{8-3}$$

We will discuss each term separately for simple polar molecules first. Later we will see that the same ideas can be applied to the situation involving an electron in a more complex fluid, such as that of a large hydrocarbon or alcohol.

The first term in Eq. (8-3) is the electronic energy for the electron interacting with molecules in the first coordination layer, the closest molecules, as well as with the remainder of the medium. It is the solution of a Schroedinger equation for the effective energy of the electron:

$$(-\tfrac{1}{2}\nabla^2 + V + V_{\mathrm{R}} + V_{\mathrm{C}})\Psi = E_{\mathrm{E}}\Psi \tag{8-4}$$

where $V$ is the classical potential between an electron and the molecule. In the case of polar molecules this is often approximated as simply the charge-dipole interaction with the first coordination shell molecules and some sort of charge continuum interaction for the interactions with molecules beyond the first layer. The first part of this term is then of the form

$$V(\text{dipole}) = \frac{-e\langle\mu\rangle}{r^2} \tag{8-5}$$

where $\mu$ is the average dipole moment along the radius vector to the center of the cavity. At zero temperature this term would simply be the dipole moment times the cosine of the angle it makes with respect to the radius vector. At finite temperatures we can approximate the average value of the cosine by using the Langevin expression if the only field orienting the molecules is the electric field of the trapped electron. In that case we assume that the fraction of the charge within the "cavity" generates the field that aligns the molecules. In this way we also obtain some measure of the effects of temperature on the energy levels. To get a better measure we need to do a complete statistical study of the inner coordination layer. Various schemes have been used to include the effects of $V_{\mathrm{R}}$, the pseudopotential correction. The best attempt was by Gaathon and Jortner in work presented at Colloque Weyl III.[12] In that work the "core" of the molecule was represented by a "soft" sphere, that is, one with a finite barrier to penetration. In other models this term has been largely ignored because the electron is so diffuse it spends very little time near any one molecule anyway. Of course, its effects are explicitly included if all of the electrons are considered as in the classic Hartree–Fock work of Newton[13] on the hydrated and ammoniated electron.

In Eq. (8-4), $V_C$ is the interaction of the electron with the medium beyond the first coordination layer. This has been approximated in most calculations as the sum of two contributions; the effective pseudopotential between the electron and the medium, and the interaction between the electron and the dipolar field of the medium. The former can be well represented as $V_0$, the energy of the bottom of the conduction band.[14] This is small for most polar liquids and rather uncertain ($\pm 0.5$ eV). Often this is arbitrarily assumed to be zero for many polar liquids. For nonpolar liquids there are good experimental data, as indicated by the previous chapter. The second contribution is normally approximated by using formulas derived from polaron theory. In these calculations the medium is treated as a polar continuum field, and the interaction is assumed to be the same as that of an electron in a polar solid (or ionic crystal). The best approach is to use the self-consistent field equations derived by Jortner:

$$\begin{aligned} V_C &= -\left(\frac{e}{2}\right)\left(1-\frac{1}{D_s}\right)f(R) \qquad r < R \\ &= -\left(\frac{e}{2}\right)\left(1-\frac{1}{D_s}\right)f(r) \qquad r > R \end{aligned} \tag{8-6}$$

where $D_s$ is the static dielectric constant of the fluid, and $f(r)$ is the solution of Poisson's equation

$$\nabla^2 f = 4\pi e \Psi^2 \tag{8-7}$$

where $\Psi$ is the wave function of the extra electron creating the field for $r > R$. It is obvious that the solution of this equation involves self-consistency because the potential $f$ depends on $\Psi$, which in turn depends on $f$. In practice, the solutions converge very rapidly because the electron is almost totally trapped in the first coordination layer and $f(r)$ is roughly $-e^2/r$. That is usually used as the initial guess. In the above equations the entire polarization contributes to the binding of the electron.

It is often true that the major factor in localizing the electron is only the orientational polarization, the so-called inertial polarization, which has a slower response than does the electronic polarization. In such a case the binding is best given by the adiabatic approximation to $V_C$:

$$\begin{aligned} V_C &= -\frac{\beta e^2}{R} \qquad r < R \\ &= -\frac{\beta e^2}{r} \qquad r > R \end{aligned} \tag{8-8}$$

with

$$\beta = \frac{1}{D_{op}} - \frac{1}{D_s}$$

These equations are easier to evaluate than Eqs. (8-6) and (8-7) but require that the electronic contributions be added separately. For ammonia the two approaches yield similar results. For water the electronic part is large enough that Eqs. (8-6) and (8-7) are to be preferred, although the general features of the results are quite similar in the two cases.

There are several important points to be made regarding these formulae. They are not exact for liquids. Many approximations are involved: the continuum approximation, the neglect of frequency and spatial dependence of the dielectric constant, and many more. The latter effect has been evaluated in a variety of ways by Dogonadze and co-workers[15] and appears to have some small but significant effects.[16] For water it tends to modify the parameter $\beta$ by about 10 to 15%; the other parameters change by about the same amount. Concerning the continuum approximation, Iguchi,[17] Kestner,[18] Carmichael,[19] and others have pointed out that the interaction energy at large distances, especially in a hydrogen-bonded fluid, is overestimated by these formulae because they assume that the only orientational influence is from the trapped electron, whereas, in fact, the strongest forces on molecules far from the electron arise from their nearest neighbors, which want to maintain the hydrogen-bonded structure of the liquid as much as possible. This provides a cutoff of the effective potential and significantly modifies the electron medium interaction. It also means that there are now a finite number of bound vertical excited electronic states, even within this model. Most importantly, however, is the observation that in this discussion we are only using the electron medium interaction for the medium beyond the first coordination layer. Thus, any uncertainties in the detailed interaction are minimized. In contrast to this model, recent work by a number of authors, most notably Brodsky and Tsarevskii[20] as well as Tachiya,[21] uses more elaborate models for the electron medium interaction but does not treat the first coordination layer explicitly. It remains to be seen if modifying their work to include the first coordination layer explictly greatly changes their results. From studies on a variety of fluids it is expected that there could be major effects. Golden and Tuttle[22] and Funabashi et al[23] have used such continuum models to indicate that the solvated anion picture is to be preferred, but their approaches are not justified by any ab initio calculations. These authors need to carry out more detailed ab initio calculations. Indications are, however, that the optimal structure is far from that of a solvated anion, as we will see shortly. Further work needs to be done to blend all of the ideas into one general model. However, that is not easy because the techniques used by the various groups differ so dramatically.

We have yet to define the two remaining terms in Eq. (8-3). The energy contribution $E_{MM}$ is the molecule–molecule interactions that must be included explicitly in the model but, of course, are automatically included in a priori calculations that include all of the electrons of all of the molecules in the first coordination layer. The model has the advantage of placing emphasis on this term explicitly. In general, the advantage of models is their ability to explore the emphasis of this term. The main contribution in the molecule–molecule term is the effect of hydrogen–hydrogen repulsions because the electron attempts to rotate the molecules in the first coordination layer in an effort to achieve the maximum interaction with the molecule's dipole moment. In doing this, however, at least in small molecules such as water, ammonia, and small-chain alcohols, this will force the hydrogens on adjacent molecules to interfere with one another. Thus there will be a strong repulsion that will raise the energy of the system unless the molecules are allowed to move outward and hence apart. For large molecules such as long-chain hydrocarbons or alcohols, the $E_{MM}$ term for the entire molecule is not as important as that part that relates to just the orientational potential of the bonds in the immediate vicinity of the trapped electron. The rest of the molecule, in that case, has very little to do with the electronic energy of the solvated electron alone.

The final contribution, $E_S$, in Eq. (8-3) includes all effects neglected in the other terms. For a very large cavity, such as that found for an electron trapped in liquid helium, this must include the pressure–volume work. In early studies the microscopic surface tension energy was also included. That term represents the energy needed to create a surface in a fluid. If the macroscopic surface tension is used in polar fluids, the effect is found to be quite small. In addition, there is a serious concern that the surface tension contribution does not include terms already present in $E_{MM}$. If the $E_{MM}$ term only includes hydrogen–hydrogen repulsions, then double counting of the effects is unlikely, but if all intermolecular interactions are considered, we are faced with possible inconsistencies. These are not serious unless the term is large. This entire matter has not been considered adequately and should be studied as part of a general statistical mechanical review of electrons bound in a cluster at a finite temperature. Such studies have not yet been done, but there are several nice papers on hydrated ions of nitrite ions by Bannerjee et al.[24]

In the discussion thus far we have concentrated primarily on the ground-state structure of the trapped electron. Of equal importance are the excited states. These are obtained in this model by using the excited-state wave function obtained from Eq. (8-4), except that the proper polarization term must be included. The expression in Eq. (8-7) is incorrect because the transition occurs with no change in the slow or inertial polarization during the transition. The detailed expressions are in papers by Jortner[25] and the review by Kevan and Feng.[26] The wave function is roughly like a *p*-orbital if the ground state is rather highly symmetrical (s-like). There is no rigorous symmetry in the problem, but if the environment of the electron is of reasonably high symmetry as expected in the ground state (at least at zero temperature?), then the electron wave function can be classified approximately by spherical harmonics. There are problems with the excited-state calculations that we must acknowledge. The inertial polarization of the medium does change with excitation, and this must be included correctly. Tachiya[21] and then Brodsky and Tsarevskii[21] have attempted to do this properly (Kuznetsov[27] has done something similar for reactions of solvated electrons in which similar considerations apply). They have found rather dramatic changes from the simple models, but again their results only apply to a purely continuum fluid with, at most, point defects. They find an unbound or continuum excited state in most cases, in accord with the Tuttle and Golden moment analysis,[28] but none of these studies includes all of the molecular effects. On the other hand, the models here explicitly include the remainder of the medium, but the treatment of the medium is less critical because most of the energy contributions are from molecules close to the electron. Nonetheless, in studies to be presented next the excited-state polarization of the medium is properly represented, as required by the Jortner analysis for vertical transitions.[25]

Equation (8-2) must now be solved. This is a typical quantum chemistry problem. As we have already indicated, the equation can be solved at the model level or more elaborately. Let us first consider the models where the solvent is approximated by point dipoles. This model has been extensively studied by several groups, notably those associated with Jortner,[12,29] and Kestner,[30–32] who mainly used the adiabatic model and concentrated on metal–ammonia solutions. Feng and Kevan[26] have used the more elaborate self-consistent field model to study not only the hydrated electron but also electrons in numerous alcohols.

## *A Priori Calculations*

In the work described above, the molecules are treated in a very approximate manner, namely, as point dipoles with some pseudopotential character. Such work is ideal for studying the general features of the trapped electrons and for discovering the role of the most important factors relating to energy and stability, but they are inherently subject to criticism because a fair amount of physical intuition is present in the model. The models have been explored regarding all of these possibilities, but there is always the chance that something was missed. For this reason Newton[13] undertook a major calculation. He calculated the energy of a system of several water molecules and an extra electron in the Hartree–Fock approximation. He found that a cluster of five water molecules as found in liquid water cannot attach an additional electron but has an instability of about 1 eV. On the other hand, he did find that four water molecules, if allowed to orient with their dipoles pointing toward a central point, and if allowed to move outward from that point somewhat, will just barely support a bound state. He then added to the calculation the long-range electron medium interaction discussed in the previous section. The electron then becomes bound with an energy of about 1 eV, and the transition energy to the first excited state is about 1.78 eV or very close to the experimental value of 1.72 eV.

The basis set used was a Gaussian 4-31G set, which has been used extensively in other studies of water and ice. It is believed to yield reasonable water properties despite its slight exaggeration of the water dipole moment.

Newton[13] also studied the system of four ammonia molecules and an electron. He found that the cavity here was larger and in fair agreement with simple model predictions as well as other experimental data. His excitation energy to the first excited state when the proper long-range term $V_C$ was added ($V_0 = 1.0$ eV) was found to be 0.8 eV, in good agreement with the experimental value of 0.80 eV.

There have been a number of other studies that have considered the solvated electron in water or ammonia, which preceded Newton. In particular, we want to mention Naleway and Schwarz,[33] who calculated water electron clusters but with less accurate basis sets. Also there is the work of Howat and Webster,[34] who considered clusters of up to four molecules of water or ammonia, using the Intermediate Neglect of Differential Overlap (INDO) molecular orbital approximation. They also calculated electron spin resonance (ESR) line widths and estimated volume expansion data. In addition, there are multiple scattering $-X\alpha$ calculations by Moskowitz et al,[35] but the results were not encouraging.

The above results were expanded by Noell and Morokuma,[36] who added to the four-water-molecule cluster additional water molecules represented by three point charges. In this way they were able to include a great many water molecules and study the role of the second and third coordination layers of water in determining the solvation energy without the need to introduce the $V_C$ term. Their results showed that the second and third coordination layers did indeed play a major role, but the convergence of the energy as a function of the number of waters included was disturbingly slow. Nevertheless, they were able to obtain heats of solvation close to the experimental value.

In the next section we will briefly mention the work of Rao and Kestner,[37] who extended Newton's work by including a few more molecules in the calculations. The Newton geometry was used for the first four water molecules, and then up to four more waters were added. These studies also show a very slow convergence with respect to second and third coordination layers and thus are more appropriate for comparison with real clusters than with the real hydrated electron in liquid water.

These studies are important in two major ways: first, they provide a check on the models; second, they can study certain properties that can never be treated in a model. With regard to the latter comment, we refer to such items as electron density at the nuclei, and so on.

## *Cluster Calculations*

There is one area in which it was—and still is—hoped that theory and experiment can both be refined sufficiently to allow detailed comparisons; namely, small clusters of small molecules, water, ammonia, and the like, containing an electron. The relevance of this to radiation chemistry and the real solvated electron is that these clusters should be similar to the first coordination layer, and if a model calculation yields the correct results for a small cluster it should be reasonable for electrons trapped in a real fluid.

The experimental data is of several major types: absorption spectral and electron mobility studies as functions of the density of the polar fluid. Each of these experiments is complicated to perform and equally complicated to explain. Typical facts are as follows: for water, even at densities 1% of the normal liquid, a typical hydrated electron absorption spectra was observed, suggesting that the electron was trapped. The spectrum shifted slightly with density but not dramatically.[38] For ammonia there appeared to be critical densities below which an absorption spectrum could not be observed.[39] Gaathon and Jortner[40] in a very rough calculation indicated that this region would typically have ammonia clusters of 13 molecules. This number is only a guideline and could have huge errors. Nevertheless, it did indicate that water and ammonia should differ. Krebs et al[41,42] have studied mobility of electrons in supercritical water and ammonia and have found a continuous shift in the mobility vs. fluid density. At higher densities the electron mobility behaves similarly to trapped electrons in the liquids. However, any analysis of the data is complicated. It is clear that mobility and spectroscopy yield different criteria for critical localization densities (as one would suspect).

Armbruster et al[43] reported the observation of $(H_2O)_8^-$ in a supersonic nozzle beam. That original study is now very suspect because of chloride ion contamination, but they have now constructed a better apparatus and eliminated all chloride ions. They now have definitive evidence for $(H_2O)_{11}^-$ and $(D_2O)_{13}^-$ as well as larger species.[43] Furthermore, they have seen large (about 35-molecule) ammonia negative-ion clusters as well but, as of now, no evidence for alcohol clusters.[45] Several research groups previously had searched for such species but had no success. Those groups attempted to add an electron to water clusters by electron bombardment of neutral clusters or by charge exchange of neutral clusters with heavy alkali metals. The models of the trapped electrons in water and ammonia, in particular, indicate that the solvent structures in the region of the electron are very distorted from these in the normal liquid (or in a neutral cluster), and thus it is reasonable that those previous methods did not observe the negative-ion species. The vertical electron affinity of a neutral cluster is very unfavorable for electron attachment. The process of binding the electron to a neutral cluster must involve large reorganizations of the molecules. The only exception would be a model to be discussed shortly or a solvated anion model.

The failure to observe small clusters is also evidence that the solvated anion model is not reasonable. What Haberland et al[43–45] did that was different was to create the hydrated electron in a fluid at a reasonable gas density by using $\beta$-emission in the first experiments and by trapping low-energy electrons photoejected from organic molecules. Thus, under the initial conditions of temperature and density they had "normal" hydrated electrons. That mixture was then exploded in a supersonic nozzle. The beam was allowed to expand, and the negative clusters were collected by a mass discriminator. Thus, it is clear that clusters of $(H_2O)_{11}^-$ and larger can exist for a finite time. Whether smaller species actually exist in the early stages of the beam but do not live long enough to be detected at the end of the apparatus is a question they cannot answer. Unfortunately, the beam experiments do not provide us with criteria to determine whether the species are stable with respect to the neutral, or whether they are a metastable state that would prefer to add or lose waters, given more time or source of other molecules. We do know that $H_2O^-$ is unstable, but we have no basis to assume that the relative numbers of various clusters formed by these explosions have any simple relationship with their relative thermodynamic stability. The experiments are extremely important, and further studies with other molecules will provide us with new ways to look at localized electrons.

Unfortunately, these beam experiments did not provide us with any information showing how the electron is trapped. Basically, we need to know if the trap depth in the cluster is as large as we suspect it is in the liquid or in the small clusters we have studied experimentally. This is more important in light of an observation made by Antoniewicz et al,[46] who have indicated that a cluster of sufficient size and polarizability can weakly trap an electron to its surface. This is a weakly bound electron and is probably not the species seen in the beam work, Or is it? We now need studies of the binding energy of the trapped electron in the small clusters, using laser photoelectron detachment studies. Haberland[45] is exploring such ideas if he can raise his beam intensity sufficiently.

The only theoretical work directly related to these negative cluster results are those of Rao and Kestner[37] in which small clusters were calculated using the Hartree–Fock method. Those studies found that clusters of size eight were still unstable relative to the neutral, but the inherent error of the 4-31G basis set in handling the molecule–molecule interactions caused the authors to be cautious in their analysis. Kestner and Jortner[47] considered a better model for the molecular interactions and reached the conclusion that species with eight waters would be unstable with respect to the neutral cluster by at least 20 kcal/mole. The studies also were able to show that the excitation energy of the electron in a small cluster should be smaller than that in the bulk liquid. Possibly this can be studied in the beam work also.

## *Electrons in Hydrocarbon Liquids*

The above methods have been applied by various authors to the trapping of electrons by large alkanes and other nonpolar molecules. In those media one assumes that there will exist regions in which the C—H bonds will point toward the center of a void, which can then act as a trapping center for the electron. Those regions can bind an electron.

Using typical geometries for the $CH_2$ groups, Feng et al[48] showed that a typical arrangement of about four $CH_2$ groups leads to a bound state of roughly 0.5 eV. The charge distribution, however, is quite diffuse relative to that found in polar media. Typically, only about 25% of the charge is in the cavity relative to the much larger amount in polar fluids. The theory predicts optical excitation energies of only 0.3–0.4 eV versus the experimental values, which are usually near 0.6 eV, but the calculation is quite sensitive to the value of $V_0$ used and to the long-range polarization energy used. Ichikawa and Yoshida[49] have extended this work by approximating the potential well formed in 3-methylpentane glass and performing a better calculation of the wave function and the energy. The results suggest that the first excited state is unbound, explaining the broad experimental absorption spectra. Nishida[50] and Kimura et al[51] evaluated the trapping well depth from exchange interactions, an approximation to the pseudopotential. Reasonable energies are obtained, despite some criticism that some integrals might be overestimated (remarks by Carmichael quoted in ref. 26).

## *Electrons in Mixed Solvents*

One of the most interesting areas of theoretical and experimental research relating to solvated electrons is the study of mixed solvents. Although such work has been done in all sorts of mixtures, we will deal only with a few binary mixtures of polar solvents.

The experimental work has been summarized in a number of review articles and in several chapters of this book. Some of the best reviews are by Dorfman et al[52,53] Depending on the two liquids selected, one can find a variety of behavior ranging from an almost linear dependence of the excitation spectrum on composition (eg, ammonia–water and ethylenediamine–ether) to one in which one component dominates (eg, water–ether in which water dominates and ethanol–hydrocarbon where ethanol dominates).

Two theoretical studies have been made of the water–ammonia system. Howat and Webster[34] calculated the energy of $[(H_2O)_n(NH_3)_{4-n}]^-$ species by their semiempirical (INDO) molecular orbital theory. They found an almost linear dependence in the excitation energy vs. $n$. This work makes an implicit assumption that the cavity species contains, in the first solvation layer, the same proportions of the component molecules as added stoichiometrically.

Logan and Kestner[54] have calculated the ground-state energy and the excitation energy for all species in all solvent compositions by using a model similar to Gaathon and Jortner,[12] including long- and short-range interactions. They found that it was necessary to assume intermolecular repulsions between water molecules (and a large cavity size) in order that all coordination species in a given bulk solvent have similar free energies. The free energy was obtained from the total energy by adding entropy effects from demixing and entropies of solvation as well as estimated hydrogen-bond-breaking contributions. With these assumptions the excitation energy vs. mole fraction of water varies almost linearly, but the water dominates more than in the experimental work of Dye et al.[55] More serious is the fact that all of the structures tend to contribute to the spectrum, and one expects to see very broad line shapes relative to the pure solvent cases. The experiments do not show great broadening at intermediate compositions, but the pure solvent spectra are also much broader than the theory predicts. If the cavity size is allowed to shrink by reducing the repulsions between water molecules, cavities with high proportions of water dominate and thus one obtains an excitation spectrum that would vary little with composition. In brief, more work is needed on this system.

Although the above calculations indicate that theoretical work in this area is rather simplistic, they point out the need for much more elaborate work, including studies that properly include the solvent structure. This probably cannot be done except as part of a detailed statistical study of liquid structure including an electron. Unless the effects of medium reorganization are properly included for all possible species that must be involved and unless their relative free energies are correctly calculated, any results on mixed solvent systems must be viewed with caution. As we have seen, at the moment no calculations satisfying these criteria exist.

In a series of papers Tuttle and Golden[56] have attempted to explain not only the line shape of an individual solvated species but also that of mixtures of polar species. They assume that the local environment of the electron is essentially a simple average of the structures in pure solvents. Their procedure, called "shape stability," assumes that one can construct from the shape of the spectrum in each pure solvent the shape in a mixed solvent and, furthermore, that there exists some fundamental shape of the solvated electron spectra in "all" solvents. The results of Stupak et al[57] on ammonia and methylamine mixtures at various compositions and temperatures are surprisingly good. It is not clear, however, what the true theoretical basis is for the procedure. It needs to be investigated further, but the high accuracy it provides suggests that there is some fundamental underlying theory. It also represents a rather different approach, which can be used to supplement model calculations.

# GENERAL COMPARISONS: THEORY AND EXPERIMENT

The general result of the model calculations is that the molecules become oriented with respect to the trapped electron if the molecule is small enough for the molecule to rotate. The rotations of small molecules such as water or ammonia, however, cause the hydrogen atoms on the molecules to be forced into close proximity. That closeness causes the solvent molecules to have repulsion interactions with each other. This can be reduced and the total energy lowered if a void volume is generated. In summary, the lowest energy state found in these model calculations (and in the a priori work of Newton[13]) for electrons in water and ammonia as well as in liquids containing small alcohol molecules is essentially a "cavity"-type model in which the solvent is effectively pushed away to varying extents from the trapped electron. There is a distortion of the liquid structure in the theory that is much larger than would be reasonable in a solvated anion model. Depending on the molecules and the situation, the size of the void can be almost any size. Nevertheless the electron density is widely distributed over many molecules. These statements are true regardless of the form of the wave function used in the studies. Often Slater or exponential functions are used for $\Psi$ because they are natural when coulombic forces are involved. Others have used Gaussian functions,[58] exponential in distance from the cavity center squared, or even numerical methods.[59] The energies in the cases of ammonia and water are similar for all wave functions, differing in energy by 10 to 15% from one another.

It is very misleading to think of these results as representing a cavity model, because one then associates this with the electron-in-helium example where a large well-defined cavity does indeed exist. As we keep emphasizing, the species in polar fluids are very different. Although the charge is localized between molecules, it is a very diffuse charge distribution, and the void volume generated can be very small. There is no good way to measure void volume in a liquid in which the trapped species are at low concentration and short lived. However, for metal–ammonia solutions the void volume is quite large (primarily because the ammonia molecule has three hydrogens that prevent an oriented molecule from getting close to another oriented molecule). In that case the theoretical results are in reasonable agreement with the experimental values as estimated by Jortner after correcting for the positive ions and their electrostructive effects.[60,61] For water the Hentz et al[62] and Schindewolf–Olinger[63] values for the pressure effects on the reaction

$$(e^-)_{aq} + (e^-)_{aq} \rightarrow H_2 + 2\,OH^- \qquad (8\text{-}9)$$

have been taken to be a measure of the size of the hydrated electron, as well as as that in ammonia. It is not clear that such identification is valid, because according to theory the electron wave function is very diffuse and the real void volume is small. Pressure effects on solvated electron reaction rates have not been studied theoretically, but they are quite complex because the major effect of pressure is on dielectric constants, compression of solvent, and many other features almost totally unrelated to void volume. Thus, although the theory predicts a void volume a factor of three larger than that needed to explain the pressure dependence of Reaction (8-9), it is unclear whether theoretical and experimental results are in serious conflict or even if they are describing the same parameter. It is interesting that the theoretical void volume predicted by simple models and the molecular calculations of Newton are in good agreement. From these arguments it is obvious that the void volume in liquids of even more complex structures cannot be explained in any simplistic way.

Another set of data that can be used to study the exact geometry of the environment of a trapped electron is the analysis of the line shapes of multiresonance ESR. Kevan and co-workers[64] have studied several systems: 10 M NaOH glass ice, and some alcohol solutions. The trapped electron in sodium hydroxide glass is quite similar to the results of theoretical models; There is a substantial void volume, the molecules are oriented, and the electron density is centered about the void. The major difference is that the water molecules in the ice appear to be oriented with their O—H bonds rather than their dipole moment (which bisects the H—O—H angle) pointing at the center of the void. The experimental studies have been criticized in two major ways: First, such a high concentration of hydroxide must modify the solid structure; second, the analysis of the resonance spectra is not unique.[65] Kevan partially answered the first objection by studying sodium–water mixtures codeposited on a cold finger.[66] Although the analysis could not be made as detailed as in the case of the hydroxide glass, the resonance spectrum did appear to be extremely similar. The second objection has not been answered in the literature, but it appears to be directed only at the most elementary spectral (moment) analyses of the Kevan group and is not relevant to their more complete work. It is clear, however, that a 10 *M* NaOH glass is not the same as pure water or pure ice. For their studies of simple alcohols[64] the solvated electron reveals a structure in complete agreement with the very general features one would expect from the model studies regarding molecular orientation and the local environment of the trapped electron except for exact orientational angles.

A discrepancy between the models and the experimental results of Kevan relates to the electron density at the hydrogens closest to the center of the trapped electron.[25,64] For water Kevan finds a positive spin density at the hydrogens, whereas all calculations, including the best work of Newton,[13] yield a negative value. Spin density is very sensitive to details of electronic wave functions, and thus it is not surprising that all of the present theoretical calculations are not sufficiently accurate; all have used quite simple basis sets, no electron correlation, and, as pointed out recently by Catterall,[67] all can include small amounts of spin contamination due to the use of the unrestricted Hartree–Fock calculations, which could give misleading results. It is unlikely that we will ever have theoretical results sufficiently accurate to agree with these experimental results unless some extremely lengthy calculations are performed.

One can get a good idea of the utility as well as the accuracy of the theoretical models by reviewing some of the results of Kevan et al[26] and other workers for the excitation energy of solvated electrons in various media. In Table 8-1 we present some results on various systems using the self-consistent field (SCF) model. All used various geometries for the molecules that were consistent with dense packing while maximizing the electron dipolar interaction. The agreement with peak position is reasonable, but, more important, the trends with change in the hydrocarbon fragment of the alcohol are very nicely reproduced. The big fragments such as *tert*-butyl prevents the cavity size from becoming small, and thus the excitation energy is large. The linear alcohols, on the other hand, can still pack closely and thus have excitation energies quite similar to one another (see ref. 68 for a summary of experimental data). Simple attempts to make such extensive calculations with continuum models in which the only parameter is the macroscopic medium dielectric constant have not been successful.

**Table 8-1.** Theoretical Results for Band Maxima of Solvated Electrons in Various Liquids

| Liquid | Theoretical Results (eV) | Experimental Results (eV) |
|---|---|---|
| Ammonia | 0.89[12] | 0.80 |
| | 0.80[13] | |
| | 1.18[26] | |
| Water | 2.00[12] | 1.72 |
| | 1.78[13] | |
| | 2.15[26] | |
| Ice | 1.84[26] | 1.9 |
| Methanol | 2.09[26] | 2.3 (77K) |
| | 1.85[26] | 1.87 (298K) |
| Ethanol | 2.15[26] | 2.3 (77K) |
| Tetrahydrofuran | 0.583[26] | 0.585 Room Temperature |
| 2-Methyl tetrahydrofuran | 1.04[26] | 1.0 (77K) |
| Triethylamine | 0.83[26] | 0.75 |

The local molecular structure of the molecule about the trap is most important. Nevertheless, for a model that has a great many known shortcomings as well as probably others not yet explored in detail, the agreement of theory and experiment is reasonable. The aim of theory in these very complex areas is simply to provide the unifying features that relate one system to another. The present models appear to do this quite well. Any theory of this simplicity is bound to have its limitations.

The major shortcoming of the models is their inability to explain the very broad line shape characteristic of the solvated electron. The line widths predicted by the theory in its simplest form are very narrow, around 0.2 eV, rather than the 0.6 to 0.8 eV observed.[69,32] Furthermore, the temperature dependence of the experimental line width is very small compared to the square root of temperature dependence one would expect from a purely bound state model. Some authors have used this failure as a reason to reject the entire semimolecular model we have been discussing. Those authors prefer to treat the solvated electron as having the spectrum of an electron in a shallow well with a continuum excited state.[22] Unfortunately, those models cannot and do not derive this bound species from first principles or any type of molecular model. This means it has no predictive value in a new system. Despite our reservations about the simple empirical continuum excited state model, it does indicate that our simple semimolecular model may have some limitations in treating the excited states. There is an additional clue from another set of purely continuum calculations, namely, those of Tachiya,[21] more recent work of Brodsky and Tsvetskov,[70] and also Kuznetsov.[27] Those authors have allowed the polarization of the medium to be represented by a parameter in the calculation, and the energy states are a function of not only the more obvious molecular parameters but also the polarization, which is a continuous function. The energy states are then rather complex functions of all of these parameters and functionals. This surface is minimized, and the various excitations allowed are calculated. When these are thermally averaged, they obtain much broader line widths, widths comparable to the experimental data. However, they use a purely continuum model, and much of the variations that they include are already in the semimolecular model when various motions of the first coordination layer are included, as was done by Logan and Kestner,[32] for example. The more molecular models appear to allow for less variation and, hence, smaller line shapes. Obviously, something is wrong with parts of the molecular model. It could be just a matter of better representing the true solvent structure.

Using general ideas of fluctuating fields, several Russian workers have been able to obtain better line shapes for the static case[71] as well as for the spectra of time-dependent electron localization.[72,73] Their models are continuum type and require some integration with the more molecular models. Recently, Zusman has carried[73] out such analysis of his models within the context of electron transfer reactions.

The molecular model is known to have some serious limitations, but these may not affect the line shape. For example, the long-range electron solvent interaction in both the semimolecular and the purely continuum models behaves as the inverse in the distance, a typical coulombic interaction. Coulombic interactions can support an infinite number of bound states, and yet the experimental data does not give any hint of higher terms that would be possibly embedded in the known experimental line shape. All known, very high-energy transitions appear to be more of the nature of charge transfer to solvent spectra (CTTS). Iguchi[17] suggested the explanation for the true spectrum, which was then expanded upon by Kestner;[18] namely, at very large distances from the electron the field due to the electron is too weak to complete with hydrogen bonding in adjacent molecules. The net result is that the long coulombic tail of the interaction is washed out and only one bound state can result. The exact nature of the absorption spectrum from such a model has not been worked out but is expected to be similar to the semimolecular model because the long-range part of the interaction is small relative to the interaction with the molecules in the first coordination layer.

Except for the absorption spectra and maybe the volume expansion in the case of metal–ammonia solutions, there is little direct comparison possible to prove or to disprove the detailed theoretical models. Theory can calculate a large number of other features, but in most cases experimental values of these features are very uncertain. This refers particularly to heats of solution and the value of the conduction-band minimum. The experimental data are uncertain in these cases by at least 0.5 eV and maybe more. There is no direct observation of photoconductive states in polar liquids that could be used to determine $V_0$. Concerning structural data, the only possible comparison are with the alkaline glass studies by Kevan. That group claims to have sixfold coordination of water around the electron, but the medium is loaded with hydroxyl ions that cloud the picture. Several groups have considered theoretically the geometry question in detail. Using INDO, Howat and Webster[34] found that the OH oriented structure had the lower energy, but the entire calculation at the four-molecule level is unstable. Noell and Morokuma[36] also found the bond-oriented model slightly lower in energy, as did Rao and Kestner.[37] However, in all of these calculations there is no accurate way to estimate the changes that must be made in the liquid structure in order to accommodate either the fourfold or the sixfold coordination. Thus, the theoretical work is of limited use in deciding between the alternatives. Both experiment and theory, however, agree that the medium is substantially distorted and certainly not representative of what one would normally call a solvated anion. Another conclusive proof of the theory would be the prediction of spin densities at the various nuclei, especially the hydrogens. Present theory predicts the wrong sign for the spin densities at the hydrogens if the results in the alkaline glasses are typical of real liquids. In ammonia the sign appears to be correct. However, as any theoretician knows, the spin density is very hard to determine correctly, and the present calculations are very rough relative to what would be needed to get detailed agreement with experiment or even to compare the two.

The present semimolecular models or their more sophisticated extensions by Newton (and even the point-charge extensions by Morokuma and Noell[35]) seem capable of explaining the gross features of the structure of the solvated electron, namely, maxima in absorption spectra and general structural details. It cannot explain the spin densities or even the line shapes that appear to be related to the proper treatment of the excited states and lifetime broadening of the solvated electron. What can we expect from the theory in the future? It would be nice to be able to say that with a larger computer we can just put in a bigger basis set and get all of the right answers. Unfortunately, that does not appear to be likely. We will probably see a great many more involved calculations—and they are needed—but it is also unlikely that they will answer many questions. There are serious problems in applying simple quantum ideas to a liquid. For a start, the quantum work ignores thermal effects (except in some simple ways used in the semimolecular models) and is rather limited usually in the size of the system it can treat. In quantum chemistry the difficulty of the problem increases as the fourth or fifth power of the number of electrons, and that does not even begin to consider the geometrical parameters that need to be varied. So what then is the hope? The best answer seems to lie in some quantum statistical approach that will have to use some approximate interaction functions.

We have three such quantum statistical calculations so far in simple systems. The first study by Hiroike et al[74] applied boson liquid theory of impurity states to electrons in helium, and this resulted in detailed density profiles around the electron in liquid and gaseous helium. It was limited to low temperatures and first-order corrections to the liquid structure factors. Recently, several groups have expanded the similarities of the Feynman path integrals and quantum statistics to treat some solvated electron problems. Some of the basic theory is in the papers by Chandler and Wolynes,[75] Wolynes and Skinner,[76] Hynes and Grate,[77] and Adelman[78] to name a few (and not all of their co-workers). The applications are being made by Chandler et al,[79] Phillips,[80] Berne,[81] and a group led by Rahman and Parrinello.[83] The first group has studied electrons in dense gaseous helium first and is now involved in studying electrons trapped in water vapor. Although no new concepts have come from these studies as of now, the techniques are being fine tuned. They have confirmed the features of the more approximate earlier work. Rahman[82] is treating a completely different area, namely, electrons in molten salts. This study is of interest because the potentials and pseudopotentials required by the calculations have been extensively studied. Furthermore, there is a great deal of experimental data on metals in molten salts to which the theory can be applied. These quantum studies are very complex in their computer needs, and the issues of convergence of the results have not been settled in all cases. Nevertheless, these studies represent the best one can do at the moment in treating all of the features of these complex problems. It is possible that some of the work of Brodsky,[70] Kuznetsov,[27] and Zusman and Helmon[72] can be merged with the quantum statistical work to simplify the calculations, but that is speculation at the moment. In summary, in regard to the static properties of solvated electrons we can expect more detailed calculations that take proper account of the liquid and the temperature dependences. We will also see more studies on small clusters with much more precision due to their use of larger basis sets, but such results must always be treated cautiously. The real problem is that the solvated electron is a very complex species, subject to the whims of parameters like temperature, pressure, and, especially, the medium. It will require the entire bag of tricks being developed by modern many-body quantum statistical mechanicians.

## REACTIONS OF SOLVATED ELECTRONS

A few comments on applying theory to the prediction of reactions of the solvated electrons are in order because that topic constitutes a major portion of the book. There are two aspects to this problem: one concerns the form of the kinetics, the time dependence, the role of the medium (diffusion controlled or rate limited as in reactions in glasses), and the problem of the microscopic quantum theory of the rate of the reaction of the solvated species with various acceptors. Because the former topic is treated elsewhere in this volume and requires a major treatment on its own, we will limit our discussion to the reaction rate process for the two species fixed at a certain separation from one another.

Marcus,[83] as in most studies of electron transfer processes, in an American Chemical Society symposium in 1965, was the first to address this problem in any detail. He applied his semiclassical theory directly to these reactions and made some generalizations. He fully appreciated, even at that time, the magnitude of the approximations he made. Basically, he assumed that the solvated electron was like any other ion, except for issues of size, and had similar physical properties; he did not consider the tenuous nature of the species. Namely, the solvated electron is very sensitive to its environment, and one cannot routinely assume that the electronic properties are independent of the vibrational components. When one studies the results of semimolecular models, as a function of the vibrations, one sees dramatic effects of motions on the electronic wave function—that is, a breakdown of the Condon factorization used so commonly in electron transfer theory. In a brief attempt to study this problem, Webman and Kester[84] reported on some preliminary results in 1979. They found a typical 10 to 15% effect on the rate of a solvated electron reacting with typical species. Webman and Kestner used the semimolecular model wave functions to obtain their estimates. Beratan and Hopfield[85] have also considered other models in which the Born–Oppenheimer principle breaks down. This is essentially the same effect as the failure of the Condon approximation in that the wave function can no longer be factored into independent electronic and vibrational portions.

Recently, Ulstrup and Kuznetsov,[27] followed by more general papers by Kuznetsov,[27] have put these failures of the Condon approximation in a more general formalism. They also are able to include the disperion of the dielectric constant of the medium as well as polarization changes that can occur during the course of the reaction. They have considered some general electron transfer reactions and found the typical 10 to 15% effect for typical cases involving solvated electrons (much smaller when ions or normal molecules are involved), though undoubtedly larger discrepancies and larger effects occur in some cases. Their studies used a continuum model of the liquid, and thus they are not identical to those of Webman and Kestner. However, the two studies complement one another because each placed emphasis on different portions of the entire problem. This does not mean that in a real case the effect would be the sum of the two contributions (molecular and bulk liquid or continuum), but that each study alone is probably somewhat incomplete. Clearly, research on the reactions of solvated electrons is in its infancy, trailing the general theory of electron transfer reactions, which is also incomplete. What do we expect for the future of the theory in these treatments of the nearest neighbor effects and the bulk medium, with very special attention to the special features of liquid structure and the effects of temperature? Reactions cannot be studied without better understanding of the species, and that, as we have seen, is still full of uncertainty. Nevertheless, there are extremely important issues to be addressed here—whole classes of new questions. For example, the solvated electron is very sensitive to pressure, magnetic fields, and other factors. Just what is the effect of pressure on a reaction of a solvated electron? The answer is unclear, but it depends as much on the effect of pressure on the dielectric constant of the fluid as it does on any property of the "void" volume. One of the most interesting problems challenging theory is, What happens when two solvated (especially two hydrated electrons) interact? We know the final products from experiment, but the mechanism is unexplained and subject to a great deal of wide-ranging speculation. Little is known about the reaction at short times from experiment, also. Therefore, although the study of trapped electrons has introduced some nice unique theoretical problems, the problems involved in studying such reactions are complex, but therefore also more challenging.

## CONCLUDING REMARKS

Although we have made much progress in the study of trapped electrons in various fluids, there are even more intriguing issues remaining. In this chapter we did not elaborate about specific models but tried to concentrate on the important features and the generalizations that one can make. Some people do not agree even with the general outline of the theoretical description, preferring instead their own constructs. What was presented, however, is based on detailed model calculations. The critics must therefore have models in at least as much detail as these if the generalizations of this chapter are to be challenged. Thus far, most critics have not had detailed models. On some issues major corrections to the model may be necessary, but they must be detailed on the molecular level. We look forward to continued debate.

## ACKNOWLEDGMENTS

Special thanks are due to the Atomic Energy Commission, then Energy Research and Development Agency, and now the Department of Energy for supporting the research on solvated electrons and their reactions. Many thanks are obviously due to co-workers Jortner, Rice, Cohen, Hiroike, Copeland, Finley, Logan, Rao, and, for many hours of discussions and fresh insights, a word of appreciation to Newton, Friedman, and Sutin, and to the atmosphere at Brookhaven National Laboratory.

## REFERENCES

1. Webster, B. C.; Howat, G. *Radiat. Res. Rev.* **1972**, *4*, 259.
2. Kenney, G. A.; Walker, D. C. In "Electroanalytical Chemistry"; Bard, A., Ed.; Marcel Dekker; New York, 1975, Vol. 5, pp. 1–65.
3. Jortner, J.; Rice, S. A.; Wilson, E. G. In "Solutions Metal-Ammoniac: Proprietes Physicochimiques, Colloque Weyl"; Lepoutre, G.; Sienko, M. J., Eds.; W. A. Benjamin: New York, 1964, pp. 222–276.
4. For general background see the original paper by Phillips, J. C.; Kleinman, L. *Phys. Rev.* **1959**, *116*, 287.
5. Cohen, M. H.; Heine, V. *Phys. Rev.* **1961**, *12*, 1821.
6. Jortner, J.; Kestner, N. R.; Cohen, M. H.; Rice, S. A. In "New Developments in Quantum Chemistry—Istanbul Lectures"; Sinanoglu, O., Ed.; Academic Press, New York, 1966.
7. Kestner, N. R.; Jortner, J.; Cohen, M. H.; Rice, S. A. *Phys. Rev.* **1956**, *140*, A56.
8. For a review see Jortner, J.; Kestner, N. R. In "Metal Ammonia Solutions—Colloque Weyl II"; Lagowski, J.; Sienko, M., Eds.; Butterworths: London, 1970.
9. Jortner, J. *Ber. Bunsenges. Phys. Chem.* **1971**, *75*, 696.
10. Edwards, P. P. *Adv. Inorg. Radiochem.* **1982**, *25*, 135.
11. Golden, S.; Tuttle, T. R. *J. Phys. Chem.* **1978**, *82*, 944. For a definitive response to this, see Symons, M. C. R. *Radiat. Phys. Chem.* **1981**, *17*, 425.
12. Gaathon, A.; Jortner, J. In "Electrons in Fluids"; Jortner, J.; Kestner, N. R., Eds.; Springer-Verlag: Berlin, 1973, pp. 429–446.
13. Newton, M. *J. Phys. Chem.* **1975**, *79*, 2795.
14. Springett, B. E.; Cohen, M. H.; Jortner, J. *Phys. Rev.* **1967**, *159*, 183.
15. For a review, with particular reference to charge transfer reactions, see Dogonadze, R. R.; Kuznetsov, A. M.; Marsagishvili, T. A. *Electrochim. Acta* **1980**, *25*, 1.
16. Examples are in Dogonadze, R. R.; Korneyshev, A. A. *J. Chem. Soc., Faraday Trans.* 2 **1974**, *70*, 1121; Kjaer, A. M.; Ulstrup, J. *Inorg. Chem.* **1979**, *12*, 3624.
17. Iguchi, K. *J. Chem. Phys.* **1968**, *48*, 1735; *Ibid.* **1969**, *51*, 3137.
18. Kestner, N. R. *Can. J. Chem.* **1977**, *55*, 1937.
19. Carmichael, I. "Abstracts of Papers", 28th Congress of the IUPAC, Vancouver, B.C.; 1981, Paper No. PH161.

20. Brodsky, A. M.; Tsarevskii, A. V. *Soviet Electrochem.* **1973**, *9,* 1571; **1978**, *14,* 151; *Adv. Chem. Phys.* **1980**, *44,* 483.
21. Tachiya, M.; Tabata, Y.; Oshima, K. *J. Phys. Chem.* **1973**, *77,* 263; Tachiya, M. *J. Chem. Phys.* **1974**, *60,* 2275.
22. Golden, S.; Tuttle, T. R. *J. Chem. Soc., Faraday Trans. 2,* **1979**, *75,* 474, 1146; *ibid.* **1982**, *78,* 1581.
23. Funabashi, K.; Carmichael, I.; Hamill, W. *J. Chem. Phys.* **1978**, *69,* 2652. Stradowskii, Cz.; Hamill, W. H. *J. Phys. Chem.* **1978**, *80,* 1054. Razem, D.; Hamill, W. H. *J. Phys. Chem.* **1977**, *81,* 1625; *ibid.* **1978**, *82,* 488.
24. Bannerjee, R.; Shepard, R.; Simons, J. *J. Chem. Phys.* **1980**, *73,* 1814.
25. Jortner, J. *Mol. Phys.* **1962**, *5,* 257.
26. Feng, D. F.; Kevan, L. *Chem. Rev.* **1980**, *80,* 1.
27. Kuznetsov, A. M. *J. Chem. Soc. Faraday Disc.* **1983**, *74,* 25; *Soviet Electrochem.* **1982**, *18,* 522; Kuznetsov, A. M.; Ulstrup, J. *J. Chem. Soc. Faraday Disc.* **1983**, *74,* 31, Itskovich, E. M.; Kuznetsov, A. M. *Soviet Electrochem.* **1982**, *18,* 824.
28. Golden, S.; Tuttle, T. R. *J. Chem. Soc., Faraday Trans. 2* **1979**, *75,* 474.
29. Copeland, D. A.; Kestner, N. R; Jortner, J. *J. Chem. Phys.* **1970**, *53,* 1189.
30. Kestner, N. R. In "Electrons in Fluids"; Jortner, J.; Kestner, M. R., Eds.; Springer-Verlag, Berlin, 1973, pp. 1–29.
31. Kestner, N. R.; Jortner, J. *J. Phys. Chem.* **1973**, *77,* 1040.
32. Logan, J.; Kestner, N. R. *J. Phys. Chem.* **1972**, *76,* 2738.
33. Naleway, C. A.; Schwarz, M. E. *J. Phys. Chem.* **1972**, *76,* 3905. Related work is to be found in Ishirmaru, S.; Kato, H.; Yamabe, T; Fukui, K. *J. Phys. Chem.* **1973**, *77,* 14, 501; *Chem. Phys. Lett.* **1972**, *17,* 264; Ishimura, S.; Tomita, H.; Yamabe, T.; Fukui, K.; Kato, H. *Chem. Phys. Lett.* **1973**, *23,* 106.
34. Howat, G.; Webster, B. C. *J. Phys. Chem.* **1972**, *76,* 3714. Earlier semiempirical (CNDO/2) work is by Weissman, M.; Cohen, N. V. *Chem. Phys. Lett.* **1970**, *7,* 455; *J. Chem. Phys.* **1973**, *59,* 1385; Cohen, N. V.; Finkelstein, G.; Weissmann, M. *Chem. Phys. Lett.* **1974**, *26,* 93.
35. Moskowitz, J. W.; Boring, M.; Wood, J. H. *J. Chem. Phys.* **1975**, *62,* 2254.
36. Noell, J. O.; Morokuma, K. *J. Phys. Chem.* **1977**, *81,* 2295.
37. Rao, B. K.; Kestner, N. R. *J. Chem. Phys.* **1984**, *80,* 1587.
38. Gaathon, A.; Czapski, G.; Jortner, J. *J. Chem. Phys.* **1973**, *58,* 2648; Michael, B. D.; Hart, E. J.; Schmidt, K. H. *J. Phys. Chem.* **1971**, *71,* 2798.
39. Olinger, R.; Schindewolf, U.; Gaathon, A.; Jortner, J. *Ber. Bunsenges. Phys. Chem.* **1971**, *75,* 690.
40. Jortner, J.; Gaathon, A. *Can. J. Chem.* **1977**, *55,* 1795.
41. Krebs, P.; Heintze, M. *J. Chem. Phys.* **1982**, *76,* 5484.
42. Krebs, P.; Bukowski, K.; Giraud, V.; Heintze, M. *Ber. Bunsenges. Phys. Chem.* **1982**, *86,* 879. For low-density water vapor studies, see also Christophorou, L. G.; Carter, J.; Maxey, D. V. *J. Chem. Phys.* **1982**, *76,* 2653.
43. Armbruster, M.; Haberland, H.; Schindler, H. G. *Phys. Rev. Lett.* **1981**, *47,* 323.
44. Haberland, H.; Langosch, M.; Schindler, H. G.; Worsnop, D. R. *J. Phys. Chem.* **1984**, *88,* 3903.
45. Haberland, H., private communication.
46. Antoniewicz, P. R.; Bennett, G. T.; Thompson, J. C. *J. Chem. Phys.* **1983**, *77,* 4573.
47. Kestner, N. R.; Jortner, J. *J. Phys. Chem.* **1984**, *88,* 3818.
48. Feng, D. A.; Kevan, L; Yoshida, H. *J. Chem. Phys.* **1974**, *61,* 4440.
49. Ichikawa, T.; Yoshida, H. *J. Chem. Phys.* **1980**, *73,* 1540.
50. Nishida, M. *J. Chem. Phys.* **1977**, *67,* 2760; *ibid.* **1977**, *67,* 2760; *ibid.* **1977**, *67,* 4786.
51. Kimura, T.; Fueki, K.; Narayana, P. A.; Kevan, L. *Can. J. Chem.* **1977.** *55,* 1940.
52. Dorfman, L.; Jou, F. Y.; Wageman, R. *Ber. Bunsenges. Phys. Chem.* **1971**, *75,* 681.
53. Dorfman, L.; Jou, F. Y.; In "Electrons in Fluids"; Jortner, J.; Kestner, N. R., Eds.; Springer-Verlag: Berlin, 1973, pp. 447–457.
54. Logan, J.; Kestner, N. R. unpublished, Ph.D. thesis of J. Logan, 1973. See also Kestner, N. R. In "Electron-Solvent and Anion-Solvent Interactions"; Kevan, L.; Webster, B. C., Eds.; Elsevier: Amsterdam, 1976, pp. 1–88.
55. Dye, J. L.; DeBacker, M. G.; Dorfman, L. M. *J. Chem. Phys.* **1970**, *52,* 6251.
56. Tuttle, T. R.; Golden, S. *J. Chem. Soc., Faraday Trans. 2* **1981**, *77,* 873; *ibid.* **1981**, *77,* 889; *ibid.* **1981**, *77,* 1421.
57. Stupak, C. M.; Tuttle, T. R.; Golden, S. *J. Phys. Chem.* **1984**, *88,* 3799.

58. Feng, D. R.; Ebbing, D.; Kevan, L. *J. Chem. Phys.* **1974**, *61*, 249.
59. Carmichael, I.; Webster, B. *J. Chem. Soc.* **1974**, *70*, 1570.
60. Jortner, J. *J. Chem. Phys.* **1959**, *30*, 839; Cohan, N. V.; Finkelstein, G.; Weissmann, M. *Chem. Phys. Lett.* **1974**, *26*, 93.
61. The best review of all data on metal ammonia solutions is Thompson, J. C. "Metal Ammonia Solutions"; Oxford University Press, Oxford, 1976.
62. Hentz, R. R.; Farhataziz; Hansen, E. M. *J. Chem. Phys.* **1972**, *57*, 1253.
63. Schindewolf, U.; Olinger, R. In "Metal Ammonia Solutions"; Lagowski, J. J.; Sienko, M., Eds.; Butterworths: London, 1970, pp. 199–215.
64. See, for example, Schlick, S.; Narayana, P. A.; Kevan, L. *J. Chem. Phys.* **1976**, *64*, 3153; Ichikawa, T.; Kevan, L.; Bowman, M. K.; Dikanov, S. A. Tsvetkov, T. D. *J. Chem. Phys.* **1979**, *71*, 1167; Narayana, M.; Kevan, L. *J. Am. Chem. Soc.* **1981**, *103*, 1618 or the excellent review of the work in ref. 26 by Kevan, L. *Radiat. Phys. Chem.* **1981**, *17*, 413.
65. Golden, S. comments at Colloque Weyl VI, July, 1982.
66. Lin, D. P.; Kevan, L. *J. Phys. Chem.* **1977**, *81*, 1498.
67. R. Catterall, comments at Colloque Weyl VI, July, 1983.
68. For a simple picture of the effect of molecular structure on absorption spectra in alcohols, see Hentz, R. R.; Kenney-Wallace, G. A. *J. Phys. Chem.* **1974**, *78*, 514.
69. Kestner, N. R. In "Proceedings of the Fifth International Congress on Radiation Research"; Adler, H. F.; Nygaard, O. F.; Sinclair, W. K., Eds. Academic Press: New York, 1975, pp. 333–334 and ref. 32.
70. Brodsky, A. M.; Tsvetkov, A. V. *J. Phys. Chem.* **1984**, *88*, 3790.
71. Ovchinnikova, M. Ya.; Ovchinnikov, A. A. *Optics and Spectroscopy* **1970**, *28*, 522.
72. Zusman, L. D.; Helman, A. B., *Chem. Phys. Lett.* **1985**, *114*, 301; *Optics and Spectroscopy* **1982**, *53*, 248.
73. Zusman, L. D. *Chem. Phys.* **1985**, *80*, 29.
74. Hiroike, K.; Kestner, N. R.; Rice, S. A.; Jortner, J. *J. Chem. Phys.* **1965**, *43*, 2625.
75. Chandler, D.; Wolynes, P. G. *J. Chem. Phys.* **1981**, *74*, 4078. Chandler, D. In "Studies in Statistical Mechanics VIII"; Montrall, E. W.; Lebowitz, J. L., Eds.; North-Holland: Amsterdam, 1982, p. 275.
76. Wolynes, P. G. *Phys. Rev. Lett.* **1981**, *47*, 968; Skinner, J. L.; Wolynes, P. G. *J. Chem. Phys.* **1978**, *69*, 2143; *ibid.* **1980**, *72*, 4913.
77. Grote, R. F.; Hynes, J. T. *J. Chem. Phys.* **1981**, *74*, 4465; *ibid.* **1981**, *75*, 2191.
78. Adelman, S. A. *J. Chem. Phys.* **1981**, *75*, 5837, *Adv. Chem. Phys.* **1980**, *44*, 143.
79. Chandler, D.; Singh, Y.; Richardson, D. M. *J. Chem. Phys.* **1984**, *81*, 1975; Nichols, A. L. III; Chandler, D.; Singh, Y.; Richardson, D. M. *J. Chem. Phys.* **1984**, *81*, 5109.
80. Phillips. P. *J. Chem. Phys.* **1984**, *81*, 6069.
81. Berne, B., private communication.
82. Parrinello, M.; Ralman, A. *J. Chem. Phys.* **1984**, *80*, 860; comments at Colloque Weyl VI, 1983 by A. Rahman.
83. Marcus, R. A. In "Solvated Electron"; Hart, E. J., Ed.; American Chemical Society: Washington, D.C., 1965, pp. 138–148.
84. Webman, I.; Kestner, N. R. Amer. Phys. Soc. Meeting, Chicago, Jan. 21, 1980.
85. Beratan, D. N.; Hopfield, J. *J. Chem. Phys.* **1984**, *81*, 5753.

# 9

# The Radiation Chemistry of Gases

**D. A. ARMSTRONG**

*Department of Chemistry, University of Calgary*

## INTRODUCTION

Earlier chapters have discussed mechanisms of energy deposition and primary species formed by ionizing radiation. Here we focus on chemical reactions and products from irradiated gases. A typical scheme of physical and chemical processes in a system where electrons do not react with the neutral molecules is shown in Table 9-1. The reader may note that the final products T, E, Z, and F arise from the parent molecule G through several different types of reactions, which involve electrons, excited molecules ($G^*$) and ions ($G^{+*}$), free radicals ($\cdot Q$ and $\cdot R$), and stable ground state ions $M^+$. The last of these may be clustered by neutral molecules, $M^+ \cdot 2G$, before they are neutralized. The time scales of the various processes at 1 atmosphere (atm) are shown in column three. Also given is the form of their dependence (if any) on dose rate I and concentration of G or other species. The time scales also depend on the chemical structure of G.

The same product may result from quite different processes in a simple system. For example, in molecular oxygen the species $\cdot Q_{ep}$, $\cdot R_{ep}$, and D are all oxygen atoms, which produce the sole product ozone by addition to $O_2$. With more complex molecules, like the higher hydrocarbons, there is a much greater range of products, but still a certain product may arise from more than one type of process. The objectives of this chapter are to

1. Explain how yields of products in gaseous systems are determined.


*Radiation Chemistry: Principles and Applications*

**Table 9-1.** Time Scale of Radiolysis Events in a Gas

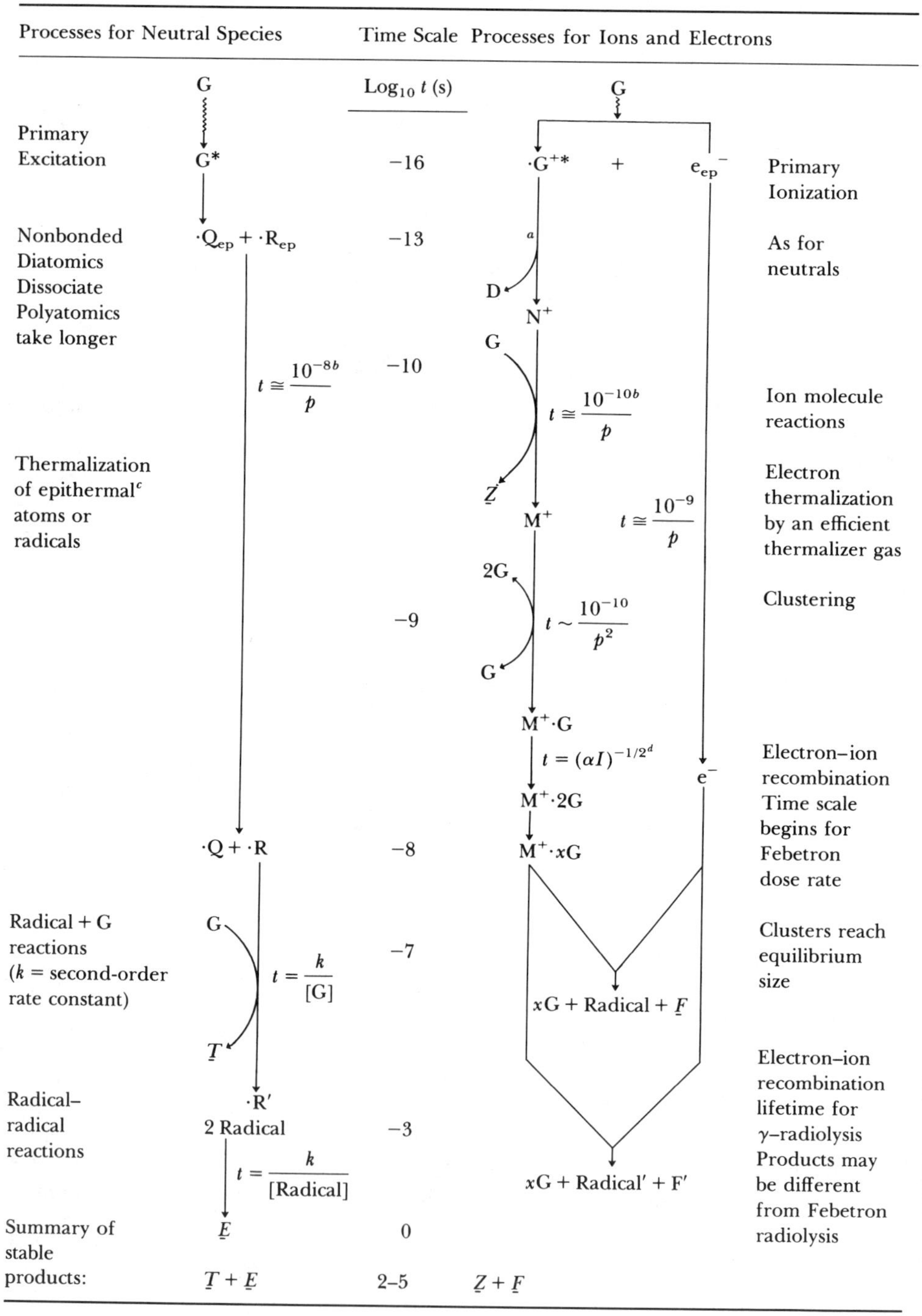

| Processes for Neutral Species | | Time Scale | Processes for Ions and Electrons | |
|---|---|---|---|---|
| | | $\log_{10} t$ (s) | | |
| | G | | G | |
| Primary Excitation | $G^*$ | −16 | $\cdot G^{+*}$ + $e_{ep}^-$ | Primary Ionization |
| Nonbonded Diatomics Dissociate Polyatomics take longer | $\cdot Q_{ep} + \cdot R_{ep}$ | −13 | $\cdot G^{+*}$ →[a] D + $N^+$ | As for neutrals |
| | $t \cong \frac{10^{-8}}{p}$[b] | −10 | $N^+$ + G → $\underline{Z}$ + $M^+$; $t \cong \frac{10^{-10}}{p}$[b] | Ion molecule reactions |
| Thermalization of epithermal[c] atoms or radicals | | | $M^+$; $e_{ep}^-$ → $e^-$: $t \cong \frac{10^{-9}}{p}$ | Electron thermalization by an efficient thermalizer gas |
| | | −9 | $M^+$ + 2G → $M^+ \cdot G$ + G; $t \sim \frac{10^{-10}}{p^2}$ | Clustering |
| | | | $M^+ \cdot G$ → $M^+ \cdot 2G$; $t = (\alpha I)^{-1/2}$[d]; $e^-$ | Electron–ion recombination Time scale begins for Febetron dose rate |
| | $\cdot Q + \cdot R$ | −8 | $M^+ \cdot xG$ | |
| Radical + G reactions ($k$ = second-order rate constant) | G → $\underline{T}$; $t = \frac{k}{[G]}$ | −7 | $M^+ \cdot xG$ + $e^-$ → $xG$ + Radical + $\underline{F}$ | Clusters reach equilibrium size |
| Radical–radical reactions | $\cdot R'$ 2 Radical | −3 | | Electron–ion recombination lifetime for γ–radiolysis Products may be different from Febetron radiolysis |
| | $t = \frac{k}{[\text{Radical}]}$ | | $M^+ \cdot xG$ + $e^-$ → $xG$ + Radical′ + F′ | |
| | $\underline{E}$ | 0 | | |
| Summary of stable products: | $\underline{T} + \underline{E}$ | 2–5 | $\underline{Z} + \underline{F}$ | |

[a] Either D or $N^+$ may have an unpaired electron.
[b] $p$ in atmospheres.
[c] ie, possessing excess kinetic or internal energy.
[d] $\alpha$ = second-order ion neutralization rate constant.
Based in part on the earlier scheme in ref. 1.

2. Demonstrate how reactions of specific intermediate species can be investigated and discuss the results of such studies.
3. Describe how mechanisms can be constructed to explain the yields of long-lived products, which are observed experimentally.

The literature concerned with gas-phase radiation chemistry is extensive, and only a few systems and reactive species can be discussed in the space available. Where possible, these have been chosen to illustrate the different types of behavior observed for different molecular structures. The reader is referred to several excellent reviews[1–8] for details of interesting topics, which cannot be covered in depth here.

# THE DETERMINATION OF YIELDS

A typical cylindrical quartz cell for studies of products from Febetron radiolysis is shown in Figure 9-1(a). The entrance–exit port may be closed

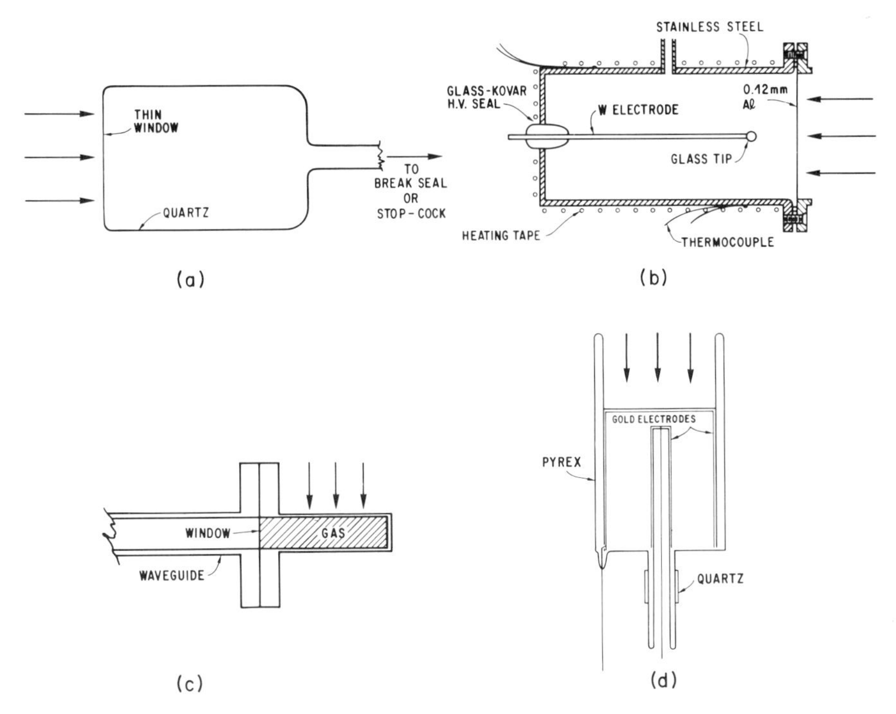


**Figure 9-1.** Cells for gas radiolysis. Direction of x-ray or electron beam is indicated by arrows: (a) Quartz cell for product yield determinations with Febetrons.[12] (b) High-temperature ionization chamber of Jones and Sworski.[21] (c) Typical x-band microwave cell used for studying electron reactions.[51,108] (d) Ionization chamber for studying rates of neutralization reactions.[110]

by a break seal or by a greaseless stopcock. The fast electron beam enters through the thin end window. Cells of a similar design are used for x-ray or $\gamma$ radiolysis, but in the latter case the thin end window is not required.[9] Also, if the cell will not be used above 600 K, the quartz may be replaced by pyrex.

After extensive baking at about 620 K and degassing on a good vacuum line,[10] the cells are filled with gas and placed in a thermostated chamber for exposure to the radiation source. When thermal equilibrium has been attained, they are irradiated. Long-lived products are analyzed by suitable procedures such as a gas chromatograph (GC), which may be coupled with a mass spectrometer detector (GCMS).[11] Occasionally, products may be absorbed in aqueous solution and titrated or determined by spectrophotometry.[12]

If the yield per unit energy absorbed (ie, *G*-value) is required, the adsorbed dose must be obtained by a method applicable to the type of radiation employed. One obvious and very common method is the use of dosimeter gases, gases for which the *G*-value of an easily measured product has been determined under standard conditions. The dose rate absorbed in the dosimeter gas can be calculated from the yield of this product, following exposure to the radiation for a measured time. The dose rate in the gas under study in the same cell with identical conditions of geometry and source intensity can then be found from the ratio of the energy absorption coefficients of the two gases. In the next section particular attention is paid to absolute yields of products from dosimeter gases. Results for different types of radiation are discussed separately.

## *α- and β-Particle Sources*

Relatively little use has been made of $\alpha$-particle sources in the past three decades. However, most of the early work in the field was done with the emissions from radium or its daughter radon.[2] The latter can be used as an internal $\alpha$-particle source in gas radiolysis, and its applications have been discussed extensively by Lind.[2] More recently tritium has found use as an internal $\beta$-particle source in the radiolysis of nitrous oxide,[13] hydrogen,[14] and water vapor.[15] With such sources the energy absorbed is calculated from the number of curies of radioactive isotope present, its decay constant, and the mean energy of the emitted particles.[2,16] A correction must be made for the fraction of energy lost by stopping a few of the particles in the wall of the container.

$\beta$-Particle sources have been of great importance in the determination of *W*-values.[17] In this application the source, which has a known activity, is located within an ionization chamber, and conditions are so designed that the average energy expended in the gas per particle emitted can be accurately calculated. The ions formed are collected by an electrode and measured with an electrometer. The $W$-value for a given set of conditions is the ratio of the energy expended by the particles to the number of ion pairs formed. A table of $W$-values has been given in Table 2-4.

A result of more direct interest to the gas-phase radiation chemist is the yield of ion pairs per 100 eV, $G$(ionization). That parameter can be determined directly from the ratio of the number of ion pairs to the energy expended in the gas, multiplied by 100. It is related to $W$ for the gas of interest—$G(\text{ionization}) = 100/W$—and can therefore be calculated from the $W$-value.

## Electron Accelerators

Willis, Boyd, and co-workers[18,19] have used adiabatic calorimetry to measure the mean dose in a cell for gas-phase radiolysis from a Febetron accelerator. By attaching thermocouples to Al or Be discs of different thickness, they were able to determine the temperature rise due to the absorbed radiation and to extrapolate the dose per gram to zero thickness of the metallic absorber. The dose for a gas in a similar cell at a given pressure was then calculated from the ratio of its stopping power to that of the metal. These stopping powers were weighted for the electron energy spectrum of the Febetron pulse. The value of $G(N_2)$ from $N_2O$ initially at 25°C and normal pressure was (12.4 ± 0.4) molecules per 100 eV for both the 1.7- and 0.7-MeV Febetron accelerators, which produce dose rates in excess of $10^{27}$ eV $g^{-1}$ $s^{-1}$.

A somewhat similar calorimetry method was used by Ghormley et al[20] to determine $G(O_3)$ for the radiolysis of oxygen with a 1.7-MeV Febetron. They observed $G(O_3) = (13.8 \pm 0.6)$ molecules per 100 eV for oxygen initially at 23°C and 750 mm of Hg pressure.

When gases for which $W$ is known are irradiated in ionization chambers, the dose rates can be calculated from saturation ionization currents determined by applying a potential between the electrodes. The relation between the dose rate and the saturation current $i_s$ is simply

$$I = i_s \times W \times e^{-1} \tag{9-1}$$

where e is the charge on the electron. The success of this technique requires that (a) the electrodes be inert to the gas used in the chamber, even in the presence of radiation; (b) all ion pairs formed in the irradiated volume be collected. Unfortunately, the technique cannot be used at the very large dose rates produced with Febetrons because the ion neutralization lifetimes are then so short that quantitative ion collection is impractical. However, the method is widely used with accelerators of lower dose rate[21,22] and with photon sources (see below). Figure 9-1(b) is a drawing of the cell employed by Jones and Sworski in their study of $N_2O$ radiolysis. The value of $G(N_2) =$ (10.0 ± 0.4) molecules per 100 eV at 24°C and 600 mm is the preferred value for this dosimeter gas at low dose rate.[23]

## *X-Ray and γ-Radiolysis*

Ionization chambers have been employed for the dosimetry of x-ray and $\gamma$-radiolysis in several laboratories.[10,24–26] The principle is the same as that described above for accelerators. However, in this case the electrodes cannot be kept out of the radiation field, and the electrode material must be selected to be compatible with the wall material of the cell. For example, platinum electrodes in a pyrex cell may produce a strongly inhomogeneous dose distribution. This problem is exacerbated as the photon energy is reduced. Thus, for 150-kV x-radiolysis aluminum or graphite electrodes are preferred in pyrex cells, provided they are inert to the gases irradiated. With photons of higher energy ($\sim$1 MeV) gold electrodes may suffice.

The most common method of dosimetry for x-ray and $\gamma$-radiolysis involves the use of dosimeter gases as described above. If $P_d$ is the dosimeter product of accurately known yield, then the dose rate $I_d$ in the dosimeter gas is given by

$$I_d = \frac{d[P_d]}{dt} \times \frac{100}{G(P_d)} \tag{9-2}$$

The dose rate $I_u$ in another gas under the same conditions of pressure, source intensity, and geometry is

$$I_u = I_d \times \rho_d{}^u \tag{9-3}$$

Here $\rho_d{}^u$ is the ratio of the energy absorption coefficients of the two gases. Because of the low stopping powers of gases, electron equilibrium is generally not attained, and the calculation of $\rho_d{}^u$ cannot be based on the same assumptions as for liquids. Indeed, a very significant fraction of the dose in the center of a typical gas-phase cell will usually arise from electrons set in motion in the walls. Actually, for gases composed of low-atomic-number atoms at normal pressure, the electron flux from this process is strongly dominant over the flux from electrons set in motion by absorption of photons in the gas itself. This is true as long as the electron energy ranges are much larger than the cell dimensions. Thus, for a typical radiolysis in a 0.3-L pyrex or quartz cell, the process of $\gamma$-radiolysis is equivalent to radiolysis with a flux of fast electrons having an energy spectrum determined by the wall material, and the Bragg–Gray cavity principle[27] can be applied to calculate $\rho_d{}^u$. In fact, that quantity becomes the ratio of stopping powers of the gases u and d weighted for the electron energy spectrum of the wall material. The calculation of weighted stopping power ratios and the use of this technique have been discussed by Davidow and Armstrong[25] and by Huyton and Woodward.[28] The latter authors have given stopping power parameters for a variety of gases.

Note that Eq. (9-3) is also applicable to the radiolysis of dosimeter and other gases by fast electron accelerators and Febetrons. In the former case the electrons are monochromatic, and calculating $\rho_d{}^u$ is especially simple.[22]

For Febetrons the spectrum of energies of the electrons can normally be obtained from the manufacturer.[18,19]

When the radiolysis cell is an ion chamber, the measured product of x-ray or $\gamma$-radiolysis may be the total yield of ion pairs. The quantity $d[P_d]/dt$ in Eq. (9-2) then becomes $i_{s(d)}$, and since $I_d = i_{s(d)} \times W_d \times e^{-1}$, it follows from (9-3) that

$$W_u \times i_{s(u)} = W_d \times i_{s(d)} \times \rho_d{}^u \quad (9\text{-}4)$$

$$W_u = W_d \times \rho_d{}^u \times \frac{i_{s(d)}}{i_{s(u)}} \quad (9\text{-}5)$$

Several laboratories have used this approach to determine $W$-values, with the inert gases, air, and nitrogen as standards.[22,24–26,28–30] Those determined with monoenergetic electrons[22] may be expected to have a somewhat smaller uncertainty.

## *Dosimeter Gases*

Only two gases, $N_2O$ and $O_2$, have found wide usage as gas-phase dosimeters. Results of several experimental determinations of $G(N_2)$ and $G(O_3)$ are summarized in Table 9-2. Oxygen has a special application in pulse experiments, because the absorbance due to $O_3$ at 225 nm may be monitored directly in the pulse radiolysis cell with the same optical setup used for the transient species in other systems. However, $G(O_3)$ in pure oxygen is dose-rate dependent[6,31] and also pressure dependent at high dose rates.[6,32] A better system with a yield independent of dose rate is $O_2$ containing approximately 0.5 mol% $SF_6$, for which $G(O_3) = 6.2$ molecule/100 eV.[6,7,31] At the present time this appears to be the preferred dosimeter for pulse radiolysis.

The yields of ozone at $10^{14}$–$10^{18}$ eV g$^{-1}$ s$^{-1}$ also tend to cluster around 6.0 (Table 9-2). The reason for this will be discussed later. Actually, the methods of dosimetry and conditions used in these studies by different laboratories were highly varied, and further, more precise determinations of $G(O_3)$ would be of value. The data for pure oxygen at Febetron dose rates above $10^{25}$ eV g$^{-1}$ s$^{-1}$ are, however, quite consistent. The agreement between the results in refs. 20 and 31 is better when the former are recalculated with the same value of $\rho_{Al}{}^{O_2}$ (Table 9-2, footnote c and ref 19). A useful list of stopping power ratios for 1–2-MeV electrons is given in ref. 19.

Nitrous oxide has been the preferred dosimeter for $\gamma$- and low-dose-rate electron radiolysis, $G(N_2)$ being taken as $10.0 \pm 0.4$ as already stated. The system is not useful for optical-pulse-radiolysis dosimetry. The nitrogen yield is generally determined by gas chromatography. Like pure oxygen the $N_2O$ system suffers from a dose-rate dependence, which causes $G(N_2)$ to rise at high dose rates. An alternative that does not exhibit this characteristic is HCl. As shown in Table 9-1, $G(H_2)$ is constant at atmospheric pressure

**Table 9-2.** Product Yields from Dosimeter Gases under Electron or Photon Radiolysis

| Product Yield (molecule/ 100 eV) | Radiation Energy/Type | Dose Rate ($eV\ g^{-1}\ s^{-1}$) | Max. mol% Conversion | Method of Dosimetry[a] | P/(mm Hg) | $T$/K | Ref. |
|---|---|---|---|---|---|---|---|
| $G(N_2)$ *from* $N_2O$ | | | | | | | |
| 10.2 | 70 kV (x-ray) | $1 \times 10^{13}$ | — | I | 200, 550 | room | 33 |
| $9.8 \pm 1.0$ | 4 MeV ($\gamma$) | $1.5 \times 10^{17}$ | 0.1 | I | 149 | 303 | 34 |
| $10.0 \pm 0.2$ | 1 MeV ($e^-$) | $1\text{–}10 \times 10^{17}$ | 0.1 | I | 200, 600 | 297 | 21 |
| $12.4 \pm 0.4$ | 1.7 MeV ($e^-$) (Febetron) | $10^{27}$ | 0.2 | C | 380–3800 | 298 | 18, 35 |
| $12.3 \pm 0.3$ | 0.6 MeV ($e^-$) (Febetron) | $2 \times 10^{28}$ | 0.5 | C | 200–1800 | 298 | 19 |
| $G(O_3)$ *from* $O_2$ | | | | | | | |
| $5.9^b$ | $Co^{60}$ ($\gamma$) | $10^{16}\text{–}10^{18}$ | 0.002 | F | $0.2\text{–}10 \times 10^4$ | 195–283 | 36 |
| $6.3^b$ | $Co^{60}$ ($\gamma$) | $10^{16}$ | 0.02 | N | 100–800 | 90, 200, 293 | 37 |
| $6.0^b$ | $Co^{60}$ ($\gamma$) | $10^{14}\text{–}10^{17}$ | 0.003 | F | 460–665 | 195–283 | 38 |
| 6.2 | $Co^{60}$ ($\gamma$) | $10^{16}$ | — | N | 600 | 77 | 31 |
| $12.9^b$ | 0.25, 0.60 MeV ($e^-$) | $10^{25}$ | — | N | 560 | 295 | 39 |
| $13.8 \pm 0.6^c$ | 1.7 MeV ($e^-$) (Febetron) | $6 \times 10^{25}$ | 0.005 | C | 750 | 295 | 20 |
| $12.8 \pm 0.6$ | 1.7 MeV ($e^-$) (Febetron) | $10^{26}\text{–}2 \times 10^{27}$ | 0.09 | N | 200–800 | 195 & 298 | 31 |
| $G(H_2)$ *from* HCl | | | | | | | |
| $8.0 \pm 0.3$ | $Co^{60}$ ($\gamma$) | $7 \times 10^{13}$ | 0.006 | I | 150–1200 | 296 | 24 |
| $8.3 \pm 0.1$ | $Co^{60}$ ($\gamma$) | $2 \times 10^{15}$ | 0.05 | I | 268–772 | 296 | 40 |
| $7.9 \pm 0.5$ | 125 kV (x-ray) | $2 \times 10^{15}$ | — | I | 200–610 | 296 | 41 |
| $8.5 \pm 0.2^d$ | $Co^{60}$ ($\gamma$) | $7 \times 10^{15}$ | — | N | 760–3000 | 296 | 42 |
| $8.1 \pm 0.2$ | 1.7 MeV ($e^-$) (Febetron) | $10^{27}$ | 0.05 | N | 700–1400 | 296 | 43, 44 |

[a] Methods of dosimetry: I = Ionization current determination
C = Calorimetry
N = Relative to $N_2O$
F = Ferrous sulfate dosimetry, assumes pressure high enough for electron equilibrium.
[b] As corrected by Willis and Boyd; see ref. 7, Table I(c).
[c] Becomes $13.2 \pm 0.7$ if $\rho_{Al}{}^{O_2} = 1.085$ is used, see ref. 19.
[d] Recalculated using $\rho_{N_2O}{}^{HCl}$ from ref. 28.
Note that in Febetron radiolysis the temperature may rise by several degrees during the pulse.

over the dose-rate range from $10^{14}\ eV\ g^{-1}\ s^{-1}$ to $10^{27}\ eV\ g^{-1}\ s^{-1}$. Also the yields are independent of temperature and pressure over wide ranges. The main deficiency is that HCl tends to adsorb on surfaces, but since cells are normally baked anyway this should not be a problem. Also, there is no possibility of a solid residue like $S_8$ from $H_2S$. Chlorine absorbance at 330 nm would have approximately the same $G_\varepsilon$ at 330 nm as ozone does at 225 nm. Thus, in theory, HCl could be used as a pulse radiolysis dosimeter as well. However, this application remains to be tested.

## REACTIONS OF INTERMEDIATES

Many different techniques have been utilized to detect intermediates and investigate their reactions. The general method of chemical scavengers is discussed first. Subsequently, physical techniques are considered, and the results obtained from studies of the kinetics of reactions of particular species are presented.

### *The Method of Chemical Scavengers*

When a particular product is believed to be formed from a given intermediate, such as an electron, hydrogen atom, or alkyl radical, confirmation may be sought by adding to the system a compound, which is known to react rapidly with the suspected intermediate. As the concentration of this chemical scavenger is increased, the yield of the product should be suppressed. At the concentrations used the scavenger should not react with any other intermediate. Additionally, it is desirable that the scavenger form a readily identifiable product, the yield of which will rise in stoichiometric proportion to the fall in the yield of the original product as the scavenger concentration rises. Unfortunately, it is seldom possible to find scavengers that fulfill all desired criteria. Nonetheless, much important information on yields, mechanisms, and rates of reactions of intermediates has been gained through their use.

Figure 9-2(a) shows the reduction in the yield of $D_2$ from the radiolysis of DCl on the addition of $SF_6$, which is an excellent electron scavenger. The net reduction in $G(D_2)$ is 4.1 molecules/100 eV, which is equal to the yield of electrons or $G$(ionization) as determined by ion chamber measurements.[45] The conclusion is that each electron produces one molecule of $D_2$.

Carbon tetrachloride and $c$-$C_4F_8$ have also been employed as electron scavengers.[31] Nitric oxide, $H_2S$, ethylene, and butenes have been used for free radicals.[3] They are, however, susceptible to the scavenging of positive ions, and one must exercise care when interpreting results obtained with them. For a discussion of these problems see ref 46, page 56.

As a further illustration of the application of scavengers, we examine the effect of chlorine on the yield of hydrogen from HCl. As in DCl, electron scavengers show that each electron forms one molecule of hydrogen. However, the evidence (see the section entitled "Electrons" below) indicates that each electron forms an H atom, which subsequently undergoes the reaction (9-6):

$$H\cdot + HCl \rightarrow H_2 + Cl\cdot \tag{9-6}$$

On the basis of photochemical studies, chlorine is expected to suppress this process by scavenging the hydrogen atoms:[24]

$$H\cdot + Cl_2 \rightarrow HCl + Cl\cdot \tag{9-7}$$

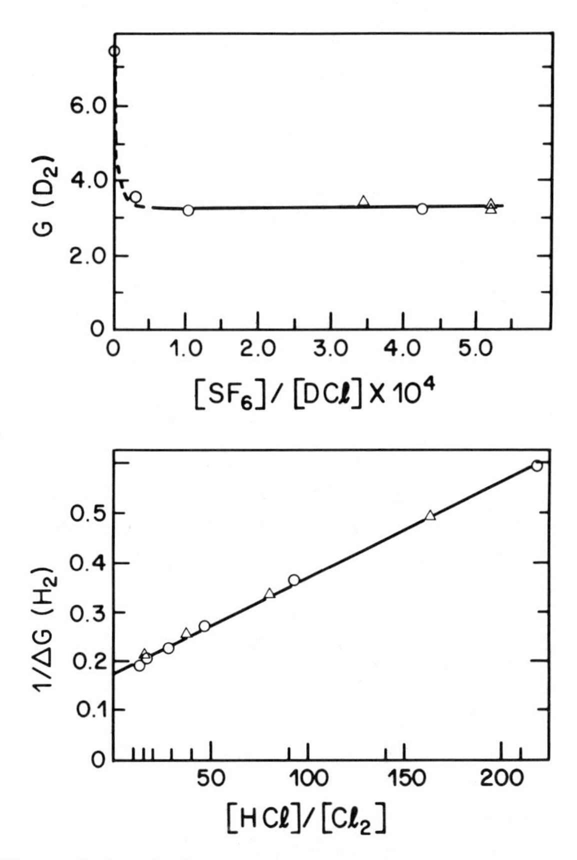


**Figure 9-2.** Effects of chemical scavengers. (a) Change in $D_2$ yield from DCl on addition of the electron scavenger $SF_6$ for Febetron radiolysis at 1200 mm Hg pressure (○) and x-radiolysis at 600 mm Hg pressure at 290 K. (b) Reciprocal plot showing the effect of the H-atom scavenger $Cl_2$ on $H_2$ production from Febetron radiolysis of HCl at 700 (△) and 1200 (○) mm Hg pressure. (Reproduced with permission.[43,45])

Applying the steady-state approximation to the concentration of hydrogen atoms, one can show that

$$G(H_2) = \frac{100 \times d[H_2]/dt}{I_{HCl}} = g_H \left\{ \frac{k_6[HCl]}{k_6[HCl + k_7[Cl_2]]} \right\} \tag{9-8}$$

Here $g_H$ is the yield of H· atoms, and $k_6$ and $k_7$ are the rate constants for Reactions (9-6) and (9-7), respectively. Taking $\Delta G(H_2)$ as the difference between $G(H_2)_0$, the hydrogen yield with no chlorine, and $G(H_2)$, we get

$$\Delta G(H_2) = g_H \left\{ \frac{k_7[Cl_2]}{k_6[HCl] + k_7[Cl_2]} \right\}$$

or

$$[\Delta G(H_2)]^{-1} = g_H^{-1}\left\{1 + \frac{k_6[HCl]}{k_7[Cl_2]}\right\} \tag{9-9}$$

A plot of experimental results from the Febetron radiolysis of HCl in accord with this expression is shown in Figure 9-2(b). The slope gives $k_6/k_7 = 1.07 \times 10^{-2}$. The agreement with $(1.08 \pm 0.15) \times 10^{-2}$ from independent photochemical experiments confirms that $\cdot$H atoms are the species producing $H_2$ in Reaction 6.[44] From the intercept of the plot in Figure 9-2(b) the yield of H$\cdot$ atoms is 5.7/100 eV. This is significantly larger than $g_e$ (or $G$(ionization) = 4.0/100 eV), demonstrating that there is an additional source of approximately 1.7 scavengeable $\cdot$H atoms per 100 eV. For a further discussion of the HCl system the reader may consult refs. 43 and 45. Our purpose here is simply to illustrate how the scavenger technique can be employed in a quantitative sense. In general, if the value of $k_7$ is accurately known, $k_6$ can be determined from the ratio $k_6/k_7$, and vice versa. This has been an important application of the method.

## Electrons

The study of electron swarms has provided a great deal of quantitative information on the interactions between electrons and molecules. The reader is directed to several treatises on this subject,[47–49] which lies in the domain of atomic and molecular physics. In the last few decades radiation chemists have used microwave conductivity methods to investigate electron thermalization, attachment, and recombination. The presence of electrons in a microwave cavity causes a shift of the cavity resonance frequency. At low electron–molecule collision frequencies this shift can be used to determine the electron concentration.[50] A second effect, which becomes more useful at higher collision frequencies, is the absorption of the microwaves or the change in microwave power reflected through the cell. This change is quantitatively related to the electron concentration.[51] Thus, either technique can be used to follow the rates of electron reactions. Figure 1(c) shows the kind of cell used by Warman in recent studies.[51]

### Thermalization

As explained in Chapter 7, electrons in irradiated systems are borne with excess kinetic energy. Those with energy less than the lowest excited state of the predominant component of a gaseous mixture are called subexcitation electrons. In Table 9-1 they have been represented as $e_{ep}^-$ to indicate the epithermal character of their spectrum of kinetic energies. Their rate of loss of energy or thermalization in a given gas may be characterized by an energy loss rate coefficient $K(\bar{u})$, which pertains to an energy distribution of mean energy $\bar{u}$ and is defined by the equation

$$\frac{d\bar{u}}{dt} = -K(\bar{u}) \times (\bar{u} - \bar{u}_{th}) \times N \qquad (9\text{-}10)$$

The number of gas molecules per unit volume is $N$, and $\bar{u}_{th}$ $(=\frac{3}{2}kT)$ represents the mean of the thermal energy distribution.[52] In the most simple zero-order treatment $K(\bar{u})$ is taken as an energy-independent coefficient $K_u$, and integration yields

$$tN = \frac{2.303}{K_u}\left\{\log\left(\frac{\bar{u}_0}{\bar{u}_{th}} - 1\right) - \log\left(\frac{\bar{u}}{\bar{u}_{th}} - 1\right)\right\}$$

If the thermalization time $\Upsilon_{th}$ is defined as the time for $\bar{u}$ to reach $1.1 \times \bar{u}_{th}$, then

$$\Upsilon_{th}N = \frac{2.303}{K_u}\left\{\log\left(\frac{\bar{u}_0}{\bar{u}_{th}} - 1\right) + 1\right\} \qquad (9\text{-}11)$$

Warman and Sauer calculated values of $\Upsilon_{th}N$ from an analysis of the time dependence of the rate of electron capture by $CCl_4$, for which the energy dependence of the electron capture rate constant was known.[52] Some of their results, which were obtained by the resonant frequency microwave method, are presented in Table 9-3. Also included are a few values of $\Upsilon_{th}N$ obtained by the electron swarm technique, by the microwave power absorption method, which does not require the presence of $CCl_4$, and values calculated from cross-section data. In each case the value of $N$ has been

**Table 9-3.** Energy Exchange Rate Coefficients $K_u$ and Thermalization Times at $N = 3.3 \times 10^{16}$ molecule cm$^{-3}$ for Several Gases, at (295 ± 5) K

| Gas | $K_u$/cm$^3$ s$^{-1}$ | $\Upsilon_{th}N/(3.3 \times 10^{16})$ s molecule cm$^{-3}$ | |
|---|---|---|---|
| *Ar* | 1.3 (−13) | 1.3 (−3) | 3.0 (−3)[b] |
| Ne | 2.5(−13) | 6.7 (−4) | 1.6 (−3)[b] |
| He | 6.4 (−12); 0.25–1.3 (−11)[a] | 2.6 (−5) | 4.0 (−5)[b] |
| $N_2$ | 2.2 (−11); 1.6 (−11)[a] | 7.6 (−6); 15 (−6)[a]; | 12 (−6)[c] |
| $H_2$ | 1.1 (−10) | 1.5 (−6) | 2.4 (−6)[c] |
| $CH_4$ | 8.6 (−10) | 2.0 (−7); 1.6 (−6)[d] | |
| $C_2H_6$ | 1.1 (−9) | 1.5 (−7) | |
| $C_2H_4$ | 4.4 (−9) | 3.8 (−8); 1.6 (−7)[d] | |
| $N_2O$ | 4.5 (−9) | 3.7 (−8) | |
| $CO_2$ | 5.8 (−9) | 2.9 (−8); 8(−8)[d] | |
| $NH_3$ | 5.9 (−9) | 2.8 (−8) | |

All data are from ref. 52, except as indicated:
[a] Microwave absorption.[53]
[b] Theoretical calculation from cross sections.[54]
[c] Theoretical calculation from cross sections.[55]
[d] Electron swarm method.[57]
Numbers in parentheses are exponents.

adjusted to $3.3 \times 10^{16}$ molecule $cm^{-3}$, which corresponds to 1 mm Hg pressure at 298 K. The agreement between the various methods is generally within a factor of two. In the case of the result for helium the microwave absorption method was used to obtain the energy dependence of $K(\bar{u})$, which increased from $0.25 \times 10^{-11}$ $cm^3$ $s^{-1}$ at approximately 0.03 eV to $1.3 \times 10^{-11}$ $cm^3$ $s^{-1}$ at approximately 1.5 eV. The result for the resonant frequency method, using $CCl_4$ (ref. 52), lies in the middle of that range.

The magnitudes of $K_u$ are smallest and those of $Y_{th}N$ largest for the inert gases, where only elastic collisions can occur. For molecular gases $K_u$ rises and $Y_{th}N$ decreases as molecular complexity and/or polarity increases. An insight into the relative importance of elastic collisions, rotational excitation, and vibrational excitation in molecular systems can be gained from the rigorous theoretical analyses of Mozumder and co-workers.[54–56] For diatomic gases, such as $N_2$, $H_2$, and CO, the contribution of elastic scattering is small but never entirely negligible. Vibrational excitation is the major contributor at the higher energies but becomes small as the electrons near thermal energy. Here rotational excitation is the largest contributor. The time course of the relative contributions to the overall rate of thermalization for $N_2$ is displayed in Figure 9-3. The practical value of the $K_u$ coefficients in Table 9-3 lies in the fact that they can be used to calculate rates of thermalization of electrons in particular systems. When reactions of electrons

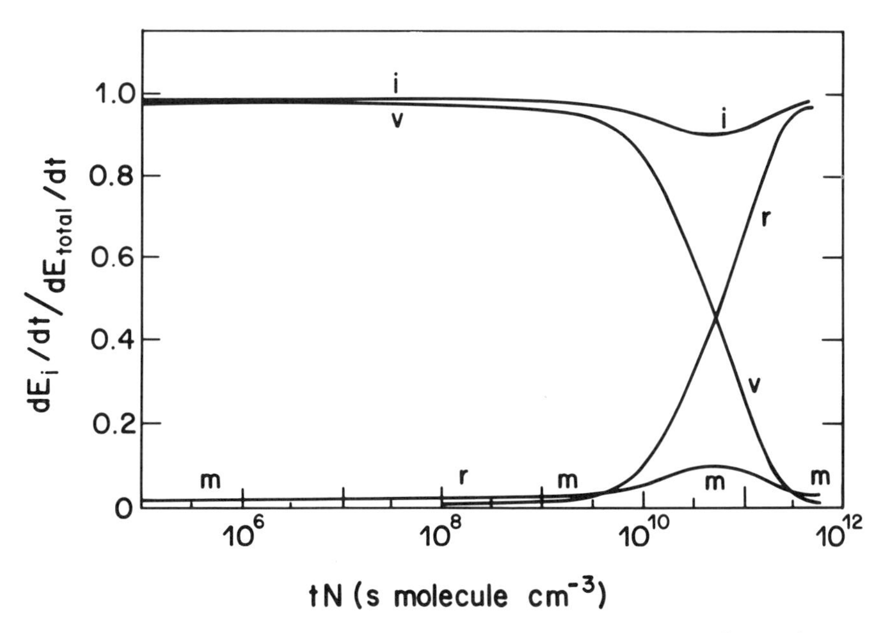


**Figure 9-3.** Ratio of energy loss rate due to different processes to the total energy loss rate as a function of $tN$ for subexcitation electrons in $N_2$ at 290 K: $m$ = elastic collisions, $i$ = total inelastic, $v$ = vibrational 0–1 excitation, and $r$ = total rotational excitation. (Reproduced with permission.[55])

are observed on similar or faster time scales, one can reasonably assume that they are reacting *before* thermalization. Mozumder has treated such reactions in the inert gases.[56] Further cases are considered below.

### Epithermal Electron Reactions

There are relatively few documented examples of these reactions. One that can be cited here is the formation of $N_2^*$ ($C^3\Pi_u$) molecules in the pulse radiolysis of helium or neon containing trace amounts ($3 \times 10^{14}$ molecule $cm^{-3}$) of molecular nitrogen.[58] The emission at 379 nm from the $C^3\Pi_u(v = 0) \rightarrow B^3\Pi_g(v = 2)$ transition was observed to grow in and decay on a time scale of about 60 ns. In helium and neon the energies of the subexcitation electrons will range up to the lowest excited atomic levels: 19.8 and 16.6 eV, respectively.[58] The growth–decay curves were interpreted in terms of the mechanism

$$X(E') + N_2 \rightarrow N_2^*(C^3\Pi_u) \qquad (9\text{-}12)$$

$$X(E') \rightarrow X(E) \qquad (9\text{-}13)$$

$$N_2^*(C^3\Pi_u) \rightarrow N_2^* + h\nu \qquad (9\text{-}14)$$

where X(E′) represents an electron with energy in excess of 11.15 eV, the level of $C^3\Pi_u$ above the ground state of $N_2$, and X(E) is an electron of lesser energy, that is, further along the course of thermalization. Analysis of the data yielded $k_{14} \cong 3 \times 10^7\ s^{-1}$, which is the spontaneous emission rate of $N_2^*(C^3\Pi_u)$, and $k_{12} = (5.0 \pm 0.8) \times 10^{12}\ M^{-1}\ s^{-1}$ in both helium and neon. This second-order rate constant is approximately equal to the product of the experimentally observed cross section for the production of $N_2^*(C^3\Pi_u)$ by impact of 14.7-eV electrons and the velocity of these electrons. It is therefore a reasonable rate constant for Reaction (9-12). The authors presented convincing arguments to show that relaxation of higher excited states of $N_2$ were not responsible for the phenomenon. Also they pointed out that the value of $k_{12}$ was an order of magnitude larger than the maximum expected for an atomic or molecular species reacting with $N_2$.

In systems containing polyatomic molecules, chemical reactions of the electrons as well as energy transfer are possible. In the following scheme $v$, $J$, $v_-$, and $J_-$ represent the thermal distributions of vibrational and rotational levels of the

$$e^-(E) + AB(v, J) \xrightarrow{(c)} AB^{*-}(v'_-, J'_-) \xrightarrow{(a)} AB(v', J') + e^-(E')$$

$$AB^{*-}(v'_-, J'_-) \xrightarrow{(d)} A + B^- \qquad (9\text{-}15)$$

$$AB^{*-}(v'_-, J'_-) \xrightarrow{(s)} AB^-(v_-, J_-) + \text{Energy}$$

diatomic AB and its molecular anion $AB^-$, respectively, and the primes indicate distributions in excited levels. The scheme may be generalized for higher polyatomics if A and B are taken to be groups of atoms and the energy levels of A and $B^-$ are also specified. It is based on concepts in refs. 48 and 59. The reaction steps (c), (a), (d), and (s) are, respectively, *attachment, autodetachment, dissociative electron attachment,* and *stabilization.* The last normally requires collision with a third body but, in principle, may also occur radiatively. It is unlikely with electrons of energy close to the subexcitation threshold, because the electron kinetic energy $E$ will usually contribute substantially to the excess energy of the compound ion in $AB^{*-}$, causing it to have relatively short lifetimes with respect to the unimolecular decay channels (a) and (d). Channel (a) contributes to the thermalization rate coefficient because $E'$ is generally less than $E$. Like (s) reaction (d) is a chemical rather than a physical process. The cross section for the combination (a) + (d), called the dissociative attachment (or dissociative capture) cross section $\sigma_d(E)$, can be determined in suitably designed swarm or beam experiments or combinations thereof.[48,59]

Many authors have pointed out that epithermal electrons may undergo dissociative capture. The following equation of Magee and Burton[61] predicts $\theta$, the fraction of electrons which *escape* capture during thermalization from energy $E_i$ to $E_{th}$.

$$\ln \theta = \int_{E_i}^{E_{th}} \frac{h(E)}{\eta(E)} \cdot \frac{dE}{E} \tag{9-16}$$

Here $\eta(E)$ is the fraction of energy lost per collision, and $h(E)$ the fraction of electrons captured per collision at energy $E$. The value of $h(E)$ is equal to the ratio of the capture cross section $\sigma_d(E)$ and the total collision cross section $\sigma(E)$, which can also be determined from swarm experiments[48] along with $\eta(E)$. As an illustration, the values of these three parameters in HCl are plotted as a function of mean electron energy in Figure 9-4. Actually, $\eta(E)$ has been multiplied by $E$ to give the mean energy lost per collision, $\Delta E(E)$.

A rigorous calculation of $\theta$ for a system under radiolysis would require numerical integration of Eq. (9-16) for each $E_i$ interval in the spectrum of initial energies of the subexcitation electrons. These values of $\theta$ would then have to be weighted in terms of that spectrum. Unfortunately, for molecules where capture is possible, values of $\sigma(E)$ and $\Delta E(E)$ are not known, certainly not with a high degree of accuracy. Also there is little quantitative information about the energy spectrum of subexcitation electrons. Detailed calculations are therefore not practical. However, a few approximations make it possible to estimate values of $\theta$, so that the important concept of capture during thermalization may be explored.

Figure 9-4 shows that $\sigma(E)$ and $\Delta E(E)$ are both to a first approximation independent of energy in the region near 1 eV. Equation (9-16) may

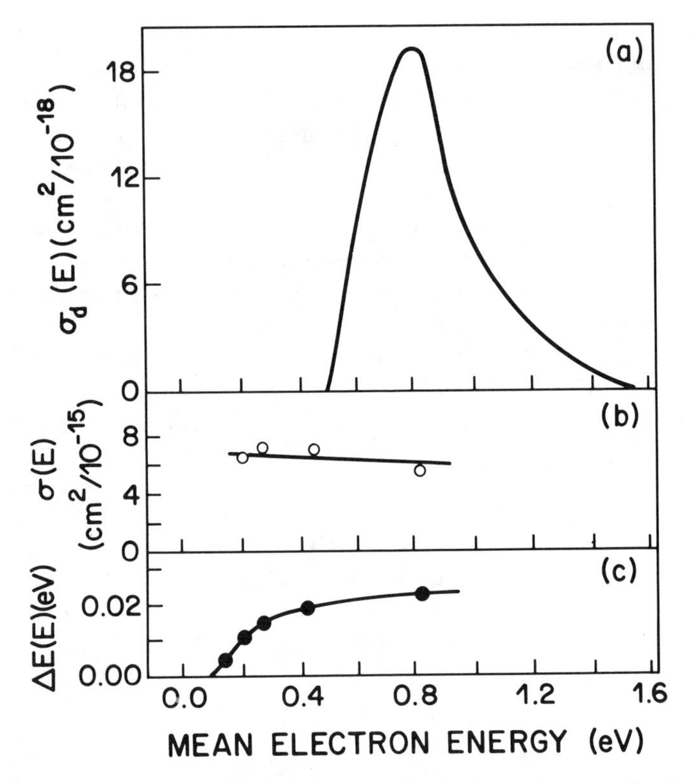


**Figure 9-4.** Energy dependence of factors affecting attachment of subexcitation electrons cooling in HCl: (a) dissociative attachment cross section; (b) total collision cross section; (c) mean energy loss per collision. (Reproduced with permission.[62])

therefore be written in the simplified form:[62]

$$\ln \theta = \frac{1}{\sigma(E) \cdot \Delta E(E)} \int_{E_i}^{E_{th}} \sigma_d(E) \cdot dE \qquad (9\text{-}17)$$

For HCl, $\sigma_d(E)$ is negligible below 0.55 eV and above 1.2 eV (Figure 9-4), and the integral in Eq. (9-17) has been calculated.[63] If one assumes that all subexcitation electrons in pure HCl have $E_i \geqslant 1.4$ eV, the value of $\theta$ is predicted to be about 92%. Because many subexcitation electrons may have $E_i$ less than 1.4 eV, this is an upper limit. Comparison of the yield of thermal electrons, determined from experiments with $SF_6$, with the total yield of electrons from $G$(ionization) indicates that $\theta = (95 \pm 5)\%$.[62] There are no values of $\sigma(E)$ and $\Delta E(E)$ for HBr, but the integral of Eq. (9-17) for this molecule is reported to be about 10 times larger than for HCl.[63] Assuming that $\sigma(E)$ and $\Delta E(E)$ are similar to the experimental values for HCl in Figure 9-4, we find $\theta$ to be $\geqslant$60%. In this system the yield of scavengeable thermal electrons indicates that $\theta$ is in fact $(73 \pm 8)\%$.[64] A further interesting point is that $\theta$ can be increased significantly when $CO_2$ is mixed with HBr.[64]

This molecule has a large value of $K_u$ (Table 9-3) but does not itself capture electrons in the 0.0 to 1.5 eV range in the gas phase at low pressure. These experiments imply that dissociative capture does occur during thermalization. However, unless the cross section is large and peaked at relatively low energies as in HBr (0.3 eV[63]), the fraction of electrons captured is small. Also it is dependent on the ratio $h(E)/\eta(E)$ for the total gas mixture. The special case of capture of epithermal electrons in inert gases has been considered by Mozumder.[56]

### Reactions of Thermal Electrons

This section is concerned with systems where the rate of electron kinetic energy exchange with the gaseous matrix is sufficiently fast that electrons attain a thermal energy distribution before reacting. The kinetics of reaction are again discussed in terms of Scheme (9-15), in which the requirement of thermalization before reaction is equivalent to $\langle E \rangle = \frac{3}{2}kT$. The total rate of capture is given by the master equation

$$-\frac{d[e^-]}{dt} = \frac{k_c(k_d + k_v + \sum k_m[M])}{k_a + k_d + k_v + \sum k_m[M]} \tag{9-18}$$

where $k_c$, $k_a$, and $k_d$ are the rate constants for steps (c), (a), and (d) in Scheme (9-15), and $k_v$ and $k_m$ are those for the radiative and collisional stabilization reactions (9-19) and (9-20), respectively:

$$AB^{*-}(v'_-, J'_-) \rightarrow AB^-(v_-, J_-) + h\nu \tag{9-19}$$

$$AB^{*-}(v'_-, J'_-) + M \rightarrow AB^-(v_-, J_-) + M \tag{9-20}$$

The sum $k_v + \sum k_m[M]$ is equal to the first-order rate constant $k_s$. The relative magnitudes of all of the above rate constants determine which terms in Eq. 18 are important and thus dictate the form of the rate expression and the order of reaction. These magnitudes in turn depend on properties of the molecule AB, including its size, symmetry, structure, and chemical makeup.[48,59]

#### *Second-Order Reactions*

For a large number of electron-reactive molecules, the kinetics have been shown to obey the empirical equation (9-21) with no effect of added non-reactive stabilizing gases:

$$-\frac{d[e]}{dt} = k_D[AB][e^-] \tag{9-21}$$

(See for example $CCl_4$ in ref. 65.) Values of $k_D$ for such systems have been listed in Table 9-4 along with the products and exothermicities of reaction, where they are known. Unless otherwise stated, the rates of reaction were followed by the microwave method.

**Table 9-4.** Rate Constants for Second-Order Electron-Attachment Reactions

| Compound | $k_D \left(\frac{cm^3}{molec}\right)$ | Product Ion | $\Delta E$ Reaction/eV | Ref. |
|---|---|---|---|---|
| $CCl_4$ | $4.1 \times 10^{-7}$ | $Cl^-$ | −0.62 | *a* |
| | $3.9 \times 10^{-7}$ | | | *b* |
| $CFCl_3$ | $2.4 \times 10^{-7}$ | $Cl^-$ | −0.4 | *c* |
| | $2.6 \times 10^{-7}$ | | | *b* |
| HI | $2.4 \times 10^{-7}$ | $I^-$ | −0.05 | *d* |
| $SF_6$ | $2.3 \times 10^{-7}$ | $SF_6^-$ | | *c* |
| | $2.2 \times 10^{-7}$ | | | *e* |
| $I_2$ | $3.1 \times 10^{-8}$ | $I^-$ | −1.56 | *d* |
| $F_2$ | $3–6 \times 10^{-9}$ | $F^-$ | −1.85 | *f* |
| $Cl_2$ | $1–4 \times 10^{-9}$ | $Cl^-$ | −1.13 | *f* |
| | $2 \times 10^{-9}$ | | | *b* |
| $CH_3Br$ | $7 \times 10^{-12}$ | ($Br^-$?) | (−0.46) | *e* |
| $CF_3Cl$ | $5 \times 10^{-14}$ | ($Cl^-$?) | (−0.14) | *e* |

[a] Ref. 65 microwave.
[b] Ref. 66 flowing afterglow.
[c] Ref. 67 Cavalleri method.
[d] Ref. 70, calculated from cross-section data.
[e] Refs. 68, 69 microwave.
[f] From refs. 71 and 72, range of values calculated from cross-section data and measurements of $k_D$ from several sources.

Dissociative attachment [ie, steps (c) + (d) of Scheme (9-15)] is in principle the simplest process to which second-order kinetics, as in Eq. (9-21), would apply. It can occur with a thermalized electron population if the energy level of $A + B^-$ lies below that of $e^-(E) + AB(v, J)$ [refer to Scheme (9-15)]. This is illustrated for the case of $Br_2$ in Figure 9-5(a). The same situation applies to the other halogens, HI, $CCl_4$ ($A = CCl_3$, $B = Cl$), and $CFCl_3$ ($A = CFCl_2$, $B = Cl$) in Table 9-4. For these systems dissociation is evidently too fast for collisional stabilization to have any effect at normal pressures (ie, $k_d \gg k_v + \sum k_m[M]$). Hence, the value of $k_D$ can be expressed as

$$k_D = \frac{k_d k_c}{k_a + k_d} \tag{9-22}$$

The absence of any reported effect of stabilizing collisions by nonreactive gases implies that the same situation applies for $CH_3Br$ and $CF_3Cl$. However, the products for these remain undetermined.

When $k_d$ is large relative to $k_a$, the overall rate constant $k_D$ approaches the attachment rate constant $k_c$. The maximum attachment cross section for an electron of velocity $v$ is $\lambda^2/4\pi$, where $\lambda$ is the de Broglie wavelength $h/mv$ of the electron.[59,65] The mean value of the maximum rate constant

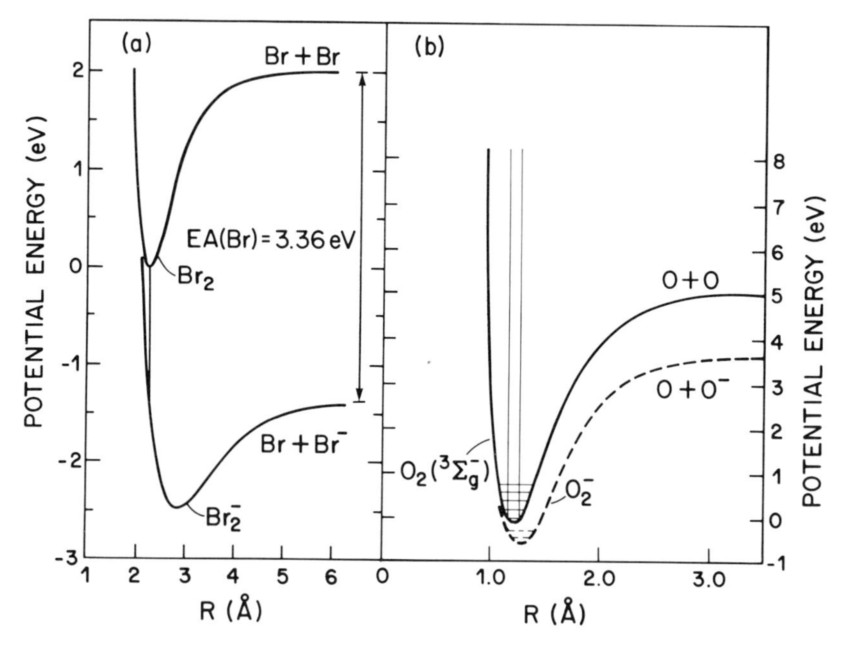


**Figure 9-5.** Potential energy diagrams for diatomic molecules and their anions: (a) $Br_2$ (Reproduced with permission.[60]); (b) $O_2$ (Reproduced with permission.[48]).

for a thermal electron energy distribution is therefore given by[59,65]

$$
\begin{aligned}
k_c(\max) &= \int_0^\infty \left(\frac{h}{mv}\right)^2 v^3 \left(\frac{m}{2\pi kt}\right)^{3/2} e^{-mv^2/2kT}\, dv \\
&= \frac{k^2}{(2\pi M)^{3/2}(kT)^{1/2}} \qquad (9\text{-}23) \\
&= 5.0 \times 10^{-7}\ \text{cm}^3\ \text{molecule}^{-1}\ \text{s}^{-1}
\end{aligned}
$$

Examination of Table 9-4 reveals that $k_D$ is close to $k_c(\max)$ for $CCl_4$, $CFCl_3$, and HI. However, it falls by a factor of $10^7$ as one proceeds down the list of molecules with exothermic dissociative attachment channels. It is evident, therefore, that exothermicity per se does not ensure a large value of $k_D$. The activation energies of a number of the reactions (eg, $CCl_4$) are negative,[65,66,73,74] but some are positive, even for certain fast reactions.[66] Some of these effects can be rationalized in terms of the symmetry, the spatial relations of the potential energy surfaces, and other properties of the negative ion and neutral molecule.[59,74] However, much remains to be done in this interesting area, and a unified picture for these systems cannot be presented here.

When the level of $[e(\langle E\rangle = \frac{3}{2}kT) + AB]$ lies between those of $AB^-$ and $A + B^-$, as illustrated by oxygen in Figure 9-5(b), channel (d) of Scheme

(9-15) is effectively closed and $AB^-$ becomes the main reaction product. The formation of $AB^-$ is called *nondissociative attachment.* The rate is then given by

$$-\frac{d[e^-]}{dt} = \frac{k_c(k_v + \sum k_M[M])}{k_a + k_v + \sum k_M[M]}[AB][e^-] \qquad (9\text{-}18a)$$

which loses its dependence on [M] if $k_a \ll (k_v + \sum k_M[M])$ or $k_v \gg \sum k_M[M]$. For these two conditions (9-18a) simplifies to

$$-\frac{d[e^-]}{dt} = k_c[AB][e^-] \qquad (9\text{-}24)$$

and, respectively,

$$-\frac{d[e^-]}{dt} = \frac{k_c k_v}{k_a + k_v}[AB][e^-] \qquad (9\text{-}25)$$

Equation (9-25) describes the kinetics of purely radiative stabilization. There is evidence for this from an ion cyclotron resonance experiment with $SF_6$ in which $\sum k_M[M]$ would have been relatively small, about $10^4$ s$^{-1}$. For the electron energies established in this ion cyclotron resonance study, the results suggested that $k_v/(k_a + k_v)$ was about 0.1.[75] However, radiative coefficients $k_v$ are not expected to be large where low-energy quanta from vibrational transitions of $AB^{*-}$ are involved. The condition that $k_v \gg \sum k_M[M]$ is therefore unlikely at normal pressures, where $\sum k_M[M]$ is about $10^{10}$ s$^{-1}$. Indeed, no confirmed cases for radiative stabilization under typical gas-phase radiolysis conditions have been reported.

The alternative condition, namely, $k_a \ll (k_v + \sum k_M[M])$, or more properly $k_a \ll \sum k_M[M]$ for normal pressures, is quite frequently encountered. Systematic studies[59] of autodetachment lifetimes have shown that $k_a$ may be $10^6$ s$^{-1}$ or less when AB is a large polyatomic ion. For example, for the series of fluorocarbons $C_4F_6$, $C_5F_8$, $C_6F_{10}$, $C_6F_{12}$, and $C_7F_{14}$, $k_a^{-1}$ is $1.3 \times 10^5$ s$^{-1}$ for the first member, decreasing to $1.3 \times 10^3$ s$^{-1}$ for the last. This decrease correlates well with the increased number of vibrational degrees of freedom in which the excess energy of electron attachment can be stored.[59] It follows that for $k_M > 10^{-10}$ cm$^3$ molecule$^{-1}$ s$^{-1}$ matrix gas densities of about $10^{16}$ molecule cm$^{-3}$ or pressures of a few millimeters of Hg and greater can ensure efficient stabilization of polyatomic ions with about 10 or more internal degrees of freedom.

Sulfur hexafluoride is the most intensively studied system of the above type.[76] As shown in Table 9-4, the magnitude of $k_D$ at normal pressure is close to $k_c$(max). The value of $k_a$ depends on the internal energy contributed by the thermal energy of $SF_6$ and the electron kinetic energy. It varies from about $10^6$ s$^{-1}$ at a few tenths of an electronvolt to $10^2$ s$^{-1}$ when the total energy is near the threshold for autoionization.[77] The structure of $SF_6^-$ is still not entirely resolved, but it appears to have a symmetry different from $SF_6$. The electron affinity is apparently in the range of 0.8 to 1.2 eV.[76] The electron-attachment rate constant at normal pressure is firmly established (Table 9-4), and the molecule has proven its worth as an electron scavenger in many systems.[7] It appears to be quite unreactive toward hydrogen atoms,[78] which is a distinct advantage in hydrogen-containing systems.

### *Third- and Higher-Order Reactions*

The rate expression for the general case of nondissociative attachment at normal pressures with $k_v$ neglected is

$$-\frac{d[e^-]}{dt} = \frac{k_c \sum k_M[M]}{k_a + \sum k_M[M]}[AB][e^-] \tag{9-26}$$

With small molecular ions (eg, $AB^-$ diatomic or triatomic) $k_a$ can be quite large $(10^{10}–10^{14})\ s^{-1}$, and the condition $k_a \gg \sum k_M[M]$ may be encountered.[48,59] The kinetic equation for the rate of capture for this case, viz,

$$-\frac{d[e^-]}{dt} = \frac{k_c \sum k_M[M]}{k_a}[AB][e^-] \tag{9-27}$$

was first derived by Block and Bradbury[79] in 1935. For a given composition and pressure of matrix gases there is an effective second-order rate constant $k_{eff}^{BB} = k_c \sum k_M[M]/k_a$, where BB means Block–Bradbury. It can be calculated from the experimental values of $(-d[e^-]/dt)/[AB][e^-]$.

### *Oxygen*

The BB mechanism was first applied to oxygen, which is still the most extensively studied diatomic molecule. Over the intervening years experimental[80] and theoretical studies[81] have greatly refined our understanding of that system. For example, it is now recognized that in process (c) [Scheme (9-15)] the electron is captured by $O_2$ in the $v = 0$ level to form $O_2^{*-}(v'_- = 4)$. For $O_2^{16}$ the $O_2^{16-}$ level lies 0.08 eV above the initial level of the neutral.[82] Thus, process (c) has an activation energy of 0.08 eV, and the rate should decrease with decreasing temperature.

Studies of the dependence of the rate of electron capture on matrix gas density [M] have been performed in several laboratories,[82–87] the microwave technique being the most commonly used in recent years. Results from different laboratories are generally in agreement, and the refined BB mechanism has been shown to be applicable at room temperature for $M = O_2$, $H_2$, $D_2$, $CO_2$ and $n\text{-}C_4H_{10}$ up to $[M] \sim 4 \times 10^{18}$ molecule $cm^{-3}$.[82,87] This conclusion was based on careful studies of the isotope effects in $O_2^{16}$ and $O_2^{18}$ and other experimental observations. Also, for a particular M in excess over $O_2$, Eq. (9-26) can be rearranged as

$$-\frac{[O_2][e^-]}{d[e^-]/dt} = \frac{k_a}{k_c k_M[M]} + \frac{1}{k_c} \tag{9-28}$$

Thus, $k_c$ can be found from the intercepts of plots of the left side of this expression vs. $[M]^{-1}$ and $k_a/k_M$ from the ratio of the slopes to intercepts. Also, because $k_M$ can be calculated from ion–molecule collision rate theory,[80] $k_a$ can be found from $k_a/k_M$. On this basis Hatano and co-workers have shown that $k_a$ is about $10^{10}$ s, which is comparable to theoretical estimates.[80] Also, $k_c$ was found to be $(4.5 \pm 1.0) \times 10^{-11}$ and $(8^{+1}_{-2}) \times 10^{-11}$ cm$^3$ molecule$^{-1}$ s$^{-1}$ for $O_2{}^{16}$ and $O_2{}^{18}$, respectively.[80,86] In the low-pressure limit (ie, $\sum [M] \rightarrow 0$) for specific matrix gases, Eq. (9-27) becomes

$$-\frac{d[e^-]}{dt} = \frac{k_c}{k_a}\{k_M[M] + k_{O_2}[O_2]\}[O_2][e^-] \qquad (9\text{-}29)$$

Using this relation, workers have found values of $k_c k_M/k_a$ for a variety of matrix gases.[86] For pure oxygen $k_c k_{O_2}/k_a = k_{eff}{}^{BB}/[O_2]$. At room temperature

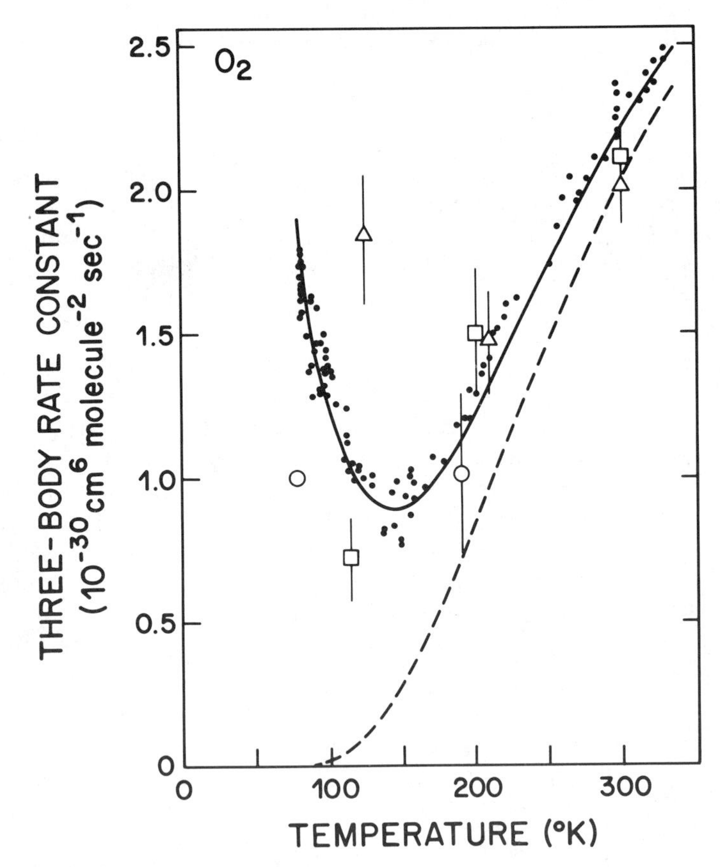


**Figure 9-6.** The temperature dependence of the three-body rate constant of $O_2$. The broken line shows the rate constant calculated from Herzenberg's refinement of the BB theory. The solid line includes the contributions of the broken line and of the electron attachment due to van der Waal's $(O_2)_2$ dimers. The points are experimental data from various laboratories. (Reproduced with permission.[82])

the value of this third-order (or three-body) rate constant is $(2.2 \pm 0.2) \times 10^{-30}$ $cm^6$ $molecule^{-2}$ $s^{-1}$.[80,82,83,87]

Despite the apparent success of the BB mechanism at room temperature and low-to-moderate pressure, strong deviations were observed from it at large (>10 atm) pressures or low temperatures. A clear illustration for the latter condition is seen in Figure 9-6, where $k_{eff}^{BB}/[O_2]$ has been plotted against temperature for the M = $O_2$[16] system. The broken line shows the steady falloff with temperature for $k_c k_{O_2}/k_a$, which was calculated from Herzenberg's equations by Shimamori and Fessenden.[82] Above 200 K the experimental results are in good agreement. However, instead of a continuing decrease below this temperature, they exhibit an increase. This has been attributed to the contribution of $O_2 \cdot O_2$ complexes or van der Waal dimers to the attachment reaction. An even more drastic deviation from theory on reduction of temperature is seen with M = $N_2$. Here, there is no falloff in the third-order rate constant at all—only a sharp rise occurring below 300 K.[82] Evidence for a strong contribution of electron capture by the $O_2 \cdot N_2$ van der Waal complex was also obtained from an examination of the isotope effect.[87]

The diagram in Figure 9-7 was taken from the review of Hatano and Shimamori.[80] It shows the variation of potential energy for the $O_2(v = 0)$—M and $O_2^-(v'_- = 4)$—interactions as a function of the intermolecular separations. They calculated the curve for the $O_2(v = 0)$—M system assuming a 12-6 Lennard-Jones potential, while for the $O_2^-(v'_- = 4)$—M system an

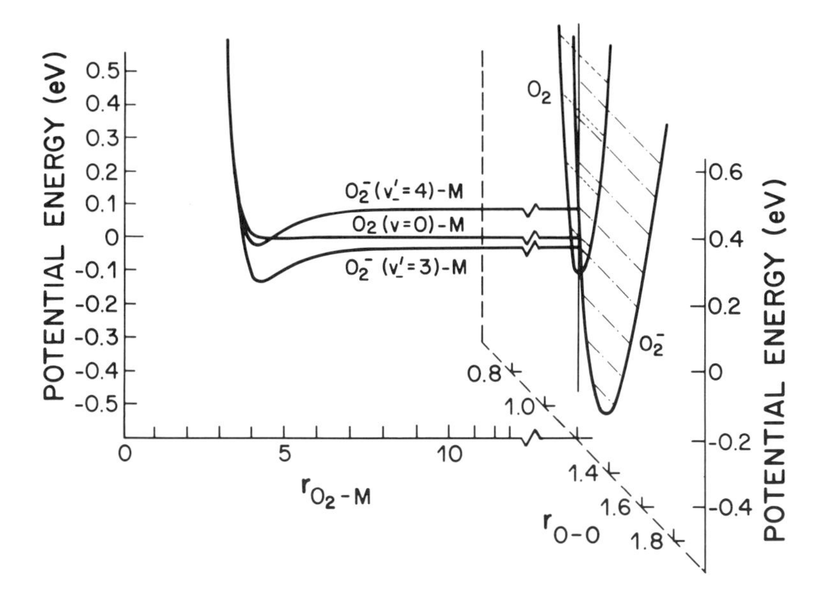


**Figure 9-7.** Variation of potential energies for $O_2(v = 0)$—M and $O_2^-$ $(v'_- = 4)$—M systems as a function of intermolecular distance $r_{O_2—M}$. (Reproduced with permission.[80])

induced dipole $-\alpha e^2/2r^4$ term was added. The value used for $\alpha$, the polarizability of M, was equivalent to that of a molecule such as $C_2H_6$ or $C_2H_4$. At infinite distance from M the separation in energy between $O_2^-(v'_- = 4)$ and $O_2(v = 0)$ is 80 meV. However, as $r_{O_2-M}$ is reduced this separation becomes steadily less until eventually there is a curve crossing at approximately 0.45 nm. Clearly, the activation energy for attachment is greatly reduced when the $O_2$ molecule is present in a van der Waal complex. However, other factors may also come into play to enhance the rate of capture in the complex.[82]

The following reactions of $O_2{\cdot}M$ complexes were postulated to occur:

$$O_2 + M \rightleftharpoons O_2{\cdot}M \tag{9-30}$$

$$O_2{\cdot}M + e^- \rightleftharpoons (O_2{\cdot}M)^{*-} \tag{9-31/-31}$$

$$(O_2{\cdot}M)^{*-} \rightarrow O_2 + M + e^- \tag{9-32}$$

$$(O_2{\cdot}M)^{*-} \rightarrow O_2^- + M \tag{9-33}$$

$$(O_2{\cdot}M)^{*-} + M \rightarrow O_2^- + 2M \tag{9-34}$$

The overall rate for capture in the presence of only one stabilizing M species in excess over $O_2$ then becomes

$$-\frac{d[e^-]}{dt} = \left\{\frac{k_c k_M[M]}{k_a + k_M[M]} + \frac{K_{30}k_{31}(k_{33} + k_{34}[M])[M]}{k_{-31} + k_{32} + k_{33} + k_{34}[M]}\right\}[O_2][e^-] \tag{9-35}$$

Values of $k_{31}$ found from studies of the dependence of the capture rate on [M] for species like $C_2H_4$, $C_2H_6$, $n$-$C_4H_{10}$, and $CO_2$ are from 5 to $20 \times 10^{-9}$ $cm^3$ $molecule^{-1}$ $s^{-1}$. The 100- to 400-fold increase over $k_c$ for $O_2$ demonstrates the enhancing effect of the complexation by M. About 20% of $(O_2{\cdot}M)^{*-}$ were estimated to decay via Reaction (9-32), and the lifetime of $(O_2{\cdot}M)^{*-}$ for $M = C_2H_6$ was found to be $\geqslant 10^{-12}$ s. The contribution of the van der Waal term relative to the $k_{eff}^{BB}$ or first term in Eq. (9-35) varies from one system to another and depends on temperature. In pure $O_2$[16] it is about 10% at room temperature, increasing at lower temperatures.[80]

It should be noted that at moderate pressures the inequalities $k_M[M] < k_a$ and $k_{34}[M] < k_{33}$ may hold, in which case (9-35) reduces to

$$-\frac{d[e^-]}{dt} = \left\{\frac{k_c k_M[M]}{k_a} + \frac{K_{30}k_{31}k_{33}[M]}{k_{-31} + k_{32} + k_{33}}\right\}[O_2][e^-] \tag{9-36}$$

The two terms now have the same dependence on [M] and will appear as a single third-order rate constant. Some of the values of $k_c k_M/k_a$, which have been reported in the literature, may in fact include contributions from van der Waal complexes.[80]

Reactions (9-30) to (9-34) accounted for the effects observed in ethane up to $3 \times 10^{20}$ molecule $cm^{-3}$. For higher values of [M] further reactions with more than one M per van der Waal complex had to be postulated to account for the observed rates, and the scheme is very complex. Fortunately, for most oxygen-containing systems, in which electron lifetimes or rates of attachment may have to be calculated, the pressures are fairly low and Eqs. (9-27) or (9-36) should hold. Thus, the tabulated third-order rate constants may be used to obtain realistic estimates of these parameters, which are needed for the establishment of mechanisms of radiolysis.

### *Hydrogen Halides*

Herzenberg[81] pointed out that when the value of $k_a$ becomes very large the distance traveled by the stabilizing collider M during the lifetime of $AB^{*-}(v'_-, J'_-)$ would approach the AB-M equilibrium separation in a van der Waal complex. Under those circumstances the two steps of the BB mechanism—electron attachment and collisional stabilization—coalesce into one. This means, in effect, that only the molecular complex mechanism need be considered. Nagra and Armstrong concluded that would be the case for the hydrogen halides HCl and HBr for which the dissociative attachment seen in HI (Table 9-4) is blocked by endothermicity.[88] In both of these HX systems (and HI) the extra electron goes into a $\sigma$-orbital, and coupling with s-wave electrons is permitted.[89] This strongly enhances both $k_c$ and $k_a$.[81,88] (Note that the same arguments apply to attachment to HI. There the dissociative channel is open, and from Table 9-4 one may note that $k_D(\cong k_c)$ is close to the maximum value.) Estimates of $k_a$ for $HCl^{*-}$ and $HBr^{*-}$ formed by attachment of thermal electrons fall in the region $>10^{-13}$ s. As one can see, this is much larger than the value of $k_a$ for oxygen ($10^{10}$ s$^{-1}$; see above), where d-wave electrons are involved.[81] Also in $10^{-13}$ s the average molecule at room temperature travels only about 0.04 nm.

Based on the above information and earlier experimental studies[42,90,91] from different laboratories, in 1976 Nagra and Armstrong proposed the following reaction mechanism, which involves the molecular dimer HX·HX:[88]

$$2HX \rightleftharpoons HX{\cdot}HX \tag{9-37}$$

$$e^- + HX{\cdot}HX \underset{k_a}{\overset{k_c}{\rightleftharpoons}} (HX{\cdot}HX)^{*-} \tag{9-38}$$

$$(HX{\cdot}HX)^{*-} \rightarrow {\cdot}H + X{-}H{-}X^- \tag{9-39}$$

$$(HX{\cdot}HX)^{*-} + HX \rightarrow (HX)_2^- + HX \tag{9-40}$$

$$(HX{\cdot}HX)^{*-} + HX \rightarrow {\cdot}H + X^-{\cdot}2\,HX \tag{9-41}$$

$$(HX{\cdot}HX)^{*-} + M \rightarrow (HX)_2^- + M \tag{9-42}$$

$$(HX)_2^- + HX \rightarrow {\cdot}H + X^-{\cdot}2\,HX \tag{9-43}$$

The HX·HX species may actually only be a transient collision complex (cf ref. 82). The value of $K_{37}$ calculated from pressure–volume–temperature data for both HCl and HBr is $3 \times 10^{22}$ $cm^3$ molecule at 298 K. From studies of the dependence of the electron capture rate on [HX] and [M], measured by competition kinetics with $SF_6$ as a scavenger, it is apparent that for HCl $k_{39}$ is $<2 \times 10^9$ $s^{-1}$ and (9-40) to (9-43) are the main product-forming reactions. The rate of capture is given by

$$-\frac{d[e^-]}{dt} = \left\{k_{40} + k_{41} + \frac{k_{42}[M]}{[HCl]}\right\} K_{37} k_c k_a^{-1} [HCl]^3 [e^-] \qquad (9\text{-}44)$$

where $k_c$ and $k_a$ are the attachment and autodetachment rate constants of Eq. (9-38). The ratio $k_{42}/(k_{40} + k_{41})$ is 0.90 for both $CO_2$ and $C_3H_8$ as M, which is in agreement with the ratio of collision rate constants from ion–molecule reaction rate theory.[88] The latter also gave $k_{40} + k_{41} = 1.2 \times 10^{-9}$ $cm^3$ $molecule^{-1}$ $s^{-1}$. The quantity $(k_{40} + k_{41})K_{37}k_c k_a^{-1}$ is found from the experimental data to be $1.0 \times 10^{-49}$ $cm^9$ $molecule^{-3}$ $s^{-1}$.[42,88] Assuming $k_c$ is close to the maximum value of $5 \times 10^{-7}$ $cm^3$ $molecule^{-1}$ $s^{-1}$ and taking the above values of $K_{37}$ and $(k_{40} + k_{41})$, it can be shown that $k_a$ must be $\leq 10^{12}$ $s^{-1}$. The fact that this is smaller than $k_a$ for the monomer is attributable at least in part to the increased number of degrees of freedom.[88]

Nagra and Armstrong also studied the HBr·HBr, HBr·$H_2S$, and HBr·HCl van der Waal systems.[92] In each case the rate of capture is independent of the concentration of stabilizing species M and obeys the equation

$$-\frac{d[e^-]}{dt} = k_T[HX][HY][e^-] \qquad (9\text{-}45)$$

The values of $k_T$ are $(1.0 \pm 0.2)$, $(1.3 \pm 0.3)$, and $(1.4 \pm 0.2) \times 10^{-28}$ $cm^6$ $molecule^{-2}$ $s^{-1}$, respectively, for the three systems. This can be explained if, for each system, $k_{39} \gg k_a$, $k_{39} > (k_{40} + k_{41})[HX]$, and $k_{39} > k_{42}[M]$. From these inequalities it follows that $k_T = K_{37}k_c$. Also $k_c$ must again be close to the maximum permissible value.

Although there are similarities between the van der Waal mechanisms for oxygen and HX molecules, there are some significant differences. The enhancement in $k_c$ for $O_2$·M complexes is considered to arise from a lowering of the activation energy and other factors,[82] whereas for the HX molecules $k_c$ may already be close to the maximum value even for the monomers.[88] For $(O_2 \cdot M)^{*-}$ the dissociative capture channel (9-33) produces $O_2^- + M$, whereas in the HBr·HX mixed dimers it actually leads to the breaking of the HBr bond and formation of $Br^-$·HX in Reaction (9-39). Actually, it is the association energy of $Br^-$—HX that couples with the electron affinity of Br to provide the driving force for dissociative attachment by the HBr·HX species. Independent evidence for the occurrence of processes of this type comes from a study of the transfer of electrons from neutral Rb atoms to $(HX)_n$ clusters.[93]

For the HCl system the main effect of the van der Waal complex formation appears to be an enhancement in the lifetime of the excited $HCl \cdot HCl^{*-}$

ion, which allows time for Reactions (9-40) and (9-41) to occur. In contrast, for the $(O_2{\cdot}M)^{*-}$complexes the lifetimes are shorter than for $O_2^{*-}$ due to the breakdown of symmetry and other effects.[82]

The independent discoveries of the roles of molecular complexes in the capture of electrons by oxygen and hydrogen halide molecules have led to the realization that such complexes may be important in many other systems. Electron beam experiments have already been conducted with neutral clusters of $H_2O$, $N_2O$, $CO_2$, and other molecules.[94] These have demonstrated formation of $(CO_2)_n^-$ ($n > 2$) and other interesting new effects. In addition, for $N_2O$ there is direct evidence from microwave studies[95] for the involvement of $N_2O{\cdot}M$ complexes at high pressures of hydrocarbon and other matrix gases, M. The system seems to resemble oxygen in that Reaction (9-46) has a larger rate constant than attachment to the isolated monomer:[80]

$$e^- + N_2O{\cdot}M \rightarrow (N_2O{\cdot}M)^{*-} \quad (9\text{-}46)$$

$$e^- + N_2O \rightarrow N_2O^{*-} \quad (9\text{-}47)$$

Clearly, recognition of the influence of molecular association on electron-attachment processes will be a major feature of studies over the next few decades. This fascinating field may provide another bridge between our understanding of gas- and liquid-phase kinetics.

## *Molecular Ions*

A few examples of cations formed in the radiolysis of gases and the reactions that they may undergo were given in Chapter 2. Negative atomic or molecular ions are normally formed by the electron-attachment processes described above. In a few systems ion pair production may occur as a result of electronic excitation and dissociation:

$$AB \rightsquigarrow AB^* \rightarrow A^+ + B^-$$

Experience suggests that it is not a very common phenomenon in radiolysis, and it will not be pursued further. However, if it does occur, the methods used for determining the yields and reaction rates of $A^+$ and $B^-$ would not differ from those discussed below.

The full importance of the contribution of ionic processes to radiolysis mechanisms was recognized in the late 1950s. The parent ions formed by ionizing radiation, $\cdot G^{+*}$ in Table 9-1, possess a distribution of internal energies, $E_i$, which may include electronic, vibrational, and rotational excitation. These ions may lose energy by radiation or collisional energy transfer, or they may undergo uni- or bimolecular reactions.[46,96] This is illustrated in the following mechanism, in which $E_0$ represents a lower internal energy:

$$AB \rightsquigarrow AB(E_i)^+ + e^- \quad (9\text{-}48)$$

$$AB(E_i)^+ \rightarrow AB(E_0)^+ + h\nu \quad (9\text{-}49)$$

$$AB(E_i)^+ + M \rightarrow AB(E_0)^+ + M \quad (9\text{-}50)$$

$$AB(E_i)^+ \rightarrow A^+ + B \quad (9\text{-}51a)$$

$$AB(E_i)^+ \rightarrow BA(E_j)^+ \text{ (Isomerization)} \quad (9\text{-}51b)$$

$$AB(E_i)^+ + M \rightarrow K^+ + Z \quad (9\text{-}52)$$

In contrast to the time sequence scheme in Table 9-1, this concentrates on reactions of a single ionic species $AB^+$, which competes on the same time scale. The rates for these reactions in a given system depend on $E_i$, the collision frequency $k_M[M]$, and the identities of AB and M. A key factor here is the gas-phase collision frequency at normal pressure—approximately $10^{10}\ s^{-1}$. This sets the time scale within which unimolecular processes must occur if they are not to be swamped by the bimolecular reactions (9-50) and (9-52). In this section a discussion of examples of Reactions (9-48)–(9-52) is given. A much more comprehensive systematic discussion of ion–molecule reactions can be found in ref. 46.

### Unimolecular Reactions of Ions and Identities of Fragments from Parent Ions

The structures and identities of the primary ionic species are, of course, of direct interest to the radiation chemist.[7,46,96] Most of this information comes from the field of mass spectrometry, although photoelectron and vacuum ultraviolet spectroscopy also make valuable contributions. Mass spectrometers with electron impact sources typically operate at $k_M[M] \leq 3000$ and detect parent $AB(E_i)^+$ ions if $k_{51} < 10^6\ s^{-1}$.[96] For small molecules $k_{51}$ is likely to be very large for dissociative electronic states—up to $10^{13}\ s^{-1}$ for diatomic hydrides[97]—and parent ions in such states are not seen in mass spectra. For such small molecular systems dissociation should also be the dominant process under radiolysis conditions, where $k_M[M]$ is $10^{10}\ s^{-1}$ at normal pressure and still too slow to compete with Reaction (9-51a). As a result the distribution of fragment and parent ions obtained from 100-eV electron impact mass spectra is often used to calculate $g$-values for the so-called primary ionic species in diatomics, most triatomics, and even some larger molecules such as $NH_3$.[7] This is justified by the fact that relative ionization cross sections $\sigma_i(E)/\sigma(E)$, where $\sigma_i(E)$ is the cross section for formation of ion $i$ at electron energy $E$ and $\sigma(E) = \Sigma\sigma_i(E)$, are only weakly dependent on $E$ above about 70 eV (refs. 96 and 46, page 10). Some examples[7,97] of ion fragmentations occurring on the sub-$10^{-10}$ s time scale are given in Eqs. (9-53) and (9-54).

$$N_2O \rightsquigarrow N_2O^+(\tilde{A}\,^2\Sigma^+) + e^- \rightarrow NO^+ + N + e^- \quad (9\text{-}53)$$

$$HCl \rightsquigarrow HCl^+(^2\Pi) + e^- \rightarrow H^+ + Cl + e^- \quad (9\text{-}54)$$

Also the distributions of ionic primary process yields calculated for molecular oxygen and water vapor by Willis and Boyd[7] are given in Table 9-5. The

**Table 9-5.** Primary Ionization Processes in Oxygen and Water Vapor

| Primary Process | | Yield ($G$ units) | Threshold Energy (eV) | Probable Energy (Ev) | Energy Used (eV/100 eV) |
|---|---|---|---|---|---|
| *Oxygen* | | | | | |
| $O_2 \rightarrow O_2^+(X^2\Pi_g) + e^-$ | (d) | 0.24 | 12.08 | 12.4 | 3.0 |
| | (a) | 0.00 | | | |
| $O_2 \rightarrow O_2^+(a^4\Pi_u) + e^-$ | (d) | 0.47 | 16.11 | 16.7 | 7.8 |
| | (a) | 0.21 | | 16.2 | 3.4 |
| $O_2 \rightarrow O_2^+(A^2\Pi_u) + e^-$ | (d) | 0.43 | 16.82 | 17.6 | 7.6 |
| | (a) | 0.18 | | 16.9 | 3.0 |
| $O_2 \rightarrow O_2^+(b^4\Sigma_g^-) + e^-$ | (d) | 0.32 | 18.17 | 18.3 | 5.9 |
| | (a) | 0.07 | | 18.2 | 1.3 |
| $O_2 \rightarrow O_2^+(\text{B state}) + e^-$ | (d) | 0.13 | 24.56 | 24.6 | 3.2 |
| $[B = (C^4\Sigma_g^-, c^2\Sigma_g^-)]$ | (a) | 0.02 | | 24.6 | 0.5 |
| $O_2 \rightarrow O(^3P \text{ or } ^1D) + O^+(^4S) + e^-$ | | 0.86 | 18.8 or 20.55 | 20.6 | 17.2 |
| $O_2 \rightarrow O(^3P) + O^+(^2D) + e^-$ | | 0.37 | 22.0 | 22.5 | 8.4 |
| $G$(ionization) | | 3.28 | Total for ionization processes | | 61.3 |
| *Water* | | | | | |
| $H_2O \rightarrow H_2O^+ + e^-$ | | 1.99 | 12.6 | 14.6 | 29.1 |
| $H_2O \rightarrow OH^+ + H + e^-$ | | 0.57 | 18.1 | 20.1 | 11.5 |
| $H_2O \rightarrow H^+ + OH + e^-$ | | 0.67 | 19.6 | 21.6 | 11.5 |
| $H_2O \rightarrow H_2^+ + O + e^-$ | | 0.01 | 20.5 | 22.5 | 0.3 |
| $H_2O \rightarrow H_2 + O^+(^2D) + e^-$ | | 0.06 | 22.4 | 34.4 | 1.4 |
| $G$(ionization) | | 3.33 | Total for ionization processes | | 56.8 |

(a) = autoionization.
(d) = direct ionization.
Based on data in ref. 7.

yield of a particular process or species i was taken as

$$g_i = \frac{\sigma_i(E)}{\sigma(E)} \times \frac{100}{W} \tag{9-55}$$

where $E$ corresponds to 100-eV electron kinetic energy. This energy is considered sufficiently large that optical selection rules are obeyed. A certain number of forbidden transitions to ionic or neutral states may occur. However, the proportion of these events is expected to be small.[7]

Also given in Table 9-5 are the threshold energies and most probable energies taken from potential energy diagrams for the two systems. Finally, in the last column is the energy for ionization. For a fuller discussion of the methods involved in these calculations, the reader is referred to refs. 7 and 97. We shall return to this subject later.

For larger molecules the ion fragmentation pattern seen in a mass spectrometer strictly only applies to radiolysis studies at the same pressure and $k_M[M]$. A priori predictions of the ions emerging from the reactions at normal pressure requires a detailed knowledge of the rate constants of

Reactions (9-48) to (9-52), which depend on $E_i$.[46,96] A large amount of very elegant work has been done on the energy dependence of both bi- and uni-molecular reaction rate constants of ions. For the latter one usually makes recourse to the quasi-equilibrium theory of mass spectra, because the energy dependences are not readily accessible by experiment. For a discussion of the research on unimolecular ionic reactions, the reader is referred to recent reviews and treatises in the field of mass spectrometry.[98]

Unfortunately, the very large number of energy-dependent rate constants, which must be known, normally makes precise predictions of ion yields from low-pressure mass spectra impractical for large polyatomic systems. The most reliable information on primary ions at high pressures and on the effects of collisions on isomerization or fragmentation of ions actually comes from analyses of long-lived chemical products obtained in the presence of scavengers—the so-called radiolytic method. This method, applied extensively to hydrocarbons, relies heavily on the use of isotopic labeling and on the neutralization of certain ions by charge transfer. The rate constants for these charge transfer reactions are often established by low-pressure mass spectrometric studies. Space permits mention of only a few examples; for a more detailed discussion the reader is directed to ref. 3, Chapter 6, and ref. 46.

As an illustration of the radiolytic method, we consider its application to the determination of the effect of pressure on the formation of $(CH_3)_3C^+$ and $CH_3CH_2C^+HCH_3$ ions in the radiolysis of $n$-hexane.[11] This compound was irradiated with $Co^{60}$ gamma rays in the presence of an excess of 3-methylpentane-$d_{14}$ ($C_6D_{14}$) and 0.4–5% $O_2$, which serves as a free radical scavenger. Under these conditions the following reactions occur:

$$n\text{-}C_6H_{14} \rightsquigarrow C_4H_9'^{+} + e^- + \text{Other Products} \tag{9-56}$$

$$C_4H_9'^{+} \rightarrow CH_3CH_2C^+HCH_3^* \tag{9-57}$$

$$CH_3CH_2C^+HCH_3^* \rightarrow (CH_3)_3C^+ \tag{9-58}$$

$$CH_3CH_2C^+HCH_3^* + M \rightarrow CH_3CH_2C^+HCH_3 + M \tag{9-59}$$

$$CH_3CH_2C^+HCH_3^* + C_6D_{14} \rightarrow CH_3CH_2CHDCH_3 + C_6D_{13}^+ \tag{9-60a}$$

$$CH_3CH_2C^+HCH_3 + C_6D_{14} \rightarrow CH_3CH_2CHDCH_3 + C_6D_{13}^+ \tag{9-60b}$$

$$(CH_3)_3C^+ + C_6D_{14} \rightarrow (CH_3)_3CD + C_6D_{13}^+ \tag{9-61}$$

The butanes formed are separated from other products by gas chromatography and then analyzed by mass spectrometry to determine the isomeric composition. Butanes from the $C_6D_{14}$ radiolysis are distinguished by their much greater degree of deuteration. The initially formed $C_4H_9'^{+}$ ions may include some $CH_3CH_2CH_2CH_2^+$, which apparently isomerize rapidly to the secondary ion. The tertiary: secondary ratios from the radiolysis experiments at 11 and 116 mm of Hg pressure are given in Table 9-6 along with ratios obtained from mass spectrometric studies at 1 and $10^{-6}$ mm Hg. The

**Table 9-6.** Extent of Isomerization of Butyl Ions from Fragmentation of *n*-Hexane as a Function of Pressure

| Pressure (mm of Hg) | Collision Interval/s | Ratio of *tert*-butyl to *sec*-butyl |
|---|---|---|
| $10^{-6}$ | $2 \times 10^{-2}$ | 1.92 |
| 1 | $2 \times 10^{-8}$ | 0.38 |
| 11 | $2 \times 10^{-9}$ | 0.22 |
| 116 | $2 \times 10^{-10}$ | 0.13 |

Based on data in ref. 11.

ratio is seen to fall off with pressure, and this can be attributed to deactivating collisions—Reaction (9-59). Rather interestingly some secondary ions are seen even at $10^{-6}$ torr. This implies that some $CH_3CH_2CH^+CH_3$ are formed from the parent ion with insufficient energy to overcome the barrier (~17 kcal mole$^{-1}$) for isomerization.[11]

As a second illustration of the utility of the radiolytic method, Figure 9-8 shows the effect of pressure on ion pair yields of several fragment ions in the radiolysis of propane. At high pressures all fragment ion yields are suppressed, suggesting that collisional deactivation of parent $C_3H_8^{*+}$ ions occurs on about a $10^{-11}$ s time scale. The increase in $C_2H_5^+$ at moderate pressure is probably due to the quenching of $C_2H_5^{*+}$, which evidently must have a longer lifetime than $C_3H_8^{*+}$.

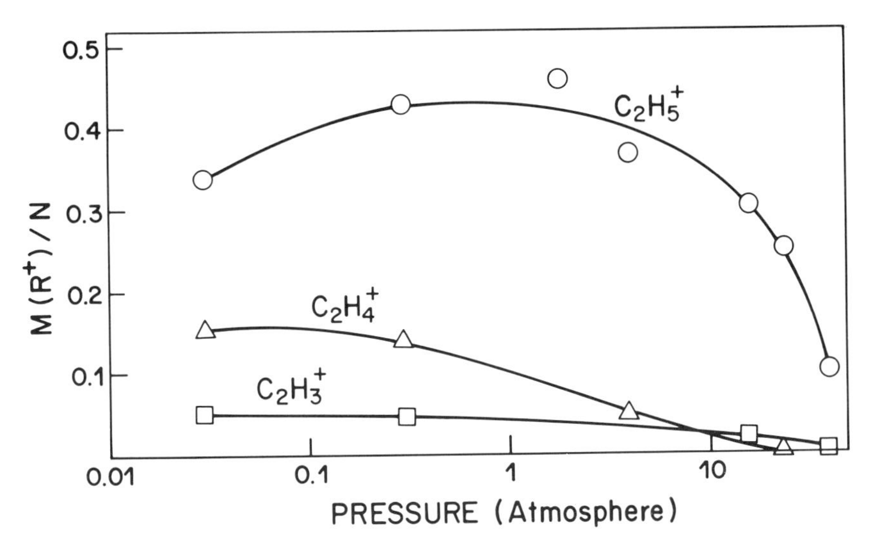


**Figure 9-8.** The effect of pressure on the yields of $C_2H_5^+$, $C_2H_4^+$, and $C_2H_3^+$ expressed as fractions of the total yield of ions from the fragmentation of $C_3H_8^+$. (Reproduced with permission.[46])

An important technique for the study of molecular ions is the photoionization–photoelectron coincidence method. Here the molecule is ionized by a photon of known energy, and the energy of the emitted electron is determined. By this method the excitation energy residing in the ion is known precisely. Hence, observed rates of reaction can be correlated with internal energy more readily. One system examined by this method and a variety of other techniques is $C_6H_6^+$. A review of the electronic states of the isomeric species and a detailed discussion of the isomerization and fragmentation pathways have been given by Rosenstock and Dannacher.[99] For some isomers radiative decay of the $\tilde{A}$ state has been found to be quite important. Thus, for $CH_3{-}C{\equiv}C{-}C{\equiv}C{-}CH_3^+$ the radiative decay constant for $\tilde{A} \rightarrow \tilde{X}$ $0 \rightarrow 0$ and $5 \rightarrow 1$ emission was $4.2 \times 10^7\ s^{-1}$, and the nonradiative decay constant was about $2 \times 10^7\ s^{-1}$.

Radiative decay of electronically excited ions has also been observed in diatomic and triatomic systems. For example, emission due to the $b^4\Sigma_g^- \rightarrow a^4\Pi_u$ transition of the $O_2^+$ parent ion at 560.2 nm has been observed in the pulse radiolysis of oxygen[6]. The radiative decay constant $k_{49}$ is $(1.2 \pm 0.4) \times 10^6\ s^{-1}$, and the quenching constant for oxygen $[k_{50}(O_2)]$ was $(7.5 \pm 2.5) \times 10^{-10}\ cm^3\ molecule^{-1}\ s^{-1}$. Similarly, for the $CO_2^+$ parent ion $A^2\Pi_u \rightarrow X^2\Pi_g$ transition at 434.1 nm, $k_{49} = (6.8 \pm 0.5) \times 10^6\ s^{-1}$ and $k_{50}(CO_2) \cong 5.3 \times 10^{-10}\ cm^3\ molecule^{-1}\ s^{-1}$.[6]

**Biomolecular Reactions of Ions**

Reactions of this kind have been investigated by mass spectrometric, flowing afterglow, absorption spectroscopic, and radiolytic methods. Exothermic ion–molecule reactions are usually rapid, and their rate constants may be calculated by means of the average dipole orientation (or ADO) theory,[46,100] which takes into account both ion–dipole and ion-induced dipole interactions. Recent developments also include repulsive terms.[101] Examples of reactions, which illustrate how parent and fragment ions in oxygen and water vapor may be converted to free radicals and more stable ion species, are

$$H_2O^+ + H_2O \rightarrow H_3O^- + \cdot OH$$

$$OH^+ + H_2O \rightarrow H_2O^{+}\cdot + \cdot OH \qquad k_{63} = 1.5 \times 10^{-9} \qquad (9\text{-}63)$$

$$O^+(^4S) + O_2 \rightarrow O_2^+(X^2\Pi_g) + O(^3P) \qquad k_{64} = 2 \times 10^{-11} \qquad (9\text{-}64)$$

$$O^+(^2D) + O_2 \rightarrow O_2^+(A^2\Pi_u \text{ or } a^4\Pi_u) + O(^3P) \qquad k_{65} = 2 \times 10^{-10} \qquad (9\text{-}65)$$

The rate constants, which are all in units of $cm^3\ molecule^{-1}\ s^{-1}$, are taken from the review of Willis and Boyd.[7] Their magnitude is such that these reactions all typically occur on a time scale of $10^{-9}$ s or less for parent molecule concentrations of $2.69 \times 10^{19}$ molecule $cm^{-3}$, that is, 1 atm at ambient temperature.

For hydride molecules for reaction pairs like $H_2O^{+\cdot}$ and $H_2O$ in Reaction (9–62), both H atom and $H^+$ transfer lead to the same products. Isotopic labeling has shown that for this pair and for $NH_3^{+\cdot}$—$NH_3$ proton transfer is more probable than H-atom transfer.[46] These reactions have been studied in some depth, and it is known that increase in the relative kinetic energy lowers the cross section of both processes. However, the cross section for atom transfer is reduced more.[46]

When the difference in ionization potentials makes charge transfer exothermic, it can compete with proton transfer. For example, studies show that for $C_4H_8^+$ ions, impacting on amines with a relative kinetic energy of 0.3 eV, charge transfer becomes more important by a factor of about 10, when the ionization potential of the amine diminishes from 8.97 eV ($CH_3NH_2$) to 7.82 eV [$(CH_3)_3N$].[46]

An interesting type of protonation occurs in hydrocarbons where ions of the form $C_nH_{2n+3}^+$ are formed. The structures and reaction of these have recently been discussed, along with thermochemical data for the equilibria:[102]

$$C_nH_{2n+3}^+ \rightleftharpoons C_nH_{2n+1}^+ + H_2 \quad (9\text{-}66)$$

A role for these species is envisaged in radiolysis.[46] For example, for $n$-pentane protonated by $H_3^+$ the following reactions were observed with the percentage yields indicated:

$$C_5D_{12}H^+ \xrightarrow{73\%} CD_3H + C_4D_9^+ \quad (9\text{-}67a)$$

$$C_5D_{12}H^+ \xrightarrow{18\%} C_2D_5H + C_3D_7^+ \quad (9\text{-}67b)$$

$$C_5D_{12}H^+ \xrightarrow{9\%} C_3D_7H + C_2D_5^+ \quad (9\text{-}67c)$$

Other types of bimolecular reaction in hydrocarbons involve $H^-$, $H_2^-$, and $H_2$ transfers. For example, for the former, one has

$$C_2H_5^+ + C_3H_8 \rightarrow C_2H_6 + C_3H_7^+ \qquad k = 6.3 \times 10^{-10} \quad (9\text{-}68)$$

$$sec\text{-}C_4H_9^+ + n\text{-}C_5H_{12} \rightarrow C_4H_{10} + C_5H_{11}^+ \qquad k = 3.7 \times 10^{-10} \quad (9\text{-}69)$$

and Reactions (9-60a), (9-60b), and (9-61). Transfers of $H_2^-$ and $H_2$ are discussed in ref. 46.

Halide ions may also be transferred. As an example, from a high-pressure mass spectrometric study of hexafluoroacetone[103] the following reaction is reported:

$$CF_3^+ + CF_3COCF_3 \rightarrow CF_4 + CF_2COCF_3^+ \qquad k = 0.84 \times 10^{-10} \quad (9\text{-}70)$$

Likewise, for the negative $SF_6^-$ ion, Reaction (9-71)

$$SF_6^- + HCl \rightarrow SF_5\cdot + F^-\cdot HCl \tag{9-71}$$

has been observed,[46] the ion product here being a hydrogen-bonded species.

From these illustrations of bimolecular reactions, the reader will begin to understand that each system is subject to its own complexities. Thus, experiment rather than theory remains the best way of determining which ionic species are active participants in product formation by radiolysis. To this end, many experiments have been made in mass spectrometers at increased pressures. Several of these are reviewed in ref. 46, and ref. 103 provides another example.

**Ion Clustering and Condensation Reactions**

As examples of clustering of small ionic species, we again cite Willis and Boyd's review:[7]

$$H_3O^+(H_2O)_n + H_2O + M \rightarrow H_3O^+(H_2O)_{n+1} + M \qquad k = 3 \times 10^{-27} \quad \text{for } n = 0\text{–}3 \tag{9-72}$$

$$O_2^+ + 2\,O_2 \rightarrow O_4^+ + O_2 \qquad k = 3 \times 10^{-30} \tag{9-73}$$

These processes are three-body reactions because the energy of ion plus molecule association must be removed. The units of the rate constants are in molecule$^{-2}$ cm$^6$ s$^{-1}$. At the normal parent molecule density, $2.69 \times 10^{19}$ molecule cm$^{-3}$, the ion lifetimes with respect to clustering are again in the region of $10^{-9}$ s or less. This is much faster than the time scale of ion–ion or ion–electron recombination, except at very high dose rates (see below). Thus, secondary ionic reactions are normally complete before the ions are neutralized. Indeed, at low dose rates, such as those obtained with $Co^{60}$ or x-ray sources, one can normally assume that the cluster sizes for a given ion core have reached a distribution close to the thermal equilibrium value. For very high dose rates detailed computer calculations based on kinetic parameters must be performed to determine the rates of neutralization of ions of each cluster size.

Tables of rate constants for clustering reactions and free energies of clustering of small ions are found in refs. 104 and 105. Condensation reactions and clustering of larger organic and inorganic species are reviewed in ref. 46, Chapter 7. As examples of these reactions for larger systems, one may cite from ref. 103:

$$CF_3COCF_2^+ + CF_3COCF_3 \rightarrow CF_3COCF_2(CF_3COCF_3)^+ \tag{9-74}$$

$$CF_3COCF_3^+ + CF_3COCF_3 \rightarrow (CF_3COCF_3)_2^+ \tag{9-75}$$

For such large molecules with many degrees of freedom, the rate of dissociation should be quite slow. Thus, stabilization may occur at relatively low values of $k_M[M]$. For this reason the reaction may appear as second order even at about a pressure of 1 mm Hg.

## *Electron–Ion and Ion–Ion Neutralization Reactions*

### Methods

The populations of ions in thermal equilibrium with gases at the temperatures and pressures of interest to radiation chemists are negligible, and therefore all ionic species produced by ionizing radiation eventually undergo neutralization, viz,

$$X^+ + Y^- \text{ or } e^- \rightarrow \text{Neutral Products} \tag{9-76}$$

The time scale for these reactions is of fundamental importance to any understanding of a mechanism of radiolysis. Because positive ions and negative ions or electrons are formed in equal numbers, homogeneous ion decay at a given pressure is a simple second-order process.[8] By convention the second-order rate constant is given the symbol $\alpha$ for ion plus ion and $\alpha_e$ for ion plus electron neutralization reactions.

Although other methods have been used,[106] the magnitude of this parameter is most easily determined in pulse experiments. If $n_0$ is the concentration of positive or negative ions at the end of pulse and $n$ the concentration after elapsed time $t$, $\alpha$ may be found from the slope of a plot of $n^{-1}$ vs. $t$. This procedure is in accord with the integrated second-order rate expression

$$n^{-1} = n_0^{-1} + \alpha t \tag{9-77}$$

Nonlinearity in such plots at low $n$ is a sign that wall diffusion, which adds a first-order component, is occurring. Corrections have to be made for this.[107] In systems where electrons remain free and undergo reaction (9-76), the large difference between the mobilities of the electron and positive ion (the electrons are more mobile) may give rise to the development of a space charge near the walls of the radiolysis cell. In those cases corrections for wall diffusion require rather detailed computer modeling.[106] Fortunately, for typical ion concentrations produced by linear accelerator (linac) or Febetron pulses space charge effects are negligible.[108]

The *microwave conductivity techniques* used for electron-attachment investigations can also be used to follow the decay of electrons due to electron–ion neutralization.[108] In electron-attaching systems ion concentrations can be monitored continuously in ionization chambers by measuring the dc conductivity in the presence of a small potential applied across the electrodes.[109] Alternatively, at prescribed times after the ionizing pulse, all ions may be swept out on a very short time scale by a strong pulsed field.[106] The total charge collected gives the ion concentration. This *charge-collection method* is restricted to fairly low ion concentrations ($\sim 10^8$ cm$^{-3}$). However, it can be used for both ion–ion and ion–electron reactions, as can the *dc conductivity method.*

Examples of plots of $n^{-1}$ vs. time for ions in sulfur hexafluoride at a given pressure have been exhibited in Figure 9-9. The ion concentrations were

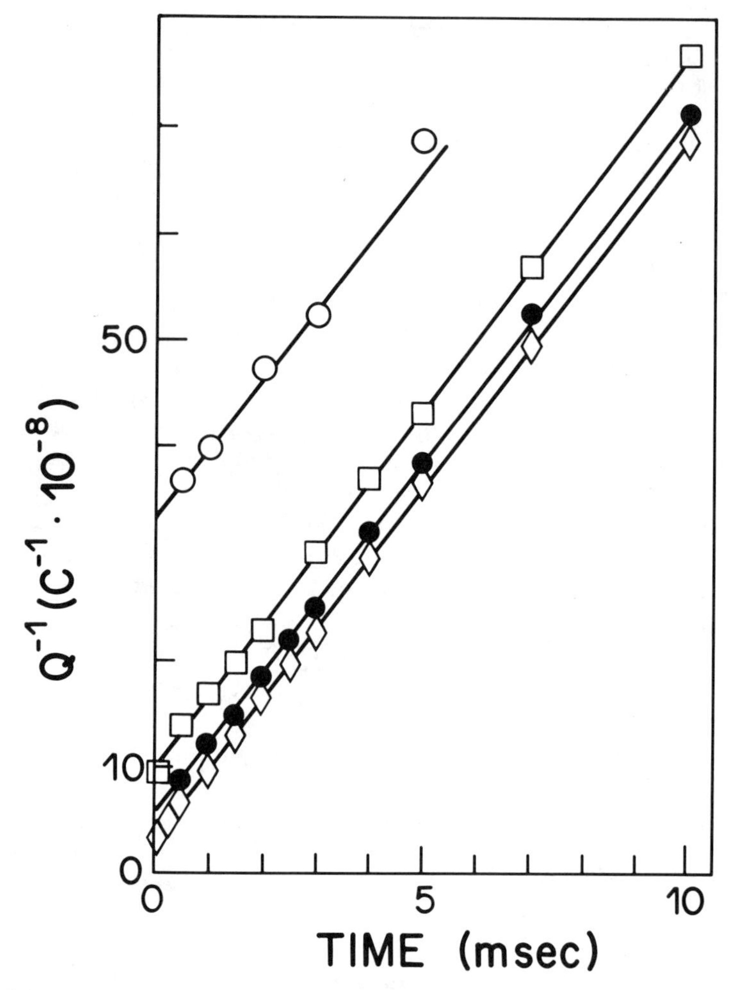


**Figure 9-9.** Plots of the reciprocal of Q, the total charge of ions remaining unneutralized in an ionization chamber, as a function of time after the ionizing electron pulse of electrons. The ion concentration is proportional to Q, and the symbols represent different initial ion concentrations obtained by using different doses per pulse. (Reproduced with permission.[109])

determined by the charge-collection method. The slopes, which are proportional to $\alpha$, are the same for the four different initial ion concentrations. An example of an ionization chamber that can be used in the charge-collection method with 1-MeV x-rays is given in Fig. 9-1(d).

**Neutralization Reactions**

In principle, the nature of the physical or chemical reaction whereby $X^+$ and $Y^-$ are converted to neutrals in Reaction (9-76) can be inferred from a detailed study of the dependence of $\alpha$ on temperature and molecular concentration or pressure of the matrix gas, and from the products of the neutralization reaction. Unfortunately, it is often very difficult to determine precisely what these are. However, some cases where specific products have been assigned to neutralization reactions on the basis of either experimental result or a theoretical analysis are listed in Table 9-7. For ion–ion neutralization, with atomic and small molecular ions, such as $O_4^+$ and $O_2^-$, the simple process of electron transfer is most important, whereas with heavily clustered hydrogen-containing species proton transfer dominates. Clustering has the important effect of lowering the potential energies of the ionic species and rendering electron transfer endothermic.[110] Much remains to be done in this area. For example, there is little known about neutralization products of organic ions and nonhydrogenic inorganic ions.

Neutralization of molecular ions by electrons normally leads to dissociation. Also, as shown in Table 9-7, one of the fragments is often in an electronically excited state. When the emission lifetime of this state is sufficiently short, the neutralization rate can be followed by observing the luminescence.[6] All of the reactions in Table 9-7 are central to the mechanisms of radiolysis of the parent matrix gases.

**Table 9-7.** Neutralization Reactions: Ion + Ion and Ion + Electron

| Matrix Gas | Neutralization Reaction | Ref. |
|---|---|---|
| | *Ion + Ion* | |
| $O_2$ | $O_4^+ + O_2^- \rightarrow O(^1S) + O(^3P)$ (25%) | 7 |
| | $\rightarrow O(^1D) + O(^3P)$ (75%) | |
| | $O_4^+ + (cyclo\text{-}C_4F_8)^- \rightarrow 2O_2\ (+\ cyclo\text{-}C_4F_8)$ | 7 |
| | $O_4^+ + O_3^- \rightarrow 2O_2 + O_3$ or $O_2 + O$ | 7 |
| $H_2O$ | $H_3O^+ \cdot nH_2O + Cl^- \cdot mH_2O \rightarrow (m + n + 1)H_2O + HCl$ | 110 |
| | $H_3O^+ \cdot nH_2O + \cdot O_2^- \cdot mH_2O \rightarrow (m + n + 1)H_2O + HO_2\cdot$ | 110 |
| | $H_3O^+ \cdot nH_2O + OH^- \cdot mH_2O \rightarrow (m + n + 2)H_2O$ | 110 |
| HCl | $H_2Cl^+ \cdot nHCl + Cl^- \cdot mHCl \rightarrow (m + n + 2)HCl$ | 62 |
| HBr | $H_2Br^+ \cdot nHBr + Br^- \cdot mHBr \rightarrow (m + n + 2)HBr$ | 62 |
| | *Ion + Electron* | |
| Ar | $Ar_2^+ + e^- \rightarrow 2Ar + h\nu$ | 111 |
| $NH_3$ | $NH_4^+ \cdot nNH_3 + e^- \rightarrow \cdot H + (n + 1)NH_3$ | 7 |
| $H_2O$ | $H_3O^+ \cdot nH_2O + e^- \rightarrow \cdot H + (n + 1)H_2O$ | 7 |
| $CO_2$ | $CO_2^+ + e^- \rightarrow CO(x^1\Sigma) + O(?)$ (44%) | 7 |
| | $\rightarrow CO(a^3\Pi) + O(?)$ (56%) | |
| | $C_2O_4^+ + e^- \rightarrow \dot{C}O + O + CO_2$ | 7 |
| $O_2$ | $O_2^+ + e^- \rightarrow O(^1S) + O(^3P)$ (10%) | 7 |
| | $\rightarrow O(^1D) + O(^3P)$ (90%) | |

## Kinetics of Neutralization

As well as being of theoretical importance, investigations of the dependences of $\alpha$ and $\alpha_e$ on the molecular concentration [M] and temperature of the

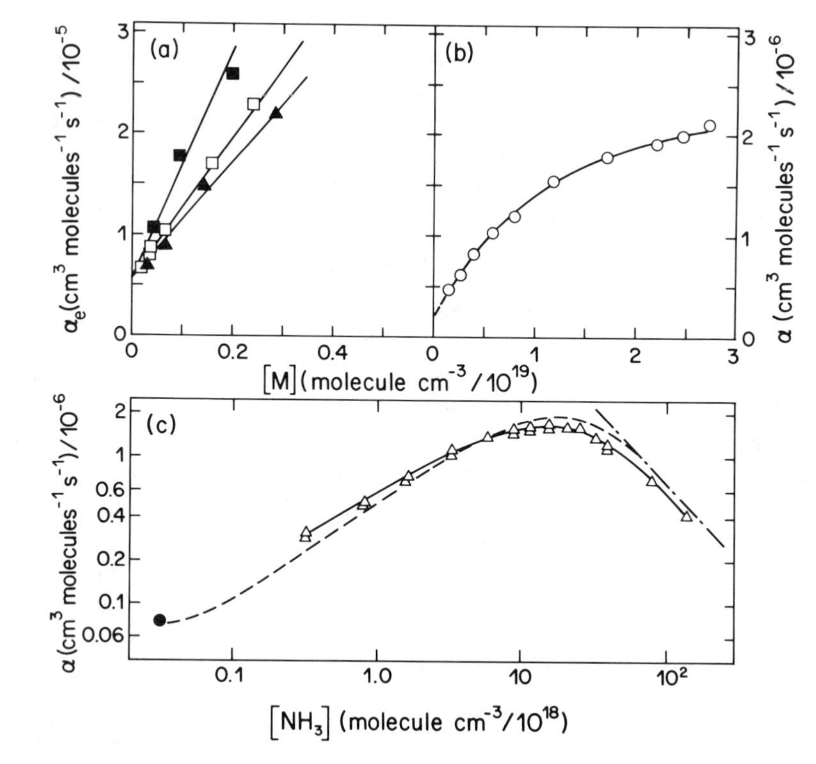


**Figure 9-10.** Dependence of the ion–electron and ion–ion neutralization rate constants, $\alpha_e$ and $\alpha$, on molecule concentration, [M]: (a) $\alpha_e$ for $NH_4^+ \cdot nNH_3$ ions in ammonia vapor at (■) 291, (□) 329, and (▲) 371 K, based on data of ref. 108; (b) $\alpha$ for ion–ion neutralization in $O_2$ at 298 K, based on data of McGowan in Figure 1-9-1 of ref. 115; (c) $\alpha$ for $NH_4^+ \cdot nNH_3$ and $Cl^- \cdot mNH_3$ in ammonia vapor, based on data in refs 8 and 110. The dashed line is calculated on the basis of Bates and Flannery's modification of Natanson's theory (see text). The line –·–·–·– represents the high-pressure limit given by Eq. (9-80b).

matrix gas are of practical value. In particular, they enable one to calculate the lifetimes of ions with respect to neutralization under a given set of conditions. Figures 9-10(a) and (b), respectively, illustrate the density dependences of $\alpha_e$ in ammonia and $\alpha$ in oxygen. Both $\alpha_e$ and $\alpha$ are seen to increase linearly with [M] in the low-density range and to extrapolate to a finite value at zero concentration, which demonstrates the presence of two- and three-body contributions in the overall rate constants. This behavior has been observed in all systems studied thus far, and in the low [M] ($<4 \times 10^{18}$ molecules cm$^{-3}$) range the linear equation

$$\alpha = \alpha_2 + \alpha_3^{M}[M] \tag{9-78}$$

is an acceptable approximation for $\alpha$[112] or $\alpha_e$.[108] The parameters $\alpha_2$ and $\alpha_3^{M}$ correspond to the two- and three-body terms, respectively. Examples of two- and three-body rate constants for a number of systems at 290 to 300 K are listed in Table 9-8. The rate constants for electron neutralizations are generally larger, because the thermal electrons have larger velocities than do the ions. Values of $\alpha_2$ and $\alpha_3$ are seen to be sensitive to the nature of the matrix gas and the ionic species.

**Table 9-8.** Rate Constants for Two- and Three-Body Neutralization Reactions[a]

| Matrix Gas | Ions | | $\alpha_2/(10^{-6}$ cm$^3$ molecule$^{-1}$ s$^{-1})$ | $\alpha_3/(10^{-25}$ cm$^6$ molecule$^{-2}$ s$^{-1})$ | Ref. |
|---|---|---|---|---|---|
| | Positive | Negative | | | |
| $O_2$ | $O_4^+$ | $O_2^-, O_4^-, O_3^{-b}$ | 0.20 | 1.91 | 115 |
| Ne | $(NO)_2^+$ | $NO_2^-, NO_3^-$ | 0.21 | 1.04 | 112 |
| Ar | $(NO)_2^+$ | $NO_2^-, NO_3^-$ | 0.21 | 1.45 | 112 |
| Xe | $(NO)_2^+$ | $NO_2^-, NO_3^-$ | 0.21 | 2.78 | 112 |
| $NH_3$ | $NH_4^+ \cdot nNH_3{}^b$ | $Cl^- \cdot mNH_3{}^c$ | 0.10 ± 0.05 | 3.70 | 110 |
| $H_2O$ | $H_3O^+ \cdot nH_2O$ | $e^-$ | 4.0 ± 1.0 | 270 | 108 |
| $NH_3$ | $NH_4^+ \cdot nNH_3$ | $e^-$ | 5.6 | 69 | 108 |
| $CO_2$ | $CO_2^+ \cdot nCO_2$ | $e^-$ | 4.0 | 6 | 108 |

[a] Unless otherwise stated, experimental uncertainties are ±20% or less.
[b] $O_3^-$ probably dominates for low-intensity pulses (see the section entitled "Mechanisms of Radiolysis").
[c] Values of $n$ vary from 4 to ~6 for $H_2O$ and $NH_3$ and 2 to 3 for $CO_2$, and $m$ is 3 to 4.

The origin of the three-body term was explained by Thomson.[113] He pointed out that if a pair of ions are within each other's sphere of coulombic attraction and one of them suffers a collision with a neutral molecule, then the loss of kinetic energy of relative motion of the pair may be sufficiently great that they become trapped in a closed orbit. This greatly increases the duration of the interaction and the probability of a transition to neutral states. The two-body rate constant corresponds to the transition taking place between unbound ions during a single pass. A similar picture has developed in recent years for electron–ion neutralization.[114] In addition, although the linear equation (9-78) with the separation of $\alpha$ into two- and three-body components is still supported by experiments at low pressure, *multiple-collision* models are now favored over the single-collision concept of Thomson for both the ion–ion and ion–electron cases. These models make calculating the pressure dependence of $\alpha$ in the low-to-moderate [M] (or pressure) regime a rather formidable task.[114–117] Space does not permit a discussion of the theory and methods of calculating $\alpha$ or $\alpha_e$ as a function of [M] here, and the interested reader should consult refs. 114–117. However, several important points remain to be discussed in this section.

The early theory of Thomson[113] predicted that the three-body stabilization would saturate; that is, the increase of $\alpha$ (or $\alpha_e$) with [M] would eventually flatten out. This is illustrated by the behavior for oxygen in Figure 9-10(b). For typical gases $\alpha$ was predicted by Thomson's theory to reach an upper limit at about 1 atm, and indeed this is seen experimentally for many gases.[118] However, $\alpha$ does not remain constant with pressure. It eventually begins to fall, because at some point the diffusion together of the two ions becomes rate limiting.[8,115] This behavior is illustrated in Figure 9-10(c) in ammonia vapor.

The first equation that described the overall pressure dependence of $\alpha$ was that of Natanson referred to in refs. 8 and 115. It was applicable only to systems where the ions and matrix molecules had identical masses.

$$\alpha = \left\{ \frac{1 - e^{-r_0/R}}{4\Pi r_0 D} + \frac{1}{k_R\, e^{r_0/R}} \right\}^{-1} \tag{9-79}$$

The first and second terms are associated, respectively, with diffusion control and the neutralization process, $r_0$ $(=e^2/4\Pi\varepsilon_0\varepsilon \boldsymbol{k}T)$ is the interionic separation at which the coulomb potential is equal to $kT$, $D$ is the sum of the individual ionic diffusion coefficients $D_+$ and $D_-$, and $R$ is a reaction radius, chosen so that the ions are still in thermal equilibrium with the matrix gas at that distance apart, given by

$$R = \frac{\beta\lambda}{2}\left\{\left(1 + \frac{5r_0}{3\beta\lambda}\right)^{1/2} + 1\right\}$$

where $\beta$ is a constant usually taken as unity and $\lambda$ is the ion-neutral mean free path. The symbol $k_R$ is the rate constant for the neutralization reaction, which would apply if the concentration of ions at the separation $R$ were as predicted by equilibrium statistics.[8] In the low-[M] $(<1 \times 10^{19}$ molecules cm$^{-3}$) regime $R$ and $D$ become very large and $\alpha = k_R$, which can be set equal to $\alpha_2 + \alpha_3{}^{M}[M]$. On the other hand, at high [M] $(>10^{20}$ molecule cm$^{-3}$ for most gases) the first term dominates, and (9-79) simplifies to

$$\alpha = 4\Pi r_0 D \tag{9-80a}$$

This equation can be recast in the form

$$\alpha = \frac{e}{\varepsilon_0 \varepsilon}(\mu_+ + \mu_-) \tag{9-80b}$$

due to the Einstein relation $D = \mu(kT/e)$.[8] Equation (9-80b) has been tested and shown to hold for ions in several electron-attaching gases by Schmidt and co-workers.[109] For these conditions it is therefore quite a straightforward process to calculate $\alpha$ from $\mu_+$ and $\mu_-$. Furthermore, it is easily seen that the temperature and molecular density dependence of $\alpha$ are the same as those of $(\mu_+ + \mu_-)$. For typical systems, therefore, $\alpha$ varies inversely with [M] and has only a small temperature dependence. A tabulation of ionic mobilities can be found in ref. 119.

The calculation of $k_R$ at low [M] or of $\alpha$ at low or intermediate [M] is a much more formidable task, for which the reader is referred to refs. 115 to 117. For clustered ion systems the distribution of ion sizes must be taken into account.[110] Agreement between experiment and theory is usually within ±40% and sometimes is better.[8,115] The dashed line in Figure 9-10(c) shows the value $\alpha$ predicted for the $NH_4^+ \cdot nNH_3 + Cl^- \cdot mNH_3$ neutralization on the basis of an improved form of Natanson's equation, developed by Bates[114] and Flannery[115]. The agreement with experiment here is within ±20% over the whole range of [M] or pressure. This newer theoretical treatment allows for the effect of different masses of the ions and matrix gas molecules, which is essential for clustered systems. Theoretical work on the dependence of $\alpha_e$ on M has also progressed significantly,[114] but there is much less than for the ionic systems. Detailed calculations of $\alpha_e$ in the low-pressure regime are formidable. However, Warman et al[108] have suggested a semiempirical formula that may be used to estimate $\alpha_e$ at low [M] from available information for polar matrix gases. This method is not as useful for nonpolar molecules like $CO_2$ and $N_2$. However, for $NH_3$ and $H_2O$ it does predict the observed negative temperature exponent for $\alpha$ fairly well.

In general, all theories predict a strong negative temperature dependence of $\alpha$ and $\alpha_e$ in the low-pressure regime, with a falloff to a relatively small dependence as [M] rises to the point where Eq. (9-80a) applies.[110,115] The negative temperature dependence of $\alpha_e$ in $NH_3$ is illustrated by the data in Figure 9-10(a). Few studies of the temperature dependence of $\alpha$ or $\alpha_e$ have thus far been made.

## *Neutral Intermediates*

### Methods

Electronically excited molecules and ground-state, excited atomic, or polyatomic free radicals are formed in almost every irradiated system. Optical absorbance techniques are frequently used in the study of the spectra and kinetics of reaction of these species. A typical experimental setup is shown in Figure 9-11. The design is referred to as a "collinear system," because the analyzing light beam lies parallel to the electron beam. It is most often employed with machines that produce electrons of energy ≥5 MeV at low or moderate beam currents.[6] Perpendicular systems, in which the analyzing light beam is at right angles to the electron beam, can be utilized with Febetrons because of the lower electron energies, larger beam currents, and the inherent tendency of these beams to diverge.[6]

The main requirements, which must be met for the observation of second-order reaction kinetics and for the determination of molar absorbance coefficients are (a) dose uniformity in the region of the cell traversed by the light beam, and (b) a sufficiently long optical path length. The second of these is often achieved by using multipass cells with reflecting mirrors. Detailed discussions of experimental problems and designs may be found in refs. 4, 6, 9, and 18–20. However, one other point should be made here. In gas-phase studies where the absorption spectra consist of discrete lines,

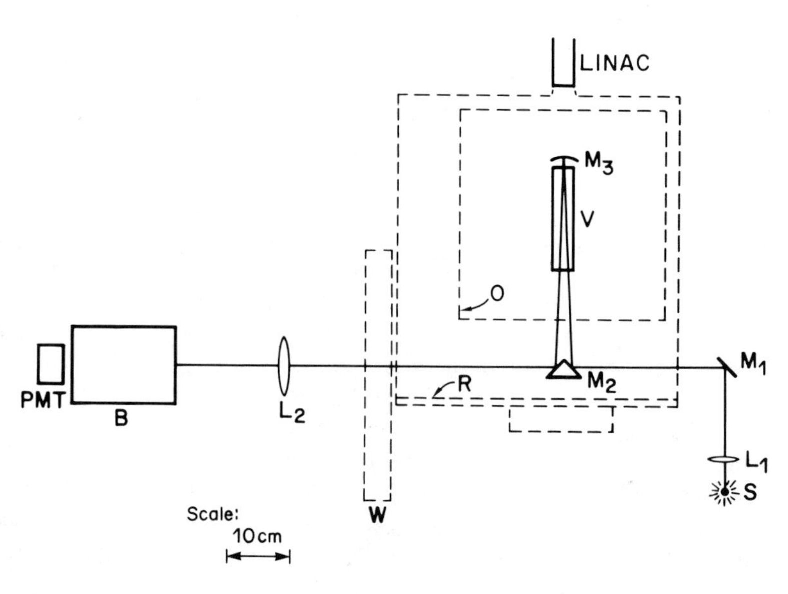


**Figure 9-11.** An optical system for pulse radiolysis of gases with a linac. V = sample vessel; O = thermostat oven; R = beam stopper and ground return to linac; W = lead brick wall; $L_1$, $L_2$ = suprasil lenses; $M_1$, $M_2$, $M_3$ = aluminized surface mirrors; S = 450-W xenon light source; B = monochromator; PMT = photomultiplier tube with lead shield. (Reproduced with permission.[120])

the Beer–Lambert law $A = \varepsilon cl$ may not be obeyed. Here $A$ is the absorbance of the intermediate, which has an absorbance coefficient $\varepsilon$ at the chosen wavelength and is present at concentration $c$ in a cell of optical path length $l$. The empirical relation $A = (\varepsilon cl)^n$, where $n$ has to be determined for the particular transient and apparatus, has proven to be an acceptable quantitative expression in several systems.[6] A detailed treatment of the problem is given in ref. 121.

A large increase in sensitivity and specificity for a particular radical can be produced by utilizing resonance absorption.[6] For example, Bishop and Dorfman[122] developed a method for detecting H atoms by employing the Lyman-$\alpha$ resonance absorbance line at 121.57 nm. The light from a hydrogen discharge lamp was passed through an oxygen gas filter to remove wavelengths outside the 121.4–122.0-nm region, through the pulse radiolysis cell, and then into a suitable detector. In more recent applications narrow band-pass interference filters are often used to select the required wavelengths. The technique has been employed for OH radicals[123] and, in principle, is applicable to a wide variety of species.

Apart from the failure of the Beer–Lambert law and the use of the resonance absorbance method, studies of the absorbance of neutral species in gases are very similar from a practical point of view to pulse radiolysis studies in liquids (see Chapter 3). This is especially true for species with broad absorption bands. Examples of absorption spectra for radicals pro-

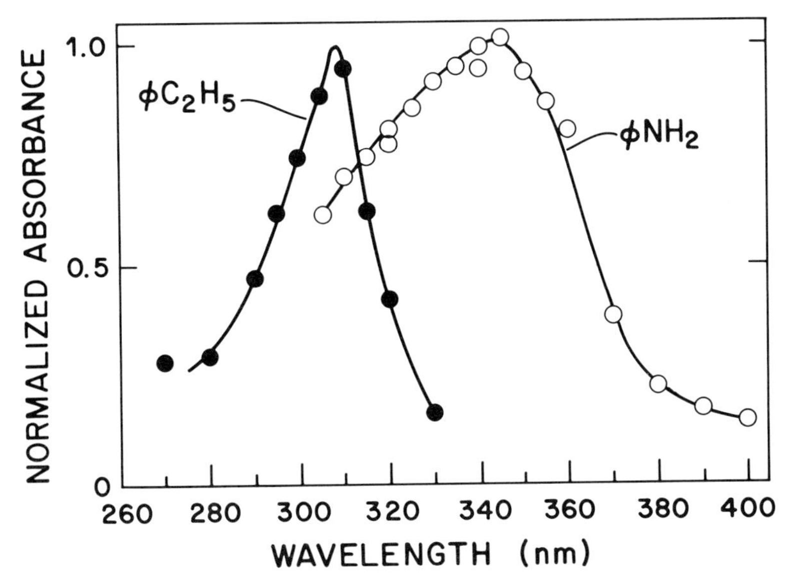


**Figure 9-12.** Spectra of radicals formed by the addition of hydrogen atoms to ethyl benzene ($\phi C_2H_5$) and aniline ($\phi NH_2$) obtained by pulse radiolysis. Conditions: (○) 20 mm Hg pressure of $\phi NH_2$, 100 mm $H_2$ and 600 mm of Ar at 330 K; (●) 8 mm pressure of $\phi C_2H_5$, 13 atm of $H_2$, and 60 atm of Ar at 298 K. (Reproduced with permission.[124])

duced by the reactions of hydrogen atoms with ethyl benzene and aniline[124] can be seen in Figure 9-12.

A second technique, which is applicable to electronically excited species, is the detection of emission. The observed pattern of wavelengths normally furnishes an unequivocal identification of the species, while the change of intensity with time allows one to follow changes in concentration.

### Excited States

As an example of the application of absorption and emission spectroscopy in gas-phase radiation chemistry, see Figure 9-13, which displays the emission (lower line) and absorbance (upper line) of 842-nm radiation for argon at 200 torr, following a 5-ns Febetron pulse.[6,125] The emission arises from the radiative decay of $Ar^3D_2(2p_8)$ to $Ar^3P_1{}^0(1s_4)$ (terminology in brackets is that of Paschen). The latter species is detected in absorption via the reverse transition. The change in its concentration as given by the absorbance reflects its initial growth and subsequent decay. Observations of this sort have shown that the lower 1s levels of argon are mainly populated from upper states by processes occurring *after* the energy has been absorbed from the radiation pulse.[6]

The data displayed in Figure 9-13 are typical of those obtained in gaseous systems. The decay portions of the curves can be utilized to calculate

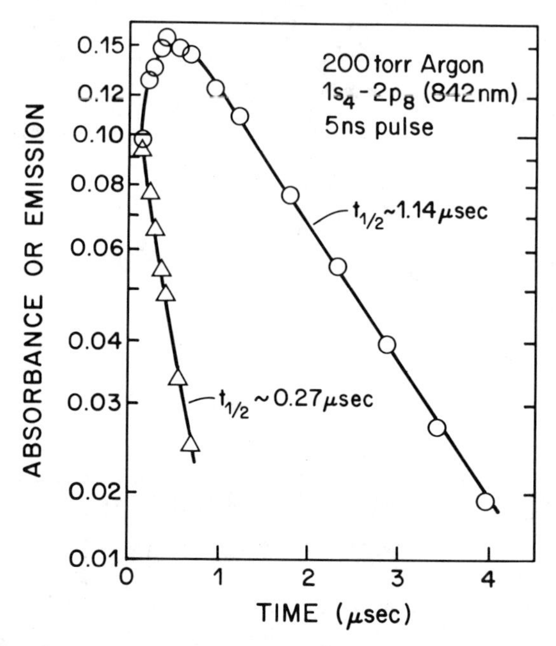


**Figure 9-13.** Absorbance and emission of Ar($^3P_1$) at 842 nm in 205 mm pressure of Ar as a function of time after pulse. △ = emission, ○ = absorbance corrected for emission. The ordinate scale for emission is in arbitrary units. Based on data of ref. 125. (Reproduced with permission.[6])

first-order rate constants, $k_I'$ (or half-lives), for the removal of the excited species. Studies of the pressure dependence are then necessary to determine whether the excited Ar* states are decaying unimolecularly or upon collision with other species. Extensive investigations of the pure rare gas systems have been made by Firestone and co-workers[121,125,126] and others (see refs. 6 and 127). The first-order rate constant $k_I'$ has been shown to contain one term that is first order in [Ar] and a second term that depends on $[Ar]^2$. The following mechanism has been postulated to explain this:[128]

$$Ar^* + Ar \rightleftharpoons Ar_2^*(v) \quad (9\text{-}81)/(9\text{-}82)$$

$$Ar_2^*(v) + Ar \rightarrow Ar_2^*(v') + Ar \quad (9\text{-}83)$$

$$Ar_2^*(v) \rightarrow 2\,Ar + h\nu \quad (9\text{-}84)$$

For the condition that $k_{82}$ is large, it can be shown that [6,128]

$$k_I' = \frac{k_{81}k_{84}}{k_{82}+k_{84}}[Ar] + \frac{k_{81}k_{83}}{k_{82}+k_{84}}[Ar]^2 \quad (9\text{-}85)$$

Values of these rate constants for various inert gas systems and quenching constants for certain additives may be found in refs. 121, 125, and 126.

The inert gas systems can be used to illustrate a further point—namely, that neutral excited states may be formed from both direct excitation and ion neutralization. Cooper et al[129] demonstrated this for the 2p levels by examining the emissions with a time resolution of a few nanoseconds. They found that the emissions consisted of two peaks, which were separated in time. The time profile of the first emission peak was independent of the concentration of ionic or neutral states produced (ie, the dose per pulse). This corresponded to the direct excitation of 2p levels or to their population by decays from higher levels. The second peak grew in more and more slowly as the dose per pulse was reduced, which is what one expects for a second-order ion neutralization reaction. It was entirely eliminated in the presence of the electron scavenger $SF_6$.

In further work,[130] Cooper's group examined the 222-nm emission from KrCl* exciplexes formed by the reactions

$$Kr^* + CF_3Cl \rightarrow KrCl^* + CF_3\cdot \quad (9\text{-}86)$$

$$e^- + CF_3Cl \rightarrow CF_3\cdot + Cl^- \quad (9\text{-}87)$$

$$Kr^+ (\text{or } Kr_2^+) + Cl^- \rightarrow KrCl^* (\text{or } +Kr) \quad (9\text{-}88)$$

Again the ionic [Reaction (9-88)] was slower and separated in time from the fast direct formation of KrCl* and Reaction (9-86). Addition of 0.1 mm Hg of $SF_6$ to a system containing 0.5 mm Hg of $CF_3Cl$ in 50 mm Hg of Kr caused the complete suppression of reaction (9-87), because $k_{89}$ is severalfold faster than $k_{87}$ (Table 9-4):

$$e^- + SF_6 \rightarrow SF_6^- \quad (9\text{-}89)$$

However, the emission from KCl* formed in (9-86) remained. In addition, a new emission from KrF*, formed via Reactions (9-89) and (9-90),

$$Kr^+ (\text{or } Kr_2^+) + SF_6^- \rightarrow KrF^* + SF_5\cdot (\text{or } +Kr) \quad (9\text{-}90)$$

could now be seen to replace the slow component of the KrCl* emission. Further details of these studies and the rate constants reported may be found in ref. 131.

Studies of excited states have also been made in $N_2$, $CO_2$, $O_2$, CO, $CH_4$, and other molecular systems.[6] A recent study of $H_2O$ reported[132] evidence for electronically excited H ($n = 3$ to 6), OH in a number of vibrational levels of the $C^2\Sigma^+$ and $A^2\Sigma^+$ states, $O(^3p^5P)$ and $O(^3p^3P)$, and $O^+3p^4D^\circ_{5/2}$, $D^\circ_{7/2}$, and $D^\circ_{1/2}$. In many cases observed lifetimes $\tau_p$ were found to depend on the concentration of water vapor in accord with the expression

$$\tau_p^{-1} = \tau^{-1} + k_{aq}[H_2O] \quad (9\text{-}91)$$

Here $k_{aq}$ is the quenching rate constant for water vapor, and $\tau$ is the natural lifetime of the species. The values of $k_{aq}$ were found to be $(12.3 \pm 2.3)$, $(10.7 \pm 0.5)$, $(9.8 \pm 0.3) \times 10^{-9}$ cm$^3$ molecule$^{-1}$ s$^{-1}$, respectively, at 293 K for H($n = 6$), H($n = 5$), and H($n = 4$), while $\tau^{-1}$ was $0 \pm 16$, $4 \pm 2$, and $15 \pm 2\ \mu s^{-1}$.

**Free Radicals**

A large number of bimolecular radical reactions have been studied by pulse radiolysis, and examples of these are given in Table 9-9. For further details the reader is referred to the reviews of Sauer[6,128] and Firestone and Dorfman.[4] In many cases activation energies have been reported. In this regard the reactions of OH and OD with CO are quite complex.[123] The normal Arrhenius equation does not hold. The dependence of the rate constant on temperature may be described by the expressions

$$k(\mathrm{OH} + \mathrm{CO}) = 4.05 \times 10^{-14}\, e^{660/T} + 3.04 \times 10^{-9}\, e^{-3801/T}$$

and

$$k(\mathrm{OD} + \mathrm{CO}) = 1.27 \times 10^{-14}\, e^{455/T} + 2.33 \times 10^{-10}\, e^{-5668/T}$$

in units of $\mathrm{cm}^3$ $\mathrm{molecule}^{-1}$ $\mathrm{s}^{-1}$, for the temperature range 340 to 1250 K in the presence of 1 atm of argon and 15 mm Hg pressure of water vapor. These contain both negative and positive terms and remain to be fully explained.

Radical–radical reactions in several systems have also been investigated. The case of $HO_2\cdot$ is interesting.[6] In hydrogen–oxygen mixtures the mechanism of radical decay can be described as

$$HO_2 + HO_2 \rightleftharpoons H_2O_4 \tag{9-92}$$

$$H_2O_4 \rightarrow H_2O_2 + O_2 \tag{9-93}$$

The negative activation energy of the overall process has been attributed to the equilibrium in the first step.[6,133] The rate of decay is greatly enhanced

**Table 9-9.** Rate Constants for Radical + Molecule Reactions

| Reactants | $k^a/\mathrm{cm}^3\ \mathrm{molecule}^{-1}\ \mathrm{s}^{-1}$ |
|---|---|
| $H + C_6H_6$ | $(9.8 \times 10^{-11})\, e^{-18{,}000/RT}$ |
| $H + C_6H_5F$ | $(1.6 \times 10^{-11})\, e^{-13{,}400/RT}$ |
| $H + C_6H_5Cl$ | $(9.1 \times 10^{-12})\, e^{-10{,}040/RT}$ |
| $O + C_6H_6$ | $(6.0 \pm 0.1) \times 10^{-14}$ |
| $O + C_6H_5CH_3$ | $(2.3 \pm 0.5) \times 10^{-13}$ |
| $CH(X^2\Pi) + CH_4$ | $(3.3 \pm 0.1) \times 10^{-11}$ |
| $CH(X^2\Pi) + C_2H_4$ | $(1.1 \pm 0.1) \times 10^{-10}$ |
| $CN(X^2\Sigma) + O_2$ | $(1.13 \pm 0.03) \times 10^{-11}$ |
| $CN(X^2\Sigma) + CH_4$ | $(7.5 \pm 0.2) \times 10^{-13}$ |
| $CN(X^2\Sigma) + C_2H_4$ | $(2.0 \pm 0.3) \times 10^{-10}$ |
| SH + cyclohexane | $7.8 \times 10^{-14}$ |
| SH + 1-butene | $<1.3 \times 10^{-14}$ |
| SH + 1,3-butadiene | $1.0 \times 10^{-10}$ |

[a] Arrhenius activation energies for H reactions are in kJ $\mathrm{mole}^{-1}$. Rate constants for all other reactions are at 295–303 K.
Based on data in ref. 6.

in the presence of $NH_3$ or $H_2O$. This feature has been attributed to the formation of a complex, for example, $H_3N—HO_2$, which then reacts rapidly with a second $HO_2\cdot$ radical.

Many additional radical reactions are described in ref. 6.

# MECHANISMS OF RADIOLYSIS

Although complete product analyses have been carried out on many irradiated gases, there are few for which detailed mechanisms of radiolysis have been established. The relatively large numbers of intermediates and the complexity of their reactions explains why this is so. Actually, the rich variety of species, which may be produced by radiolysis, has caused studies of the reactions of intermediates per se to be of greater interest than mechanisms. However, a knowledge of overall mechanisms of radiolysis is important in some systems for practical as well as theoretical reasons. For example, the mechanism of radiolysis of water vapor is relevant to an understanding of the decomposition of steam in nuclear reactors. Second, many reactions in the radiolysis of oxygen occur in the upper atmosphere. Attention is therefore confined to these two systems, which are also simple enough to be amenable to a meaningful analysis.

A mechanism of radiolysis must (i) account for the *G*-values of all stable products and (ii) predict an acceptable energy balance in terms of the equation:[7,134]

$$\sum_{i} g_i E_i + \sum_{ex} g_{ex} E_{ex} + G_i \bar{E}_{se} = 100\ \text{eV} \tag{9-94}$$

Here $g_i$ and $g_{ex}$ are the yields per 100 eV of primary ionizations and excitations, and $\bar{E}_{se}$ is the average energy of subexcitation electrons. In principle, it is possible to compute the loss of energy from the fast electrons and the yields of primary ionic and neutral states from cross sections for electron impact ionization and excitation.[134] This requires a detailed knowledge of the energy dependences of cross sections for *all* transitions and of the degradation spectrum, that is, the spectrum of energies of the secondary electrons produced in ionizing events. Unfortunately, these data are available for only a few very simple systems, such as $H_2$ and He. Thus, for the present at least, one is constrained to use relative cross sections for 100-eV electrons. This procedure can only be accurate to within about 10% for allowed transitions and may be subject to much greater error for forbidden transitions with low-energy thresholds.[7] It is justified by the fact that ratios of cross sections do not change appreciably above 100 eV[7,97] and by theoretical studies that show that the major part of the energy is imparted to gas molecules by electrons of energy in the optical range.[134,135]

The present treatment is drawn largely from the review of Willis and Boyd.[7] For each system there are separate tabulations of primary processes, reactions and rate constants for intermediates, and radiation chemical data. The values of $E_i$ and $E_{ex}$ for each type of transition were usually derived from detailed potential energy curves and are the most probable energies for Franck–Condon excitation from ground-state molecules. Where detailed potential energy curves were not available, they were arbitrarily taken to be "threshold energy" plus 2 eV.[7] Equation (9-55) was used to calculate *g*-values for each primary process. The discussion here is brief because we assume the reader is familiar with complex kinetic mechanisms.

## *Oxygen*

Primary excitations to neutral states and reactions of intermediates are listed in Tables 9-10 and 9-11. Radiolysis yields and primary ionization processes may be found in Tables 9-2 and 9-5.

A consideration of the rate constants in Table 9-11 leads to the following conclusions for oxygen at normal pressure and 298 K. Because [O] is normally ≪ [$O_2$], each O atom reacts with $O_2$ to form $O_3$. (Some of this is initially in excited states and cools by collisions.[136]) Even at Febetron dose rates Reaction (9-73) is fast enough that almost all $O_2^+$ become clustered before neutralization. Most electrons are captured to form $O_2^-$. Also all $O^+$ ions should charge transfer with $O_2$ to form $O_2^+$.

Some information must be derived from the radiation chemical data per se. For example, Willis and Boyd conclude from the absence of a "fast" yield of $O_3$ that the formation of $O_3^+$,

$$O_2^+(A^2\Pi_u \text{ or } a^4\Pi_u) + O_2 \rightarrow O_3^+ + O(^3P) \qquad (9\text{-}95)$$

**Table 9-10.** Primary Excitation of Neutral States in Oxygen

| Primary Process | Yield (*G* units) | Threshold Energy (eV) | Probable Energy (eV) | Energy Used (eV/100 eV) |
|---|---|---|---|---|
| $O_2 \rightarrow O_2(a^1\Delta_g)$ | 0.012[a] | 1.0 | 1.0 | 0.012 |
| $O_2 \rightarrow O_2(b^1\Sigma_g^+)$ | 0.002[a] | 1.6 | 1.7 | 0.014 |
| $O_2 \rightarrow O_2(A^3\Sigma_u^+)$ and $(C^3\Delta_u)$ | 0.063[a] | 4.3 | 5.0 | 0.31 |
| $O_2 \rightarrow O_2(B^3\Sigma_u^-)$ (dissociates to $O(^3P) + O(^1D)$) | 1.82 | 5.2 | 9.7 | 17.7 |
| $O_2 \rightarrow O + O^*$ (dissociates to give O above $^1S$ level) | ~0.18 | | ~12 | 2.2 |
| Total dissociative excitation | 2.00 | Total for neutral excitation | | 20.2 eV |

[a] Do not contribute to dissociation.
Based on data in ref. 7.

**Table 9-11.** Relevant Kinetic Data for Oxygen

| Reaction | Rate Constant |
|---|---|
| *Ion–molecule reactions* | |
| $O^+(^4S) + O_2 \rightarrow O_2^+(X^2\Pi_g) + O(^3P)$ | $2 \times 10^{-11}$ |
| $O^+(^2D) + O_2 \rightarrow O_2^+(A^2\Pi_u$ and/or $a^4\Pi_u) + O(^3P)$ | $2 \times 10^{-10}$ |
| $O^+(^2D) \rightarrow O^+(^4S) + h\nu$ | $\tau_{1/2} \approx 3.6$ h |
| $O_2^+ + 2\,O_2 \rightarrow O_4^+ + O_2$ | $2.8 \times 10^{-30\dagger}$ |
| $O_4^+ + O_2 + He \rightarrow O_6^+ + He$ | $\approx 5 \times 10^{-30\dagger}$ |
| $O^+(^4S) + SF_6 \rightarrow SF_5^+ + OF$ | $1.5 \times 10^{-9}$ |
| $O_2^+(X^2\Pi_g) + SF_6 \rightarrow SF_5^+ + FO_2$ | $<10^{-11}$ |
| $e^- + 2\,O_2 \rightarrow O_2^- + O_2$ | $2.0 \times 10^{-30\dagger}$ |
| $e^- + O_2 \rightarrow O_2^- + h\nu$ | $10^{-19}$ |
| $O_2^- + 2\,O_2 \rightarrow O_4^- + O_2$ | $3 \times 10^{-31}$ |
| $O_2^- + SF_6 \rightarrow SF_6^- + O_2$ | $7 \times 10^{-11}$ |
| $e^- + SF_6 \rightarrow SF_6^-$ | $2.4 \times 10^{-7}$ |
| $SF_6^- + O_3 \rightarrow O_3^- + SF_6$ | $7 \times 10^{-11}$ |
| $O_2^- + O_3 \rightarrow O_3^- + O_2$ | $3 \times 10^{-10}$ |
| *Neutral reactions* | |
| $O(^3P) + O(^3P) + O_2 \rightarrow 2\,O_2$ | $7.2 \times 10^{-33\dagger}$ |
| $O(^3P) + 2\,O_2 \rightarrow O_3 + O_2$ | $3.6 \times 10^{-34\dagger}$ |
| $O(^3P) + O_3 \rightarrow 2\,O_2$ | $7.5 \times 10^{-15}$ |
| $O(^1D) \rightarrow O(^3P) + h\nu$ | $\tau_{1/2} = 110$ s |
| $O(^1D) + O_2 \rightarrow O(^3P) + O_2$ | $4 \times 10^{-11}$ |
| $O(^1D) + O_3 \rightarrow 2\,O_2$ | $3 \times 10^{-10}$ |
| $O(^1S) \rightarrow O(^1D) + h\nu$ | $\tau_{1/2} = 0.74$ s |
| $O(^1S) + O_3 \rightarrow 2\,O_2$ | $6 \times 10^{-10}$ |
| $O_2(a^1\Delta_g) \rightarrow O_2(X^3\Sigma_g^-) + h\nu$ | $\tau_{1/2} = 45$ s |
| $O_2(a^1\Delta_g) + O_2 \rightarrow 2\,O_2$ | $2 \times 10^{-18}$ |
| $O_2(a^1\Delta_g) + O_3 \rightarrow 2\,O_2 + O(^3P)$ | $3 \times 10^{-15}$ |
| $O_2(b^1\Sigma_{g+}) \rightarrow O_2(X^3\Sigma_g^-) + h\nu$ | $\tau_{1/2} \cong 12$ s |
| $O_2(b^1\Sigma_{g+}) + O_2 \rightarrow 2\,O_2$ | $1 \times 10^{-16}$ |
| $O_2(b^1\Sigma_g) + O_3 \rightarrow 2\,O_2 + O(^3P)$ | $7 \times 10^{-12}$ |

Rate constants have units $cm^3$ $molecule^{-1}$ $s^{-1}$ except for three-body reactions, which have units $cm^6$ $molecule^{-2}$ $s^{-1}$ and are indicated by $^\dagger$, and unimolecular reactions for which lifetimes have been given.
Based on data in ref. 7.

and its neutralization by electron transfer from $O_2$ is unimportant. Second, at high dose rates the reactions

$$O_4^+ + O_2^- \rightarrow 2\,O_2 + 2\,O \quad (9\text{-}96a)$$

$$O_4^+ + O_4^- \rightarrow 3\,O_2 + 2\,O \quad (9\text{-}96b)$$

are considered to be the main neutralization reactions. Addition of the electron scavengers $SF_6$ or $c$-$C_4F_8$ suppresses this process, replacing it with

$$O_4^+ + SF_6^- \rightarrow \text{No } O_3 \text{ or O atoms}$$

or the equivalent for $c$-$C_4F_8$ negative ions.[31] These reactions account for the fact that $\Delta G(O_3) = 6.6$:

$$\Delta G(O_3) = G(O_3)_{\text{pure}\,O_2} - G(O_3)_{O_2+SF_6} = 12.8 - 6.2 = 6.6$$

is equal to $2 \times G$(ionization).

The reader will find it instructive to examine the concentration-time curves for important intermediates and the product ozone shown in Figure 9-14 for a typical Febetron pulse irradiation. The time scale for the ionic processes is seen to be well separated from that for $O_3$ production. Also, as stated above, $O_4^+$ is the dominant positive ion species. The electron concentration is appreciable. However, the computer calculations show that Reactions (9-96a) and (9-96b) are more important than $e^- +$ ion neutralization,[31] because the latter has a lower rate constant. At lower pressures electron–ion neutralization becomes more important. The concentration of $O^+$ is always small.[31]

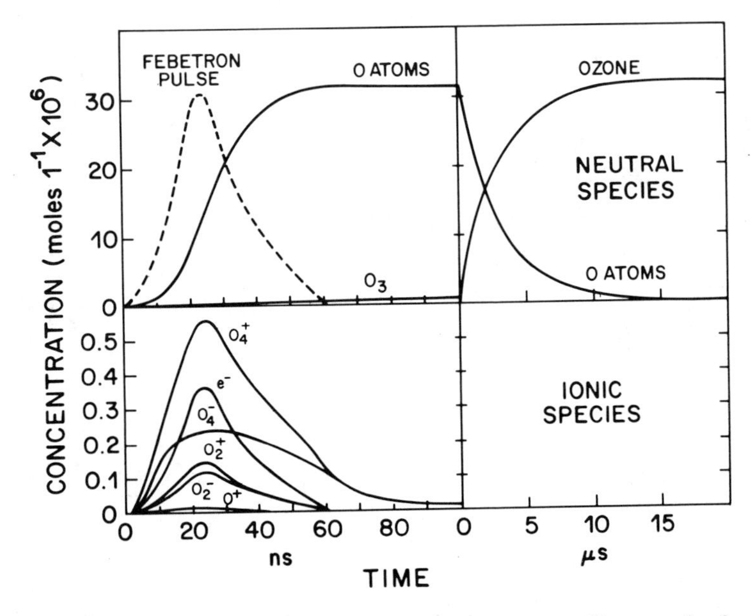


**Figure 9-14.** Concentrations of ionic and neutral intermediates and of product ozone during and following a Febetron pulse irradiation of 830 mm of Hg pressure of $O_2$ at 298 K. The pulse intensity profile is shown by the dashed line. The maximum intensity is $6.16 \times 10^{27}$ eV $L^{-1}$ $s^{-1}$, and 10 units on the ordinate scale correspond to $2 \times 10^{27}$ eV $L^{-1}$ $s^{-1}$. (Reproduced with permission.[31])

In contrast to the above behavior it can be shown that, at dose rates of $10^{16}$–$10^{18}$ eV $g^{-1}$ $s^{-1}$ and typical gamma or x-ray radiolysis periods of several minutes, all intermediates will attain steady concentrations. The ozone concentration will quickly reach approximately $10^{13}$ molecule $cm^{-3}$, and it can be demonstrated from the rate constants that Reaction (9-97) will then dominate over (9-96).[31]

$$O_2^- + O_3 \rightarrow O_2 + O_3^- \quad (9\text{-}97)$$

Therefore $O_3^-$ ions will react with $O_4^+$ in the neutralization reaction, viz,

$$O_4^+ + O_3^- \rightarrow 2\,O_2 + O_3 \text{ (or } O_2 + O) \quad (9\text{-}98)$$

The observation (Table 9-2) that $G(O_3)$ for these conditions is unaffected by electron scavengers and the same as at high dose rates in the presence of electron scavengers is taken as evidence for the absence of more than one O atom in the products of (9-98).

Given the reactions in Table 9-11 and the above neutralization processes at high dose rate (HDR) and low dose rate (LDR), the mechanism can be seen to predict

$$G(O_3)_{HDR} = 2\sum G(O_2^+) + 4\sum G(O^+) + 2G(\text{Neutral Dissociation})$$
$$= 13.06$$

and

$$G(O_3)_{LDR} = 2\sum G(O^+) + 2G(\text{Neutral Dissociation})$$
$$= 6.3$$

where the yields are calculated from the yields for primary processes in Tables 9-5 and 9-10. These results are in reasonable agreement with the experimental yields from Table 9-2:

$$G(O_3)_{HDR} = 13.3 \text{ (Average of results from refs. 20 and 31)}$$

and

$$G(O_3)_{LDR} = 6.1 \text{ (Average of all low-dose-rate data)}$$

Using a calculated value of $\bar{E}_{se} = 7.8$ eV, Willis and Boyd[7] found a total energy based on Eq. (9-94) of 89.3 eV. This is much less than the theoretical 100 eV. A reason for the discrepancy could be that the estimates of $E_i$ for ions from autoionizing neutral states are too low.[7] However, as stated earlier, one has to expect some uncertainty in the present simplified method.

## Water Vapor

The primary ion yields in Table 9-5 and the kinetic data in Table 9-12 can be used to calculate the yields of species formed by ionizing events before neutralization or free-radical reactions take place. These are shown in Table 9-13, column two.

The radicals ·OH and H· do not react with $H_2O$ at the temperatures of concern here. In the presence of an olefinic scavenger, H atoms are

**Table 9-12.** Relevant Kinetic Data for Water Vapor

| Reaction | Rate constant |
|---|---|
| *Ion–Molecule reactions* | |
| $H_2O^+ + H_2O \rightarrow H_3O^+ + OH$ | $1.2 \times 10^{-9}$ |
| $D_2O^+ + D_2O \rightarrow D_3O^+ + OD$ | $1.5 \times 10^{-9}$ |
| $OH^+ + H_2O \rightarrow H_3O^+ + O$ | $1.5 \times 10^{-9}$ |
| $OD^+ + D_2O \rightarrow D_3O^+ + O$ | $3.0 \times 10^{-9}$ |
| $H^+ + 2\,H_2O \rightarrow H_3O^+ + H_2O$ | |
| $H_3O^+(H_2O)_n + H_2O + M \rightarrow H_3O^+(H_2O)^{n+1} + M$ | $3 \times 10^{-27\dagger}$ |
| $n = 0–3$ | |
| $O^+(^4S) + H_2O \rightarrow H_2O^+ + O$ | $2.3 \times 10^{-9}$ |
| *Radical reactions* | |
| $H + H + H_2O \rightarrow H_2 + H_2O$ | $2 \times 10^{-32\dagger}$ |
| $H + OH + H_2O \rightarrow 2H_2O$ | $1 \times 10^{-30\dagger}$ |
| $OH + OH + H_2O \rightarrow H_2O_2 + H_2O$ | $1 \times 10^{-31\dagger}$ |
| $O(^3P) + OH \rightarrow O_2 + H$ | $5 \times 10^{-11}$ |
| $H + O_2 + H_2O \rightarrow HO_2 + H_2O$ | $4 \times 10^{-31\dagger}$ |
| $H + HO_2 \rightarrow H_2 + O_2$ | $\sim 1 \times 10^{-11}$ |
| $H + HO_2 \rightarrow OH + OH$ | $\sim 2 \times 10^{-12}$ |
| $H + HO_2 \rightarrow H_2O + O$ | $\sim 1 \times 10^{-21}$ |
| $HO_2 + HO_2 \rightarrow H_2O_2 + O_2$ | $3.6 \times 10^{-12}$ |
| $H + O + H_2O \rightarrow OH + H_2O$ | $2.5 \times 10^{-32\dagger}$ |
| $H + H_2O_2 \rightarrow H_2O + OH$ | $1 \times 10^{-13}$ |
| $OH + H_2O_2 \rightarrow H_2O + HO_2$ | $2 \times 10^{-12}$ |
| $OH + H_2 \rightarrow H_2O + H$ | $9 \times 10^{-14}$ |
| $OH + OH \rightarrow H_2O + O$ | $2.3 \times 10^{-12}$ |
| $O + O + H_2O \rightarrow O_2 + H_2O$ | $2.7 \times 10^{-33\dagger}$ |
| $O + H_2 \rightarrow H + OH$ | $3 \times 10^{-16}$ |

Rate constant have units $cm^3$ $molecule^{-1}$ $s^{-1}$ except for three-body reactions, which have units $cm^6$ $molecule^{-2}$ $s^{-1}$ and are indicated by $^\dagger$, and unimolecular reactions for which lifetimes have been given.
Based on data in ref. 7.

**Table 9-13.** Yields of Species in Water Vapor Prior to Neutralization and Radical Reactions

| Species | From Ionization[a] | From Neutral Excitation[b] | Total[c] |
|---|---|---|---|
| $e^-$ | 3.3 | — | 3.3 |
| $H_3O^+ \cdot nH_2O$ | 3.3 | — | 3.3 |
| OH | 2.7 | 3.5 | 6.2 |
| H | 0.57 | 3.5 | 4.1 |
| O | 0.63 | 0.45 | 1.08 |
| $H_2$ | 0.06 | 0.45 | 0.51 |

[a] Calculated from the yields in Table 9-5.
[b] Deduced from the effects of scavengers on $G(H_2)$.
[c] Uncertainties are ±0.2 for OH and H and ±0.1 for other species.

**Table 9-14.** Radiation Yields from Water Vapor at 390–430 K

| Dose Rate ($eV\ g^{-1}\ s^{-1}$) | Additives | Yields $G(H_2)$ |
|---|---|---|
| $\sim 10^{16}$ | RH ($=CH_3OH, C_3H_7OH, C_3H_8$) | 7–9 |
| $\sim 10^{16}$ | $RH + SF_6$ | 5–6 |
| $\sim 10^{16}$ | Ol ($=C_3H_6, C_5H_{10}, C_6D_6$) | 0.51 |
| $10^{27}$ | RH ($=HCl, HBr$) | 7.9 |
| $10^{27}$ | $HCl, HBr + SF_6$ | 4.6 |

Based on data in ref. 7 and 137.

consumed:

$$H\cdot + C_3H_6 \rightarrow C_3H_7\cdot \qquad (9\text{-}99)$$

$$2\,C_3H_7\cdot \rightarrow C_3H_6 + C_3H_8 \text{ or } C_6H_{14} \qquad (9\text{-}100)$$

Therefore, the $H_2$ formed under these conditions, $G(H_2)_{Ol} = 0.51$ (Table 9-14) must be formed directly by dissociation of $H_2O^{+*}$ or $H_2O^*$. The yield from $H_2O^+$ is given in Table 9-5 (0.06 molecules/100 eV). Hence, the yield from neutral molecules must be 0.45. We turn now to the use of hydrogen atom-donating scavengers RH ($CH_3OH$, $C_2H_5OH$, HBr, HCl, . . . , etc: Table 9-14), which form the specific easily identified product $H_2$ on reaction with $H\cdot$ atoms:

$$H\cdot + RH \rightarrow H_2 + R\cdot \qquad (9\text{-}101)$$

In the presence of these the hydrogen yield $G(H_2)_{RH}$ is about 8.0 molecules/100 eV at both low and high dose rates. Electron scavengers reduce $G(H_2)_{RH}$ by an amount close to $G$(ionization) (=3.3 ionizations/100 eV). From this it is concluded that the main neutralization reaction is

$$H_3O^+\cdot nH_2O + e^-(\text{or } O_2^-) \rightarrow H\cdot + (n+1)H_2O(+O_2) \qquad (9.102)$$

The total hydrogen atom yield must be $G(H_2)_{RH} - G(H_2)_{Ol} = 8.0 - 0.5 = 7.4$ atoms/100 eV. A yield of 0.57 $H\cdot$/100 eV is estimated to come from dissociative ionization (Table 9-5). Therefore, the yield from neutral excited states is $7.4 - 3.3 - 0.57 = 3.5$ atoms/100 eV. These yields are summarized in Table 9-13.

Willis and Boyd[7] estimate $\bar{E}_{se} = 6.6$ eV and give probable energies of 8.8 and 9.0 eV for the processes:

$$H_2O \rightsquigarrow H\cdot + \cdot OH \qquad (9\text{-}103)$$

and

$$H_2O \rightsquigarrow H_2 + O \qquad (9\text{-}104)$$

respectively. Using the latter two energies and the $G$-values in Table 9-13, we see that the total energy going into neutral excitation per 100 eV is 35.8 eV. The total energy expenditure according to Eq. (9-94) is then 99 eV, in good agreement with the theoretical value. Although it is possible to account for the radiation yields from water vapor in these simple terms, it is apparent that more extensive data on cross sections for neutral processes especially would provide more insight into the details of the mechanism of radiolysis for this important molecule.

## REFERENCES


1. Hart, E. J.; Platzman, R. L. "Mechanisms in Radiobiology"; Errera, M.; and Forssberg, A. Eds.; Academic Press: New York, 1961, p. 93.
2. Lind, S. C. "Radiation Chemistry of Gases"; Reinhold: New York, **1961**.
3. "Fundamental Processes in Radiation Chemistry"; Ausloos, P., Ed.; Wiley Interscience: New York, 1968.
4. Firestone, R. F.; Dorfman, L. M. In "Actions Chimiques et Biologiques des Radiations"; Haissinsky, M. Ed.; Masson, Paris, 1971, Vol. 15, p. 76.
5. Hurst, G. S.; Klots, C. E. *Adv. Radiat. Chem.* **1976**, *5*, 1.
6. Sauer, M. C., Jr. *Adv. Radiat. Chem.* **1976**, *5*, 97.
7. Willis, C.; Boyd, A. W. *Radiat. Phys. Chem.* **1976**, *8*, 71.
8. Armstrong, D. A. *Radiat. Phys. Chem.* **1982**, *20*, 75.
9. Willis, C.; Boyd, A. W.; Miller, O. A. *Can. J. Chem.* **1969**, *47*, 3007.
10. Back, R. A.; Woodward, T. W.; McLauchlan, K. A. *Can. J. Chem.* **1962**, *40*, 1380.
11. Shold, D. M.; Ausloos, P. *J. Am. Chem. Soc.* **1978**, *100*, 7915.
12. Willis, C.; Boyd, A. W.; Miller, A. O. *Can. J. Chem.* **1969**, *47*, 3007; Boyd, A. W.; Willis, C.; Cyr. R. *Anal. Chem.* **1970**, *42*, 670.
13. Hearne, J. A.; Hummel, R. W. *Radiat. Res.* **1961**, *15*, 254.
14. Jones, W. M.; Dever, D. F. *J. Chem. Phys.* **1974**, *60*, 2900.
15. Firestone, R. F. *J. Am. Chem. Soc.* **1957**, *79*, 5593.
16. Spinks, J. W. T.; Woods, R. J. "An Introduction to Radiation Chemistry", 2nd ed., Wiley: New York, 1976, p. 88.
17. Jesse, W. P. *Phys. Rev.* **1958**, *109*, 2002; Jesse, W. P.; Sadauskis, J. *ibid.* **1957**, *107*, 766.
18. Willis, C.; Miller, O. A.; Rothwell, A. E.; Boyd, A. W. *Adv. Chem. Ser.* **1968**, *81*, 539.
19. Willis, C.; Boyd, A. W.; Miller, O. A. *Radiat. Res.* **1971**, *46*, 428.
20. Ghormley, J. A.; Hochanadel, C. J.; Boyle, J. W. *J. Chem. Phys.* **1969**, *50*, 419.
21. Jones, F. T.; Sworski, T. J. *J. Phys. Chem.* **1966**, *70*, 1546.
22. Meisels, G. G. *J. Chem. Phys.* **1964**, *41*, 51.
23. Johnson, G. R. A. "Radiation Chemistry of $N_2O$ Gas: Primary Processes, Elementary Reactions and Yields"; U.S. Dept. of Commerce: Washington, D.C., 1973, Report NSRDS-NBS45.
24. Lee, R. A.; Davidow, R. S.; Armstrong, D. A. *Can J. Chem.* **1964**, *42*, 1906.
25. Davidow, R. S.; Armstrong, D. A. *Radiat. Res.* **1966**, *28*, 143.
26. Leblanc, R. M.; Herman, J. A. *J. Chim. Phys.* **1966**, *7*, 1055.
27. Spencer, L. V.; Attix, F. H. *Radiat. Res.* **1955**, *3*, 239.
28. Huyton, D. W.; Woodward, T. W. *Radiat. Res. Rev.* **1970**, *2*, 205.
29. Hunter, L. M.; Johnson, R. H. *J. Phys. Chem.* **1967**, *71*, 3228.
30. Cooper, R.; Mooring, R.-M. *Australian J. Chem.* **1972**, *25*, 2125.
31. Willis, C.; Boyd, A. W.; Young, M. J.; Armstrong, D. A. *Can J. Chem.* **1970**, *48*, 1505.
32. Johnson, G. R. A.; Wilkey, D. D. *J. Chem. Soc. D.* **1969**, 1455.
33. Burtt, B. P.; Kircher, J. F. *Radiat. Res.* **1958**, *9*, 1.
34. Gordon, R.; Ausloos, P. *J. Res. Nat. Bur. Std.* **1965**, *69A*, 79.
35. Willis, C.; Boyd, A. W.; Binder, P. E. *Can. J. Chem.* **1972**, *50*, 1557.
36. Kircher, J. F.; McNutty, J. S.; McFarling, J. L.; Levy, A. *Radiat. Res.* **1960**, *13*, 452.

37. Johnson, G. R. A.; Warman, J. M. *Discuss. Faraday Soc.* **1964**, *37*, 87.
38. Sears, J. T.; Sutherland, J. W. *J. Phys. Chem.* **1968**, *72*, 1166.
39. Meaburn, G. M.; Perner, D.; LeCalvé, J.; Bourène, M. *J. Phys. Chem.* **1968**, *72*, 3920.
40. Lee, R. A. Ph.D. Dissertation, University of London, 1967; *Nature* **1967**, *216*, 57.
41. Chen, J. D. M.Sc. Thesis, University of Calgary, 1967.
42. Johnson, G. R. A.; Redpath, J. L. *Trans. Faraday Soc.* **1969**, *66*, 861.
43. Willis, C.; Boyd, A. W.; Armstrong, D. A. *Can. J. Chem.* **1969**, *47*, 3783.
44. Boyd, A. W.; Armstrong, D. A.; Willis, C.; Miller, O. A. *Radiat. Res.* **1969**, *40*, 255.
45. Jardine, D. K.; Armstrong, D. A.; Boyd, A. W.; Willis, C. *Can. J. Chem.* **1971**, *49*, 187.
46. Lias, S. G.; Ausloos, P. "Ion Molecule Reactions, Their Role in Radiation Chemistry"; American Chemical Society: Washington, D. C., 1975.
47. McDaniel, E. W. "Collision Phenomena in Ionised Gases"; Wiley: New York, 1964.
48. Christophorou, L. G. "Atomic and Molecular Radiation Physics"; Wiley: New York, 1971.
49. Huxley, L. G. H.; Crompton, R. W. "The Diffusion and Drift of Electrons in Gases"; Wiley: New York, 1974
50. Fessenden, R. W.; Warman, J. M. *Adv. Chem.* **1968**, *82*, 222.
51. Warman, J. M., In "The Study of Fast Processes and Transient Species by Electron Pulse Radiolysis"; Baxendale, J. H.; and Busi, F. Eds.; Reidel: Dordrecht, 1982, pp. 129, 433.
52. Warman, J. M.; Sauer, M. C. *J. Chem. Phys.* **1970**, *52*, 6428 and *ibid.* **1975**, *62*, 1971.
53. Warman, J. M.; de Haas, M. P. *J. Chem. Phys.* **1975**, *63*, 2094.
54. Mozumder, A. *J. Chem. Phys.* **1980**, *72*, 6289.
55. Tembe, B. L.; Mozumder, A. *J. Chem. Phys.* **1983**, *78*, 2030.
56. Mozumder, A. *J. Chem. Phys.* **1981**, *74*, 6911.
57. Christophorou, L. G.; Gant, K. S.; Baird, J. K. *Chem. Phys. Lett.* **1975**, *30*, 104.
58. Cooper, R.; Denison, L.; Sauer, M. C., Jr. *J. Phys. Chem.* **1982**, *86*, 5093.
59. Compton, R. N.; Huebner, R. H. *Adv. Radiat. Chem.* **1970**, *2*, 281; Compton, R. N. In "NATO Conference on Photophysics and Photochemistry in Vacuum Ultraviolet", 1982.
60. Tang, S. Y.; Leffert, C. B.; Rothe, E. W. *J. Chem. Phys.* **1975**, *62*, 132.
61. Magee, J. L.; Burton, M. *J. Am. Chem. Soc.* **1951**, *73*, 523.
62. Wilson, D. E.; Armstrong, D. A. *Radiat. Res. Rev.* **1970**, *2*, 297.
63. Christophorou, L. G.; Compton, R. N.; Dickson, H. W. *J. Chem. Phys.* **1968**, *48*, 1949.
64. Nagra, S. S.; Armstrong, D. A. *Can. J. Chem.* **1975**, *53*, 3305.
65. Warman, J. M.; Sauer, M. C., Jr. *Radiat. Phys. Chem.* **1971**, *3*, 273.
66. Smith, D.; Adams, N. G.; Alge, E. *J. Phys. B. Atom Mol. Phys.* **1984**, *17*, 461.
67. Crompton, R. W.; Haddad, G. N.; Hegerberg, R.; Robertson, A. G. *J. Phys. B. Atom. Mol. Phys.* **1982**, *15*, L483.
68. Fessenden, R. W.; Bansal, K. M. *J. Chem. Phys.* **1970**, *53*, 3468.
69. Bansal, K. M.; Fessenden, R. W. *Chem. Phys. Lett.* **1972**, *15*, 21.
70. Christophorou, L. G.; Blaunstein, R. P. *Chem. Phys. Lett.* **1971**, *12*, 173.
71. Kurepa, M. V.; Babić, D. S.; Belić, D. S. *Chem. Phys.* **1981**, *59*, 125.
72. Sides, G. D.; Tiernan, T. O.; Hanrahan, R. J. *J. Chem. Phys.* **1976**, *65*, 1966.
73. Chen, E.; George, R. D.; Wentworth, W. E. *J. Chem. Phys.* **1973**, *49*, 1968.
74. Wentworth, W. E.; George, R.; Keith, H. *J. Chem. Phys.* **1969**, *51*, 1791.
75. Foster, M. S.; Beauchamp, J. L. *Chem. Phys. Lett.* **1975**, *31*, 482.
76. Lifshitz, C. *J. Phys. Chem.* **1983**, *87*, 3474.
77. Odem, R. W.; Smith, D. L.; Futrell, J. *J. Phys. B. Atom. Mol. Phys.* **1975**, *8*, 1349.
78. Fenimore, C. P.; Jones, G. W. *Combust. Flame* **1964**, *8*, 231.
79. Bloch, F.; Bradbury, N. E. *Phys. Rev.* **1935**, *48*, 689.
80. Hatano, Y.; Shimamori, H. In "Electron and Ion Swarms"; Christophorou, L. G., Ed.; Pergamon Press, New York, **1981**.
81. Herzenberg, A. *J. Chem. Phys.* **1969**, *51*, 4942.
82. Shimamori, H.; Fessenden, R. W. *J. Chem. Phys.* **1981**, *74*, 453.
83. Pack, J. L.; Phelps, A. V. *J. Chem. Phys.* **1966**, *44*, 1870.
84. Truby, F. *Phys. Rev.* **1972**, *A6*, 671.
85. McCorkle, D. L.; Christophorou, L. G.; Anderson, V. E. *J. Phys. B. Atom. Mol. Phys.* **1972**, *5*, 1211; Goans, R. E.; Christophorou, L. G. *J. Chem. Phys.* **1974**, *60*, 1036.
86. Shimamori, H.; Hatano, Y. *Chem. Phys.* **1977**, *21*, 187; Kokaku, Y.-I.; Hatano, Y. *J. Chem. Phys.* **1979**, *71*, 4883; Kokaku, Y.-I.; Toriumi, M.; Hatano, Y. *J. Chem. Phys.* **1980**, *73*, 6167; Toriumi, M.; Hatano, Y. *J. Chem. Phys.* **1983**, *79*, 3749.
87. Shimamori, H.; Hotta, H. *J. Chem. Phys.* **1983**, *78*, 1318.

88. Nagra, S. S.; Armstrong, D. A. *Can. J. Chem.* **1976**, *54*, 3580.
89. Crawford, O. H.; Koch, B. J. D. *J. Chem. Phys.* **1974**, *60*, 4512.
90. Chen, J. D.; Armstrong, D. A. *J. Chem. Phys.* **1968**, *48*, 2310.
91. Davidow, R. S.; Armstrong, D. A. *J. Chem. Phys.* **1968**, *48*, 1235.
92. Nagra, S. S.; Armstrong, D. A. *J. Phys. Chem.* **1977**, *81*, 599; *Radiat. Phys. Chem.* **1978**, *11*, 305.
93. Quitevis, E. L.; Bowen, K. H.; Liesegang, G. W.; Herschbach, D. R. *J. Phys. Chem.* **1983**, *87*, 2076.
94. Klots, C. E.; Compton, R. N. *J. Chem. Phys.* **1978**, *69*, 1636, *ibid.* **1978**, *69*, 1644.
95. Shimamori, H.; Fessenden, R. W. *J. Chem. Phys.* **1978**, *68*, 2757; *ibid.* **1978**, *69*, 4732; *ibid.* **1979**, *70*, 1137; *ibid.* **1979**, *71*, 3009.
96. Meisels, G. G. *Radiat. Phys. Chem.* **1982**, *20*, 7.
97. Armstrong, D. A.; Willis, C. *Radiat. Phys. Chem.* **1976**, *8*, 221.
98. Rosenstock, H. M.; Wallenstein, M. B.; Wahrhaftig, A. L.; Eyring, H. *Proc. Natl. Acad. Sci. U.S.* **1952**, *38*, 667.
99. Rosenstock, H. M.; Dannacher, J. *Radiat. Phys. Chem.* **1982**, *20*, 7.
100. Chesnavich, W. J.; Su, T.; Bowers, T. *J. Chem. Phys.* **1980**, *72*, 2641 and prior references.
101. Celli, F.; Weddle, G.; Ridge, D. P. *J. Chem. Phys.* **1980**, *73*, 801; Ridge, D. P. In "Kinetics of Ion Molecule Reactions"; Ausloos, P., Ed.; Plenum, New York, 1979.
102. Hiraoka, K.; Kebarle, P. *Radiat. Phys. Chem.* **1982**, *20*, 41.
103. Chang, C.; Hanrahan, R. J. *Radiat. Phys. Chem.* **1982**, *20*, 111.
104. Meot-Ner, M. In "Gas Phase Ion Chemistry"; Bowers, M. T., Ed.; Academic Press: New York, 1979, Vol. 1, p. 197.
105. Kebarle, P. *Ann. Rev. Phys. Chem.* **1977**, *28*, 445.
106. Sennhauser, E. S.; Armstrong, D. A. *Radiat. Phys. Chem.* **1978**, *11*, 17; *ibid.* **1977**, *10*, 25.
107. Gray, E. P.; Kerr, D. E. *Ann. Phys.* **1962**, *17*, 276.
108. Warman, J. M.; Sennhauser, E. S.; Armstrong, D. A. *J. Chem. Phys.* **1979**, *70*, 995; Sennhauser, E. S.; Armstrong, D. A.; Warman, J. M. *Radiat. Phys. Chem.* **1980**, *15*, 479.
109. Schmidt, W. F.; Jungblut, H.; Hansen, D.; Tagashira, H. *Proceedings of the Second International Symposium on Gaseous Dielectrics*, March 9–13, Knoxville, Tenn., 1980, Pergamon Press: New York, 1982.
110. Sennhauser, E. S.; Armstrong, D. A. *Radiat. Phys. Chem.* **1978**, *12*, 115; *J. Phys. Chem.* **1980**, *84*, 123.
111. Shiu, Y.-J. Biondi, M. A. *Phys. Rev. A.* **1978**, *17*, 868.
112. Mahan, B. H.; Person, J. C. *J. Chem. Phys.* **1984**, *40*, 392.
113. Thomson, J. J. *Philos. Mag.* **1924**, *47*, 337.
114. Bates, D. R. *J. Phys. B. Atom. Mol. Phys.* **1981**, *14*, 3525.
115. Flannery, M. R. *Case Studies Atom. Phys.* **1972**, *2*, 3.
116. Bates, D. R.; Mendas, I. *Proc. Roy. Soc. London* **1978**, *A359*, 275; *ibid.* **1978**, *A359*, 287.
117. Bates, D. R. *J. Phys. B. Atom. Mol. Phys.* **1980**, *13*, L623; *Proc. Roy. Soc. London* **1980**, *A369*, 327; *Chem. Phys. Lett.* **1980**, *75*, 409.
118. Wilson, D. E.; Armstrong, D. A. *Can. J. Chem.* **1970**, *48*, 598.
119. Ellis, H. W.; McDaniel, E. W.; Albritton, D. L.; Viehland, L. A.; Lin, S.; Mason, E. A. *Atom. Data Nucl. Data Tables* **1978**, *22*, 179 and references therein.
120. Sauer, M. C., Jr.; Mulac, W. A. *J. Chem. Phys.* **1972**, *56*, 4995.
121. Firestone, R. F.; Oka, T.; Takao, S.; Beno, M. F. *J. Phys. Chem.* **1978**, *82*, 2183.
122. Bishop, W. P.; Dorfman, L. M. *J. Chem. Phys.* **1970**, *52*, 3210.
123. Jonah, C. D.; Mulac, W. A.; Zeglinski, P. *J. Phys. Chem.* **1984**, *88*, 4100.
124. Sauer, M. C.; Mani, I. *J. Phys. Chem.* **1970**, *74*, 59.
125. Arai, S.; Firestone, R. F. *J. Chem. Phys.* **1969**, *50*, 4575.
126. Firestone, R. F.; Chen, M. *J. Chem. Phys.* **1978**, *69*, 2943; Firestone, R. F.; Oka, T.; Takao, S. *J. Chem. Phys.* **1979**, *70*, 123; Loeb, D. W.; Chen, M.; Firestone, R. F. *J. Chem. Phys.* **1981**, *74*, 3270.
127. Golde, M. F. *Gas Kinet. Energy Transfer* **1977**, *2*, 123.
128. Sauer, M. C., Jr. In "The Study of Fast Processes and Transient Species by Electron Pulse Radiolysis"; Baxendale, J. H.; and Busi, F., Eds.; Reidel, Dordrecht, **1982**. p. 601.
129. Cooper, R.; Grieser, F.; Sauer, M. C., Jr.; Sangster, D. F. *J. Phys. Chem.* **1977**, *81*, 2215.
130. Cooper, R.; Denison, L. S.; Zeglinski, P.; Roy, C. R.; Gillis, H. *J. Appl. Phys.* **1983**, *54*, 3053.
131. Cooper, R.; Grieser, F.; Sauer, M. C., Jr. *J. Phys. Chem.* **1976**, *80*, 2138; Sauer, M. C., Jr.; Mulac, W. A.; Cooper, R.; Grieser, F. *J. Chem. Phys.* **1976**, *64*, 4587; Cooper, R.; Grieser, F.; Sauer, M. C., Jr. *J. Phys. Chem.* **1977**, *81*, 1889.

132. Freeman, C. G.; Quickenden, T. I.; Sangster, D. F. *J. Chem. Phys.* **1984**, *80*, 2336.
133. Lii, R. R.; Gorse, R. A., Jr.; Sauer, M. C., Jr.; Gordon, S. *J. Phys. Chem.* **1980**, *84*, 813.
134. Platzman, R. L. *Radiat. Res.* **1962**, *17*, 419; *Vortex* **1962**, *23*, 1.
135. Santar, I.; Bednář *Czech, Chem. Commun.* **1969**, *34*, 1.
136. Ramirez, J. E.; Bera, R. K.; Hanrahan, R. J. *Radiat. Phys. Chem.* **1984**, *23*, 685.
137. Dixon, R. S. *Radiat. Res. Rev.* **1970**, *2*, 237.

# 10

# Radiation Chemistry of the Liquid State: (1) Water and Homogeneous Aqueous Solutions

G. V. BUXTON

*Cookridge Radiation Research Center, The University of Leeds*

## INTRODUCTION

More is known about the radiation chemistry of water than any other liquid.[1–6] From a practical viewpoint our knowledge is virtually complete, and water radiolysis now provides a very convenient way of generating an enormous variety of unstable species under well-defined conditions. This facility, coupled with the techniques of pulse radiolysis, has opened up new areas in aqueous inorganic, organic, and biochemistry that cannot be readily studied by thermal or photochemical methods. This chapter is aimed, therefore, at those who wish to use radiolytic methods to generate and study unstable species in aqueous solution. The basic features of the radiation chemistry of water are described first to show how the primary radical and molecular products evolve with time and to delineate the bounds of useful experimental conditions. Next, the properties of the primary radicals are summarized, and examples are given to show how the primary radicals can


*Radiation Chemistry: Principles and Applications*

be converted into secondary radicals, often of a single kind. This is an important aspect of the radiation chemistry of aqueous solutions. Lastly, the impact of our knowledge of the radiation chemistry of water on advances in general chemistry is illustrated by examples from the fields of inorganic and organic chemistry.

## THE RADIATION CHEMISTRY OF WATER

The time scale of events initiated by the absorption of energy by water from the incident ionizing radiation is shown in Scheme 10-1.

Initially, a water molecule is ionized by an energetic photon, for example, a gamma photon from $^{60}Co$, or by a charged particle, such as an electron or proton generated by a particle accelerator, or an α-particle from a suitable radioactive nuclide. Generally, the electron liberated in this ionization event has sufficient energy to ionize further water molecules, and this leads to the formation of clusters of ions along the track of the ionizing particle.

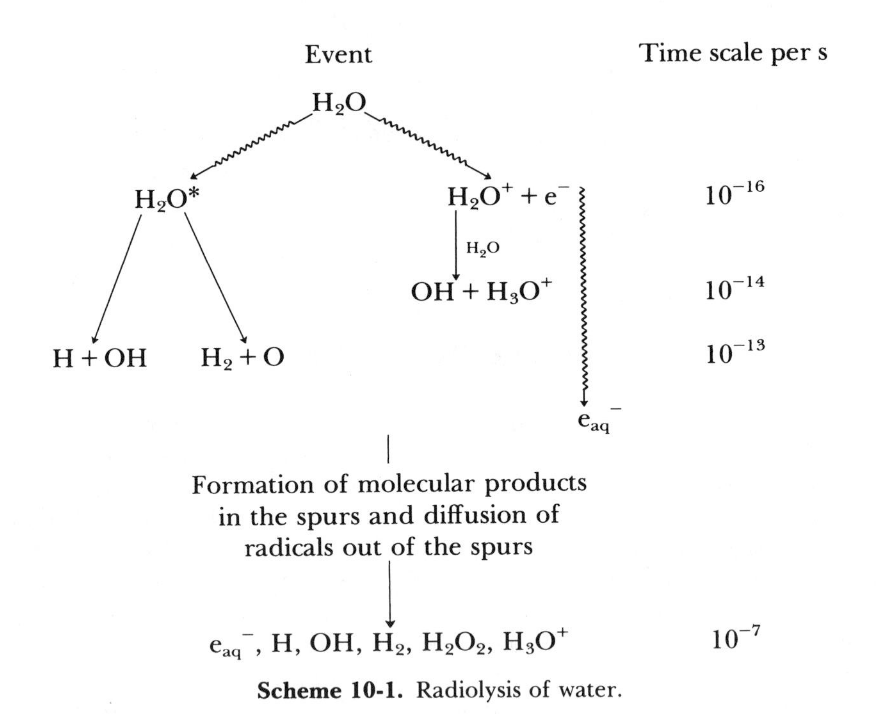


**Scheme 10-1.** Radiolysis of water.

These clusters are called spurs, and a detailed description of their structure and distribution along the particle's track is given in Chapter 6. Some water molecules, instead of being ionized, may be excited to upper electronic states from which they can autoionize, dissociate, or simply fall back to the ground electronic state. Although excited-state water molecules have sometimes been invoked to explain certain observations in the radiation chemistry of aqueous solutions, there is no unequivocal evidence to show that they play any significant role.

The ionization event occurs on the time scale of an electronic transition. The positive ion $H_2O^+$ is known to undergo the ion–molecule reaction in the gas phase with a rate constant of $8 \times 10^{12}$ $M^{-1}$ $s^{-1}$,[7] which, when extrapolated to liquid water, sets the lifetime of $H_2O^+$ in this medium at less than $10^{-14}$ s. However, Hamill[8] has pointed out that initially $H_2O^+$ will have the structure of a neutral water molecule so that it may migrate rapidly over distances of a few molecular diameters by resonance electron transfer with a succession of neighboring water molecules. Some support for this idea is provided by data obtained for solutions containing fairly high (ca. 1 *M*) concentrations of halide ions[6] that, because of their low ionization potentials, could be expected to trap $H_2O^+$ efficiently.

$$H_2O^+ + X^- \rightarrow X + H_2O$$

The electron released in the ionization event can, as mentioned above, cause further ionization to occur if it has sufficient kinetic energy. Eventually its energy falls below the ionization threshold of water, and it loses the rest of its energy by exciting vibrational and rotational modes of the solvent molecules; and finally it is localized in a potential energy well long enough to become solvated as a result of the molecular dipoles rotating under the influence of the negative charge. Recent experimental evidence from a study using a subpicosecond laser pulse to generate the excess electron by photoionization[9] shows that the electron is solvated in less than $10^{-12}$ s. This is much shorter than the dielectric relaxation time of water ($10^{-11}$ s) and may mean that the electron finds a preformed site in the medium that largely satisfies its solvation requirements. It should be noted, however, that the intense electric field due to the excess electron is calculated to shorten the dielectric relaxation time by an order of magnitude.[10] The electron is often termed "dry" in its presolvation state, but this is a rather nebulous description because it does not refer to any well-defined state. Nevertheless, there is good evidence that the electron in its presolvation state can be captured by solutes at concentrations of about 1 *M*.

The generally accepted view is that the physicochemical processes described above are complete by about $10^{-12}$ s after the ionization event, and that the species $e_{aq}^-$, OH, $H_3O_{aq}^+$ and any radical and molecular fragments resulting from dissociation of excited-state molecules are then in thermal equilibrium with the bulk medium. At this time, which may be regarded as the beginning of the chemical stage, these species are clustered in spurs, as indicated earlier. Next, the radiolysis products begin to diffuse, with the result that a fraction of them encounter one another and react together to form molecular or secondary radical products, while the remainder escape into the bulk solution and effectively become homogeneously distributed throughout the medium. This so-called "spur expansion" is complete by about $10^{-7}$ s, at which time the radiolysis products are as shown in Scheme 10-1. This time can be thought of as marking the end of the radiation chemical stage. The reactions occurring while the spurs expand are listed in Table 10-1.

**Table 10-1.** Spur Reactions in Water

| Reactions | $10^{-10}\ k/M^{-1}\ s^{-1}$ |
|---|---|
| (10-1) $e_{aq}^- + e_{aq}^- \rightarrow H_2 + 2OH^-$ | 0.54 |
| (10-2) $e_{aq}^- + OH \rightarrow OH^-$ | 3.0 |
| (10-3) $e_{aq}^- + H_3O^+ \rightarrow H + H_2O$ | 2.3 |
| (10-4) $e_{aq}^- + H \rightarrow H_2 + OH^-$ | 2.5 |
| (10-5) $H + H \rightarrow H_2$ | 1.3 |
| (10-6) $OH + OH \rightarrow H_2O_2$ | 0.53 |
| (10-7) $OH + H \rightarrow H_2O$ | 3.2 |
| (10-8) $H_3O^+ + OH^- \rightarrow 2\ H_2O$ | 14.3 |

# EXPERIMENTAL EVIDENCE FOR SPURS

The brief and qualitative description just presented of the spur diffusion model for water and which is described in quantitative detail elsewhere has been built up from numerous experimental observations and models to explain them. It is appropriate to examine the main pieces of evidence on which the spur diffusion model is based because they bear directly on the quantitative interpretations of radiolytically induced chemical processes in general. For example, it is obvious that when the product of the rate constant and the concentration of a scavenger, S, for a particular free radical, R, exceeds $10^7\ s^{-1}$, Reaction (10-9) will compete with the spur reactions of that radical, so that more product will be formed than would be the case if $k_9[S] < 10^7\ s^{-1}$.

$$R + S \xrightarrow{k_9} P \qquad (10\text{-}9)$$

A knowledge of the effects of spur processes is important, therefore, in interpreting correctly the radiation chemistry of concentrated ($\geqslant 10^{-2}\ M$) aqueous solutions.

The original evidence for spurs was deduced from measurements of stable radiolysis products in water and aqueous solutions. For pure water it was noted that, if no gas space is available above the liquid, scarcely any decomposition products are formed, save for traces of hydrogen, when the ionizing radiation is of low mean linear energy transfer (LET; Chapter 5), but $H_2$, $O_2$, and $H_2O_2$ are formed when gas space is available. In contrast, when high-LET radiation is used, these products are formed under all conditions.

Further experiments with aqueous solutions revealed three characteristic features:

1. A fraction of the radicals $e_{aq}^-$, H, and OH are readily scavenged by quite low concentrations of solute ($\leqslant 10^{-3}$ $M$) in yields that are independent of solute concentration. Similarly, the molecular products $H_2$ and $H_2O_2$ are formed in constant yield.
2. When the solute concentration is raised above about $10^{-3}$ $M$, the yields of scavenged radicals increase and the yields of $H_2$ and $H_2O_2$ decrease, the changes in the yields being approximately proportional to the change in log [scavenger].
3. The yields of radical products decrease, and the yields of molecular products increase, as the LET of the radiation increases.

These features are readily explicable in terms of the spur diffusion model. First, only those radicals that diffuse out of the spurs into the bulk solution are scavenged by solutes in low concentration. The only requirement is that there should be sufficient solute present to scavenge all the escaping radicals. Second, at sufficiently high solute concentrations Reactions (10-10) and/or (10-11) occur in the spur;

$$e_{aq}^- + S_1 \rightarrow P_1 \qquad (10\text{-}10)$$

$$OH + S_2 \rightarrow P_2 \qquad (10\text{-}11)$$

where $S$ represents scavenger, and $P$ product. Reactions (10-10) and (10-11) compete with the spur reactions of $e_{aq}^-$ and OH (Table 10-1), and consequently the yields of $H_2$ and $H_2O_2$ diminish. Third, with increasing LET the ionization events become more closely spaced, and the spurs are formed closer together and eventually coalesce into a continuous cylindrical shape. In this case it is easy to see, qualitatively, that the number of radicals diffusing out of the track will decrease, and, correspondingly, the number reacting to form molecular products will increase, because (i) the concentration of radicals in overlapping spurs is higher than in isolated spurs of a given size, and (ii) radicals diffusing along the track will have less chance to escape reaction with other radicals in the track.

These features are illustrated by experimental results shown in Figures 10-1 and 10-2 and Table 10-2 for yields of radical and molecular products obtained in steady-state experiments under a variety of conditions.

Radiation chemical yields are expressed as $G$-values, where $G$ is the number of species created (or destroyed) per 100 eV of absorbed dose.* For historical reasons the $G$-value of a species X may be written as $G$(X),

* This definition of $G$-value does not conform to SI units. A new definition expresses the units of $G$-values in mol $J^{-1}$, which is equivalent to $9.65 \times 10^6$ molecules $(100\ eV)^{-1}$, so that the $G$-value as originally defined can be converted to SI units by multiplying by $1.036 \times 10^{-7}$. The vast majority of $G$-values quoted in the literature use the original definition given in the text.

$G_X$, or $g(x)$. $G(X)$ refers to an experimentally measured yield, and $G_X$ or $g(x)$ refer to the so-called primary yield of X, which is derived from the experimental data. The primary yield, however, is the term given to the yield of X that remains when the spur reactions are complete; that is, it pertains to dilute solutions. This term is something of a misnomer now because it would be more logical to refer to yields of the species formed in the initial event as primary yields. However, we shall refer to these yields as initial $G$-values, which are sometimes written as $G_X{}^0$. The yields listed in Table 10-2 are primary yields (the designation $g(x)$ is not commonly used) and have, of course, been derived from experimental measurements made in dilute solutions.

In the absence of solutes, reactions occur between the radical and molecular products in the bulk liquid; they are listed in Table 10-3, from which it is evident that they ultimately reform water. Table 10-2 shows that for low-LET radiation $G_{radicals} > G_{molecules}$, so that the molecular species are almost completely decomposed, whereas for high-LET radiation $G_{radicals} < G_{molecules}$, so that molecular products remain when the reactions listed in Table 10-3 are complete. Thus, the effect of LET on the extent of the radiolytic decomposition of pure water, described earlier, is explained by the spur diffusion model.

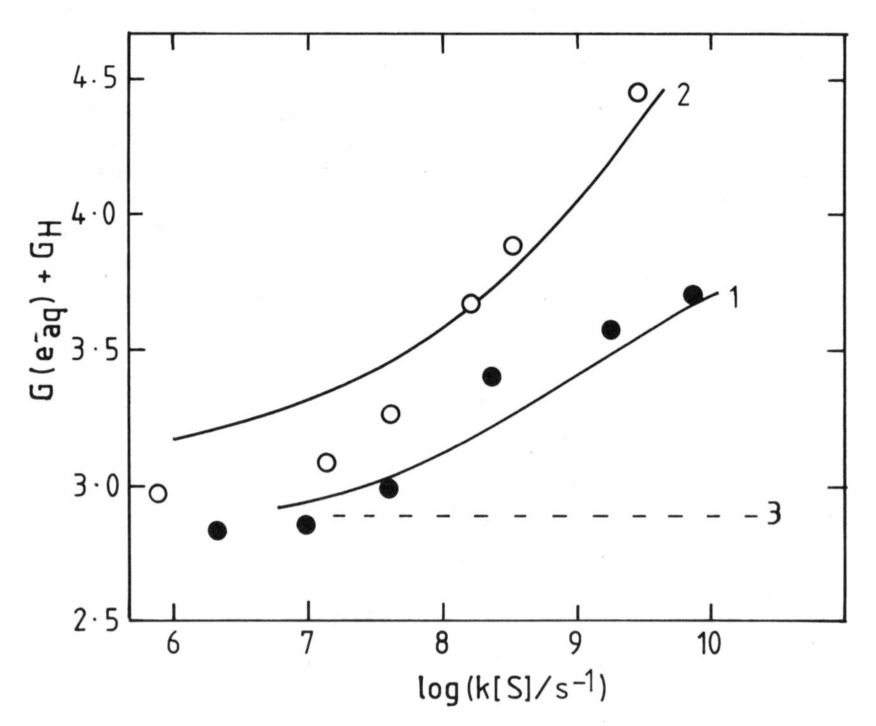


**Figure 10-1.** Dependence of the yield of scavenged hydrated electrons on the scavenging power of solutes. (○, $N_2O$; ●, $NO_3^-$; curve 1, $H_3O^+$; curve 2, Eq. (10-27); curve 3 is for no net scavenging in the spur.) (Reproduced with permission.[39])

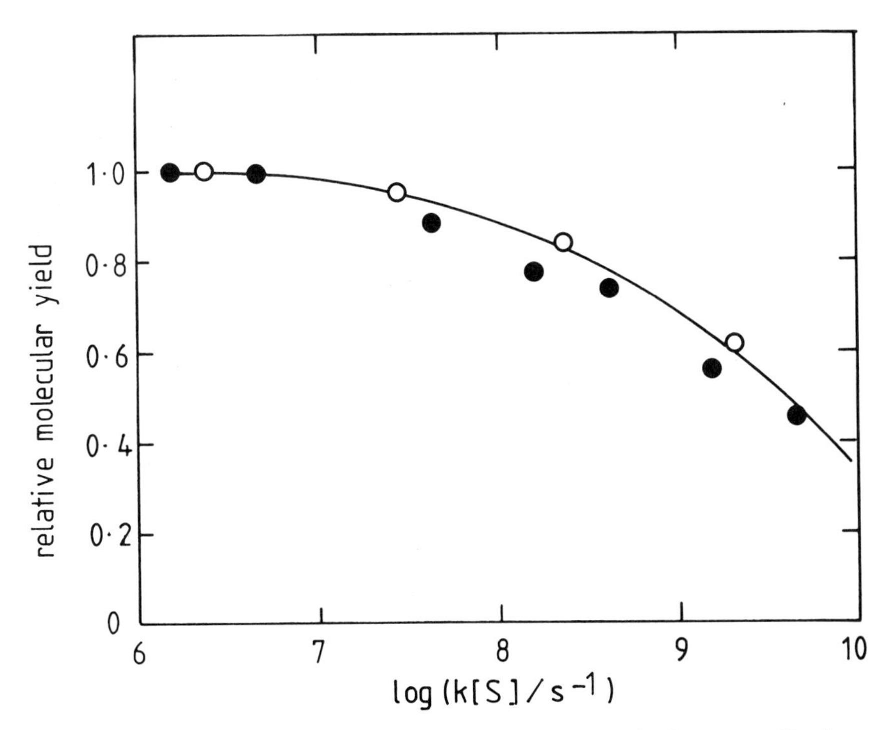


**Figure 10-2.** Effect of scavengers on the yields of molecular hydrogen and hydrogen peroxide. (●, $C_2H_5OH$; ○, $H_2O_2$; the solid line is the averaged data for several scavengers.) (Adapted from Draganic and Draganic.[3])

**Table 10-2.** Dependence of the Yields of Radiolysis Products from Neutral Water on LET[1,11]

| | *G* | | | | | | |
|---|---|---|---|---|---|---|---|
| LET/eV $nm^{-1}$ | $-H_2O$ | $e_{aq}^-$ | OH | H | $H_2$ | $H_2O_2$ | $HO_2$ |
| 0.23 | 4.08 | 2.63 | 2.72 | 0.55 | 0.45 | 0.68 | 0.008 |
| 12.3 | 3.46 | 1.48 | 1.78 | 0.62 | 0.68 | 0.84 | |
| 61 | 3.01 | 0.72 | 0.91 | 0.42 | 0.96 | 1.00 | 0.05 |
| 108 | 2.84 | 0.42 | 0.54 | 0.27 | 1.11 | 1.08 | 0.07 |

**Table 10-3.** Reactions Between Radical and Molecular Products

| | Reaction | $10^{-7}\,k/M^{-1}\,s^{-1}$ |
|---|---|---|
| (10-12) | $e_{aq}^- + H_2O_2 \rightarrow OH + OH^-$ | 1200 |
| (10-13) | $OH + H_2 \rightarrow H + H_2O$ | 4.9 |
| (10-14) | $OH + H_2O_2 \rightarrow HO_2 + H_2O$ | 2.7 |
| (10-15) | $HO_2 + HO_2 \rightarrow H_2O_2 + O_2$ | 0.25 |
| (10-16) | $O_2^- + HO_2 \rightarrow HO_2^- + O_2$ | 4.4 |
| (10-17) | $H + O_2 \rightarrow HO_2$ | 1900 |
| (10-18) | $e_{aq}^- + O_2 \rightarrow O_2^-$ | 1900 |

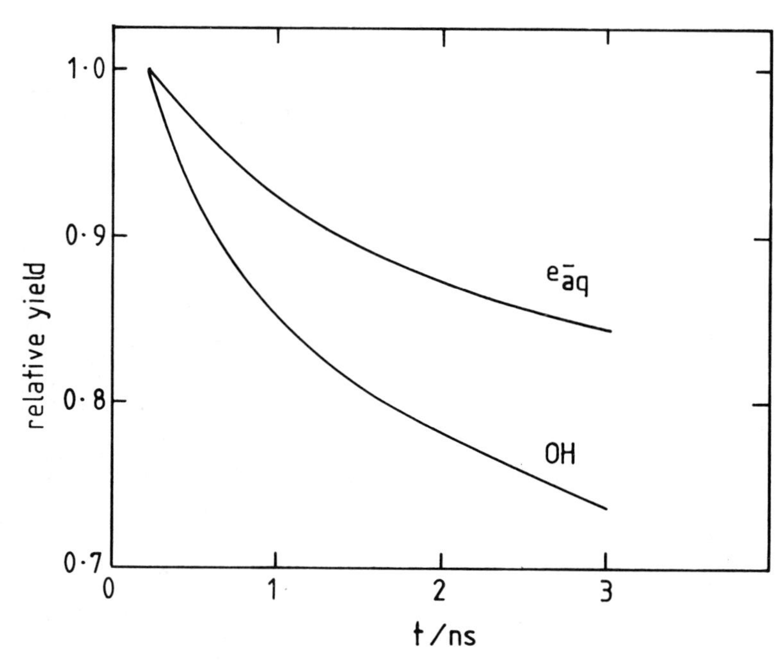


**Figure 10-3.** Decay of $e_{aq}^-$ and OH produced by a 30-ps pulse of 20-MeV electrons in water. (Reproduced with permission.[39])

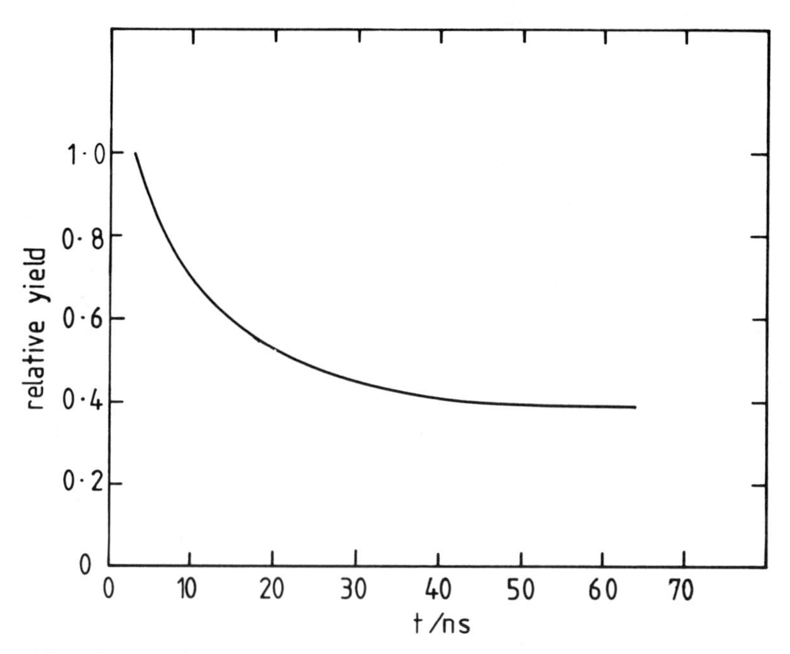


**Figure 10-4.** Decay of $e_{aq}^-$ produced by a 1-ns pulse of 3-MeV protons in water. (Reproduced with permission.[39])

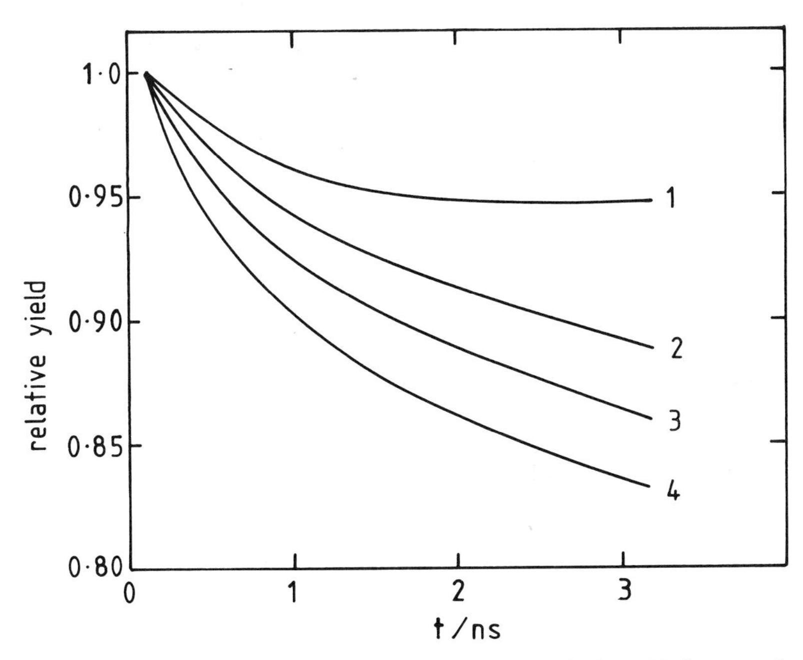


**Figure 10-5.** Decay of $e_{aq}^-$ produced by a 30-ps pulse of 20-MeV electrons in 1 *M* NaOH + 3 *M* ethanol solution (curve 1); 1 *M* NaOH solution (curve 2); 3 *M* ethanol solution (curve 3); and water (curve 4). (Reproduced with permission.[39])

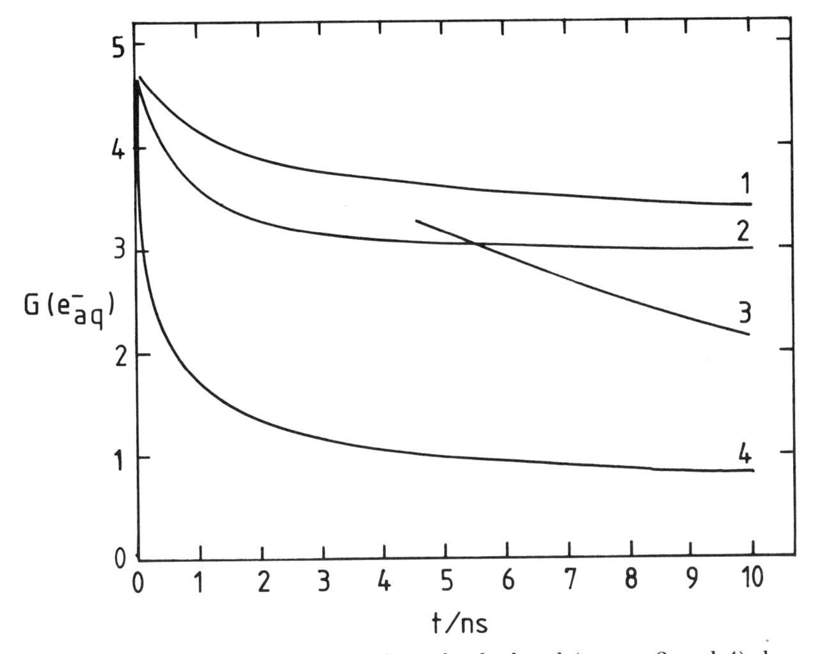


**Figure 10-6.** Observed (curves 1 and 3) and calculated (curves 2 and 4) decays of $e_{aq}^-$ for 20-MeV electrons (curves 1 and 2) and 3-MeV protons (curves 3 and 4). (Reproduced with permission.[39])

Direct observations of spur reactions, made possible by the development of subnanosecond pulse radiolysis facilities in the last decade or so, have fully confirmed the main framework of the spur diffusion model,[6,12–14] as demonstrated by the data shown in Figures 10-3 to 10-5. At the same time, however, they have also shown that some adjustments need to be made to the spatial and number distributions of the initial species $e_{aq}^-$, OH, H, and $H_3O_{aq}^+$ for the time profiles of the spur reactions calculated using the spur diffusion model to match the experimental ones.[15,16] This is particularly true in the case of higher-LET radiation such as 3-MeV protons, as Figure 10-6 shows, but for low-LET radiations, as commonly used in general chemical studies, the discrepancy between model and experiment is not very great.

## PRIMARY YIELDS

As already indicated, primary yields are the yields of the species remaining when all the spur reactions are complete, that is, some $10^{-7}$ s after the ionization event. At this time the radiolytic change in water is represented by

$$H_2O \rightsquigarrow e_{aq}^-, H, OH, HO_2, H_2O_2, H_3O^+ \qquad (10\text{-}19)$$

and the material balance equations are

$$G_{-H_2O} = 2G_{H_2} + G_H + G_{e_{aq}^-} - G_{HO_2} = 2G_{H_2O_2} + G_{OH} + 2G_{HO_2} \qquad (10\text{-}20)$$

or, if $HO_2$ is neglected, which is justified for low-LET radiation (Table 10-2),

$$G_{-H_2O} = 2G_{H_2} + G_H + G_{e_{aq}^-} = 2G_{H_2O_2} + G_{OH} \qquad (10\text{-}21)$$

Numerous measurements have been made of these primary yields, particularly for low-LET radiations such as $^{60}Co$ $\gamma$-rays and fast electrons. Most measurements have been carried out in dilute solutions under steady-state conditions where solutes are employed to scavenge each particular species and form an identifiable stable product. It is generally not possible to measure the yields of all of the primary species by using a single system; instead, the measured yields are substituted into the material balance Eqs. (10-20) or (10-21) in order to calculate the yield of the unmeasured species. It is possible, in principle, to measure the yields of the radicals directly by using pulse radiolysis, but in practice only $e_{aq}^-$ is measured in this way because H and OH have only weak absorption spectra in the ultraviolet that are generally masked by those of added solutes. However, H and OH can be measured by adding solutes that react to give observable products under pulse radiolytic conditions. Some examples of these will be described later when reaction rate constants are discussed.

### *Yields in Neutral Solution*

For low-LET radiations the generally accepted values of the primary yields in the pH range 3 to 11 are the following:

$$G_{e_{aq}^-} = G_{OH} = G_{H_3O^+} = 2.7 \quad G_H = 0.6 G_{H_2} = 0.45, \quad G_{H_2O_2} = 0.7$$

They have been derived from measurements made in many different laboratories on many different solutions and represent a consensus view of the results. Naturally, individual measurements vary from system to system, and this is illustrated by a few examples in Table 10-4, but the variation is not very great, and for most work in dilute solution the yields listed above are appropriate.

Some care must be exercised, however, in the choice of $G$-values appropriate to more concentrated solutions, as is apparent from an examination of Figure 10-1. There we see that the dependence of the $G$-value on concentration can vary markedly from one solute to another. These variations can be rationalized in terms of the spur diffusion model in the following way. The spurs contain approximately equal numbers of oxidizing and reducing radicals so that if $P_1$ formed in

$$e_{aq}^- + S_1 \rightarrow P_1 \qquad (10\text{-}10)$$

is less reactive with OH than $e_{aq}^-$ is,

$$OH + P_1 \rightarrow S_1 + OH^- \qquad (10\text{-}22)$$

then $G(P_1)$ will be larger than $G_{e_{aq}^-}$ when the spur processes are complete. On the other hand, if Reaction (10-22) is as efficient as

$$OH + e_{aq}^- \rightarrow OH^- \qquad (10\text{-}2)$$

there will be little or no difference between $G(P_1)$ and $G_{e_{aq}^-}$. The analogous situation occurs in the case of the hydroxyl radical through Reactions (10-11) and (10-23):

$$OH + S_2 \rightarrow P_2 \qquad (10\text{-}11)$$

$$e_{aq}^- + P_2 \rightarrow S_2 \qquad (10\text{-}23)$$

**Table 10-4.** Some Values of $G_{e_{aq}^-}$ and $G_H$ in Neutral Solution[17]

| System | $G_{e_{aq}^-}$ | $G_H$ | $G_{e_{aq}^-} + G_H$ |
|---|---|---|---|
| 2-propanol + acetone | 2.65 | 0.55 | 3.2 |
| 2-propanol + $N_2O$ | 2.8 | 0.6 | 3.4 |
| $N_2O + NO_2^-$ | 2.6 | — | — |
| $HCO_2^- + O_2$ | 2.3 | 0.75 | 3.05 |
| $NO_3^- + HPO_3^-$ | 2.65 | 0.55 | 3.2 |
| $NO_3^-$ + 2-propanol | 2.8 | 0.6 | 3.4 |
| $C(NO_2)_4$, pulse radiolysis | 2.6 | — | — |

However, all reactive solutes $S_1$ and $S_2$ decrease $G_{H_2}$ and $G_{H_2O_2}$, respectively, because $P_1$ and $P_2$ replace the radicals in Reactions (10-1) and (10-6) in Table 10-1. In fact, as Figure 10-2 demonstrates, $G_{H_2}$ and $G_{H_2O_2}$ depend only on the scavenging power, that is, the product of rate constant and concentration of the solute, in contrast to the much more varied dependences of the radical yields exemplified in Figure 10-1. Some examples of solutes that increase the yields of scavenged radicals are listed in Table 10-5. Solutes such as $Br^-$, $I^-$, and $SCN^-$ are examples of scavengers that do not enhance the radical yields because Reaction (10-23) is as efficient as Reaction (10-2).

**Table 10-5.** Solutes That Increase Radical Yields at High Concentrations[17]

| Solute | Scavenged radical | Product |
|---|---|---|
| $NO_3^-$ | $e_{aq}^-$ | $NO_3^{2-}(NO_2 + 2OH^-)$ |
| $N_2O$ | $e_{aq}^-$ | $N_2 + O^-$ |
| $ClCH_2CO_2^-$ | $e_{aq}^-$ | $Cl^- + CH_2CO_2^-$ |
| $HCO_2^-$ | OH | $CO_2^-$ |

## *pH Dependence of the Primary Yields*

This is illustrated in Figure 10-7 for low-LET radiation. Above pH 3 the yields are essentially independent of pH, although it should be noted that Reaction (10-24) occurs at high pH with $k_{24} = 2.3 \times 10^7\ M^{-1}\ s^{-1}$, so that it may compete with the reaction of H with other solutes under this condition.

$$H + OH^- \rightarrow e_{aq}^- \qquad (10\text{-}24)$$

Below pH 3 Reaction (10-3) occurs and $G_{e_{aq}^-} + G_H$ increases as the proportion of H increases through Reaction (10-3), competing with other reactions of $e_{aq}^-$ in the spurs. This is probably the result of the diffusion coefficient of H ($D_H = (7.0 \pm 1.5) \times 10^{-5}\ cm^2\ s^{-1}$ [18]) being somewhat greater than that of $e_{aq}^-$ ($D_{e_{aq}^-} = 4.9 \times 10^{-5}\ cm^2\ s^{-1}$). The changes in the individual values of $G_{e_{aq}^-}$ and $G_H$ with pH have been measured with chloroacetic acid as the solute, which reacts differently with $e_{aq}^-$ and H to yield $Cl^-$ and $H_2$, respectively, and these are shown in Figure 10-8.

$$e_{aq}^- + ClCH_2CO_2H \rightarrow Cl^- + \dot{C}H_2CO_2H \qquad (10\text{-}25)$$

$$H + ClCH_2CO_2H \rightarrow H_2 + Cl\dot{C}HCO_2H \qquad (10\text{-}26)$$

$G_{OH}$ and $G_{H_2O_2}$ also increase with pH below 3; an increase in the yield of oxidizing equivalents is expected, of course, from considerations of material balance, but the increase in $G_{H_2O_2}$ does suggest that the spur reaction (10-6) becomes more important at low pH at the expense of the spur reaction (10-7) in Table 10-1.

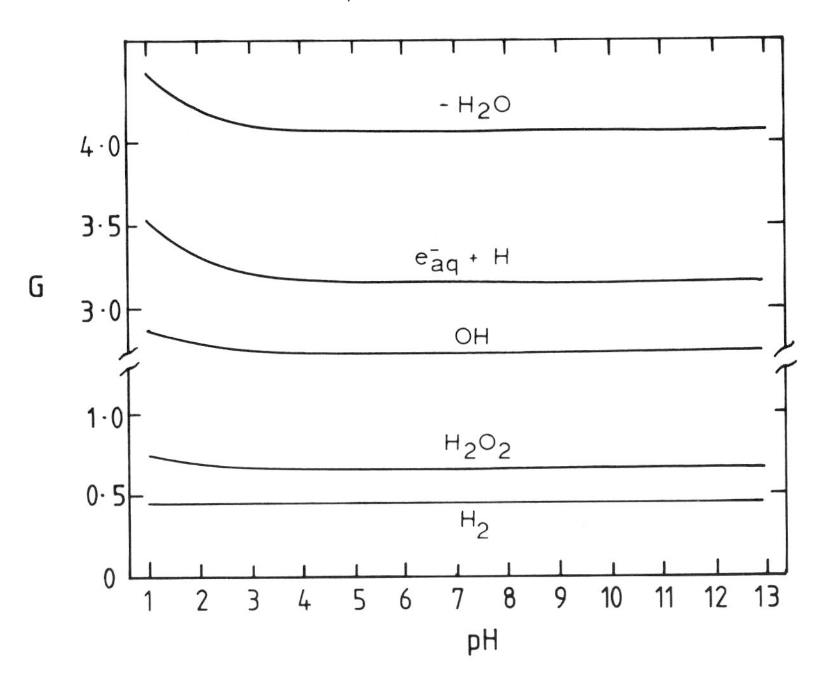


**Figure 10-7.** Dependence of the primary radical and molecular yields from the $\gamma$-radiolysis of water on pH. (Reproduced with permission.[3])

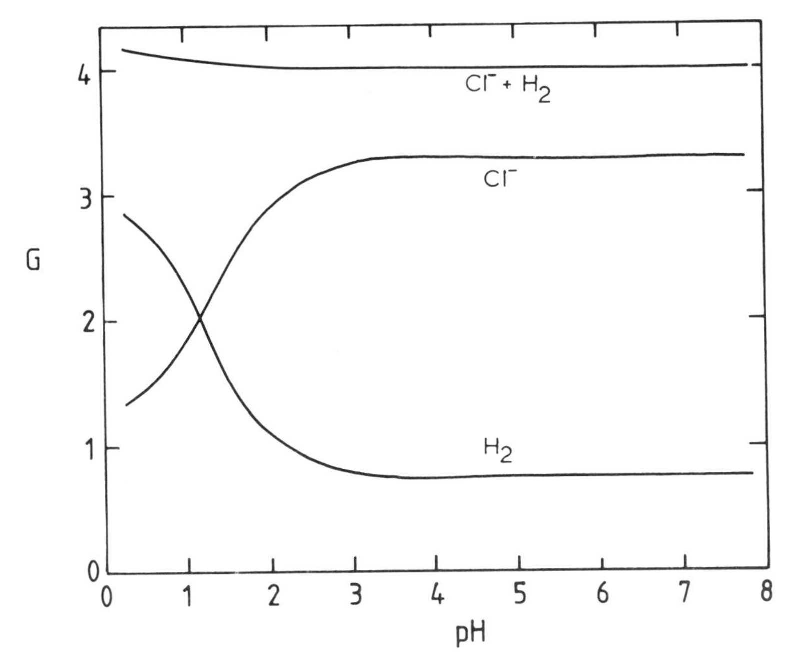


**Figure 10-8.** Effect of pH on the yields of $Cl^-$ and $H_2$ formed on irradiation of $10^{-1}$ *M* $ClCH_2CO_2H$ solution. (Reproduced with permission.[3])

# INITIAL YIELDS

The initial yields of $e_{aq}^-$ and OH are formed in $10^{-12}$ s and therefore are difficult to measure. Before the advent of pulse radiolysis equipment capable of subnanosecond time resolution, initial yields were calculated with the spur diffusion model. This was achieved by choosing a set of parameters that, when substituted into the diffusion equation, gave a good match between the experimental and calculated primary yields. The set of parameters that was found to give excellent results is listed in Table 10-6.

**Table 10-6.** Spur Parameters Used to Fit Primary Yields[19]

| Species | Initial *G*-Value | Mean Initial Distribution Radius/nm |
|---|---|---|
| $e_{aq}^-$ | 4.78 | 2.3 |
| H | 0.62 | 0.75 |
| $H_2$ | 0.15 | — |
| OH | 5.6 | 0.75 |
| $H_3O^+$ | 4.78 | 0.75 |

The most important feature of these parameters is that the radius of the initial (Gaussian) distribution of $e_{aq}^-$ is required to be considerably larger than those of the other species. The initial species also include atomic and molecular hydrogen formed from excited water molecules (Scheme 10-1).

Subsequent direct measurements of the yields of $e_{aq}^-$ and OH at 100 ps by pulse radiolysis at low LET show that $G(e_{aq}^-) = 4.7 \pm 0.2$ and $G(\text{OH}) = 5.9 \pm 0.2$, in good agreement with the diffusion model calculations. However, as Figure 10-9 shows, the calculated and experimentally observed time profiles of $G(e_{aq}^-)$ differ by about one decade of time. The discrepancy between the diffusion model calculation and experiment appears to be even greater at high LET (Figure 10.6).

A useful empirical equation, which allows the yield of scavenged $e_{aq}^-$ to be calculated at high solute concentrations is[20]

$$G(\text{P}) = 2.55 + \frac{2.23(k[\text{S}_1]/\lambda)^{1/2}}{1 + (k[\text{S}_1]/\lambda)^{1/2}} \qquad (10\text{-}27)$$

Here $G(\text{P}_1)$ is the yield of product from Reaction (10-10), $k$ is the rate constant for Reaction (10-10), $\lambda$ is a constant estimated to be $8 \times 10^8\ \text{s}^{-1}$, and the initial and primary yields of $e_{aq}^-$ are taken as 4.78 and 2.55, respectively. The initial yield was chosen to agree with Schwarz's value.[19] The fit of Eq. (10-27) to the data for $N_2O$ is shown in Figure 10-1, from

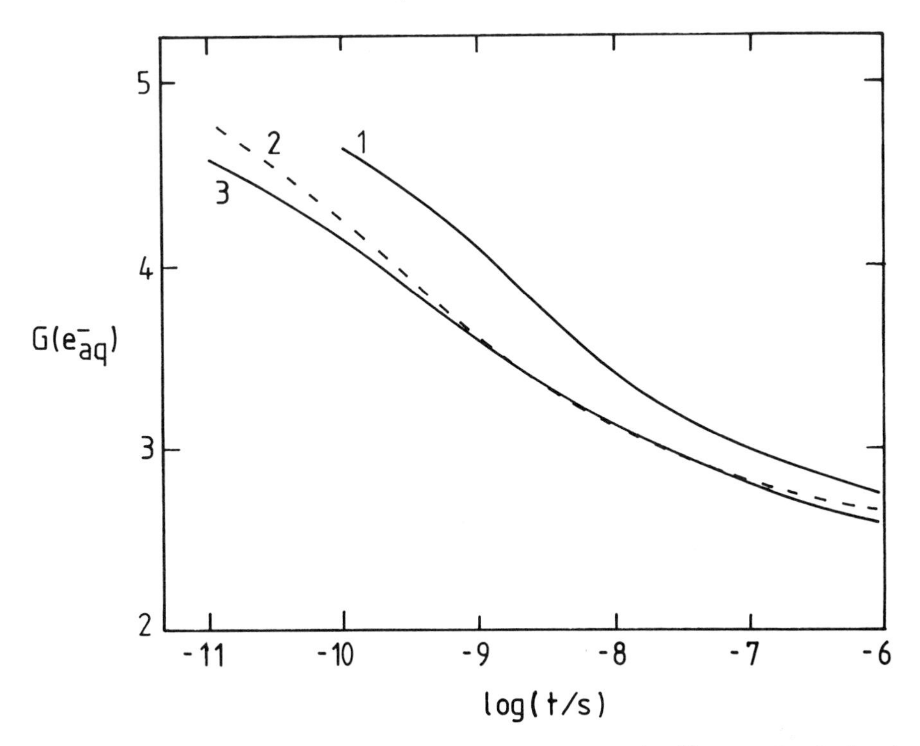


**Figure 10-9.** Time profile of $e_{aq}^-$. Curve 1, pulse radiolysis data[12]; curve 2, calculated using Schwarz's model[19]; curve 3, calculated using Eq. (10-28).[20] (Reproduced with permission.[39])

which it is seen that Eq. (10-27) may be used to calculate $G(P_1)$ for values of $k_{10}[S_1]$ up to $2 \times 10^9\ s^{-1}$ with an accuracy sufficient for most chemical applications. However, Eq. (10-27) will only apply to scavengers that form products that are unreactive towards OH. Because of its wider spatial distribution in the spur (see Table 10-6), $e_{aq}^-$ is scavenged more easily than OH in the spur, so that generally higher concentrations of reactive solutes are required to cause an increase in the scavenged yield of OH.

The Laplace transform of Eq. (10-27) is Eq. (10-28), which gives the time profile of $e_{aq}^-$. This compares very well with the profile derived from the diffusion model, as Figure 10-9 shows.

$$G(e_{aq}^-) = 2.55 + 2.23\, e^{\lambda t}\, \mathrm{erfc}(\lambda t) \qquad (10\text{-}28)$$

At very high solute concentrations (ca. 1 $M$) there is evidence to indicate that some electrons can be scavenged in their presolvation state. For example, it has been shown that $G(P_1) \simeq 5.4$ at $3 \times 10^{-11}$ s when $S_1$ is $Cd^{2+}$ or cysteamine,[6,12] which implies that $G \simeq 0.8$ is the yield of electrons that do not become solvated but instead undergo geminate recombination with $H_2O^+$ in more dilute solution.

# PROPERTIES OF THE PRIMARY RADICALS

## *The Hydrated Electron*

The main properties of $e_{aq}^-$ are summarized in Table 10-7, and its absorption spectrum is given in Figure 10-10. Because of its intense absorption band, which is well removed from the spectra of most other chemical species, it is a simple matter to follow the reaction of $e_{aq}^-$ by using pulse radiolysis combined with kinetic spectroscopy.[21] Several hundreds of its reaction rate constants have been measured in this way.[22]

**Table 10-7.** Properties of the Hydrated Electron

|  |  |
|---|---|
| Radius of charge distribution | 0.25–0.3 nm |
| Diffusion constant | $4.9 \times 10^{-5}$ cm$^2$ s$^{-1}$ |
| Wavelength of maximum absorption | 715 nm |
| Extinction coefficient at 715 nm | $1.85 \times 10^4$ $M^{-1}$ cm$^{-1}$ |
| Standard reduction potential | −2.9 V |
| Half-life in neutral water | $2.1 \times 10^{-4}$ s |

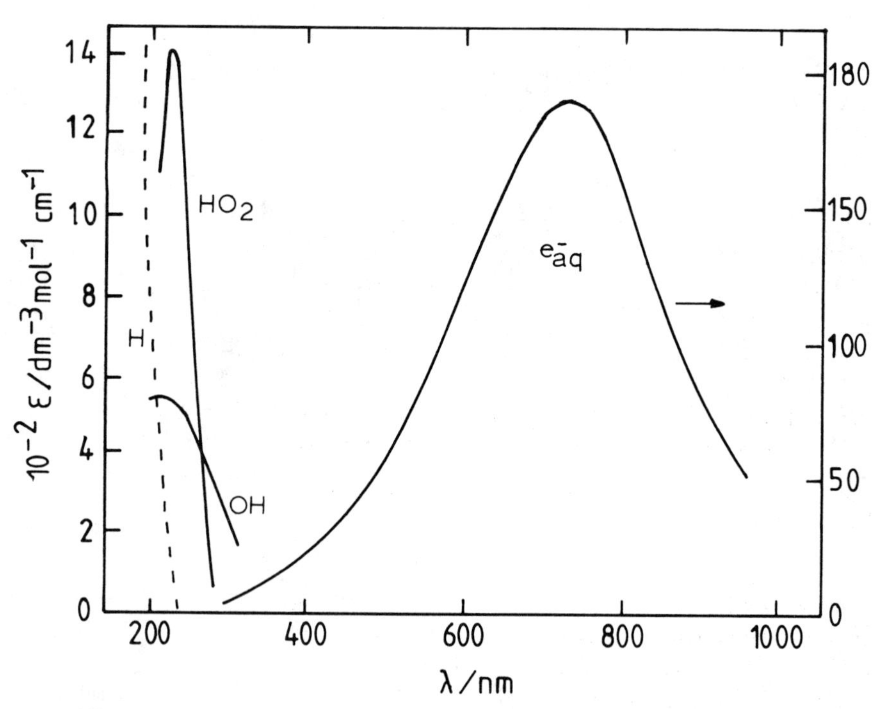


**Figure 10-10.** Absorption spectra of $e_{aq}^-$, H, OH, and $HO_2$. (Adapted from Fig. 7.3 of Spinks and Woods.[1])

The hydrated electron may be visualized as a localized electron surrounded by oriented water molecules, although the precise details of its structure in liquid water are not yet settled. It is a powerful reducing agent, and its reactions are one-electron transfer processes that can be written as

$$e_{aq}^{-} + S^{n} \rightarrow S^{n-1} \qquad (10\text{-}29)$$

where $n$ is the charge on the solute S. Rate constants for Reaction (10-29) range from $16\ M^{-1}\ s^{-1}$ ($S = H_2O$) up to the diffusion-controlled limit, although the measured activation energy is invariably small (6–30 kJ $mol^{-1}$), which suggests that the entropy of activation is the dominant kinetic parameter. This can be understood in terms of the availability of a suitable vacant orbital on S for the electron to enter. Molecules such as water, simple alcohols, ethers, and amines have no low-lying vacant orbitals, and this explains why solvated electrons can be observed in these liquids. Organic molecules with low-lying vacant orbitals—for example, most aromatics, halides, carbonyl compounds, thiols, disulfides, and nitro compounds—react rapidly, $e_{aq}^{-}$ acting as a nucleophile. Thus, its reactivity is greatly enhanced by electron-withdrawing substituents adjacent to double bonds or attached to aromatic rings. Although the first step in the reaction is electron addition, in some instances, notably with halides, bond breakage occurs rapidly, and the reaction can be considered as a dissociative electron capture process,

$$e_{aq}^{-} + RX \rightarrow (RX)^{-} \rightarrow R\cdot + X^{-} \qquad (10\text{-}30)$$

Some features of the reactivity of $e_{aq}^{-}$ are illustrated in Table 10-8.

**Table 10-8.** Rate Constants for Some Reactions of the Hydrated Electron[22]

| Solute | $10^{-10}\ k/\ M^{-1}\ s^{-1}$ | Solute | $10^{-10}\ k/M^{-1}\ s^{-1}$ |
|---|---|---|---|
| | *Inorganic* | | *Organic* |
| $O_2$ | 1.9 | $C_6H_6$ | $1.2 \times 10^{-3}$ |
| $H_3O^+$ | 2.3 | $C_6H_5Cl$ | $5 \times 10^{-2}$ |
| $NH_4^+$ | $< 2 \times 10^{-4}$ | $C_6H_5I$ | 1.2 |
| $Ag^+$ | 3.6 | $CH_2{=}CH_2$ | $<2.5 \times 10^{-4}$ |
| $Cd^{2+}$ | 5.0 | $CH_2{=}CCl_2$ | 2.3 |
| $In^{3+}$ | 5.6 | $CH_2{=}CHCONH_2$ | 1.8 |
| $Fe(CN)_6^{3-}$ | 0.3 | $CH_4$ | $<10^{-3}$ |
| $CrO_4^{2-}$ | 1.8 | $CH_3I$ | 1.7 |
| $NO_3^-$ | 1.0 | $CH_3OH$ | $<10^{-6}$ |
| $N_2O$ | 0.87 | | |
| $H_2O$ | $1.6 \times 10^{-9}$ | | |

As might be expected, $e_{aq}^{-}$ reacts rapidly with many species having reduction potentials more positive than −2.9 V. In some cases, for example, $CrO_4^{2-}$, the rate constant is higher than predicted by the Debye–Smoluchowski equation for diffusion-controlled reactions, and it is suggested[4] that the reaction radius is larger than normal because the electron tunnels from its solvent trap to the acceptor. Tunneling between solvent traps may also explain the mobility of $e_{aq}^-$ because it is much higher than expected for a singly charged anion of radius 0.3 nm.

## *The Hydrogen Atom*

The hydrogen atom is not an important species in the radiation chemistry of neutral and alkaline solution, but it is the major reducing radical in acidic media. It is a less powerful reductant ($E° = -2.3$ V) than $e_{aq}^-$ and can be thought of as a weak acid with a p$K$ of 9.6, obtained from the rate constants for Reactions (10-24) and (10-31):

$$e_{aq}^- + H_2O \rightarrow H + OH^- \qquad (10\text{-}31)$$

It readily reduces cations having more positive reduction potentials than itself but at a slower rate than $e_{aq}^-$. In some cases it reacts effectively as an oxidant via the formation of a hydride; for example,

$$H + Fe^{2+} \rightarrow Fe^{3+}\cdot H^- \xrightarrow{H^+} Fe^{3+} + H_2 \qquad (10\text{-}32)$$

With organic compounds it reacts more like OH (see below) than $e_{aq}^-$, adding to centers of unsaturation and abstracting H from saturated molecules. Many of its rate constants have been measured by using competition methods.[23]

## *The Hydroxyl Radical*

The hydroxyl radical has a standard reduction potential of 2.72 V in acidic solution[24] and is therefore a powerful oxidant. In neutral solution, where the free energy of neutralization of $OH^-$ by $H^+$ is not available, the reduction potential is calculated to be 1.89 V.[24]

The hydroxyl radical readily oxidizes inorganic ions. The reaction is often represented as a simple electron transfer process

$$OH + S^n \rightarrow S^{n+1} + OH^- \qquad (10\text{-}33)$$

but, although Reaction (10-33) describes the stoichiometry of the reaction, in most cases it is unlikely that the reaction proceeds by simple outer-sphere electron transfer because of the large solvent reorganization energy involved in forming the hydrated hydroxide ion. In the oxidation of halide ions, for example, intermediate adducts can be observed by pulse radiolysis,

$$OH + X^- \rightarrow HOX^- \qquad (10\text{-}34)$$

and it has been suggested[25] that this is the reaction path for most inorganic anions.

There are several examples of OH reacting with anions at the diffusion-controlled rate,[26] but rate constants for oxidation of aquated metal cations seem to have an upper limit of approximately $3 \times 10^8\ M^{-1}\ s^{-1}$—that is, a 100-fold lower than the diffusion-controlled rate. A suggested reason for this is that oxidation of these cations involves abstraction of H from a coordinated water molecule followed by electron transfer from the metal to the oxidized ligand.[27]

$$OH + M^{n+}\cdot H_2O \rightarrow M^{n+}\cdot OH \rightarrow M^{(n+1)+}\cdot H_2O \qquad (10\text{-}35)$$

Pulse radiolysis experiments confirm that $M^{n+}\cdot OH$ is indeed the initial product in several cases ($M^{n+} = Tl^+$, $Ag^+$, $Cu^{2+}$, $Sn^{2+}$, $Fe^{2+}$, $Mn^{2+}$). There is no correlation between the measured rate constants and the rates of exchange of coordinated water molecules, which rules out formation of $M^{n+}\cdot OH$ by ligand substitution as the general mechanism.

In strongly alkaline solution, OH is rapidly converted to its anionic form $O^-$, with $k_{36} = 1.2 \times 10^{10}\ M^{-1}\ s^{-1}$ and $k_{-36} = 9.3 \times 10^7\ s^{-1}$.

$$OH + OH^- \rightleftharpoons O^- + H_2O \qquad (10\text{-}36)$$

$O^-$ reacts more slowly than OH with several inorganic anions. For example, $Br^-$, $CO_3{}^{2-}$, and $Fe(CN)_6{}^{4-}$ react immeasurably slowly with $O^-$, although they all react rapidly with OH. On the other hand, unlike OH, $O^-$ reacts rapidly with $O_2$ to form the ozonide ion, which, because of its characteristic absorption spectrum ($\lambda_{max} = 430$ nm, $\varepsilon_{max} = 2000\ M^{-1}\ s^{-1}$), provides a good way of measuring $O^-$ in alkaline solution ($k_{37} = 3.6 \times 10^9\ M^{-1}\ s^{-1}$ and $K_{37} = 6 \times 10^5\ M^{-1}$):

$$O^- + O_2 \rightleftharpoons O_3{}^- \qquad (10\text{-}37)$$

The hydroxyl radical behaves as an electrophile in its reactions with organic molecules, whereas $O^-$ is a nucleophile. Thus, OH readily adds to unsaturated bonds, but $O^-$ does not. Both forms of the radical abstract H from C—H bonds. This can result in the formation of different products when the pH is raised, so that $O^-$, rather than OH, is the reactant. For example, if an aromatic molecule carries an aliphatic side chain, OH adds preferentially to the aromatic ring, whereas $O^-$ abstracts H from the side chain. Although OH and H undergo similar types of reaction with organic molecules, OH is less selective, and more reactive, than H in abstraction reactions because formation of the H—OH bond is 57 kJ mol$^{-1}$ more exothermic than the formation of H—H. The reactivities of H and OH with some organic molecules are compared in Table 10-9.

**Table 10-9.** Rate Constants for Selected Reactions of OH and H[26,23]

| Solute | Reaction Type | $10^{-7}\ k/M^{-1}\ s^{-1}$ | |
|---|---|---|---|
| | | OH | H |
| $CH_2{=}CHCONH_2$ | Addition | 450 | 1800 |
| $C_6H_6$ | Addition | 530 | 53 |
| $C_6H_5NO_2$ | Addition | 340 | 170 |
| $C_2H_5OH$ | Abstraction | 180 | 1.7 |
| $CH_3OH$ | Abstraction | 84 | 0.16 |

The hydrogen atom and the hydroxyl radical both absorb only weakly in the UV, as shown in Figure 10-10, so that rate constants for their reactions must be determined by measuring the rate of product formation or by using the competition method, where two solutes compete for the radicals and one of the products can be determined. Ferricyanide ion is a suitable solute for measuring H-atom rate constants because Reaction (10-38) can be followed from the decrease in absorbance at 420 nm due to $Fe(CN)_6^{3-}$ ($\varepsilon_{420} = 1040\ M^{-1}\ cm^{-1}$). When a second competing solute is present, $G(-Fe(CN)_6^{3-})$ is given by Eq. (10-40):

$$H + Fe(CN)_6^{3-} \rightarrow Fe(CN)_6^{4-} + H^+ \quad (10\text{-}38)$$

$$H + S \rightarrow P \quad (10\text{-}39)$$

$$\frac{1}{G(-Fe(CN)_6^{3-})} = \frac{1}{G(H)}\left(1 + \frac{k_{39}[S]}{k_{38}[Fe(CN)_6^{3-}]}\right) \quad (10\text{-}40)$$

Rate constants for reactions of OH are measured in the same manner; suitable solutes for the competition method include $Fe(CN)_6^{4-}$, $CO_3^{2-}$, and $HCO_3^-$. Thiocyanate ion is often used, but this can lead to erroneous results because the measured product $(SCN)_2^-$ is a secondary one, viz,

$$OH + SCN^- \rightleftharpoons HOSCN^- \quad (10\text{-}41)$$

$$HOSCN^- \rightarrow SCN + OH^- \quad (10\text{-}42)$$

$$SCN + SCN^- \rightleftharpoons (SCN)_2^- \quad (10\text{-}43)$$

and the possibility exists that $HOSCN^-$ or SCN may also react with the solute that competes with $SCN^-$ for OH. Many of the details of Reactions (10-41)–(10-43) have been worked out[28] by using pulse radiolysis methods. If the measured product is stable, relative rate constants may also be obtained by steady-state methods, but again it is important to establish that no interfering secondary reactions take place.

## *The Perhydroxyl Radical*

The perhydroxyl radical $HO_2$ is not a significant primary radical at low LET, but it is an important secondary radical in oxygenated solution,[29] where it is formed from other primary radicals:

$$e_{aq}^- + O_2 \rightarrow O_2^- \quad (10\text{-}18)$$

$$H + O_2 \rightarrow HO_2 \quad (10\text{-}17)$$

$$HO_2 \rightleftharpoons O_2^- + H^+ \quad (10\text{-}44)$$

The p$K$ of $HO_2$ is 4.7. The standard reduction potentials are −0.05 V for $HO_2$ (acidic solution) and −0.33 V for $O_2^-$, referred to oxygen gas at atmospheric pressure. Thus, $O_2^-$ is a stronger reducing agent than $HO_2$.

Conversely, $HO_2$ is a stronger oxidant than $O_2^-$; $E°$ for the couple ($HO_2$, $H^+/H_2O_2$) = 1.45 V and $E°$ for the couple ($O_2^-$, $H^+/HO_2^-$) = 1.03 V.

$HO_2$ and $O_2^-$ have characteristic absorption spectra with $\varepsilon_{max}$ = 225 nm and 245 nm, respectively, which are intense enough ($\varepsilon_{225} = 1400\ M^{-1}\ cm^{-1}$ and $\varepsilon_{245} = 2300\ M^{-1}\ cm^{-1}$) to allow the reactions of the radicals to be followed by direct observation. Both species are relatively unreactive toward organic molecules, abstracting only weakly bonded hydrogen atoms (eg, in ascorbic acid, cysteine, and hydroquinone), but they react readily with metal ions to often form complexes; for example,

$$O_2^- + Mn^{2+} \rightarrow MnO_2^+ \tag{10-45}$$

In the absence of metal ions $HO_2$ and $O_2^-$ decay by second-order kinetics. Careful work[29] has shown that the disproportionation reactions are

$$HO_2 + HO_2 \rightarrow H_2O_2 + O_2 \tag{10-15}$$

$$HO_2 + O_2^- \rightarrow HO_2^- + O_2 \tag{10-16}$$

with $k_{14} = 8.6 \times 10^5\ M^{-1}\ s^{-1}$ and $k_{15} = 1.02 \times 10^8\ M^{-1}\ s^{-1}$. The third possible reaction,

$$O_2^- + O_2^- \xrightarrow{H_2O} HO_2^- + O_2 + OH^- \tag{10-46}$$

does not occur at all. However, it is catalyzed by proteins containing copper, manganese, or iron, the superoxide dismutases, and by some simple metal complexes such as $Cu_{aq}^{2+}$ and $Fe^{II}EDTA$.

## GENERATION OF SECONDARY RADICALS

The radiolysis of water produces approximately equal numbers of powerful oxidizing and reducing radicals, but for chemical applications it is usually desirable to have totally oxidizing or totally reducing conditions. This can be achieved by interconversion of the primary radicals, or by converting the primary radicals into a single kind of secondary radical, or by converting the unwanted primary radical into a relatively unreactive secondary radical. Some useful systems that meet these requirements are described below.

### *Oxidizing Conditions*

The most convenient way of obtaining almost totally oxidizing conditions is to saturate the solution with $N_2O$, which converts $e_{aq}^-$ to OH.

$$e_{aq}^- + N_2O \rightarrow N_2 + O^- \xrightarrow{H_2O} OH + OH^- \tag{10-47}$$

Under these conditions the concentration of $N_2O$ is $2.5 \times 10^{-2}\ M$ and $k_{47}[N_2O]$ is $2.5 \times 10^8\ s^{-1}$; steady-state product measurements show that

$G(N_2) = 3.2$. However, the available yield of OH under these conditions has been shown[30] to fit an empirical equation analogous to Eq. (10-27), viz,

$$G(\text{OH}) = 5.2 + \frac{3.0(k[\text{S}]/\lambda)^{1/2}}{1 + (k[\text{S}]/\lambda)^{1/2}} \tag{10-48}$$

where $\lambda = 4.7 \times 10^8\ \text{s}^{-1}$, and $k$ is the rate constant for Reaction (10-11). Thus, the OH that replaces $e_{aq}^-$ in the spur through Reaction (10-47) also participates in spur reactions. Values of $G$(OH) calculated from Eq. (10-48) are shown in Figure 10-11. Another important quantity is the sum of $G(\text{OH}) + G(\text{H})$, that is, $G(\text{OH}) + 0.6$, which is a measure of $G$(reducing radicals) formed when appropriate organic solutes—for example, 2-propanol, formate ion, and others—are used to generate reducing conditions (see below). In $N_2O$-saturated 0.1 $M$ 2-propanol solution, for example, where $k[\text{S}] = 2 \times 10^8\ \text{s}^{-1}$, the yield of $(CH_3)_2\dot{C}OH$ is 7.0 radicals/100 eV.

Nitrous oxide reacts only slowly with the hydrogen atom ($k = 2.3 \times 10^6\ M^{-1}\ \text{s}^{-1}$),[23] so that approximately 10% of the radicals available in $N_2O$-saturated neutral solution are H atoms and 90% are OH radicals. At pH > 11, where Reaction (10-24) becomes increasingly important, $G$(OH) can

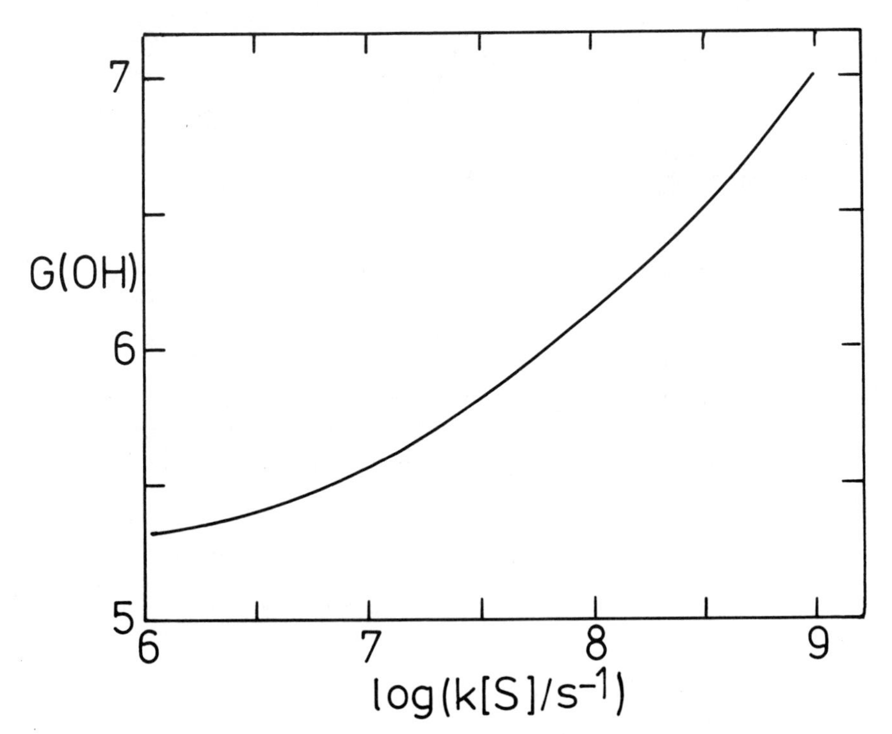


**Figure 10-11.** Dependence of the yield of OH on the scavenging power, $k[\text{S}]$, of a solute in neutral aqueous solution saturated with $N_2O$.

increase by up to 0.6 (ie, $G_H$), whereas at pH < 3, where Reaction (10-3) competes with (10-47), the fraction of oxidizing radicals decreases.

It should be remembered that the product of Reaction (10-47) is $O^-$ rather than OH.[31] Thus, if a solute is present that reacts with $O^-$ in competition with the protonation Reaction (10-36), that is, $k[S] > 9 \times 10^7\ s^{-1}$, then because $O^-$ can react differently from OH, different products may be obtained in $N_2O$-saturated solutions containing hydroxyl radical scavengers in high and low concentrations.

It was mentioned above that OH is rather unselective in its reactions with organic molecules. More selective oxidizing radicals can be produced by converting OH into another inorganic radical: for example, $Br_2^-$, $I_2^-$, $(SCN)_2^-$, $CO_3^-$, $N_3$, and so on. This strategy is particularly useful for studying redox changes in metalloproteins and organometallic complexes because the secondary inorganic radicals are more likely to react at the metal center, whereas OH will also attack the organic moiety and introduce there, by H-atom abstraction, a reducing radical center.

## *Reducing Conditions*

Although it is possible to convert OH to $e_{aq}^-$ through Reactions (10-13) and (10-24), this can only be achieved usefully under the rather severe conditions of high pH and hydrogen at a pressure of 100 atm.

$$OH + H_2 \rightarrow H + H_2O \qquad (10\text{-}13)$$

$$H + OH^- \rightarrow e_{aq}^- \qquad (10\text{-}24)$$

A much more convenient method of obtaining totally reducing conditions is to convert all the primary radicals to the same secondary reducing radical by adding an organic solute to $N_2O$-saturated water. For example, in the presence of 2-propanol Reactions (10-49) and (10-50) take place to produce $(CH_3)_2\dot{C}OH$:

$$OH + (CH_3)_2CHOH \rightarrow (CH_3)_2\dot{C}OH + H_2O \qquad (10\text{-}49)$$

$$H + (CH_3)_2CHOH \rightarrow (CH_3)_2\dot{C}OH + H_2 \qquad (10\text{-}50)$$

This technique provides a simple way of generating organic radicals in aqueous solution and has been exploited widely in the study of their chemistry. When the organic radical is formed by H-atom abstraction, the method can be used over the whole pH range, but one should bear in mind that the secondary radical may contain a dissociable proton so that the ionized form of the radical will be formed when pH > p*K*. Generally, the basic form of the radical is more strongly reducing than the acidic form.

Simple alcohols are often used as the source of the secondary radicals, and it is the hydrogen atom on the $\alpha$-carbon atom that is most readily abstracted by OH, as is indicated above for Reactions (10-49) and (10-50).

Nevertheless, H is also abstracted from other positions so that a mixture of radicals is produced where only one is desired. The distributions of radicals formed from a number of simple alcohols are given in Table 10-10. The properties of some of the more commonly used organic radicals are summarized in Table 10-11.

**Table 10-10.** Percentage Abstraction by OH Radicals from Various Positions in Alcohols[a]

| Compound | $\alpha$ C—H | Other C—H | OH |
|---|---|---|---|
| $CH_3OH$ | 93.0 | — | 7.0 |
| $CH_3CH_2OH$ | 84.3 | 13.2 | 2.5 |
| $CH_3CH_2CH_2OH$ | 53.4 | 46.0 | <0.5 |
| $(CH_3)_2CHOH$ | 85.5 | 13.3 | 1.2 |
| $CH_3CH_2CH_2CH_2OH$ | 41.0 | 58.5 | <0.5 |
| $(CH_3)_3COH$ | — | 95.7 | 4.3 |
| $(CH_2OH)_2$ | 100 | — | <0.1 |
| $CH_3CH(OH)CH_2OH$ | 79.2 | 20.7 | <0.1 |
| $CH_3CH(OH)CH(OH)CH_3$ | 71.0 | 29.0 | <0.1 |

[a] Asmus, K.-D.; Möckel, H.; Henglein, A. *J. Phys. Chem.* **1973**, *77*, 1218. Reproduced with permission.

**Table 10-11.** Acid Dissociation Constants and Standard Reduction Potentials of Some Organic Radicals

| Radical | $pK_a$[a] | $E°/V$[b] $\dot{R}OH$ | $E°/V$[b] $\dot{R}O^-$ |
|---|---|---|---|
| $\dot{C}H_2OH$ | 10.7 | −1.0 | −1.6 |
| $CH_3\dot{C}HOH$ | 11.5 | −1.5 | −1.8 |
| $(CH_3)_2\dot{C}OH$ | 12.2 | −1.5 | −2.2 |
| $\dot{C}O_2H$ | 1.4 | — | −2.0 |

[a] Toffel, P.; Henglein, A. *Ber. Bunsenges. Phys. Chem.* **1976**, *80*, 525. Reproduced with permission.
[b] Butler, J.; Henglein, A. *Radiat. Phys. Chem.* **1980**, *15*, 603. Reproduced with permission.

The low p$K$ of $\dot{C}O_2H$ means that $CO_2^-$ can be used as an electron transfer reducing agent under conditions where $e_{aq}^-$ would react predominantly with $H_3O^+$. $CO_2^-$ reacts rapidly with oxygen ($k_{51} = 4.2 \times 10^9\ M^{-1}\ s^{-1}$), so that radiolysis of oxygenated formate solutions provides a good method for generating $O_2^-$ and $HO_2$.

$$CO_2^- + O_2 \rightarrow O_2^- + CO_2 \qquad (10\text{-}51)$$

As Table 10-11 shows, the organic radicals are less powerful reductants than $e_{aq}^-$. This property can be exploited in pulse radiolysis studies by saturating the solution within an inert gas instead of $N_2O$, so that only OH and H are converted to the less reactive organic radical and $e_{aq}^-$ remains. Tertiary butanol is a particularly good solute for this purpose because the radical $\cdot CH_2(CH_3)_2COH$ is rather unreactive. It has the additional advantage of absorbing only weakly in the UV, so it interferes to a minimal extent in the measurement of the absorption spectra of other products.

# APPLICATIONS IN GENERAL CHEMISTRY

Radiation chemistry techniques can be applied very widely as a way of generating and observing free radicals and unusual oxidation states of metals, measuring very fast reaction rates, and elucidating the role of short-lived intermediates in reaction mechanisms.

## *Inorganic Free Radicals*

These radicals are readily formed by oxidation of inorganic anions with the hydroxyl radical: for example,

$$OH + CO_3^{2-} \rightarrow CO_3^- + OH^- \qquad (10\text{-}52)$$

or by reduction of peroxo-compounds with $e_{aq}^-$:

$$e_{aq}^- + S_2O_8^{2-} \rightarrow SO_4^- + SO_4^{2-} \qquad (10\text{-}53)$$

$SO_4^-$ is a very powerful oxidizing agent and will oxidize $OH^-$ to OH; it will also oxidize $Cl^-$ in neutral and alkaline solutions. The phosphate radical $PO_4^{2-}$ and its protonated forms $H_2PO_4$ (p$K$ = 5.7) and $HPO_4^-$ (p$K$ = 8.9) are also good oxidants, and $H_2PO_4$ will oxidize $OH^-$.

The halide and pseudohalide radicals $X_2^-$ (X = Cl, Br, I, SCN, etc) are readily formed by oxidation of the halide, as indicated above. Their reactivities are generally in the order $Cl_2^- > Br_2^- > (SCN)_2^- > I_2^-$, and $Cl_2^-$ can abstract hydrogen from aliphatic compounds. The rate constants are usually less than, but show a similar pattern to, those for the H atom, so that C—H bond strength is a controlling factor. Hydrogen-atom abstraction reactions by other $X_2^-$ radicals are barely detectable in pulse radiolysis experiments. Also, $Cl_2^-$ reacts with unsaturated organic compounds to form Cl adducts, and this may be a general reaction of $X_2^-$, but the main reaction of $X_2^-$ is one-electron oxidation, which occurs with many aromatic and heterocyclic compounds and certain inorganic ions. However, $X_2^-$ radicals are weaker oxidants than $SO_4^-$ and $H_2PO_4$, although $Cl_2^-$ is a stronger oxidant than $HPO_4^-$.

Other inorganic radicals can be prepared from sulfite, selenite, nitrite, nitrate, and so forth, but the reactions of these radicals have not been studied to any great extent. A compilation of the rate constants for the reactions of inorganic radicals is available.[32]

## *Organic Free Radicals*

Organic radicals can readily be prepared by one of the following reactions:

$$OH + RH \rightarrow R\cdot + H_2O \tag{10-54}$$

$$e_{aq}^- + RX \rightarrow R\cdot + X^- \tag{10-55}$$

$$OH + \phi H \rightarrow (HO\phi)\cdot \tag{10-56}$$

$$OH + HR{=}R'H \rightarrow H(OH)R{-}R'H \tag{10-57}$$

where R is an alkyl or substituted alkyl group, X is halogen, and $\phi$ is an aryl group. There are several hundred papers dealing with organic free radicals in solution. Their reactions are described in a very comprehensive review,[33] and rate constants for reactions of aliphatic carbon-centered radicals have been compiled.[34]

The radicals studied most extensively in aqueous radiation chemistry are the $\alpha$-hydroxyalkyl radicals (see Table 10-10), which react with many organic and inorganic compounds by reduction, although abstraction and addition reactions can also take place. There are several examples known of addition reactions with metal ions to form metal–carbon bonds. Alkyl radicals that do not contain an —OH or —$NH_2$ group on the carbon atom carrying the unpaired electron are generally weaker reductants and less reactive than the $\alpha$-substituted ones. Alkoxy radicals, and alkyl radicals substituted with a carbonyl group, are oxidizing radicals. The latter are formed by the reaction sequence

$$OH + HOCH_2CH_2OH \rightarrow HOCH_2\dot{C}HOH \tag{10-58}$$

$$HOCH_2CHOH \rightarrow H_2O + \dot{C}H_2CHO \tag{10-59}$$

and react by one-electron transfer. An analogous reaction sequence involving addition of OH to phenol followed by acid- or base-catalyzed elimination of water from the dihydroxy species produces the phenoxyl radical. The rate of water elimination depends on the position of the added OH; the *p*-adduct undergoes the fastest elimination. This radical can also be generated by oxidation of the phenoxide ion in an alkaline solution by one-electron oxidants such as $\dot{C}H_2CHO$, $Br_2^-$, and others.

Organic sulfur compounds such as thiols and sulfides are interesting, because they can be oxidized to form dimeric radical anions and cations respectively, viz,

$$OH + RSH \rightarrow R\dot{S} + H_2O \tag{10-60}$$

$$RS\cdot + RS^- \rightleftharpoons RSSR^{-}\cdot \tag{10-61}$$

$$OH + R_2S \rightarrow R_2S^{+}\cdot + OH^- \tag{10-62}$$

$$R_2S^+ + R_2S \rightleftharpoons R_2SSR_2^{+}\cdot \tag{10-63}$$

where Reaction (10-61) comprises a complex sequence of steps. $RSSR^{-}\cdot$ can also be formed by one-electron reduction of RSSR, for example, by the hydrated electron. The sulfur–sulfur bond in $R_2SSR_2^+$ is a three-electron bond[35] formulated as

$$\begin{matrix} R & & + & & R \\ & \diagdown & & \diagup & \\ & S & \therefore & S & \\ & \diagup & & \diagdown & \\ R & & & & R \end{matrix}$$

with two electrons in the $\sigma$-bonding orbital and one in the $\sigma^*$-antibonding orbital. Analogous species are also formed in the oxidation of organic sulfides by $X_2^-$ radicals.

## *Reactions of Radicals with Metal Ions and Their Complexes*

Primary and secondary radicals generated by radiolysis of aqueous solution react with metal complexes by the following mechanisms: one-electron oxidation and reduction, addition, and atom transfer. Reaction may occur at the metal center or at the ligand and be followed by intramolecular electron transfer, ligand exchange, ligand degradation, and loss. Examples of all of these processes are known and have been reviewed.[36,37]

The hydrated electron reacts with simple aquo cations to produce hyper-reduced oxidation states that are not accessible by other methods: for example,

$$e_{aq}^- + Ni^{2+} \rightarrow Ni^+ \tag{10-64}$$

The monovalent ions $Cd^+$, $Co^+$, $Ni^+$, and $Zn^+$ have been studied in depth. They are strong reductants, and the general order of reactivity with inorganic oxidants is $Zn^+ \simeq Co^+ \geqslant Cd^+ > Ni^+$. These reactions occur by one-electron transfer, but other types of reactions are known: for example,

$$M^+ + N_2O \rightarrow MO^+ + N_2 \tag{10-65}$$

$$M^+ + O_2 \rightarrow MO_2^+ \tag{10-66}$$

$$M^+ + \text{olefin} \rightarrow (M - \text{olefin})^+ \tag{10-67}$$

$$M^+ + R\cdot \rightarrow MR^+ \tag{10-68}$$

Reaction (10-68) is a general example of the formation of a metal–carbon bond between the reduced metal ion and a carbon-centered free radical. Such reactions occur when there is an accessible higher oxidation state of the metal; for example,

$$\dot{C}H_2CH_2OH + Cu^{2+} \rightarrow (Cu^{III} - CH_2CH_2OH)^{2+} \tag{10-69}$$

A compilation of rate constants for reactions of metal ions in unusual oxidation states is available.[38]

Reactions of free radicals with transition-metal complexes, including bioinorganic molecules have been extensively investigated. Radicals that selectively react by electron transfer usually attack the metal center directly, and both outer-sphere and inner-sphere mechanisms have been identified. When the ligand contains abstractable hydrogen, the hydroxyl radical also reacts there to produce a ligand radical or, in the case of metalloproteins, a radical center on the protein moiety. The ligand radical can then transfer its unpaired electron to the metal center with the result that the ligand is oxidized by two equivalents and the metal center is reduced by one. Other processes that can occur with transition metal complexes include aquation, in which reduction of the metal center results in ligand exchange with the solvent, and expansion of the coordination sphere, although few examples of this last process are known.[39]

## REFERENCES


1. Spinks, J. W. T.; Woods, R. J. "An Introduction to Radiation Chemistry"; Wiley Interscience: New York, 1976.
2. Swallow, A. J. "Radiation Chemistry, An Introduction"; Longman: London 1973.
3. Draganic, I. G.; Draganic, Z. D. "The Radiation Chemistry of Water"; Academic Press: New York, 1971.
4. Hart, E. J.; Anbar, M. "The Hydrated Electron"; Wiley Interscience; New York, 1970.
5. Thomas, J. K. In "Advances in Radiation Chemistry"; Burton, M.; and Magee, J. L., Eds.; Wiley Interscience: New York, 1969, Vol. 1, pp. 103–198.
6. Hunt, J. W. In "Advances in Radiation Chemistry"; Burton, M.; and Magee, J. L., Eds.; Wiley Interscience: New York, 1976, Vol. 5, pp. 185–315.
7. Lampe, F. W.; Field, F. H.; Franklin, J. L. *J. Am. Chem. Soc.* **1957**, *79*, 6132.
8. Hamill, W. H. *J. Phys. Chem.* **1969**, *73*, 1341.
9. Wiesenfeld, J. M.; Ippen, E. P. *Chem. Phys. Lett.* **1980**, *73*, 47.
10. Mozumder, A. *J. Chem. Phys.* **1969**, *50*, 3153.
11. Appleby, A.; Schwarz, H. A. *J. Phys. Chem.* **1969**, *73*, 1937.
12. Jonah, C. D.; Matheson, M. S.; Miller, J. R.; Hart, E. J. *J. Phys. Chem.* **1976**, *80*, 1267.
13. Jonah, C. D.; Miller, J. R. *J. Phys. Chem.* **1977**, *81*, 1974.
14. Burns, W. G.; May, R.; Buxton, G. V.; Wilkinson-Tough, G. S. *J. Chem. Soc., Faraday Trans. 1*, **1981**, *77*, 1543.
15. Kupperman, A. In "Physical Mechanisms in Radiation Biology "; Technical Information Center, Office of Information Services, U.S. Atomic Energy Commission: Washington, D.C., 1974, p. 155.
16. Trumbore, C. N.; Short, D. R.; Fanning, J. E.; Olson, J. H. *J. Phys. Chem.* **1978**, *82*, 2762.
17. Buxton, G. V. *Radiat. Res. Rev.* **1968**, *1*, 209.
18. Benderskii, V. A.; Krivento, A. G.; Rukin, A. N. *High Energy Chem.* **1980**, *14*, 303.
19. Schwarz, H. A. *J. Phys. Chem.* **1969**, *73*, 1928.
20. Balkas, T. I.; Fendler, J. H.; Schuler, R. H. *J. Phys. Chem.* **1970**, *74*, 4497.
21. Matheson, M. S.; Dorfman, L. M. "Pulse Radiolysis"; M.I.T. Press; Cambridge, Mass., 1969.
22. Anbar, M.; Bambenek, M.; Ross, A. B. *Natl. Stand. Ref. Data Ser., Natl. Bur. Stand.* (*U.S.*) **1973**, *43*.
23. Anbar, M.; Farhataziz; Ross, A. B. *Natl. Stand. Ref. Data Ser., Natl. Bur. Stand.* (*U.S.*) **1975**, *51*.
24. Schwatz, H. A.; Dodson, R. W. *J. Phys. Chem.* **1984**, *88*, 3643.
25. Meyerstein, D. *J. Chem. Soc., Faraday Disc.* **1977**, *63*, 203.
26. Farhataziz; Ross, A. B. *Natl. Stand. Ref. Data Ser., Natl. Bur. Stand.* (*U.S.*) **1977**, *59*.

27. Berdnikov, V. M. *Russ. J. Phys. Chem.* **1973**, *47*, 1547.
28. Behar, D.; Bevan, P. L. T.; Scholes, G. *J. Phys. Chem.* **1972**, *76*, 1537.
29. Allen, A. O.; Bielski, B. H. J. In "Superoxide Dismutase"; Oberley, L. W., Ed.; CRC Press: Boca Raton, Flo., 1982, Vol. 1, Chapter 6, pp. 125–141.
30. Schuler, R. H.; Hartzell, A. L.; Behar, B. *J. Phys. Chem.* **1981**, *85*, 192.
31. Buxton, G. V. *J. Chem. Soc., Faraday Trans. 1* **1970**, *66*, 1656.
32. Ross, A. B.: Neta, P. *Natl. Stand. Ref. Data Ser., Natl. Bur. Stand.* (*U.S.*) **1979**, *65*.
33. Swallow, A. J. *Prog. React. Kinet.* **1978**, *9*, 195.
34. Ross, A. B.; Neta, P. *Natl. Stand. Ref. Data Ser., Natl. Bur. Stand.* (*U.S.*) **1982**, *70*.
35. Asmus, K.-D. *Acc. Chem. Res.* **1979**, *12*, 436.
36. Buxton, G. V.; Sellers, R. M. *Coord. Chem. Rev.* **1977**, *22*, 195.
37. Meyerstein, D. *Acc. Chem. Res.* **1978**, *11*, 43.
38. Buxton, G. V.; Sellers, R. M. *Natl. Stand. Ref. Data Ser., Natl. Bur. Stand.* (*U.S.*) **1978**, *62*.
39. Buxton, G. V. In "The Study of Fast Processes and Transient Species by Electron Pulse Radiolysis"; Baxendale, J. H.; and Busi, F., Eds.; Reidel: Dordrecht, 1982, pp. 241–266.

# 11

# Radiation Chemistry of the Liquid State: (2) Organic Liquids

A. J. SWALLOW

*Paterson Laboratories,*
*Christie Hospital and Holt Radium Institute*

## INTRODUCTION

The consequences of ionization and excitation depend on the physical state of the material irradiated and on the molecular composition. Thus, the radiation chemistry of organic liquids may be compared, on the one hand, with that of organic compounds in the gaseous or solid state and, on the other hand, with that of inorganic liquids, of which water is by far the most important. Among the significant differences between organic liquids and organic gases are the following:

1. Organic liquids are denser than gases, so that the electrons produced in the ionization events will lose their energy before they get far from the associated positive ions. In fact, many electrons will never leave the coulombic field of their positive ions ("geminate ions"). This has a profound influence on the kinetics of subsequent processes and, among other things, means that it is never possible to collect all the positive and negative ions by applying a clearing electric field.
2. Every molecule in the liquid is in contact with other molecules. Hence, the energy of an excited molecule may be taken away before it decomposes ("collisional deactivation") or if decomposition into fragments does occur,


*Radiation Chemistry: Principles and Applications*

the fragments may recombine within the "cage" of nearby molecules. The existence of these processes means that fewer decomposition products are formed in the liquid phase than in the gas phase.

There are not so many significant differences between organic liquids and organic solids. The main one is that diffusion processes are much faster in the liquid state, so that reactive species such as free radicals have a very much shorter lifetime.

The principal features that distinguish organic liquids from water include the following:

1. Most organic liquids have a lower dielectric constant than that of water, which is about 80 at ambient temperature. Exceptions such as *N*-methylacetamide (about 180), hydrocyanic acid (about 115), and formamide (about 110) are rare. The lower dielectric constants mean that the coulombic field of the positive ions extends much further out than with water, so that the reactions of isolated species play a smaller part in the radiation chemistry.
2. Some organic molecules possess nondissociative singlet and triplet excited states that have been characterized photochemically. For other molecules excited states have been demonstrated by radiation chemistry. Of course, all molecules produce excited states on irradiation, but those of water probably dissociate very rapidly to give free radicals, resembling those produced by ionization, or, alternatively, may simply dissipate their energy as heat, and this is by no means always the case for organic molecules.
3. Organic molecules have more atoms than water. This and the presence of the carbon atom allow a much greater diversity of decomposition processes to take place than with water, resulting in a greater variety of reactions and products.

This chapter gives an overall view of the radiation chemistry of organic liquids, the most thoroughly studied of which have been hydrocarbons and alcohols. Emphasis is placed on radiations of low linear energy transfer (LET). Several aspects are dealt with in greater detail in other chapters. No attempt has been made to survey the literature comprehensively, but a few references are given to provide a starting point for further reading.

## IONIZATION: ELECTRONS AND NEGATIVE IONS

The average amount of energy dissipated per unit length of track depends on the energy of the ionizing particle and the number of electrons per cubic centimeter. When organic liquids are irradiated with $\gamma$-rays or fast electrons, the value will be in the region of 0.15 eV/nm. The average energy to produce an ion pair cannot be measured directly but might be expected to be comparable to that in the gas phase, about 25 eV, so there will be an average of around six ion pairs per micrometer of track; that is, the ion pairs will be on average about 170 nm apart. The distance $r_c$ at which the potential energy of an ion pair is equal to the thermal energy $kT$, is given by an expression derived from the work of Onsager.[1] In SI terms $r_c$ is given by

$$r_c = \frac{e^2}{4\pi\varepsilon_0\varepsilon_r kT} \tag{11-1}$$

where $r_c$ is in meters, e is the charge on the electron ($1.602 \times 10^{-19}$ C), $\varepsilon_0$ is the permittivity of free space $8.854 \times 10^{-12}$ $Fm^{-1}$), $\varepsilon_r$ is the dielectric constant of the medium (a pure number), $k$ is the Boltzmann constant ($1.381 \times 10^{-23}$ $JK^{-1}$), and $T$ is the temperature of the medium in kelvins. Values of $r_c$ for room temperature calculated from Eq. (11-1) are given in Table 11-1. The values are seen to be less than the average distance separating the ion pairs, so the particle track can be thought of as a series of isolated positive charges, each with its own coulombic field around it. Some of the associated electrons will escape, yielding free ion pairs, while others will be retained within the coulombic field ("geminate ion pairs"). Actually, some of the ionizations occur in groups, so the average distance apart of the ionization *events* will be greater than 170 nm. Within a group of ionizations the last ion pair to recombine may be thought of in terms of the Onsager theory. The Onsager theory does not, however, apply to high-LET radiation, for which the tracks should be treated as cylinders.

**Table 11-1.** Onsager Radii

| Liquid | $r_c$ (nm) |
|---|---|
| Neopentane | 32 |
| Tetramethylsilane | 31 |
| *n*-Hexane | 30 |
| Cyclohexane | 28 |
| Cyclohexene | 26 |
| Dioxan | 26 |
| Benzene | 25 |
| Carbon tetrachloride | 25 |
| Toluene | 24 |
| Chloroform | 12 |
| Ethylamine | 8.1 |
| Tetrahydrofuran | 7.7 |
| Benzyl alcohol | 4.4 |
| *n*-Propanol | 2.8 |
| Acetone | 2.8 |
| Ethanol | 2.3 |
| Benzonitrile | 2.2 |
| Methanol | 1.7 |
| Acetonitrile | 1.5 |
| Water | 0.7 |

The probability of an electron escaping its positive ion to become a free ion is equal to $e^{-r_c/r}$, where $r$ is the distance at which the electron becomes thermalized. There is no a priori way of determining $r$, but reliable measurements of the yield of free-ion pairs can be made in non-polar liquids with a very low electrical conductivity in the absence of radiation. In such liquids the conductance at very low applied voltages is governed by the rate at which free ion pairs are produced and the rate at which they are removed by second-order recombination. Yields of free ion pairs can be obtained from conductance measurements, provided the sum of the mobilities of the positive and negative ions, the recombination coefficient, and the dose rate are known. Alternatively, very high applied voltages can be used to prevent recombination. In this condition the geminate ions begin to be collected too, but extrapolation of the current to zero field strength can be used to obtain free ion yields. In a third method, "the clearing field method," a pulse of radiation is given and all of the free ions are collected before significant second-order recombination has had time to take place. Some values obtained by such methods for the yield of free ion pairs, $G_{fi}$, in hydrocarbons at room temperature[2] are shown in Table 11-2. Yields in liquids like alcohols, acetone, and amines may be deduced from a variety of steady-state and pulse radiolysis experiments to be more than the yields for nonpolar liquids, but less than the yield of hydrated electrons in water (2.7 electrons/100 eV), in qualitative agreement with expectations based on Eq. (11-1).

A striking feature of the results in Table 11-2 is that the yields, although all fairly small, show a considerable variability from compound to compound. An extreme example is neopentane, where the yield is several times greater than that in other pentanes. This cannot be accounted for on the basis of dielectric constant because all of the liquids in the table have similar values, giving an Onsager radius near 30 nm.

**Table 11-2.** Yields of Free Ions in Hydrocarbons

| Liquid | $G_{fi}$ (ion pairs/100 eV) | Mobility (cm$^2$ V$^{-1}$ s$^{-1}$) |
|---|---|---|
| Neopentane | 0.86–1.09 | 55 |
| Tetramethylsilane | 0.74 | 90 |
| 2,2,4-Trimethylpentane | 0.332 | 7 |
| 2,2-Dimethylbutane | 0.304 | 10 |
| *n*-Butane | 0.193 | 0.4 |
| Isopentane | 0.170 | — |
| Cyclohexene | 0.15–0.20 | — |
| Cyclohexane | 0.15–0.19 | 0.35 |
| *n*-Hexane | 0.10–0.18 | 0.09 |
| Cyclopentane | 0.155 | 1.1 |
| *n*-Pentane | 0.145 | 0.16 |
| Toluene | 0.093 | — |
| Benzene | 0.053–0.081 | — |

The reason can be found by examining electron mobility.[3] The mobilities are quite high (Table 11-2) for compounds whose molecules can be seen to be more spherical (more isotropic) and quite low for the less spherical (more anisotropic). The higher yields of free ion pairs in liquids composed of more spherical molecules are attributable to the electrons finding it easier to move away from their positive partners, so that a smaller fraction remains within the coulombic field of the positive ion. The high mobility with the more spherical molecules also means that thermalization will take longer, perhaps in the region of $10^{-11}$ s, compared with about $10^{-12}$ s with nonspherical, nonpolar molecules and about $10^{-13}$ s with polar molecules. More details of these areas can be found in Chapter 7.

It is possible for electrons to react chemically before they become thermalized.[4] Reaction with the solvent is most likely in the case of electron-attaching liquids like $CCl_4$ and $C_6F_6$. With other liquids, electrons formed close to their geminate partners may become neutralized before they become thermalized. If electrons do attain thermal energies, whether in the coulombic field of their geminate partners or not, they may become trapped in the solvent. In nonpolar liquids the traps are not expected to be deep, and there will be an equilibrium between the population of electrons in the "quasi-free" state (analogous to the conduction band in solids) and those in the trapped state, the proportion depending on temperature and sphericity. In polar liquids the traps are deeper, although when an electron is first trapped in a polar medium the surrounding molecules are unlikely to be in the correct equilibrium configuration for molecules surrounding a negative charge. Solvent relaxation would be expected to give rise to fully solvated electrons.

Solvated electrons possess the property of absorbing light. The energy of maximum absorption, being inversely proportional to $\lambda_{max}$, the wavelength of maximum absorption, is an indication of trap depth. Values of $\lambda_{max}$ for room temperature are given in Table 11-3.[5] The absorptions

**Table 11-3.** Absorption Maxima of Solvated Electrons

| Liquid | $\lambda_{max}$ (nm) |
|---|---|
| Ethylene glycol | 580 |
| Methanol | 640 |
| Ethanol | 720 |
| *n*-Propanol | 740 |
| Isopropanol | 830 |
| Ethylene diamine | 1300 |
| *N,N*-Dimethylformamide | 1700 |
| *N,N*-Dimethylacetamide | 1800 |
| Diethylamine | 1900 |
| Tetrahydrofuran | 2100 |
| Diethyl ether | 2100 |

are all very broad. Molar extinction coefficients are quite high, in the region of 1.5 to $4 \times 10^4\ M^{-1}\ \mathrm{cm}^{-1}$. In aliphatic hydrocarbons the wavelengths of maximum absorption are greater than 2000 nm.

The later stages in the history of electrons in organic liquids can be observed pulse radiolytically. Experiments with alcohols are particularly informative because there are approximately equal yields of geminate and free ion pairs, and the alcohols can be irradiated at low temperatures where processes are slow enough to be experimentally observable. Observations on *n*-propanol at 152 K are shown in Figure 11-1.[6] Immediately after a

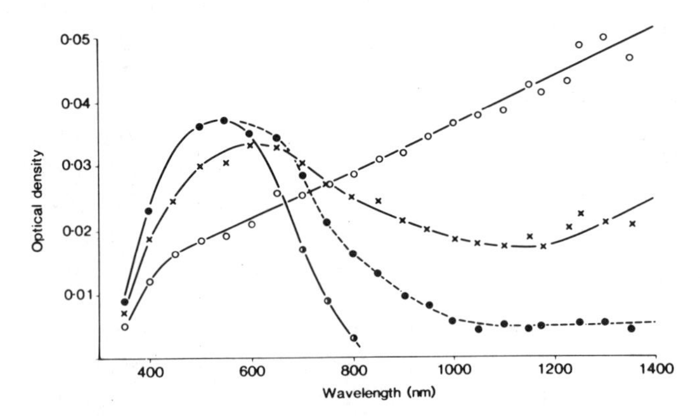


**Figure 11-1.** Absorption spectra after giving a 5-ns pulse of radiation to *n*-propanol at 152 K; ○——○-end of pulse; × —— × -65 ns; ●- - -●-200 ns; ◑——◑-1 $\mu$s.

5-ns pulse there is a broad absorption extending into the infrared. The energies corresponding to these absorption wavelengths are small, so the absorption band is attributed to electrons trapped in shallow traps, resembling those in hydrocarbons and other nonpolar molecules. In the case of alcohols it is assumed that the alcohol molecules have not fully relaxed to form ordinary solvated electrons because these should absorb at shorter wavelengths (cf water). As time goes by, the long-wavelength absorption decays (half-life for the experiment of Figure 11-1 about 60 ns), and simultaneously a new absorption with a peak at 550 nm appears, attributable to solvated electrons. The energy of the absorption is greater than that in Table 11-3 because of the lower temperatures. The process probably consists mainly of a simple reorientation of alcohol molecules around the trapped electrons, but it may also include migration of electrons from the shallower to the deeper traps. There appears to be some loss of electrons during this time, perhaps due to geminate neutralization. After the solvated electrons have been formed in alcohols, they disappear in two stages. The first, with typical half-lives of a few microseconds at low temperatures, or a fraction of a microsecond at room temperature, corresponds to geminate neutralization or reaction with nearby neutral species. The second, typical half-lives >1 ms at low temperatures and a few microseconds at room temperature, corresponds to reaction of the free electrons with solvent, probably in a reaction analogous to that assumed for hydrated electrons in water;

$$e_{solv}^{-} + ROH \rightarrow RO^{-} + H \tag{11-2}$$

As in water, thermalized electrons in organic liquids are able to react with appropriate solutes. In contrast to water, however, there is significant competition between reaction with the solute and reaction with positive ions over the whole of the accessible range of solute concentrations. In such a situation the yield of a product P formed by reaction of electrons with a solute S may be fitted to an empirical equation of the form[7]

$$G(\mathrm{P}) = G_{\mathrm{fi}} + \frac{G_{\mathrm{gi}}\sqrt{\alpha[S]}}{1 + \sqrt{\alpha[\mathrm{S}]}} \tag{11-3}$$

where $G_{gi}$ is a parameter numerically not very different from what the yield of geminate ions might be expected to be, and $\alpha$ is a factor that depends on the reactivity of the electrons with the solute. The equation was originally applied to solutions of alkyl halides ($10^{-4}$–0.5 $M$) in cyclohexane, where alkyl radicals were formed by reactions such as

$$e^{-} + CH_3Cl \rightarrow CH_3\cdot + Cl^{-} \tag{11-4}$$

but has since been applied to a variety of organic solutions, with appropriate modifications where necessary to take into account reactions such as concurrent positive ion scavenging and reaction of electrons with solvent.

Rate constants for reactions of solvated electrons with solutes can be determined optically or conductimetrically, but the measurements are more difficult than in aqueous solution because the electron lifetimes are generally much shorter. With optical measurements on nonpolar liquids, another problem is that the electrons absorb only at rather long wavelengths. For different solutes in a given solvent, the values correlate reasonably well with values for $\alpha$ needed for a fit of Eq. (11-3). In nonpolar solvents the rate constants can be very much higher than found in aqueous solution: values as high as $10^{12}\ M^{-1}\ s^{-1}$ are common, and values more than $10^{14}\ M^{-1}\ s^{-1}$ have been found for good electron acceptors in spherical nonpolar liquids. A partial explanation of such high rate constants can be found in high rates of electron diffusion, which are also reflected in measurements of mobility. The expression for the rate constant $k_D$ of a diffusion-controlled reaction

$$k_D = 4\pi RD \tag{11-5}$$

where $R$ is the reaction radius and $D$ is the diffusion constant, may be combined with

$$D = \frac{\mu kT}{e} \tag{11-6}$$

where $\mu$ is the mobility, to obtain

$$k_D = 1.9 \times 10^{13}\ \mu R\ M^{-1}\ s^{-1} \tag{11-7}$$

for room temperature, where $\mu$ is in $cm^2\,V^{-1}\,s^{-1}$ and $R$ is in nm. For $\mu = 0.1\,cm^2\,V^{-1}\,s^{-1}$ and $R = 0.5\,nm$, $k_D \simeq 10^{12}\,M^{-1}\,s^{-1}$. For $\mu = 50\,cm^2\,V^{-1}\,s^{-1}$, $k_D = 5 \times 10^{14}\,M^{-1}\,s^{-1}$. Although the reactions can be thought of in terms of diffusion, other considerations play a part in attempting to explain the full range of experimental values of rate constants as a function of nature of solvent and solute, temperature, and other factors.

Reaction of electrons generated from solvents such as alcohols or amines provides a convenient method of producing solute radical anions and carbanions in isolation from any counterion. The method has been used with aromatic solutes such as biphenyl, anthracene, pyrene, and others to study electron transfer equilibria of the type[8]

$$\text{arene(1)}^{-}\cdot + \text{arene(2)} \rightleftharpoons \text{arene(1)} + \text{arene(2)}^{-}\cdot \qquad (11\text{-}8)$$

The dependence of the rates of such reactions on reduction potential and nature of solvent is in reasonable agreement with the Marcus theory of electron transfer rates. The anions become protonated on reaction with alcohols, and this reaction too can be observed:

$$\text{arene}^{-}\cdot + \text{ROH} \rightarrow \text{areneH}\cdot + \text{RO}^{-} \qquad (11\text{-}9)$$

Carbanions are conveniently formed in solution in tetrahydrofuran by dissociative electron capture; for example,

$$e_{solv}^{-} + (C_6H_5CH_2)_2Hg \rightarrow C_6H_5CH_2^{-} + C_6H_5CH_2Hg\cdot \qquad (11\text{-}10)$$

Rates of protonation can be measured by adding fairly low concentrations (eg, $10^{-2}$–$10^{-3}\,M$) of proton donors such as water or alcohols.[9] For carbanions the values found fall in the range $10^7$ to $10^{10}\,M^{-1}\,s^{-1}$.

## POSITIVE IONS: RECOMBINATION

In the first instance the positive ions produced by radiation are radical cations possessing various amounts of energy of electronic excitation. Three different types of processes—unimolecular fragmentation or rearrangement, ion–molecule reaction, and hole migration—may take place within a picosecond or so, depending on the molecules concerned.

Fragmentation is observable in the gas phase by means of mass spectra. Less fragmentation is to be expected in the liquid phase because the ions are in contact with the molecules of the medium, but mass spectra provide a guide to the type of fragmentation that might be expected. Ion–molecule reactions take place when ions are close enough to neutral molecules for attractive forces to bring the two species into a favorable position for any rearrangement to a more stable configuration. Among the types of reaction to be expected between parent or fragment ions and parent molecules are proton or hydrogen atom transfer, such as

$$C_2H_5OH^{+}\cdot + C_2H_5OH \rightarrow C_2H_5O\cdot + C_2H_5OH_2^{+} \qquad (11\text{-}11)$$

hydride ion transfer, such as

$$C_4H_9^+ + C_6H_{14} \rightarrow C_4H_{10} + C_6H_{13}^+ \qquad (11\text{-}12)$$

and with olefins, condensation, such as

$$R\dot{C}H\overset{+}{C}H_2 + CH_2{=}CHR \rightarrow R\dot{C}HCH_2{-}CH_2\overset{+}{C}HR \qquad (11\text{-}13)$$

The nature of the fragmentation, rearrangement, and ion–molecule reactions taking place in particular cases is deduced from analysis of the irradiation products as a function of variables such as temperature, dose rate, and presence of scavengers, taking into account other reactions occurring in the system (electrons, excited molecules, free radicals, etc).

Where a parent radical cation does not fragment, rearrange or react with the solvent, it may transfer its charge to a neighboring molecule by a resonant charge transfer process (hole transfer) in which it captures an electron, thus restoring itself to its original condition while the formerly neutral molecule acquires the charge.[4] This process would be expected to be favored when the geometry of the radical cation is close to that of the parent molecule. It would lead to a very high ion mobility. Various lines of evidence, of which conductivity measurements are the most direct, have demonstrated this process in acetone, chlorinated hydrocarbons, and certain hydrocarbons including cyclohexane, although with other liquids it does not take place.

Positive ions may react with solutes if they encounter them before neutralization. The reactions include ion–molecule reactions such as those represented by Eq. (11-11)–(11-13), but an additional and important class of reaction is simple charge transfer, which can occur if the electron affinity of the positive ion is greater than the ionization potential of the solute

$$A^{+\cdot} + B \rightarrow A + B^{+\cdot} \qquad (11\text{-}14)$$

The solute cation radical may in turn hand on its charge to another solute and so forth.

Competition between reaction with solutes and ion recombination means that the yields of products formed by reaction with solute must depend on solute concentration. The dependence takes a similar form to that noted above for reaction of electrons with solutes [Eq. (11-3)]. An illustration of the dependence can be seen in the irradiation of cyclohexane solutions of deuterated ammonia,[10] where the positive ions formed from the cyclohexane can donate a proton to the more basic ammonia, giving $ND_3H^+$ ions. When these ions are eventually neutralized they most probably give D atoms, which abstract H from cyclohexane to give HD. Hydrogen formed from the cyclohexane by other routes will consist of $H_2$, so that the proportion of HD in the total hydrogen gas provides a measure of the extent of the ion–solute reaction. Figure 11-2 shows the low yield up to about $5 \times 10^{-3}$ *M*, due to reaction of free positive ions, and the increasing yield at higher and higher concentrations as the geminate positive ions are increasingly able to react with solute before they become neutralized. It is generally found that

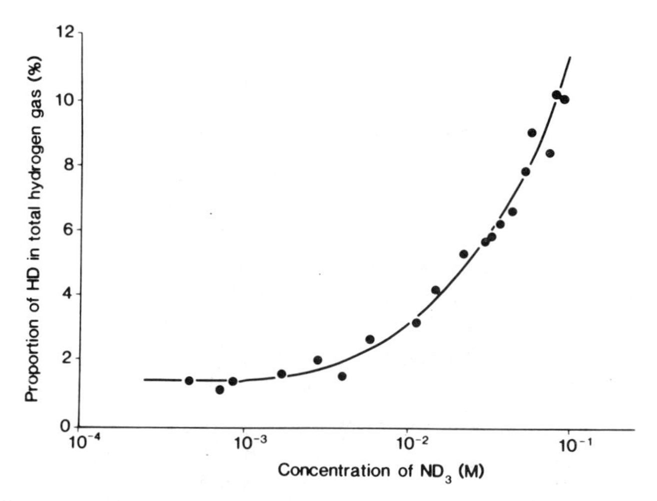


**Figure 11-2.** Proportion of HD in hydrogen gas formed by irradiation of solutions of $ND_3$ in $C_6H_{12}$.

higher solute concentrations are needed to obtain significant reaction than are needed with electrons, and this is reflected in $\alpha$ being smaller, typically by a factor of 10. There are comparatively few direct measurements of rate constants for reactions of positive ions with solutes, but the ones obtained tend to be lower than for electrons, in agreement with the values of $\alpha$.

Reactions of radical cations and carbocations[11] can be studied by pulse radiolysis of appropriate substances dissolved in chlorinated hydrocarbons, where the electrons are removed by the dissociative attachment

$$e^- + RCl \rightarrow R\cdot + Cl^- \qquad (11\text{-}15)$$

The radical cations are formed by charge transfer from the solvent radical cations; for example,

$$RCl^{+\cdot} + (C_6H_5)_2 \rightarrow RCl + (C_6H_5)_2^{+\cdot} \qquad (11\text{-}16)$$

The carbocations are formed by a dissociative process; for example,

$$RCl^{+\cdot} + (C_6H_5CH_2)_2Hg \rightarrow RCl + C_6H_5CH_2^+ + C_6H_5CH_2Hg\cdot \qquad (11\text{-}17)$$

Among the reactions of the carbocations is that with alkenes, which is probably a condensation [cf Reaction (11-13)]. The rate constants for the reactions of the benzyl cation and the benzhydryl cation with a variety of olefins range from below $10^5$ to $10^9$ $M^{-1}$ $s^{-1}$. Such reactions represent the initiation step in cationic polymerization.

The final stage in the lifetime of the ions is recombination. The kinetics of the recombination fall into two parts. The first is the recombination of geminate ions, and the second is the homogeneous recombination of free ions. Within the geminate region electrons that have not moved far from their geminate partners recombine the fastest, whereas with greater separations geminate recombination is also fast where the electron mobility is high. The time scale for geminate recombination of electrons with positive ions can be described empirically by the Laplace transform of Eq. (11-3):[12]

$$G(t) = G_{fi} + G_{gi}\, e^{\lambda t} \mathrm{erfc}(\lambda t)^{1/2} \tag{11-18}$$

where $G(t)$ is the yield of ions surviving at time $t$ and $\lambda$ is an empirical parameter equal to $\alpha$ divided by the rate constant for the reaction of the electron with the solute. The inverse of $\lambda$ is equal to the time $\tau_{gi}$ at which 57% of the ions that will finally undergo geminate recombination have already done so. Values of $\tau_{gi}$ depend on the rates of diffusion of the electrons and the positive ions and are in the region of picoseconds for nonpolar liquids at room temperature. They can also be considerably slower, as with alcohols at low temperatures[6] where recombination took place over microseconds. Geminate recombination of negative molecular ions with positive ions is slower than that between electrons and positive ions and can be readily observed by fast-response pulse radiolysis over tens or hundreds of nanoseconds, even in nonpolar liquids.

The homogeneous recombination must follow second-order kinetics. The rate constants are of the order of magnitude of those expected for reactions controlled by diffusion [or mobility, cf Eq. (11-6)]. For recombination of electrons with cations in liquids of high sphericity and low dielectric constant, the rate constants are extremely high, in the region of $10^{16}$ to $10^{17}$ $M^{-1}$ s$^{-1}$. This is consistent with expectations based on Eq. (11-7), where the mobilities are those of the electrons that have the high values given in Table 11-2 and the reaction radius $R$ is the Onsager radius $r_c$ expressed in nanometers. For example, for tetramethylsilane with $\mu = 90$ cm$^2$ V$^{-1}$ s$^{-1}$ and $r_c = 31$ nm, Eq. (11-7) gives $k_D = 5.3 \times 10^{16}$ $M^{-1}$ s$^{-1}$, which may be compared with an experimental value[13] of about $5 \times 10^{16}$ $M^{-1}$ s$^{-1}$.

# EXCITED STATES

The excited states produced *directly* by energetic ionizing particles are optically allowed states with energies up to and above those needed to enable ionization to take place. For molecules that are singlets in the ground state (most molecules), these excited states are therefore singlets. Once the energy of the particles (including the electrons ejected by the original particles) drops below 100 eV or so, optically forbidden processes can also take place so that triplets are produced too. Excited states (singlets) must also be produced when Čerenkov radiation is absorbed. As well as the excited states produced directly, the energy liberated in ion recombination is often sufficient to enable excited singlet and triplet states of the recombining entities to be produced. The excited states produced by radiation are liable to have energies that are quite high, but unless dissociation takes place, internal conversion might be expected to reduce the level to that of the lowest excited state within a picosecond or so. After that, various processes familiar from photochemistry take place; intersystem crossing, excimer formation, energy transfer, luminescence, and so on.

Yields of solvent singlet excited states are obtained most directly from measurements of fluorescence. Yields of triplets are obtained by measuring the yields of chemical reactions known from photochemistry to be originated by triplets or by measuring solute triplet–triplet absorptions pulse radiolytically. Allowance must be made for triplets arising from intersystem crossing from singlets. Yields determined by such methods are given in Table 11-4.[14]

**Table 11-4.** Yields of Excited States

| Liquid | Singlet (Number/100 eV) | Triplet (number/100 eV) |
|---|---|---|
| Benzene | 1.6 | 4.2 |
| Toluene | 1.35 | 2.8 |
| Benzonitrile | 0.7 | 1.4 |
| Benzyl alcohol | 0.7 | 1.1 |
| Dioxan | 1.4 | 0 |
| Cyclohexane | 1.45 | — |
| Acetone | 0.34 | 1.0 |
| Tetrahydrofuran | 0.1 | <0.1 |
| Ethanol | <0.1 | <0.1 |
| Acetonitrile | — | <0.1 |

The excited states of solutes are formed by three main mechanisms

1. Excitation by subexcitation electrons or Čerenkov radiation
2. Energy transfer from solvent excited states
3. Ion recombinations involving solute anions and/or cations.

The contribution made by the various processes varies with the nature of the solvent and the solute. It is not always easy to establish. With aromatic hydrocarbons as solvents, however, energy transfer from solvent excited states is certainly a major process, whereas with aliphatic hydrocarbons as solvents, ion recombination dominates.

An important kinetic treatment, Stern–Volmer kinetics, relates the first-order rate constant for decay of an excited state

$$A^* \xrightarrow{k_I} A \qquad (11\text{-}19)$$

to the rate constant for interaction with a solute; for example,

$$A^* + B \xrightarrow{k_{II}} A + B^* \qquad (11\text{-}20)$$

If $E$ is the emission from $B^*$, then

$$\frac{1}{E} = \frac{1 + k_I/k_{II}[B]}{E_{max}} \tag{11-21}$$

where $E_{max}$ is the emission when all the $A^*$ reacts with B. Lifetimes of singlet states are typically of the order of nanoseconds. Lifetimes of triplet states are usually much longer (milliseconds to seconds), so that in the liquid phase bimolecular processes often take place before triplet emission (phosphorescence) has time to occur.

The rate constants for diffusion-controlled reactions of neutral species can be conveniently calculated in terms of solvent viscosity by combining the expression relating rate constants to diffusion constants [cf Eq. (11-5)] with the Stokes–Einstein equation relating diffusion constant and viscosity. This gives

$$k_D = \frac{2.22 \times 10^4\ T}{\eta} M^{-1}\ s^{-1} \tag{11-22}$$

where $T$ is the temperature in kelvins and $\eta$ is the viscosity in kg $m^{-1}$ $s^{-1}$. Rate constants for reactions of triplets are always diffusion controlled or slower, but rate constants for transfer of singlet excitation energy are often found to be substantially higher than diffusion controlled. For example, the excited singlet state of *cis*-decalin is quenched by $CCl_4$ with $k = 1.4 \times 10^{10}\ M^{-1}\ s^{-1}$, whereas the diffusion-controlled rate is only $3.2 \times 10^9\ M^{-1}\ s^{-1}$. Resonant energy transfer and abnormally high-reaction radii are among the possible explanations.[15]

The emission of fluorescence from solutes excited by ionizing radiation is put to important practical use in the scintillation counter. Another practical application of radiation excitation is in the determination of certain properties of excited states and their complexes with ground state molecules (excimers). One example is the use of radiation to determine triplet–triplet extinction coefficients. In a typical determination a concentrated cyclohexane, hexane, or benzene solution of biphenyl (eg, $10^{-1}\ M$) containing the solute of interest at lower concentration (eg, $10^{-4}\ M$) is given a short pulse of radiation. The triplet of the biphenyl is first produced. The mechanism is irrelevant to the determination, although, as mentioned above, it is likely to be mainly ion recombination or mainly energy transfer, depending on the solvent. The absorption spectrum of the biphenyl triplet in cyclohexane[16] based on an extinction coefficient at 361 nm of 42,800 $M^{-1}\ cm^{-1}$ is shown in Figure 11-3. During a few microseconds the biphenyl triplet transfers its energy to the solute of interest, resulting in the replacement of the biphenyl triplet by the solute triplet, the shape of whose spectrum can now be measured. The biphenyl extinction coefficient is used to calculate the absolute value of the extinction coefficient of the solute of interest, based on the assumption that the transfer is 100% efficient. Triplet energy levels can be measured by making use of standards whose triplets have known

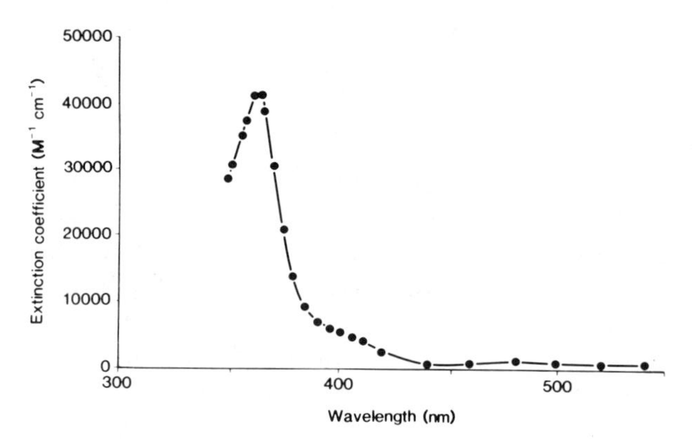


**Figure 11-3.** Absorption spectrum of biphenyl triplet in cyclohexane.

energy levels; for example, naphthalene (255 kJ $mol^{-1}$), anthracene (176 kJ $mol^{-1}$), or naphthacene (123 kJ $mol^{-1}$). If transfer takes place from standard to substrate, the triplet energy of the substrate is less than that of the standard, and vice versa. Attainment of equilibria between substrate and standard enables the precise energy level to be determined.

## FREE RADICALS

The entities formed by loss or gain of an electron are free radicals as well as ions. These radical ions can give rise to neutral free radicals through unimolecular scission or through bimolecular processes like deprotonation or protonation. Free radicals can also be formed by dissociation of excited molecules.

Some free radicals are formed in close proximity to other radicals and would react so rapidly that they could not be detected either by conventional fast-reaction techniques or by letting them react with radical scavengers. Those radicals that escape fast recombination diffuse freely through the liquid and are available to react with radical scavengers like diphenylpicrylhydrazyl or iodine. Measurement of the change produced in the reaction enables the yield of radicals to be determined. Low concentrations of scavengers must be used ($< \sim 10^{-3}$ $M$) to minimize complicating reactions. Yields of radicals obtained by such methods are included in Table 11-5.[17] It may be noted that few radicals are formed from benzene, which is typical of aromatic liquids. This is consistent with a high proportion of the radiation energy being degraded by one mechanism or another into that of the lowest singlet and triplet states, which are formed in good yield (Table 11-4) and do not give rise to free radicals. Table 11-5 also indicates a degree of stability of other compounds containing $\pi$-electrons.

**Table 11-5.** Yields of Free Radicals

| Liquid | Radicals/100 eV |
|---|---|
| Ethyl ether | 9–12 |
| Ethanol | 5–8 |
| Methyl acetate | 7 |
| Carbon tetrachloride | 3–7 |
| Methanol | 6 |
| Ethyl iodide | 4–6 |
| *n*-Hexane | 4.8–5.9 |
| Cyclohexane | 4.8–5.5 |
| Cyclopentane | 4.6–5.5 |
| Cyclohexene | 5.2 |
| Chloroform | 5 |
| *Trans*-2-butene | 4.0 |
| Acetone | 4 |
| Acetic acid | 4 |
| Ethylene | 2.6 |
| Benzene | 0.7–1.0 |

The nature of the radicals can be deduced from studies of stable products formed by irradiating the liquid, both pure and in the presence of appropriately chosen solutes. For instance, the distribution of the various alkyl iodides formed from aliphatic hydrocarbons in the presence of iodine can provide a basis for estimating the yields of alkyl radicals in the system. Another method is "radical sampling" in which $^{14}C_2H_5\cdot$ radicals are formed by action of H atoms on labeled ethylene, or $^{14}CH_3\cdot$ radicals are formed by action of electrons on $^{14}CH_3I$, and then react randomly with all the other radicals in the system to give labeled hydrocarbons that can be analyzed. Such studies provide evidence that more radicals tend to be formed by scission of weak bonds than of strong bonds.

Reactions undergone by radiolytically produced free radicals are of the same type as those undergone by radicals produced in other ways. The principal categories of reaction include electron transfer, hydrogen atom transfer, such as

$$CH_3O\cdot + CH_3OH \rightarrow CH_3OH + \cdot CH_2OH \qquad (11\text{–}23)$$

and addition to oxygen or to molecules containing double bonds, such as

$$C_6H_5{-}\dot{C}HCH_2R + C_6H_5CH{=}CH_2 \rightarrow C_6H_5\dot{C}HCH_2CH(C_6H_5)CH_2R \qquad (11\text{-}24)$$

Atom transfer and addition reactions lead to the formation of further radicals, which must react further. Chain reactions of the normal type can readily occur, for example, with vinyl monomers or in the irradiation of mixtures containing reactive solutes such as chlorine.

The final step in the life of free radicals is usually mutual reaction by disproportionation (dismutation),

$$2RH\cdot \rightarrow RH_2 + R \qquad (11\text{-}25)$$

or dimerization (combination),

$$2RH\cdot \rightarrow (RH)_2 \qquad (11\text{-}26)$$

The ratio of the rates of Reactions (11-25) and (11-26) is not very different from that in the gas phase. It depends on the structure of the radicals. For example, with secondary hydrocarbon radicals, the two reactions proceed at comparable rates, whereas with primary hydrocarbon radicals dimerization is preferred, and with tertiary hydrocarbon radicals disproportionation predominates.

Electron spin resonance (ESR) provides the most direct information about radiolytically produced free radicals. Molecular tumbling reduces the magnetic anisotropy in liquids, so the interpretation of the spectra is more certain than with solids. Radicals tend to disappear by reacting with each other at rates approaching the diffusion controlled, so it is necessary to irradiate inside the cavity of the spectrometer in order to obtain an adequate signal. The doses required are large enough for radicals to be formed by secondary processes as well as by direct radiolysis of the liquid.

The ESR spectrum of irradiated liquid ethane[18] is typical and shown in Figure 11-4. The 12 principal lines are attributed to the ethyl radical on the basis of ESR theory and agreement with previous work in which the radical had been produced from various substances in the solid phase. Weaker lines are seen in Figure 11-4 between the fifth and sixth and the seventh and eighth lines from the left. These are attributable to the methyl radical. It is also possible to discern even weaker lines attributable to vinyl radicals ($C_2H_3\cdot$). The greater part of the ethyl radical probably derives from direct action of the radiation on the ethane, but a substantial fraction is likely to derive from a secondary reaction between radiolytically generated hydrogen atoms and ethylene. The vinyl radicals must be entirely of secondary origin, because it is not possible to envisage any likely process by which they could be formed directly.

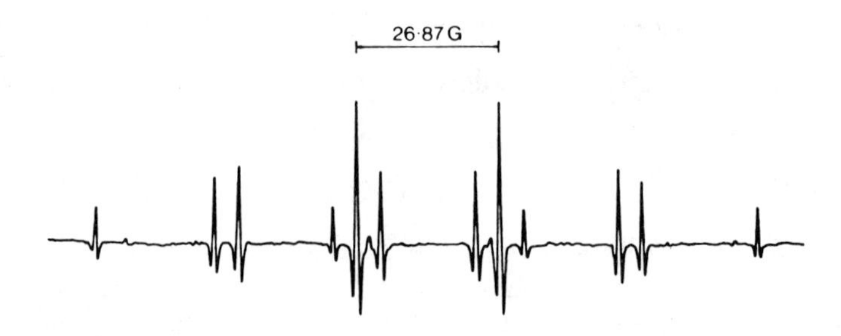


**Figure 11-4.** ESR spectrum of radicals in liquid ethane at 93 K.

By measuring the steady-state concentration of radicals produced by a given dose rate, it is possible for one to use ESR to measure rate constants for mutual reaction of free radicals. Activation energies can also be determined. The results are not very accurate but are generally in agreement with reactions proceeding at a diffusion-controlled rate.

The free radicals formed by pulse irradiation of pure liquids or solutions can be observed by means of optical spectrophotometry. The absorption spectra of the radicals formed from pure liquids do not, in general, show strong individual features. That of the cyclohexyl radical,[19] calculated on the basis of an extinction coefficient at 240 nm of 1700 $M^{-1}$ $cm^{-1}$, is shown in Figure 11-5. The spectra of radicals formed from solutes are more characteristic, but fewer studies have been made than in aqueous solution, partly because the radiolytic processes are not as simple as in well-composed aqueous systems and partly because recent interest in the radiation chemistry of organic liquids has focused on ions and excited states rather than radicals.

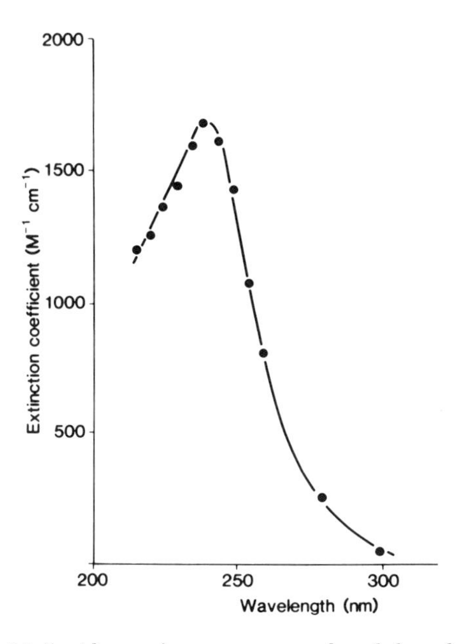


**Figure 11-5.** Absorption spectrum of cyclohexyl radical.

# REACTION PRODUCTS

The ionic, excited-state, and free-radical species discussed so far have lifetimes in the range $10^{-12}$–1 s. After they have disappeared they leave behind longer-lived species with lifetimes in the region of 10-$10^9$ s or more. These are the "stable" irradiation products, which can be studied by conventional chemical techniques. Examination of the products provided the foundation that enabled fast-reaction experiments to build up our present understanding of the active species. In fact, the products can provide information about active species that is unobtainable in any other way. The irradiation products are also of legitimate interest in themselves as well as in a number of applications of radiation in industrial and other fields.

An important general point is that all atoms that have formed part of molecules that have been destroyed by the radiation should be found in the products. When a pure liquid has been irradiated to a low-percentage conversion, it is not practical to measure the $G$-values for loss of starting material, but the proportion of C, H, O atoms, and so on, in the totality of the products should be the same as in the starting material. Such material balance considerations can provide useful pointers to the existence of previously unidentified processes. For example, hydrogen ($H_2$) is normally produced when hydrogenous organic liquids are irradiated, and it is sometimes found that the yield is greater than can be accounted for by the volatile organic products. In such cases a nonvolatile hydrogen-deficient species would be suspected.

Another generalization is that although the concentration of certain products may build up linearly with dose, it is also common for product $G$-values to increase or decrease as the irradiation proceeds. This is generally because of reactions of intermediates (positive and negative species, excited states, or free radicals) with the products formed in the first instance. Product yields depend on the presence of additives for similar reasons. Dose rate often has an influence on product yields, because at high dose rates the active species are present at higher steady-state concentrations than at low dose rates, so that they are more likely to react with each other than with the "stable" molecules in the system. Dose rate becomes extremely important in free-radical polymerization, where the kinetics resemble those of "conventional" free-radical polymerization. Another experimental variable is temperature, which affects product yields where the mechanism involves steps with significant activation energies. The LET affects product yields because of the effect of the concentration of active species in the track.

Cyclohexane has been one of the most exhaustively studied of all organic liquids, being chosen because it is very simple (all the C—H bonds and all the C—C bonds are almost identical), fairly easily purified, and inexpensive. Table 11-6 gives one of the most complete lists of initial product yields[20] obtained by making measurements at doses up to about 50 kGy and extrapolating to zero dose. Measurements of yields by other workers agree quite well with those in Table 11-6. A complete material balance would require the sum of the yields times the number of H atoms in product to be twice the sum of the yields times the number of C atoms in product, that is, 92.92. At 94.02 it is distinctly greater. Furthermore, it could not be brought into equality by any variation of the yields within the stated experimental errors. Thus cyclohexane illustrates the production of unidentified hydrogen-deficient species. In the present instance these are most likely to be oligomeric in nature.

**Table 11-6.** Material Balance in the $\gamma$-Radiolysis of Cyclohexane

| Product | Initial yield (molecules/100 eV) | Yield × no. C atoms in product | Yield × no. H atoms in product |
|---|---|---|---|
| $H_2$ | $5.6 \pm 0.1$ | 0 | 11.2 |
| $c$-$C_6H_{10}$ | $3.2 \pm 0.2$ | 19.2 | 32.0 |
| $(c$-$C_6H_{11})_2$ | $1.76 \pm 0.05$ | 21.12 | 38.72 |
| $CH{=}CH(CH_2)_3CH_2$ | $0.40 \pm 0.05$ | 2.40 | 4.80 |
| $c$-$C_5H_7{-}CH_3$ | $0.15 \pm 0.01$ | 0.90 | 1.8 |
| $c$-$C_6H_{11}{-}(CH_2)_4CH{=}CH_2$ | $0.12 \pm 0.02$ | 1.44 | 2.64 |
| $n$-$C_6H_{14}$ | $0.08 \pm 0.02$ | 0.48 | 1.12 |
| Unidentified C12 | ~0.05 | ~0.60 | 1.1[a] |
| $c$-$C_6H_{11}{-}C_2H_5$ | ~0.04 | ~0.32 | ~ 0.64 |
| $c$-$C_6H_{11}{-}cC_6H_9$ | 0 | 0 | 0 |
| | | 46.46 | 94.02 |

[a] Assuming $C_{12}H_{22}$.

Examination of Table 11-6 shows that, unlike the gas phase, there is little C—C rupture. Collisional deactivation and the cage effect account for the difference. Hydrogen is a principal product, as it is in the gas phase. Some of it is likely to be formed by reaction of thermalized H atoms with cyclohexane,

$$H + C_6H_{12} \rightarrow H_2 + C_6H_{11}\cdot \qquad (11\text{-}27)$$

where some of the hydrogen atoms are formed when the cyclohexane is excited directly, and some when the positive ions are neutralized by electrons. The larger part is probably formed without the intermediate formation of thermalized hydrogen atoms, either by direct detachment of $H_2$ from the molecule or by reaction of hot hydrogen atoms. This part also derives both from direct excitation and from neutralization. The cyclohexene is likely to derive mainly from unimolecular decomposition of cyclohexane and disproportionation of cyclohexyl radicals:

$$C_6H_{12}^* \rightarrow C_6H_{10} + H_2 \text{ or } 2H \qquad (11\text{-}28)$$

$$2C_6H_{11}\cdot \rightarrow C_6H_{12} + C_6H_{10} \qquad (11\text{-}29)$$

Some of the cyclohexyl radicals dimerize, thus accounting for the bicyclohexyl

$$2C_6H_{11} \rightarrow (C_6H_{11})_2 \qquad (11\text{-}30)$$

The ratio of the rate constants for Reactions (11-29) and (11-30) has been found from a number of photochemical and radiation-chemical experiments to be about 1.1, so that if the whole of the bicyclohexyl comes from Reaction

(11-30), $G = 1.76 \times 1.1 = 1.9$ of the cyclohexene would come from Reaction (11-29).*

The products formed from cyclopentane[22] are analogous to those from cyclohexane, but those from cyclopropane and cyclobutane contain a substantially smaller yield of hydrogen ($G \sim 1$ and 1.7, respectively) and contain ethylene as a major product ($G \sim 1.5$ and 4.9, respectively). The change in product distribution is associated with the high strain energy of the three- and four-carbon rings. The ethylene derives largely from decomposition of the excited parent

$$c\text{-}C_3H_6^{*\cdot} \rightarrow C_2H_4 + CH_2 \tag{11-31}$$

$$c\text{-}C_4H_8^{*} \rightarrow 2C_2H_4 \tag{11-32}$$

Straight- and branched-chain hydrocarbons resemble cyclohexane in giving hydrogen, the congruent olefins, and the paraffins with twice the number of carbon atoms as the parent. But they differ from cyclohexane in giving substantial yields of both saturated and unsaturated hydrocarbons of lower molecular weight and also saturated hydrocarbons of molecular weight intermediate between the parent and the dimer. In contrast to the cyclic hydrocarbons, ion–molecule reactions of fragment ions play a significant part in the mechanism.

Olefins give rise to appreciably smaller yields of hydrogen than saturated hydrocarbons do, partly because the $\pi$-electrons help to stabilize the molecule against radiolysis and partly because hydrogen atoms can add to the double bond. However, yields of oligomeric material are higher than with saturated hydrocarbons. This is accounted for by short-chain reactions initiated by ion–molecule condensation [Reaction (11-13)]. With highly purified olefins ionic chain polymerizations can occur in very high yield indeed ($G = 10^3$–$10^6$ molecules of monomer converted to polymer per 100 eV), as first seen in the formation of polyisobutylene from isobutylene (at low temperature).[23]

Aromatic hydrocarbons are characterized by being rather stable to radiation, which can be seen by comparing the magnitude of the yields from benzene (Table 11-7)[24] with those from cylohexane (Table 11-6). This must be due to the efficient conversion of energy into that of the lowest excited states. With high-LET radiation, however, the aromatic hydrocarbons are less stable, perhaps because the upper excited states are able to react with each other in the tracks. Polycyclic aromatic hydrocarbons such as polyphenyls were at one time thought to be sufficiently stable for use as coolant moderators in nuclear reactors; but although adequately stable to low-LET radiation, they proved too unstable to higher LET radiation, especially at the temperatures encountered in the reactor. Furthermore, the high-

* Reference 21 erroneously states that it is the ratio of dimerization to disproportionation which is 1.1, so that the yield of cyclohexene formed by disproportionation would be $G = 1.6$.

**Table 11-7.** Products formed by $\gamma$-Irradiation of Benzene

| Product | Initial yield (molecules/100 eV) |
|---|---|
| $C_6$ units incorporated in polymer | 0.8 |
| $(C_6H_5)_2$ | 0.065 |
| Phenyl-2,5-cyclohexadiene | 0.045 |
| $H_2$ | 0.039 |
| $C_2H_4$ | 0.022 |
| 1,4-Cyclohexadiene | 0.021 |
| Phenyl-2,4-Cyclohexadiene | 0.021 |
| $C_2H_2$ | 0.020 |
| 1,3-Cyclohexadiene | 0.008 |

molecular-weight material produced was exceptionally troublesome because it tended to precipitate on the heat transfer surfaces.

Those olefin compounds whose double bonds are activated by substitution at the vinyl position (vinyl monomers) readily polymerize in good yield on irradiation. The mechanism depends on the nature of the monomer and the experimental conditions. Certain monomers such as vinyl halides can polymerize only by free radical mechanisms. Others, like isobutylene, polymerize only by ionic mechanisms. Some monomers can polymerize either way (eg, styrene). Free-radical polymerization is favored by increasing temperatures (typical activation energies are in the region of 20 kJ $mol^{-1}$) and by the absence of free-radical inhibitors like oxygen and benzoquinone. Ionic polymerization tends to predominate at lower temperatures and is inhibited by a different range of inhibitors, including water, methanol, and acetone. Polymerization by radiation has been of considerable interest to industry, and a few processes have passed into industrial practice.

In the irradiation of chlorides, bromides, and iodides, the carbon–halogen bond tends to break most easily. Yields from ethyl iodide[25] are listed in Table 11-8. Ion–molecule, electron-attachment, and free-radical reactions

**Table 11-8.** Products Formed by $\gamma$-Irradiation of Ethyl Iodide

| Product | Yield at dose of $1.6 \times 10^4$ Gy (molecules/100 eV) |
|---|---|
| $C_2H_4$ | 2.20 |
| $I_2$ | 2.12 |
| $C_2H_6$ | 1.92 |
| $n$-$C_4H_{10}$ | 0.33 |
| HI | 0.33 |
| $H_2$ | 0.23 |
| $C_2H_2$ | 0.09 |
| $CH_4$ | 0.01 |

all play a part in the mechanism. One of the most interesting processes is seen with alkyl chlorides and bromides, where isomerization can occur in very high yield (eg, $G = 60$ or more). For example, irradiation of *n*-propyl or butyl chloride gives high yields of secondary propyl or butyl chloride, and irradiation of isobutyl chloride or bromide gives *t*-butyl chloride or bromide. The mechanism is likely to be a free-radical chain reaction and may perhaps involve an equilibration with halogen atoms in the essential propagation step. For isobutyl bromide[26] the $(CH_3)_2\dot{C}CH_2Br$ radical could, by means of isobutylene and the bromine atom, yield $(CH_3)_2CBr\dot{C}H_2$:

$$(CH_3)_2\dot{C}CH_2Br \rightarrow (CH_3)_2CBr\dot{C}H_2 \quad (11\text{-}33)$$

which would then abstract a hydrogen atom to form *t*-butyl bromide and another radical to continue the chain. Reaction (11-33) could also take place intramolecularly, or, indeed, there may be only a single form of the bromoalkyl radical rather than the two in Eq. (11-33). Another interesting halide is carbon tetrachloride, which produces only hexachloroethylene and chlorine in quite small yield ($G < 1$ molecule/100 eV). This is not due to any intrinsic stability, but because $\cdot CCl_3$ radicals formed with normal yield react with chlorine within a sequence that adds up to zero effect:

$$CCl_4 \rightarrow CCl_3\cdot + Cl\cdot \quad (11\text{-}34)$$

$$CCl_3\cdot + Cl_2 \rightarrow CCl_4 + Cl\cdot \quad (11\text{-}35)$$

$$2Cl\cdot \rightarrow Cl_2 \quad (11\text{-}36)$$

The small amount of hexachloroethylene and chlorine must be formed by minor processes.

Yields of products obtained by irradiating pure ethanol to doses in the region of $10^3$ to $10^4$ Gy are shown in Table 11-9.[27] The yields of hydrogen and acetaldehyde are significantly more dose dependent at relatively low

**Table 11-9.** Products Formed by $\gamma$-Irradiation of Ethanol

| Product | Yield at dose of $10^3$–$10^4$ Gy (molecules/100 eV) |
|---|---|
| $H_2$ | 4.6 |
| $(CH_3CHOH)_2$ | 2.2 |
| $CH_3CHO$ | 1.9 |
| $CH_4$ | 0.6 |
| $H_2O$ | 0.5 |
| $C_2H_6$ | 0.24 |
| $C_2H_4$ | 0.14 |
| HCHO | 0.13 |
| $CH_3CH(OH)CH_2OH$ | 0.13 |
| $CH_3CH(OH)C_2H_5$ | 0.08 |
| CO | 0.06 |
| $CH_3CH(OH)CH_2CH_2OH$ | 0.05 |

doses than are the yields from the other liquids so far discussed. For example, at less than about 2 Gy the yields (in molecules/100 eV) are $G(H_2) = 5.8$ and $G(CH_3CHO) = 3.7$. The yields are also more dependent on impurities. The reason is that much, if not all, of the hydrogen derives from the reaction of solvated electrons: product acetaldehyde competes with the hydrogen-generating steps by taking up solvated electrons:

$$e_{alc}^{-} + CH_3CHO \rightarrow CH_3CHO^{-}\cdot \quad (11\text{-}37)$$

leading to a reduction in $G(H_2)$ while giving the deprotonated form of the hydroxyalkyl radical $CH_3\dot{C}HOH$. These radicals subsequently react by dimerizing and by disproportionating, resulting in a net loss of acetaldehyde as well as of hydrogen whenever Reaction (11-37) takes place.

Other alcohols resemble ethanol in giving as major products hydrogen, carbonyl compound (aldehyde or ketone), and the diol containing twice as many carbon atoms as the original alcohol. Ethers have not been as extensively studied as alcohols but appear to resemble alcohols in giving hydrogen and the dimeric compound with twice as many carbon atoms as the parent. Substantial fission of the C—O bond also seems to take place, leading in the case of noncyclic ethers to the formation of alcohols and carbonyl compounds.

Results of one determination of the radiolysis products of acetone are given in Table 11-10.[28] It can be seen that in contrast to aliphatic hydrocarbons or alcohols, but as with halides, little hydrogen is formed. This is understandable on the basis that the electrons are captured by acetone

**Table 11-10.** Products Formed by $\gamma$-Irradiation of Acetone

| Product | Initial yield (Molecules/100 eV) |
|---|---|
| $CH_4$ | 2.00 |
| $CO$ | 0.75 |
| $H_2$ | 0.54 |
| $(CH_3CO)_2$ | 0.47 |
| $CH_3COOH$ | 0.31 |
| $C_2H_6$ | 0.29 |
| $(CH_3COCH_2)_2$ | 0.28 |
| $CH_3COC_2H_5$ | 0.21 |
| $CH_3CH(OH)CH_3$ | 0.2 |
| $CH_3CHO$ | 0.14 |
| $CH_3COCH_2COCH_3$ | 0.11 |
| $C_2H_4$ | 0.027 |
| $C_3H_8$ | 0.018 |
| $C_3H_6$ | 0.015 |
| $C_2H_2$ | 0.012 |
| $C_3H_4$ | 0.006 |

molecules, and the various excitations, decompositions, and recombinations give rise to C-C scission. Carboxylic acids similarly give fairly low yields of hydrogen, but give $CO_2$ as a major product, together with hydrocarbons. Lists of products formed from other compounds and further data on those discussed above are given elsewhere.[29]

It will be clear from what has gone before that radiation would be expected to give quite different products from organic liquids when substances are present that could scavenge intermediates such as free radicals. Oxygen, for example, causes oxidation products like $H_2O_2$ and/or organic peroxides to appear. Irradiation of appropriate mixtures has led to the discovery of numerous interesting reactions.[30,31] Here the radiation chemistry of organic liquids, having started with purely physical considerations, enters the field of organic chemistry.

## ACKNOWLEDGMENTS

The author is grateful to his colleagues, especially Drs. Hoey and Land, for helpful discussions, and to Drs. Buxton and Warman for reviewing the manuscript. Some of the work on which this chapter is based was supported by grants from the Cancer Research Campaign.

## REFERENCES

1. Onsager, L. *Phys. Rev.* **1938**, *54*, 554.
2. Allen, A. O. "Yields of Free Ions Formed in Liquids by Radiation"; U.S. Department of Commerce: Washington, D.C., 1976, Report NSRDS-NBS 57.
3. Schmidt, W. F., Allen, A. O. *J. Chem. Phys.* **1970**, *52*, 4788.
4. Warman, J. M. In "The Study of Fast Processes and Transient Species by Electron Pulse Radiolysis"; Baxendale, J. H.; and Busi, F. Eds.; Reidel: Dordrecht, 1983, pp. 433–533.
5. Based on values given in Dorfman, L. M.; Jou, F. Y.; Wageman, R. *Ber. Bunsenges. Phys. Chem.* **1971**, *75*, 681, but more recent determinations from several laboratories have been incorporated.
6. Baxendale, J. H.; Wardman, P. *J. Chem. Soc., Faraday Trans. 1* **1973**, *69*, 584.
7. Warman, J. M.; Asmus, K.-D.; Schuler, R. H. *J. Phys. Chem.* **1969**, *73*, 931.
8. Dorfman, L. M. *Acc. Chem. Res.* **1970**, *3*, 224.
9. Dorfman, L. M.; Sujdak, R. J.; Bockrath, B. *Acc. Chem. Res.* **1976**, *9*, 352.
10. Williams, F. *J. Am. Chem. Soc.* **1964**, *86*, 3954.
11. Dorfman, L. M. In "Fast Reactions in Energetic Systems"; Capellos, C.; and Walker, R. F. Eds.; Reidel: Dordrecht, 1981, pp. 95–109.
12. Rzad, S. J.; Infelta, P. P.; Warman; J. M.; Schuler, R. H. *J. Chem. Phys.* **1970**, *52*, 3971.
13. Allen, A. O.; Holroyd, R. A. *J. Phys. Chem.* **1974**, *78*, 796.
14. Value for cyclohexane from Choi, H. T.; Askew, D.; Lipsky, S. *Radiat. Phys. Chem.* **1982**, *19*, 373; other values from Rodgers, M. A. J. In "The Study of Fast Processes and Transient Species by Electron Pulse Radiolysis"; Baxendale, J. H.; and Busi, F., Eds.; Reidel: Dordrecht, 1983, pp. 535–550.
15. Luthjens, L. H.; Codee, H. D. K.; DeLeng, H. C.; Hummel, A. *Chem. Phys. Lett.* **1981**, *79*, 444.
16. Land, E. J. *Proc. Roy. Soc.* **1968**, *A305*, 457.

17. Holroyd, R. A. In "Fundamental Processes in Radiation Chemistry"; Ausloos, P., Ed. Wiley Interscience: New York, 1968, pp. 413–514.
18. Fessenden, R. W.; Schuler, R. H. *J. Chem. Phys.* **1963**, *39*, 2147.
19. Simic, M.; Hayon, E. *J. Phys. Chem.* **1971**, *75*, 1677.
20. Ho, S. K.; Freeman, G. R. *J. Phys. Chem.* **1964**, *68*, 2189.
21. Swallow, A. J. "Radiation Chemistry"; Longman: London, **1973**, p. 175.
22. Wojnárovits, L. In "Radiation Chemistry of Hydrocarbons"; Földiák, G. Ed.; Elsevier: Amsterdam, 1981, pp. 177–251.
23. Davison, W. H. T.; Pinner, S. H.; Worrall, R. *Chem. Ind.* (London) **1957**, 1274.
24. Based on references listed by Hoigné, J. In "Aspects of Hydrocarbon Radiolysis"; Gäumann, T.; and Hoigné, J., Eds.; Academic Press: London, 1968, pp. 61–151.
25. Gillis, H. A.; Williams, R. R.; Hamill, W. H. *J. Am. Chem. Soc.* **1961**, *83*, 17.
26. Bakale, D. K.; Gillis, H. A. *J. Phys. Chem.* *74*, 2074.
27. Freeman, G. R. "Radiation Chemistry of Ethanol; A Review of Data on Yields, Reaction Rate Parameters, and Spectral Properties of Transients"; U.S. Department of Commerce: Washington, D.C., 1974, Report NSRDS-NBS 48.
28. Barker, R. *Trans. Faraday Soc.* **1963**, *59*, 375.
29. Tabata, Y. (ed.) CRC Handbook of Radiation Chemistry"; in preparation.
30. Wagner, C. D. In "Advances in Radiation Chemistry"; Burton, M.; and Magee, J. L., Eds.; Wiley Interscience: New York, 1969, Vol. 1, pp. 199–244.
31. Vereschchinskii, I. V. In "Advances in Radiation Chemistry"; Burton, M.; and Magee, J. L. Eds; Wiley Interscience: New York, 1972, Vol. 3, pp. 75–123.

# 12

# Radiation Chemistry of Colloidal Aggregates

J. K. THOMAS

*Department of Chemistry,*
*University of Notre Dame*

## INTRODUCTION

Full descriptions of the organized assemblies discussed in this chapter have been given elsewhere,[1] and it suffices here to state briefly our terminology. The systems are colloidal and consist of small particles (2 nm to 20 nm in radius) suspended either in aqueous or organic bulk media. In many systems of micelles or microemulsions, the colloidal species result from a specific gathering of surfactant molecules into an organized assembly. In bulk aqueous media the interior of the assembly is alkane or hydrophobic in nature, whereas in bulk oil media the interior of the assembly is hydrophilic (aqueous) in character. The specific details of the systems are known (as shown in other chapters), and we can use this information with confidence in discussing chemical reactions. The terms *suspended particle, colloidal particle, micelle, vesicle, microemulsion,* and *organized assembly* will be used interchangeably in this discussion.

## OVERVIEW OF REACTIONS IN ORGANIZED ASSEMBLIES

Figure 12-1 shows schematically several possible reactions that might occur in colloidal systems as a consequence of the organized and assembled colloid.


*Radiation Chemistry: Principles and Applications*

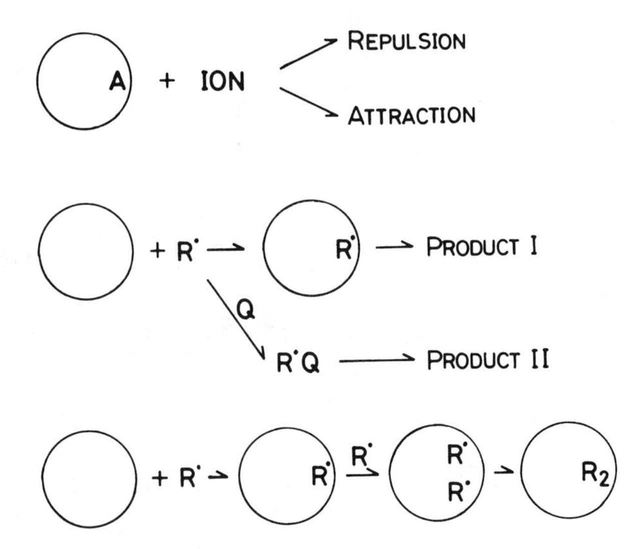


**Figure 12-1.** Possible reactions in micellar system. (Large circle = micelle; R· = radical; A and Q = reactant; $R_2$ = product.)

This brief consideration is useful to bear in mind when considering the detailed kinetics in later sections.[2,3]

The surfaces of all aggregates, and in particular micelles, may be divided into three types: positively charged, negatively charged, or uncharged. These situations usually arise from ionization of groups on the particle surfaces. Situation (a) in Figure 12-1 illustrates electrostatic attraction or repulsion between an ionic species, such as $e_{aq}^-$ or $Cu^{++}$, and a charged particle surface, thus dramatically affecting the mode of reaction of these species with a solute S contained in the particle. Quite a large enhancement or retardation of reaction can result from these simple electrostatic effects.

Situation (b) illustrates the attraction of two reactive species to a micelle either via electrostatic attraction, for example, $Br_2^-$ on a cationic micelle, or by hydrophobic affiliation of R· in the aqueous phase. Typically, R· could be large hydrocarbon residues such as heptyl, dodecyl, and others.

Finally, it is important to draw attention to the energy of the charged micellar surface, which may produce a field of $10^5$–$10^6$ V/cm at the surface. It is anticipated that such fields can influence the transition state of a reaction and, thence, the outcome of this reaction. The redox potential of the reactions may also be altered significantly by the surface, again changing the reaction.

It is now opportune to consider specific examples of the influence of organized assemblies on high-energy radiation-induced reactions, bearing in mind the above broad generalizations.

# REACTIONS OF HYDRATED ELECTRONS

There are several accounts already available that discuss the reactions of $e_{aq}^-$ in micellar systems.[4,5] The main feature of interest is the strong electrostatic repulsion of $e_{aq}^-$ by anionic micelles and the strong attraction to cationic micelles that can strongly influence its reaction with a solute located in the micelle. Table 12-1 lists rate constants for several reactants of interest.

## *Reactions in Anionic Micelles*

Solutes such as naphthalene, biphenyl, and others, react at diffusion rates with $e_{aq}^-$ in homogeneous media.[6] The rate constants for reaction are severely reduced in anionic micelles such as sodium lauryl sulfate (NaLS), where the rate constant for reaction of $e_{aq}^-$ with pyrene is reduced from $10^{10}$ to less than $10^7\ M^{-1}\ s^{-1}$. The reduction in rate is not as marked with biphenyl[6] and naphthalene,[7] and it is suggested that $e_{aq}^-$ reacts with these solutes in the aqueous bulk. Both naphthalene and biphenyl are more soluble in water than pyrene, and hence a small but significant aqueous concentration of the solute exists. In fact, this situation is readily shown by means of an electron transfer reaction.[6] The anion of biphenyl, $\phi_2^-$, readily transfers an electron to pyrene in homogeneous solutions:

$$\phi_2^- + \text{pyrene} \rightarrow (\text{pyrene}^-) + \phi_2 \qquad (12\text{-}1)$$

In a NaLS micellar system of $\phi_2$ and pyrene, pulse radiolysis leads to $\phi_2^-$, but no subsequent transfer to pyrene is observed. This is because $\phi_2^-$ in the aqueous phase experiences difficulty in approaching pyrene in the negatively charged micelle. Such effects are common in micellar systems, and care regarding the partition of solute between the micelle and aqueous phase has to be taken. The partition data for many solutes between micelles and water have been reported and rationalized in terms of thermodynamic properties of the solutes.[8,9]

The rates of reaction of $e_{aq}^-$ with solutes in anionic micelles increase if the ionic strength of the solution is increased, and if a polar group is placed on the solute to bring it further into the micelle–water interface. Both conditions lead to a reduction in the repulsion experienced by $e_{aq}^-$ on approaching the surface: Increased ionic strength decreases the surface charge of the micelle and also decreases the repulsion effect, and location of the solute further towards the aqueous phase reduces the distance into the micelle that $e_{aq}^-$ has to travel. Two explanations have been put forward to explain the decreased reaction rates. The mechanism of electron reaction with the solute may involve tunneling to the solute while still some distance from the micelle. Gratzel and co-workers[10] have published many data to substantiate this mechanism. The alternative explanation concerns the conventional electrostatic effect on reaction rates as expressed by the Debye

**Table 12-1.** Rate Constants for Reactions of $e_{aq}^-$ ($10^9\ M^{-1}\ s^{-1}$)

| Solute | Rate constant in aqueous ethanol solution | Micelle | | | | | | |
|---|---|---|---|---|---|---|---|---|
| | | NaLS | NaTC | CTAB | Igepal | Triton | Lyso-lecithin | Distearyl lecithin |
| Pyrene | 10 | $<10^{-4}$<br>$<10^{-2}$<br>$4.8 \times 10^{-2}$<br>(0.2 *M* NaCl) | 0.13 | $<10^{-3}$<br>$>10^{-2}$ (ISS) | $1.7 \times 10^{-2}$<br>1.7 | $3.8 \times 10^{-2}$<br>2.0 | 3.9 (86)<br>3.8 (86) | $<10^{-3}$<br>$<10^{-2}$<br>fast |
| Biphenyl | 5.0 | 0.13 | | | 3.8 | 4.0 | | 0.005 |
| Pyrene carboxaldehyde | 16 | 2.5 | | | | | | |
| Amino pyrene | 14 | 1.1 | | | | | | |
| Benzophenone | 10 | 1.5 | 20 | | 2.0 | | | |
| Nitro anthracene | 10 | 1.5 | | 90 | | | | |

formulation[11]:

$$k = \frac{4\pi\gamma DN}{1000}\left\{\left|\frac{z_1 z_2 e^2}{\gamma EkT}\right| \left(\exp\left\{\frac{z_1 z_2 e^2}{\gamma EkT}\right\}^{-1}\right)\right\} \tag{12-2}$$

where $k$ is the rate constant, $z_1$ and $z_2$ the charges of $e_{aq}^-$ and micelle, $\gamma$ the separation for reaction, and $D$ the dielectric constant. This type of calculation quantitatively fits the data. It is difficult to decide between the two mechanisms at this stage, and this review will adhere to the more conventional description in terms of electrostatic repulsion.

The assumption that hydrated electrons react extremely slowly with solutes located well into the Stern layer can be used to assess binding constants of ions to micellar surfaces. For example, $Cd^{++}$ and $Cu^{++}$ bind to NaLS micelles with $K = 820$ and $1400\ M^{-1}$, respectively.[12] The association equilibrium

$$M^{++} + \text{micelle} \rightleftharpoons M_{\text{micelle}}^{++} \tag{12-3}$$

follows a Langmuir-type equation

$$\frac{1}{[M^{++}]_{\text{micelle}}} = \left(\frac{1}{KS}\right)\left(\frac{1}{[M^{++}]_{\text{micelle}}} + \frac{1}{S}\right) \tag{12-4}$$

Similar data were observed[13] for doubly charged methyl viologen, $MV^{++}$. The binding of doubly charged ruthenium bipyridyl ions to polyvinyl sulfate also leads to a dramatic decrease in its rate of reaction with $e_{aq}$[14].

## *Reactions in Cationic Micelles*

The strongly positive electrostatic field of cationic micelles such as cetyl trimethyl ammonium bromide (CTAB) attracts the electron and greatly enhances the rate of reaction of $e_{aq}^-$ with solutes located in the micelle.[15] In all cases the anion of the solute $S^-$ is observed as the direct product of the reaction.

$$S_{\text{micelle}} + e_{aq}^- \rightarrow S_{\text{micelle}}^- \tag{12-5}$$

The strong catalysis (10-fold or greater) is readily explained in terms of either theory indicated earlier for anionic micelles. The rate of encounter of $e_{aq}^-$ with the cationic micelles is readily measured by using a reactive micelle such as cetyl pyridinium chloride.[16] The monomer form of this surfactant reacts with $e_{aq}^-$ with $k \sim 10^{10}\ M^{-1}\ s^{-1}$—in other words, at every encounter. The reaction rate with the micellar form of the surfactant is greater than $10^{12}\ M^{-1}\ s^{-1}$, a catalytic increase of about 100-fold.

## *Special Systems*

### Proteins

It is well established that $e_{aq}^-$ reacts with many proteins. In particular, bovine serum albumin (BSA) has received much attention, and it is suggested that

$e_{aq}^-$ reacts with —S—S linkages in the protein. Altering the charge of the protein, by adding surfactant, dramatically affects the rate of $e_{aq}^-$ capture.[17] Figure 12.2 shows typical data for absorption of NaLS and CTAB on BSA, on the yield of the $e_{aq}^-$ reaction observed by means of absorption at $\lambda$ = 410 nm, and on the reaction rate. Small amounts of NaLS first increase the reaction rate, followed by a steady decrease with further addition; the yield of the reaction product shows similar effects. CTAB immediately increases the rate of reaction up to the critical micelle concentration of the surfactant ($\sim 10^{-3}$ *M*). The yield of product shows a corresponding increase with [CTAB]. The effects are readily explained by simple electrostatic repulsion or attraction of $e_{aq}^-$ by the superficial charge on the protein. Photochemical studies where pyrene is incorporated into the protein and subsequently

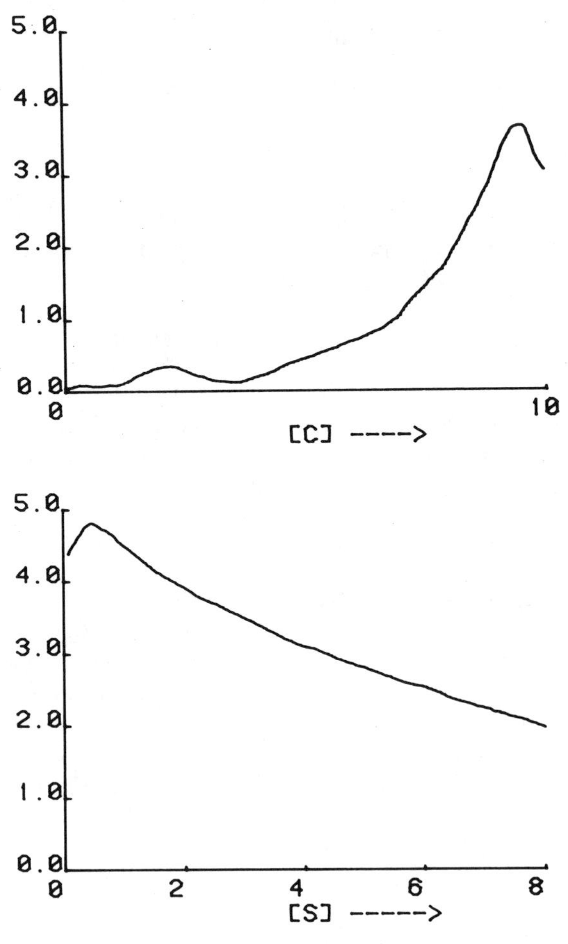


**Figure 12-2.** Effect of CTAB, and NaLS on the relative rate of reaction of $l_{aq}^-$ with BSA. [CTAB] – [C] in $10^{-3}$ *M*; [NaLS] – [S] in $10^{-4}$ *M*.

quenched by various molecules (for example, $I^-$) show directly comparable data to that reported for $e_{aq}^-$. It is noteworthy that this approach, that is, observations of the effects of additives on the rates of $e_{aq}^-$ reaction with proteins, has been successfully used to study the binding of several drugs to proteins.[18]

**Carrageenan Gels**

Quite rigid aqueous polysaccharide gel matrices may be formed with $\kappa$-carrageenan and agarose.[19] The mixtures, apparently presenting a rigid microstructure to the external viewer, nevertheless show a fluid microstructure when viewed by means of reaction kinetics or fluorescence polarization. For example, the rate constant of the reaction

$$e_{aq}^- + NO_3^- \rightarrow NO_3^{2-} \tag{12-6}$$

is $2.08 \times 10^{10}\ M^{-1}\ s^{-1}$ in a $\kappa$- carrageenan solution and $1.61 \times 10^{10}\ M^{-1}\ s^{-1}$ in a $\kappa$-carrageenan gel. Similar effects are noted with hydroxyl radical reactions, suggesting that small species can readily diffuse in the gelled form of the polysaccharide.

Anionic and neutral surfactants in micellar form can be incorporated into these gel systems by interaction with the polysaccharide network. The critical micelle concentrations of surfactants in the gels are lower than those obtained in water alone, and the micellar environments are more fluid to aromatic probes, although the polarities of the environments are similar in all systems.

**Nonionic Micelles**

Nonionic micelles, such as those formed with Triton X100, also reduce the rate of reaction of $e_{aq}^-$ with solutes located therein.[6,20–22] For the most part the rate constants are some three to five-fold lower than those observed in a homogeneous solution. Incorporation of charged surfactants into these micelles either leads to further decrease in reactivity, as with NaLS, or an increased reactivity, as with CTAB. The environment of the solutes in these micelles is quite polar,[23] but the molecule is probably surrounded by the large ethylene oxide mantle of the nonionic surfactant. This could lead to the decreased reactivity of $e_{aq}^-$ in these systems. Some molecules, such as $\beta$-carotene, may lie further into the core of the micelle because no reactivity with $e_{aq}^-$ is observed.[24] A mediator molecule such as biphenyl may be used to facilitate transport of electrons to the $\beta$-carotene.

**Vesicles**

Vesicles are usually formed by sonication of double-stranded surfactant-type molecules, such as dialkyldimethyl ammonium halides or lecithin, a naturally occurring component of biosystems.[25] The geometry of these molecules requires that they form closed bilayers rather than micelles. However, these entities still possess many of the properties associated with micelles; in particular, hydrophobic molecules can be placed in the bilayer region. The large size of vesicles dictates that only low vesicle concentrations ($\sim 10^{-7}$–$10^{-6}$ $M$) can be used in optical experiments before sample turbidity spoils the experiments. Hence, it is usual to have several solute molecules per vesicle in pulse radiolysis, or laser photolysis, experiments. The kinetics of radical reactions are thus modified, as the rate-controlling step becomes the rate at which the radical species encounters the vesicle, regardless of the solute concentration in that vesicle. Such effects have been observed in vesicles and micellar systems[3,26,27]. Below the phase transition temperature of the vesicles, the reactivity of $e_{aq}^-$ with pyrene is very low;[26] however, above the phase transition temperature appreciable reactivity is restored.[27] At high pyrene concentration (about 20 per vesicle) the electron capture rate constant is $10^9$ $M^{-1}$ $s^{-1}$, much like that in nonionic micelles.

Hydrated electrons react rapidly with solutes such as pyrene located in vesicles constructed from surfactants with quaternary ammonium head group.[28] The rate constants are similar to those observed in cationic micelles, and the kinetics are similar to those observed in lecithin vesicles.

It can be stated with some confidence that, to date, radiation-induced processes in the more complex vesicular systems can be quantitatively interpreted in terms of the structure of the vesicle and kinetic patterns learned from studies in micellar systems; no new features have to be introduced.

## REACTIONS OF HYDROGEN ATOMS

Little is known regarding reactions of H atoms in micellar systems. The only reports available[6,20] indicate that micelles do not affect reactions of H

**Table 12-2.** Rate Constants for Some Radiation-Induced Free-Radical Reactions in Micelles

| | $10^{-9}ko$ $M^{-1}$ $s^{-1}$ | | | | |
|---|---|---|---|---|---|
| Radical reactions | Water or alcohol | NaLS | CTAB | Bilayer distearyl-lecithin | Lyso lecithin micelle |
| OH | | 7.6 (monomer) 0.5 (micelle) | | | |
| OH + pyrene | | | | | |
| OH + benzene | | ~2 | ~3 | 1.7 | |
| H | | $1.2 \times 10^{-2}$ (monomer) (& micelle) | $1.6 \times 10^{-2}$ (monomer) (& micelle) | | |
| H + biphenyl | 6.0 | 6.0 | 6.0 | | |
| $O_2^*$ + $\alpha$ tocopherol (singlet $O_2$) | | 0.64 | | | |

atoms with molecules solubilized within them (Table 12.2). The surfactants themselves react with H atoms presumably by means of H-atom abstraction producing $H_2$. Thus, the molecule may be protected from H-atom attack due to this effect.

# REACTIONS OF HYDROXYL RADICALS

## *Micellar Systems*

Hydroxyl radicals are very reactive with micellar systems,[6,20,29] a result that is not unexpected from what has been observed in simple homogeneous systems. Counter ions can also be quite reactive with OH radicals: for example, the CTAB system where $Br^-$ reacts to produce $Br_2^-$;

$$OH + Br^- \rightarrow OH^- + Br \tag{12-7}$$

$$Br + Br^- \rightleftharpoons Br_2^- \tag{12-8}$$

Pulse radiolysis of nitrous-oxide–saturated CTAB solution produces $Br_2^-$; however, the decay of $Br_2^-$ is quite unusual[30,31] (Figure 12-3). Under conditions where less than one $Br_2^-$ is produced per micelle, that is, at high [CTAB], the decay rate of $Br_2^-$ is much slower than that observed in surfactant-free aqueous $Br^-$ solution. This is due to the adsorption of $Br_2^-$ on the cationic micellar surface. Hence, the approach of two $Br_2^-$ entities to each other for reaction to occur is diminished due to strong repulsion between the micelles. Reaction is only achieved when $Br_2^-$ exits a micelle and is attached to another micelle containing a $Br_2^-$, when rapid reaction ensues. The residence time of $Br_2^-$ on a micelle has been measured[30] as $1.5 \times 10^{-5}$ s.

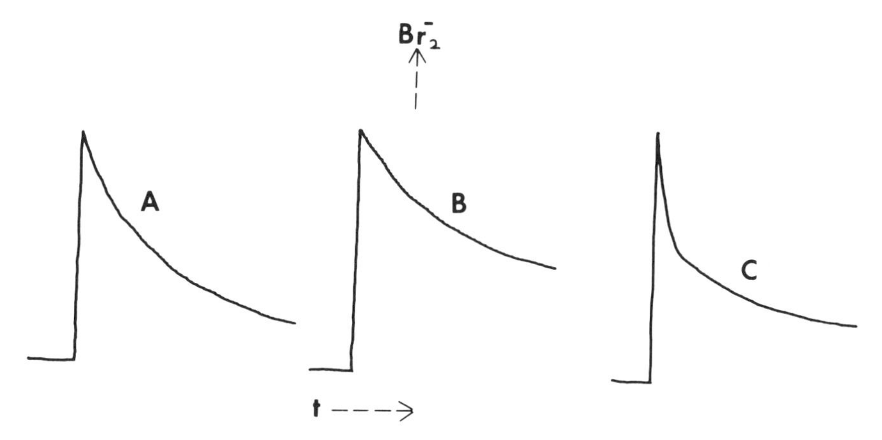


**Figure 12-3.** Representation of the effect of [CTAB] on the dismutation of two $Br_2^-$ radicals in the pulse radiolysis of $N_2O$ saturated solutions. A. [CTAB] = 0; B. [CTAB]: large (eg, 0.1 *M*); C. [CTAB]: small (eg, $10^{-2}$ *M*).

**Table 12-3.** Bimolecular Rate Constants[a] For Radiation-Induced Reactions in the Reversed Micellar System, 3% AOT/$H_2O$/Heptane

| Reactions | | $k(M^{-1}\,S^{-1})$ | | | |
|---|---|---|---|---|---|
| | $H_2O$% (v/v) | 6 | 3 | 2 | 1 |
| $b_{\phi_2}^{-} + PSA \rightarrow \phi_2 + PSA^{-}$ | | $1.05 \times 10^{10}$ $(6.3 \times 10^{8})$ | $1.84 \times 10^{10}$ $(5.5 \times 10^{8})$ | $3.7 \times 10^{10}$ $(7.4 \times 10^{8})$ | |
| $c_{\phi_2}^{T} + PSA \rightarrow \phi_2 + PSA^{T}$ | | $6.10^{9}$ $(3.6 \times 10^{8})$ | $7.5 \times 10^{9}$ $(2.25 \times 10^{8})$ | | $(6.5 \times 10^{9})$ $6.5 \times 10^{7}$ |
| $c_{\phi_2}^{T} + PBA \rightarrow \phi_2 + PBA^{T}$ | | $6.4 \times 10^{9}$ $(3.8 \times 10^{8})$ | | | |
| $\phi_2^{T} + Py \rightarrow \phi_2 + Py^{T}$ | | $2.7 \times 10^{10}$ | | | |
| $e_{\phi_2}^{T} + PTSA \rightarrow \phi_2 + PTSA^{T} < 2 \times 10^{9}$ | | | | | |
| $H_3O^{+} + \phi_2^{-} < \phi_2H + H_2O$ | | $5 \times 10^{9}$ $(3 \times 10^{8})$ | $5.8 \times 10^{9}$ $(1.78 \times 10^{8})$ | $4.6 \times 10^{9}$ $(9.2 \times 10^{7})$ | $4.7 \times 10^{9}$ $(4.7 \times 10^{7})$ |
| $Cu^{++} + \phi_2 - Cu^{+} + \phi_2$ | | $6.3 \times 10^{9}$ $(3.78 \times 10^{8})$ | $3.0 \times 10^{9}$ $(1.1 \times 10^{8})$ | $3.4 \times 10^{9}$ $(7.9 \times 10^{7})$ | $1.7 \times 10^{9}$ $(1.7 \times 10^{7})$ |
| $e^{-}_{aq} + \phi_2 \rightarrow \phi_2^{-}$ | | $2.0 \times 10^{10}$ | | | |
| $l^{-} + \text{micelle} \rightarrow l^{-}_{aq}$ | | $>10^{13}$ | | | |

[a] Rate constants are calculated from the bulk solute concentration over the whole solution. The rate constants in the parentheses are calculated from the local solute concentration in the micelle. From refs. 29, 32.

[b] $\phi_2$ anion of biphenyl.

[c] $\phi_2^{T}$ triplet excited state of biphenyl.

At lower [CTAB], for example $1 \times 10^{-2}$ $M$, the possibility exists for two $Br_2^-$ to be formed instantaneously on a micelle. These pairs of $Br_2^-$ react rapidly on the micelle surface due to the proximity effect. A slower reaction follows due to isolated $Br_2^-$. At very low [CTAB], for example $2 \times 10^{-3}$, $M$ the rapid dismutation of $Br_2^-$ on individual micelles is the predominant reaction.

OH radical reactions with solutes are not significantly affected by the micelles; a few rate constants are given in Table 12-3.

### *Vesicle Systems*

Hydroxyl radicals react readily with lecithin vesicles, electron paramagnetic resonance (EPR) and other spectroscopic data indicating an attack on the choline region of the molecule.[26] This is not unexpected, because the OH radical encounters this part of the molecule as it diffuses from the aqueous phase into the vesicle. Solutes such as pyrene, solubilized in the vesicle, also react with OH radicals, the kinetics of the process being similar to those already discussed for reactions of $e_{aq}^-$ with solutes in vesicles.

Extensive irradiation of the vesicle suspension (doses of several megarads) leads to damage and modified vesicles.

## REACTIONS OF SECONDARY RADICALS

Micelles and other organized assemblies can also influence the reactions of the secondary radicals formed by $e_{aq}^-$, H, and OH radicals. Figure 12-4 illustrates one such situation where radicals of the surfactant NaLS or sodium octylsulfate (NaOS) are produced by radiolysis of $N_2O$-saturated solutions of these surfactants. The spin trap $(CH_3)_3C{-}NO$ is used to trap the surfactant radicals, forming in turn a stable nitroxide radical.[32]

$$OH + SH \rightarrow S\cdot + H_2O \tag{12-9}$$

$$S\cdot(CH_3)_3C{-}NO \rightarrow (CH_3)_3{-}C\overset{\displaystyle S}{\overset{|}{N}}{-}O\cdot \tag{12-10}$$

No trapping of S· is observed below the critical micelle concentration (cmc) of these surfactants ($\sim 10^{-2}$ $M$ for NaLS and $\sim 0.35$ for NaOS), whereas extensive trapping is observed above the cmc. The data clearly show that under homogeneous conditions the reaction of S· and $(CH_3)_3C{-}NO$ is too slow to compete with radical–radical reactions while the micelles promote the trapping reaction as they hold S· and the nitroso-compound in close proximity.

Reactions of surfactant radicals have also been studied by pulse radiolysis, the OH adduct of sodium H-($6^1$-dodecyl)-benzene sulfonate,[33] the radical

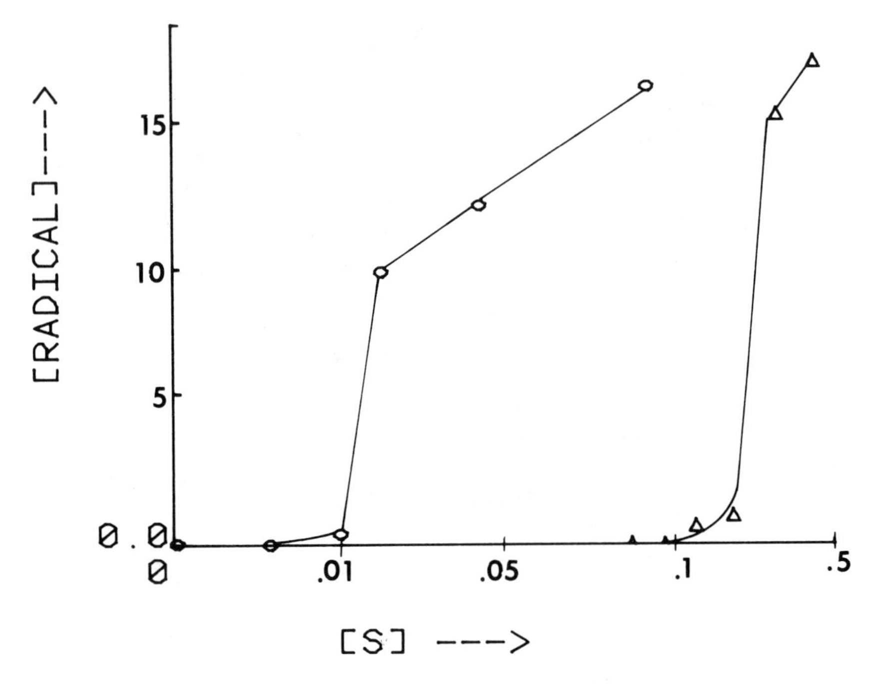


**Figure 12-4.** Effect of surfactant concentration [S] denoted as NaLS (○) and NaOS (△) on the yield of nitroxide radicals in the radiolysis of $N_2O$-saturated solution of these surfactants and nitroso spin trap.

formed by H-atom abstraction from $C_{16}H_{33}(OCH_2CH_2)OH$, and $C_{14}H_{29}(OCH_2CH_3)_3—SO_3Na$[34]; and radicals of NaLS[35] were observed by means of their reduction of ferricyanide. In all cases the surfactant radical could exist under two conditions: (a) as a monomeric surfactant radical S·, or (b) as a radical in a micelle M·:

$$S\cdot + \text{micelle} \rightarrow M\cdot \qquad (12\text{-}11)$$

The reactivity of S· was greatly affected by the environment of the radical. For example, $Fe(CN)_6^{3-}$ reacts rapidly with S· but not with M·, due to repulsion of $Fe(CN)_6^{3-}$ from the site of S· on the micelle by the strongly negative micellar field. The radical dimerization reaction was also affected by the micellar environment, being promoted if two S· occupied one micelle and decreased if they occupied different micelles. Analysis of the data give the rate constants for the equilibrium of a surfactant radical with a micelle, the absolute numbers comparing favorably with those in the literature. In some instances prolonged irradiation led to extensive cross-linking of the micelles.[34]

The so-called harvesting of reaction products by micelles is a general and useful process. Figure 12-5 shows the rate of growth of the benzoquinone anion $BQ^-$, following reaction of benzoquinone BQ and R·, R· being formed

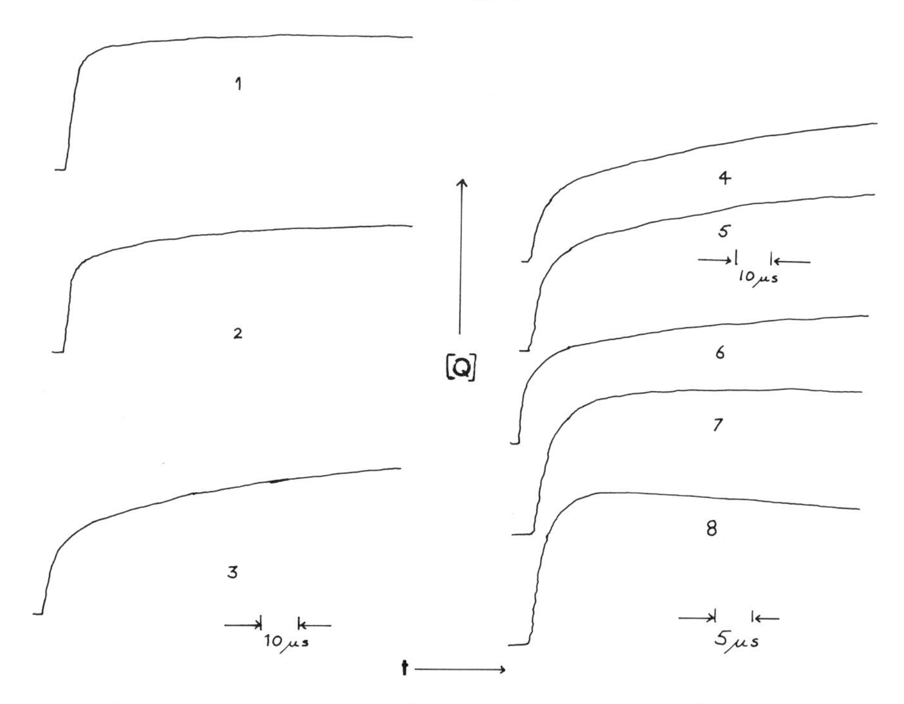


**Figure 12-5.** Effect of [NaLS] and [benzoquinone] on the yield of benzoquinone anion $[Q^-]$ in the pulse radiolysis of $N_2O$ saturated solution.

| | [NaLS] *M* | [Benzoquinone] $10^{-3}$ *M* |
|---|---|---|
| 1 | 0.1 | 5.0 |
| 2 | 0.1 | 3.0 |
| 3 | 0.1 | 1.0 |
| 4 | 0.3 | 1.0 |
| 5 | 0.1 | 1.0 |
| 6 | 0.06 | 1.0 |
| 7 | 0.03 | 1.0 |
| 8 | 0.01 | 1.0 |

by OH-attack on monomeric or micellar NaLS.[35] At [NaLS] below the cmc ($8.1 \times 10^{-3}$ *M*) the rate is proportional to both [NaLS] and to [BQ]. Above the cmc the growth of $BQ^-$ no longer shows simple kinetics. The traces clearly show two-step kinetic processes fast growth and a slow growth. On increasing the concentration of BQ, the extent of the fast growth increases, whereas increasing the concentration of NaLS causes a decrease in intensity of the fast growth; but the total intensity of absorption remains constant. These observations lead to the assumption that the fast growth is the electron transfer reaction from R· to BQ in the water phase, and the slow growth is due to the electron transfer reaction from R· attached to the micelles to

BQ in the water phase. This assumption is in accord with the fact that if [BQ] increases, then both the rate and the yield of reaction increase. However, on increasing the concentration of micelles, the aqueous quinone concentration $[BQ]_{aq}$ will be reduced according to equilibrium (12-13) below, and the rate of the first reaction will be increased. These processes cause a slowing down of the rate and a reduction of the yield to the R· + BQ reaction.

$$R\cdot + (Mic) \rightarrow (Mic,R\cdot) \quad (12\text{-}12)$$

$$BQ + (Mic) \rightleftharpoons [BQ;Mic] \quad (12\text{-}13)$$

$$R\cdot + BQ \rightarrow BQ^- + R^+ \quad (12\text{-}14)$$

## POLYMERIZATION

It was shown that extensive irradiation of surfactants containing ethylene oxide units leads to cross-linking of the material.[24] Similar effects are noted with unsaturated fatty acids[36] such as sodium 11-undecenoic acid, which, in micellar form, polymerizes to form a polymer with 10 surfactant units. The radiation doses required are in the several megarad range. A recent report outlines the polymerization of a vesicle form of lecithin containing a methacrylate group.[37] The polymerization was initiated thermally. However, there is no reason to believe that radiolysis of these systems would not produce similar results.

## ELECTRON TRANSFER REACTIONS

The rates of electron transfer from ethanol radical to several molecules solubilized in micelles have been measured by pulse radiolysis.[38] For the acceptors tetranitromethane, 1,2,4,5-tetracyanobenzene, and 1,2,4,5-tetrachlorobenzoquinone, in both CTAB, and NaLS micelles, the rate constants for electron and transfer from ethanol radicals were in the range $(3 \pm 1) \times 10^9\ M^{-1}\ s^{-1}$ and close to those observed in homogeneous solution. The anion $CO_2^-$ produced by

$$e_{aq}^- + CO_2 \rightarrow CO_2^- \quad (12\text{-}15)$$

or

$$HCOO^- + OH \rightarrow CO_2^- + H_2O \quad (12\text{-}16)$$

reacts rapidly with the triplet state of pyrene solubilized in CTAB micelles, The ground state of pyrene is apparently nonreactive, and the reverse reaction

$$(\text{pyrene})^- + CO_2 \rightarrow CO_2^- + \text{pyrene} \quad (12\text{-}17)$$

has a rate constant of near $10^7\ M^{-1}\ s^{-1}$.

The hydrated electron does not react efficiently with β-carotene solubilized in nonionic micelles, although other solutes/(e.g., biphenyl) react quite rapidly to produce solute anions. However, the yield of β-carotene anion can be improved if biphenyl is present in the micelle. In this case $e_{aq}^-$ first reacts to produce the biphenyl anion, which subsequently transfers an electron to β-carotene.[24] The biphenyl here acts as an intermediary in the electron transfer reaction. Electron transport through vesicles has also been demonstrated[39] by utilizing the β-carotene anion. Vesicles were prepared with $Eu^{3+}$ in the aqueous interior compartment and β-carotene in the lipid phase. Pulsed radiolysis produces $e_{aq}^-$ in the bulk aqueous phase, 40% of which react to produce β-carotene anions. Subsequently, these anions decay by electron transfer to $Eu^{3+}$ in the vesicle interior. In the absence of $Eu^{3+}$ the β-carotene anions decay slowly; in the absence of β-carotene the $e_{aq}^-$ in the bulk fail to react with $Eu^{3+}$ in the vesicle interior.

The micellar surface potential can significantly alter the equilibrium setup between two anions.[40] Pulse radiolysis has been used to study the equilibrium between anthraquinone sulfonate (AQS), and its anion $AQS^-$, and duroquinone (DQ), and its anion $DQ^-$. The equilibrium constant is changed in a manner predicted by solubilization of DQ at the micelle and AQS, $AQS^-$, and $DQ^-$ in the aqueous phase. The kinetics of the equilibration show that electron transfer at the micellar surface is important; that is, AQS and $DQ^-$ are also somewhat associated with the micelle. It was concluded that the association of a redox agent with a micelle affects the redox potential in much the same way as the p$K$ value of an acid is changed at a micelle surface.

## REVERSED MICELLES

Radiolysis of reversed micellar systems, where the bulk phase is hydrocarbon, initially leads to nonaqueous chemistry: that is, if the bulk phase is benzene or toluene, then excited states are formed; if the bulk phase is an alkane (eg, heptane), then both ions and excited states are formed. The most widely studied system is Aerosol OT (AOT) in heptane or isooctane and the rate constants for several reactions in this system have been determined. Pulse radiolysis leads to the observation of hydrated electrons. These come from electrons that are initially produced in the alkane phase but which are rapidly captured by the water bubbles.[41,42] The ratio of water to surfactant, $\omega_0$, defines the efficiency of electron capture. To see appreciable yields of $e_{aq}^-$ ($G > 0.1$), we must have $\omega_0$ greater than 10 when [AOT] ~ 0.1 $M$. The rate constants for reaction of $e_{aq}^-$ with solutes in the water pools increase with increasing $\omega_0$,[42,43] in a manner reminiscent of the variation of other physical properties of the water bubble, that is, NMR,[44] and fluorescence studies, and so on,[45] with water content.

Ions formed in the alkane phase by electron capture (eg, biphenyl anion) are also captured by the water bubbles and may react with polar solutes (eg, $Cu^{++}$ or $H^{+}$) located in the water pools.[46] Similar effects are noted with energy transfer from biphenyl triplet to various pyrene derivatives in the pools.

The chemistry of these systems is readily described in terms of what is already known about the radiation chemistry of alkanes or arenes with modifications of the subsequent chemistry by the water pools, which is understood from photochemical studies. Table 12-3 lists rate constants for various reactions in reversed micelles.

## INORGANIC COLLOIDS: METAL CLUSTERS

Colloids of silver atoms stabilized by surfactants are made by radiolysis of silver nitrate solutions.[47] These particles act as strong electron acceptors and organic-free radicals of high negative redox potential (eg, $\alpha$-alcohol radicals) transfer electrons to the silver clusters. The electrons stored in the Ag clusters subsequently reduce water to form $H_2$, or they may react with suitable solutes. Similar effects are noted with gold colloids.[48] Colloidal platinum may also be used[49] to catalyze $H_2$ formation by electron transfer. In particular, the reduced form of methyl viologen, $MV^{+}$ forms $H_2$ and methyl viologen in the presence of surfactant-stabilized colloidal platinum.

Early data on the radiation chemistry of colloidal particles is somewhat sparse, although it is reported[50–52] that radiolysis of colloidal systems may lead to stabilization or sensitization with regard to coagulation by electrolytes. A recent study[53] of colloidal silver bromide in aqueous suspension reports that the colloid is reduced to silver metal and $Br^{-}$ in the presence of ethanol, isopropanol, and acetone. The rate of this reduction is proportional to the surface area of the particles, and *G*-values of about unity are reported for particles with radius of 54 nm at a pH of 3.5 in 0.15 *M* ethanol. It is suggested that the reduction reaction site is on the particle surface rather than with homogeneously distributed $Ag^{+}$ ions.

It has recently been shown that $\gamma$-irradiation of heterogeneous mixtures of sulfur and water lead to the formation of sulfuric acid by OH radical attack on the colloid.[54].

## ACKNOWLEDGMENT

The author wishes to thank the NSF (grant CHE 78-24867) and the ARO (grant DAAG-80-K-0007) for research support.

# REFERENCES


1. Fendler, J. H. "Membrane Mimetic Chemistry"; Wiley Interscience: New York, 1982.
1b. Thomas, J. K. "Chemistry of Excitation at Interfaces"; ACS Symposium Series No. 181, Washington, 1984.
2. Thomas, J. K.; Almgren, M. In "Solution Chemistry of Surfactants"; Mittal, K. L.; Ed.; Plenum Press: New York, 1979, p. 559.
3. Thomas, J. K.; Chen, T. S. *J. Chem. Ed.* **1973**, *58*, 140.
4. Henglein, A.; Gratzel, M. "Solar Power and Fuels"; Academic Press: New York, 1977.
5. Thomas, J. K. *Acc. Chem. Res.* **1976**, *10*, 133.
6. Wallace, S. C.; Thomas, J. K. *Radiat. Res.* **1973**, *54*, 49.
7. Evers, E. L.; Jayson, G. G.; Robb, I. D.; Swallow, A. J. *J. Chem. Soc., Faraday Trans. 1* **1980**, *76*, 528.
8. Almgren, M.; Grieser, F.; Thomas, J. K. *J. Am. Chem. Soc.* **1979**, *101*, 279.
9. Almgren, M; Grieser, F.; Powell, J.; Thomas, J. K. *J. Chem. Eng. Data* **1979**, *24*, 205.
10. Gratzel, M.; Henglein, A.; Janata, E. *Ber. Bunsenges. Phys. Chem.* **1975**, *79*, 475.
11. Debye, P. *Trans. Electrochem. Soc.* **1942**, *82*, 265.
12. Gratzel, M.; Thomas, J. K. *J. Phys. Chem.* **1974**, *78*, 2248.
13. Rodgers, M. A. J.; Foyt, D. C.; Zimek, Z. A. *Radiat. Res.* **1978**, *75*, 296.
14. Matheson, M. S.; Meisel, D.; Rabani, J. *J. Am. Chem. Soc.* **1978**, *100*, 117.
15. Gratzel, M.; Kozak, J.; Thomas, J. K. *J. Chem. Phys.* **1975**, *62*, 1632.
16. Gratzel, M.; Patterson, L.; Thomas, J. K. *Chem. Phys. Lett.* **1974**, *29*, 393.
17. Cooper, M.; Thomas, J. K. *Radiat. Res.* **1977**, *70*, 312.
18. Phillips, G. O.; Power, D. M.; Davies, J. V. In "Fast Processes in Radiation Biology and Chemistry"; John Wiley, London, 1975, p. 180.
19. Wedlock, D. J.; Phillips, G. O.; Thomas, J. K. *Polymer J.* **1979**, *11*, 681.
20. Fendler, J. H.; Fendler, E. J. "Catalysis in Micellar and Macromolecular Systems"; Academic Press; New York, 1975.
21. Kalyanasundaram, K.; Thomas, J. K. In "Micellization, Solubilization, and Microemulsions"; Mittal, K., Ed.; Plenum Press; New York, 1977, Vol. 2, 569.
22. Proske, T. L.; Fischer, Cl-H; Gratzel, M.; Henglein, A. *Ber. Bunsenges. Phys. Chem.* **1977**, *81*, 816.
23. Kalyanasundaram, K.; Thomas, J. K. *J. Am. Chem. Soc.* **1977**, *99*, 2039.
24. Almgren, M; Thomas, J. K. *Photochem. Photobiol.* **1989**, *31*, 329.
25. Israelachvili, J. N.; Mitchell, D. J.; Ninham, B. W. *J. Chem. Soc., Faraday Trans. 2*, **1976**, *72*, 1525.
26. Barber, D. J. W.; Thomas, J. K. *Radiat. Res.* **1978**, *74*, 51.
27. Schnecke, W.; Gratzel, M. *Ber. Bunsenges. Phys. Chem.* **1977**, *81*, 821.
28. Henglein, A.; Proske, T. L.; Schnecke, W. *Ber. Bunsenges. Phys. Chem.* **1978**, *82*, 956.
29. Patterson, L.; Hasegawa, K. *Ber. Bunsenges. Phys. Chem.* **1978**, *82*, 951.
30. Proske, T. L.; Henglein, A. *Ber. Bunsenges, Phys. Chem.* **1978**, *82*, 711.
31. Frank, A. J.; Gratzel, M.; Kozak, J. J. *J. Am. Chem. Soc.* **1976**, *98*, 3317.
32. Bakalik, D.; Thomas, J. K. *J. Phys. Chem.* **1977**, *81*, 1905.
33. Henglein, A.; Proske, T. L. *J. Am. Chem. Soc.* **1978**, *100*, 3706.
34. Henglein, A.; Proske, T. L. *Maksonol. Chem.* **1978**, *179*, 2279.
35. Almgren, M.; Grieser, F.; Thomas, J. K. *J. Chem. Soc. Faraday Trans. 1* **1979**, *75*, 1674.
36. Larrabee, C. E.; Sprague, E. D. *J. Pol. Sci. Lett.* **1979**, *17*, 749.
37. Regen, S. L.; Bronislaw, C.; Singh, A. *J. Am. Chem. Soc.* **1980**, *102*, 6630.
38. Frank, A. J.; Gratzel, M.; Henglein, A.; Janata, E. *Ber. Bunsenges. Phys. Chem.* **1976**, *80*, 548.
39. Frank, A. J.; Gratzel, M.; Henglein, A. *Ber. Bunsenges. Phys. Chem.* **1976**, *80*, 393.
40. Almgren, M.; Grieser, F.; Thomas, J. K. *J. Phys. Chem.* **1979**, *83*, 3232.
41. Wong, M.; Gratzel, M.; Thomas, J. K. *Chem. Phys. Lett.* **1975**, *30*, 329.
42. Wong, M.; Griesser, F.; Thomas, J. K. *Ber. Bunsenges Phys. Chem.* **1978**, *82*, 937.
43. Beck, G.; Bakale, G.; Thomas, J. K. *J. Phys. Chem.* **1981**, *85*, 1062.
44. Wong, M.; Nowak, T.; Thomas, J. K. *J. Am. Chem. Soc* **1977**, *99*, 4730.
45. Wong, M.; Gratzel, M.; Thomas, J. K. *J. Am. Chem. Soc.* **1976**, *98*, 2391.
46. Wong, M.; Thomas, J. K. In "Micellization, Solubilization and Microemulsions"; Mittal, K. L., Ed.; Plenum Press; New York, 1977, Vol. 2.
47. Henglein, A. *J. Phys. Chem.* **1979**, *83*, 2209.

48. Meisells, D.; Kopple, K.; Meyerstein, D. *J. Phys. Chem.* **1980**, *84,* 870.
49. Gratzel, M.; Kalyanasundaram, M. *Helv. Chim. Acta* **1980**, *63,* 478.
50. Anretts, M. *J. Phys. Chem.* **1935**, *39,* 509.
51. Crouther, J. A.; Liebman, H.; Jones, R. *Philos. Mag.* **1940**, *29,* 339.
52. Spivak, M. A.; Linford, H. B.; Odian, G.; Cropper, W. H. *J. Colloid Interface Sci.* **1967**, *23,* 358.
53. Johnston, F. *Radiat. Res.* **1978**, *75,* 286.
54. Dellaguardia, R. A.; Johnston, F. J. *Radiat. Res.* **1980**, *84,* 259.

# 13

# The Radiation Chemistry of Organic Solids

**J. E. WILLARD**

*Department of Chemistry,*
*University of Wisconsin*

## INTRODUCTION

The yields of primary products (ions, electrons, and excited states) produced by exposure of an organic compound to ionizing radiation are essentially independent of whether it is in the gas, liquid, or solid state. However, the nature and yields of the final products are often dependent on the state. This is the result of the effects of density and temperature on the relative probabilities of competing reactions of the primary species and of the radicals which they produce.

The density effects are of two types. First, the close proximity of neighboring molecules in the solid favors deactivation rather than decomposition of excited molecules and favors prompt recombination in the parent cage of the fragments of any that do decompose. Second, since the distance traveled by an energetic electron in depositing its energy is inversely proportional to the density of the medium, the tracks are shorter and the spur radii smaller in the solid than in the liquid (and in great contrast to the gas, where spur effects are negligible). The increased role of intraspur reactions of radicals, electrons, and cations in solids is clearly shown by the results to be discussed in this chapter.

The low temperatures required for solid-state studies, coupled with the high densities, make diffusion rates extremely slow, thus allowing the storage


*Radiation Chemistry: Principles and Applications*

of electrons, ions, and radicals for long times (sometimes even years). This allows the leisurely examination of the optical and electron spin resonance (ESR) absorption, electrical conductivity, and luminescence properties of species, which in other phases would be accessible only by very fast pulse techniques.[1] Use of fast pulse techniques at cryogenic temperatures allows observation of processes even in the time domain of the solvation of electrons by molecular orientation.[1d] The low temperatures also preclude any chemical reactions with the solvent or solutes that have activation energies of greater than about 1 kcal $mole^{-1}$. (Average thermal energies are 153 cal $mole^{-1}$ at 77 K and 8 cal $mole^{-1}$ at 4 K.)

This chapter is designed to give an overview of some of the unique features of organic solid-state radiation chemistry.[1] Selected references, which provide further references, will be given.

## ELECTRONS

### *Effect of Phase on Recombination of Geminate Ion Pairs*

When a molecule in a gas is ionized by a high-energy electron, the positive ion formed has negligible probability of recapturing the electron lost, because the mean free path of the latter is too long. For an identical ionization event in a liquid, there is a high probability that the electron will return to its "own" positive ion, because the higher density medium reduces the kinetic energy to thermal energies at a distance where the energy of coulombic attraction of the electron for the cation remains greater than thermal energies (Chapter 1). The *G*-value for escape of electrons from combination with their parent cations, and other siblings in the parent spur in organic liquids, is commonly only 0.1 to 0.2,[2a] whereas it is 3 to 5 in the gas phase.

In the solid state, electrons are thermalized at even shorter distances from the parent cation than in liquids, because of the higher density. However, this does not lead to an increase in prompt electron–cation recombination in organic glasses, because physical trapping of the electrons is possible. $G(e_t^-)$, the trapping yield, which depends on the medium and the temperature, is sometimes as high as 2 to 3. The $e^-$ may become stabilized for long times by interaction with permanent and induced dipoles of the molecules.

Trapping of $e^-$ does not normally occur in crystalline organic media. There is limited knowledge of the *G*-values for escape of electrons from geminate combination or of their mobilities and thermalization path lengths in crystals.[2b]

### *Observation of $e_t^-$*

Trapped electrons in organic glasses are most readily observed by their optical and ESR spectra. Glassy alkanes, alkenes, amines, ethers, and alcohols exposed to ionizing radiation at cryogenic temperatures all show broad (>2000 nm) spectra in the visible-IR region,[3] with extinction coefficients at $\lambda_{max}$ of $10^3$ to $>10^4$ $M^{-1}$ $cm^{-1}$. The values of $\lambda_{max}$ at 77 K range from about 1600 nm for the alkanes to about 500 nm for some alcohols and more polar compounds. The ESR spectra of $e_t^-$ are singlets centered at or near $g = 2.0023$. The peak-to-peak line widths are a few gauss and generally increase with the polarity of the matrix molecules and with temperature.[4,5] The mechanistic reasons for the variations of the optical and ESR spectra with the conditions of the matrix will be considered in later sections.

## *Mechanism of Trapping of $e_t^-$*

Varied types of experimental results support the conclusion that trapping occurs when a thermalized electron finds itself at a spot in the amorphous organic matrix, where potential fluctuations cause it to tarry long enough to orient molecules around it by interaction with molecular, bond, and induced dipoles, a process often referred to as "digging its own hole." Examples of the evidence are the following:

1. The $\lambda_{max}$ values of the optical spectra of $e_t^-$ in organic glasses move to the blue, and the widths of the ESR lines increase, as the polarity of the medium increases.[4,5] Both trends are to be expected if stronger trapping occurs in the more polar matrices as a result of stronger dipole interaction between the electron and neighboring molecules.
2. The optical spectra of $e_t^-$ produced in a matrix at 77 K are blue shifted and narrower relative to the spectra of those injected at 4 K,[6] consistent with greater ease of movement of the molecules in the field of the electron (and hence stronger trapping) at the higher temperature.
3. The $\lambda_{max}$ value of the stable spectrum of $e_t^-$ produced in $C_2H_5OH$ at 4 K is at 1400 nm but shifts to 540 nm on warming the sample to 77 K, where the ease of molecular rotation is increased.[7] As expected, the process is not reversible.
4. The value of $G(e_t^-)$ is usually higher at 77 K than 4 K,[8] and higher in polar media than in nonpolar media.[5]
5. The initial spectra of $e_t^-$ produced in alcohols at 77 K by a <40-ns pulse of 13-MeV electrons undergo a continuous blue shift of $\lambda_{max}$ from 1300 nm to 540 nm from $10^{-7}$ to 2 s after production. The rate of shifting (ie, of molecular orientation around electrons) is $10^5$ times faster in $CH_3OH$ than in $C_2H_5OH$ and $n$-$C_3H_7OH$.[9] The initial spectra produced in the glassy hydrocarbons 3-methylpentane (3MP) and 3-methylhexane (3MHx) at 76 K by short electron pulses also show major blue shifts with time, but on a time scale of many seconds.[10]
6. The spectra of trapped electrons in propyl, butyl, and pentyl alcohols at times of approximately 20 ns after production are essentially independent of temperature from 6 K to 115 K.[1d]

The probability of electron trapping depends not only on the polarity of the molecules of the medium but also on their structure. This is illustrated by selected $G(e_t^-)$ values:[5,8] for matrices at 77 K, 3MP = 0.70; 3MP-$d_{14}$ = 1.1; 3-ethylpentane (3EP) = 0.33; methylcyclohexane (MCHx) = 0.23; MCHx-$d_{14}$ = 0.38; 3-methylheptane (3MHp) = 0.47; 3MHp-$d_{18}$ = 0.63; ethylcyclohexane (EcHx) = 0.19; isopropylcyclohexane (*i*-PrcHx) = 0.03; 2-methyltetrahydrofuran (MTHF) = 2.6; ethyl alcohol = 2.4; *n*--propyl alcohol = 1.5; isopropyl alcohol = 1.1.

Two rationalizations of the effects of molecular structure on $G(e_t^-)$ are the following:

1. The probability of trapping rather than prompt return to the geminate cation is determined by differences in local potential fluctuations experienced by the electrons in matrices of different structure.
2. In all of the matrices, most of the electrons are initially trapped, and the rate of their removal by tunneling to the matrix cations is determined by Franck–Condon factors, which are a sensitive function of the structure of the cations. The latter seems to be the only plausible explanation of the difference in $G(e_t^-)$ values always observed between the protiated and deuterated forms of the same compound.[11]

## *Total Ionization Yield; Role of Scavengers*

Even the highest $G(e_t^-)$ values observed in pure organic solids account for only a fraction of the electrons ejected by the ionizing radiation. The *total* ionization yield in 3MP at 77 K has been shown to be greater than or equal to $(5.4 \pm 0.5)$ $e^-$/100 eV by measuring the yield of $SF_6^-$ in 3MP glass containing both the electron scavenger $SF_6$ (0.3 mol%) and the positive charge scavenger 2-methylpentene-1 (2MP-1) (0.75 mol%).[12] Thus, in the absence of scavengers, only 0.7/5.4 = 0.13 of the electrons produced are trapped, whereas 0.87 of the total are neutralized. The dual scavengers complement each other in eliminating the neutralization. A good working hypothesis is as follows: All of the electrons ejected are initially weakly trapped. In pure 3MP 87% tunnel rapidly to $3MP^+$ cations. When 2MP-1 is present, all of the positive charge is lost from $3MP^+$ to form 2MP-$1^+$ before the return of electrons can occur. The probability of $e_t^-$ tunneling to 2MP-$1^+$ is much lower than the probability of tunneling to $3MP^+$, because of unfavorable Franck–Condon factors; it is also much lower than the probability of tunneling to $SF_6$, which consequently captures all of the electrons.

In general, additives with a lower ionization potential than that of the solvent enhance $e_t^-$ yields, and those with a positive electron affinity capture electrons to produce molecular anions.[1a] Radiolytically produced radicals have an electron affinity of 0.5 to 1 eV and thus may act as electron scavengers ($R\cdot + e^- \rightarrow R{:}^-$).[12]

## *Origin of the Shape of the Spectra of $e_t^-$*

It is consistent with the known facts to interpret the broad spectra of $e_t^-$ in organic glasses as resulting from ejection of $e_t^-$ from their traps into the conduction band with varying energies equal to the difference between the trapping energy and the photon energy. Consistent with this conclusion, photon energies on the extreme red tails of the spectra of $e_t^-$ in alkane glasses induce bleaching of the electrons (ie, their removal by combination with cations or radicals).[13] In the somewhat more polar MTHF, the quantum yield for bleaching at 77 K, is 0.9 at 532 nm, but drops to $8 \times 10^{-3}$ at 1064 nm, and $<7 \times 10^{-6}$ at 1338 nm.[13] This implies that, at the longer wavelengths, the energy given to the $e_t^-$ in excess of that necessary to reach the conduction band is insufficient to prevent formation of a new trap, without appreciable migration (or prompt return to the oriented coulomb well of the parent trap). The greater probability of prompt trapping in MTHF as compared to hydrocarbons is reflected by its $G(e_t^-)$ value of 2.6 compared to that of 0.7 for 3MP glass.

At one time it was suggested[14] that photon absorption at wavelengths where the quantum yield of bleaching is low promotes electrons to an excited state of the trap rather than to the conduction band. According to this view, any photoconductivity or bleaching at such wavelengths must be due to biphotonic absorption and thus be proportional to the second power of the light intensity. More recent tests indicate that first-order dependence is observed.[15]

If light absorption by $e_t^-$ ejects into the conduction band, the spectrum would be expected to be broad because of the continuum of energies possible in the latter. It is also broad because the electrons are trapped with varied energies as a result of being born in a glass with random molecular orientation.

Exposure of $\gamma$-irradiated $C_2H_5OH$,[6c] MTHF,[6a,13] and alkane[6a,13,16] glasses to monochromatic light results in selective bleaching ("hole burning") of the $e_t^-$ spectrum in the region of the activating wavelengths.[13] The data have been interpreted[17] to indicate that the extinction coefficient of each electron rises rapidly with increasing photon energy on the red end of its spectrum, passes through a maximum, and has a long, low tail on the blue end. Superposition of the spectra of different $e_t^-$, with many different trapping energies, produces the observed envelope. Monochromatic light at the red end of the $e_t^-$ spectrum in alkane glasses bleaches completely at that wavelength while "peeling off" a little of the absorption at shorter wavelengths.[13]

## *Mechanism of Electron Decay*

Spontaneous combination of trapped electrons with cations, radicals, and additives occurs in $\gamma$-irradiated organic glasses even when the reactant species are separated by several molecular diameters and the temperature is far below the glass transition temperature. Movement to a reaction partner by thermal detrapping and hopping appears improbable because the photo-detrapping thresholds are approximately 0.5 eV in hydrocarbons and much higher in more polar matrices, whereas $kT$ at 77 K is 0.006 eV.

There is convincing evidence[18] that electrons trapped within a few tens of angstroms of cations, radicals, or scavenger molecules, and that react with them on the time scale of $10^{-3}$–$10^{2}$ s do so by tunneling. Electrons at greater distances may react by tunneling over longer times. At higher temperatures the decay of electrons not close to reaction partners seems to involve diffusion-assisted tunneling.

Readily observable decay[13] of $e_t^-$ on a time scale of 0 to 250 min occurs in neat alkane glasses even at 20 to 40 K. It is temperature independent, consistent with a tunneling mechanism. The most weakly trapped electrons, which account for the red end of the optical spectra, decay the most rapidly. The rates of decay are "composite first order" (ie, the apparent first-order rate constant is independent of the initial concentration but decreases as decay proceeds). In pulse radiolysis experiments the rate of tunneling transfer of $e_t^-$ to scavengers at 77 K decreases linearly with increase in the log of time over many decades.[18]

The rates of decay of $e_t^-$ in different alkane glasses differ greatly and are always slower in the perdeuterated than in the protiated form of the same compound.[11] When an additive of lower ionization potential, which captures the positive charge, is present, the decay rate is slower than in the neat alkane. All of these differences have been ascribed to changes in the tunneling probability dependent on the extent of matching of spectroscopic states between the $e_t^-$ and the cationic acceptor.[11]

The rates of decay of $e_t^-$ in organic glasses depend on the length of time the glass has been annealed below the glass transition temperature before or during irradiation.[19,20] For example, the half-life at 77 K of $e_t^-$ present following a 5-min $^{60}Co$ irradiation of 3MP, which has just been cooled to 77 K, may be less than 10 min, whereas if the 3MP is allowed to stand for 200 hr before irradiation, $t_{1/2}$ becomes 60 min. Annealing causes a decrease in free volume and enthalpy and an increase in the viscosity. The rate of annealing depends on the size and shape of the container, which determines the ease with which the sample can contract in volume.[20]

## *Population Dynamics of Growth of $e_t^-$*

During exposure of an alkane glass to ionizing radiation, the concentration of $e_t^-$ increases linearly with dose at first and then, more slowly, passes through a maximum at about $5 \times 10^{-4}$ mole fraction at a dose of about $1 \times 10^{20}$ eV gm$^{-1}$: it then decreases, approaching zero at about $3 \times 10^{20}$ eV gm$^{-1}$.[21,22] The linear rise occurs at doses where the probability of spur overlap is negligible. In this region electrons react only with species in the parent spur, because diffusion is negligible. At higher doses, where spur overlap occurs, the probability for $e^-$ to be born close enough to a cation or radical to be lost by rapid tunneling to it increases, and $G(e_t^-)$ decreases. If all such reactions were with cations (which are produced in equal number to the electrons), the $e_t^-$ and cations would achieve a steady-state concentration with the differential $G(e_t^-)$ for continuing radiation being zero. The observed *decrease* in $[e_t^-]$ at higher doses is attributable to the continuing increase in the concentration of radicals. These are produced with $G(R)$ about five times larger than $G(e_t^-)$ and react with $e_t^-$ to form carbanions ($R\cdot + e^- \rightarrow R{:}^-$).

In matrices more polar than alkanes, the maxima of the $[e_t^-]$-vs.-dose curves occur at higher doses ($\sim 2.4 \times 10^{20}$ eV $gm^{-1}$ for MTHF[23] and $6 \times 10^{20}$ eV $gm^{-1}$ for $C_2H_5OH$),[24] because the spur radii are smaller, and higher concentrations are required for spur overlap. The $e_t^-$ spur radii for 3MP, triethylamine (TEA), and MTHF have been estimated[23] to be >130 Å, 101 Å, and 63 Å, respectively, by measurement of the dose at which the ESR relaxation time $(T_1 T_2)^{1/2}$ starts to decrease as a result of increased spin–spin relaxation resulting from the onset of spur overlap.

Further evidence of the changing competition between radicals and cations for electrons as the radiation dose increases is given by the heat liberated when $e_t^-$ are detrapped by light and combine with the trapped radicals and cations[20] following different $\gamma$-doses to a glass. For 3MP the heat evolved decreases from 150 kcal $mole^{-1}$ of $e_t^-$ bleached at zero dose (extrapolated), to 80 kcal $mole^{-1}$ at $1.25 \times 10^{20}$ eV $gm^{-1}$. The decrease is ascribed to the increasing ratio of reaction with radicals to reaction with cations. Some gas-phase heats of reaction ($\Delta H$, kcal $mole^{-1}$) are

$$C_6H_{14}^+ + e^- \rightarrow C_6H_{14} - 230 \quad (13\text{-}1)$$

$$C_6H_{14}^+ + e^- \rightarrow C_6H_{12} + H_2 - 200 \quad (13\text{-}2)$$

$$C_6H_{13}^+ + e^- \rightarrow C_6H_{13} - 170 \quad (13\text{-}3)$$

$$C_6H_{13} + e^- \rightarrow C_6H_{13}^- - 20 \quad (13\text{-}4)$$

Assuming $\Delta H$ for reaction with cations in the glassy state to be $-150$ kcal $mole^{-1}$, and $\Delta H$ of Reaction (13-4) to be $-20$ kcal $mole^{-1}$, then 54% of the photobleached $e_t^-$ react with cations, and 46% with radicals at a dose of $1.2 \times 10^{20}$ eV $gm^{-1}$, to give an average of $\Delta H$ of $-80$ kcal $mole^{-1}$. If Reaction (13-1) were the only neutralization reaction and no $e^-$ reacted with $R\cdot$ at low dose, the estimated heat of solvation of the cations would be $-230$ kcal $mole^{-1} - (-150$ kcal $mole^{-1}) = -80$ kcal $mole^{-1}$.

## Information from ESR

The singlet of a few gauss width, which is the ESR signal of trapped electrons, is useful for their detection, determination of their concentration, and as a source of information about their spatial distribution relative to neighboring molecules and paramagnetic species.

The area under the ESR absorption curve is proportional to the concentration of the $e_t^-$. It is determined by double integration of the first-derivative signal typically given by the experimental equipment. Absolute concentrations are determined by comparison with a sample of a stable free radical containing a known number of spins. The concentrations of $e_t^-$ determined by ESR agree well with determinations by optical absorption.[8] For the latter the extinction coefficients ($\varepsilon$) are determined by comparing the optical densities with those of an anion of known $\varepsilon$, such as biphenylide, when the $e_t^-$ are quantitatively converted to the latter by photodetrapping.[25] Direct proportionality of the area under the ESR absorption curve to that under the optical absorption spectrum holds even for selective partial bleaching.[13]

The width of the ESR line of $e_t^-$ in a deuterated glass is narrower than in the protiated form of the same compound by a factor approximating the ratio predicted if hyperfine interaction with deuterons replaces that with protons. This effect has provided evidence that the molecules surrounding trapped electrons in alcohol glasses have the hydroxyl groups oriented toward the $e_t^-$; for example, the line width of $e_t^-$ in $C_2H_5OD$ is 6 G, compared to 14 G in $C_2H_5OH$.[26]

As noted in a preceding section, spur sizes have been estimated from the change in the microwave power threshold for ESR saturation as a function of dose.[23]

A quantitative model of the organization of molecules around trapped electrons in MTHF glass at 77 K, based on electron spin echo studies, indicates that three or four MTHF molecules are oriented around each $e_t^-$, with the plane of each molecule facing the $e_t^-$ at a distance of approximately 3.7 Å and with the $CH_3$ group away from the $e_t^-$.[27] The model requires the rather surprising conclusion that the coulombic field of the $e_t^-$ can organize the MTHF molecules into the suggested configuration even at a temperature well below the glass transition temperature, and that the $e_t^-$ is centered in a cage of approximately 6 Å inner diameter, the walls of which have large gaps.

# CATIONS

### *Formation, Migration, and Stabilization*

Direct knowledge of the properties of the cations formed in pure organic solids by ionizing radiation is much more limited than that of electrons, because they often do not have readily observable optical or ESR spectra. The primary cations are molecules of the matrix from which an electron has been removed by the radiation. These become solvated by polarization interaction with the surrounding molecules. The electron affinity of the solvated cation in its ground electronic state is less than the ionization potential of neighboring neutral molecules of the same kind, precluding electron transfer from the latter, which would cause migration of the positive hole. If the cation is formed in an excited state, tunneling transfer, which would be endoergic for the ground state, may occur. If a solute with lower ionization potential is present, electron tunneling transfer from it may occur over many molecular diameters, producing a solute cation that can often be characterized by its optical spectrum. In alkanes such solutes include biphenyl, various amines, and alkenes.[1a] The yield of solute cations is increased by electron scavengers that stabilize the negative charge.[1a] Good evidence has been advanced for the formation of dimeric cations of solutes.[28] The yields of solute cations are, in general, much higher than can be accounted for by direct interaction of the radiation with the solute, confirming that charge is transferred from solvent cations. In the absence of additives the possible fates of the original cations include neutralization, $RH^+ + e^- \rightarrow RH$ (or $R\cdot + H$), proton transfer $RH^+ + RH \rightarrow R + RH_2^+$ and dimerization $RH^+ + RH \rightarrow (RH)_2^+$. Proton transfer, demonstrated by mass spectrometric evidence, has for many years been known to occur in the gaseous $C_1$ to $C_4$ alkanes.[29a] Much more recently it has been confirmed for several alkanes in solid-state matrices.[29b]

## *Optical Spectra of Cations*

In pioneering work on ionic species in organic solids, Hamill and co-workers[1a] found no optical spectra of the cations in a series of single-component alkanes following $\gamma$-radiation at 77 K with or without $CO_2$ present to scavenge electrons. However, when alkanes with seven or more carbon atoms are present as solute in 3MP containing $CO_2$, at 77 K, ionizing radiation produces a broad (400-nm half-width) absorption band.[30] The $\lambda_{max}$ increases from 560 nm for $C_7H_{16}$ solute to about 910 nm for $C_{15}H_{32}$. The bands are attributed to $RH^+$. More recently, cation absorption has been observed in neat 3-methyloctane (3MO), with $\lambda_{max}$ at 600 nm,[31] and in neat squalane (SQ), with $\lambda_{max}$ at 1400 nm,[32] at 100 ns after a 40-ns, 35-MeV electron pulse. These bands are attributed to the $3MO^+$ and $SQ^+$ ions. In 3MO containing 1 vol% SQ and the electron scavenger $N_2O$, pulse radiolysis experiments showed a decrease in $[3MO^+]$ and concomitant increase in $[SQ^+]$ from $10^{-7}$ to $10^{-5}$ s after the pulse as charge was transferred (presumably by electron tunneling) to SQ.[32] Still more recently, convincing evidence for a stable cation absorption peak at 490 nm in 3MP glass has been reported.[12] $SF_6$ was used to stabilize the electrons as $SF_6^-$ and free the 490-nm region of interference from the spectra of the $e_t^-$ and carbanions. The optical spectra of the cation radicals formed from the series $n$-$C_4H_{10}$ through $n$-$C_9H_{20}$ by radiolysis in $CCl_3F$ at 77 K have been determined and confirmed by correlation with the ESR spectra.[33]

### ESR Spectra of Cations

No ESR spectra of cations are found in neat irradiated organic solids. The initial ion formed in alkanes, $RH^+$, is paramagnetic, but $RH_2^+$, which would be formed by proton transfer, is not. The possibility of line-broadening effects and the overlapping of the spectrum with that of the prominent radical signal prevent the conclusion that stable trapped $RH^+$ is not present.

Despite these negative results with single-component systems, radiation chemistry in solids has proved useful in preparing and stabilizing many cationic species that cannot be readily prepared by chemical oxidation. This is done by positive charge transfer from an irradiated matrix to a solute that is the neutral parent of the desired cation. Good matrices for such studies are $SF_6$ and $CCl_3F$, each of which has a high ionization potential (15.69 and 11.9 eV, respectively), captures electrons to form stable anions, and shows negligible ESR signals from its radiation products.[33] The tetramethylethylene cation ($TME^+$) and the dimer $(TME)_2^+$ have been investigated in 3MP at 77 K.[28] Using 1 mol% solutions of the simple alkanes, $C_2H_6$, $C_3H_8$, and $C_4H_{10}$ in $SF_6$, workers have been able to obtain and characterize well-defined spectra of their cations.[34a]

## RADICALS[34b]

### Production and Observation of Radicals

Exposure of any organic solid to ionizing radiation produces trapped radicals, with half-lives of minutes or longer if the temperature is near or below the glass transition temperature or, for crystalline samples, less than about 0.8 of the absolute melting point. The most useful method of studying these is ESR spectroscopy. It is sensitive (to $\sim 10^{-9}$ *M* radicals) and provides a theoretically predictable "fingerprint" for each type of radical. The line broadening that occurs in solids and the overlapping of spectra of different radicals are limitations, but they ordinarily do not prevent the identification and study. Infrared analysis for organic radicals in organic matrices is usually precluded by the interfering absorption bands of the matrix. The UV absorption spectrum of the 3MP radical in the irradiated glass at 77 K has been characterized,[12] but neither this nor other radical spectra have been used for analysis because of interference from overlapping absorptions due to $e_t^-$, carbanions, and olefinic reaction products.

### Mechanisms of Formation

Possible mechanisms of radical formation, for which there are varying degrees of evidence, include the following:

$$RH \rightsquigarrow RH^* \rightarrow R + H \tag{13-5}$$

$$\rightarrow R' + R'' \tag{13-5a}$$

$$RH^+ + e^- \rightarrow RH^* \rightarrow R + H \tag{13-6}$$

$$\rightarrow R' + R'' \tag{13-6a}$$

$$RH^+ + RH \rightarrow RH_2{}^+ + R \tag{13-7}$$

$$RH_2{}^+ + e^- \rightarrow R + H_2 \tag{13-8}$$

$$\rightarrow RH + H \tag{13-8a}$$

$$RH + H \rightarrow R + H_2 \tag{13-9}$$

$$RH^+ + R^- \rightarrow R + RH \tag{13-10}$$

$$\rightarrow 2R + H \tag{13-10a}$$

$$RX + e^- \rightarrow R + X^- \tag{13-11}$$

$$CH_3 + RH \rightarrow R + CH_4 \tag{13-12}$$

Excited molecules (RH*) that may decompose to radicals by rupture of a C—H or C—C bond [Reactions (13-5)–(13-6a)] are produced in irradiated systems both directly (13-5) and by the neutralization of cations (13-6). The first excited singlet state, to which higher states decay rapidly by fluorescence and from which any decomposition is expected to occur,[35] typically has an energy of greater than or equal to 7 eV. This is higher than either C—H bond energies ($\leq$4.3 eV) or C—C bond energies ($\leq$3.7 eV), so rupture of either type of bond is energetically possible. Although C—C bonds are weaker than C—H bonds, C—H rupture predominates. The yields from C—C rupture are usually too low to be observed by ESR in the presence of the radicals from C—H rupture. Some low-yield stable products, detectable by gas chromatography[36] following melting, are of the type expected from combination of radicals formed by C—C bond splitting.

For reasons as yet unknown, the C—H bond rupture is neither random nor localized at the weakest bond. In singly branched alkanes with branching in the 3 or 4 position it occurs selectively at a secondary C—H bond of the $-CH_2CH_3$ group of the longest branch of the molecule.[37,38] When branching is in the 2 position, the major rupture is the same, but there is also significant rupture of the tertiary hydrogen.[37] Surprisingly, H-atom attack[37,38] and photosensitization by Hg[39] and aromatic solutes[37] cause selective rupture at the same point in the molecule as does ionizing radiation.

There is experimental evidence consistent with the conclusion that at least a part of the radical yield results from neutralization reactions.[12,28,40] $G$(R) is reduced by electron scavengers. For example, in 3MP there is a reduction from 3.0 to 2.2–2.4[40] by biphenyl or $C_3H_7I$; in 3MP-$d_{14}$ from 2.0 to 0.7 by $CH_3I$;[40] in 3MHp from 3.5 to 3.06 by $CO_2$.[28] Each of the scavengers was present at less than or equal to 1 mol% in the matrix irradiated at 77 K.

An upper limit of the fraction of the neutralization events in 3MP at 77 K that produce radicals may be estimated from the evidence[12] that $G_0(e^-)$ = $G_0$ (cations) = 5.4 (where $G_0$ is the primary product yield before any interaction with the medium), $G(R^-) = 0.3$, $G(e_t^-) = 0.7$, and $G(R) = 3.0$. From these data it follows that 4.4 $G$-units (5.4 − 0.7 − 0.3) of electrons are neutralized rather than forming stable $e_t^-$ or $R^-$. If all of the positive charge was in the form of $RH_2^+$ (ie, every $C_6H_{14}^+$ had undergone proton transfer before neutralization occurred), the neutralization would produce 4.4 $G$-units of radicals by (13-8) and/or (13-8a) + (13-9). Because one R is produced in the formation of $RH_2^+$, the total $G(R)$ would then be 8.8, and since the observed $G(R) = 3.0$, no more than 3/8.8 = 0.34 of the neutralizations can produce radicals by this mechanism. Similar reasoning indicates that not more than 3/8.8 of the neutralization events produce radicals by Reactions (13-6a) or (13-6) + (13-9). Thus, at least 66% of the electrons are neutralized without producing radicals.

When $e_t^-$ in 3MP are detrapped by light, the ratio of R formed to $e_t^-$ neutralized is 0.27.[12] If the same ratio holds for the electrons formed by ionizing radiation, $G(R)$ formed by neutralization must be $0.27 \times 4.4 = 1.2$. Since the observed $G(R)$ is 3.0, this would indicate that only 1.2/3.0 = 0.40 of the observed radicals are formed by neutralization processes, and 0.60 are formed by Reactions (13-5) + (13-9), and (13-5a). If this is true, then only (1.2/2)/4.4 = 0.14 of the electrons neutralized in 3MP produce radicals (noting that 2 R are produced for each neutralization that produces any R). It follows from this reasoning that no more than (1.2/2)/5.4 = 0.11 of the original $C_6H_{14}^+$ ions undergo proton transfer, because such transfer inevitably results in the production of two radicals. As noted, these conclusions involve the assumption that the radical yield (% of events) from the prompt neutralization of electrons during irradiation is the same as that from the neutralization of $e_t^-$ detrapped by light in an irradiated 3MP matrix.

### *Mechanism of $CH_3$ Formation in $CH_4$ at 4 K*

Methane differs from all other organic substances tested at greater than or equal to 4 K in that it does not react with H atoms but traps them, thus eliminating Reaction (13-9) as a contributor to R production. The data indicate that $G(CH_3) = G(H_t) = 3.3$.[41] This equality implies either that all $CH_3$ and H formation occurs by Reactions (13-5) and (13-6) or that if proton transfer (13-7) occurs, it is always followed by Reaction (13-8a) (since (13-7) + (13-8) would produce two $CH_3$ and no H). Using the reasoning of the previous section and assuming that $G_0$ ($e^-$) is the same in $CH_4$ at 4 K as in 3MP at 77 K, one can estimate that at least 39% of the electrons are neutralized without observable bond dissociation and that no more than 61% of the $CH_4^+$ undergo proton transfer. Since no electrons are trapped in $CH_4$ and only one $CH_3$ is formed for each electron that causes bond rupture on neutralization, the calculation is $G_0(e^-) - G(CH_3) = 5.4 - 3.3 = 2.1$; $2.1/5.4 = 0.39$. Although the $CH_4^+ + CH_4 \rightarrow CH_5^+ + CH_3$ reaction has a large cross section and zero activation energy in the gas phase,[29a] it is not surprising if the transfer of the relatively heavy nucleus does not compete effectively with the return of the relatively light electron to its geminate cation.

## *Formation of Radicals by Dissociative Electron Capture*

When an electron is captured by a molecule that has an electron affinity greater than the energy of a bond in the capturing molecule, dissociation may occur [Reaction (13-11)], leaving a trapped radical in close proximity to the anion of the group removed. Examples of dissociative capture reactions include $CH_3I + e^- \rightarrow CH_3 + I^-$;[42,43] $CH_3CN + e^- \rightarrow CH_3 + CN^-$; $C_6H_5Cl + e^- \rightarrow C_6H_5 + Cl^-$.

It is of interest that the radicals trapped following dissociative electron capture by alkyl halides in hydrocarbon glasses are always the radical of the alkyl halide, whereas photolysis of the halide with 254-nm light ($RI \xrightarrow{h\nu} R + I$) always produces the radical[37,40] of the solvent. Apparently, the radicals from the photolysis are born with sufficient energy to insure hot abstraction from the matrix molecules (eg, $CH_3 + C_6H_{14} \rightarrow CH_4 + C_6H_{13}$). When a pure alkyl halide matrix is irradiated, the radical formed by splitting off the halogen atom is usually the only one observed.[44] This may occur wholly by dissociative electron capture, or may also include dissociation of excited molecules.

The yields of alkyl radicals from dissociative electron capture of alkyl halide solutes in 3MP increase with increasing concentration of the halide up to approximately 1 mol%, above which they remain constant.[45a] In the more polar MTHF, where the electrons are more strongly trapped, the plateau is reached at approximately 2 mol%. The reported[45a] $G(R)$ of unity is now known[46] to be low by a factor of at least 2 because of ESR saturation effects. It has been reported[45b] that the plateau concentrations of $CH_3$ radicals produced by radiolysis of $CH_3I$ and $CH_3Cl$ in 3MHx glass at 77 K are achieved at halide concentrations of approximately 0.1 mol%, and also that $CH_3F$ undergoes dissociative capture in this matrix, even though the process would be endothermic in the gas phase.

## *Radical Pairs: Temperature-Dependent H Diffusion*

Electron spin resonance measurements have demonstrated[47] that a few percent of the radicals formed by the $\gamma$-irradiation of organic glasses at 77 K are trapped as pairs, with interradical distances of 5 to 25 Å. At 4 K the pair yield may be as high as 100%.[47b]

Pair formation is to be expected if H atoms produced in Reaction (13-5) promptly abstract from an adjacent molecule by (13-9). It is also probable if the reaction sequences (13-7) + (13-8) and (13-7) + (13-8a) + (13-9) occur without significant diffusion. The H atoms from Reactions (13-5), (13-6), and (13-8a) may be born with kinetic energies much in excess of thermal energies (ie, "hot"), and thus have a high probability of abstracting on an early collision. In all alkanes except methane, thermal H atoms abstract from C—H and C—D bonds by rapid quantum mechanical tunneling even at 4 K.[48] Their inability to diffuse at 4 K[49] (see later section on hydrogen atoms), as contrasted to 77 K, may account for the much higher pair yields at the lower temperatures.

## *Radical Spurs*

Upper limits of 34 Å and 37 Å for the radii of radical spurs produced in MTHF and TEA by irradiation at 77 K are suggested by ESR evaluation of the $(T_1 T_2)^{1/2}$ relaxation times of the radical spectra as a function of dose.[50] Tests on 3MP give similar results, but radical decay and spectral shape changes during the long irradiation times required for high doses cause some uncertainty.[47d]

It is not surprising that the radii of the radical spurs are smaller than those of the electron spurs (>130 Å in 3MP, 101 Å in TEA, and 63 Å in MTHF),[23] because each electron trapped must be ejected from its parent molecule with sufficient kinetic energy to escape prompt geminate recombination, and may travel many molecular diameters, whereas the radicals formed by the mechanisms discussed above must all be trapped close to their sites of formation along the paths of the secondary electrons.

Independent evidence that the radicals are produced in spurs of high local concentration is given by the kinetics of their decay. In 3MP at 77 K, for example, 57% decay by a time-dependent ("composite") first-order process.[40] This is evidenced by the superimposability of the decay curves for different initial concentrations, produced by different doses, when plotted as fraction decayed vs. time. Such kinetics are typical of intraspur reactions, which are not influenced by the presence of radicals in other spurs. The plots are convex toward the origin; that is, there is a progressive decrease in the magnitude of the first order rate constant as decay proceeds. This reflects random distribution of interradical distances in the spurs and the decreasing probability of intraspur radical encounters with time. The radicals that escape intraspur combination or disproportionation (~43%), decay by reaction with radicals from other spurs encountered after random diffusion, and exhibit clean second-order decay kinetics.[40]

Exposure of $\gamma$-irradiated glassy 3MP, 3EP, 3MHp,[51] or polycrystalline *n*-alkanes[52] to 254-nm light at 77 K removes 23 to 35% of the trapped radicals. The remainder are not affected by continued exposure. Those removed must be those in closest proximity in the spurs. There are two mechanisms by which the light may promote intraspur radical–radical encounters. One is by dissipation of the photon energy as heat that warms a small volume of the matrix sufficiently for radical diffusion to occur. For the photon energies and matrices used, this is plausible at 77 K[51] but not at 4 K. The second mechanism is excitation of the radicals in such a way as to induce migration of the vacant bond by intermolecular or intramolecular hydrogen hopping. Photoremoval at 4 K of 23% of the radicals produced in polycrystalline $n$-$C_6H_{14}$[52] seems to require the latter. However, the former may be responsible in glassy 3MP, where 30% of the radicals are removed by light at 77 K but none at 15 K (footnote 8(a) of ref. 12).

## *Radical Yields*

If the energy deposited in organic solids by ionizing radiation was used 100% efficiently in producing the net reaction $2RH \rightarrow 2R + H_2$ by the processes of Eqs. (13-5) to (13-11), approximately 24 radicals would be formed per 100 eV absorbed. The actual yields are less than or equal to six radicals per 100 eV (Table 13-1),[52,53] indicating that 75% of the energy is dissipated as light or heat or in reactions that do not involve trapped radical intermediates.

There are striking systematic similarities and differences in the radical yields from glassy, as compared to polycrystalline, and from deuterated, as compared to protiated, $\gamma$-irradiated hydrocarbons (Table 13-1).

1. $G$(R)-values for $n$-alkanes from $C_6$ through $C_{10}$, which are all polycrystalline, are, within experimental error, independent of carbon number.
2. $G$(R)-values for the branched alkanes, nearly all of which are glassy (including methyl and ethyl branching, single branching, double branching, and different branching positions), are also independent of carbon number but are all about 40% lower than for the polycrystalline compounds.
3. $G$(R)-values for the crystalline cyclo and branched alkanes tested (Table 13-1) do not fit in either category (1) or (2).
4. $G$(R) for completely deuterated hydrocarbons is always 40 ± 15% less than that for the protiated form of the same compound, regardless of molecular structure and the glassy or crystalline state (except for benzene, where the change is greater).

Are the lower yields from the glasses a result of the amorphous state or of the branched molecular structure? The lower yield for MCHx glass compared to crystalline MCHx (Table 13-1) indicates that, for this compound at least, it is the amorphous state. (It is rare that a compound can be prepared in each state.)

**Table 13-1.** Yields of Trapped Radicals in $\gamma$-Irradiated Hydrocarbons at 77 K[52,53]

| Compound | $G$ (radicals)[a] | | $G(R\cdot)_H/G(R\cdot)_D$[b] |
|---|---|---|---|
| | Perprotiated | Perdeuterated | |
| (1) *Glasses* | | | |
| 3-Methylpentane | 3.0[c] | 2.0[c] | 1.5 |
| Methylcyclohexane | 3.4 ± 0.4 (6) | 2.6 ± 0.2 (4) | 1.3 |
| 3-Methylheptane | 3.5 ± 0.1 (6) | 2.7 ± 0.3 (4) | 1.3 |
| 3-Ethylpentane | 3.2 ± 0.4 (4) | | |
| 3-Methylhexane | 3.0 ± 0.2 (4) | | |
| 2-Methylhexane | 3.3 ± 0.2 (4) | | |
| 2,4-Dimethylpentane | 3.2 ± 0.3 (4) | | |
| 2,4-Dimethylhexane | 3.2 ± 0.3 (4) | | |
| 2,5-Dimethylhexane | 3.6 ± 0.3 (4) | | |
| Ethylcyclohexane | 3.7 ± 0.3 (4) | | |
| Isopropylcyclohexane | 3.6 ± 0.1 (3) | | |
| (2) *Polycrystals* | | | |
| *n*-Hexane | 5.3 ± 0.4 (12) | 3.9 ± 0.1 (5) | 1.4 |
| *n*-Heptane | 5.6 ± 0.2 (3) | 3.6 ± 0.2 (2) | 1.6 |
| *n*-Octane | 5.7 ± 0.3 (3) | 4.5 ± 0.4 (2) | 1.3 |
| *n*-Nonane | 5.9 ± 0.1 (2) | | |
| *n*-Decane | 6.0 ± 1.0 (3) | | |
| Neopentane | 2.5 ± 0.3 (2) | | |
| Cyclohexane | 3.8 ± 0.3 (7) | 3.0 ± 0.2 (2) | 1.3 |
| Methylcyclohexane | 4.3 ± 0.3 (3) | 3.0 (1) | 1.4 |
| Benzene | 0.15 ± 0.02 (4) | 0.065 ± 0.007 (3) | 2.3 |

[a] The number of determinations made on each compound is shown in the parentheses. The error limits given are the average deviation from the average.
[b] Ratio of the $G$(radiccals) for the protiated form to that for the deuterated form.
[c] Values from ref. 40, to which determinations in the present work have been normalized.

An early summary[54] of $G(R)$ for 16 aromatic compounds and 19 other compounds, including some with nonconjugated double bonds, alcohols, bicyclic and monocyclic alkanes, and branched and linear hydrocarbons, found $G \leqslant 0.3$ for all of the aromatics and $G \geqslant 3$ for most of the others. For cyclohexane, the only compound also reported in Table 13-1, the agreement is good. The authors[54] suggest that $G(R)$ is related to the energy of the lowest excited state of the molecule relative to the C—H bond energy. The estimated first excited-state energies are higher than C—H energies in the nonaromatic compounds, but about the same or lower than the bond energies in the aromatics. This hypothesis cannot explain the glass–crystal and deuteration effects illustrated in Table 13-1.

The remarkable systematic correlations of the data of Table 13-1, with respect to crystallinity and deuteration, suggest that Mother Nature is seeking to convey a clear message about the controlling mechanistic variables, but her language still awaits interpretation.

A source of the observed isotope effects on radical yields may be the influence of deuteration on the branching ratios between decomposition of excited molecules and stabilization by radiative decay or collisional deactivation.[53] The change in vibrational energy-level spacing, caused by deuteration, with accompanying less favorable Franck–Condon factors for intersystem crossing, has been shown[35] to decrease the probability of radiationless loss of excitation, relative to radiative loss. This shift would be expected to decrease the probability of bond rupture, although little is known about what factors determine the ratio between bond rupture and loss of energy to the vibrational modes of the matrix.

## *Radical Decay*

### Types of Reaction

When an organic solid containing trapped radicals is held at a temperature where diffusion can occur, the radicals disappear. The reactions responsible and their kinetics depend on the identity and spatial distribution of the radicals and the nature of the matrix. The reactions that may occur are radical–radical combination and disproportion, abstraction of H from the C—H bonds of the matrix by quantum mechanical tunneling, reaction with a solute, reaction with an electron to produce a carbanion, and reaction with a cation to produce a carbonium ion. These are illustrated by

$$C_6H_{13} + C_6H_{13} \rightarrow C_{12}H_{26} \quad (13\text{-}13a)$$

$$\rightarrow C_6H_{12} + C_6H_{14} \quad (13\text{-}13b)$$

$$CH_3 + C_6H_{14} \rightarrow CH_4 + C_6H_{13} \quad (13\text{-}14)$$

$$C_6H_{13} + HI \rightarrow C_6H_{14} + I \quad (13\text{-}15)$$

$$C_6H_{13} + e^- \rightarrow C_6H_{13}^- \quad (13\text{-}16)$$

$$C_6H_{13} + C_6H_{14}^+ \rightarrow C_6H_{13}^+ + C_6H_{14} \quad (13\text{-}17)$$

### Intraspur and Random Radical–Radical Reactions

Reactions of type (13-13) are composite first order in their kinetics when they are intraspur (see the section entitled "Radical Spurs" above), and are second order when they result from random radical encounters throughout the medium.[40] (Early conclusions that all $C_6H_{13}$ radicals produced by the $\gamma$-irradiation of $C_6H_{14}$ decay by second-order kinetics were in error. In one experiment[55] the composite first-order intraspur decay was overlooked because it occurred during warming of the sample to 87 K prior to monitoring the decay. In another, where monitoring was at 77 K, the length and intensity of the radiation dose resulted in both significant intraspur decay during the irradiation and in spur–spur overlap.)

A special case of intraspur reaction is the combination of geminate pairs of radicals. It is difficult to isolate this type of decay from other intraspur encounters in radiolyzed systems, but it seems to be clearly illustrated by the rapid decay of $C_6H_{13}$ in a system where hot H atoms from the photolysis of HI solute have produced the $C_6H_{13}$

$$\left(HI \xrightarrow[C_6H_{14}]{h\nu} C_6H_{13} + I + H_2\right)$$

in the same cage with an I atom with which it can combine.[55]

### Tunneling Abstraction of H from C—H Bonds by Radicals

The activation energy for abstraction of H from a hydrocarbon molecule by a radical [Reaction (13-14)] is typically greater than or equal to 6 kcal mole$^{-1}$, as measured in the gas and liquid phases. Such an energy requirement precludes observable reaction in less than months or years at cryogenic temperatures ($RT = 153$ cal mole$^{-1}$ at 77 K and 8 cal mole$^{-1}$ at 4 K). However, such reactions of $CH_3$ radicals have been observed. They must occur by quantum mechanical tunneling. The most-studied examples include abstraction by $CH_3$ from $CH_3CN$[56] and $CH_3NC$[57] in the polycrystalline state and from $CH_3OH$[58] and 3MP[40,59–61] in the glassy state. In each the reaction has been confirmed by observation of the growth of the product radical at a rate equal to the decay of the $CH_3$. In the case of 3MP a yield of $CH_4$ equivalent to the $CH_3$ reacted has been found.[61] The activation energies for the abstraction reaction in the four matrices have been reported as 0.9, 1.4, 1.4, and 0.8 kcal mole$^{-1}$, respectively.

The rate of abstraction by $CH_3$ from the perdeuterated forms of the matrices is less than $10^{-3}$ of that from the protiated forms, consistent with a tunneling mechanism. $CH_3$ produced by the $CH_3X + e^- \rightarrow CH_3 + X^-$ process in 3MP-$d_{14}$ decays not by abstraction,[60] but by a competing process thought to be combination with its geminate partner ($CH_3 + X^- \rightarrow CH_3X^-$). The rate of this process depends on whether $X^-$ is $Cl^-$, $Br^-$, or $I^-$, whereas the tunneling decay in 3MP-$h_{14}$, it is independent of the identity of the partner species.[40,60] $CD_3$ radicals decay at the same rate as $CH_3$ in both 3MP-$h_{14}$[55,60] and $CH_3OH$.[63] Decay is slower in $CD_3OH$ than in $CH_3OH$.[63] A low yield ($G = 0.7$) of $CH_3D$ is produced by D abstraction by hot $CH_3$ radicals when $CH_3I$ undergoes dissociative electron capture in 3MP-$d_{14}$.[60]

The kinetics of the tunneling abstraction are composite first order rather than first order, indicating that the "tunneling rate constant" is sensitive to steric factors in the positioning of $CH_3$ relative to the matrix molecules. When the matrix temperature is raised above the glass transition temperature,[55] allowing relatively rapid diffusion, the rates of decay of $CH_3$ in protiated and deuterated matrices approach equality, apparently because the $CH_3 + I^-$ and radical–radical reactions can compete with the abstraction process.

There are insufficient data to show whether radicals more complex than $CH_3$ can abstract H from C—H bonds at cryogenic temperatures. For reaction to occur it must be exothermic, or at least thermoneutral, thus limiting the possibilities.

### Reaction with Solute Molecules

Solutes with which radicals can react with low activation energy compete effectively by reactions of type (13-15) for randomly diffusing radicals that would otherwise react with each other (13-13). The kinetics are pseudo first order, giving a linear log[R]-vs.-*t* plot.[40]

### Reactions with Electrons and Cations

Free radicals have electron affinities of the order of 1 eV, and are therefore able to undergo reactions of type (13-16) (see section entitled Carbanions and Anion Radicals).

The ionization potentials of most molecules used as trapping matrices ($\geqslant$10 eV) are higher than the ionization potentials of radicals ($\sim$8 eV).[64] Thus, it is probable that radicals lose electrons to matrix cations that are in sufficient proximity for tunneling to occur.

### Effects of Radical and Matrix Structure on Decay Rates

The rates of radical decay depend on the nature of the radicals and the nature of the matrix. This has been extensively demonstrated for radicals produced from alkyl halides in alkane glasses and in matrices of the pure compound.[44a] In all cases the decay kinetics are composite first order, implying either tunneling abstraction or combination with the geminate partner.

The rates of decay of $CH_3$ in 3MP, 2MP, and MCHx at 77 K decrease in that order.[55] Presumably the decay is by tunneling abstraction. The differences in rate imply differences in the average proximity and orientation of the $CH_3$ relative to the vulnerable C—H bonds.

Radical decay rates are slower in samples that have been annealed near their glass transition temperature,[40] producing a decrease in free volume.[20]

### Radical Decay in Crystalline Alkanes

Typically, radical decay in irradiated polycrystalline alkanes becomes fast enough for observation at approximately 0.8 of the absolute melting point. At each higher temperature to which the sample is raised, the radical concentration decreases rapidly at first and then levels off to a plateau, giving a stepwise profile of concentration vs. time. In contrast, the radicals in an uncracked single crystal of *n*-hexane showed second-order decay kinetics at each temperature,[52] suggesting that the decay pattern in the polycrystals is the result of varied softening points of crystallites, and surface irregularities of different size and/or strain.[52]

**Radical Growth; Limiting Concentrations**

When an organic solid is exposed to a constant source of radiation, the concentration of trapped radicals grows linearly at first, and then at a progressively slower rate until a steady-state plateau is reached where the rate of removal of radicals by decay and/or radiation-induced events is equal to the rate of production.[65,66] Plateau concentrations of radicals observed include 0.4 mol% $C_6H_{13}$ in 3MP at 77 K ($2 \times 10^{21}$ eV $gm^{-1}$ $\gamma$-irradiation at $3 \times 10^{18}$ eV $gm^{-1}$ $min^{-1}$);[47d] 0.05 mol% $C_2H_5$ in $C_2H_5I$ at 77 K ($3 \times 10^{20}$ eV $gm^{-1}$ $\gamma$-irradiation at $2 \times 10^{18}$ eV $gm^{-1}$ $min^{-1}$);[65] 0.1 mol% N in $N_2$ at 4 K;[67] 0.5 mol% for the radicals in polar aromatic compounds and 5 mol% for those in nonpolar aromatics.[68] Factors controlling the maximum achievable concentrations of radicals have been discussed.[69] It appears that the trapping of radiation-produced radicals, for subsequent release of the energy of combination on warming, will not prove useful in powering space ships or exploiting the waste radiation energy from spent reactor fuels.

When a steady-state plateau of radical concentration is reached during continued irradiation, the rate at which radicals are being removed must equal the rate at which they are being produced, which is equal to the observed growth rate at the start of exposure. Thus, it would be predicted that the decay rate when irradiation is stopped must equal the initial growth rate. In actuality, this rate is slower than the initial growth for some compounds (including alkyl halides,[65,66] glassy and polycrystalline ethylene glycol,[70] polycrystalline $CBr_4$,[71] and a variety of biological compounds).[72] This indicates that during irradiation radicals are removed by a process or processes other than those which occur in the absence of radiation. Mechanisms that may, in principle, contribute to the effect include the following:

1. Tunneling electron transfer ($R\cdot + e_t^- \rightarrow R:^-$ and $R\cdot + RH^+ \rightarrow R^+ + RH$)
2. Reaction of newly formed H atoms with radicals in competition with other reactions
3. Stimulation of recombination of trapped geminate partners by migrating phonons
4. Diffusional encounters of radicals within small volumes warmed by energy from neutralization and deexcitation events
5. Photoremoval of radicals by luminescence from deexcitation processes.

Evidence on luminescence yields,[66,73–75] extinction coefficients,[12,66] and decomposition quantum yields of radicals[12,66,76] is too limited to exclude 5, but it seems improbable that it is a major contributor. The formation of hot spots of sufficient volume and temperature to enhance diffusive combination (4) is not ruled out at 77 K[66,51] but cannot be important at 4 K.[52] The stimulation of combination by phonons is most plausible in matrices such as $C_2H_5I$ glass, where 98% of the radicals appear to decay by geminate recombination.[66]

For there to be significant probability for a radical produced by irradiation to be formed in contact with one already present, the radical concentration must be several mol%.[70] For a concentration of 0.16 mol%, the average distance between radicals, if evenly spaced, would be about 8.5 molecular diameters (~40 Å[22]). The average distance from a new radical, $e_t^-$, or cation introduced into this system to a radical already present would be about 3 molecular diameters. This scale of distances is such as to afford a high probability that radicals will be removed by tunneling capture of $e_t^-$, or loss of $e^-$ to cations, or reaction with mobile H atoms.

## Optical Spectra and Photoeffects

Pulse radiolysis studies indicate that radicals in gaseous and liquid alkanes have optical spectra with $\lambda_{max}$ values in the region of 217 to 245 nm (refs. 15 and 16 of ref. 12). Resolution of the spectrum of the radicals produced in 3MP glass at 77 K by $\gamma$-irradiation from the overlapping spectra of R:$^-$, $e_t^-$, and stable products indicates that $\lambda_{max} = 218$ nm and $\varepsilon = 1675\ M^{-1}\ cm^{-1}$, with a long wavelength onset of absorption at about 260 nm.[12]

As noted above, when $\gamma$-irradiated samples of alkane glasses and crystals are exposed to light absorbed by the radicals, the concentration is reduced by approximately 30%. Continued illumination causes no further reduction in concentration but produces thermally reversible changes in the ESR spectra[51] that reflect either isomerization, changes in rotamer states, or abstraction from adjacent molecules. Photochemical changes in the structure of trapped radicals produced from other types of compounds have also been reported[46,52] (see also refs. 12–16 of ref. 51).

The photolytic decompositions of several simple radicals in $CH_4$, $CD_4$, and, Xe matrices at 5 K have been observed by ESR.[76] These include

$$CH_3 + h\nu\ (185\ nm) \rightarrow CH_2 + H$$

$$CD_3 + h\nu\ (185\ nm) \rightarrow CD_2 + D$$

$$C_2H_5 + h\nu\ (185\ nm) \rightarrow C_2H_4 + H$$

$$C_2H_3 + h\nu\ (185\ nm) \rightarrow C_2H_2 + H$$

$$HCO + h\nu\ (\sim 500\ nm) \rightarrow H + CO$$

$$DCO + h\nu\ (\sim 500\ nm) \rightarrow D + CO$$

$$HO_2 + h\nu\ (254\ nm) \rightarrow H + O_2$$

$$DO_2 + h\nu\ (254\ nm) \rightarrow D + O_2$$

$$CH_3O_2 + h\nu\ (254\ nm) \rightarrow CH_3 + O_2$$

(and HCO + evidence for H + OH).

# CARBANIONS AND ANION RADICALS

## *Carbanions*

The presence of carbanions formed by the $R\cdot + e^- \rightarrow R:^-$ reaction in $\gamma$-irradiated organic solids is revealed by their optical spectra and by the generation of $e_t^-$, radicals, luminescence, and electrical conductivity when they are photoionized.[12,22,77] Since they have no unpaired electrons, they cannot be detected by ESR. In the branched alkane glasses the near-UV optical spectrum has a low-wavelength limit of about 210 nm, $\lambda_{max}$ at about 240 nm[12] with $\varepsilon = 6300\ M^{-1}\ cm^{-1}$, and a long red tail extending to greater than 800 nm.[77d] The spectrum is overlapped by the spectra of both the radicals and trapped electrons. The value of $G(R:^-)$ from the $\gamma$-irradiation ($\leq 6 \times 10^{19}$ eV gm$^{-1}$) of 3MP glass is 0.3.[12] When the $R:^-$ are photoionized at 15 K with light of less than 285 nm, 100% of the electrons ejected become trapped rather than reacting with cations. In contrast, when the $e_t^-$ population is completely bleached with IR light, 88% become neutralized and 12% react with radicals to form $R:^-$.[12] This results because the UV light is not absorbed by $e_t^-$, whereas the IR light can repeatedly mobilize any $e_t^-$ retrapped after initial detrapping.

## *Anion Radicals*

Early work reported the optical spectra of the radical anions formed by electron capture by biphenyl and naphthalene when these were present as solutes in $\gamma$-irradiated MTHF or 3MP at 77 K.[1a] Subsequently, this type of radiation chemistry has been extensively used for the preparation of radical anions for the purpose of characterizing their spectra.[78]

# HYDROGEN ATOMS

Gamma-irradiation of $CH_4$ at 5 K produces trapped H atoms with $G(H_t) = G(CH_3) = 3.3$.[41,79] The atoms, which can be readily monitored by their two-line ESR spectrum with 504 $G$ splitting, are stable for weeks. In contrast, no trapped hydrogen atoms have been found in other radiolyzed solid hydrocarbons (with the exception of a low yield of $D_t$ from $C_2D_6$),[80a] despite an intensive search.[80b] As noted in an earlier section, their absence must result from their rapid reaction with the matrix, either by hot abstraction or tunneling[48b,81,82] abstraction by thermalized atoms.

Conclusive evidence that thermal abstraction by H can occur at cryogenic temperatures is given by experiments in which $CH_4$[41,48a] and Xe matrices[82,83] containing dilute $C_2H_6$ and $H_t$ or $D_t$ at 4 K are warmed to a temperature where the atoms can diffuse and encounter the $C_2H_6$ (~12 K in $CH_4$ and

45 K in Xe). Under these conditions, $C_2H_5$ radicals appear while the atoms disappear at an equal rate.

The stability of $H_t$ in $CH_4$, in contrast to higher hydrocarbons, is attributed to the greater C—H bond strength, which reduces the probability of tunneling abstraction. Similarly, the reduction in the tunneling rate by the deuterium isotope effect explains why $H_t$ and $D_t$ produced in deuterated alkane glasses by the photolysis of hydrogen halides have an observable lifetime (seconds to minutes) while they decay too rapidly to be seen in protiated matrices.[48d] The apparent activation energies for tunneling abstraction by these atoms in 3MP-$d_{14}$, 3MHp-$d_{16}$, and MCHx-$d_{14}$ at less than 30 K are all less than 100 cal $mole^{-1}$.[48d] The decay kinetics are composite first order, ruling out H removal by the second-order $H + H \rightarrow H_2$ process.

Although H atoms are not observable in radiolyzed matrices of alkanes other than methane, there is convincing indirect evidence that they are formed and abstract. For example, in solid *neo*-$C_5H_{12}$ containing less than 2% of a solute with weaker C—H bonds, high yields of the radical of the solute are produced by irradiation at 77 K. These are presumed to be formed by H abstraction by atoms ejected from the solvent.[49] At 4 K, diffusion of H cannot occur, and abstraction is almost entirely from the solvent.[49] The data suggest that 20 to 30% of the atoms undergo hot reaction. Qualitatively similar results have been obtained for the reaction of D atoms from the radiolysis of *n*-$C_{10}D_{22}$ with *n*-$C_{10}H_{22}$ present as dilute solute.[49d]

Investigations in this field have provided an interesting example of the effect of the duration of encounter times on competitive reaction rates. Although the theory of tunneling reactions predicts a much smaller rate constant for tunneling abstraction from C—D than from C—H bonds,[48b,81] the ratio $k_H/k_D$ for abstraction by H atoms from $C_2H_6$, as compared to $C_2D_6$ in Xe matrices containing low concentrations of $C_2H_6$—$C_2D_6$ mixtures, is approximately unity at temperatures near the onset of H diffusion (~45 K).[82a] This is interpreted to indicate that the residence time of an H atom at each site encountered in its hopping diffusion is sufficient to give approximately 100% probability for reaction if either a $C_2H_6$ or $C_2D_6$ molecule is present. If the true $k_H/k_D$ ratio for very short encounter times is much greater than 1, as predicted from knowledge of isotope effects, it would be expected that at higher temperatures, where faster hopping rates result in shorter encounter times, the observed $k_H/k_D$ ratio would be higher. The value at 50 K is $k_H/k_D = 60$, which is consistent with this expectation.[82a]

## ORGANIC SOLUTES IN RARE GAS MATRICES

### *Properties and Uses*

Xe, Kr, and Ar (melting points 161, 116, and 84 K, respectively) are useful as matrices for the study of the radiolytic decomposition or organic molecules because they do not themselves yield trapped intermediates, since they are relatively unreactive toward reactive species produced from the solutes and they insulate these species from reactions with other solute molecules. For example, H atoms produced by the radiolysis of organic solutes in Xe, Kr, and Ar are stabilized and readily observed, whereas they would immediately react in an organic glass.[82,83,84]

In both rare gas and $CH_4$ matrices, the temperature threshold for significant diffusion of H atoms is well below that for radicals (Table 13-2), making it relatively easy to distinguish their reactions when the samples are warmed.

Sometimes the ESR signal of a radical that is undetectable in an organic matrix is observable in a rare gas (eg, $CH_2$ in Xe).[85a]

**Table 13-2.** Approximate Temperatures (K) of Onset of Rapid Diffusion of H and Radicals in $CH_4$ and Rare Gas Matrices[a],[41,84]

| | Matrix | | | |
|---|---|---|---|---|
| | $CH_4$ | Xe | Kr | Ar |
| $H(D_t)$ | 15 | 45 | 35 | 20 |
| $CH_3$ | 45 | 85 | | |
| $CH_2$ | 35 | 70 | | |
| $C_6D_{13}$ | | 100 | | |

[a] As indicated by measurable decrease in concentration in a few minutes.

## *Energy Transfer to Solutes*

Gamma-irradiations of Xe containing organic solutes provide good examples of transfer of energy from an irradiated matrix to a solute. The $G(H_t)$ values from the irradiation of 0.5 mol% 3MP in Xe and of 0.5 mol% $CH_4$ in Xe are approximately 100 and 1000, respectively, if calculated on the basis of energy absorbed only by the solutes, but are 0.5 and 1.35 if calculated from the energy absorbed by the matrix.[84] Since the higher values are energetically impossible, decomposition of the solutes must result from energy absorbed by the matrix. For 3MP, charge transfer from $Xe^+$ to 3MP by tunneling or electron hopping, followed by the $RH^+ + e^- \rightarrow R + H$ reaction, may account for the decomposition. This mechanism is precluded for $CH_4$ because its ionization potential (12.99 eV in the gas) is higher than that of Xe (12.13 eV). (If transfer occurred from the excited $Xe^+(^2P_{1/2})$ state (13.43 eV),[85b] the $CH_4$ would be neutralized without bond rupture by electron transfer from an adjacent Xe.) Thus, the production of $H_t$ and $CH_3$ by irradiation of 0.5 mol% $CH_4$ in Xe implies that the decomposition can occur by photolysis by light produced in the process of neutralization of $Xe^+$ by electrons. The luminescence spectrum of Xe has a maximum at 1482 Å. The vacuum-UV absorption spectrum of solid $CH_4$ extends up to 1575 Å, and the extinction coefficient at 1482 Å is $2 \times 10^{-1}$ $M^{-1}$ $cm^{-1}$. It is assumed here that differences in the energy of "solvation" of $CH_4^+$ and $Xe^+$ in the Xe matrix do not reduce the ionization potential of $CH_4$ to less than that of Xe.

### *Dosimetry*

The dose rate to Xe matrices exposed to $^{60}$Co $\gamma$-rays (~1.1 MeV) cannot be determined directly from Fricke dosimetry because the high atomic number results in significant absorption by the photoelectric effect, which is negligible with the low-atomic-number dosimeter solutions. For x-rays, which normally are of much lower energy, the correction for the photoelectric effect depends on the energy and may be very large (eg, in a typical case the true x-ray dose to a Xe-3MP matrix was 18 times that measured by Fricke dosimetry).[84]

## ESR SATURATION EFFECTS

Electron spin resonance saturation effects in the measurement of trapped radicals and electrons may be a source of both error and information. For trapped electrons in $\gamma$-irradiated 3MP at 77 K, the power at which the onset of saturation occurs increases with increasing $\gamma$-dose (and hence $e_t^-$ concentration) but decreases as the electrons decay.[19] The saturation thresholds for trapped H, D, $CH_3$, and $CD_3$ in $CH_4$ and $CD_4$ depend on the method of their production (and hence their geminate partners), on the isotopic nature of the matrix, on the temperature, and sometimes on the extent to which the initial population of trapped species has decayed.[41] Saturation thresholds decrease with increase in the spin–spin relaxation time, which depends on the nature and proximity of neighboring paramagnetic species.

## STABLE PRODUCTS

The yields of stable products from the $\gamma$-irradiation of organic solids are usually obtained after melting, by gas chromatographic or mass spectrometric analysis, or, in the case of inorganic halide products, wet assay. The products include those from the combination and disproportionation of the trapped radicals observed by ESR, lower-yield products formed from trapped radicals whose ESR signals are obscured by the stronger signals of the predominant radicals, and products formed by excited molecules, prompt radical–radical, and charge neutralization processes.

### Yields from 3MP-$h_{14}$ and 3MP-$d_{14}$ in Glass, Liquid, and Gas

Thirty different products of the radiolysis of 3MP-$h_{14}$ at 77 K, with a total $G$-value of 9.02,[86] have been detected by gas chromatography. Individual $G$-values range from 0.01 for 2-methylbutane to 3.2 for $H_2$. (Later work has shown that at lower $\gamma$-doses than the $1.2 \times 10^{21}$ eV gm$^{-1}$ used in ref. 86, $G(H_2) = 4.3$.[60]) Many of the products, such as $CH_4$, $C_2H_6$, $C_3H_8$, $C_4H_{10}$, $C_5H_{12}$, require C—C bond rupture for their formation. Comparison of the yields of the individual products from 3MP-$d_{14}$ with those from 3MP-$h_{14}$ indicates that there is no significant isotope effect.[86] The total yields from radiolysis of 3MP-$h_{14}$ in the liquid and gas states are $G = 12.7$ and 16.6, as compared to 9.0 in the solid state, the increases resulting predominantly from greater C—C bond rupture.[86] This suggests that such rupture is reduced by deexcitation and caging effects in the solid. The yields of products that require isomerization without change in chain length (eg, 2-methylpentane) and concerted $H_2$ elimination (3-methyl-1-pentene) are lower in the liquid than in the solid, and lower in the gas than in the liquid.[86]

### Effects of Crystal Structure and Scavengers in Alkanes

In $\gamma$-irradiated alkane solids:

1. $G$-values are different for the glassy and polycrystalline states of a compound.[87]
2. Dissolution of the $\gamma$-irradiated crystals in liquid propane containing dissolved oxygen, rather than melting them, allows $O_2$ to scavenge the trapped radicals and thus reduces the yields of dimers. The remaining dimer yield, attributed to processes completed during irradiation, is systematically different for the $n$-alkanes of odd C number than for those of even C number, indicating that the radiation chemistry is sensitive to crystal structure.[89]
3. The scavenging by $O_2$ also reduces the yields of olefins by eliminating the disproportionation of trapped radicals.[89]
4. At high doses the yield of dimers of the type that can be formed by the reaction of radicals with olefins increases as a result of accumulation of olefins.[90]

### Alkyl Iodides

The sum of the yields of HI and $I_2$ from the $\gamma$-radiolysis of $C_2H_5I$, $n$-$C_3H_7I$, or $n$-$C_4H_9I$ is similar in the liquid, glassy, and crystalline states, but the ratio of HI to $I_2$ is much higher in the solid phases than in the liquid phases.[91]

In general, the same products, including ethylene, methyl-, ethyl-, propyl-, and butyl-halides, some dihalides, and some still less volatile compounds, are produced from the radiolysis of $C_2H_5I$, $C_2H_5Br$, and $C_2H_5Cl$[44b] in the polycrystalline and glassy states and at 77 K and 4.2 K. (The ESR spectra of trapped radicals in the alkyl halides show systematic differences dependent on phase (glass or crystal) and C number.[44b])

### $CCl_3Br$

The value of $G(Br_2)$ from the radiolysis of solid $CCl_3Br$ is constant (0.12) from 83 K to 153 K, above which it increases, reaching 3.5 in the liquid at 371 K. The increase gives a linear Arrhenius plot (2 kcal $mole^{-1}$) from 195 to 371 K with no discontinuities at solid-state transitions at 238 and 260 K or at the melting point (267 K).[92] In contrast, alkyl iodides show a positive temperature coefficient for $I_2$ production between 83 K and 153 K, but $G(I_2)$ is temperature independent in the liquid phase.[93] The mechanisms responsible for these effects have not yet been elucidated.

### Choline Chloride

The radiolysis at approximately 65°C of crystalline choline chloride $[(CH_3)_3NCH_2CH_2OH]^+Cl^-$ to yield $(CH_3)_3NHCl + CH_3CHO$ holds the all-time record of 55,000 for the highest $G$-value observed for a decomposition reaction.[94] A chain-reaction mechanism involving electrons and H as chain-carrying intermediates has been proposed. Choline bromide also gives a high, but lower, $G$-value. Choline iodide gives a normal value. For the chloride and bromide the values are normal below 20°C, rise to maxima at about 65° and 75°C, and fall to zero above polymorphic phase transitions, which occur at about 80° and 90°C, respectively.[94]

## MATRIX PROPERTIES

### Criteria for Selection

Ideally, an organic matrix for the study of radiation-produced intermediates should have the following properties:

1. The ability for stable trapping of electrons, cations, radicals, carbanions, and H atoms at a convenient temperature.
2. Transparency in the UV, visible, and IR.
3. Irradiation should not generate species that produce ESR signals other than those of the intermediates to be studied.
4. The ability to dissolve desired additives.

5. When intended as a solvent for the study of solute cations, it should have an ionization potential higher than that of the solute and an electron affinity enabling it to capture electrons and thus prevent neutralization of the cations.

Comments will be made here on the properties of some of the most commonly used types of matrix. Detailed added information on the physical properties of those particularly useful for the study of trapped electrons is given in ref. 5.

## *Hydrocarbons*

### Glassy and Polycrystalline States

All of the noncyclic branched chain alkanes and alkenes form clear glasses when rapidly cooled to 77 K, whereas the *n*-alkanes are invariably polycrystalline. Methylcyclohexane (MCHx) can be solidified as the glass by rapid cooling or, in the polycrystalline state, by slow cooling or by warming the glass to 125 K,[95] thus allowing comparison of the radiation chemistry of the glassy and crystalline states using a single compound.

### Spectral Considerations

The pure alkanes are transparent from less than 200 nm to greater than 1100 nm. The overtone and combination bands of their vibrational spectra in the 1100–2500-nm region overlap the spectra of trapped electrons (see Figure 2 of ref. 11), but the latter can be resolved by subtraction. The interference is much smaller in perdeuterated than in protiated matrices.[11]

When radiolyzed, all of the hydrocarbons produce radicals that give ESR spectra centered at approximately $g = 2.0023$. In the perprotiated matrices these are about 160 G wide; in the perdeuterated matrices, about 50 G. The latter are used when it is desired to minimize interference with the spectrum from a perprotiated solute radical.

### Viscosities

The viscosities of a series of glassy hydrocarbons, measured to the lowest temperature possible by the pressure extrusion method and extrapolated to 77 K, range from $2.2 \times 10^{12}$ P (about the viscosity of window glass) for 3MP to $1.2 \times 10^{31}$ P for 3-methyloctane.[96] Except for 3MP, the 77 K values are probably all low because the required extrapolation is to a temperature below their glass transition temperatures. Within experimental error, the viscosity of 3MP-$d_{14}$ is equal to that of 3MP-$h_{14}$[96a] and so does not explain the slower decay rates of $e_t^-$ and radicals in the former.

### Glass Transition Temperature; Annealing

As the viscosity of a glass increases on cooling, the glass transition temperature ($T_g$) is reached (typically $10^{12}$–$10^{13}$ P). As cooling is continued, the rate of molecular relaxation becomes too slow for the decreasing enthalpy and free volume to reach their equilibrium values except with long periods of annealing (Figure 1 of ref. 17). Glasses cooled to below $T_g$ anneal toward the equilibrium state with time but at rates that are much slower at temperatures considerably below $T_g$ than near $T_g$.[20] The changes caused by annealing decrease the decay rates of both trapped radicals[40] and trapped electrons[19] but do not affect the $G$-values.[19] Surprisingly, the rate of annealing toward the equilibrium condition is dependent on the size, shape, and wall thickness of the sample container. Thin, flexible, walls or a large surface area allow a more rapid volume contraction, as does cracking of the glass.[19,20]

### Molecular Orientation by Electric Fields

Partial orientation of the molecules in 3MP, MCHx, and MTHF glasses at 77 K by electric fields as low as $10^4$ V cm$^{-1}$ have been demonstrated.[95] The orientation in MTHF, which has a permanent dipole, is much greater than in 3MP, where the interaction must be with bond dipoles. These observations are consistent with the conclusion that the intense field in the vicinity of a trapped electron aids it in producing molecular orientations that stabilize the trap.

### Density

Cooling of a hydrocarbon from the liquid state at room temperature to the glassy state at 77 K is accompanied by a significant increase in density, and, hence, in the molarity of any solute, and in the energy absorbed per mL during irradiation. The density of 3MP-$h_{14}$ is 0.676 gm mL$^{-1}$ at 298 K and 0.866 gm mL$^{-1}$ at 77 K. The values for 3MP-$d_{14}$ are 0.766, 0.877, and 0.994 gm mL$^{-1}$ at 298, 195, and 77 K, respectively. The molar volumes of the protiated and deuterated species are equal.[46]

## $CCl_3F$, $CBr_2FCBr_2F$, sec-$C_4H_5Cl$, $SF_6$

Certain freons and $SF_6$ have ionization potentials, electron affinities, and spectral properties particularly favorable for the formation and stabilization of solute cations (as noted in the section on cations). A 50:50 vol% mixture of $CCl_3F$ (Freon 11) (IP 11.78 eV) and $CBr_2FCBr_2F$ (Freon 114B2) (IP 11.40) forms a clear glass when rapidly cooled to 77 K[97] (but becomes polycrystalline if cooled slowly[98]); the glass is transparent from 270 to 2700[98] nm. The ESR signal of its radicals is sufficiently broad so that it does not interfere seriously with that of solute cations. However, in some cases (eg, pyridine solute) the cation ESR spectrum is not observed, although it is in $CCl_3F$,[99] which is an excellent solvent for ESR study of irradiation-produced cations but is less desirable for optical studies because it is polycrystalline and opaque. The background ESR signal from its radical is negligible. $SF_6$ has given better resolution of some solute cation ESR spectra (eg, $C_2H_6^+$) than obtained with $CCl_3F$ as the solvent.[100a] $SF_6^-$ gives a 1500-G wide anisotropic ESR spectrum at 4 K that converts to an isotropic spectrum at 100 K,[100a] similar to that observed when $SF_6$ is present as a solute in *neo*-$C_5H_{12}$ irradiated at 77 K.[100b]

Examples of the many types of compounds whose radical cations have been investigated in irradiated matrices include olefins and dienes in $CCl_3F$,[101] hexamethylethane in a mixture of $CCl_3F$ and $CBr_2FCBr_2F$ (designated "FM" for "freon mixture"),[102] aliphatic and aromatic hydrocarbons in *sec*-$C_4H_9Cl$ and FM,[103] 1-3-cyclohexadiene in FM,[98] pyrene–biphenyl and other solute pairs in *sec*-$C_4H_9Cl$ and in MTHF,[104] and others.

## Alkyl Halides

All of the straight-chain alkyl iodides[44b] and bromides[105a] from $C_2$ through $C_6$ and the chlorides from $C_3$ through $C_6$ [105a] can be prepared as transparent glasses by rapid cooling to 77 K. If slowly cooled, they all become polycrystalline. Attempts to prepare *i*-$C_3H_7I$, the $C_7$, $C_8$, and $C_{10}$ iodides, *n*-$C_7H_{15}Br$, $CH_2BrCH_2Br$, $C_2H_5Cl$, and $CCl_4$ as glasses have been unsuccessful.[105a] The radiolytic products of the solid iodides and bromides give intense optical spectra in the visible and near-UV,[105a,b,c] and all of the irradiated solid halides give strong ESR signals.[44b,105a] In most cases these include the spectrum of the parent alkyl radical in the $g = 2.0023$ region. In the polycrystalline iodides of even C number, however, the signals are very broad (350–1000 G), many-line (15–30) spectra.[44b,105d] All of the $\gamma$-irradiated iodides and bromides exhibit additional spectra extending to less than 1600 G on the low side of the central signal (3200 G) and to greater than 5000 G on the high side.[105e]

## Methyltetrahydrofuran

The radiation chemistry of glassy MTHF, with and without solutes, has probably been more extensively studied than that of any organic solid other than 3MP. The glass is transparent. It has a glass transition temperature of about 91 K[20] and an extrapolated viscosity at 77 K of $3.7 \times 10^{20}$ P.[96a] With a static dielectric constant of 2.9,[5] it is slightly more polar than 3MP. As a result, the spectrum of trapped electrons is blue shifted relative to that in 3MP ($\lambda_{max}$ at 77 K ~ 1200 nm as compared to 1600 nm), although not as much as in the more polar alcohols. The $e_t^-$ spectrum at 77 K has three peaks, suggesting three distinctly different types of trapping center,[6a] in contrast to most other matrices studied. The ESR spectrum of $\gamma$-irradiated

MTHF consists of the $e_t^-$ singlet superimposed on a seven-line spectrum of the radical formed by H atom removal.

### *Alcohols*

Most alcohols studied, except $CH_3OH$ and $t$-$C_4H_9OH$, readily form glasses on cooling to 77 K and do not easily crystallize. $CH_3OH$ forms a clear transparent glass if greater than 2 mol% of $H_2O$ is added. Gamma-irradiation of the glassy alcohols produces trapped electrons and, typically, the radical formed by removal of a beta hydrogen atom. The properties of alcohol matrices and their radiation chemistry are discussed in ref. 1(c).

### *Adamantane*

The polycyclic hydrocarbon adamantane has been extensively used as a matrix for the ESR examination of radicals produced from solutes by radiolysis.[106] Advantages over investigation of the radicals in the solid parent compounds at low temperatures include freer rotation (and hence better resolved spectra) and the possibility of studies over a wide temperature range without excessive decay due to softening of the matrix.

### *Other Matrices*

Information on the matrix characteristics and radiolytic properties of other matrices, including amines, cyanides, ketones, and ethers has been summarized in ref. 1(c).

## TECHNIQUES

### *Cryogens and Cryogenic Equipment*

The techniques used for study of the radiation chemistry of organic solids are the same as those for the liquid and gas phases, with the added requirement that it be possible to cool samples at desired rates to desired cryogenic temperatures and to maintain them there during both irradiation and measurement.

Liquid $N_2$ (BP 77 K) and liquid He (BP 4.2 K) are the most commonly used cryogens (other possibilities are discussed in ref. 107). Each can be used for a wide range of temperatures. Immersion of a sample in the liquid at atmospheric pressure maintains it at the boiling point of the cryogen. Lower temperatures (to ~64 K with $N_2$ and ~1.5 K with He) can be obtained by lowering the pressure, with a mechanical pump, to a controlled value equal to the vapor pressure of the liquid at the temperature desired. Temperatures less than 77 K with liquid $N_2$ can also be obtained by bubbling He gas through it at a vigorous rate.

Continuously variable temperatures can be achieved by bathing the sample in a stream of the boil-off gas from He or $N_2$, with provision for electrical heating to the desired temperature before it reaches the sample. With liquid $N_2$ it is common to pass gaseous $N_2$ from a tank through a heat-exchanging coil immersed in the liquid, then over the sample.

Types of equipment used for variable temperature studies in the range of 4 to 77 K include[108]

1. Systems that deliver liquid or gaseous He directly from its shipping container through a flexible Dewar-walled hose to the sample Dewar in the measuring instrument.
2. Refrigerators using the Joule–Thomson effect, with sequential expansion of $N_2$, $H_2$, and He to cool a metal tail that holds the sample in the analysis position.
3. Metal Dewars containing liquid He that can be released at a controlled rate to a built-in sample chamber.

Each of these types of equipment is useful for optical studies on irradiated samples. Only the first allows easy placement and cooling of the sample in an ESR cavity. For ESR studies at 77 K the sample in a 3-4-mm OD quartz tube is typically held under liquid $N_2$ in an 11-mm or less OD Dewar-walled tail of a liquid $N_2$ reservoir. The sample position in this tail is centered in the ESR cavity. With liquid He such equipment has a high boil-off rate. However, using an ESR cavity with an opening large enough to accommodate a 15-mm OD sample finger,[109] with four Suprasil walls, and liquid $N_2$ shielding of the liquid He, workers have achieved excellent performance, with only 1 L of liquid He consumption per 30 hr (as contrasted to ~1.5 L/hr for cooling with He boiling off the liquid). Equipment for the variable temperature control of samples in the range above 5 K during ESR and optical analysis has been described.[110]

## *Radiation Sources*

Gamma-irradiation of organic solids is typically done with a $^{60}Co$ source, with the sample under liquid $N_2$ or He in a heavily shielded irradiation chamber. Following irradiation, the sample must be transferred to the analyzing Dewar and carried to the analyzing instrument. The need for such transfer from the irradiator can be eliminated by x-irradiating the sample while it is positioned in the analyzing instrument by using an x-ray tube attached to its power supply by a flexible cable. Only a few millimeters of Pb shielding are required to protect personnel. The sample is irradiated in a portion of the Dewar above the analysis region to avoid producing paramagnetic and color centers in the latter. It is then lowered to the analyzing position. Measurements can be started within seconds of the end of the irradiation. This procedure avoids the risks of warming, and exposure to humidity with resultant condensation, that are present when the sample must be transferred through air from a $\gamma$-irradiation Dewar to the analysis Dewar. Dewars in which the sample is immersed in liquid He[109] and those in which it is cooled by a flowing He stream[85a,110] have been used for x-irradiations. The lower energy of the x-rays as opposed to $^{60}Co$ $\gamma$-rays has two disadvantages:

1. Their absorption cross sections are very sensitive to atomic number, thus making the dosimetry more tedious when samples of different composition are being compared.[85a,41]
2. Their penetrating power is too low to allow them to be effectively used with thick-walled cryostats.

Samples synthesized to include tritium ($^3H$, the $\beta$-emitting isotope of hydrogen) at high specific activity allow continual observance of the growth of trapped radiation intermediates during self-irradiation.[66,111] Because the weak $\beta$-particles are all absorbed within the sample, the dose rate is known if the specific activity is known. A disadvantage, in addition to the problems of synthesis, is that the radiation cannot be turned off, so the sample deteriorates on standing.

## *Sample Preparation*

### Purification

A few special considerations for purification of samples for solid-phase studies will be noted. For investigations of $e_t^-$ it is important to remove electron scavenging impurities. These include $O_2$ and $CO_2$, which can be eliminated by flushing the liquid with pure $N_2$ or Ar or by vacuum line techniques. To remove $CO_2$ by the latter, one must pump on the liquid, rather than on the solid, because the $CO_2$ has negligible vapor pressure at 77 K. Storage of hydrocarbons under vacuum over molecular sieve (degassed under vacuum at ~290°C) has proved effective in removing olefins, $H_2O$, and the last traces of $O_2$. Hydrocarbons containing traces of $O_2$ invariably show a rising optical absorption in the region from about 250 to 200 nm, attributed to a charge transfer complex. A different type of impurity effect is illustrated by the observation that $C_2H_5I$ dried by passing through $P_2O_5$ readily forms a glass, whereas untreated samples always crystallize.

### Sample Cells and Freezing Procedure

Ionizing radiation darkens Pyrex and produces paramagnetic centers. For this reason samples for optical or ESR analysis are usually prepared in synthetic high-purity fused silica (eg, "Suprasil"), except when trapped H atoms are to be studied. Ordinary fused quartz gives a much weaker radiation-induced H-atom signal. Suprasil, which is made from hydrolyzed $SiCl_4$, contains residual OH groups that produce H on irradiation. The heights of the lines of the H-atom ESR doublet in empty sample tubes made of Suprasil, ordinary quartz, and Pyrex that had received a $\gamma$-dose of $5 \times 10^{19}$ eV $gm^{-1}$ were in the ratio of 30:1:3. In the same tubes the heights of the signal of the impurity center at approximately 3200 G, where free-radical signals typically appear, were in the ratio of 1:15:30.

Cells of 1 cm × 1 cm or 0.5 cm × 0.5 cm square-cross-section Suprasil tubing are commonly used for optical spectrometry, and 2-mm ID, 3-mm OD or 3-mm ID, 4-mm OD Suprasil tubes for ESR examination. For samples with high-absorbancy, optical cells of less than or equal to 1-mm thickness can be made by rapid drawing out of heated 2-cm ID tubing over a rectangular piece of Mo or stainless steel, machined to the desired width and thickness. Alternatively, polished glass plates, separated by a spacer, can be sealed around the edges with the aid of a quartz cane bead.

As a liquid cools and freezes, it becomes denser. Therefore, it expands when warmed and often cracks cells. Varying success is achieved in avoiding this by plunging the cold cell into hot water to rapidly liquify the surface of the sample before the center warms, or by allowing warming to take place slowly from the top downward. The tendency of samples to crack when lowered to cryogenic temperatures is greater for large samples than for small, and greater in square cells than in round. The tendency to crystallize, for samples that can form either a glass or crystal, varies similarly.

When a quartz ESR sample tube containing an organic sample frozen to an uncracked glass at 77 K is rapidly immersed in He gas or liquid at a much lower temperature, both the organic glass and the quartz usually fracture. If, however, the organic glass is first cracked at a higher temperature (eg, by cycling back and forth between Dewars of $N_2$ at 77 K and 65 K, or by moving it slowly into progressively colder regions in a stream of flowing He), fracturing of the quartz can be avoided.

When a gaseous mixture (eg, $C_4H_{10}$ in Xe or HI in $CH_4$) is to be solidified, particular care is necessary to avoid segregated deposition of the solute. Two proven successful methods are

1. Mixing the gas in a large volume bulb and condensing it at cold He gas temperatures in the tip of a long ESR sample tube attached directly to the bulb.[41]
2. Introducing the mixture to the cold surface in small puffs through a stopcock,[100a] thus minimizing softening of the initially condensed sample by the heat of condensation of subsequent portions.

There is some question about the time required to assure complete cooling of condensed samples. It has been reported[49c] that polycrystalline samples in evacuated ESR tubes require as long as 1 hr to achieve temperature equilibrium at 4.2 K when immersed in liquid He from 77 K, but that equilibrium is achieved in a few minutes when 500 mm of He gas are present in the ESR sample tube to serve as a heat transfer medium. In contrast, the cooling of samples in evacuated tubes to 5 K by flowing He gas has been found to be complete in less than 3 min.[41] The reason for the difference is not clear.

## *Electrical Conductivity*

The temperatures of onset of migration of electrons, anions, and cations when a $\gamma$-irradiated organic solid is warmed can be determined by measuring the current passing through a thin (eg, 0.13-mm) sample in a volume defined by a Mylar spacer between two glass plates coated to make them electrically conducting.[95,112] An electrometer with a sensitivity of about $10^{-14}$ A and a voltage supply of 1000 V are needed.

## *Differential Thermal Analysis*

The thermochemistry of reactions of the electrons, ions, and radicals produced by the $\gamma$-irradiation of organic solids at 77 K can be studied. The equipment used[20] employs a copper cylinder, which serves as a heat sink, wound with heating wire and suspended in an evacuated Pyrex cylinder surrounded by liquid $N_2$. Differential thermal analysis (DTA) curves are obtained by amplifying the output from opposing thermocouples embedded in a sealed sample and a reference tube positioned in holes in the copper block, as the block is heated. A quartz light pipe allows illumination of the sample cell to rapidly detrap electrons, for measurement of their heats of reaction with radicals and cations. Gamma-irradiations are made by suspending the block in liquid nitrogen and lowering a $^{60}$Co source to a position adjacent to it.

## *Luminescence Techniques*

### Isothermal Luminescence

When an organic solid is exposed to ionizing radiation, it emits visible and UV light during and for minutes or hours after irradiation.[73] At temperatures where the diffusion of radicals and ions is negligible, this results from the tunneling of electrons to cations, producing excited states that lose their energy (in part) by radiative transitions. The intensity decreases with time, as the electrons are removed, and as the remaining electron–cation separations become greater. The wavelengths emitted depend on the cation, which in a pure hydrocarbon may be of the type $C_6H_{14}^+$, $C_6H_{13}^+$, $C_6H_{15}^+$, or $(C_{12}H_{26})_2^+$. If positive charge transfer to a solute is complete before neutralization occurs, the luminescence has a wavelength determined by the excited states of the solute species.

The spectra of the excited states and the probability of light emission relative to radiationless decay may be different in the glassy phase than in the gas or liquid phases. The glass provides a dielectric medium which

1. Reduces the total energy of neutralization.
2. Constrains free rotation, with the result that the excited molecules formed by neutralization may luminesce from the geometry of the cations rather than the lowest excited state of the neutral molecules.
3. May reduce the decomposition that accompanies neutralization in the gas or liquid, and hence increase the probability of luminescence from the undecomposed form.

For these reasons there is considerable uncertainty in the assignment of the isothermal luminescence peaks to specific processes in pure compounds; the uncertainty is less when the cations are produced from well-characterized solute molecules.

For pure hydrocarbons there is evidence[74] that only a small fraction of the neutralization events (eg, $10^{-3}$) results in the emission of photons, but also that the low emission yield in liquids[75] is due to a low fluorescence yield from the first excited singlet state rather than to a low $G$-value for production of the state.

### Photostimulated Luminescence

The spontaneous luminescence on standing at a fixed temperature, described above, is referred to as *isothermal luminescence* (ITL). The emission of luminescence can be greatly accelerated by exposing the sample to IR light, which repeatedly detraps the electrons and allows rapid neutralization by cations. If monochromatic light is used and scanned from the red end of the $e_t^-$ spectrum in the IR into the UV, a photostimulated luminescence (PSL) spectrum is obtained, with the intensity of the luminescence as a function of wavelength being proportional to $I\Phi\varepsilon C\alpha$ (the intensity of the incident light × the quantum yield for free electron production from the absorbing species × the extinction coefficient × the concentration × the relative sensitivity of the detector for the luminescent wavelength observed). The PSL spectrum is thus a reflection of the spectrum of the electrons and carbanions,[73a,113] plus any positive holes that are caused to migrate and become neutralized.

### Electrophotoluminescence

Some of the electrons produced by the irradiation of hydrocarbon glasses are trapped sufficiently close to cations so they can be caused to combine by applying an electric field across the cell.[114] The intensity of this "electrophotoluminescence" (EPL) dies off with time of application of the field in one direction but can be activated to about the same initial intensity by reversal of the field direction.

### Warmup Luminescence

When a $\gamma$-irradiated alkane is slowly warmed, there is a burst of luminescence in the temperature range just above the glass transition temperature. This results from neutralization of the electrons, as in isothermal decay, but at a faster rate because the tunneling process is assisted by diffusion. On further warming, one or more added emissions occur. In 3MP these are at approximately 86 K and 91 K.[73b] The first is attributed to the reaction of cations with carbanions, because it can be completely eliminated by bleaching in the carbanion band before warming. The second is, in part, resistant to such bleaching and may result from radical–radical combination. This emission is enhanced by the presence of $O_2$ in the matrix, indicating that it may involve peroxy radicals and/or ions.

## REFERENCES


1. Reviews dealing with this subject include (a) Hamill, W. H. In "Radical Ions"; Kaiser and Kevan, Eds.; Wiley Interscience; New York, 1968, pp. 321–416; (b) Willard, J. E. In "Fundamental Processes in Radiation Chemistry"; Ausloos, P. Ed.; Wiley Interscience; New York, 1968, pp. 599–650; (c) Kevan, L. In "Actions Chimiques et Biologiques des Radiations"; Haissinsky, M. Ed.; Masson: Paris, 1971, Vol. 15, pp. 5–143. (d) For an example and references see Klassen, N. V. *J. Phys. Chem.* **1983**, *87*, 3894.
2. (a) Schmidt, W. F.; Allen, A. O. *Science* **1968**, *30*, (b) Tezuka, T.; Namba, H.; Nakamura, Y.; Chiba, M.; Shinsaka, K.; Hatano, Y. *Int. J. Radiat. Phys. Chem.* **1983**, *21*, 197.
3. Shida, T.; Iwata, S.; Watanabe, T. *J. Phys. Chem.* **1972**, *76*, 3683.
4. Ekstrom, A. In "Radiation Research Reviews"; Elsevier: Amsterdam, 1970, pp. 381–409.
5. Kevan, L. In "Advances in Radiation Chemistry"; Burton, M.; and Magee, J. L. Eds.; Wiley Interscience: New York, 1974, Vol. 4, pp. 181–306.
6. (a) Hager, S. L.; Willard, J. E. *J. Chem. Phys.* **1974**, *61*, 3244; (b) Perkey, L. M.; Farhataziz; Hentz, R. R. *J. Chem. Phys.* **1971**, *54*, 2975; (c) Namiki, A.; Noda, M.; Higashimura, T. *Chem. Phys. Lett.* **1973**, *23*, 402.
7. Hase, H.; Noda, M.; Higashimura, T. *J. Chem. Phys.* **1971**, *54*, 2975.
8. Kimura, T.; Bremer, N.; Willard, J. E. *J. Chem. Phys.* **1977**, *66*, 1127.
9. Miller, J. R.; Clifft, B. E.; Hines, J. J.; Runowski, R. F.; Johnson, K. W. *J. Phys. Chem.* **1976**, *80*, 457.
10. Klassen, N. V.; Gillis, H. A.; Teather, G. G. *J. Phys. Chem.* **1972**, *76*, 3847.
11. Wang, H. Y.; Willard, J. E. *J. Chem. Phys.* **1978**, *69*, 2964.
12. Bhattacharya, D.; Willard, J. E. *J. Phys. Chem.* **1980**, *84*, 146.
13. Paraszczak, J.; Willard, J. E. *J. Chem. Phys.* **1979**, *70*, 5823.
14. (a) Huang, T.; Eisele, I.; Lin, D. P.; Kevan, L. *J. Chem. Phys.* **1972**, *56*, 4702; (b) Kevan, L. *J. Phys. Chem.* **1972**, *76*, 3830; (c) Huang, T.; Kevan, L. *J. Am. Chem. Soc.* **1973**, *95*, 3122.
15. Dismukes, G. C.; Hager, S. L.; Morine, G. H.; Willard, J. E. *J. Chem. Phys.* **1974**, *61*, 426.
16. Hager, S. L.; Willard, J. E. *Chem. Phys. Lett.* **1974**, *24*, 102.
17. Willard, J. E. *J. Phys. Chem.* **1975**, *79*, 2966.
18. For examples and references see (a) Calcaterra, L. T.; Closs, G. L.; Miller, J. R. *J. Am. Chem. Soc.* **1983**, *105*, 670; (b) Miller, J. R.; Beitz, J. W. *J. Chem. Phys.* **1981**, *74*, 6746;

(c) Miller, J. R.; Clifft, B.; Runowski, R. F.; Johnson, K. W. *J. Phys. Chem.* **1976**, *80*, 457; (d) Miller, J. R. *J. Chem. Phys.* **1972**, *56*, 5173.
19. Shooter, D.; Willard, J. E. *J. Phys. Chem.* **1972**, *76*, 3167.
20. Hager, S. L.; Willard, J. E. *J. Chem. Phys.* **1975**, *63*, 942.
21. Shirom, M.; Willard, J. E. *J. Am. Chem. Soc.* **1968**, *90*, 2184.
22. Ekstrom, A.; Suenram, R.; Willard, J. E. *J. Phys. Chem.* **1970**, *74*, 1888.
23. Lin, D. P.; Kevan, L. *J. Chem. Phys.* **1971**, *55*, 2629.
24. Fujii, S.; Willard, J. E. *J. Phys. Chem.* **1970**, *74*, 4314.
25. Gallivan, J. B.; Hamill, W. H. *J. Chem. Phys.* **1966**, *44*, 1279.
26. (a) Blandamer, M. J.; Shields, L.; Symons, M. C. R. *J. Chem. Soc.* **1967**, 1127; (b) Smith, D. R.; Pieroni, J. J. *Can. J. Chem.* **1967**, *45*, 2723.
27. Kevan, L.; Bowman, M. K.; Narayana, P. A.; Boeckman, R. K.; Yudanov, V. F.; Tsvethov, Y. D. *J. Chem. Phys.* **1975**, *63*, 409.
28. Ichikawa, A.; Ohta, M.; Kajioka, H. *J. Phys. Chem.* **1979**, *83*, 284.
29. (a) Lias, S. G.; Ausloos, P. "Ion-Molecule Reactions, Their Role in Radiation Chemistry"; American Chemical Society, Washington, D.C., **1975**; (b) Iwasaki, M.; Toriyama, K.; Nunome, N. *Int. J. Radiat. Phys. Chem.* **1983**, *21*, 147.
30. Louwrier, P. W. F.; Hamill, W. H. *J. Phys. Chem.* **1968**, *72*, 3878.
31. (a) Klassen, N. V.; Teather, G. G. *J. Phys. Chem.* **1979**, *83*, 326; (b) Cygler, J.; Teather, G. G.; Klassen, N. V. *J. Phys. Chem.* **1983**, *87*, 455.
32. Teather, G. G.; Klassen, N. V. *J. Phys. Chem.* **1981**, *84*, 3044.
33. For examples and references see (a) Kubodera, H.; Shida, T.; Shimokoshi, K. *J. Phys. Chem.* **1981**, *85*, 2583; (b) Toriyama, K.; Nunome, K.; Iwasaki, M. *J. Phys. Chem.* **1981**, *85*, 2149.
34. (a) Iwasaki, M.; Toriyama, K.; Nunome, J. *J. Am. Chem. Soc.* **1981**, *103*, 3591. (b) For additional discussion of the reactions of radicals produced by radiation in the solid state, see Willard, J. E. In "Chemical Kinetics of Small Organic Radicals"; Alfassi, Z. Ed.; CRC Press, Boca Raton, Fla, in press.
35. (a) Henry, B. T.; Siebrand, W. In "Organic Molecular Photophysics"; Birks, J. B. Ed.; Wiley: New York, **1976**, Vol. 1, p. 153; (b) Rothman, W.; Hirayama, F.; Lipsky, S. *J. Chem. Phys.* **1973**, *58*, 1300.
36. Mainwaring, D. D.; Willard, J. E. *J. Phys. Chem.* **1973**, *77*, 2864.
37. Henderson, D. J.; Willard, J. E. *J. Am. Chem. Soc.* **1969**, *91*, 3014.
38. Ichikawa, T.; Nobuaki, O. *J. Phys. Chem.* **1977**, *81*, 560.
39. Bremer, N.; Brown, B. J.; Morine, G. H.; Willard, J. E. *J. Phys. Chem.* **1975**, *79*, 2187.
40. Neiss, M. A.; Willard, J. E. *J. Phys. Chem.* **1975**, *79*, 783.
41. Bhattacharya, D.; Wang, H. Y.; Willard, J. E. *J. Phys. Chem.* **1981**, *85*, 1310.
42. Skelly, D. W.; Hayes, R. G.; Hamill, W. H. *J. Chem. Phys.* **1965**, *43*, 2795.
43. Claridge, R. F. C.; Willard, J. E. *J. Am. Chem. Soc.* **1965**, *87*, 4992.
44. (a) Roy, C. R.; Willard, J. E. *J. Phys. Chem.* **1972**, *76*, 1405; (b) Fenrick, H. W.; Filseth, S. V.; Hanson, A. L.; Willard, J. E. *J. Am. Chem. Soc.* **1963**, *85*, 3731.
45. (a) Shirom, M.; Willard, J. E. *J. Phys. Chem.* **1968**, *72*, 1702; (b) Harada, K.; Irie, M.; Yoshida, H. *Int. J. Radiat. Phys. Chem.* **1976**, *8*, 331.
46. Perkey, L.; Willard, J. E. *J. Chem. Phys.* **1974**, *60*, 2732.
47. For examples and other references see (a) Toriyama, K.; Iwasaki, M.; Nunome, K. *J. Chem. Phys.* **1979**, *71*, 1698; (b) Iwasaki, M.; Toriyama, K.; Muto, H.; Nunome, K. *J. Chem. Phys.* **1976**, *65*, 596; (c) Gillbro, T.; Lund, A. *Chem. Phys. Lett.* **1975**, *34*, 375; (d) Lin, D. P.; Willard, J. E. *J. Phys. Chem.* **1974**, *78*, 2233.
48. (a) Iwasaki, M.; Toriyama, K.; Muto, H.; Nunome, K. *Chem. Phys. Lett.* **1978**, *56*, 494; (b) Toriyama, K.; Iwasaki, M. *J. Phys. Chem.* **1978**, *82*, 2056; (c) Aditya, S.; Wilkey, D. D.; Wang, H. Y.; Willard, J. E. *J. Phys. Chem.* **1979**, *83*, 599; (d) Wang, H. Y.; Willard, J. E. *J. Phys. Chem.* **1979**, *83*, 2585.
49. (a) Miyazaki, T.; Hirayama, T. *J. Phys. Chem.* **1975**, *79*, 566; (b) Iwasaki, M.; Toriyama, K.; Nunome, K.; Fukaya, M.; Muto, H. *J. Phys. Chem.* **1977**, *81*, 1410; (c) Iwasaki, M.; Muto, H.; Toriyama, K.; Fukaya, M.; Nunome, K. *J. Phys. Chem.* **1979**, *83*, 1590; (d) Claesson, O.; Lund, A. *Chem. Phys. Lett.* **1977**, *47*, 155; (e) Miyazaki, T.; Wakahara, A.; Kimura, T.; Fueki, K. *J. Phys. Chem.* **1981**, *85*, 564: A summary of references to earlier work by this group on this phenomenon is given in their ref. 1.
50. Lin, D. P.; Hamlet, P.; Kevan, L. *J. Phys. Chem.* **1972**, *76*, 1226.

51. Sprague, E. D.; Willard, J. E. *J. Chem. Phys.* **1975**, *63*, 2603.
52. Wilkey, D. D.; Fenrick, H. W.; Willard, J. E. *J. Phys. Chem.* **1977**, *81*, 220.
53. Wilkey, D. D.; Fenrick, H. W.; Hager, S. L.; Bremer, N.; Willard, J. E. *J. Chem. Phys.* **1977**, *66*, 1170.
54. Voevodskii, V. V.; Molin, Y. N. *Radiat. Res.* **1962**, *17*, 366.
55. French, W. G.; Willard, J. E. *J. Phys. Chem.* **1968**, *72*, 4604.
56. Sprague, E. D.; Williams, F. *J. Am. Chem. Soc.* **1971**, *93*, 787.
57. Wang, J. T.; Williams, F. *J. Am. Chem. Soc.* **1972**, *94*, 2930.
58. Campion, A.; Williams, F. *J. Am. Chem. Soc.* **1972**, *94*, 7633.
59. Sprague, E. D. *J. Phys. Chem.* **1973**, *77*, 2066.
60. Neiss, M. A.; Sprague, E. D.; Willard, J. E. *J. Chem. Phys.* **1975**, *63*, 1118.
61. Early results on the decay of $CH_3$ in 3MP at 77 K indicated that the $C_6H_{13}$ radical did not grow during $CH_3$ decay[45] and that the $CH_4$ yield[45,62] did not equal the $CH_3$ loss. Improved knowledge of the change in shape of the $C_6H_{13}$ ESR spectrum with time and improved analytical techniques showed these conclusions to be in error.[40,59,60]
62. Claridge, R. F. C.; Willard, J. E. *J. Am. Chem. Soc.* **1967**, *89*, 510.
63. Sullivan, P. J.; Koski, W. S. *J. Am. Chem. Soc,* **1963**, *85*, 384.
64. Williams, J. T.; Hamill, W. H. *J. Chem. Phys.* **1968**, *49*, 4467.
65. Fenrick, H. W.; Nazhat, N. B.; Ogren, P. J.; Willard, J. E. *J. Phys. Chem.* **1971**, *75*, 472.
66. Ogren, P. J.; Willard, J. E. *J. Phys. Chem.* **1971**, *71*, 3359.
67. (a) Brown, D. W.; Florin, R. E.; Wall, L. A. *J. Chem. Phys.* **1962**, *66*, 2602; (b) Fontana, B. J. *ibid.* **1959**, *31*, 148.
68. Trofimov, V. I.; Chkheidze, I. I.; Buben, N. Ya. *Russ. J. Phys. Chem.* **1965**, *39*, 881.
69. Pimentel, G. C. In "Formation and Trapping of Free Radicals"; Bass, A. M. and Broida, H. P. Eds.; Academic Press: New York, 1960, Chapter 4, pp. 105–107.
70. Ermolaev, V. K.; Voevodskii, V. V. In "Proceedings of the Second Tihany Symposium on Radiation Chemistry"; Akademiai Kiado: Budapest, 1967, pp. 211–224.
71. Iyer, R. M.; Willard, J. E. *J. Chem. Phys.* **1967**, *46*, 3501.
72. For examples and further references, see Snipes, W.; Horan, P. K. *Radiat. Res.* **1967**, *30*, 307.
73. For examples and references see (a) Kieffer, F.; Magat, M. In "Actions Chimiques et Biologiques de Radiation"; Haissinsky, M. Ed.; Masson: Paris, 1970, Vol. 14, pp. 135–189; (b) Morine, G. H.; Willard, J. E. *J. Phys. Chem.* **1977**, *81*, 2668.
74. (a) Brocklehurst, B.; Robinson, J. *Chem.. Phys. Lett.* **1971**, *10*, 277; (b) Brocklehurst, B. *Int. J. Radiat. Phys. Chem.* **1974**, *6*, 483.
75. Walter, L.; Lipsky, S. *Int. J. Radiat. Phys. Chem.* **1975**, *7*, 175.
76. Bhattacharya, D.; Willard, J. E. *J. Phys. Chem.* **1982**, *86*, 962.
77. (a) Lin, D. P.; Willard, J. E. *J. Phys. Chem.* **1974**, *78*, 1135; (b) Wilkey, D. D.; Fenrick, H. W.; Willard, J. E. *J. Phys. Chem.* **1977**, *81*, 220; (c) Paraszczak, J. R.; Willard, J. E. *J. Chem. Phys.* **1979**, *70*, 5823. References with evidence from photoluminescence and photo- and thermally excited conductivity include (d) Bernas, A.; Truong, T. B.; Roncin, J. *J. Phys. Chem.* **1974**, *78*, 867; (e) Wiseall, B.; Willard, J. E. *J. Chem. Phys.* **1967**, *46*, 4387; Willard, J. E. *Int. J. Radiat. Phys. Chem.* **1972**, *4*, 405; (f) Tal'rose, V. L. *Izv. Akad. Nauk. SSSR, Odtel. Khim. Nauk.* **1959**, No. 2368; (g) Nikol'skii, V. J.; Tochin, V. A.; Buben, N. *Ya. Fiz. Tverd. Tela (Kharkov)* **1963**, *5*, 2248; (h) Takovlav, B. S.; Frankavich, E. L. *Izv. Akad. Nauk. SSSR, Ser. Khim.* **1966**, *3*, 402.
78. For examples and references see Kato, T.; Shida, T. *J. Am. Chem. Soc.* **1973**, *95*, 3473.
79. (a) Smaller, B.; Matheson, M. S. *J. Chem. Phys.* **1958**, *28*, 1169; (b) Brown, D. W.; Florin, R. E.; Wall, L. A. *J. Phys. Chem.* **1962**, *66*, 2602; (c) Gordy, W.; Morehouse, R. *Phys. Rev.* **1966**, *151*, 207.
80. (a) Toriyama, K.; Iwasaki, M.; Nunome, K. *J. Chem. Phys.* **1979**, *71*, 1698; (b) Timm, D.; Willard, J. E. *J. Phys. Chem.* **1969**, *73*, 2403.
81. Le Roy, R. J.; Murai, H.; Williams, F. *J. Am. Chem. Soc.* **1980**, *102*, 2325.
82. (a) Toriyama, K.; Nunome, K.; Iwasaki, M. *J. Phys. Chem.* **1980**, *84*, 2374; (b) Muto, H.; Nunome, K.; Iwasaki, M. *J. Phys. Chem.* **1980**, *84*, 3402; (c) Iwasaki, M.; Toriyama, K.; Muto, H.; Nunome, K.; Fukaya, M. *J. Phys. Chem.* **1981**, *85*, 1326.
83. Fenrick, H. W.; Bhattacharya, D.; Willard, J. E. *J. Phys. Chem.* **1981**, *85*, 1324.
84. Bhattacharya, D.; Willard, J. E. *J. Phys. Chem.* **1981**, *85*, 154.

85 (a) Bhattacharya, D.; Willard, J. E. *J. Phys. Chem.* **1982**, *86*, 967; (b) Lindholm, E. In "Ion Molecule Reactions"; Franklin, J. L. Ed.; Plenum Press: New York, 1972, p. 465.
86. Mainwaring, D. D.; Willard, J. E. *J. Phys. Chem.* **1973**, *77*, 2864.
87. Tilquin, B.; Louveaux, C.; Bombaert, C.; Claes, P. *Radiat. Effects* **1977**, *32*, 37.
88. Tilquin, B.; Tilman, P.; Claes, P. *Radiat. Phys. Chem.* **1980**, *16*, 321.
89. Tilman, P.; Claes, P.; Tilquin, B.; *Radiat. Phys. Chem.* **1980**, *15*, 465.
90. Tilquin, B.; Bombaert, C.; Claes, P. *Radiat. Res.* **1981**, *85*, 262.
91. Arnikar, H. J.; Willard, J. E. *Radiat. Res.* **1967**, *30*, 204.
92. Firestone, R. F.; Willard, J. E. *J. Am. Chem. Soc.* **1961**, *83*, 3551.
93. Hornig, E. O.; Willard, J. E. *J. Am. Chem. Soc.* **1957**, *79*, 2429.
94. Petrouleas, V.; Nath, A.; Lemmon, R. M. *Radiat. Phys. Chem.* **1980**, *16*, 113.
95. Ling, A. C.; Willard, J. E. *J. Phys. Chem.* **1969**, *73*, 2408.
96. (a) Ling, A. C.; Willard, J. E. *J. Phys. Chem.* **1968**, *72*, 1918; *ibid.* **1968**, *72*, 3349; (b)Hutzler, J. S.; Cotton, R. J.; Ling, A. C. *J. Chem. Eng. Data* **1972**, *17*, 324.
97. (a) Sandorfy, C. *Can. Spectry.* **1965**, *10*, 85; (b) Grimison, A.; Simpson, G. A. *J. Phys. Chem.* **1968**, *72*, 1776.
98. Shida, T.; Kato, T.; Nosaka, Y. *J. Phys. Chem.* **1977**, *81*, 1095.
99. Shida, T.; Kato, T. *Chem. Phys. Lett.* **1979**, *68*, 106.
100. (a) Iwasaki, M.; Toriyama, K.; Nunome, K. *J. Am. Chem. Soc.* **1981**, *103*, 3591. (b) Iwasaki, M.; Muto, H.; Toriyama, K.; Fukaya, M.; Nunome, K. *J. Phys. Chem.* **1979**, *83*, 1590.
101. Shida, T.; Egawa, Y.; Kubodera, H. *J. Chem. Phys.* **1980**, *73*, 5963.
102. Wang, J. T.; Williams, F. *J. Phys. Chem.* **1980**, *84*, 3157.
103. Shida, T.; Nosaka, Y.; Kato, T. *J. Phys. Chem.* **1978**, *82*, 695.
104. Kira, A.; Nakamura, T.; Imamura, M. *J. Phys. Chem.* **1978**, *82*, 1961; *ibid.* **1978**, *82*, 1966.
105. (a) Egland, R. J. Ph.D. dissertation, University of Wisconsin-Madison, 1968; available from University Microfilms, Ann Arbor, Mich.; (b) Claridge, R. F. C.; Willard, J. E. *J. Am. Chem. Soc.* **1966**, *88*, 2404; (c) Bertin, E. P.; Hamill, W. H. *J. Am. Chem. Soc.* **1964**, *86*, 1301; (d) Fenrick, H. W.; Willard, J. E. *J. Am. Chem. Soc.* **1966**, *88*, 412; (e) Egland, R. J.; Willard, J. E. *J. Phys. Chem.* **1967**, *71*, 4158.
106. For examples and references see Dismukes, G. C.; Willard, J. E. *J. Phys. Chem.* **1976**, *80*, 1435.
107. Meyer, B. "Low Temperature Spectroscopy"; Elsevier: New York, 1971, Chapters 7, 8.
108. Suppliers of such equipment include Air Products and Chemicals, Inc., APD Cryogenics, Box 2802, Allentown, Pa. 18105; The Oxford Instrument Co., Ltd., Osney Mead, Oxford OX2 0DX, England; Janis Research Co., Inc., 22 Spencer St., Stoneham, Mass. 02108.
109. Nunome, K.; Muto, H.; Toriyama, K.; Iwasaki, M. *Chem. Phys. Lett.* **1976**, *39*, 542.
110. Willard, J. E. *Cryogenics* **1982**, July, 361.
111. Long, M. A.; Willard, J. E. *J. Phys. Chem.* **1970**, *74*, 1207.
112. (a) Wiseall, B.; Willard, J. E. *J. Chem. Phys.* **1967**, *46*, 4387; (b) Willard, J. E. *Int. J. Radiat. Phys. Chem.* **1972**, *4*, 405.
113. (a) Dèroulède, A. *J. Luminescence* **1971**, *3*, 302; (b) Denian, C.; Dèroulède, A.; Kieffer, F.; Rigaut, J. *J. Luminescence* **1971**, *3*, 325.
114. (a) Doheny, A. J.; Albrecht, A. C. *Can. J. Chem.* **1977**, *55*, 2065; (b) Charlesby, A. *Radiat. Phys. Chem.* **1981**, *17*, 399.

# 14

# Radiation Chemistry of the Alkali Halides

V. J. ROBINSON AND M. R. CHANDRATILLAKE

*Department of Chemistry,*
*University of Manchester*

## INTRODUCTION

By far the most thoroughly investigated group of compounds in solid-state radiation chemistry are the alkali halides. Some of the reasons are undoubtedly practical: large single crystals of high purity are readily prepared. The crystals are transparent over a wide range of wavelengths. They are more sensitive to radiation damage than most other ionic solids. The crystals have simple well-defined structures, and the products of radiolysis have also in many cases been clearly identified by a variety of experimental techniques, the most important being optical methods and electron paramagnetic resonance (EPR). In recent years the application of pulse techniques—radiolysis and laser photolysis—has yielded a wealth of information concerning the mechanisms of the primary processes of radiation damage, on the one hand, and of thermal and photolytic reactions that the radiolysis products undergo, on the other.

What relevance have these studies to the bulk of radiation chemistry, carried out largely in fluid or glassy media? In many cases the species are similar: the trapped electron, now so well established in aqueous media, has been known longer in solid alkali halides as the F center; and the $X_2^-$ ion (X = halogen), which is a product of radiolysis of halide ion solutions, also plays a vital role in the primary processes of radiation damage in the solid halides. There are also important differences, though. The F center,


*Radiation Chemistry: Principles and Applications*

for example, has a much more clearly defined structure than an electron trapped in amorphous media. Also, the nature of the solid lattice is such that atomic movement is difficult, whereas electron transfer processes are relatively facile. Reactions involving translocation of atoms through the lattice are therefore of particular interest and may be studied in a degree of detail not possible in less well-defined media.

From a practical point of view there are several handicaps facing the experimenter. The requirement for purity is as great as in solution studies, especially with respect to multivalent cations. Much of the early work in this field is of questionable validity because of the unavailability of pure materials. By the same token one cannot, for example, easily add reactants (eg, scavengers) to check a reaction scheme, and, in general, the techniques available are much more limited than in solution studies.

Most of the species produced by radiolysis fall into the category known as "color centers." Originally these were species that absorbed in the visible region, but now the term is a much more general one, and a more suitable definition would be "species in which a lattice defect and an electronic defect are associated," In this chapter the structures and properties of the known color centers are outlined first, followed by accounts of the mechanisms of the primary processes of radiation damage, and then of subsequent reactions that these species may undergo.

## STRUCTURES OF COLOR CENTERS

### *The F-Center*

This was the first color center to be recognized experimentally: It is responsible for the brilliant colors that many alkali halides acquire when exposed to radiation: the purple of KCl, yellow of NaCl, sky blue of KBr, and so on. The structure, postulated as long ago as 1937 by de Boer,[1] has been shown correct by electron resonance experiments:[2] it consists of an electron trapped by a vacant anion lattice site. Since a negative ion vacancy is a positively charged species in the lattice, the F center itself is electrically neutral. The simplest model of an F center compares it to a hydrogen atom,[3] with the positive charge not localized on a single nucleus but spread over the lattice in the region of the vacancy—mainly associated with the neighboring cations (6 in NaCl structures, 8 in CsCl). According to this model, the intense absorption responsible for the color is a transition from the ground state (1s) to an excited state (2p) (see Figure 14-1). The F center is very unlike a hydrogen atom in that the absorption is a broad band rather than a narrow line. This arises from strong interaction between the electron and the surrounding lattice; this is explained in more detail later.

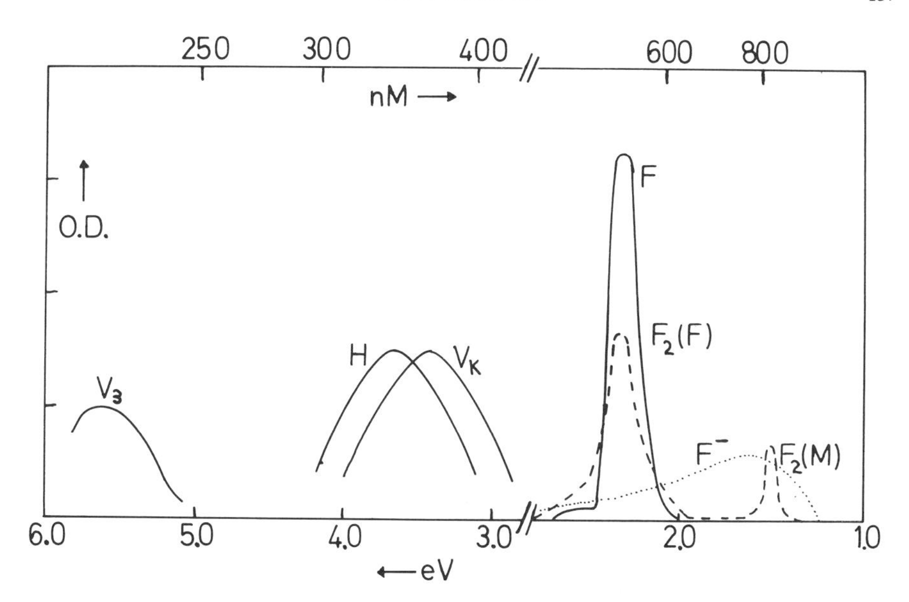


**Figure 14-1.** Absorption bands of color centers produced by radiolysis in alkali halides. The positions and widths correspond to KCl at low temperature, but the ODs of the different bands are not to scale.

In addition to radiation, F centers can be produced by a process called *additive coloration.*[4] The crystals are exposed at a high temperature to the vapor of the alkali metal: metal ions are incorporated at new cation sites, and the compensating electrons occupy anion vacancies to give F centers. Thus, there is one F center for every excess metal ion in the lattice, and the number of F centers present can be "counted" by dissolving the crystal and measuring the alkalinity of the solution. Additive coloration provides a controlled method of producing specific centers, but only a few kinds (mainly F type) can be obtained in this way.

## *Other F-Type Centers*

The F center is the most important of a family of centers, sometimes called "electron centers." The other group is called "hole centers" or V centers. The electron–hole nomenclature has arisen by analogy with semiconductors, but it is likely to produce confusion, because the centers in alkali halides do not necessarily carry a net charge, as has already been noted in the case of the F center.

The $F^-$ center contains two electrons trapped at a single anion vacancy. Only the spin-paired form of $F^-$ is known, and the second electron is rather weakly bound. The center is not EPR active, but its structure has been well established by optical experiments,[5] described later. There appears to be no bound excited state, and the absorption band corresponds to a transition resulting in photoionization of the center. $F^-$ centers can be produced both by radiolysis and by photolysis of F centers.

The $F_2$ center is an F center dimer, with two electrons trapped by two adjacent anion vacancies. Although the ground state is spin paired and not EPR active, its structure has been well established by optical methods.[4] It has a complex absorption spectrum with two principal bands. The higher energy band is situated in almost exactly the same position as the F band in most alkali halides, a phenomenon that caused considerable confusion in the early literature. The lower energy absorption, called the M band, is a much less intense peak. The $F_2$ center can be produced by additive coloration, as they equilibrate with F centers at higher temperatures,[4] and also directly by radiation.[6] A third, unusual, method in some alkali halides is by photolysis of F centers at temperatures around room temperature and above.[4] This reaction has been known for more than 20 years.

Further aggregate centers, such as $F_3$ and $F_4$, can be produced by extended photobleaching of F centers: The process is best termed *photoaggregation.* The F center aggregates are also produced by intense radiolysis—for example by heavy charged particles—and analyses of these processes have been made.[7]

## V Centers

In contrast to the electron centers, much less is known of the hole or V centers. The two best-established species are both based on the $X_2^-$ ion, which exists in different environments in the H and $V_K$ centers, resulting in very different properties of the two species. The structures have both been established by EPR.[8]

In the H center the $X_2^-$ ion occupies a single anion position. It can be thought of as a halogen atom interstitial. The center is electrically neutral, and it is in fact the antimorph of the F center, because the F center is an anion site with a missing halogen atom, whereas the H center is an anion site with an extra halogen atom. The formation of F–H pairs, an important primary process in radiolysis, is discussed below. It may be noted that the direct analogue of the F center, a hole associated with a cation vacancy, is not an established structure in the alkali halides.

The $V_K$ center, also known as the "self-trapped hole," has the $X_2^-$ ion straddled across two adjacent anion lattice sites,[9] so that the structure has an overall positive charge. The free hole (a halogen atom in an anion site) is not known, which suggests that it would form a bond to produce $V_K$ very quickly. The $V_K$ centers are not produced directly by radiation in pure crystals, but they are produced with high efficiency in crystals containing small amounts of multivalent cations, such as $Pb^{2+}$.[8]

Because H and $V_K$ are the same molecular species, they have very similar absorption spectra, which are determined mainly by the halogen present, and depend relatively little on the alkali metal. For all the halides there is a strong band between 300 and 400 nm, and the chlorides, bromides, and iodides have, in addition, a less intense absorption in the near-IR. The internuclear distance is different in H and $V_K$ because the H is compressed and the $V_K$ is stretched compared with the free $X_2^-$ ion. This leads to slight shifts in the absorption peaks: For example, in KCl the maxima are at 340 nm and 360 nm for H and $V_K$, respectively (Figure 14-1).

A species related to the $V_K$ is the self-trapped exciton $(X_2^=)^*$. This is analogous to the excimers and exiplexes observed in solution radiation chemistry. In all such cases chemical bond formation only occurs when one of the partners is in an excited state, because when both are in the ground state the interaction is repulsive. The trapped exciton has been extensively investigated and is thought to play an important role in the primary processes of radiation damage in several different solids.[10]

In many alkali halides none of the hole centers discussed above is stable around room temperature and above. The nature of the hole species that accompanies the F center as a stable radiolysis product at these temperatures is not known at present. An absorption known as the $V_3$ band (Figure 14-1) is produced by radiolysis, but the species responsible has not so far been clearly identified.

## PRODUCTION OF COLOR CENTERS BY RADIATION

### *The Formation of F–H Pairs*

Much effort has been devoted to the task of measuring the efficiency of production of color centers by radiation,[10] but it is really the advent of pulse techniques that has led to a clearer understanding of the processes involved. The use of pulse radiolysis was pioneered by Ueta.[11] The production of F centers following a radiation pulse in KCl is shown in Figure 14-2. The increase in OD follows the pulse, and this is followed by a fast decay in which about one-half of the F centers disappear. Subsequent behavior depends on the crystal. In a well-annealed sample there is no alteration in F concentration after about 300 ns [Figure 14-2(iii)], but in a crystal containing a significant concentration of anion vacancies (as a result of previous radiation treatment), the fast decay is followed by a relatively slow increase in OD [Figure 14-2(iv)]. This growth has the same time constant of about 600 ns as the radiative decay of the F excited state (see later), so this was interpreted[12] as F centers produced by trapping electrons into preexisting anion vacancies: The electrons are initially trapped into the 2p excited state, which then decays with the characteristic lifetime. The implication is that the F centers produced during the pulse are formed in their ground states,

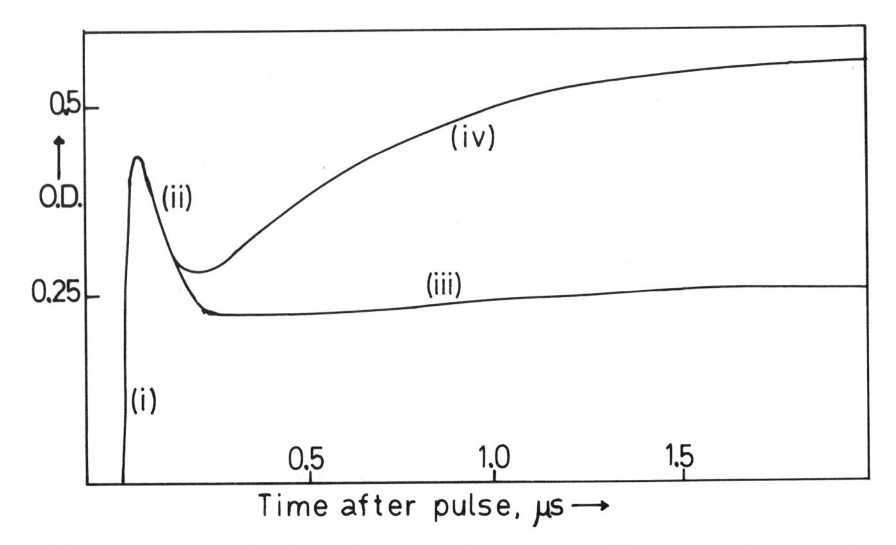


**Figure 14-2.** Change in OD at 540 nm, following a radiation pulse in KCl at 77 K. (i) Rapid increase in F center concentration, which follows the pulse on this time scale. This is due to production of F–H correlated pairs, which has been shown to be complete in a few tens of picoseconds (ref. 14). (ii) Recombination of the correlated pairs. Both the fraction that recombines and the rate of recombination increase with temperature. (iii) At 77 K in a pure annealed crystal, about half the F centers remain after F–H recombination. (iv) When the crystal contains anion vacancies prior to the pulse, some F centers are produced by electron capture into the excited F(2p) state: This then undergoes radiative decay with a characteristic lifetime (~600 ns), resulting in a growth of the ground-state absorption. Data from ref. 12.

not at preexisting vacancies, and on a time scale less than about $10^{-8}$ s. More recent work has confirmed these suggestions and shown that H centers are produced as partners of the F center:[12] the process involves ejection of a halogen atom into an adjacent site. There is a considerable probability of rapid recombination at this stage, which accounts for the sharp decrease in OD immediately following the pulse.

There has been much discussion concerning the mechanism by which the halogen atom moves out of its site,[13] but in recent years there has been increasing agreement that the process is associated in some way with the decay of the exciton. An authoritative account of the current situation is given by Williams.[14] On both energetic grounds and from kinetic arguments, it appears to be excited states of the exciton that decay to give the F–H pair, although there is still uncertainty about the specific states involved. In any event the process is very fast, with F–H pair formation being complete within a few tens of picoseconds.

The efficiency of F center production has an unusual dependence on temperature, as shown in Figure 14-3 for KCl. The main features can be interpreted in terms of the above model as follows. The decrease up to 100 K is due to temperature-dependent recombination of F–H correlated

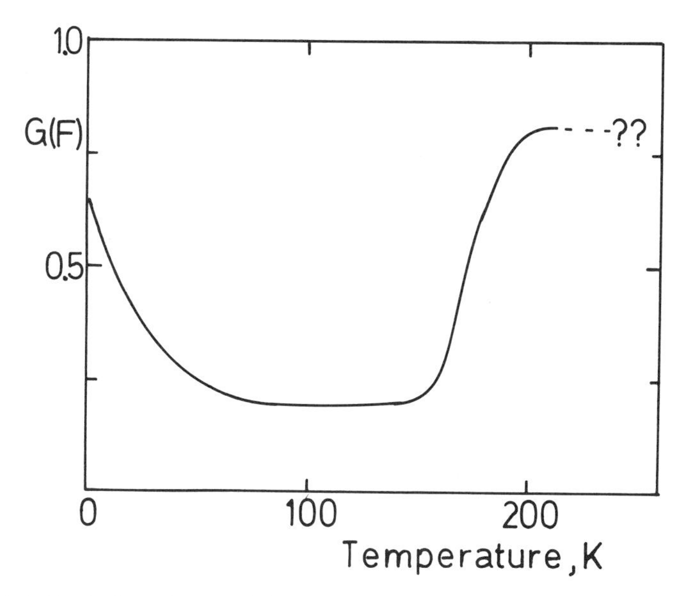


**Figure 14-3.** Variation of the efficiency of production of stable F centers in KCl with temperature. The decrease between 0 and 100 K arises from the temperature-dependent recombination of correlated F–H pairs, and the increase between 150 K and 200 K is associated with the onset of H center mobility. At higher temperatures $G$(F) is dose dependent, probably because reactions such as formation of $F^-$ or $F_2$ can remove F centers. Adapted from ref. 14.

pairs. The sharp rise between 150 and 200 K has been interpreted as an increase in the probability of the F–H pairs diffusing apart and consequent stabilization of the F centers. It appears to be H and not F that is the mobile species, and indeed the H center lifetime decreases rapidly over this temperature range. It is interesting to note that, although the H centers disappear completely, there appears to be no corresponding decrease in the F concentration. Although F–H recombination between correlated pairs may take place with high probability immediately after their formation, the reaction between uncorrelated pairs does not appear to take place at all. Part of the reason for this could be that when H centers do become mobile, they are rapidly removed by another process that does not involve F centers.

## *The Formation of $F^-$–$V_K$ Pairs*

Up to now the production of F–H pairs is the only primary process recognized in pure alkali halide crystals. Other color centers can be produced by secondary reactions, which involve reactions with species already present. A good example is the formation of $F^-$–$V_K$ pairs in crystals that already contain F centers.

The formation of $V_K$ centers results from ionization of a halide ion, which is almost certainly a primary process, although the free halogen atom as such has not been observed—presumably because immediate reaction with a neighboring halide ion gives $X_2^-$ as a $V_K$ center. Stabilization of the $V_K$ requires a trap to be available for the liberated electron; otherwise recombination takes place rapidly:

$$2X^- \rightsquigarrow V_K + e^- \tag{14-1(a)}$$

$$e^- + \text{trap} \rightarrow \text{trapped electron} \tag{14-1(b)}$$

$$e^- + V_K \rightarrow 2X^- \tag{14-1(c)}$$

The trap can be a multivalent cation, such as $Pb^{2+}$, since there is very efficient production of $V_K$ in $Pb^{2+}$-doped crystals. The F centers produced by earlier radiation damage will also act as efficient traps, giving $F^-$ centers, and this is the method of both $F^-$ and $V_K$ production in pure crystals. It has been noted[15] that the probability of formation of $F^-$ and $V_K$ both increase with F center concentration—that is, with the availability of electron traps. In the absence of such traps formation of $V_K$ by radiation is rapidly followed by recombination [Eq. 14-1(c)], and it is probably this process that is responsible for production of the excited exciton states that are the precursors of F–H pair formation.[14]

## *Other Radiation-Produced Centers*

The $F_2$ center can be produced directly by radiation in crystals already containing F centers. At low temperatures the process appears to be straightforward because the probability of $F_2$ production is comparable with the statistical probability of producing a second anion vacancy adjacent to a preexisting F center.[6] At higher temperatures $F_2$ production efficiency is much higher, and there are two possible explanations. Either the nascent F centers could be mobile in the lattice until they reach and combine with an F center already present, or the anion positions surrounding an F center may be sensitized and more likely to undergo damage than normal sites. In view of the fact that F centers can be mobilized by light around ambient temperature, it seems quite possible that radiation can produce the same effect, supporting the first explanation.

There is evidence that other defects are also produced by radiation, but little or nothing is known of the detailed mechanisms involved. The phenomenon of thermoluminescence, for example, is widespread among crystalline solids: Crystals exposed to radiation will emit visible light when subsequently heated to temperatures characteristic of the compound under study. The phenomenon has been utilized by archaeologists as a method of dating ceramic materials.[16] There is much work still required before we have a comprehensive understanding of the effects of radiation in even the simplest and most accessible crystal structures.

# REACTIONS OF COLOR CENTERS

Color centers produced by radiolysis are always metastable in the crystal lattice, and given suitable treatment, such as heating to just below the melting point, damage centers can be removed and the crystal returned to its original form. These processes are called annealing reactions. In addition to these, however, there are a number of thermal and photolytic reactions that can take place at lower temperatures that may produce other species.

## *Reactions of F Centers*

### Low Temperature Fluorescence

The absorption band of the F center is an intense, allowed transition with an oscillator strength approaching 1.0, the maximum possible for a one-electron transition.[17] On the basis of the H-atom model of the center, the transition is labeled 1s–2p. The properties of the excited state have been examined in some detail since Swank and Brown observed fluorescence from crystals following light absorption in the F band at low temperatures.[18] They measured the lifetime and intensity of the fluorescence at temperatures up to about 180 K in KCl and noted several features of interest. Below 100 K the fluorescence quantum yield was 1.0, indicating that only radiative decay was taking place. The lifetime was almost independent of temperature and was unexpectedly long, about 600 ns in KCl. Above 100 K the lifetime and quantum yield decreased with temperature, consistent with a competing temperature-dependent nonradiative decay route for the excited state. Similar phenomena occur in most of the alkali halides, but in a few (all the lithium halides, NaBr, and NaI) no fluorescence has been observed. The nonradiative decay route has been shown in several cases to be due to thermal ionization of the center, and the thermal ionization energy has been obtained from the temperature dependence of the radiative lifetime.[19] The process releases an electron, leading to destruction of the F center, as discussed below.

The lifetime of the excited state is much longer than would be expected for such an intense absorption. Absorption and emission probabilities are connected, because both depend on the transition moment, which in turn depends on the wave functions of the lower and upper states. For a transition in the visible region with an oscillator strength of 1.0, the expected lifetime would be about 10 ns. The long radiative lifetime has been ascribed to changes in the surrounding lattice as a result of the absorption process. This is also consistent with other phenomena, such as the large Stokes shift—the difference between absorption and emission energies—which is about a factor of two, compared with typical values of approximately 10% found in covalent molecules. The processes of absorption and emission can be visualized as follows. In the F center ground state the electron is bound tightly in the anion vacancy. Because the wave function confines the negative charge largely within the vacancy, the surrounding positive ions are drawn inward compared with their normal positions. This in turn reinforces the positive potential holding the electron, giving a "bootstrap" effect and further increasing the binding of the electron. When the electron is raised into the excited state, its charge is diffused over a much larger volume of the crystal, with the result that the positive ions relax back. Because of the Franck–Condon principle, this relaxation takes place *after* the electronic transition, so the excited state is initially highly compressed. Subsequent outward movement of the positive ions results in a decrease in the potential and therefore even greater diffuseness of the electron density in the excited state. The process is illustrated schematically in Figure 14-4. The emission then occurs from the relaxed excited state to a highly distorted ground

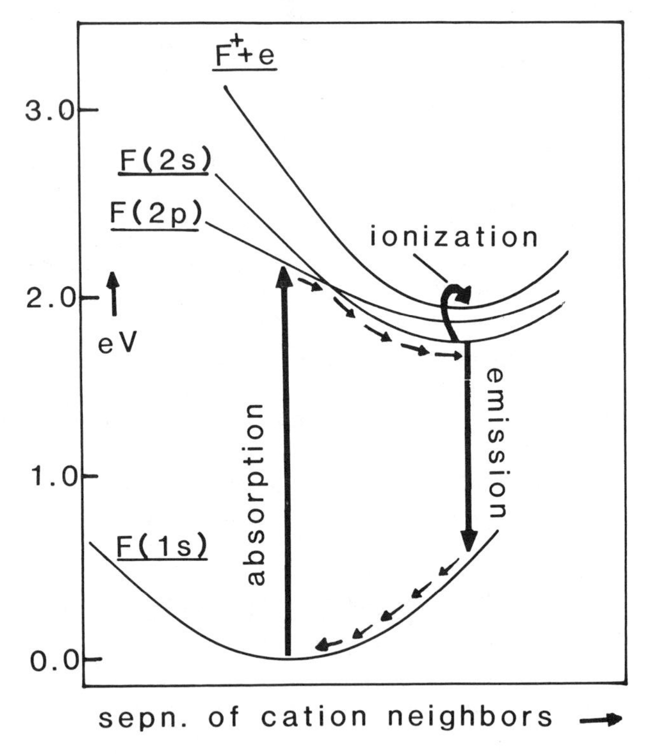


**Figure 14-4.** Configuration coordinate diagram of the F center. F band absorption corresponds to an allowed transition to the F(2p), which then undergoes relaxation to an equilibrium position with a much greater separation of the neighboring cations. (It has also been postulated that relaxation involves a change in the electronic state from 2p to 2s, as shown in the diagram.) The relaxed excited state can either undergo ionization or undergo radiative decay to the ground state, depending on the temperature.

state, which also has a diffuse electron density distribution compared to the normal ground state. Thus, the absorption process is a transition between states of relatively concentrated electron densities, whereas emission is not. This in itself may be sufficient to account for the reduced probability of emission compared to absorption. An alternative, or possibly additional, explanation is that, although the excited state in absorption is largely 2p in character, the relaxation process brings the 2s state, initially above the 2p, down below it, so the relaxed state has much more 2s character.[20] The 2s–1s transition is forbidden by the Laporte selection rule ($\Delta l = \pm 1$) and so is reduced in probability. Confirmation that the wave function in the relaxed excited state is indeed diffuse has come from EPR measurements,[21] but it has not been possible to distinguish between the possibilities of 2s or 2p wave functions.

Figure 14-4 is an example of the configuration-coordinate model of the F center. Although this must be a gross oversimplification of the complex potential energy surfaces characterizing the ground and excited states, it is nevertheless a useful qualitative method of understanding the processes involved. The lack of radiative decay in the lithium salts and NaBr and NaI can, for example, be explained in terms of the relatively free movement of the small cations, so that intersystem crossing can occur between upper and lower states during relaxation, which eliminates the formation of the relaxed excited state.

### Formation of $F^+$–$F^-$ Pairs

With increasing temperature, thermal ionization from the F(2p) excited state results in a decrease in fluorescence quantum yield and increased efficiency of F center bleaching. This latter process was studied quantitatively by Pick.[2] He found a broad absorption on the red side of the F band appearing as the F center bleached and also noted that bleaching this new band restored the F band. The quantum yield for F center destruction increased with temperature to a limiting value of 2.0, and this suggested that the new band was associated with the $F^-$ center. The reactions involved are as follows:

$$F \underset{h\nu'}{\overset{h\nu}{\rightleftharpoons}} F^* \rightarrow F^+ + e^- \tag{14-2(a)}$$

$$F + e^- \underset{h\nu''}{\rightleftharpoons} F^- \tag{14-2(b)}$$

Unlike the F center, $F^-$ appears to have no bound excited states. Photon absorption anywhere in the absorption band at any temperature results in destruction of the center by photoionization with a quantum yield of 1.0. This, and its relatively low thermal stability, gives the $F^-$ a somewhat transitory and ephemeral nature, which makes it difficult to study.

At somewhat higher temperatures (eg, above 250 K in KCl), the $F^-$ undergoes thermal decay. There has been some discussion of this process and it has been suggested that it is involved in the formation of $F_2$ centers by F center photobleaching.[22] The balance of evidence, however, is now rather against this, and the thermal decay of $F^-$ appears to be simply an ionization process.[23] The net reaction that occurs depends on the species that traps the electron; that is,

$$F^- \rightarrow F + e^- \qquad (14\text{-}3(a))$$

$$e^- + F^+ \rightarrow F \quad (F^- \text{ from photolysis of F}) \qquad (14\text{-}3(b))$$

$$e^- + V_K \rightarrow 2X^- \quad (F^- \text{ from radiolysis}) \qquad (14\text{-}3(c))$$

The activation energies for thermal decay of $F^-$ are about one-half of the energies of the absorption peak.[24] This is because the Franck–Condon principle does not apply to the thermal process.

The thermal decay of $F^-$ to give F centers explains one unusual feature of F center bleaching: In steady-state experiments the efficiency of permanent photobleaching of F centers increases rapidly between 100 and 200 K in KCl [due to increasing probability of ionization of F(2p)] but then decreases above 250 K. Around 270 to 290 K the F center in KCl appears remarkably photostable but becomes unstable again as the temperature rises further. The apparent photostability between 250 and 300 K is the result of the thermal decay of $F^-$ to reform F. The subsequent rise in bleaching rate is connected with $F_2$ formation (see below).

## Photoaggregation of F Centers

A clear recognition of the reactions taking place during photobleaching of F centers near ambient temperatures is mainly due to van Doorn,[4] who showed that the initial processs was conversion of F to $F_2$ centers; this was followed by further aggregation to $F_3$, $F_4$, and probably even higher aggregates. Many of the difficulties in earlier studies arose from not fully appreciating that $F_2$ (and $F_3$) centers have absorption bands underlying the F center, so even complete conversion of F to $F_2$ does not bleach the F band totally. The formation of $F_2$ centers is easily recognized by the appearance of the relatively weak "M band" on the red side of the F band [the two $F_2$ bands are sometimes called $F_2$(M) and $F_2$(F)]. The properties of the $F_2$ center have been examined in some detail. Absorption of light in either the $F_2$(F) or $F_2$(M) band gives rise to fluorescence, which has the same lifetime[25] for excitation in either band. The lifetime is relatively short and independent of temperature in most halides,[19] with a value in accordance with estimates of the oscillator strength of the M band. The presumption is that absorption in the $F_2$(F) band is followed by a very fast radiationless transition to the lower excited state, which then undergoes radiative decay.[26]

It has long been known that crystals containing $F_2$ centers become dichroic on exposure to light in the F band region. That is, the optical density is different along different crystal axes. This is due to $F_2$ centers taking up preferred orientations. The absorption of an $F_2$(F) photon can take place only along the axis of the center, and it can also produce reorientation by a vacancy jump.[27] Prolonged exposure to $F_2$(F) light therefore results in $F_2$ centers becoming aligned with their axes perpendicular to the incident light. As a result the OD for $F_2$(F) absorption is less along the original axis of the incident light than along axes perpendicular to it. The opposite is true for the $F_2$(M), which corresponds to a transition for which the light beam must be perpendicular to the axis of the center.

The photostimulated jumping of vacancies involved in reorientation must be similar to that occurring in the conversion of F to $F_2$ centers, although the jumps in this latter case must be over greater distances. If the F centers are initially randomly distributed, the average distance apart is between 10 and 50 lattice units at the concentrations normally encountered in these studies. The mechanism by which jumps over such large distances occur is not established at present, but the high quantum yield of the process would appear to rule out a "random walk" with the F center moving one unit at a time.

Investigations that have thrown some light on the process of photomigration of F centers have been carried out by Lüty and co-workers,[22] using alkali halides in which subtle changes in the lattice were produced by deliberate doping with another alkali cation—for example, KCl crystals containing about 0.1% NaCl. In such a crystal around 1% of the anions will occupy anomalous sites, with one impurity cation and five normal cations as nearest neighbors. Lüty showed that for F centers at such sites the absorption band split in two in the ratio 2:1 because of the reduction in the degeneracy of the 2p levels arising from the decreased site symmetry. These he called $F_A$ centers. Normal F centers can be produced in such a crystal by radiation or additive coloration. When photolyzed they are efficiently converted into $F_A$ centers, even at temperatures below that at which $F_2$ formation occurs. Lüty suggested that $F^-$ centers are involved in the photomigration, but the experimental evidence on this point is not now completely convincing.

## *Reactions of Hole Centers*

Hole centers have been much less thoroughly studied than F centers. The thermal instability of the H center has been known for a long time, and it was observed, for example, that in a KCl crystal irradiated at 4 K the H band disappeared in about 20 min on warming to 77 K.[27] A more detailed study of H center decay kinetics in KCl by using pulse radiolysis has recently been carried out,[15] and calculations of the stabilities and possible reactions of hole centers in alkali halides are also available.[28] The pulse radiolysis experiments showed the KCl H center decayed by a first-order process, with an Arrhenius activation energy of 0.24 eV. No other centers with optical absorptions in the UV-visible region appeared as the H center decayed, so that although the kinetic behavior is straightforward, the mechanism of the reaction is by no means established. There was no evidence of any reaction between F and H centers because the F center band showed no decay on the time scale of the H band decay.

The calculations of Catlow et al[28] are consistent with a low activation energy, which is associated with the jump energy of the H center: The motion of the interstitial halogen atom from a lattice position ($X_2^-$) to an intermediate position on the ridge between two halide ions ($X_3^=$) apparently requires little energy. However, the calculations also suggest that the disappearance of the H center is due to a dimerization process forming an interstitial halogen molecule, $X_2$, or a species such as $X_3^-$. Dimerization does not seem to be consistent with the observation of first-order decay, and this discrepancy has yet to be resolved.

Reactions of the $V_K$ center have also been examined. The $V_K$ center is an efficient electron trap, but in the absence of species such as $F^-$ that release electrons readily, the center has considerable stability. Photoreorientation of the center leading to dichroism occurs similarly to that of the $F_2$ center.[8] Motion of the center by means of photothermally activated jumps has also been studied both theoretically[3] and experimentally.[5]

## SUMMARY OF THE PRESENT SITUATION

Our current understanding of the radiation chemistry of the alkali halides is reasonably good at low temperature. The nature of the primary species (F–H pairs) is established, as is the mechanism of their formation by decay of excited states of the self-trapped exciton, although the specific excited state or states involved have not yet been exactly identified. The recombination of correlated F–H pairs and its temperature dependence accounts for the decrease of F center production efficiency with temperature, up to about 100 K. The increase above 150 K is likely to be associated with the onset of mobility of the H center.

At higher temperature there is considerably more uncertainty. The H center is highly unstable, and its decay process is not well understood. As a result, the stable "partner" of the F center produced by radiolysis at around ambient temperature is not known, nor is its presence very obvious in any of the reactions that F centers undergo. (F centers produced by additive coloration or by radiation seem to have the same response to photolysis, for example.) At temperatures a few hundred degrees above ambient, radiolytically produced F centers do undergo thermal bleaching, but the kinetics are complex and do not admit any simple interpretation.[29] For additively colored crystals such bleaching is the result of diffusion of the excess metal and its eventual evaporation from the surface.

Little has been said concerning centers associated with impurities or with crystalline solids other than the alkali halides. There are undoubtedly color centers produced by radiation in these materials, but studies are yet in their infancy. Before mechanisms of production and reactions of such centers can be studied, it is necessary to establish their structures. In this context it is chastening to realize that 50 years of intensive research in the alkali halides has produced only some half-dozen or so well-established structures, and there still exist large gaps concerning the nature of hole species at ambient temperature.

## ACKNOWLEDGMENTS

We should like to thank our many colleagues and collaborators over the years for stimulating discussions, especially Drs. I. Hamblett, G. W. A. Newton, S. F. Patil, and M. A. J. Rodgers.

## REFERENCES


1. de Boer, J. H. *Rec. Trav. Chim. Pays-Bas* **1937**, *56*, 301.
2. Markham, J. J. "*F*-Centers in Alkali Halides"; Academic Press: New York, 1966.
3. Stoneham, A. M. "Theory of Defects in Solids"; Oxford University Press, 1975.
4. van Doorn, C. A. Ph.D. Dissertation, University of Utrecht **1962**; published as Phillips Research Reports Suppl. **1962**, 4.
5. Fowler, W. B. In "Physics of Color Centers"; Fowler, W. B., Ed.; Academic Press: New York, 1968, Chapter 2.
6. (a) Faraday, B. J.; Rabin, H.; Compton, W. D. *Phys. Rev. Lett.* **1961**, *7*, 57; (b) Compton, W. D.; Rabin, H. *Solid State Phys.* **1964**, *16*, 121.
7. Jain, U.; Lidiard, A. B. *Philos. Mag. A* **1977**, *35*, 245.
8. Seidel, H.; Wolf, H. C. In ref. 5, Chap. 8.
9. Castner, T. G.; Känzig, W. *Phys. Chem. Solids* **1957**, *3*, 178.
10. Townsend, P. D.; Agullo-Lopez, F. *J. Phys. C6* **1980**, *41*, 279.
11. Ueta, M. *J. Phys. Soc. Japan* **1967**, *23*, 1265.
12. Ueta, M.; Kondo, Y.; Hirai, M.; Yoshinari, T. *ibid.* **1969**, *26*, 1000.
13. (a) Varley, J. H. O. *Nature* **1954**, *174*, 886. (b) Pooley, D. *Proc. Phys. Soc.* **1966**, *87*, 245, 257. (c) Hersh, H. N. *Phys. Rev.* **1966**, *148*, 928.
14. Williams, R. T. *Semiconductors and Insulators* **1978**, *3*, 251.
15. Chandratillake, M. R.; Hamblett, I.; Newton, G. W. A.; Robinson, V. J. *J. Chem. Soc., Faraday Trans. 2* **1981**, *77*, 2319.
16. Aitken, M. J.; Fleming, S. J. "*Topics in Radiation Dosimetry*", Academic Press; New York, 1972, suppl. 1, p. 1.
17. Calvert, J. G.; Pitts, J. N., Jr. "Photochemistry", John Wiley: New York **1966**, pp. 170 seq.
18. Swank, R. K.; Brown, F. C. *Phys. Rev.* **1963**, *130*, 34.
19. Ref. 5, Appendix B.
20. Ref. 3, p. 560.
21. Baldacchini, G.; Mollenauer, L. F. *J. Phys. C9* **1973**, *34*, 141.
22. Lüty, F. In ref. 5, Chap. 3.
23. Chandratillake, M. R.; Newton, G. W. A.; Robinson, V. J.; Rodgers, M. A. J. *J. Chem. Soc., Faraday Trans. 2* **1977**, *73*, 1739.
24. Chandratillake, M. R.; Hamblett, I.; Newton, G. W. A.; Patil, S. F.; Robinson, V. J.; Rodgers, M. A. J. *J. Chem. Soc., Faraday Trans 2* **1978**, *74*, 1342.

25. Bosi, L.; Bussolati, C.; Spinolo, G. *Phys. Rev. B* **1970**, *1*, 890.
26. Delbecq, C. J. *Z. Phys.* **1963**, *171*, 560.
27. Duerig, W. H.; Markham, J. J. *Phys. Rev.* **1952**, *88*, 1043.
28. Catlow, C. R. A.; Diller, K. M.; Hobbs, L. W. *Philos Mag. A* **1980**, *42*, 123.
29. (a) Pikaev, A. K.; Ershov, B. G.; Makarov, I. E. *J. Phys. Chem.* **1975**, *79*, 3025. (b) Chandratillake, M. R.; Newton, G. W. A.; Patil, S. F.; Robinson, V. J., unpublished.

15

# Radiation Chemistry of Polymers

A. CHARLESBY

*Department of Physics,*
*Royal Military College of Science*

## INTRODUCTION

For a number of decades the study of radiation effects in materials has been moving in different directions. In radiation chemistry one may have as an objective the study of products and intermediates, the kinetics of radiation-initiated reactions, the early stages of reactions including the chemistry of the electron, the difference between reactions in solids and liquids, as well as in gases, and so on. As a different objective one may search and develop commercial products using radiation processes based on earlier fundamental radiation chemistry.

Basically, the reactions involved in the irradiation of polymers should not differ from those involved in the radiation chemistry of low molecular-weight compounds, especially as the basic unit involved, the monomer, is usually of simple structure. Yet the radiation of polymers as a subject of research and ultimately of applications has been growing away from those usually involved in radiation chemistry. Most basic studies of radiation chemistry are concerned with low-molecular-weight compounds in the liquid state, and it is the changes in chemical structure and the intermediate states and mechanisms that are of major concern. Recently, there have been changes in these approaches, and the behavior of the electron at times shorter than those involved in atomic/molecular rearrangement is becoming of increasing interest. In the irradiation of polymers the objectives are


*Radiation Chemistry: Principles and Applications*

basically very different, and it is usually the modification of physical properties, such as flow and mechanical behavior, conductivity, light emission, melting, and solubility that are of major interest. In fact, the chemical changes involved in the radiation of polymers are relatively very small, and it is because such chemical modifications on a very minor scale produce such major physical modifications that the radiation treatment of polymers has acquired its present industrial importance, far greater indeed than those resulting from the radiation chemistry of low-molecular-weight compounds to which far more academic research has been devoted. It is, of course, true that radiation research reveals many new aspects of basic chemistry. The same can be claimed for the radiation-induced modification of polymers that provides an extremely valuable method for the study of polymer science, where the mechanical and electrical properties can be readily linked to minor changes in the chemical structure, as determined quantitatively by the radiation dose.

The same is true of radiobiology, where doses are so low that chemical changes can only be minute; nevertheless, they result in major biological effects. The explanation both for polymers and biopolymers is usually that they involve macromolecules in which minute chemical changes at the appropriate places have great influence on molecular morphology. A single bond altered in about $10^5$ may convert a free-flowing polymer from a fluid to an elastic solid. An even smaller change may be lethal in biological systems. A basic question, therefore, is why are these properties so sensitive to minute modification in molecular structure? To clarify this problem, we must summarize the relationship between structure and physical properties. A more detailed review would not be appropriate in a volume concerned with radiation chemistry as a subject on its own, and reference is therefore made to more specialized texts.

There are several and quite distinct major fields of interest. The first deals with the modification of existing polymers either by main-chain scission or by cross-linking; the theories of these patterns of behavior are summarized here, and typical examples of polymers that show these alternative reactions are given. These are not comprehensive to avoid this presentation becoming a catalog of similar patterns of behavior.

The second field deals with the cases when a chain reaction is involved and which radiation only intervenes in the initiation step; from then onward the subject becomes of major interest primarily to polymer science. Often these reactions can also be obtained with chemical initiation. Nevertheless, radiation initiation of ionic or radical polymerization has some unique advantages over the more conventional chemical process.

Finally, attention is drawn to the special condition that may prevail when the irradiated polymer is in an aqueous environment. Other aspects of the radiation treatment of polymers, although of considerable interest, may not be considered of direct importance to radiation chemistry and are therefore not dealt with here. Reference to them will be found in the texts concerned specifically with the radiation of polymers. In addition to the production or modification of polymers are such topics as radiation-induced conductivity, both during and after radiation, thermoluminescence (the emission of light by irradiated polymers), the trapping of electrons and their subsequent release, the nature of these traps and of luminescent centers, and diffusion in the glassy or crystalline state. Of even greater interest, at least to the author, are the considerable analogies between the radiation treatments of simple polymeric systems and biomaterials. Many of the phenomena observed in polymers can be used to interpret some radiobiological effects, but for this one needs scientists reasonably familiar with polymer and biopolymer, as well as with radiation science, collaborating closely.

## PLASTICS, RUBBERS, NATURAL AND SYNTHETIC POLYMERS

Early distinctions made between plastics and rubbers, naturally occurring and man-made polymers have been shown not to be of fundamental importance in terms of their behavior at appropriate temperatures. In what follows, some of the basic parameters are described, especially those readily modified by radiation.

The simplest polymer structure consists of a linear array of identical units (monomers) forming a molecule of high molecular weight. If the arrangement of successive units is regular, the individual molecules may form a polycrystalline structure at a sufficiently low temperature, the precise arrangement depending on conditions during crystallization. If the monomer arrangement is less regular (eg, —MMWWMWMM—), crystallization is no longer possible, and on cooling the polymer forms a glass. By orientation during cooling or under stress, largely parallel polymer chains may be obtained as in fibers.

Well above the melting point or the glass temperature, polymer chains may be very flexible, forming very viscous fluids. At sufficiently high molecular weight these chains can be entangled, giving elastic properties of a temporary character (viscoelasticity). If the chains are permanently linked together (as in vulcanization of rubber), the polymer chain may be readily deformed, but for thermodynamic reasons, revert to a random configuration when external stresses are removed.

Not all polymer molecules will be of the same size, and many properties depend not only on the average but also on the molecular-weight distribution. The average may be defined in several ways, depending on the property and on its method of measurement; a complete molecular-weight distribution [number $n$ of molecules of molecular weight $M$; $n(M)$] cannot always be obtained. Osmotic pressure depends on the number average

$$M_n = \frac{\sum n(M)M}{\sum n(M)}$$

in which all molecules, large or small, are equally important; light scattering gives the weight average

$$M_w = \frac{\sum n(M)M^2}{\sum n(M)M}$$

biased in favor of the larger molecules. Other averages include viscosity average $M_v$ and z-average

$$M_z = \frac{\sum n(M)M^3}{\sum n(M)M^2}$$

When a specimen consists of linear polymer molecules linked together to form a network structure, the most characteristic parameters are the fraction of the total specimen present in the network and the average molecular weight $M_c$ between successive links. The molecular weight $M_c$ serves to determine such important properties as elastic modulus and swelling. A property not always adequately taken into account is the physical entanglement between adjacent flexible polymer molecules; the density of such entanglements may be represented by a somewhat similar quantity to $M_c$ and denoted by $M_e$. All these properties and many others can be profoundly and often advantageously modified by radiation, and it is our objective to illustrate these changes in a few typical polymers.

Other polymeric properties influenced by radiation are the electrical conductivity, fluorescence, phosphorescence, and chemical structure. Long-term effects arise due to the radicals and charges trapped in the solid structure, especially in the crystalline or glassy regions. These may react subsequently with each other, with oxygen, and so forth, to produce an improved or a weakened material. Many of these physical and chemical changes are, of course, not restricted to polymers but also occur to some extent in related low-molecular-weight compounds, but they are then of relatively minor importance and require much higher doses. Some properties such as mechanical structure are of major importance for macromolecules and can be readily influenced in a quantitative manner by relatively low doses. The initial stages in the radiation process intervene directly in such chain reactions as ionic or radical polymerization, and even earlier stages may be involved when dealing with electrical conductivity, radiation protection, and energy transfer under conditions where molecular diffusion is not possible.

# MOLECULAR-WEIGHT CHANGES IN DEGRADING POLYMERS

## *Main Chain Scission*

Many widely used polymers, when exposed to radiation, suffer main-chain scission[1] and a loss in mechanical properties such as strength;[2,3] for this reason they are termed "degrading" polymers. Side chains may also be broken off, but this has relatively little effect on physical behavior. The radiation dose required is a relatively modest one, so that the fraction of bonds modified is correspondingly small. Thus a dose of 10 kGy (1 Mrad) will only raise the temperature of a typical degrading polymer by about 7°C, with a modification of only $10^{-5}$ of all chemical bonds, but could reduce the average molecular weight from $10^6$ to $0.3 \times 10^6$, leading to a drastic change in mechanical strength.

The effect of main-chain scissions depends on how they are distributed. In depolymerization by heat, for example, the scission of one bond allows unzipping to occur with the release of many monomer units. This is presumably because, to allow thermal decomposition, the average temperature is everywhere high, and the whole polymer chain is in a state of quasi-unstable equilibrium. With radiation the absorbed energy is very localized (apart from energy transfer, etc) and so will be the effects.

The simplest case to consider is one where the scissions occur at random along polymer chains. The exact point of scission within each monomer or its neighbor is largely irrelevant to the effect on overall molecular-weight changes. One may, of course, also consider the cases where scissions tend to occur in clusters (eg, due to track effects along the high-energy particle) or end effects where the polymer end is particularly sensitive. Neither case will be considered here; indeed they have not been reported to any significant extent.

Changes due to main-chain scission can be followed in a number of ways. We first consider $M_n$ and denote its value initially and after a dose $r$ by $M_n(0)$, $M_n(r)$. With Avogadro's number $N = 6.02 \times 10^{23}$ the initial number of polymer molecules per gram is $N/M_n(0)$, and the final number per gram is $N/M_n(r)$.

A dose of $r$ kGy ($0.1r$ Mrad) deposits $(0.624 \times 10^{19})$ $r$ eV/g and causes $(0.624 \times 10^{17})$ $Gr$ scissions/g, where $G$ is the number of scissions per 100 eV absorbed energy. Each scission increases the number of polymer molecules by unity.

$$\frac{N}{M_n(r)} = \frac{N}{M_n(0)} + 0.624 \times 10^{17} Gr \tag{15-1}$$

and

$$\frac{1}{M_n(r)} = (1.04 \times 10^{-7}) G(r + r_0) \tag{15-2}$$

where $r_0$ is a virtual dose, namely, that needed to scission an infinite molecule to obtain the initial distribution

$$\frac{1}{M_n(0)} = (1.04 \times 10^7)Gr_0$$

If $G$(scission) is independent of molecular weight, this relation requires that $1/M_n(r)$ be linear with dose. This has been found repeatedly to be the case, whether it be electron, $\gamma$-, or reactor radiation, and whether the characterization of the irradiated polymer is by osmotic pressure by light scattering or by pulsed NMR. This agreement holds over a decrease in $M_n$ over a very wide range, for at least a 100-fold decrease in $M_n$. It must therefore be accepted that the scission rate, as determined by $G$, is independent of molecular weight. Obviously, this conclusion does not apply to short end fragments or side chains unless these are revealed separately by a measurement technique.

To obtain any idea of the repartition of these main-chain scissions requires a technique, such as viscosity or light scattering, that involves the size or degree of polymerization of the individual molecules. It may also be necessary to know the initial molecular-weight distribution or at least its degree of polydispersity. This is best judged from such ratios as $M_w/M_n$, $M_z/M_n$, and so on. For a uniform distribution, with all molecules equal in molecular weight, these ratios equal 1, because $M_n = M_w = M_z$; for a random distribution of molecular weights, as would be obtained by random scission of a polymer, initially of infinite molecular weight, $M_w = 2M_n$; $M_z = 3M_n$. The viscosity average $M_v$ is related to $M_n$ by

$$M_v = M_n[(1 + a)\Gamma(a + 1)]^{1/a} \tag{15-3}$$

where $a$ is the parameter in the equation

$$[\eta] = kM_v^{\,a} \tag{15-4}$$

and $\Gamma$ is the gamma function. Thus, if $a = 0.5$, $M_v/M_n = 1.77$; if $a = 0.8$, $M_v/M_n = 1.91$; and if $a = 1$, $M_v = 2M_n = M_w$.

An important feature of the random or exponential distribution is that further random scission does not alter the distribution; $M_w(r)/M_n(r)$ remains equal to 2, whatever $r$ is. Thus, if $1/M(r) = (1.04 \times 10^{-7})G(r + r_0)$, $1/M_w(r) = (0.52 \times 10^{-7})G(r + r_0)$, and, similarly, $1/M_z(r) = (0.35 \times 10^{-7})G(r + r_0)$. A plot of $1/M_w(r)$ vs. $r$ should reveal a linear relation, and the slope gives $G$ directly. The same will be true of $1/M_v(r)$ and $1/M_z(r)$ with the appropriate coefficient. For any other distribution this will no longer be true until the dose is sufficiently high and, consequently, the number of scissions adequate to degrade the distribution sufficiently for it to approximate a random one.

Table 15-1 shows that for an initially random distribution the ratio $M_n(0)/M_w(r)$ increases uniformly with dose $r$ (expressed as $r/r_0$, the number of scissions per initial molecule). For a distribution consisting of two uniform parts ($M_n$, $2M_n$ originally), this random distribution equivalent is attained

**Table 15-1.** Effect of Radiation-Induced Scissions on Molecular Weight Distribution $[M_n(0)/M_w(r)]$

| Radiation-induced scissions per initial (number average) molecule | 0 | 1 | 2 | 3 | 4 | 6 | 8 | $r/r_0$[a] |
|---|---|---|---|---|---|---|---|---|
| (a) initially random | 0.5 | 1 | 1.5 | 2 | 2.5 | 3.5 | 4.5 | linear |
| (b) initially uniform | 1 | 1.35 | 1.76 | 2.195 | 2.685 | 3.598 | 4.571 | |
| (c) $M_n(0) + 2M_n(0)$ | 0.7 | 1.05 | 1.5 | 2 | | | | |
| Difference between initially uniform and random distribution | 100 | 35 | 18 | 10 | 7 | 3 | 1.6 | % |

[a] $r/r_0$ represents average number of radiation-induced scissions per initial molecule.

very rapidly, but for an initially uniform distribution much higher scission densities (and hence doses) are needed.

The ability to determine the *G*-values for scissions for any degrading polymer, independent of its initial molecular weight, can be of considerable value in assessing quantitatively technological and scientific possibilities in the radiation processing of polymers. It offers a simple method of assessing these effects by using conventional polymer characterization techniques and of finding how these effects may be modified—for example, by the presence of additives, of crystalline material, or a change in temperature. It may also be extended to the characterization and analysis of biomolecules studied in radiobiology. Although several common polymers—polymethyl methacrylate, polyisobutylene, polytetrafluoroethylene, and cellulose—all undergo radiation degradation, only one, polyisobutylene, is considered in detail, because of space limitations.

## *Polyisobutylene*

Polyisobutylene (PiB), with a repeat unit

$$\begin{array}{cc} CH_3 & H \\ | & | \\ -C & -C- \\ | & | \\ CH_3 & H \end{array}$$

has an amorphous structure at room temperature and, like natural rubber, crystallizes on stretching. However, unlike the latter, it suffers main-chain scission on irradiation with a corresponding reduction in molecular weight as predicted by the theory of random scission outlined above.[4–6] There is a corresponding reduction in mechanical strength, so that by measuring certain physical properties (eg, stress–strain, flow rate, moduli, etc) for various doses, one can readily relate these quantitatively to molecular weight.[3]

Plots of $10^6/M_v(r)$ for three samples of polyisobutylene, subjected to $\gamma$ radiation at four different intensities and to electron radiation at 20°C,[1] show linearity of log $M_v(r)$ vs. log $(r + r_e)$ over a 100-fold range of molecular weights, confirming the random location of radiation-induced scissions, and their independence of molecular weight at least down to $M_v = 5000$. Later measurements using pulsed NMR confirm this behavior to even lower molecular weights. Furthermore, the results for gammas and fast electrons agree quantitatively over this range of doses and molecular weights, in spite of a radiation intensity ranging by a factor of perhaps $10^3$ or more. Thus $G$(scission) is independent of molecular weight and radiation intensity. Each scission is an independent event.

On the other hand, there is a noticeable dependence on temperature of scissions per dose. Over the temperature range 0 to 100°C this can be represented by an Arrhenius relation between $G$(scissions) and temperature ($G \propto e^{-E/kT}$), with $E = 0.03$ eV $= 0.8$ kcal mole$^{-1}$. At much lower temperature the dependence of $G$ on temperature is even less marked ($E = 0.005$ eV or 0.1 kcal mole$^{-1}$), possibly because of reduced molecular mobility in the frozen (glassy or crystalline) state.

It is of considerable interest to note that this reduction in scission efficiency at lower temperatures is paralleled quantitatively by the temperature dependence of inactivation of several biological systems such as *E. coli* and red cell catalase. Since these must also contain very high-molecular-weight structures, more complex admittedly than polyisobutylene, one might perhaps expect a somewhat similar temperature dependence in scission effects, as is in fact found.[4] It is therefore not necessary to interpret the temperature effect in radiation inactivation in purely biological terms—simple main-chain scission would explain the observed data at least equally well.

In addition to the chemical changes associated directly with main-chain scission, other chemical effects include the splitting off of $H_2$ and $CH_4$ from the side chains, and at much higher doses the release of isobutylene monomer. Some such evolution is to be expected when, purely for statistical reasons, two neighboring scissions occur in the main chain. Whether the release of $H_2$ and $CH_4$ (from H and $CH_3$) occurs in the same reaction as main-chain scission has not yet been determined. The same problem arises in polymethyl methacrylate in which both main-chain and side-chain scissions occurs. Another question of basic interest is whether the $G$-values for scission are affected if, during radiation, the polymer chains are stressed. One would perhaps expect this to be the case, since main-chain scission leaves two chains with radical ends in very close proximity, and there would be a likelihood of recombination before the chains move apart. Under stress, the time available would be greatly reduced. Similarly, one could anticipate a greater probability of permanent-chain scission at higher temperature and, hence, more rapid separation of these highly reactive chain ends.

Main-chain scission (or degradation) of polyisobutylene by radiation is inherently different from thermal degradation. In the latter, which occurs at a higher temperature, a chain reaction of depolymerization is involved, releasing considerable amounts of monomer. Indeed, we would expect a limit to $G$(scission) at high temperature, in that each scission is an independent radiation-induced event, and, therefore, their number cannot exceed the number of primary ionizations and excitations. The temperature effect may then possibly be little more than the recoupling of these separate but adjacent chain ends. Many factors can be expected to influence these radiation-induced scissions, and this influence can be readily assessed by the decrease in molecular weight (viscosity, light scattering, etc).

# CROSS-LINKING AND NETWORK FORMATION

## *Theoretical Aspects*

The use of high-energy radiation to promote cross-linking between adjacent polymer molecules is one of the major industrial applications of radiation. To understand adequately the influence of such radiation-induced cross-links on the physical behavior of polymers, it is necessary to fall back on the theory of changes in molecular weight due to cross-links, the condition for network formation, and the theory of rubber-like elasticity of polymers.

An original theory of network formation due to Flory[7] and to Stockmayer[8] calculates the weight changes, but assumes polymer molecules originally all of the same size and molecular weight $M_1$. By random cross-linking (and ignoring links between monomers in the same original polymer) the number of final molecules was calculated of molecular weight $M_1$, $2M_1$, $3M_1, \ldots,$ $iM_1$ due to combination of the appropriate number of the original molecules. The number of polymer molecules left unaffected decreases exponentially, and the number of cross-linked molecules of a certain size, $iM$, increases to a maximum, then decreases as the cross-link density increases, and eventually a network is formed. A later theory by Charlesby[9,10] dealt with the case when an arbitrary molecular-weight distribution is initially involved. It transpires that the important parameter in network formation is not $\gamma$, the average number of cross-linked units per number average molecule $M_n$, but $\delta$, the average number per weight average molecule. The physical properties of the network are, of course, quite different from those of the original, unirradiated polymer with separate molecules.

We may distinguish the following major aspects, depending on the initial molecular weight distribution and the density of cross-links, that is, on $\delta$:[11]

1. The modified molecular weight distribution up to formation of an incipient network
2. The cross-link density needed for this incipient network structure

3. The separation of the network (gel) fraction that is insoluble, and the residual soluble (sol) fraction at higher cross-link densities; the dependence of sol fraction on cross-link density depends on molecular weight distribution
4. The properties of the network fraction ($g$), for example, elastic behavior
5. Properties of the residual soluble (sol) fraction ($s$): $s + g = 1$

Up to the point at which a network is formed, it can be shown that the average molecular weight increases as follows:

$$M_n(r) = \frac{M_n(0)}{1 - 1/2\gamma} \quad M_w(r) = \frac{M_w(0)}{1 - \delta} \quad M_z(r) = \frac{M_z(0)}{(1 - \delta)^2}$$

This is true whatever the initial molecular weight distribution, and

$$\frac{\delta}{M_w(0)} = \frac{\gamma}{M_n(0)}$$

It is therefore possible to produce a series of polymers by cross-linking, of any desired higher weight, viscosity, and $z$-average molecular weight. It will be realized that starting from a linear polymer, this increase in molecular weight produces a branched material, although below the gel point there is an average of less than one link per average molecule so that the degree of branching is low. At $\delta = 1$, $M_w$ and $M_z$ become infinite, and this is the cross-link density at which an incipient gel network first appears. A critical change in many properties occurs when $\delta$ exceeds 1, and this occurs whatever the initial molecular weight distribution.

For a polymer of weight average molecular weight $M_w$, the average energy deposited per molecule for a dose of $r$ kGy (or $0.1r$ Mrad) amounts to $(0.624 \times 10^{19})rM_w/(6.02 \times 10^{23}) = (1.04 \times 10^{-5})rM_w$ eV. If $G$ is the number of cross-linked units per 100 eV of absorbed energy, then

$$\delta = (1.04 \times 10^{-5})rM_w\left(\frac{G}{100}\right) = (1.04 \times 10^{-7})rGM_w$$

At the gel point of incipient gelation, $\delta = 1$, and the corresponding dose is $r_g$.

$$1 = \delta = (1.04 \times 10^{-7})r_gGM_w$$

This determines $r_g$. At any dose, assuming random cross-linking proportional to dose $r$,

$$\delta = (1.04 \times 10^{-7})rGM_w = \frac{r}{r_g}$$

Thus, if $M_w = 10^5$ and $r_g = 30$ kGy, the $G$-value for cross-links is $G(X)$, where

$$G(X) = 0.5G(\text{cross-linked units}) = 1.6$$

since each cross-link involves two cross-linked units, one on each molecule.

To determine the gel or sol fractions in an irradiated polymer and, hence, the $G$-value for cross-linking, various techniques have been developed. The most well established is the determination of the insoluble fraction in Soxhlet equipment with the irradiated polymer suspended in a boiling and recirculating solvent for a lengthy period—after 24 hours for polyethylene—in the absence of oxygen to avoid oxidation reactions. This separates sol and gel fractions. Another promising new method is the use of pulsed NMR.

The progress of cross-linking can also be followed—even when the fraction of nonnetwork material(s) has become too small to be a reliable measure—by determination of the swelling ratio[12] or the elastic modulus $Y$ at a temperature when the chains between cross-links are flexible. The density of cross-links is then best determined not by the number of cross-linked units per weight average molecule ($\delta$) but by the average molecular weight between cross-links $M_c$. According to Flory the swelling ratio ($V/V_0$) is proportional to $M_c^{0.6}$ (if $\delta \gg 1$), the constant or proportionality depending on the interaction of polymer with solvent.

Much research has been carried out on the properties of polymers cross-linked to varying extents by varying doses. Such radiation experiments can be helpful by offering well-calibrated polymer networks for polymer research, as well as for radiation research, by providing methods for evaluating radiation constants and for interpreting radiation effects. Conversely, the use of varying radiation doses for a cross-linking polymer can be used to verify the validity of these more theoretical analyses. Two of the many documented systems are briefly described below.

## *Polydimethylsiloxanes*

Although the radiation-induced cross-linking of polyethylene has the greatest industrial importance and was observed before the corresponding behavior in the polydimethylsiloxanes (PDMS), it is more convenient to begin with the presentation of the latter.[13–15] The results are not confused by the presence of two phases (crystalline and amorphous) at room temperature; a wide range of molecular weights is readily available and can be deduced from the published viscosity values.

The structure of PDMS is analogous to that of polyisobutylene (PiB)

$$\left( -\underset{\substack{| \\ CH_3}}{\overset{\substack{CH_3 \\ |}}{Si}}-O- \right)_n \quad \text{compared to} \quad \left( -\underset{\substack{| \\ CH_3}}{\overset{\substack{CH_3 \\ |}}{C}}-CH_2- \right)_n$$

but the former cross-links very readily, whereas the latter scissions. This basic difference must be ascribed to the configuration of the chains, which in both cases are flexible at room temperature. However, in PiB there is steric hindrance between methyl groups on alternating carbons, whereas in

PDMS this is not the case due to the more open structure with the divalent oxygen.

This relationship between bulk viscosity $\eta$ and the weight average molecular weight may be expressed in the form

$$\log_{10} \eta = A \log_{10} M_w + C$$

provided that $M_w$ exceeds $3 \times 10^4$ or $4 \times 10^4$, with $A = 3.7$ or 3.64. Below this critical value the parameter $A$ sinks to about 1.4 or less. Other flexible polymers, both those which cross-link or scission under radiation, show the same pattern of behavior of viscosity. This can best be explained as due to entanglements between polymer chains. When the chains move due to their thermal energy, these entanglements disappear and reform elsewhere, their number and lifetime depending on temperature. For sufficiently short periods these entanglements therefore behave as temporary cross-links and, by analogy with the theory presented above, a minimum of one entangled unit (0.5 entanglement) per weight average molecule (ie, $\delta \geq 1$) is needed to form a temporary network. The transition in $A$ from about 3.7 to 1.4 occurs at this point.

The minimum dose to form an incipient network $r_g$, has been determined for a range of PDMS materials of widely different viscosity and, hence, molecular weight. Early data, making use of mixed reactor radiation, showed that $r_g$ varied inversely as weight average molecular weight (deduced from bulk viscosity). Unit reactor radiation is equivalent to approximately 450 kGy, so that the $G$-value for cross-link formation can be deduced (Table 15-2).

**Table 15-2.** PDMS Incipient Network Formation by Reactor Radiation

| Bulk Viscosity $\eta$ (cs[a] at 25°C) | $M_w$ | $r_g$ (reactor units) | $r_g M_w$ |
|---|---|---|---|
| 100,000 | $100 \times 10^3$ | $3.7 \times 10^{-2}$ | $3.9 \times 10^3$ |
| 30,000 | $80 \times 10^3$ | $5.1 \times 10^{-2}$ | $4.1 \times 10^3$ |
| 12,500 | $62 \times 10^3$ | $7 \times 10^{-2}$ | $4.3 \times 10^3$ |
| 1,000 | $26.4 \times 10^3$ | $12 \times 10^{-2}$ | $3.2 \times 10^3$ |
| 50 | $3.9 \times 10^3$ | $72 \times 10^{-2}$ | $2.8 \times 10^3$ |
| 10 | $1.2 \times 10^3$ | 3.2 | $3.8 \times 10^3$ |
| | | mean | $3.7 \times 10^3$ |

[a] cs = centistokes.
Counting unit reactor dose as equivalent to 450 to 500 kGy, the mean value of $r_g M_w = 1.67 \times 10^6$ to $1.85 \times 10^6$. For random cross-linking the theoretical formula outlined above, $G r_g M_w = 0.96 \times 10^7$, where $G$ is the number of cross-linked units per 100 eV, so that $G$(c.u.) = 5.7 to 5.2; $G$(X) = 0.5$G$(c.u.) = 2.8 to 2.6. Each cross-link X requires two cross-linked units, one on each molecule.

The gelation dose may be determined in a number of ways, for example, from solubility determinations or from the minimum dose at which the polymer ceases to flow (a small network fraction can hold the soluble fraction within its pores). The *G*-value for cross-linking can also be deduced from the elastic modulus at a range of higher doses or from the analysis of low-molecular-weight silicones (eg, pentadimethyl siloxane), from swelling data, or more recently from pulsed NMR measurements of a series of samples of varying viscosity or radiation dose.[16–18]

The theory of cross-linking predicts the decrease in soluble fraction as a function of cross-link density (ie, dose and molecular weight distribution). The shape of the sol/dose curve depends on initial molecular weight distribution, and this can be calculated and compared with experimental results. Good agreement is found between the sol/dose curves observed for PDMS of viscosities 50, 1000, 12,500, and 100,000 centistokes (cs) and that calculated for a random initial distribution. It may be concluded that $G(X)$ is substantially independent of molecular weight over more than two decades, as well as independent of prior cross-linking.

The elastic modulus is also proportional to radiation dose, that is, to the number of cross-links or chain segments able to partake in an elastic deformation. The modulus is independent of molecular weight except at low doses, when there is a significant fraction of the irradiation polymer in the sol or tied onto the network at only one point. Various values of $G(X)$, varying from 2.3 to 3.4, have been derived by different authors. It may be expected that when rapid measurements are made, sufficient chain entanglements will remain to increase the effective modulus and, hence, give an apparently high *G*-value.

Temperature influences the *G*-value, changing little from −78°C ($G(X) =$ 1.3) to 0°C (1.4), but increasing more rapidly from 20°C (1.55) to 150°C (2.35). Most of this change must be related to a true radiation chemical effect and not to a change in structure (as in polyethylene) or to radical traps, because the chains are mobile. The increase may be due to the increased free volume available for chain motion which is needed to allow two chains to link together.

The nature of the cross-link may be surmized from the gases released. These consist primarily of hydrogen and methane, indicative of primarily ethylene and methylene linkages, with a ratio of 6:1 in the H/C ratio.

## Polyethylene

The radiation of polyethylene and, more recently, its copolymers has been the most commercially successful of all applications of radiation in the polymer field. The fully saturated nature of the polymer (apart from small terminal and pendant groups in certain polyethylenes) initially caused objection to the possibility of a reaction such as cross-linking, which in rubber involves polymer unsaturation; such objections have long been overruled, but the precise mechanism by which cross-links are formed is still not generally agreed, although the basic reactions have been very fully studied.[19]

Shortly after the discovery of polyethylene, experiments to determine its structure were carried out in the 1940s using x-ray and fast electron diffraction, and abnormal melting effects were observed in the latter.[20] Fuller investigations were made in the late 1940s and early 1950s with reactor radiation, gammas from radioactive sources, and fast electrons from accelerators, and it was soon established that beyond a certain dose new properties emerged, notably little or no fluid flow above the usual melting point.[21–24] There was no evidence that the type of radiation was important, except insofar as low-intensity radiation, as usually obtained from Co-60 $\gamma$-sources, led to increased oxidation effects. These were not due to the nature of the radiation, however, but to the longer exposure time, allowing increased diffusion of oxygen into the specimen.

In conventional low-density polyethylene (as available in the 1950s) several regions may be distinguished: a crystalline region melting at about 125 to 130°C; an amorphous region consisting of chains linking the crystalline regions and accounting for its flexibility; folds and chain fragments near the crystalline surfaces; and regions near defects or special features such as residual unsaturated groups. Each of these regions will behave differently when exposed to radiation, although there is every reason to believe that the incident energy is absorbed at random. To explain these differences, workers have proposed various theories, including energy transfer, ionic and electron reactions, mobile radicals, and H abstraction–addition reactions. Some will be discussed below.

The main chemical reactions induced by radiation are

a. The formation of tetrafunctional links (H or X links) between adjacent chains.

```
  H  H  H
—C—C—C—
  H  |  H
  H  |  H
—C—C—C—
  H  H  H
```

Internal links must also be expected to form between neighboring parts of the same polymer molecule, as, for example, in the folded regions on crystalizing surfaces.

b. Increase in transvinylene unsaturation in the main chain.
c. Rapid destruction of pendant and terminal unsaturation.
d. Slow destruction of crystallinity, possibly by reactions a–c within the bulk of the crystal.
e. Production of alkyl radicals, especially at low temperature, which then react to provide the more stable allylic radicals (readily observed with EPR).
f. Evolution of $H_2$ and possibly traces of other gases (eg, $CH_4$).

When irradiations are carried out on polyethylene at room temperature, no obvious changes in polymer behavior are seen until very high doses of the order of 10 MGy ($10^3$ Mrad) are reached, when there is a marked decrease in crystallinity.[25,26] At even higher doses the crystallinity is largely destroyed, leaving a glassy, brittle material, brown in color due to conjugated unsaturation. This corresponds roughly to a very highly vulcanized rubber, such as ebonite, but achieved without the introduction of additives such as sulfur. These doses are very considerably higher than those used in commercial irradiation processes.

If the polymer, irradiated at room temperature to only a few Mrad, is subsequently heated above the usual melting point, some obvious changes then take place. For zero dose the polymer becomes transparent (no light scattering from crystals) and can flow slowly under external stress. This is often measured by the melt index, which decreases with increase in molecular weight. With small radiation doses—up to about 30 kGy, 3 Mrad—there is less than one cross-linked unit per weight average molecule ($\delta < 1$), and no network is formed. The viscosity is increased, but not greatly because many molecules are unchanged, and those linked together form "stars" with lower viscosity than a linear molecule of the same molecular weight. At somewhat higher doses, typically greater than or equal to 50 kGy, a network is formed ($\delta > 1$) that can hold the nonnetwork fraction in its pores.[27] In the absence of any external stress it appears that the polymer has not melted because it does not flow, although it becomes transparent. This is an error; the crystalline polymer has indeed melted and is readily deformable, but the cross-links prevent it from viscous flow. On cooling it recrystallizes. In the molten state it behaves as an elastomer with modulus proportional to radiation dose, as required by the theory of random cross-linking (with appropriate end-effect corrections), and on cooling it reverts to its original shape for thermodynamic reasons of equilibrium. It is only at very high doses, when the polymer begins to acquire some of the properties of a glass, that the theory of elastic deformation no longer applies, and there is little effect of temperature.

Thus, in polyethylene above the melting point, we can study the change in physical behavior from a viscous linear liquid to a viscous branched liquid, to a weak rubbery material, to a strong rubber of high modulus, to an overcured cheesy material, to a leathery type material, and eventually to a glass by merely adjusting the radiation dose.

One of the great advantages in using irradiated polyethylene, and indeed other polymers showing similar behavior, lies in the so-called memory effect. This depends on the thermodynamic equilibrium of a flexible cross-linked network that takes up a favored or equilibrium configuration in the absence of an external stress. This behavior is, of course, typical of conventionally cured rubber and is also valid for cross-linked polyethylene above its melting point. The difference from rubber is that polyethylene is largely crystalline at room temperature, and these crystals can lock polymer chains into a configuration quite different from their thermodynamic equilibrium. This configuration can be retained indefinitely until the crystals are melted when, in the absence of external stresses, it reverts to its shape during irradiation, as typified by the distribution of cross-links engendered by the radiation dose.

In industrial practice, polyethylene is irradiated, possibly at room temperature, then heated above the melting point, when it is converted to an elastomer. If stressed or otherwise distorted into a new shape and then cooled, crystals reform and hold the polyethylene in this new configuration. When subsequently it is reheated to remove the crystals that locked the polymer chains, the cross-linked polyethylene remembers its shape during irradiation and, if possible, resumes the original configuration.

Practical use of this "memory" is in shrink film, used for food packaging and encapsulation, for shrinkable connectors in cable and other electrical installations, and for many other devices in which a controlled shape change is desired, triggered by a simple temperature change. The critical temperature change is directly related to the polymer melting point and, hence, to its chemical structure, and the retractive force increases with radiation dose.

## POLYMERIZATION[28]

Polymerization of a monomer comprises three major steps—initiation, propagation, termination—possibly modified by other reactions such as transfer. In radiation-induced polymerization attention is essentially confined to the first step, where radiation can often be considered as an attractive alternative to chemical initiation by a thermally activated catalyst. Major advantages include the following:

1. Absence of any residue as from the decomposition of a chemical catalyst.
2. An immense range of intensities and, hence, initiation rates.
3. Little or no temperature effect, so that the choice of an operating temperature may be fixed by other considerations (eg, propagation rate).
4. There is no danger of runaway reaction due to the energy released in the propagation step reacting back to enhance the initiation rate of the catalyst.
5. Unusual features such as the possibility of initiation in the solid state, polymerization of oriented monomers, and so on.

The radiation intensities available range from very low intensities (leading to very high molecular weights) to intensities exceeding 10 kGy/s, or radical production of about 0.01 mole $kg^{-1}\ s^{-1}$, yet with a temperature rise due to initiation of only a few degrees Celsius. Any exotherm due to the propagation step will have little effect in increasing initiation.

Radiation can produce both radicals and ions, and therefore it initiates both types of polymerization. Radical initiation has been far more closely studied, and, in the early years, ion initiation[24] was not considered likely.

At low intensities the rate of radical-initiated polymerization is determined primarily by initiation rate, itself determined by radiation intensity $I$, and overall polymerization varies as $I^n$, with $n = 0.5$ under steady-state radical conditions. This is because termination occurs when two growing chains meet, and their concentration and encounter rate increases with the increase in $I$. Thus, higher radiation intensities produce polymers of lower molecular weight, but the number of polymer chains is proportional to dose. At much higher intensities (>10 Gy/min) there is a further decrease because termination may occur by radiation-induced primary radicals of different mobility and reactivity than growing polymer chains.

Non-steady-state conditions will also lead to a departure from this dependence on intensity $I$: for example, when polymer is insoluble in monomer or in solvent, when there may be a long post-irradiation effect; in certain monomers these lead to a conversion rate $I^n$ with $n = 0.8$. Another feature is attributable to the Trommsdorf effect, whereby the overall polymerization reaction speeds up with conversion, due to the higher viscosity and, hence, lower termination rate of the longer growing polymer chains by natural combination; monomer addition is less affected, being largely by individual more mobile monomers. At high degrees of conversion, branched polymer can also be formed by means of the radicals created along the length of already formed polymer chains being subjected to further radiation.

When the monomer is irradiated in a solvent such as water or alcohol, the radiolytic products of the solvent can themselves initiate polymerization of the monomer present or, indeed, serve to terminate the reaction.

When monomer is irradiated in the solid state, high degrees of conversion may be reached.[29–33] It can be argued that the radicals formed by radiation are trapped in a glassy or crystalline lattice and initiate polymerization when the specimen is subsequently heated, but there is very strong evidence that polymerization does occur at much lower temperatures (eg, a large exotherm from the propagation reaction). In some cases the rate of polymerization is far greater than when the monomer is irradiated as a liquid, pointing to some unusual kinetics for molecular motion in crystalline organic solids. It would appear unlikely for monomers in a crystalline solid to be located in such an unusually favorable orientation to undergo linkages with their neighbors, without which considerable dislocations would have to occur during the conversion to a crystalline, oriented polymer.

Ionic polymerization has been studied;[39] for isobutene at low temperature (−77°C), the $G$(initiation) value is low—0.1 to 0.2. This would agree with the fraction of the total number of electrons that escape following irradiation from their geminate ion in an $n$-alkane. Of the total $G(\text{ion}) \simeq 3$, as in a gas, almost all the electrons (2.9) are rapidly recaptured to give highly excited and eventually radical species. In ionic polymerization low temperatures are favored (compared with the radical reaction), and trace water must be eliminated. Inorganic additives greatly enhance the apparent yield,[30,35] but it has been claimed that this is due to the removal of trace water; another explanation is that electrons are trapped which otherwise would be recaptured to give an excited species.[36]

## CROSS-LINKING OF POLYMERS IN SOLUTION

Irradiation of long-chain polymers in solution, especially in aqueous solution, reveals some unexpected features:[37–39] for example, in the solid-state polystyrene cross-links, although because of the presence of the aromatic group, the doses needed are high. In solution in chloroform, however, it suffers main-chain scission even in the absence of oxygen, which often causes scission in irradiated solid polymers. Polyvinyl chloride, whose behavior on irradiation is very complex, cross-links readily when dissolved in tetrahydrofuran. This may be partly due to the greater free volume offered to polymer motion in the solvent.

The behavior of water-soluble polymer such as polyvinyl pyrrolidone (PVP) is of interest not only as a research subject of radiation of polymers but also as a model for radiobiology, where many of the same basic reactions occur in the more complex systems and are difficult to elucidate directly.

Aqueous solutions of PVP cross-link readily, and the change is readily observed. From a free-flowing liquid it is transformed to a jelly-like material that trembles on agitation and breaks into small fragments under small stress but does not flow. This behavior is to be expected from a very lightly cross-linked gel, swollen in the soluble fraction. But the most interesting aspect is the very small gel fraction needed to incorporate the sol fraction fully. Thus, the change from a completely fluid solution ($\delta < 1$) to one that behaves as a weak solid is remarkably sharp and offers an extremely simple and quite accurate method of determining the gelation dose $r_g$ ($\delta = 1$) to within a few percent or better; one just inverts the glass vessel and sees whether the specimen flows or remains in place.

At doses well above the gelation dose, the cross-link density increases, and the pore size, determined by $M_c$, shrinks in proportion. It can no longer incorporate all the residual polymer solution, excluding the excess. The irradiated sample then shows two phases; a swollen gel and an aqueous solution with a diminishing amount of polymer.

An effect known from the early days of radiation chemistry and radiobiology is readily seen in the irradiation of dilute aqueous solutions of polymers. Energy from the incident radiation is absorbed mainly in the water, whose radiolytic products are H, OH, $e_{aq}^-$ as well as $H_2O_2$, $H_2$, and $H_3O^+$. Some of these can attack polymer molecules to produce reactive species such as polymer radicals in molecules that have, purely on statistical grounds, not been modified directly by radiation. This indirect effect can therefore lead to an apparently enhanced effect, and the minimum radiation dose needed for network formation is decreased in more dilute solution. In other words, the further apart the polymer molecules, the easier it is to link them together. This apparent contradiction is due to the simple fact that as the solution becomes more dilute, the polymer molecules available to be affected by the increased radiolytic water products decreases. In many polymer solutions this increased sensitivity continues down to a concentration of about 1%. There is every reason to believe that the same behavior occurs in many radiobiological systems that will be especially sensitive to the presence of low concentrations of additives able to intervene by reacting with the water products before the polymer or biological macromolecules are affected.

This sensitization (eg, reduced dose for network formation) offered by the presence of water may not persist at low concentrations, typically below about 1%. As the concentration is further reduced, the dose for gel formation no longer decreases but increases very rapidly, perhaps by a factor of 10 when the concentration is reduced from 1% to 0.5%. This sudden change, however, is not due to a change in the radiation process but results from polymer chain configuration. For a cross-link to form, two radicals, one on each polymer molecule, must be in close proximity, as can readily occur in a solution with high polymer mobility. However, these radicals are distributed at random, and there is a good possibility of a pair being formed on the same molecule. With mobile chains, therefore, there will be direct competition between external and internal links, and only the former lead to network formation. Internal links draw different parts of a single polymer chain together in that they convert a temporary proximity into a permanent one. The originally linear polymer chain is thereby converted to a two- or even three-dimensional microgel of molecular dimensions. The concentration at which the transition from macro- to microgel occurs must depend on the configuration of the original polymer in solution, and its evaluation may therefore be of value in polymer science. One would expect this critical configuration to depend on molecular weight, chain flexibility, and especially the pH of the solution.

The ability to form network systems in such hydrophilic polymers, with pore sizes readily controlled by the radiation dose, is a promising technique for controlling, for example, diffusion rates of small molecular structures, especially those of medical and biological interest.

## GRAFTED COPOLYMERS

Radiation can be readily used to produce grafted copolymers,[40,41,42] whereby a polymer consisting of monomer units A has attached to it branches, each consisting of a chain formed from monomer units B.

```
—AAAAA—  ⇝  —AAAAA—
+B+B+B+          B
                 B
                 B
                 |
```

The permanent combination of two polymer chains A and B with very different properties may provide a copolymer with very useful new properties—rather like an alloy of two dissimilar metals but far more permanent and indissoluble in nature. Of course, there is always the possibility that the copolymer may present the disadvantages of A and of B. It is up to the researcher to predict, if possible, the properties of the new product. For example, by grafting a hydrophilic monomer (such as acrylic acid) to hydrophobic fiber (such as nylon), we can obtain a far more hydrophilic fibrous material—even more hydrophilic than cotton. In general, the properties of a grafted system depend not only on the chemical nature of A and B units and their relative numbers but also on the disposition of branches B within the copolymer (eg, in possible crystalline or amorphous components), on the length of these branches, and on their number (the product equals the total number of monomers B).

Several methods have been proposed for the preparation of grafted systems by radiation:

(a) Polymer A is dipped in monomer B and subsequently irradiated so that the radicals formed on A initiate polymerization of trapped B as side chains. The degree of grafting is therefore limited to at most the absorption capacity of polymer A for monomer B. Some degree of homopolymerization of B and of cross-linking or scission of A must be envisaged. An effect of dose rate as well as dose is to be expected because there will be competition between adjacent growing chains for available monomer.

(b) Polymer A is irradiated, resulting in the formation of radicals on the backbone chain. If the irradiated polymer is now dipped in monomer B, these trapped radicals react with B and provide the initiating centers for the branches. The extent of the reaction is restricted to the concentration of trapped radicals on A, when these can also react—for example by cross-link formation or scission. One might also expect a difference in graft concentration from outside in.

(c) Simultaneous irradiation of polymer A immersed in monomer B occurs. This provides a good density of grafted material but also produces homopolymerization of B—a very undesirable feature, because the homopolymer B must be removed from the surface of A before the copolymer can be used as such. However, this behavior can be eliminated by introducing into the solution of monomer B (in which polymer A, usually in the form of films, is immersed) a metallic salt such as an iron sulphate. This has the effect of avoiding homopolymerization of B in the solution but not of grafting within polymer A. A number of explanations of this behavior have been advanced, but no really convincing theory is generally accepted. It is difficult to see why, if B enters A with difficulty, the metallic salt cannot penetrate equally well. It may be that the salt can prevent initiation by monomer B, but not radical formation on A, and these radicals provide the anchor for the B side chains.

The penetration of branch formation with A can be readily traced at various stages by using suitable dyes that react readily with B. In this way it is possible to show that the grafted portion penetrates on a remarkably sharp front, indicating that, once initiated at a point, further grafting occurs at a rapidly increasing rate. However, it soon reaches a limiting value, as in acrylic acid in polyethylene, so that once the two advancing grafting fronts meet a surprising degree of uniformity in graft is obtained. The reason for this limitation, in spite of the continuous availability of monomer B, is not fully settled.

(d) When polymer A is irradiated in air, peroxide groups are formed. These are somewhat unstable, and decompose on heating to provide radicals. Then the specimen A, following irradiation, is placed in monomer B and heated to produce a grafted system. The time and location of grafting need not be close to the radiation site or process.

It will be appreciated that with these alternative techniques, the use of radiation for graft formation should be of considerable promise. In fact the industrial applications have been relatively sparse—membranes for battery separators, biocompatible surfaces on polymeric structures, and so on. The main difficulty is to apply such composite systems, taking full advantage of their properties, as distinct from those of a simple intimate mixture. In addition to the need for discovering new fields to fully use the unique properties of grafts, scientists still must answer some very intriguing questions concerning, for instance, the mobility and reactivity of monomers within a solid, the transfer of radicals, the length of side chains, and the resultant influence on physical properties.

## ENHANCED CROSS-LINKING

In the conventional theory of cross-linking, one assumes that each cross-link is formed independently of all the others and that each requires at least one or possibly two radiation events—one on each chain. To provide even the minimum cross-link density to produce an incipient network then requires a radiation dose $r_g$ such that there is an average of one cross-linked link per molecule. Thus,

$$r_g G(\text{c.u.}) M_w = 0.96 \times 10^6$$

if $r_g$ is in Mrad. For a typical polymer with, say, $M_w = 2 \times 10^5$, and $G$(c.u.) = 2, $r_g \sim 2.4$ Mrad. For a low-molecular-weight polymer with, say, $M_w = 10^4$, the dose needed per initial network formation would be very high, about 48 Mrad at the very least. This would not be economically feasible in an industrial process. In certain systems containing, for example, highly unsaturated polymers of low molecular weight, a fully cured (highly cross-linked) system giving a glassy product can be obtained at far lower doses, 1 Mrad or even less.[43,44,45]

The explanation resides in the fact that in such systems one is dealing with a cross-linking reaction and a polymerization reaction. Polymer A carries a number of unsaturated groups on each molecule by which they can be linked together as if they were monomers. If molecule $A_1$ becomes linked to molecule $A_2$, it forms a new radical on $A_2$, which then allows it to be linked to a third molecule $A_3$, and so on. Radiation only intervenes in the very first step, although it can eventually result in the linking together of a number of polymer molecules. Here the cross-links are no longer distributed at random—if $A_2$ is linked directly to $A_1$, it is very likely to be linked directly to $A_3$, and indirectly to $A_0 \ldots, A_4 \ldots$. The conditions for network formation are then no longer those deduced for a random distribution of cross-links, and, in particular, the gelation dose is greatly reduced. The mathematics of this form of "enhanced cross-linking" have been derived and tested experimentally.

It is not necessary for the unsaturated groups to be located on the polymer A itself. A similar reaction can be obtained with a saturated polymer A by using polyfunctional monomer B serving as a link and chain propagator.

This type of cross-linking can result in a highly cross-linked structure and even a glassy rigid network at relatively small doses. This has led to a growing interest in the radiation curing of thin surface films. In this it is in direct competition with curing by ultraviolet radiation. The latter has the advantage of low capital cost (and corresponding low output) but requires chromophores and polymeric systems of suitable optical transmission. It cannot be used for surface films with coloring matter (solid particles such as aerosil, for example). With the advent of lower-voltage electron radiation sources (of a few hundred thousand volts) and low penetration, the capital cost of high-energy radiation for surface treatment can be greatly reduced, and the shielding requirement is also lessened. However, it is still necessary to allow for the cost of the polyfunctional monomer to serve as a linking agent. Although this type of system is being rapidly explored, and indeed already utilized on a wide scale as in the cross-linking of polyvinyl chloride (PVC), the reactions involved have been inadequately studied, and there is still some confusion as to the difference between conventional and enhanced cross-linking. The characterization of these highly cross-linked systems also leaves many questions open—such as, What is the minimum cross-link density needed to provide a hard glossy material?

# CONCLUSIONS

In this chapter an attempt was made to emphasize at least briefly those aspects of the radiation chemistry of polymers insofar as they differ from those typical of the radiation chemistry of low-molecular-weight compounds. These differences relate particularly to such facts as molecular orientation, the reaction possibilities in solids, the large mechanical changes resulting from very small chemical changes, and long-term stored reactivity. It has not proven possible to go into any detail about the behavior of electrons trapped in solids, the transfer of some form of reactivity in polymers to additives to provide radiation protection, or even enhancement. Thus, polymers usually of very high electrical resistivity can show a considerable loss in insulating properties not only during but also for some time after radiation. This must be due to radiation-induced charged species, probably electrons, which also account for subsequent fluorescence and phosphorescence at temperatures, but not wavelengths, determined by the polymer. These effects can be enhanced by external fields. A highly desirable feature is the close analogies existing between the radiation behavior of polymers and biopolymers: obvious examples are the effects of oxygen, of temperature, of molecular configuration, of additives and of concentration. This again calls for close collaboration between radiobiologists and polymer radiation scientists.

# BIBLIOGRAPHY

The radiation effects in polymeric systems cover at least two widely divergent subjects and are best described in specialized texts. A number of volumes specifically devoted to radiation modification of polymers and to radiation polymerization provide excellent sources for further reading and full lists of references classified by topic. There are also a few shorter summaries and overall reviews in books covering radiation chemistry in general and, therefore, not particularly concerned with the unusual physical behavior of polymers following irradiation, although it is these properties that are the major source of interest. Up-to-date information on application is best found in published conference proceedings devoted to the subject. Only a few references are given here to illustrate and elaborate some of the fundamental reactions involved.

## *Texts*


Charlesby, A. "Atomic Radiation and Polymers"; Pergamon Press: Oxford, 1960.

Chapiro, A. "Radiation Chemistry of Polymeric Systems"; Wiley Interscience: New York, 1962.

Platzer, N. In "Irradiation of Polymers (Advances in Chemistry)"; Platzer, N., Ed.; American Chemical Society: Washington, D.C., 1967.

Dole, M. "Radiation Chemistry of Macromolecules I and II"; Academic Press: New York, 1972.
Makhlis, F. A. "Radiation Physics and Chemistry of Polymers" (translated from Russian); Wiley: New York, 1975.
Wilson, J. E. "Radiation Chemistry of Monomers, Polymers and Plastics"; Marcel Dekker: New York, 1974.
Spikes, J. W. T.; Woods, R. J. "Introduction to Radiation Chemistry"; Wiley: New York, 1976.
Swallow, A. J. "Radiation Chemistry—An Introduction"; Longman, London, 1973.

## *Conferences on Radiation Processing*

Puerto Rico, 1976 (*Radiat. Phys. Chem. 4* **1977**)
Miami, 1978 (*Radiat. Phys. Chem. 14* **1979**)
Tokyo, 1980 (*Radiat. Phys. Chem. 18* **1981**)
Dubrovnik, 1982 (*Radiat. Phys. Chem. 22* **1983**)
San Diego, 1984 (*Radiat. Phys. Chem. 29* **1985**)

# REFERENCES

1. Charlesby, A. *Proc. Roy. Soc.* **1954**, *A224*, 120.
2. Sisman, O.; Bopp, C. D. *ORNL* **1951**, 928.
3. Bopp, C. D.; Sisman, O. *Nucleonics* **1955**, *13*(7), 28; *ibid*, **1955**, *13*(10), 51.
4. Alexander, P.; Black, R. M.; Charlesby, A. *Proc. Roy. Soc.* **1955**, *A232*, 31.
5. Charlesby, A.; Steven, J. H. *Radiat Phys. Chem.* **1976**, *8*, 719.
6. Charlesby, A.; Bridges, B. *Radiat. Phys. Chem.* **1982**, *19*, 155.
7. Flory, P. J. *J. Am. Chem. Soc.* **1941**, *63*, 3096.
8. Stockmayer, W. H. *J. Chem. Phys.* **1944**, *12*, 125.
9. Charlesby, A. *Proc. Roy. Soc.* **1954**, *A222*, 542.
10. Charlesby, A. *J. Polym. Sci.* **1953**, *11*, 513, 521.
11. Charlesby, A. "Atomic Radiation and Polymers"; Pergamon Press: Oxford, 1960, Chapter 9.
12. Flory, P. J. "Principles of Polymer Chemistry"; Cornell University Press: New York, 1953.
13. Charlesby, A. *Proc. Roy. Soc.* **1955**, *A230*, 120.
14. Folland, R.; Charlesby, A. *Radiat. Phys. Chem.* **1977**, *10*, 61.
15. Bueche, F. *J. Polym. Sci.* **1956**, *19*, 297.
16. Charlesby, A. *Radiat. Phys. Chem.* **1979**, *14*, 919.
17. Charlesby, A.; Folland, R. *Radiat. Phys. Chem.* **1980**, *15*, 393.
18. Charlesby, A.; Folland, R. Fifth Symposium Tihany, Akademai Kiado, Budapest, 1982.
19. Charlesby, A. "Atomic Radiation and Polymers", Pergamon Press: Oxford, 1960, Chapter 13.
20. Charlesby, A. *Proc. Phys. Soc. (London)* **1945**, *57*, 496, 510.
21. Charlesby, A. *Proc. Roy. Soc.* **1954**, *A222*, 60.
22. Dole, M.; Keeling, C. D. *J. Amer. Chem. Soc.* **1953**, *75*, 6082.
23. Lawton, E. J.; Balwit, J. S.; Bueche, A. M. *Ind. Eng. Chem. (Anal.)* **1954**, *46*, 1703.
24. Dole, M.; Keeling, C. D.; Rose, D. G. *J. Amer. Chem. Soc.* **1954**, *76*, 4304.
25. Ahmad, S. R.; Charlesby, A. *Radiat. Phys. Chem.* **1976**, *8*, 497, 585.
26. Charlesby, A.; Callaghan, A. *J. Phys. Chem. Solids* **1958**, *4*, 306.
27. Charlesby, A.; Pinah, S. H. *Proc. Roy. Soc.* **1959**, *A249*, 367.
28. Chapiro, A. "Radiation Chemistry of Polymer Systems"; Wiley Interscience: New York, 1962, Chapter IV.
29. Sobue, H.; Tabata, Y. *J. Polym. Sci.* **1960**, *43*, 459.
30. Bensasson, R.; Marx, R. *J. Polym. Sci.* **1960**, *48*, 53.
31. Schmitz, S. V.; Lawton, E. J. *Science* **1951**, *113*, 718.
32. Mesrobian, et al. *J. Chem. Phys.* **1954**, *22*, 565.
33. Henglein, A.; Schultz, R. *Z. Naturforsch.* **1954**, *9B*, 617.
34. Davisson, W. H. T.; Pinner, S. H.; Worrall, R. *Chem. Ind. (Res.)* **1957**, 1274.

35. Worrall, R.; Charlesby, A. *Int. J. Appl. Radiat. Isotopes* **1958**, *4*, 84.
36. Charlesby, A.; Pinner, S. H.; Worrall, R. *Proc. Roy. Soc.* **1959**, *A259*, 386.
37. Charlesby, A.; Alexander, P. *J. Chem. Phys.* **1955**, *52*, 699.
38. Berkowitch, J.; Charlesby, A.; Desreux, V. *J. Polym. Sci.* **1957**, *25*, 490.
39. Alexander, P.; Charlesby, A. *J. Polym. Sci.* **1957**, *23*, 355.
40. Chapiro, A. *Chim. Ind. (Milan)* **1956**, *76(4)*, 754.
41. Chapiro, A. *J. Polym. Sci.* **1957**, *23*, 377.
42. Charlesby, A.; Pinner, S. H. *Ind. Mat. Plast.* **1957**, *9(9)*, 30.
43. Charlesby, A. "Atomic Radiation and Polymers", Pergamon Press: London, 1960, Chapter 24.
44. Wycherley, V.; Greenwood, T. T. *Proc. Roy. Soc.* **1958**, *A244*, 54.
45. Charlesby, A.; Fukada, E. "Rheology of Elastomers"; Pergamon Press: Oxford, 1958, p. 150.

# 16

# Radiation Chemistry of Biopolymers

**L. K. MEE**

*Department of Radiology,
Harvard Medical School*

## INTRODUCTION

The study of the radiation chemistry of biopolymers has been directed mainly toward an understanding of events in vivo. The exposure of the genetic material DNA to radiation may result in cell replication failure or in chromosome aberrations leading to mutagenesis and carcinogenesis. A knowledge of the radiation damage to proteins, lipids, and carbohydrates is relevant to effects such as enzyme inactivation and also has applications in the radiation sterilization of foods and drugs where the concern is for the toxicity of radiation products.

Studies have been made on biopolymers in the solid state (direct effect) and in dilute aqueous solution (indirect effect). In vivo the state of biopolymers lies somewhere between fluid and solid, and to understand the radiation effects, one must interpolate between the two extremes. Evidence is quite strong that hydroxyl radicals are involved in mammalian and bacterial cell killing. The structure of DNA and many proteins is now clearly defined. With this knowledge and with the development of fast reaction techniques, the sites of reaction of the primary aqueous radicals and the reaction mechanisms in these biopolymers are well understood. The identification of the radiation products has been hampered by lack of sensitive analytical methodologies. Recent developments in analytical techniques, such as capillary gas chromatography (GC), mass spectrometry (MS), and high-perform-


*Radiation Chemistry: Principles and Applications*

ance liquid chromatography (HPLC), have provided means of monitoring small changes in amino acids and DNA bases, and of detecting radiation products formed in low yields.

The focus of this chapter will be on indirect effects of the primary aqueous radicals in forming organic radicals in biopolymers and on the mechanisms of termination of these radicals to produce damage and radiation products.

# NUCLEIC ACIDS

## *Structure of DNA*

DNA is a long-chain polymer built up by extended repetition of deoxyribonucleotides (Figure 16-1). Each of these is composed of a purine or pyrimidine base, a 5-carbon sugar (2′-deoxy-D-ribose), and a phosphate group. The base is joined to the C-1′ position of the sugar by a $\beta$-glycosidic linkage, and the phosphate group forms a link between the 3′-OH of one

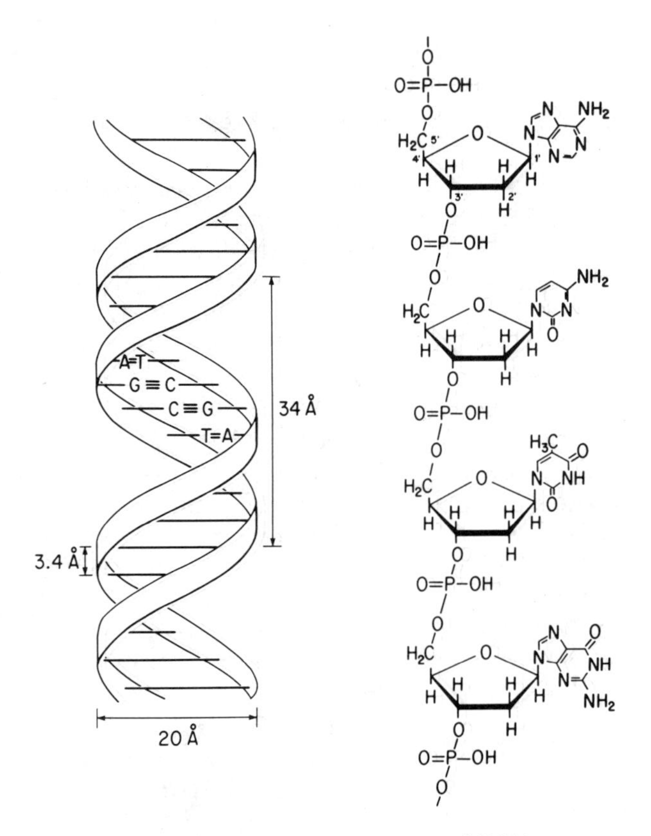


**Figure 16-1.** Primary structure of DNA.

deoxyribose residue and the 5′-OH of the adjacent residue. Four heterocyclic bases occur predominantly in DNA; these are the purines, adenine and guanine, and the pyrimidines, cytosine and thymine. In the "DNA double helix" the two polynucleotide strands that have opposite polarities (one 5′-3′, the other 3′-5′) are held together by hydrogen bonding between their bases with the sugar–phosphate chain on the outside of the helix. DNA can be categorized in three distinct forms; the right-handed helical forms, A and B, differing in the pitch and tilt of the helix, and the left-handed Z-form. In the mammalian genome most of the DNA is in the typical Watson–Crick right-handed B-form with the bases perpendicular to the helical axis; 5–10% of the DNA is in the Z-form, and another small fraction may be in the A-form.

## *Reactions of DNA and Its Constituents with the Primary Aqueous Radicals*

The rates of reaction of OH· radical, $e_{aq}^-$, and H· atom with free bases, nucleosides, nucleotides and with single- and double-stranded polynucleotides determined from pulse radiolysis studies[1–3] have given information on the relative efficiencies of attack by the primary species and on the influence of conformation of the polynucleotides upon their reactivities (Table 16-1). The values for the rate constants predict that the site of attack of $e_{aq}^-$ is exclusively the base components of the various monomeric compounds and that H· also preferentially attacks the heterocyclic bases rather

**Table 16-1.** Rate Constants for Reaction of OH·, $e_{aq}^-$, and H· with Nucleic Acid Constituents

| Compound | OH· | $e_{aq}^-$ | H· |
|---|---|---|---|
| *Bases* | | $k$ ($\times 10^9\ M^{-1}\ s^{-1}$) | |
| Adenine | 4.4 | 9.0 | 0.14 |
| Cytosine | 4.7 | 13.2 | 0.09 |
| Thymine | 5.2 | 18.0 | 0.38 |
| *Pentose* | 1.6 | <0.01 | 0.02 |
| *Nucleosides* | | | |
| Deoxyadenosine | 4.0 | 9.2 | 0.23 |
| Deoxycytidine | 4.8 | 13.2 | |
| Thymidine | 4.8 | | 0.38 |
| *Nucleotides* | | | |
| 5′-deoxyadenylic acid | 3.5 | | |
| 5′-deoxycytidylic acid | 4.4 | | |
| 5′-thymidylic acid | 5.3 | 1.5 | 0.38 |
| *Polynucleotides* | | | |
| Poly U | 1.25 | 0.75 | 0.42 |
| Poly A | 0.9 | 0.25 | |
| Poly (U + A) | 0.5 | 0.13 | |
| DNA | 0.3 | 0.14 | 0.08 |

than the sugar moiety. The rate of reaction of OH· with the sugar moiety is consistent with the conclusion from steady-state radiolysis[4] that about 25% of the reaction is with the sugar, although the major reaction is still with the base moieties in polynucleotides. Comparison of the reaction rate constants of the single-stranded polyuridylic acid (poly U) and polyadenylic acid (poly A) with double-stranded poly (U + A) and with DNA shows a lowered reactivity of the nucleotide in the bound state. This lowered reactivity has been attributed to changes in collision frequency; alternatively, it has been suggested that in the double-stranded state the bases are shielded from radical attack.

The interpretation of the data obtained by fast-reaction techniques, for example, optical absorption and electron spin resonance (ESR) spectra of transients, often depends on a knowledge of the nature of the final radiolysis products. Thus, the information from pulse and steady-state radiolysis is complementary.

Because the radical adduct is still a free radical, it must further react before becoming a stable compound. Three types of radical decay are possible: (1) unimolecular, (2) bimolecular with another radical, and (3) bimolecular with a molecule. In pulse radiolysis a high concentration of radicals is produced that favors radical–radical reactions. At lower dose rates unimolecular and radical–molecule reactions are more likely to occur. A unimolecular reaction has been inferred for 5-bromouracil.[5,6] Interest in this reaction arises because of the radiosensitization of cells in which bromouracil, a thymine analogue, is incorporated in the DNA. By dissociative electron capture, 5-bromouracil reacts with $e_{aq}^-$ to give the uracilyl free radical and bromide:

$$\text{5-bromouracil} + e_{aq}^- \longrightarrow \text{uracilyl radical} + Br^- \qquad (16\text{-}1)$$

The uracilyl radical is highly reactive and can immediately abstract hydrogen atoms from organic molecules, such as a neighboring sugar. This reaction may produce a strand break and introduce a uracil moiety into the DNA;

$$\text{uracilyl radical} + RH \longrightarrow \text{uracil} + R\cdot \qquad (16\text{-}2)$$

Of most significance to the in vivo situation are radical–molecule reactions. One of the most important is the reaction with oxygen, because it is well known that in the presence of oxygen cells are more radiosensitive. The reaction between the DNA—OH· adduct and oxygen has not been observed upon pulse radiolysis. That it exists is inferred from the formation of peroxides in DNA irradiated in oxygen-saturated solutions.[7]

The subsequent reactions of adducts of DNA with $e_{aq}^-$ have not been the subject of extensive investigation; it is generally accepted that they do not lead to significant damage. DNA—OH· adducts are the most damaging species resulting in multiple lesions; these include DNA–DNA cross-linking, base damage, and sugar damage, the latter leading to scission of the sugar–phosphate backbone (strand breaks) and release of bases.

## *Radiation-Induced Damage to the Bases in DNA*

Early work established the occurrence of chemical modifications, deamination, and ring fission in the heterocyclic bases of DNA. The extent of destruction of bases in irradiated DNA depends upon several factors, including concentration, ionic strength, pH, and the conformational state of the molecule.[7] The effects of polymerization and organization upon the radiosensitivity of the bases in the macromolecular structure have been studied in synthetic polynucleotides. The destruction of adenine occurring when polyadenylic acid is irradiated as a random coil is the same as that for the free base, $G = 0.65$, whereas in the double helical state the yield of base destruction is $G = 0.34$. These $G$-values correlate with the values for the reaction rate constants (Table 16-1). It has been suggested that the bases are sterically protected when inside the double helix. Support for this hypothesis comes from pulse radiolysis studies. The size of the optical absorption signal of radiation-produced DNA—OH· radicals is much lower than that expected for a mixture of deoxyribonucleotides. However, when the DNA has been preirradiated, the intensity of the absorption increases. This has been attributed to the opening of the double-stranded helical structure permitting more extensive attack of OH· radicals on the base moieties.

Two types of damage to the heterocyclic bases in DNA irradiated in dilute solution are recognized;[8]

1. Unaltered bases and their radiation-induced products that are released from the polynucleotide chain by rupture of the *N*-glycosidic bond (leading to strand breakage); this represents about one-fourth of the total bases affected.
2. Altered bases that remain attached to the DNA chain.

A major difficulty in identifying this second type of damage is the removal of the altered base from the DNA chain without further destruction.

The main site of attack by OH· radicals on pyrimidines is the 5,6 double bond.[9] In aerated solution, thymine ring saturation products of the 5,6-dihydroxy-5,6-dihydrothymine type are the major class of damage in DNA and have been detected in irradiated bacterial and mammalian cells. Other radiation products from the thymine moiety in DNA include 5-hydroxy-5-methyl hydantoin, formamide, and *N*-formyl urea.

A mechanism for the formation of hydroxyhydroperoxide from thymine has been postulated:[4]

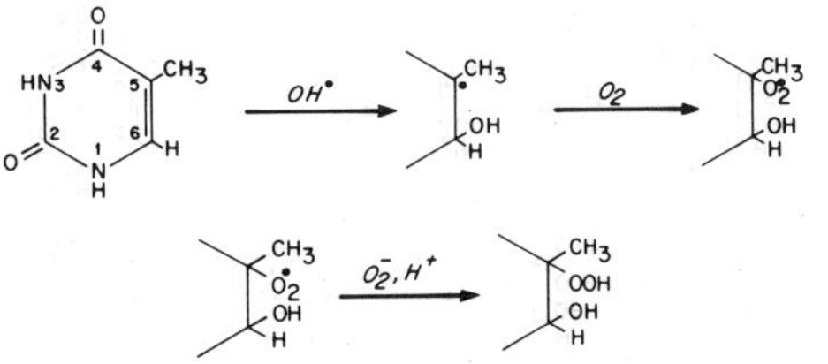

(16-3) and (16-4)

(16-5)

Various lines of evidence support this mechanism.[7] From the ESR spectra of the transients produced in pulsed solutions, the OH· adducts at the C-5 and C-6 positions of thymine have been positively identified. Oxygen reacts rapidly with the thymine—OH· radical with a rate constant for Reaction (16-4) of $1.9 \times 10^9\ M^{-1}\ s^{-1}$. The final reaction (16-5) can be inferred from the stoichiometry of the peroxide yields. Chromatographic comparison with synthesized thymine hydroxyhydroperoxides confirmed the identity of the final product.

In the case of cytosine it appears to be more difficult to release the radiation products from the DNA chain. No cytosine glycols could be detected, probably because they are likely to be dehydrated to isobarbituric acid. In early studies on the formation of peroxides in DNA, the rate of decay of the hydroperoxide activity showed two distinct species that were assigned to the two pyrimidine bases. Further studies, using the energetics of decay rather than simple kinetics, identified the fast-decaying peroxide with the cytosine product and the slow-decaying with that of thymine.

The major product from attack on the purine bases in DNA is 4,6-diamino-5-*N*-formamidopyrimidine, formed as a result of scission of the bond between C-8 and N-9 in the imidazole ring;

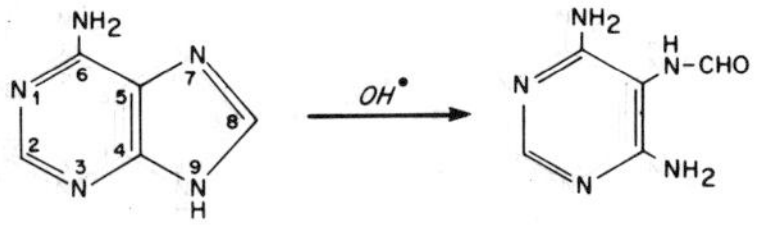

(16-6)

In the case of adenine, 7,8-dihydro-8-oxoadenine is also formed in low yield.[10] Irradiation in the presence of various radical scavengers confirmed that OH· is the effective radical in converting adenine to the formamido product, whereas $e_{aq}^-$ is ineffective.[11]

The potential for a variety of base products in DNA is high, but few of those detected in the free-base system have been pursued to the macromolecular or to the in vivo situation. If products are formed in very low yields, existing chemical methods may not be sufficiently sensitive to detect them, although these products may be significant in terms of toxicity. The application of the recently developed radioimmune assay techniques, as well as GC–MS, may be useful in detecting radiation products in very low yields.

## *Strand Breaks in DNA*

Only a small percentage of the aqueous radicals, about 20% OH·, react with the sugar moiety in DNA. Although this may seem negligible, scission of the sugar–phosphate bond (strand breaks) and release of base moieties are induced by the sugar radicals. Interruption of the polynucleotide chain is one of the most serious kinds of damage to the macromolecular structure of DNA leading to inactivation of template activity and loss of transforming activity. Strand breaks may concern either one strand of the double helix (single-strand breaks) or both strands (double-strand breaks). Although strand breaks, especially single-strand breaks, may be repaired rapidly by the cell, knowledge of the nature of the radiation-induced strand breaks is necessary for an understanding of these radiation effects and the enzymatic reactions involved in their repair.

Loss of intrinsic viscosity in DNA irradiated in solution was observed in early experiments, suggesting radiation-induced fragmentation of the DNA. The first chemical measurements of strand breaks were made by assaying with the enzyme alkaline phosphatase the number of radiation-produced monophosphate end groups.[7] Sedimentation of DNA on sucrose gradients with ultracentrifugation has become widely used for the detection and quantitation of strand breaks, especially for DNA in irradiated whole cells. On neutral sucrose gradients DNA retains its double-stranded structure, so only double-strand breaks are measured; on alkaline sucrose gradients the two DNA strands are separated, so a measure of single-strand breaks is obtained. In addition to radiation-induced strand scission, damage occurs at other sites on the DNA chain, which on alkaline hydrolysis results in scission of the sugar–phosphate bond (alkaline-labile damage). Thus, sedimentation on alkaline sucrose gradients measures both types of damage. The yields of strand breaks in DNA determined from these various techniques are in the range $G = 0.2–0.8$ for single-strand breaks, and $G = 3–9 \times 10^{-2}$ for double-strand breaks. A detailed mechanism for the formation of radiation-induced strand breaks has been proposed based on the results of chemical and enzymatic analyses.[12–14] Whereas chemical analyses identify the altered sugar moieties, enzymatic analyses determine the nature of the end groups on the strand break.

After irradiation of DNA in nitrous oxide-saturated aqueous solution, three sugars have been isolated: 2,5-dideoxypentose-4-ulose **1**, 2,3-dideoxypentose-4-ulose **2**, and 2-deoxypentose-4-ulose **3**.[15]

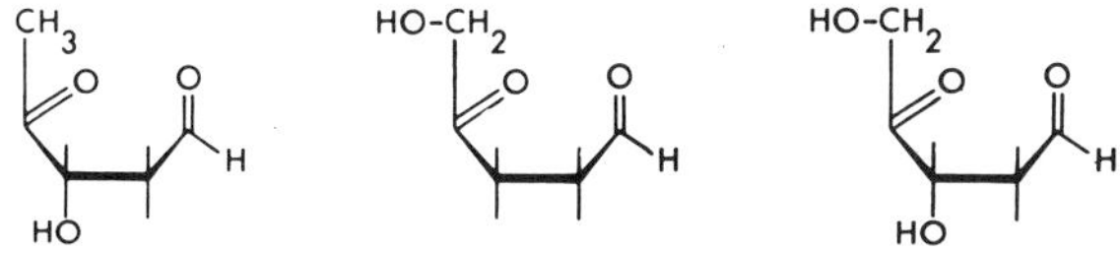

1 2 3

These sugars may be released from the chain during irradiation or, by a different mechanism, may remain bound to the DNA strand by a phosphate ester linkage; in both cases the DNA chain is broken.

Model studies on sugars indicate that hydrogen-atom abstraction by OH· radicals can occur at all five positions. The sugars **1**, **2**, and **3**, share the property that all are oxidized at the C-4′ position. Thus, the radical **4** (Figure 16-2) formed by abstraction at this position is the most likely precursor. Reaction mechanisms have been postulated for the cleavage of the phosphate ester bond and formation of the altered sugar moiety. Radical **4** first eliminates the phosphate ester anion at either the C-3′ or C-5′ position, thus breaking the chain; a radical cation is formed as an intermediate. Figure 16-2 shows reaction schemes for the radical cation **5**, formed by elimination of the phosphate at the C-3′ position, leading to the formation of radicals **6** and **7**. For radical **6** the second phosphate can be eliminated, again with a radical cation as an intermediate, leading to the formation of radical **8**. This radical is terminated in disproportionation reactions with other radicals, followed by ring opening and the elimination of the unaltered base forming sugar **1**. For radical **7** the second phosphate cannot be eliminated,

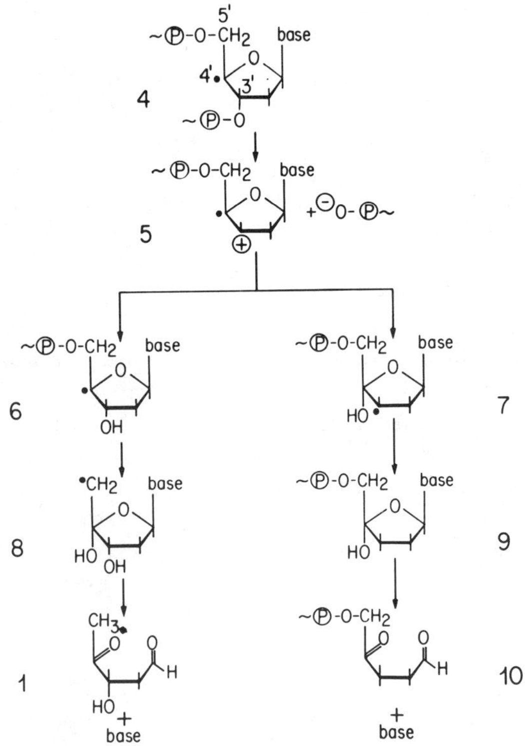

**Figure 16-2.** Reaction pathways for the formation of altered sugars in DNA leading to strand breakage.

but in an analogous series of reactions, the ring opens and an unaltered base is released. However, the altered sugar **10** remains bound to the DNA chain. For the formation of sugar **2**, radical **4** first eliminates the phosphate at the C-5′ position and this is followed by corresponding reactions. In the presence of oxygen, the carbon-centered radicals are rapidly converted into the corresponding peroxyl radicals. Based on model studies on sugars, a typical reaction of DNA peroxyl radicals is the fragmentation of the C-4′—C-5′ bond, a reaction not observed in the absence of oxygen.

Support for these mechanisms comes from other experimental evidence. The release of unaltered bases is observed after irradiation of DNA in solution.[16] Model studies on sugars predict that phosphate elimination at the C-3′ position should be faster than at the C-5′ position. In agreement with this, the $G$-value for **1** is 0.065 compared to $G = 0.01$ for **2**.

Although OH· radicals may be the important direct precursors of sugar radicals, experiments with poly U suggest an alternative mechanism.[14] Higher values for the formation of single-strand breaks are obtained than can be accounted for solely by reaction of OH· with the sugar moieties in poly U. Because 80% of the reaction of OH· is on the base moieties, the base radicals may also lead to strand breaks by radical transfer to the sugar moieties.

Several enzymes are available that react specifically with the hydroxyl or phosphate groups at the 3′ or 5′ end of the broken strand. The 5′ terminals of broken DNA strands can be determined by means of the combined action of phosphatase and polynucleotide kinase. The latter enzyme transfers the $\gamma$-phosphate of ATP to a 5′ hydroxyl group. With DNA irradiated in solution only a few end groups are phosphorylated with kinase, either before or after alkaline treatment. After dephosphorylation most of the end groups become reactive. This means that most of the 5′ ends in the actual strand breaks and in the alkali-induced strand breaks carry phosphate groups. These observations lead to the conclusion that in aqueous solutions radical attack at the sugar moiety splits the phosphoester bond predominantly at the 5′ position, consistent with the chemical data.

## CHROMATIN

### *Structure*

In eukaryotic cells nuclear DNA is associated with proteins, both histone and nonhistone, in the chromatin complex. The histones, which contain a high proportion of basic amino acids with positive charges, associate with negatively charged groups in the DNA, constraining the DNA into a regular repeating pattern.[17] In the basic structural unit, the nucleosome, two and one-half turns of DNA are wound around an octameric histone complex containing two each of the histones H2A, H2B, H3, and H4. These core regions are joined by linker DNA, which is associated with H1 histone. Extracted together with the DNA and histones in purified chromatin are nonhistone proteins, which are mainly nuclear enzyme proteins, with only a few being structural elements of chromatin. Isolated chromatin is a dilute aqueous gel, so the effects of radiation are due primarily to the indirect action of the aqueous radicals rather than to direct action.

## *Radiation-Induced Damage in Chromatin*

Irradiation of isolated chromatin results in DNA damage similar to that observed with free DNA, but the yields are generally lower. Single- and double-strand breaks, base damage, formation of intramolecular linkages, and loss of template activity are all strongly suppressed in chromatin relative to free DNA in solution. The lower yields are best accounted for by the protective effect of the histone and nonhistone proteins and the influence of DNA structure on accessibility to radical attack.

Single-strand breaks in the DNA of irradiated chromatin have been measured by sedimentation on alkaline sucrose gradients with ultracentrifugation.[18] Molecular size distributions are obtained as shown in Figure 16-3 for DNA from unirradiated and 5-krad irradiated chromatin. The DNA is degraded to lower molecular weights with irradiation. The shape of the

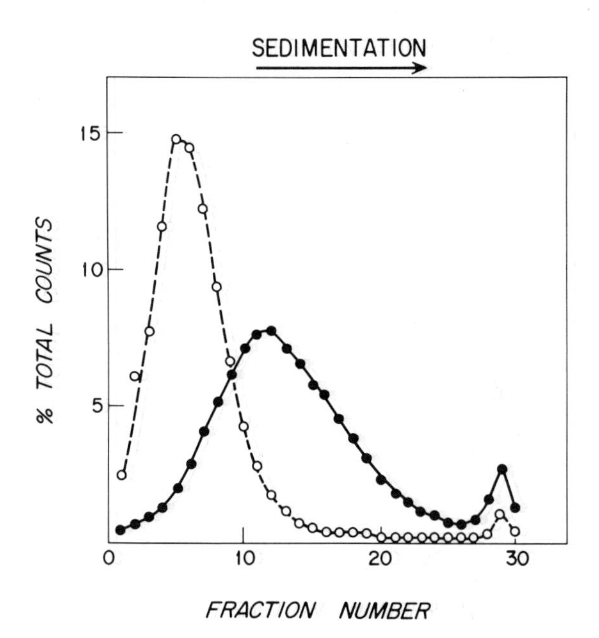


**Figure 16-3.** Sedimentation distributions of chromatin DNA on alkaline sucrose gradients: ○ = unirradiated; ● = 5 krad. (Reproduced with permission.[18])

profiles shows the DNA to be polydisperse, having a range of molecular weights, which is typical for mammalian DNA. From the molecular weights of the initial and irradiated DNA size distributions, the average number of breaks per molecule is determined and, hence, the efficiency of strand breakage. For irradiated chromatin the efficiency is $0.5–1.5 \times 10^{-2}$ single-strand breaks per krad in $10^6$-dalton DNA; the value for free double-stranded DNA in solution is 0.2–0.3,[19,20] indicating a difference of more than an order of magnitude.

The formation of double-strand breaks, determined by sedimentation on neutral sucrose gradients, is strongly suppressed in chromatin compared to free DNA in solution.[21] For irradiated chromatin the efficiency is $0.6–1.7 \times 10^{-4}$ double-strand breaks per krad in $10^6$-dalton DNA; this compares to a value for free DNA of $3–9 \times 10^{-2}$.[20,22]

The formation of single-strand breaks in the DNA of isolated chromatin and in free DNA in solution is a linear function of radiation dose, implying that single-strand breaks are caused by a single radiation event. In the case of double-strand breaks the relationship is less clear. At low doses the data fit a linear function for both chromatin DNA and free DNA in solution, although at high doses the data for free DNA fit more closely to the sum of a linear term and a quadratic term. There is no mechanism by which a single radical could produce a break in both strands of DNA. It has been postulated that double-strand breaks may be caused by multiple-radical processes in which two or more radicals are formed in a volume with dimensions similar to the DNA interstrand distance.

In isolated chromatin the reactivity of thymine to $\gamma$-radiation, as determined by the formation of ring-saturated products of the 5-hydroxy-6-hydroperoxydihydrothymine type, is 0.3 thymine destroyed per kilorad in $10^6$-dalton DNA.[23] This compares to a value of about 1.8 for free DNA in solution. In the low-dose region the dose-response curves are linear, implying the reaction is caused by a single radiation event.

Comparing the efficiency of strand damage and of thymine damage in chromatin (Table 16-2) and taking into account that damage is not restricted to thymine but also occurs at other DNA bases, we see it follows that base damage is the predominant type of radiation-induced damage in the DNA of isolated chromatin. Thus, the distribution of damage in chromatin DNA

**Table 16-2.** Radiation-Induced Damage in DNA

| Type of Damage | Chromatin DNA | Free DNA |
|---|---|---|
| | Damage per krad in $10^6$-dalton DNA | |
| Single-strand breaks | $0.5–1.5 \times 10^{-2}$ | 0.2–0.3 |
| Double-strand breaks | $0.6–1.7 \times 10^{-4}$ | $3–9 \times 10^{-2}$ |
| Thymine destroyed | 0.3 | 1.8 |

is similar to that found in free DNA in solution, but quantitatively the yields are considerably suppressed.

Covalent cross-links are formed between the DNA and chromosomal proteins upon irradiation of isolated chromatin.[24,25] Irradiation in the presence of radical scavengers shows the OH· radical to be the most effective aqueous intermediate for the promotion of cross-linking; $e_{aq}^-$ and $O_2^-$ are essentially ineffective. DNA-protein cross-links are also formed in the chromatin of irradiated intact cells, and irradiation in the presence of radical scavengers indicates that the OH· radical may be involved in the formation of cross-links in vivo. The core histones, H2A, H2B, H3, and H4, appear to be the predominant proteins involved in cross-links in isolated chromatin that would be predicted from their close proximity to DNA in the nucleosome structure. In whole cells, however, other chromosomal proteins have been implicated in cross-linking.

# PROTEINS

## *Structure*

The basic structural units of proteins are amino acids. Many amino acids are joined by peptide bonds, linking the $\alpha$-carboxyl group of one amino acid and the $\alpha$-amino group of the next one, to form a polypeptide chain. The chain spontaneously folds on itself in a manner dictated precisely by the sequence of amino acids. Many proteins, such as myoglobin and lysozyme, consist of a single polypeptide chain, whereas others contain two or more chains. The structure is held together by noncovalent bonds, including ionic, hydrogen and hydrophobic linkages, and by covalent disulfide bonds.

## *Reactivity of Proteins with the Primary Aqueous Radicals*

Proteins generally have very high rate constants for reaction with $e_{aq}^-$, H· atom, and OH· radical, in the range $10^{10}$–$10^{11}$ $M^{-1}$ $s^{-1}$.[1-3] The possible sites of attachment in proteins can be predicted from pulse radiolysis studies on amino acids and on smaller peptides. However, the intrinsic reactivities of these sites are influenced by the structure of the protein governing the accessibility of the sites to radical attack.

In reaction with $e_{aq}^-$ the main reducible sites in proteins are disulfide bridges, the protonated imidazole ring of histidine, and peptide carbonyl groups; less reducible sites are the aromatic side chains, as in tyrosine, phenylalanine, and tryptophan.[26] Based on known protein structural factors and the kinetic constants for radical attachment, we estimate that most hydrated electrons should react with the peptide carbonyl groups, and only about 20% should react directly with disulfide bridges and imidazole rings.[27] In several proteins the yields obtained from the observed transient spectra agree closely with theoretical estimates; the proportion of $e_{aq}^-$ reacting with disulfide bridges in ribonuclease is about 25%, in bovine serum albumin about 24%.[28,29] However, much higher yields of disulfide radicals are found in lysozyme (62%).[28] To account for these observations, scientists have suggested that the main sites of electron attachment are the surface peptide groups, and this is followed by internal electron migration into a potential sink, either histidine or a disulfide bridge.

In redox proteins, such as cytochrome C, in which the heme-iron group is in the higher oxidation state, $e_{aq}^-$ reacts with the metal center, resulting in reduction of the iron from the ferric to the ferrous state.

The OH· radical is a strong oxidizing agent and, in contrast to $e_{aq}^-$, can abstract hydrogen atoms. In proteins the reaction of OH· radicals will be distributed in the following ways:

1. Addition to aromatic rings:

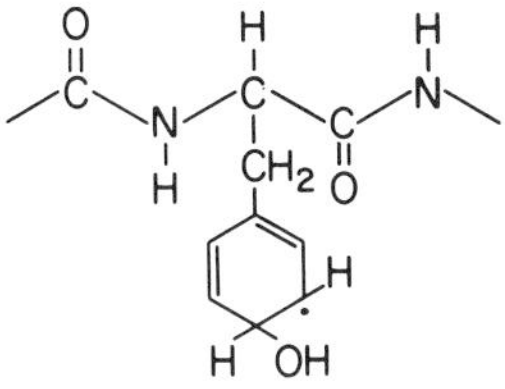

**11**

2. Abstraction of a hydrogen atom from the peptide backbone:

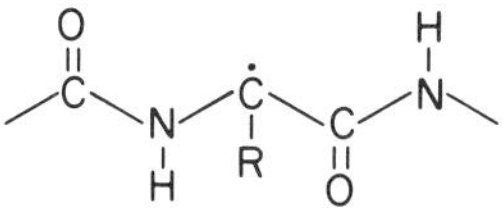

**12**

3. Abstraction of a hydrogen atom from the side chains:

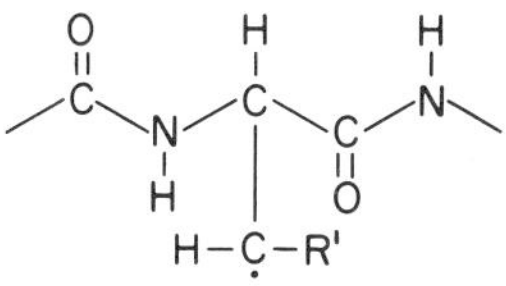

**13**

Addition can occur at the *o*-, *m*-, or *p*-position in the benzene ring to form hydroxycyclohexadienyl radicals. The abstraction of hydrogen atoms from the aromatic ring is negligible, and in compounds with aliphatic side chains, such as phenylalanine, tyrosine, tryptophan, and histidine, the main reaction is addition to the aromatic ring.

In spite of their strongly oxidizing character, OH· radicals lead to reduction of the metal center in redox proteins. Most OH· radicals react with the outer protein coat, and only a small fraction react directly with the metal. Some of these radicals on the protein coat can apparently transfer an electron intramolecularly to the metal center, resulting in oxidation of the protein radical and reduction of the metal.

In most cases the reactions of H· atom are similar to those of either $e_{aq}^-$ or OH· radical. Whereas abstraction of hydrogen atom may be defined as an oxidation process, addition to double bonds or aromatic rings may be considered a reduction process.

Spectral changes observed after pulse radiolysis of some proteins have established that radical migration may occur within the molecule. In the reaction of H· atom, OH· radical, or $e_{aq}^-$ with ribonuclease, intramolecular radical transfer is clearly established.[30] The primary addition of the reactive intermediates to sites on the enzyme molecule is followed by an intramolecular chain of events in which one electron equivalent radical transfer occurs from one amino acid residue to another, involving selectively only divalent sulfur and aromatic residues. This is consistent with the results of steady-state radiolysis showing that similar amino acid residues are chemically altered. In an investigation of the hydrolytic enzyme papain, the time-resolved radical sequences are interpreted in terms of a similar mechanism of intramolecular radical transfer involving tyrosine and disulfide linkages. In simple peptides containing tryptophan and tyrosine, intramolecular electron transfer occurs between the two residues.[31] Similarly, the radical sequences obtained from pulse radiolysis of alcohol dehydrogenase and lysozyme also indicate charge transfer between tyrosine and tryptophan.

Termination of the highly unstable secondary free radicals depends upon the conditions of the medium. In aqueous media at 25°C the most common reactions for carbon radicals are

$$\cdot RH + \cdot RH \nearrow HR{-}RH \quad \text{(Dimerization)} \qquad (16\text{-}7)$$

$$\cdot RH + \cdot RH \searrow RH_2 + R \quad \text{(Disproportionation)} \qquad (16\text{-}8)$$

Radicals formed by the addition of OH· radical to aromatic rings most likely terminate by disproportionation. The radiation products are *o*-, *m*-, and *p*-hydroxy derivatives of the original molecules, the isomeric distribution varying from system to system.

Secondary radicals can also terminate by radical–molecule reactions. These reactions are less likely in pulse radiolysis, where the initial concentration of free radicals is higher, but they occur under steady-state conditions, where the free-radical concentration is much lower. One important radical–molecule reaction involving S—H bonds is the basis of the radiation protection effect (ie, repair mechanism) in radiation biology. The S—H bond is much weaker than the C—H bond, and thus hydrogen atom abstraction can occur:

$$\cdot R' + RS{-}H \rightarrow R'H + RS\cdot \qquad (16\text{-}9)$$

If the RS· radicals are mobile, Reaction (16-10) takes place:

$$RS\cdot + RS\cdot \rightarrow RSSR \qquad (16\text{-}10)$$

## *Chemical Changes*

The effects of the highly reactive aqueous intermediates on proteins result in irreversible chemical changes. Although a great number of sites are available in proteins, due to the 20 different amino acids and several possible positions for attack within each amino acid, the action of the radicals has been shown to be specific and selective in many instances.[32–34] Amino acid analyses of proteins irradiated in dilute solutions show a significant decrease mainly in sulfur-containing, aromatic, and heterocyclic residues, correlating with the higher reaction rate constants for these residues (Table 16-3).

**Table 16-3.** Destruction of Amino Acids in Irradiated Proteins

| Protein | Amino Acids Destroyed |
|---|---|
| Insulin | cys, his, phe, tyr |
| Oxytocin | cys, phe, tyr |
| Cytochrome C | cys, his, met, phe |
| Hemoglobin | his, met, phe |
| Ovalbumin | cys, his, met, phe |
| Alkaline phosphatase | arg, cys, his, met, trp, tyr |
| Amylase | arg, his, lys, phe, tyr |
| Ribonuclease | cys, his, lys, met, phe, tyr |
| Subtilisin | his, met, trp, tyr |

Radiation products from these specific amino acids in proteins have been identified, especially with the availability of sensitive analytical techniques, such as HPLC and GC–MS.[35] By comparison with the known radiation products from the irradiation of free amino acids in solution, we can postulate reaction pathways.

2-Aminobutanoic acid has been identified as a product from OH· radical attack on lysozyme and from H· atom attack on ribonuclease. It is most likely an irradiation product of methionine, because it has been identified in the radiolysis of free methionine. The major site of attack of OH· on methionine is the sulfur atom; about 80% of OH· add to the sulfur and 20% abstract hydrogen atoms from the carbon atoms. The products of OH· attack on methionine predict that cleavage of the C—S—C bond must occur to produce thiyl and alkyl radicals:[36]

$$OH\cdot + \underset{H_3N^+}{\overset{^-O_2C}{}} \rangle CH-CH_2-CH_2-S-CH_3 \rightarrow \begin{cases} \rangle CH-CH_2-\dot{C}H_2 & \mathbf{14} \\ \dot{S}-CH_3 & \mathbf{15} \\ \rangle CH-CH_2-CH_2-\dot{S} & \mathbf{16} \\ \dot{C}H_3 & \mathbf{17} \end{cases}$$

2-Aminobutanoic acid, which has been identified as a radiation product of methionine-containing proteins, may result from a disproportionation reaction of radical **14**.

*o*-Tyrosine, *m*-tyrosine, 3-hydroxytyrosine (dopa), and 2-hydroxytyrosine have been identified as products of OH· radical attack on lysozyme; their formation can be explained by OH· addition to various aromatic rings of the constituent amino acids. By analogy with the radiolysis of phenylalanine, OH· addition to the aromatic ring followed by disproportionation reactions yields *o*-, *m*-, and *p*-tyrosine [see Structure **11** and Eq. (16-8)]. Because *p*-tyrosine is a constituent of lysozyme, only the *o*- and *m*-isomers could be identified with certainty. The formation of dopa and 2-hydroxytyrosine can be explained by OH· addition to the aromatic ring of tyrosine in lysozyme followed by disproportionation reactions.

Allo-threonine has been identified as a product of OH· radical attack on lysozyme. Its formation can be explained by hydrogen-atom abstraction reactions at the peptide backbone and at the side chain of threonine (see Structures **12** and **13**). Disproportionation and repair reactions involving both radicals by hydrogen atom transfer would yield L-threonine and its diastereomers, L-allo-threonine and D-allo-threonine.

## *Conformational Changes*

Both protein cross-linking (aggregation) and scission of the peptide backbone occur in proteins irradiated in the solid state and in solution, the effects depending upon the conditions during irradiation.[34] In the absence of oxygen intermolecular cross-linking of protein molecules predominates, whereas in the presence of oxygen few aggregates are formed, but scission of the peptide backbone as well as intramolecular cross-linking occurs.

In early studies with peptides the formation of amide and carbonyl groups was reported after irradiation in oxygenated solutions, but not under anaerobic conditions; peptide cleavage was enhanced in the presence of oxygen, but aggregation was suppressed. Recent studies on the enzyme lactate dehydrogenase and on bovine serum albumin agree with these early results.[37,38] Under anaerobic conditions the main reaction is aggregation, but the presence of oxygen prevents the intermolecular interaction of protein molecules. Fragments could be detected by gel electrophoresis after reduction of proteins irradiated in the presence of oxygen; some of the fragments are apparently held together by intramolecular disulfide bonds. Gel electrophoresis also reveals isomers produced by intramolecular disulfide interchange. At low dose rates the radiation-induced disulfide interchange is probably caused by a radical–molecule reaction:

$$R{-}S\cdot + R'{-}S{-}S{-}R \rightarrow R{-}S{-}S{-}R + R'{-}S\cdot \qquad (16\text{-}11)$$

This means that a single radical formation can alter several disulfide bonds.

A mechanism has been formulated for the fragmentation of biomolecules in general [Reactions (16-12)–(16-15)]. The radical R· formed by attack of OH· on the biomolecule reacts with oxygen, yielding peroxyradicals. The biomolecular reaction of the peroxyradicals leads to oxyradicals, which finally decompose into fragments ($F_1 + F_2$).

$$RH + OH\cdot \rightarrow R\cdot + H_2O \qquad (16\text{-}12)$$

$$R\cdot + O_2 \rightarrow RO_2\cdot \qquad (16\text{-}13)$$

$$2RO_2\cdot \rightarrow 2RO\cdot + O_2 \qquad (16\text{-}14)$$

$$RO\cdot \rightarrow F_1 + F_2 \qquad (16\text{-}15)$$

In the absence of oxygen the biomolecule radical can react with the formation of aggregates.

The specific effects of the individual radical intermediates in producing changes in molecular size have been demonstrated by separation of the radiation products of ribonuclease irradiated in solution under various environmental conditions.[32,33] In the elution profiles of ribonuclease from Sephadex chromatography (Figure 16-4), the native enzyme elutes at posi-

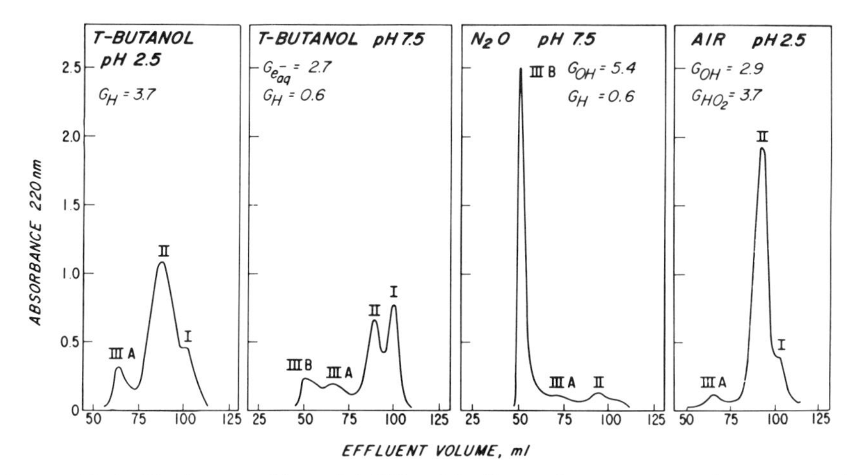


**Figure 16-4.** Elution profiles of 0.073 m*M* ribonuclease separated on Sephadex G-75. Dose: 100 krad. (Reproduced with permission.[33])

tion I, peak II is identified as a denatured monomer product, IIIA a dimer, and IIIB an aggregated trimer. The main product of H· atom attack is the denatured monomer. The hydrated electron is the least efficient radical for producing changes in molecular configuration because some native enzyme still remains after a dose of 100 krad. The most effective radical for the formation of aggregates is OH·; the formation of dimers, trimers, and higher polymers proceeds by means of a stepwise mechanism. Few aggregates are observed after irradiation of ribonuclease in the presence of oxygen, which reacts rapidly with protein radicals preventing intermolecular reactions.

The nature of the bonds linking radiation-induced protein aggregates has been studied by gel filtration and electrophoresis in the presence and absence of sodium dodecyl sulfate (which disturbs electrostatic and hydrophobic interactions) and reducing agents (β-mercaptoethanol). Such separations demonstrate that aggregation of ribonuclease[39] and of lactate dehydrogenase[37] involves both covalent and noncovalent bonds, but only some of the covalent bonds consist of disulfide bridges. Radicals **14** and **16** could be the precursors of cross-links between methionine-containing proteins. Evidence for the formation of covalent bonds between tyrosine moieties in irradiated proteins has been obtained from insulin, ribonuclease, papain, and collagen dimers which exhibit fluorescence spectra similar to bityrosine.[40]

## *Enzyme Inactivation*

Enzymes irradiated in dilute aqueous solution are inactivated very efficiently by the reactive intermediates, and in many cases the inactivation has been related to a specific and selective attack by individual radicals. The attack of H· atom and OH· radical on lysozyme leads to loss of activity, but that of $e_{aq}^-$ does not; the efficient inactivation by the OH· radical is attributed to its reaction with the tryptophan residue in the active center of the enzyme.[41] Lysozyme may be a special case because of the location of a tryptophan residue in the active center and its high reactivity with OH· radical. Lactate dehydrogenase is also efficiently inactivated by OH· radical, and the primary target appears to be the cysteine residue in the active center.[42] Trypsin, on the other hand, is inactivated more efficiently by H· atom and $e_{aq}^-$ than by OH· radical; this is attributed to reaction with the disulfide bonds disrupting the active center.[43] Ribonuclease that does not contain tryptophan is inactivated very efficiently by H· atom; approximately 1 out of 4 H· atoms inactivates a ribonuclease molecule compared to 1 out of 15 OH· radicals or $e_{aq}^-$.[33]

## *Protein–Amino Acid Cross-linking*

Cross-linking between proteins and free amino acids occurs when they are irradiated in dilute aqueous solution and is mediated predominantly by

OH· radical.[34,44] Protein–protein and amino acid–amino acid adducts, as well as protein–amino acid adducts, are formed, the relative yields depending on the molar ratio of amino acid: protein in solution. The separation of irradiated mixtures of glucagon and phenylalanine by Sephadex gel chromatography shows that at low ratios of phenylalanine: glucagon predominantly aggregated products of glucagon are formed, both glucagon–glucagon aggregates and glucagon–phenylalanine adducts within the aggregated substrate, whereas at high ratios, glucagon–phenylalanine adducts are formed that are not aggregated, and dimers and higher polymers of phenylalanine predominate.

Specific amino acids are involved in cross-linking, and the efficiency of binding generally correlates with their reactivity toward the aqueous radicals. In a comparison of the radiation-induced binding of various amino acids to bovine serum albumin, the aromatic residues, phenylalanine and tryptophan, cross-link very efficiently, the sulfur-containing methionine and cystine slightly less so, whereas the aliphatic residues, alanine and leucine, have very low binding efficiencies. The sulfur radicals **15** and **16** are probably involved in the binding of methionine to bovine serum albumin. Only a limited number of binding sites on the protein molecule is involved in cross-linking with free amino acids. Aromatic amino acids bind to serum albumin but not to polyglutamic acid, indicating that aliphatic side chains and peptide bonds are not involved in the cross-linking mechanism. Amino acid analysis of monomeric glucagon–phenylalanine adducts shows a loss of specific amino acid residues, namely, the aromatic, basic, and sulfur-containing residues. Presumably, these residues are specifically attacked by OH·, forming free radicals that are sites for several possible cross-linking reactions.

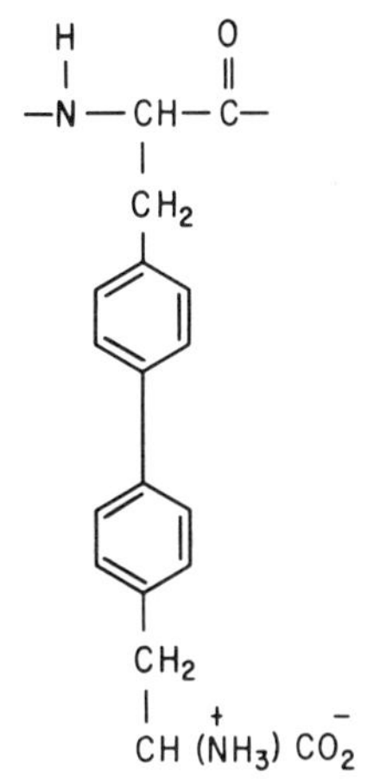

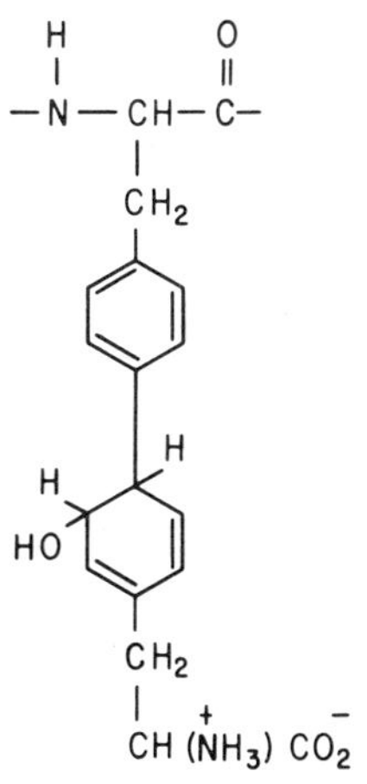

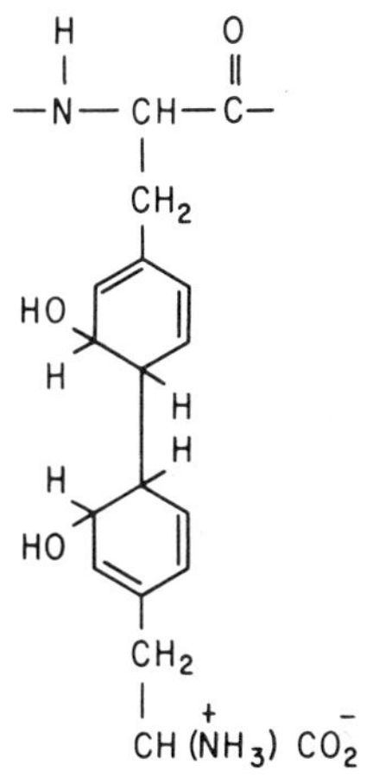

Radiation-induced cross-linking between phenylalanine and the tetrapeptide, Gly-Gly-Phe-Leu, mediated by the OH· radical, involves specifically the aromatic ring of phenylalanine.[45] The complexity of the cross-linking reaction is evident from the separation of several adduct species by gel filtration and HPLC. Examination of tetrapeptide–phenylalanine adducts with fast atom bombardment mass spectrometry identified several molecular species corresponding to adducts with zero, one, and two hydroxyl groups attached (structures **18**, **19**, and **20**). More than one phenylalanine-derived species is probably involved in peptide–phenylalanine adducts. Because OH· can attack the ring at the *o*-, *m*-, or *p*-position, structures **19** and **20** are only two of several possible configurations.

# LIPIDS

Interest in the irradiation of fats is related mainly to the use of radiation in the sterilization of foods. It is well established that the treatment of fats with high doses of ionizing radiation causes the formation of lipid peroxides, the extent depending upon the chemical composition of the fat and the conditions during irradiation.[46] Of primary importance is the fatty-acid composition of the irradiated fat, especially the degree of unsaturation (ie, the presence of one or more double bonds). Saturated and monounsaturated fatty acids, such as oleic acid, are very resistant to radiation. The polyunsaturated fatty acids containing three or more double bonds are very sensitive and are readily destroyed by radiation.

The initial event is the formation of fatty-acid free radicals, either directly or, if water is present, by abstraction of a hydrogen atom from the fatty-acid chain by OH· radical. Peroxy radicals are formed by reaction with oxygen. By abstraction of a hydrogen atom from another fatty-acid molecule, peroxy radicals are converted to hydroperoxides, thus promoting a chain reaction:

$$R\cdot + O_2 \rightarrow RO_2\cdot \qquad (16\text{-}16)$$

$$RO_2\cdot + RH \rightarrow ROOH + R\cdot \qquad (16\text{-}17)$$

Lipid hydroperoxides are very unstable and spontaneously decompose to form a mixture of mono- and dialdehydes and ketones of various chain lengths. Thus, foods containing unsaturated fats and subjected to irradiation will contain these radiation products and a decreased content of unsaturated, especially polyunsaturated, fatty acids. Hydroperoxides are known to be very toxic when injected into animals. The inclusion of high concentrations of peroxidized fats in the diet of animals restricts their growth and can be lethal.[47]

Antioxidants, such as vitamin E ($\alpha$-tocopherol) and 2,6-bis(1,1-dimethylethyl)-4-methylphenol (BHT), which are very effective in preventing oxidative deterioration in unsaturated fats, are also effective in preventing peroxidation of fats if present during irradiation.[48] The chain reaction (16-16), (16-17) can be inhibited either by retarding the formation of free radicals or by the introduction of radical acceptors. The majority of effective antioxidants is believed to act through a radical acceptor mechanism, the fatty-acid radical, R·, or the peroxyradical, $RO_2$·, being accepted. Additionally, lipid peroxidation is inhibited when proteins, such as casein or ovalbumin, are present during irradiation.

## POLYSACCHARIDES

The basic monomeric units, carbohydrates, are joined by glycosidic linkages to form oligo- and polysaccharides. The polysaccharide cellulose is the most abundant organic compound, comprising 50% or more of all carbon in vegetation; the purest source is cotton, which is at least 90% pure cellulose. There is considerable interest in the $\gamma$-radiolysis of cellulose from the viewpoints of food sterilization and textile improvement.

The predominant effect of irradiation of cellulose is degradation, which is most likely due to the splitting of the glycosidic bond. The mechanism has been investigated in a small model compound, cellobiose, which contains two glucose molecules joined by a glycosidic linkage.[49] In reactions similar to those on the deoxysugars in DNA (see above), OH· radicals can abstract hydrogen atoms at every C—H bond in the cellobiose molecule. However, only radicals with free spin at the C-1′, C-4′, and C-5′ positions lead to splitting of the glycosidic bond. Thus, the initial product resulting from hydrolytic scission of the glycosidic bond followed by hydrogen atom transfer is glucose. Other radiation products, the most abundant being gluconic acid, have been identified and are degradation products of glucose. Radicals formed at other positions on the ring will cause alteration of the cellobiose molecule without scission of the glycosidic bond. On further irradiation, altered glucose units may be released. The disaccharide, $\alpha,\alpha$-trehalose has also been studied as a model compound.[50] In similar reactions the OH· radical abstracts hydrogen atom from the sugar ring, which may result in hydrolytic cleavage of the glycosidic bond. By hydrogen-atom transfer glucose and 5-deoxyxylohexodealdose are formed as the main initial products.

The effects of radiation have been studied in model systems of various complexity containing carbohydrates, fats, amino acids, and proteins in mixtures approaching the composition in foods.[51] The radiation-induced damage in the various components may be modified, the nature and degree of alteration depending upon the composition of the mixture. The decomposition of $\alpha,\alpha$-trehalose is reduced if equimolar amounts of amino acids (alanine, leucine, phenylalanine, methionine, or cysteine) are present during irradiation; the extent of protection generally correlates with their reaction rate constants. However, cysteine protects to a greater extent than can be accounted for by its ability to act as a radical scavenger; this enhancement can be explained by a repair mechanism of the $\alpha,\alpha$-trehalose radicals [Eqs. (16-9) and (16-10)]. Similarly, proteins having high cystine–cysteine content, such as bovine serum albumin, provide a greater degree of protection against radiation damage than proteins, such as myoglobin, which lack these sulfur-containing amino acid residues. Addition of emulsified lipids (palmitic acid methylester or linoleic glycerol) has no effect on the amount or nature of the radiation products from irradiated $\alpha,\alpha$-trehalose. On the other hand, aggregation of proteins, which occurs when they are irradiated alone (see above), is suppressed in the presence of carbohydrates. However, if both carbohydrates and lipids are present in the irradiation mixture with proteins, the lipids counteract the effects of carbohydrates in decreasing protein aggregation.

## REFERENCES


1. Anbar, M.; Bambenek, M.; Ross, A. B. *Natl. Stand. Ref. Data Ser., U.S. Natl. Bur. Stand.* **1973**, *43.*
2. Anbar, M.; Farhataziz; Ross, A. B. *Natl. Stand. Ref. Data Ser., U.S. Natl. Bur. Stand.* **1975**, *51.*
3. Farhataziz; Ross, A. B. *Natl. Stand. Ref. Data Ser., U.S. Natl. Bur. Stand.* **1977**, *59.*
4. Scholes, G.; Ward, J. F.; Weiss, J. *J. Mol. Biol.* **1960**, *2*, 379.
5. Adams, G. E. *Current Topics Radiat. Res.* **1967**, *3*, 35.
6. Zimbrick, J. D.; Ward, J. F.; Myers, L. S., Jr. *Int. J. Radiat. Biol.* **1969**, *16*, 505.
7. Ward, J. F. *Adv. Radiat. Biol.* **1975**, *5*, 181.
8. Téoule, R.; Cadet, J. In "Effects of Ionizing Radiation on DNA: Physical, Chemical, and Biological Aspects"; Hütterman, J.; Köhnlein, W.; Téoule, R., Eds.; Springer-Verlag; Berlin and New York, 1978, Section II, Chapter 2, pp. 171–203.
9. Scholes, G. In "Effects of Ionizing Radiation on DNA: Physical, Chemical and Biological Aspects"; Hütterman, J.; Köhnlein, W.; Téoule, R., Eds.; Springer-Verlag: Berlin and New York, 1978, Section II, Chapter 1, pp. 153–170.
10. Bonicel, A.; Mariaggi, N.; Hughes, E.; Téoule, R. *Radiat. Res.* **1980**, *83*, 19.
11. Chetsanga, C. J.; Grigorian, C. *Int. J. Radiat. Biol.* **1983**, *44*, 321.
12. Von Sonntag, C.; Schulte-Frohlinde, D. In "Effects of Ionizing Radiation on DNA: Physical, Chemical, and Biological Aspects"; Hütterman, J.; Köhnlein, W.; Téoule, R., Eds.; Springer-Verlag: Berlin and New York, 1978, Section II, Chapter 3, pp. 204–226.
13. Von Sonntag, C.; Hagen, U.; Schön-Bopp, A.; Schulte-Frohlinde, D. *Adv. Radiat. Biol.* **1981**, *9*, 109.
14. Von Sonntag, C. *Int. J. Radiat. Biol.* **1984**, *46*, 507.
15. Dizdaroglu, M.; Von Sonntag, C.; Schulte-Frohlinde, D. *J. Am. Chem. Soc.* **1975**, *97*, 2277.
16. Hems, G. *Nature (London)* **1960**, *186*, 710.
17. McGhee, J. D.; Felsenfeld, G. *Ann. Rev. Biochem.* **1980**, *49*, 1115.
18. Mee, L. K.; Adelstein, S. J.; Stein, G. *Int. J. Radiat. Biol.* **1978**, *33*, 443.
19. Hagen, U. *Biochim. Biophys. Acta* **1967**, *134*, 45.
20. Ward, J. F.; Kuo, I. *Radiat. Res.* **1978**, *75*, 278.
21. Mee, L. K.; Adelstein, S. J. In "Proceedings of the Seventh International Congress of Radiation Research"; Broese, J. J.; Barendsen, G. W.; Kal, H. B.; Van Der Kogel, S. J., Eds.; Martinus Nijhoff: Amsterdam, 1983, pp. A3–29.
22. Frey, R.; Hagen, U. *Radiat. Environ. Biophys.* **1974**, *11*, 125.
23. Roti Roti, J. L.; Stein, G S.; Cerutti, P. A. *Biochemistry* **1974**, *13*, 1900.

24. Mee, L. K.; Adelstein, S. J. *Int. J. Radiat. Biol.* **1979**, *36*, 359.
25. Mee, L. K.; Adelstein, S. J. *Proc. Natl. Acad. Sci. U.S.A.* **1981**, *78*, 2194.
26. Simic, M. G. *J. Agric. Food Chem.* **1978**, *26*, 6.
27. Faraggi, M.; Klapper, M. H.; Dorfman, L. M. *Biophys. J.* **1978**, *24*, 307.
28. Bisby, R. H.; Cundall, R. B.; Redpath, J. L.; Adams, G. E. *J. Chem. Soc., Faraday Trans. 1* **1976**, *72*, 51.
29. Schuessler, H.; Davies, J. V. *Int. J. Radiat. Biol.* **1983**, *43*, 291.
30. Shafferman, A.; Stein, G. *Biochim. Biophys. Acta* **1975**, *416*, 287.
31. Prütz, W. A.; Land, E. J. *Int. J. Radiat. Biol.* **1979**, *36*, 513.
32. Jung, H.; Schuessler, H. *Z. Naturforsch.* **1967**, *22b*, 614.
33. Mee, L. K.; Adelstein, S. J.; Stein, G. *Radiat. Res.* **1972**, *52*, 588.
34. Yamamoto, O. In "Protein Crosslinking"; Friedman, M., Ed.; Plenum: New York, 1977, Part A, pp. 509–547.
35. Dizdaroglu, M.; Gajewski, E.; Simic, M. G.; Krutzsch, H. C. *Int. J. Radiat. Biol.* **1983**, *43*, 185.
36. Gajewski, E.; Dizdaroglu, M.; Krutzsch, H. C.; Simic, M. G. *Int. J. Radiat. Biol.* **1984**, *46*, 47.
37. Schuessler, H.; Herget, A. *Int. J. Radiat. Biol.* **1980**, *37*, 71.
38. Schuessler, H.; Schilling, K. *Int. J. Radiat. Biol.* **1984**, *45*, 267.
39. Hajós, Gy.; Delincée, H. *Int. J. Radiat. Biol.* **1983**, *44*, 333.
40. Boguta, G.; Dancewicz, A. M. *Int. J. Radiat. Biol.* **1983**, *43*, 249.
41. Adams, G. E.; Willson, R. L.; Aldrich, J. E.; Cundall, R. B. *Int. J. Radiat. Biol.* **1969**, *16*, 333.
42. Buchanan, J. D.; Armstrong, D. A. *Int. J. Radiat. Biol.* **1976**, *30*, 115.
43. Masuda, T.; Ovadia, J.; Grossweiner, L. I. *Int. J. Radiat. Biol.* **1971**, *20*, 447.
44. Mee, L. K.; Kim, H.-J.; Adelstein, S. J.; Taub, I. A. *Radiat. Res.* **1984**, *97*, 36.
45. Kim, H.-J.; Mee, L. K.; Adelstein, S. J.; Taub, I. A.; Carr, S. A.; Reinhold, V. A. *Radiat. Res.* **1984**, *100*, 30.
46. Wills, E. D. *Int. J. Radiat. Biol.* **1980**, *37*, 383.
47. Horgan, V. J.; Philpot, J. St. L.; Porter, B. W.; Roodyn, D. B. *Biochem. J.* **1957**, *67*, 551.
48. Wills, E. D. *Int. J. Radiat. Biol.* **1980**, *37*, 403.
49. Von Sonntag, C.; Dizdaroglu, M.; Schulte-Frohlinde, D. *Z. Naturforsch.* **1976**, *31b*, 857.
50. Adam, S. *Int. J. Radiat. Biol.* **1982**, *42*, 531.
51. Diehl, J. F.; Adam, S.; Delincée, H.; Jakubick, V. *J. Agric. Food Chem.* **1978**, *26*, 15.

# 17

# Application of Radiation Chemistry to Studies in the Radiation Biology of Microorganisms

D. EWING

*Department of Radiation Oncology and Nuclear Medicine, Hahnemann University*

## INTRODUCTION

Our awareness of ionizing radiation stretches back to 1895. Since that time, radiation biologists have directed considerable effort at learning how relatively small amounts of energy have such profound effects on living organisms. We now know a great deal about how radiation damages cells, but our knowledge is still far from complete.

Almost without exception the changes radiation effects in living cells are harmful. One of the most important applications we have made of this destructiveness is in cancer therapy. In fact, Grubbé reported using x-rays for treating a mammary carcinoma in 1896,[1] only one year after Roentgen had announced his discovery. As long as cancer occurs, it is reasonable to expect that we will continue to need to improve our ability to treat cancers with radiation.

Our explicit goal in applying radiation biology to cancer therapy is to manipulate the cell's sensitivity to radiation. We seek to destroy one cell while its neighbor—which received the same dose—remains unaffected. It


*Radiation Chemistry: Principles and Applications*

seems logical to believe that an understanding of exactly how cells are damaged by radiation should lead to our increased ability to control the cell's sensitivity to radiation.

One of the characteristics of radiation biology is the breadth of its component disciplines. The field cuts across both physical and biological sciences. This characteristic is one of our strengths, because it allows us to approach our goals from many different experimental directions.

The past 25 years has brought astonishing improvements in chemists' abilities to monitor fast, radiation-induced chemical reactions. Exposure times can now be in the nano- to picosecond ranges, and we speak with confidence about the biological relevance of reactions occurring in submicrosecond intervals, even when we know many years may pass before we can observe the effects of such reactions.

These improvements in mechanical techniques have been the basis for a virtual explosion of information in radiation chemistry; and, again, just as with Grubbé and Roentgen, radiation biologists have moved quickly to apply this new information to their own field.

Microorganisms have been especially valuable for such experiments. Early in this century the techniques for handling microorganisms were reasonably well established, and microorganisms were relatively resistant to the harsh experimental treatments needed to control radiation-induced chemical reactions. We should remember also that experimental techniques for using in vitro mammalian cells have become "reliable" only in the last few decades; indeed, the classic paper by Puck and Marcus[2] in 1956 is widely acknowledged as the beginning of mammalian cell radiation biology.

This chapter will focus on a few basic principles in radiation biology, with some emphasis on how data are collected and analyzed. We will then discuss several examples to show how microorganisms have been used for applying radiation chemistry techniques and information to specific problems in radiation biology. The area on which we will direct attention concerns how radiolysis products contribute to cell death.

## SURVIVAL CURVES

When organisms are exposed to ionizing radiation, one common biological endpoint concerns cell survival. In this context "survival" has a somewhat unique meaning. The term does not distinguish "living" from "nonliving" cells, but it refers to reproductive capacity. Survival, as an endpoint, is based on a cell's ability to divide until a visible clone (a colony) is formed. Under this definition a cell that functions (ie, lives) but cannot divide is classed as "dead."

Suppose a population of genetically identical microorganisms is uniformly suspended in a fluid. If the fluid is then divided into equal volumes, each volume should contain the same number of cells ($N_0$). We set one of these volumes aside and expose each of the others to a different x-ray dose. We make serial dilutions from each volume, including the nonirradiated control that was set aside, and then inoculate samples from each dilution step into Petri dishes containing a nutrient medium. Agar or a similar supporting substrate is used to fix the position of the cells in these plates. Then we allow time for the development of macroscopic colonies and examine the plates. Some will contain no colonies, and others will have far too many for accurate counting. If we made enough dilutions and platings, however, we will have countable numbers of colonies on some plates from each of the volumes of the original cell suspension. Each colony represents a single cell that survived the radiation exposure, and, with a knowledge of the dilution steps and $N_0$, it is a simple matter to calculate the number of cells ($N_D$) that survived each x-ray dose. Table 17-1 gives data from a hypothetical experiment.

**Table 17-1.** Hypothetical Data for a Survival Curve

| Dose | $\bar{x}$ | Dilution Factor | $N_D$ | $S$ |
|---|---|---|---|---|
| 0 | 153.2 | $1 \times 10^{-5}$ | $1.5 \times 10^{7}$ | 1.0 |
| 40 | 137.1 | $1 \times 10^{-5}$ | $1.37 \times 10^{7}$ | $9.0 \times 10^{-1}$ |
| 60 | 122.7 | $1 \times 10^{-5}$ | $1.23 \times 10^{7}$ | $8.0 \times 10^{-1}$ |
| 120 | 79.4 | $1 \times 10^{-5}$ | $7.9 \times 10^{6}$ | $5.2 \times 10^{-1}$ |
| 160 | 147.3 | $2 \times 10^{-5}$ | $7.4 \times 10^{6}$ | $4.8 \times 10^{-1}$ |
| 250 | 138.7 | $6 \times 10^{-4}$ | $2.3 \times 10^{6}$ | $1.5 \times 10^{-1}$ |
| 360 | 142.4 | $2 \times 10^{-4}$ | $7.1 \times 10^{5}$ | $4.6 \times 10^{-2}$ |
| 480 | 140.6 | $1 \times 10^{-3}$ | $1.4 \times 10^{5}$ | $9.2 \times 10^{-3}$ |
| 570 | 311.9 | $7 \times 10^{-3}$ | $4.5 \times 10^{4}$ | $2.9 \times 10^{-3}$ |
| 670 | 189.4 | $2 \times 10^{-2}$ | $9.5 \times 10^{3}$ | $6.2 \times 10^{-4}$ |
| 780 | 127.3 | $7 \times 10^{-2}$ | $1.8 \times 10^{3}$ | $1.2 \times 10^{-4}$ |
| 880 | 76.9 | $1 \times 10^{-1}$ | $7.7 \times 10^{2}$ | $5.0 \times 10^{-5}$ |
| 950 | 103.2 | $3 \times 10^{-1}$ | $3.4 \times 10^{2}$ | $2.2 \times 10^{-5}$ |

Dose is given in arbitrary units. The dilution factors are calculated from the dilutions that were made and the volume plated from each aliquot of irradiated cell suspension. The mean number of colonies per plate is divided by the dilution factor to obtain the viable number of cells per mL ($N_D$). The ratio of $N_D$ to $N_0$ (the number of viable cells per mL in the unirradiated control) gives the fractional survival after each radiation dose, $S$.

The relationship between survival and radiation dose is not linear, and survival curves are usually plotted on log-linear axes. For the type of illustrations we draw, the logarithmic base we use is unimportant; most often, we simply let the paper make this manipulation for us. Our calculations and analysis, however, are always based on natural logarithms. The process is analogous to the way radioactive curves are plotted.

Figure 17-1 illustrates the survival curve formed by the data in Table 17-1. Most survival curves have this shape. A low-dose region (the "shoulder"), where the line connecting the data points has a constantly changing curvature, is followed by a high-dose region (the terminal portion), where the line connecting the data points is straight. (Nothing, of course, is really "straight" on semilog paper; the illusion of "straightness" is, however, intentional and is caused by the paper manufacturer's spacing on the logarithmic axis.)

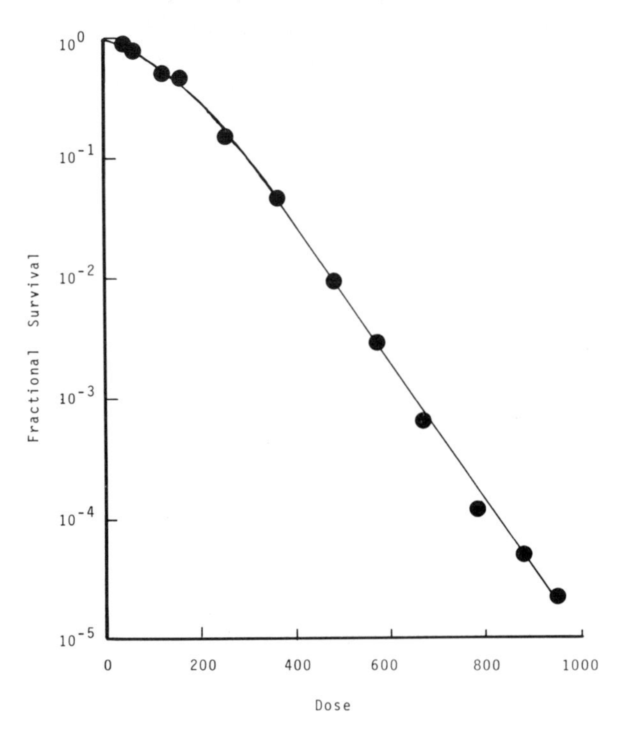


**Figure 17-1.** A semilog plot of the hypothetical survival curve data given in Table 17-1.

Most survival curves can be fitted by the general expression

$$S = 1 - (1 - e^{-kD})^n \tag{17-1}$$

where $S(=N_D/N_0)$ is the fractional survival after radiation dose $D$. The extrapolation number ($n$) is the value of $S$ when the terminal (straight) portion of the curve is extrapolated back to $D = 0$. The inactivation rate constant ($k$) is the absolute value of the slope of the straight portion of the curve. (All survival curves have negative slopes.) The value of $k$ or its reciprocal ($1/k = D_0$) is used as an index for radiation sensitivity.

# RADIATION SENSITIVITY

Organisms show a wide variation in their sensitivities to radiation. A particular radiation dose may be relatively unimportant to one kind of organism while to another it may reduce the likelihood of survival to only a fraction of a percent. In addition, radiation sensitivity itself is not a constant for a given organism; instead, an organism's response will depend on its repair capabilities as well as its environment and its physiological status at the time of irradiation. In the preceding section the steps for generating survival curve data were summarized. Each of these steps, in fact, must be carefully controlled. The composition and temperature of the suspending fluid, the cell concentration, the energy of the photons used for irradiation, the time that elapses between irradiation and dilution, the composition and temperature of the dilution fluid, the composition of the nutrient medium on which colony formation occurs, and the temperature of incubation are only a few of the parameters that must be controlled; any of these can affect the response of the organism being studied.

Figure 17-2 illustrates survival curves for three microorganisms. Note the wide variation in doses needed to reduce the survival to the same level.

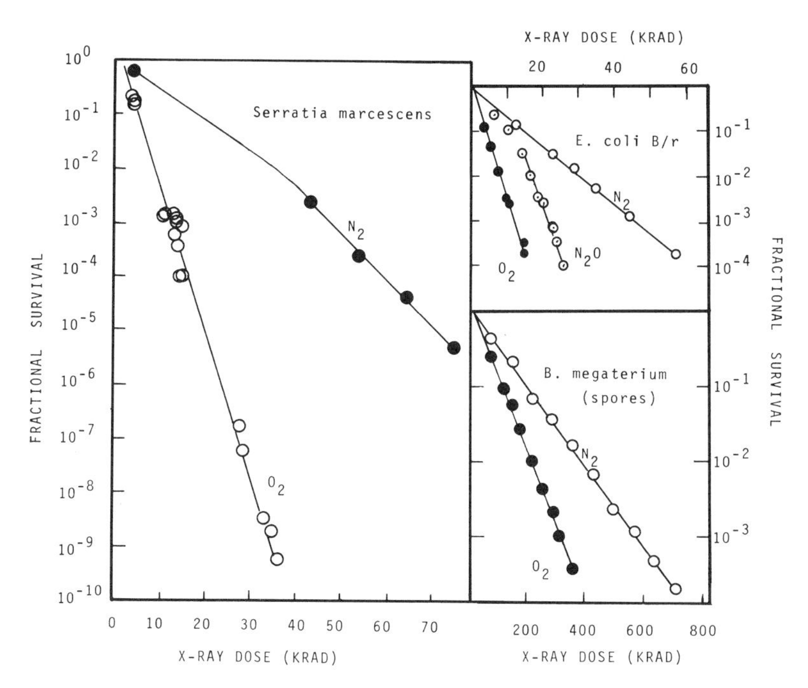


**Figure 17-2.** Survival curves for *Serratia*, *E. coli* B/r, and *B. megaterium* spores. In each panel $O_2$ means 100% $O_2$. (Data for *Serratia*[44] by permission of the author and Academic Press; data for *E. coli* B/r and *B. megaterium* spores by the author, unpublished.)

In general, eucaryotic cells are more sensitive to radiation than these procaryotic cells, and bacteriophages and viruses are less sensitive. Dugan and Trujillo[3] have noted that a relationship exists between radiation sensitivity and the nucleic acid content of different kinds of cells. The existence of such a relationship is one line of evidence that DNA is a major target for radiation-induced damage.

Figure 17-2 also illustrates another important point: when all other experimental parameters are kept constant, the gas during irradiation affects the sensitivity. We see specifically that survival curves in $O_2$ are steeper (ie. they show a more sensitive response) than those in 100% $N_2$. Abundant experimental evidence has allowed us to identify $O_2$ as a radiation sensitizer; it increases the response to ionizing radiation of almost all types of cells. (A few repair-deficient mutants exist that show virtually the same response in $N_2$ and in air but these are clearly exceptions.[4]) The oxygen enhancement ratio (OER) is frequently used to express the amount of sensitization by $O_2$. The OER is defined as the absolute value of the ratio of the inactivation rate constants, taken from the slopes of the survival curves:

$$\mathrm{OER} = \left| \frac{k_{\mathrm{air}}}{k_{\mathrm{N_2}}} \right|$$

## RADIATION SENSITIZATION, DESENSITIZATION, AND PROTECTION

Many chemical agents have been found that alter radiation sensitivity. This generalization has virtually no exceptions: Oxidizing agents are radiation sensitizers, whereas reducing agents are usually radiation protectors. We should remember, however, that two ways of decreasing the response are possible; the ability of an additive to act as a protector may be mechanistically quite different from an ability to reduce the amount of sensitization from a second additive. For example, $Ag^+$ is an excellent radiation sensitizer of bacterial spores irradiated in $N_2$.[5] *t*-Butanol, which reacts rapidly with hydroxyl radicals (·OH) can prevent this sensitization, but, in the absence of $Ag^+$, *t*-butanol itself has no effect on the spores' response in anoxia.[6,7]

One other precaution needs emphasis. We must always use great care in reaching our conclusions. If we observe a good correlation between radiation protection and additives' efficiencies at scavenging OH radicals, then we are justified in concluding that these additives protect because they scavenge ·OH. What appears to be a similar conclusion, that ·OH themselves are damaging, is not justified, and, in some instances, it may actually be wrong. For example, a reaction involving ·OH may be important only because the reaction produces the agent that actually causes damage. In that case we might observe a good correlation between protection and ·OH removal, but ·OH would actually be necessary to form the damaging product; they would not themselves cause damage. We cannot evaluate whether or not ·OH are damaging until we identify both the biological target and the specific reactions that destroy it.

## THE ANOXIC RESPONSE

Studies focusing on how various additives affect radiation sensitivity obviously need a survival curve "baseline." $N_2$ gas appears completely passive with regard to causing damage, and most biologists use the response in 100% $N_2$ as their reference. Thus, "anoxic" usually means not simply "no $O_2$," but it specifically means 100% $N_2$.

## MODIFYING THE ANOXIC RESPONSE

Sanner and Pihl[8] used *E. coli* B to judge whether or not any of the radiolytic products causes anoxic damage. They selected additives that had widely different rates of reaction with ·OH and $e_{aq}^-$ and then measured the response at a particular additive concentration. Some of these additives were excellent anoxic protectors; others had only small effects on radiation sensitivity. They presented their results (Figure 17-3) by plotting the relative response against

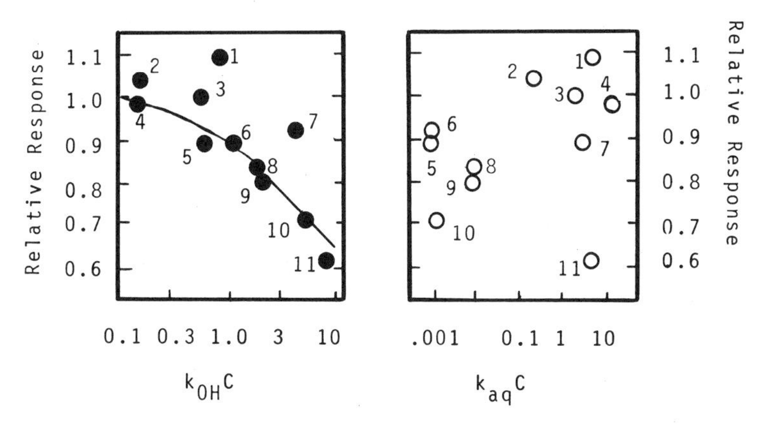


**Figure 17-3.** Ability of different compounds to protect *E. coli* B at 0°, as a function of their rate of interactions with ·OH (left panel) and with $e_{aq}^-$ (right panel). The relative sensitivity observed in the presence of the compounds is plotted vs. the product of the protector concentration and the respective second-order constants for the interaction of the different protectors with ·OH and with $e_{aq}^-$. The curve in the left panel is a theoretical curve, calculated on the assumption that radiolysis products account for 45% of the total damage. The same numbering applies to both panels: 1-cytosine; 2-glycylglycine; 3-adenine; 4-acetone; 5-thymine; 6-methanol; 7-sodium formate; 8-glucose; 9-glycerol; 10-ethanol; 11-cysteamine. (Reproduced with permission.[12])

the efficiency for scavenging either ·OH or $e_{aq}^-$. Sanner and Pihl pointed out that these compounds' protective ability increases with increasing efficiency for ·OH removal; no similar relationship occurs between response and $e_{aq}^-$ scavenging.

With complementary experiments, Sanner and Pihl found that the response at 0° was reduced about 50% when the temperature during irradiation was decreased to −15°. Further reductions in the irradiation temperature (to −196°) caused only a small reduction in sensitivity. Making the assumption that the reduction in response as the temperature decreased (0° to −15°) was based on the elimination of damage of radiolytic products Sanner and Pihl derived a hypothetical relationship between response and ·OH scavenging, shown as the solid line in the left panel of Figure 17-3. The data points show a fairly wide scatter around this theoretical, expected line. (More recent, ie, better, values of the reaction rate constants do not improve the fit.) If, however, protection must be attributed to either ·OH or $e_{aq}^-$ removal, then certainly Sanner and Pihl's conclusions are correct: The amount of protection is a function of the additives' efficiencies at scavenging ·OH. Under anoxic conditions, ·OH account for 45 to 50% of the total amount of damage in irradiated bacteria.[8]

However, it is not actually required that anoxic protection must be assigned to either ·OH or $e_{aq}^-$ removal, and a similar later study with bacterial spores reached another conclusion.[9] Several agents were tested for their effects on the anoxic sensitivity of *B. megaterium* spores (Figure 17-4, lower panel). Some of these (*t*-butanol, *t*-amyl alcohol, and benzyl alcohol) had no effect, whereas others (1-propanol and 2-propanol, in addition to those named in the legend to Figure 17-4) could reduce the response. The use of several concentrations of each additive provides enough data to judge whether or not the additives all protect to a common response level. We would expect a common maximum effect if the additives all protect through the same chemical pathways. The data may show a minimum response level for some of the compounds, but not all of them can reach it. It was observed that the additives do not protect equally well when they are used at the same efficiencies for ·OH removal, and the additives cannot all protect to the same minimum level. From this, it was concluded that ·OH scavenging is not the single process on which protection is based.

A correlation exists between protection and the ability of the additive to react with a water-derived radical to form a reducing radical ($R\dot{C}OH$, an $\alpha$-hydroxy radical). Only those additives capable of forming this radical protect.[9]

·OH competition experiments (cf. Table 17-2) between a protector (methanol) and a nonprotector (*t*-amyl alcohol) tested the validity of this correlation. The two alcohols were tested first alone and then simultaneously. The response in $N_2$ with no additive is shown. Methanol can reduce that response; *t*-amyl alcohol cannot. When 0.5 *M* methanol and 1 *M* *t*-amyl alcohol are tested together, the response is not as low as that observed when that methanol concentration is tested alone. At the two alcohol concentrations, ·OH scavenging by *t*-amyl alcohol is 4.3 times more efficient than ·OH scavenging by methanol. When the ratio of efficiencies is reversed (0.5 *M* methanol and $5.5 \times 10^{-2} M$ *t*-amyl alcohol), the response is the same as that methanol concentration tested alone. The results show that *t*-amyl

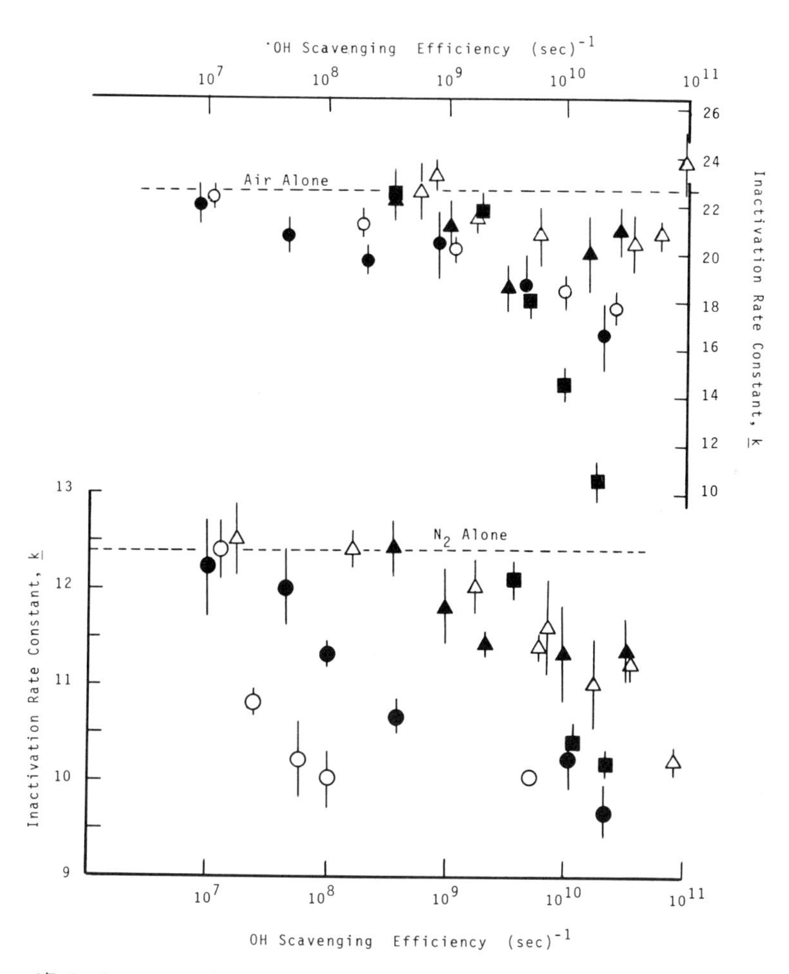


**Figure 17-4.** *Lower panel*: Anoxic protection of *B. megaterium* spores suspended in various concentrations of different additives. The response (*k*) is plotted on the ordinate against the ·OH scavenging efficiency on the abscissa. *Upper panel*: Protection of *B. megaterium* spores irradiated in air-equilibrated suspension with different concentrations of various additives present. The response (*k*) is plotted on the ordinate against the ·OH scavenging efficiency on the abscissa. The symbols are the same for both panels: methanol (●); sodium formate (○); glycerol (■); allyl alcohol (▲); and ethanol (△). Note that the ordinate scales are not the same in the upper and lower panels. The data do not show a clear relationship between response and ·OH scavenging efficiency. (Reproduced with permission.[9,15])

**Table 17-2.** An ·OH Competition Experiment with *B. megaterium* Spores

| Additive | $k$ |
|---|---|
| None | $12.4 \pm 0.3$ |
| 1 $M$ $t$-amyl alcohol | $12.7 \pm 0.3$ |
| 0.5 $M$ methanol | $10.4 \pm 0.2$ |
| 1 $M$ $t$-amyl + 0.5 $M$ methanol | $11.8 \pm 0.3$ |
| $5.5 \times 10^{-2}$ $M$ $t$-amyl alcohol | $12.6 \pm 0.1$ |
| $5.5 \times 10^{-2}$ $M$ $t$-amyl + 0.5 $M$ methanol | $10.6 \pm 0.2$ |

Reproduced with permission.[9]

alcohol, by reducing the frequency of ·OH scavenging by methanol, can partially block protection. When $t$-amyl alcohol is present at a concentration that does not affect ·OH scavenging by methanol, the full amount of protection from methanol is seen. Thus, ·OH scavenging by methanol is a necessary step for methanol to protect, but ·OH removal itself is not the single step which protects.

The results of the ·OH competition experiments support the validity of the correlation between protection and the formation of a reducing radical by the additive. If simple ·OH removal were responsible for protection, then the amount of protection when two scavengers were used together should be at least as great as, or greater than, the amount of protection from the most effective additive alone. Instead of this result, we found that one additive, itself without an effect on radiation sensitivity, could partially block protection from another.

The result with $CO_2$ gas is especially important, since it demonstrates that a removal of ·OH is not required for protection to occur. After $CO_2$ reacts with $e_{aq}^-$, the radical formed ($\cdot CO_2^-$) is a reducing radical; this could be the basis for the biological observation that $CO_2$ is a protector.

Thus, from studies with bacterial spores, it was concluded that, although ·OH scavenging may be a necessary step for protection to occur, protection itself does not rest simply on removing this radical. In at least one case protection can occur even though the levels of ·OH are undisturbed.[9]

At the present time we cannot resolve the difference in conclusions regarding the importance of ·OH in anoxic experiments with the two kinds of organisms. With bacteria about 45–50% of the total amount of damage appears to come from reactions of ·OH, whereas with bacterial spores there is no evidence that ·OH cause any damage in anoxia at all. This difference might indicate that anoxic damage in bacteria and bacterial spores occurs through different mechanistic pathways. Further study should clarify this point.

# RADIATION SENSITIZATION BY AIR

$O_2$ sensitizes virtually all organisms to ionizing radiation, and, although major studies have been directed toward understanding this phenomenon, our knowledge is not yet complete. Understanding how $O_2$ functions as a sensitizer is central not only to the question of how radiation damages cells but also to our application of this knowledge to cancer therapy.

Using very dry bacterial spores, Powers and his colleagues[10] were first to show that $O_2$ has at least two separable effects on radiation sensitivity. Later studies with cells in aqueous suspension have confirmed that observation.[11–13] $O_2$ acts in more than one way to sensitize cells.

# MODIFYING THE RESPONSE IN AIR

Johansen and Howard-Flanders[14] used *E. coli* to investigate the involvement of radiolytic products in radiation sensitization by high concentrations of $O_2$. They selected and tested a variety of compounds over a range of concentrations. For analysis they separated the protectors from the additives that had no effect on the response in air. Interestingly, they included nitric oxide (NO) gas among the protectors. Although NO has complex effects on the response, at all concentrations the sensitivity was greater in the presence of NO than in its absence. Because of this, it seems difficult to rationalize their assigning NO to the group of protectors.

Interesting also is their observation that the protectors did not all reduce the response to a common minimum level. Since this is expected to occur if all the protectors act in the same way, they took this as evidence that two protective processes were operating.

Their results are given in Figure 17-5. As they pointed out, the relationship between protector concentration (for half the maximum effect) and rate of scavenging by the two reducing radicals ($e_{aq}^-$ and $\cdot H$) is very poor indeed. On the other hand, a relationship between concentration and rate of $\cdot OH$ scavenging appears clearer. Johansen and Howard-Flanders termed the fit of these data to their hypothetical, expected line "reasonably good," and they concluded that in air $\cdot OH$ account for 40 to 60% of the total amount of damage.[14]

The results of a similar series of tests with bacterial spores[15] do not allow the same conclusion regarding the importance of OH radicals. For the data in Figure 17-4, upper panel, spores were irradiated in air with different concentrations of selected additives present. Three additives (*t*-butanol, *t*-amyl alcohol, and benzyl alcohol) did not protect over the range of concentrations tested. Other additives, as the figure shows, could reduce the response. The complexity of these data is readily apparent. The additives do not protect equally well when they are tested at the same efficiencies for

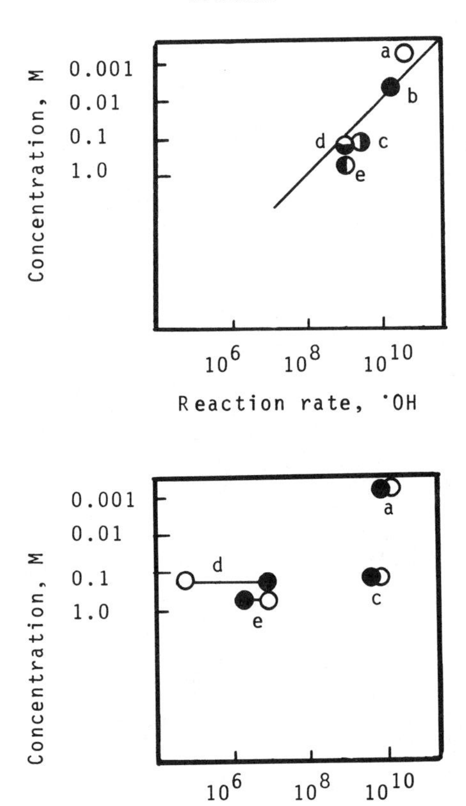


**Figure 17-5.** Protection of *E. coli* irradiated in air-equilibrated suspension. The concentration for half the maximum amount of protection is plotted on the ordinate against the reaction rates of these substances with ·OH (upper panel) or $e_{aq}^-$ and ·H (lower panel). *Upper panel symbols*: a-nitric oxide (○); b-mercaptoethanol (●); c-nitrite (◑); d-ethanol (◒); and e-methanol (◐). *Lower panel symbols*: (○)-reaction rates with $e_{aq}^-$; (●)-reaction rates with ·H. a-nitric oxide; c-nitrite; d-ethanol; e-methanol. The reaction rate for mercaptoethanol and the reducing radicals is too low to be shown in the lower panel. (Data[14] redrawn by permission of the author and Academic Press.)

·OH removal, and, in addition, the agents do not all protect to a common minimum response level. In this series of tests it seems very clear that protection is not a function of the additives' efficiencies at scavenging OH radicals.

For the anoxic protection of bacterial spores, a correlation exists between protection and an additive's ability to react with a water-derived radical to form a reducing radical. This same correlation also exists with the additives tested in air. Those additives that cannot form $\alpha$-hydroxy radicals cannot reduce the response of air-equilibrated spores. An ·OH competition experiment using a protector (methanol) and a nonprotector (*t*-amyl alcohol) gave results identical to those from the test with anoxic spores (see Table 17-2). The protector has a smaller effect if the nonprotector is present to significantly reduce the frequency of ·OH scavenging by the protector. Thus, although ·OH scavenging may be crucial, it is not the single, specific step on which protection is based.

## RADIATION SENSITIZATION BY NITROUS OXIDE

Nitrous oxide ($N_2O$) has been tested for effects on radiation sensitivity in several biological systems, although the results have not always been easy to interpret. With bacterial spores, $N_2O$ is a potent radiation sensitizer.[16] When tested with other organisms, however, $N_2O$ sometimes appears to have no effect.[17–19]

Studies have indicated that cellular repair systems may play a critical role in establishing whether or not we observe sensitization by $N_2O$. Brustad and Wohl,[20] using repair-deficient mutants of *E. coli*, and Mitchel,[21] with similar mutants of the yeast *Saccharomyces*, both noted that sensitization by $N_2O$ is observed only in strains competent for recombination repair. Strains deficient in recombination repair showed the same survival curves for $N_2$ and $N_2O$. (Recombination repair occurs in procaryotic cells as DNA is being synthesized. A damaged region in one strand of parental DNA causes a gap in the complementary daughter strand being replicated. Repair occurs when this gap is replaced by the identical region on the second parental strand. The gap, now in the parental complementary strand, can be filled by DNA synthesis because the correct sequence of bases can be read from the complementary, newly synthesized daughter strand.[22])

If it seems peculiar that we might observe sensitization by $N_2O$ only in organisms *capable* of repairing DNA damage, remember we never judge sensitization by the steepness of a single survival curve. Instead, we always compare that curve with the position of the reference survival curve in $N_2$. If that reference curve should be "too steep" because of damage that cannot be repaired, then we no longer have a proper reference for making a judgment about sensitization. Although the survival curves for $N_2$ and for $N_2O$ may be identical, this does not necessarily mean that $N_2O$ does not cause damage.

## MODIFYING THE RESPONSE IN NITROUS OXIDE

In their studies with bacterial spores, Powers and his colleagues noted that sensitization by $N_2O$ requires both ·OH and radiolytic $H_2O_2$.[16,23] From their results it is clear that both products must be present simultaneously for this kind of damage to occur.

In two closely related studies almost 10 years later, Watanabe and his colleagues found that the sensitization of the bacterium *Pseudomonas* by $N_2O$ required the same two radiolytic products.[24,25] Some of their studies showed that sensitization by $N_2O$ could be prevented by lowered dose rates and/or by high cell concentrations. In their dose-rate study they attributed this protection in $N_2O$ to an insufficient production of $H_2O_2$ at low dose rates. In their cell concentration studies they proposed that the high levels of catalase decomposed radiolytic $H_2O_2$ before it could be effective. Although they presented evidence to support this hypothesis, including a study of radiation-dependent leakage of intracellular catalase into the suspending fluid, another explanation is equally valid. Their observation that high concentrations of cells could prevent the sensitization from $N_2O$ could be taken as evidence that the origin of this kind of damage is extracellular. If an organism's survival during irradiation depended entirely on intracellular events, then the response of an individual cell should not depend on the number of its neighboring cells. Several studies have now shown that this is not always true; in some instances the cell concentration will influence the result. A number of recent studies have used cell concentration effects to distinguish between intracellular and extracellular damaging processes.[26–28]

Results of parallel tests[29] with *B. megaterium* and *E. coli* B/r are given in Table 17-3. The responses in $N_2$, air, a low concentration of $O_2$, and $N_2O$ are measured with and without various additives. The amount of sensitization by oxygen (in air) is different for the two organisms, and the OER values are 1.7 and 3.7, respectively. However, the amount of sensitization by $N_2O$ is about 60% of either organism's OER.

Data in Table 17-3 show that sensitization by $N_2O$ can be prevented by *t*-amyl alcohol, catalase, or superoxide dismutase (SOD). These three protect equally well in $N_2O$, and a test with all three present gives no more protection than any one alone. At the concentrations tested, none of the three has an effect on the sensitivity in $N_2$ or in air. Heat-activated forms of the enzymes have no effect, and further work[30] has attributed the protection from the alcohol to $\cdot OH$ removal.

Overall, the results in Table 17-3 indicate that sensitization by $N_2O$ requires three radiolytic products: $\cdot OH$, $H_2O_2$, and $\cdot O_2^-$ (or $\cdot HO_2$). Other tests showed very clearly that all three products must be present simultaneously; for example, added $H_2O_2$ will not sensitize (in $N_2$ or in $N_2O$) in the presence of a sufficiently high concentration of an $\cdot OH$ scavenger or in the presence of added SOD.[31]

The data in Table 17-3 also show identical patterns for protection when the equilibrating gas is 0.8% $O_2$ (with 99.2% $N_2$). This similarity is evidence that a single chemical pathway is responsible for at least one kind of damage in $N_2O$ or in this low concentration of $O_2$. This mechanism is discussed in the following sections.

**Table 17-3.** Parallel Experiments with Bacteria and Bacterial Spores[a]

| Gas | Additive | Radiation sensitivity, $k$[b] | |
|---|---|---|---|
| | | *Bacillus megaterium* | *E. coli* B/r |
| $N_2$ | None | 11.0 ± 0.4 | 9.4 ± 0.3 |
| | $10^{-3}$ *M t*-Amyl alcohol | 11.3 ± 0.5 | 9.2 ± 0.6 |
| | 10 μg CAT/mL | 10.7 ± 0.3 | 10.0 ± 0.7 |
| | 10 μG H-I CAT/mL | 11.0 ± 0.6 | 9.6 ± 0.4 |
| | 15 μg SOD/mL | 10.9 ± 0.2 | 9.8 ± 0.4 |
| | 15 μg H-I SOD/mL | 11.4 ± 0.6 | 8.9 ± 0.5 |
| | $10^{-3}$ *M* Histidine | 11.7 ± 0.8 | 9.2 ± 0.4 |
| AIR | None | 20.2 ± 0.6 | 34.9 ± 0.9 |
| | $10^{-3}$ *M t*-Amyl alcohol | 19.8 ± 0.4 | 33.6 ± 0.7 |
| | 10 μg CAT/mL | 21.8 ± 1.0 | 36.0 ± 1.2 |
| | 10 μg H-I CAT/mL | 19.6 ± 0.8 | 35.1 ± 0.8 |
| | 15 μg SOD/mL | 20.0 ± 0.4 | 33.7 ± 0.6 |
| | 15 μg H-I SOD/mL | 19.6 ± 0.4 | 34.2 ± 0.3 |
| | $10^{-3}$ *M* Histidine | 20.7 ± 0.6 | 35.3 ± 0.5 |
| 0.8% $O_2$ | None | 18.5 ± 0.6 | 17.6 ± 0.5 |
| | $10^{-3}$ *M t*-Amyl alcohol | 13.8 ± 0.4 | 9.3 ± 0.7 |
| | 10 μg CAT/mL | 13.2 ± 0.5 | 9.0 ± 0.5 |
| | 10 μg H-I CAT/mL | 18.7 ± 0.3 | 18.7 ± 0.2 |
| | 15 μg SOD/mL | 12.9 ± 0.6 | 9.7 ± 0.4 |
| | 15 μg H-I SOD/mL | 19.0 ± 0.7 | 18.2 ± 0.9 |
| | $10^{-3}$ *M* Histidine | 13.0 ± 0.5 | 9.0 ± 0.1 |
| | CAT + SOD + Histidine[c] | 12.6 ± 1.3 | 8.9 ± 0.5 |
| $N_2O$ | None | 16.7 ± 0.4 | 24.8 ± 0.6 |
| | $10^{-3}$ *M t*-Amyl alcohol | 13.2 ± 0.6 | 9.2 ± 0.3 |
| | 10 μg CAT/mL | 13.5 ± 0.5 | 8.8 ± 0.5 |
| | 10 μg H-I CAT/mL | 17.2 ± 0.7 | 25.0 ± 0.8 |
| | 15 μg SOD/mL | 13.1 ± 0.2 | 9.4 ± 0.7 |
| | 15 μg H-I SOD/mL | 17.0 ± 0.3 | 23.7 ± 1.3 |
| | $10^{-3}$ *M* Histidine | 13.7 ± 0.8 | 9.7 ± 0.8 |
| | CAT + SOD + Histidine[c] | 13.0 ± 0.4 | 9.1 ± 0.1 |

[a] Reproduced with permission.[29]
[b] Values of $k$ for spores were multiplied by $10^4$ Gy; values of $k$ for the bacteria were multiplied by $10^3$ Gy.
[c] For these experiments, CAT was at 10 μg/mL, SOD at 15 μg/mL, and histidine at $10^{-3}$ *M*.

# RADIATION SENSITIZATION BY LOW CONCENTRATIONS OF OXYGEN

The mechanisms for damage in low concentrations of $O_2$ are different from those at high concentrations.[12,31] Figure 17-6 illustrates this point. The left panel shows the response of *E. coli* B/r tested in $N_2$ or in air. The presence

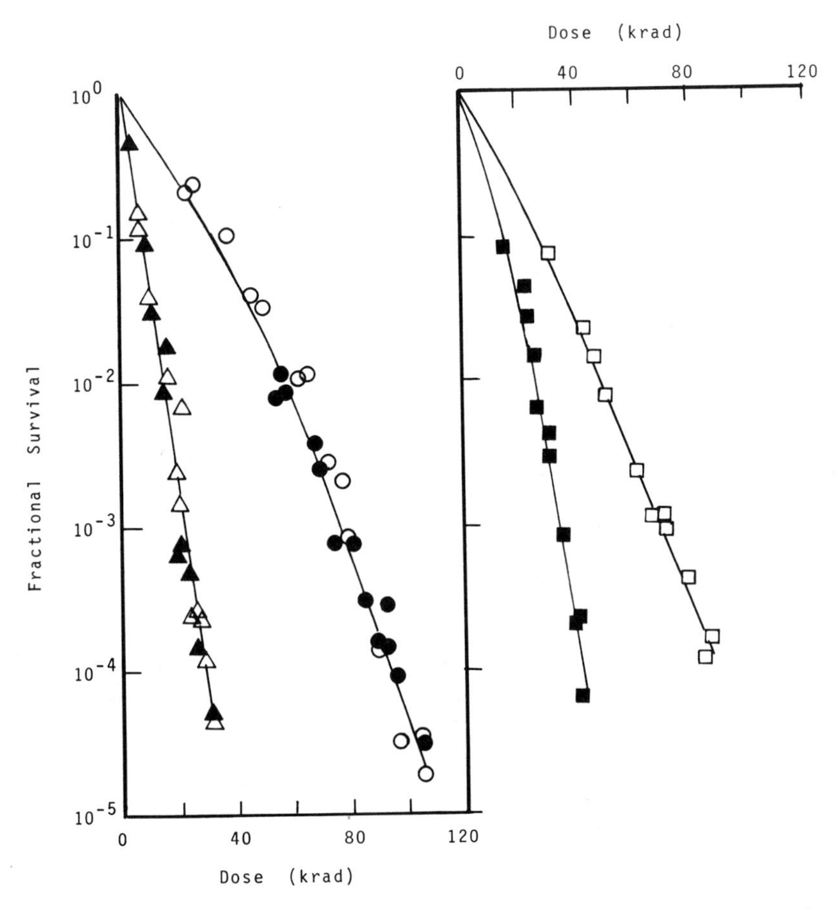


**Figure 17-6.** Survival of *E. coli* B/r. The left panel shows survival curves in air and in 100% $N_2$, with (○) and without (●) $10^{-2}$ *M t*-butanol. The presence of the alcohol has no effect on the response. The right panel shows the response in 1.3% $O_2$, with (□) and without (■) this same *t*-butanol concentration. There is considerable protection from the alcohol in this low concentration of $O_2$. (Reproduced with permission.[31])

of $10^{-2}M$ *t*-butanol has no effect on either of these responses. In contrast (right panel) when $O_2$ is present at about $10^{-5}$ *M* (about 1% $O_2$ with 99% $N_2$), the same alcohol concentration now has a large protective effect.

## MODIFYING THE RESPONSE IN LOW CONCENTRATIONS OF OXYGEN

Tests[30–32] for an involvement of ·OH in damage from low concentrations of $O_2$ are shown in Figure 17-7. In these the $O_2$ concentration was held

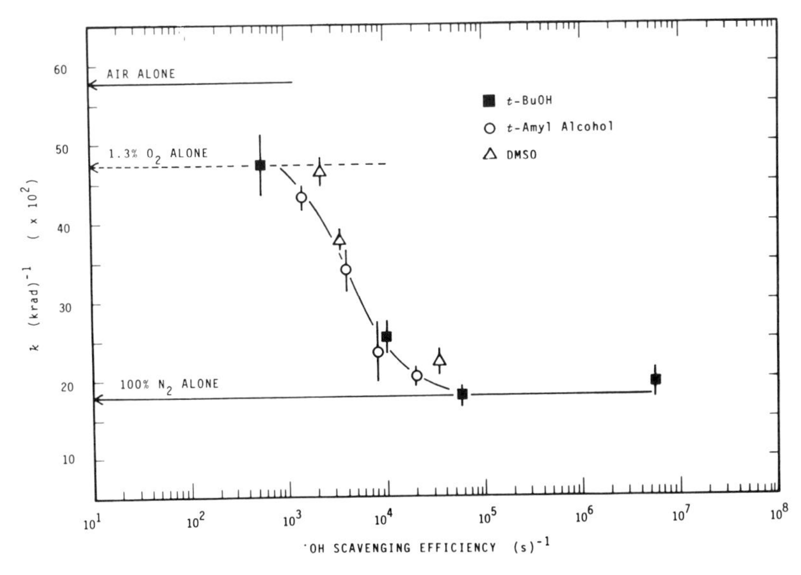


**Figure 17-7.** Tests to determine whether or not there is a relationship between response (on the ordinate) and ·OH scavenging efficiency (on the abscissa). The upper horizontal line shows the response level in 1.3% $O_2$ with no additive present. The lower horizontal line shows the response in 100% $N_2$ with no additive present. The additives were dimethyl sulfoxide (△), *t*-butanol (■), and *t*-amyl alcohol (○). The data show there is a good relationship between response and ·OH scavenging efficiency. (Reproduced with permission.[31])

constant, and the response was measured when different concentrations of several ·OH scavengers were present. The data show a good correlation between response and the additives' efficiencies at ·OH removal. From these data in low concentrations of $O_2$, we conclude that ·OH scavenging is the basis for the protection by *t*-butanol.

Bacterial spores[33] and bacteria[32] have been used to measure the ·OH scavenging efficiency needed to protect at different $O_2$ concentrations. These tests were run to ensure that $O_2$ does not somehow affect the production or equilibrium concentration of OH radicals during irradiation. The results of these tests, illustrated with *E. coli* in Figure 17-8, show clearly that the ·OH scavenging efficiency needed to protect is independent of $O_2$ concentration and independent also of the specific magnitude of the damage from ·OH.

Figure 17-9 compares results from similar experiments with bacteria and bacterial spores. The response was measured over a wide range of $O_2$ concentrations with and without *t*-butanol. The two upper panels show that sensitization of both kinds of cells begins as the dissolved $O_2$ concentration exceeds about $10^{-6}$ *M*. $O_2$ reaches its maximum effectiveness at about $10^{-4}$ *M*, and higher concentrations, up to about $1.3 \times 10^{-3}$ *M* (ie, 100% $O_2$ gas for equilibration), do not increase the response further. The results with spores

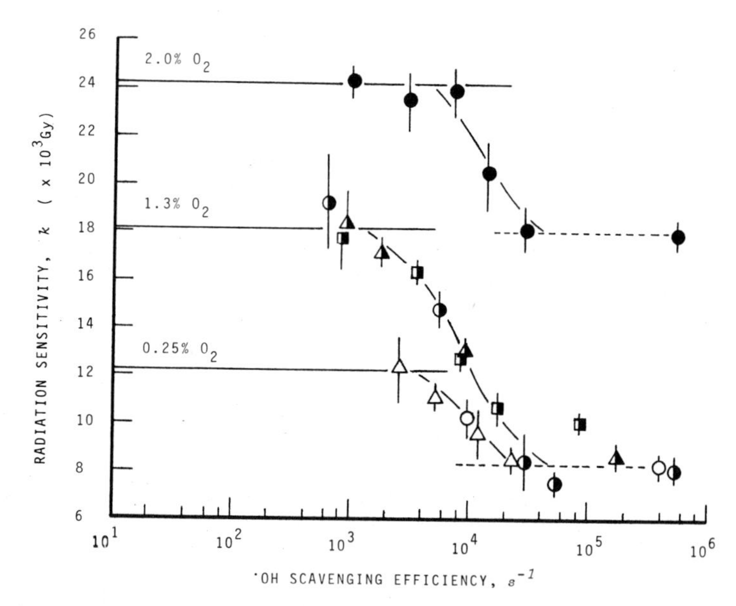


**Figure 17-8.** The radiation sensitivity of *E. coli* B/r in low concentrations of $O_2$ with and without additives present. The figure plots the response on the ordinate against the ·OH scavenging efficiency on the abscissa. For these data, *t*-butanol was used in 1.3% $O_2$ (▲) and in 0.25% $O_2$ (△); *t*-amyl alcohol was used in 2% $O_2$ (●), in 1.3% $O_2$ (◑), and in 0.25% $O_2$ (○); dimethyl sulfoxide was used in 1.3% $O_2$ (◨). The results show that the ·OH scavenging efficiency needed to protect is independent of $O_2$ concentration and independent of the magnitude of the damage from OH radicals. (Reproduced with permission.[32])

(left panel) show very clearly that at least three kinds of oxygen-dependent damage exist. With *E. coli* only two are apparent with these experiments. One kind of damage involving ·OH occurs in both organisms. (Experiments with spores,[30] like those with *E. coli* B/r in Figure 17-7, have established ·OH removal as the basis for protection by the alcohols in low concentrations of $O_2$.) An ·OH scavenging efficiency of about $10^4\ s^{-1}$ gives one-half the maximum amount of protection in either organism.

The two bottom panels in Figure 17-9 show how this kind of oxygen-dependent damage varies in magnitude as the dissolved oxygen concentration changes. (The points in the lower panels were obtained by subtracting the responses at a given $O_2$ concentration; ie, $k - k_{t\text{-BuOH}\cdot}$)

The data in Figures 17-7 and 17-8 indicate that scavenging OH radicals will protect in low concentrations of $O_2$ but will not protect in air or in $N_2$. (Note that the data do not allow the conclusion that ·OH are damaging in low concentrations of $O_2$ but not in air or in $N_2$.)

These data should be compared with the earlier data by Sanner and Pihl.[8] Those authors concluded that OH radicals cause about 45% of the damage when bacteria are irradiated in $N_2$. In that study Sanner and Pihl found that very high additive concentrations were required for an effect, and they

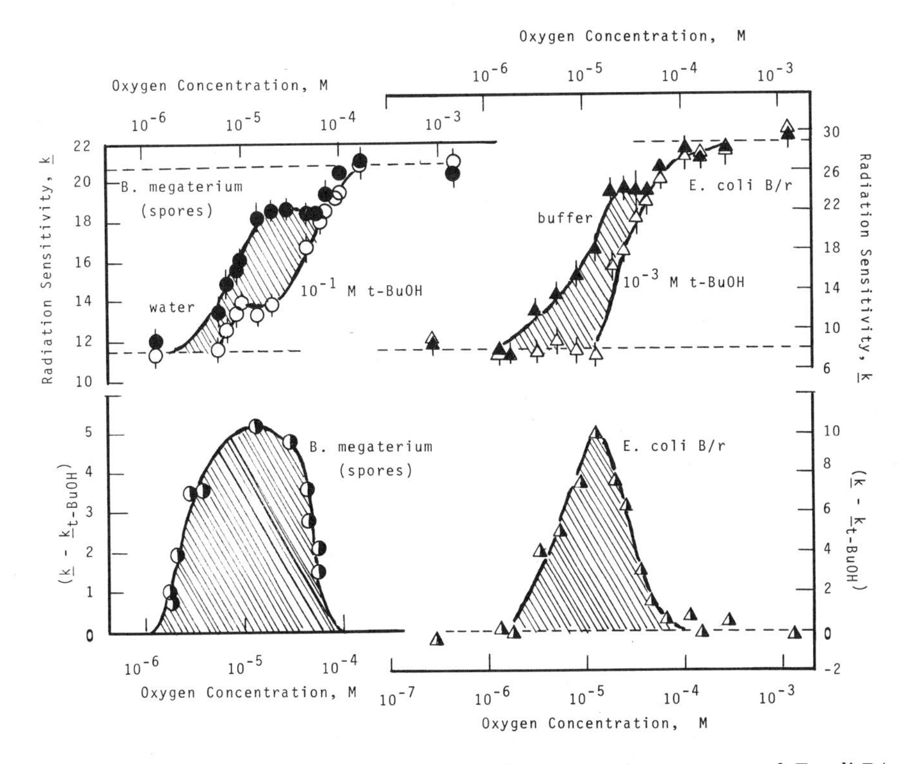


**Figure 17-9.** A comparison of the responses of *B. megaterium* spores and *E. coli* B/r over a wide range of $O_2$ concentrations with and without *t*-butanol. The response (*k*) is plotted on the ordinate against $O_2$ concentration of the abscissa in the two upper panels. Values of *k* for spores were multiplied by $10^3$ krad for entry on the figure; for *E. coli* the values of *k* were multiplied by $10^2$ krad. The points in the two lower panels were obtained by subtracting the values of *k* at a particular $O_2$ concentration. The same two multiplying factors were used for spores and bacteria as noted above. (Reproduced with permission.[32,33])

saw protection with ·OH scavenging efficiencies of about $10^8$ $s^{-1}$. In contrast, we found, using *E. coli* B/r,[32] (Figure 17-8), *B. megaterium* spores,[33] and the bacterium *Serratia marcescens*,[34] an ·OH scavenging efficiency for protection in about 1% $O_2$ of about $10^4$ $s^{-1}$. It is unclear why ·OH removal must be $10^4$ times more efficient to protect in $N_2$ than in 1% $O_2$. We cannot explain this difference by proposing two cellular targets that are destroyed by ·OH at two widely different efficiencies. This proposal is reasonable, but, if it were correct, we would observe that protection occurred in stages as the ·OH scavenging efficiency is raised; the data do not show this.

Tests with different concentrations of spores[35] have revealed another important characteristic of this ·OH-dependent damage. When the dissolved $O_2$ concentration is held constant at about $10^{-5}$ *M*, protection is observed simply by using high concentrations of cells. It was found that protective effects of *t*-butanol and high cell concentrations are additive, although the maximum amount of protection from both "treatments" is not greater than either could give alone.

It had been previously concluded that protection in 1% $O_2$ by *t*-butanol is based on $\cdot OH$ removal (cf. Figures 17-7 and 17-8), and it had also been noted that protection by high concentrations of cells is based on blocking extracellular reactions.[28] These two observations taken together indicate that the damage from OH radicals occurring over a restricted range of $O_2$ concentrations originates outside the spore. (Comparable experiments with bacteria have not yet been performed.)

The results[28] of tests with mixtures of live and dead spores support this conclusion. In $10^{-5}$ *M* $O_2$ we found the same amount of protection by raising the spore concentration to $10^{10}$/mL as by using $10^7$/mL in a suspension of $10^{10}$ dead spores/mL. In these experiments the dead spores were used as $\cdot OH$ scavengers, which obviously cannot have access to the interior of the living cells whose response is being measured.

In Table 17-3 we summarized the results of tests that showed how various additives affect the responses in $N_2O$ and in 1% $O_2$. These tests used both bacteria and bacterial spores. It was found that sensitization from either of these gases could be blocked by $\cdot OH$ removal, catalase, or SOD. These tests also showed that the heat-inactivated enzymes had no effect on the response, and all three additives used together gave no more protection than one alone. It was therefore concluded that $\cdot OH$, $H_2O_2$, and $\cdot O_2^-$ (and/or $\cdot HO_2^-$) all play an essential role in sensitization of both organisms by $N_2O$ and 1% $O_2$. One can prevent a single kind of damage from occurring by removing any one or any combination of the three radiolytic products.

There are two obvious ways in which the three radiolytic products might be related:

$$H_2O_2 + \cdot O_2^- \rightarrow O_2 + OH^- + \cdot OH \tag{17-2}$$

and/or

$$H_2O_2 + \cdot OH \rightarrow H^+ + H_2O + \cdot O_2^- \tag{17-3}$$

The first, usually called the Haber–Weiss reaction, is extremely slow,[36] although catalysis by iron salts[37] within the cell could greatly accelerate the reaction. If Reaction (17-2) is relevant to radiation biology, then $\cdot OH$ would probably be the agent that actually causes damage, and the two reactants would best be thought of as a means of generating additional $\cdot OH$. Reaction (17-3) is the relatively rapid ($k = 3 \times 10^7\ M^{-1}\ s^{-1}$)[38] decomposition of $H_2O_2$ by reaction with $\cdot OH$. If this reaction is relevant, $\cdot O_2^-$ (and/or $\cdot HO_2$) would probably be the agent that actually causes damage. Our observations that, in $N_2O$ and low concentrations of $O_2$, protection can be achieved by catalase, SOD, or an $\cdot OH$ scavenger would apply equally well to either equation.

Experiments[39] have been run that allow a choice between the two alternatives. $H_2O_2$ is a potent sensitizer of bacterial spores under anoxic conditions.[23,40] Over a range of $H_2O_2$ concentrations, added $\cdot OH$ scavengers can prevent damage from $H_2O_2$. In addition, anoxic sensitization by reagent $H_2O_2$ can be prevented by adding either catalase or SOD. The result with SOD is especially important, because, under anoxic conditions, the radiolytic production of $\cdot O_2^-$ is negligible. Thus, we add $H_2O_2$, but we can protect by enzymatically degrading $\cdot O_2^-$. The most likely source of $\cdot O_2^-$ is the $H_2O_2$ that we added.

Data in Table 17-4 show that the $\cdot OH$ scavenging efficiency needed to block sensitization by reagent $H_2O_2$ is linearly related to the concentration

**Table 17-4.** Anoxic Tests with *B. megaterium* Spores[a]

| Initial $H_2O_2$ concentration ($M$) | Ratios | $\cdot OH$ scavenging efficiency for 50% protection ($s^{-1}$) | Ratios |
|---|---|---|---|
| $2 \times 10^{-4}$ | 1.0 | $2.6 \times 10^4$ | 1.0 |
| $6 \times 10^{-4}$ | 3.0 | $8.0 \times 10^4$ | 3.1 |
| $9.7 \times 10^{-4}$ | 4.9 | $1.3 \times 10^5$ | 5.1 |

[a] Reproduced with permission.[39]

of $H_2O_2$ added. The concentration of reagent $H_2O_2$ is held constant and the $\cdot OH$ scavenging efficiency required to block half the amount of sensitization by that $H_2O_2$ concentration is measured. Three concentrations of $H_2O_2$ were used. The ratios of $H_2O_2$ concentrations and $\cdot OH$ scavenging efficiencies are the same. The results show that $H_2O_2$ and $\cdot OH$ must react with each other for anoxic sensitization by $H_2O_2$ to occur. If we double the concentration of reagent $H_2O_2$, we also double the $\cdot OH$ scavenging efficiency needed to prevent this kind of damage. This result is good evidence that $H_2O_2$ and $\cdot OH$ are reactants. This information, taken with the observation that SOD can block sensitization by reagent $H_2O_2$, allows the conclusion that, in some instances, biological damage requires a reaction between $H_2O_2$ and $\cdot OH$ to form $\cdot O_2^-$ (or $\cdot HO_2$). The experiments do not actually identify $\cdot O_2^-$ as the agent that causes damage.

The results from the anoxic experiments with $H_2O_2$ allow us to conclude that $\cdot O_2^-$ can play a critical role in radiation-induced damage. It is conceivable that those conclusions reached with reagent $H_2O_2$ might not apply to situations in which $H_2O_2$, $\cdot OH$, and $\cdot O_2^-$ are present as radiolytic products. However, no evidence suggests that such an application would be incorrect. We believe that a hypothesis will ultimately be developed that relates damage in $N_2$, air, low concentrations of $O_2$, and $N_2O$ to the formation and decomposition of those three radiolytic products.

At this time the studies by Sanner and Pihl[8] and Johansen and Howard-Flanders[14] appear to stand apart from the studies with *E. coli* B/r and bacterial spores discussed in this chapter. Those studies assigned a major damaging role to $\cdot OH$ when bacteria are irradiated in $N_2$ or in air. Until those experiments are repeated with the same organism and then run with another organism, we will not know how to incorporate that information with the hypothesis presented below.

## OVERVIEW: RADIOLYSIS PRODUCTS AND CELL DEATH

*Under anoxic conditions* neither the presence of added catalase[23] nor SOD,[29,39] nor the process of ·OH removal[9] will protect bacterial spores when they are irradiated. (Sanner and Pihl, working with *E. coli* B, concluded that ·OH removal did protect these cells, although relatively high scavenger concentrations were required.) Some, but definitely not all, ·OH scavengers can protect spores irradiated under anoxic[9] or well-oxygenated[15] conditions. This protection, however, is not based on simple ·OH removal.

The oxygen-independent formation of $H_2O_2$,

$$\cdot OH + \cdot OH \rightarrow H_2O_2 \tag{17-4}$$

involves a rapid ($k = 5.5 \times 10^9\ M^{-1}\ s^{-1}$)[38] but inefficient reaction between OH radicals. Under the experimental conditions used for most of the work discussed here, the maximum amount of $H_2O_2$ formed during anoxic irradiation is about $10^{-5}\ M$.[4] Because the enzymatic elimination of this $H_2O_2$ does not protect in anoxia, we can conclude that this is not enough $H_2O_2$ to be important. Under other conditions in which larger amounts of $H_2O_2$ might be formed, catalase should be able to protect.

*When low concentrations of $O_2$ are present,* the formation of $H_2O_2$ can proceed through two additional pathways:

Pathway I
$$O_2 + e_{aq}^- \rightarrow \cdot O_2^- \tag{17-5}$$
$$\cdot O_2^- + H^+ \rightleftharpoons \cdot HO_2 \tag{17-6}$$
$$\cdot HO_2 + \cdot HO_2 \rightarrow H_2O_2 + O_2 \tag{17-7}$$

or

$$\cdot HO_2 + \cdot O_2^- \rightarrow HO_2^- + O_2 \tag{17-8}$$

Pathway II
$$O_2 + \cdot H \rightarrow \cdot HO_2 \tag{17-9}$$

Reaction (17-9) would then be followed by Reactions (17-7) or (17-8). In addition, Reaction (17-5) would also block the decomposition of $H_2O_2$ by reaction with $e_{aq}^-$ and, therefore, the concentration of $H_2O_2$ will increase.

The observations with bacteria[31,32] and bacterial spores[33] are that the ·OH scavenging efficiency needed to protect in low concentrations of $O_2$ is constant at about $10^4\ s^{-1}$. This probably means that ·OH removal protects by blocking the oxygen-independent formation of $H_2O_2$ in Eq. (17-4).

Although the maximum concentration of $H_2O_2$ formed in $N_2O$-saturated water is about $2 \times 10^{-5}$ $M$,[41] about twice the level found in anoxia, the kinetics for damage and for protection in $N_2O$ appear the same as those in low concentrations of $O_2$. The presence of catalase will enzymatically decompose the extracellular $H_2O_2$ that is formed before it can react with $\cdot OH$.* Scavengers of $\cdot OH$ will protect in $N_2O$ and in low concentrations of $O_2$ by removing $\cdot OH$ formed either inside or outside the cell. This $\cdot OH$ removal will block the formation of $H_2O_2$ [Eq. (17-4)] or block the reaction between $H_2O_2$ and $\cdot OH$ [Eq. (17-3)].

*In air* the maximum concentration of radiolytic $H_2O_2$ is about $2 \times 10^{-4}$ $M$;[41] OH radicals are present; and the levels of $\cdot O_2^-$ (and $\cdot HO_2$) are high [reaction (17-3)], and the pathways that form $\cdot O_2^-$ directly from dissolved $O_2$, Eqs. (17-5–17-9) are available.

The relevant biological observations with *E. coli* B/r and bacterial spores are that added catalase and SOD will not protect, and neither will the process of $\cdot OH$ removal, at least at efficiencies that protect in low concentrations of $O_2$.[29] Johansen and Howard-Flanders[14] concluded that $\cdot OH$ removal in air provides large amounts of protection, although high scavenger concentrations are required.

We cannot explain these contradictory conclusions. It is possible that we have not always run the experiments we planned, and it is also possible that we have not always understood what our results mean. We must also be aware that different kinds of cells will have different capabilities for repairing radiation-induced damage. If we observe that added SOD has no effect on radiation sensitivity, we might conclude that $\cdot O_2^-$ does not react to damage cells. This conclusion is reasonable, but it is not necessarily correct. It could be that $\cdot O_2^-$ damages cells through many different reactions, but the cells we used were able to recognize and repair all this damage. Our conclusion that $\cdot O_2^-$ does not damage cells might be true in one sense, but literally it is incorrect, and the conclusion will mislead us as we try to understand how radiation damages cells.

## DAMAGE FROM $\cdot O_2^-$

A number of studies have tested for an involvement of superoxide anion radicals in damage, but all of these have not indicated that $\cdot O_2^-$ (and/or $\cdot HO_2$) is important in causing damage. Many researchers believe that SOD must protect cells irradiated in air if $\cdot O_2^-$ is important. This is not the case at all, as the data in Table 17-3 show. Superoxide dismutase does not protect those two organisms when they are irradiated in air, but we believe the

* Catalase and SOD are very large molecules, and the penetration of these into the spore interior is unlikely.[42,43] Thus protective effects by these added enzymes are most probably through extracellular degradation of the particular substrate.

evidence is convincing that $\cdot O_2^-$ (and/or $\cdot HO_2$) can play a major role in causing damage under some experimental conditions. We must also realize that if exogenous SOD cannot enter the cell, then we cannot use SOD to study intracellular damage. Clearly, we must use great caution in reaching our conclusions and in testing them from as many different ways as practicable.

Recent work by van Hemmen and Meuling[45] concerns the cellular target for damage from superoxide and hydroperoxy radicals. In their experiments they altered the pH to shift the equilibrium

$$H^+ + \cdot O_2^- \rightleftharpoons \cdot HO_2 \qquad pK_a = 4.9^{46} \tag{17-10}$$

and measured the amount of DNA damage as the pH changed. They found that $\cdot HO_2$ and $H_2O_2$ act together to damage DNA, but they found no evidence that DNA damage involves $\cdot O_2^-$. These results can be explained by the hypothesis that $H_2O_2$ functions as a source for $\cdot O_2^-$.

The concept that the cell membrane may be a likely target for damage from $\cdot O_2^-$ has developed over the last decade. Kellogg and Fridovich,[47] in an early study, followed the peroxidation of linolenic acid by $\cdot O_2^-$ (produced enzymatically by the action of xanthine oxidase on acetaldehyde). They found protection under aerobic conditions from either SOD or catalase, and they concluded that $\cdot O_2^-$ and $H_2O_2$ were both essential for damage to occur. Other evidence that membranes may be a target for superoxide radical reactions has been discussed in detail in several recent sources.[48–50]

## SUMMARY

The reactions in Figure 17-10 summarize the present position on how radiolytic products are involved in cellular damage from ionizing radiation.

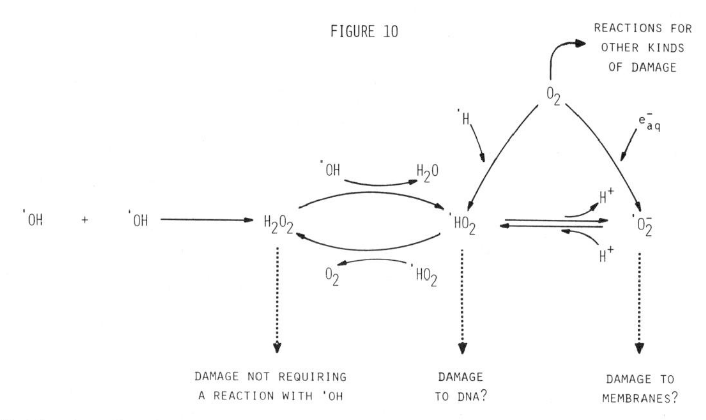


**Figure 17-10.** A collection of reactions involving $O_2$ and various radiolytic products. (Reproduced with permission.[39])

Certainly, the evidence for the importance of some of these steps is stronger than for others. The overall scheme, however, is reasonable and valid from the view of both radiation chemistry and radiation biology. There are many ways to test the relevance of the relationships in Figure 17-10, and we will undoubtedly discover others as we proceed. Studies since 1895 have told us that the answers we seek will be complex, but the value of those answers justifies our work.

# REFERENCES


1. Grubbé, E. H. *Radiology* **1933**, *21*, 156.
2. Puck, T. T.; Marcus, P. I. *J. Exp. Med.* **1956**, *103*, 653.
3. Dugan, V.; Trujillo, R. *J. Theor. Biol.* **1974**, *44*, 397.
4. Alper, T. *Nature* **1968**, *217*, 862.
5. Richmond, R. C.; Powers, E. L. *Radiat. Res.* **1974**, *58*, 471.
6. Cross, M.; Simic, M.; Powers, E. L. *Int. J. Radiat. Biol.* **1973**, *24*, 207.
7. Ewing, D. *Int. J. Radiat. Biol.* **1975**, *28*, 165.
8. Sanner, T.; Pihl, A. *Radiat. Res.* **1969**, *37*, 216.
9. Ewing, D. *Radiat. Res.* **1976**, *68*, 459.
10. Powers, E. L.; Webb, R. B.; Ehret, C. F. *Radiat. Res. (Suppl. II)* **1960**, 94.
11. Tallentire, A.; Jones, A. B.; Jacobs, G. P. *Isr. J. Chem.* **1972**, *10*, 1185.
12. Ewing, D.; Powers, E. L. *Science* **1976**, *194*, 1049.
13. Michael, B. D.; Harrop, H. A.; Maughan, R. L.; Patel, K. B. *Brit. J. Cancer* **1978**, *37 (Suppl. III)*, 29.
14. Johansen, I.; Howard-Flanders, P. *Radiat. Res.* **1965**, *24*, 184.
15. Ewing, D. *Int. J. Radiat. Biol.* **1976**, *30*, 419.
16. Powers, E. L.; Cross, M. *Int. J. Radiat. Biol.* **1970**, *17*, 501.
17. Mullenger, L.; Singh, B. B.; Ormerod, M. G. *Nature* **1967**, *216*, 372.
18. Redpath, J. L. *Radiat. Res.* **1973**, *55*, 109.
19. Tilby, M. J.; Loverock, P. S.; Fielden, E. M. *Radiat. Res.* **1982**, *89*, 488.
20. Brustad, T.; Wohl, E. *Radiat. Res.* **1976**, *66*, 215.
21. Mitchel, R. E. J.; Morrison, D. P.; Unrau, P. *Radiat. Res.* **1982**, *89*, 481.
22. Kornberg, A. "DNA Replication"; W. H. Freeman: San Francisco, 1974.
23. Cross, M.; Simic, M.; Powers, E. L. *Int. J. Radiat. Biol.* **1973**, *24*, 207.
24. Watanabe, H.; Ilzuka, H.; Takehisa, M. *Radiat. Res.* **1981**, *88*, 577.
25. Watanabe, H.; Ilzuka, H.; Takehisa, M. *Radiat. Res.* **1982**, *89*, 325.
26. Agnew, D. A.; Skarsgard, L. D. *Radiat. Res.* **1974**, *57*, 246.
27. Misra, H. P.; Fridovich, I. *Arch. Biochem. Biophys.* **1976**, *176*, 577.
28. Ewing, D. *Int. J. Radiat. Biol.* **1982**, *41*, 197.
29. Ewing, D. *Int. J. Radiat. Biol.* **1983**, *43*, 565.
30. Ewing, D. *Radiat. Res.* **1978**, *73*, 121.
31. Ewing, D. *Int. J. Radiat. Biol.* **1982**, *41*, 203.
32. Ewing, D. *Int. J. Radiat. Biol.* **1982**, *42*, 191.
33. Ewing, D. *J. Radiat. Res.* **1980**, *21*, 288.
34. Ewing, D. In "Oxygen and Oxy-Radicals in Chemistry and Biology"; Rodgers, M. A. J.; Powers, E. L., Eds.; Academic Press: New York, 1981, pp. 269–275.
35. Ewing, D. *Radiat. Res.* **1980**, *83*, 374.
36. Willson, R. L. In "Oxygen Free Radicals and Tissue Damage"; The Ciba Foundation, Excerpta Medica: Amsterdam, 1979, pp. 19–42.
37. Cohen, G. In "Superoxide and Superoxide Dismutases"; Michelson, A. M.; McCord, J. M.; Fridovich, I., Eds., Academic Press: New York, 1977, pp. 317–321.
38. Dorfman, L. M.; Adams, G. E. "Reactivity of the Hydroxyl Radical in Aqueous Solutions"; U.S. Department of Commerce, National Bureau of Standards: Washington, D.C., 1973.
39. Ewing, D. *Radiat. Res.* **1983**, *94*, 171.

40. Ewing, D. *Radiat. Res.* **1982**, *92*, 604.
41. Ewing, D. *Radiat. Res.* **1983**, *96*, 275.
42. Gerhardt, P.; Black, S. H. *J. Bacteriol.* **1961**, *82*, 750.
43. Black, S. H.; Gerhardt, P. *J. Bacteriol.* **1961**, *82*, 743.
44. Dewey, D. L. *Radiat. Res.* **1963**, *19*, 64.
45. van Hemmen, J. J.; Meuling, W. J. A. *Arch. Biochem. Biophys.* **1977**, *182*, 743.
46. Behar, D.; Czapski, G.; Rabani, J.; Dorfman, L. M.; Schwarz, H. A. *J. Phys. Chem.* **1970**, *74*, 3209.
47. Kellogg, E. W. III; Fridovich, I. *J. Biol. Chem.* **1975**, *250*, 8812.
48. Fridovich, I. In "Oxygen Free Radicals and Tissue Damage"; The Ciba Foundation, Excerpta Medica: Amsterdam, 1979, pp. 77–93.
49. Petkau, A.; Chelack, W. S. *Biochim. Biophys. Acta* **1976**, *443*, 445.
50. Bors, W.; Saran, M.; Lengfelder, E.; Spottl, R.; Michel, C. In "Current Topics in Radiation Research Quarterly"; Ebert, M.; Howard, A., Eds.; North-Holland: Amsterdam, 1977, pp. 247–309.

18

# The Effects of Ionizing Radiation on Mammalian Cells

J. E. BIAGLOW

*Department of Radiology,*
*Case Western Reserve University*

## INTRODUCTION

Human and animal cells, both normal and tumorigenic, can be grown in culture in vitro and studied for their responses to ionizing radiation.[1–8] In vitro studies with cell cultures may involve cytotoxic and mutagenic effects of radiation. There is also a good deal of current interest in the effects of drug and radiation combinations on cell cytotoxicity. In addition, the effects of hormones and radioprotective agents are being actively pursued. For these studies the cells can be grown under various conditions such as log phase, plateau phase, or growth arrested, and in multicellular arrays such as spheroids. Cells may also be grown at lower density and concentrated before irradiation in order to duplicate tissue-like densities. Cells may be collected when they round up for division, or they may be chemically synchronized in order to study the effects of radiation alone or with drugs at various stages of the cell cycle.

An important advancement in understanding the nature and severity of radiation damage to mammalian cells was accomplished by the introduction of several techniques to measure the survival of single cells.[1–4] These techniques provided the first opportunity to quantify the biological effects of


*Radiation Chemistry: Principles and Applications*

radiation on a cellular basis. The introduction of these survival techniques also hastened the understanding of effects of various biological, chemical, and physical modifications on the survival of cells after radiation.

Cell survival techniques are limited to dividing or potentially dividing cells. Although this limitation may seem severe, the dividing cells of an organism are important because they compose the stem-cell compartment and are more sensitive to radiation than the differentiated cells. In whole body exposure to radiation, the ability of the stem cells to divide determines the life or death status of the individual. The dividing cells are also the ones of interest in the treatment of mammalian tumors with radiation.

In this chapter we are concerned with the effects of radiation on dividing cells and the factors that influence the division. In addition, the radical mechanism for radiation damage is briefly reviewed. The mechanism of action of hypoxic cell radiosensitizing drugs is discussed as well as the ability for thiol oxidizing agents to enhance the radiation response. A model for the influence of oxidation-reduction reactions on radiation and damage is examined. The ability of oxygen to increase radiation damage, and the effect of oxygen-sparing drugs on the radiation response of multicellular systems is described. A good deal of emphasis is given in this chapter to oxygen, oxygen-mimicking drugs, and oxygen-sparing drugs because of the central importance of oxygen in radiation effects.[1–3,5,6]

## ASSAYS FOR RADIATION EFFECTS

In radiation biology the basic assay for the effect of radiation involves the survival of cells after treatment (the survival curve). Cells may be treated with a single or multiple dose of radiation in the presence or absence of drugs and under different growth conditions. This survival depends on the ability of the initial cell to divide and produce a stainable colony after 7–14 days of growth in a chemically defined medium. The in vitro colony determination was made possible by advances in cell culture technique[1–6] to allow for the growth of single cells. The colony formation assay was developed by Puck and Marcus for mammalian cells in 1956.[4] The technique has not changed much since that time. However, some cells that are not easily cultured in liquid medium have been found to produce colonies if agar is included in the growth medium.[7,8] Agar techniques permit the assay of a number of additional cell lines, particularly those derived from human tumor tissue,[7] that cannot be assayed in the conventional liquid medium. Both techniques involve plating of single cells into a Petri dish or a flask containing a cell culture media with serum or protein supplement.[1–7] The cells are treated either in suspension or after plating onto the assay flask. Different series of replicate plates, or the suspensions, are treated with a particular radiation dose. The survival is measured by scoring the number

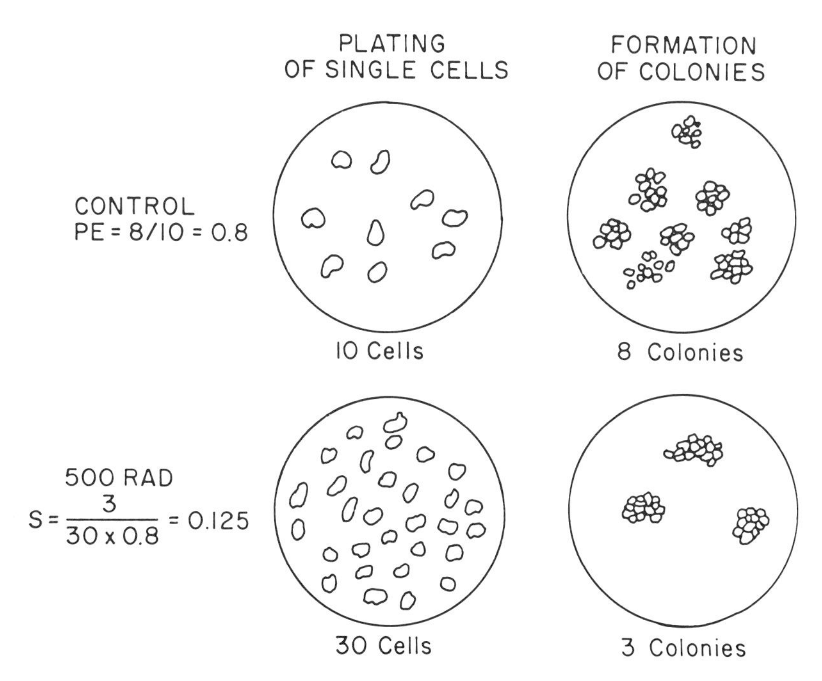


**Figure 18-1.** The colony-forming assay: in vitro cells are plated as single cells and scored 1–2 weeks later as colonies.

of single cells that develop colonies (Figure 18-1). Since some unirradiated cells fail to form colonies, a correction has to be made in terms of the plating efficiency. Thus, survival is calculated by

$$S = \frac{\text{Number of Colonies Formed}}{(\text{Number of Single Cells Plated}) \times (\text{PE})_c}$$

where $(\text{PE})_c$ is the number of colonies formed per number of single cells plated without irradiation. A simplified example showing only one plate for control and 500 rad of radiation is given in Figure 10-1. In practice it is difficult to obtain pure single-cell populations, so a correction is made generally for higher multiplicities (average number of cells per group at the time of irradiation). Other complications arise from the difficulty of obtaining surviving colonies of uniform size. Therefore, when counting colonies for clonogenic assays, an arbitrary number of cells per colony, usually 7 to 8 doublings (128–256 cells), is chosen, and colonies below that number are excluded or counted as a special group.

The effects of irradiation on cells, as initially determined by Puck and Marcus with HeLa cells, have been verified by scores of investigators using a variety of animal and human cell types in vitro.[1–7]

Figure 18-2 shows the shape and parameters of a typical survival curve obtained from target theory. The slope of the exponential portions of the

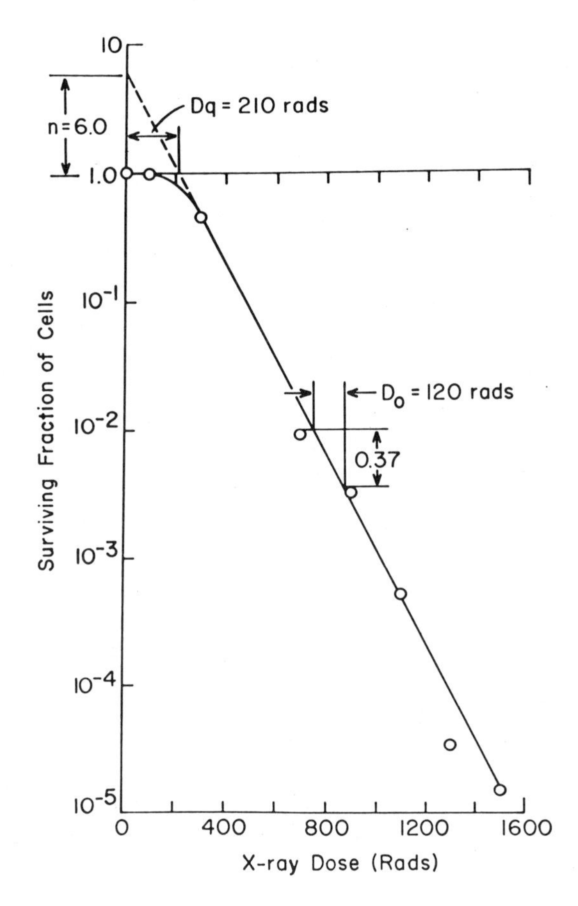


**Figure 18-2.** A diagrammatic representation of a typical dose-response curve.

curve is given in terms of $D_0$, the dose to give an average of one hit per target and, thereby, to reduce survival to 0.37 of any survival level of the exponential portion of the curve. The shoulder region of the curve at low doses suggests that the cells have the ability to accumulate sublethal damage before additional radiation causes the sublethal damage to become lethal. In terms of the target theory, the shoulder region is indicative of either several targets in the cell that must be inactivated or target(s) that must be hit several times to kill the cell. The exponential portion of the curve can be extrapolated to zero dose to obtain the extrapolation number $n$. This number is highly variable and, in practice, is not a reliable index.

Another useful term to describe the shape of the survival curve is $D_q$, the dose in rad where the exponential portion of the curve intercepts 100% survival. The quantity $D_q$ is a convenient number because it measures the shoulder width in units (rad), unlike $n$. The three parameters are related from derivation of multitarget theory:

$$D_q = D_0 \ln n$$

# SURVIVAL CURVE DIFFERENCES

There is a vast amount of literature on dose-survival curves for numerous cell lines; however, most of these have been obtained under different growth conditions. Therefore, we have obtained information on a number of cell lines irradiated and assayed under the same conditions (Figure 18-3) and have determined the survival curves for the mouse lymphocytic leukemic L5178Y (L suspended culture), L5178Y (L-A grown attached), Chinese hamster ovary (CHO), and $CHO_2$, a morphological variant. We have also determined the radiation response of V79-171 hamster lung cells and an isolated variant of this line, the V79-171B. All of these cell lines vary in their response to radiation. The most sensitive lines are the L5178Y (suspended) and the L5178Y (attached). The difference between these two lines is

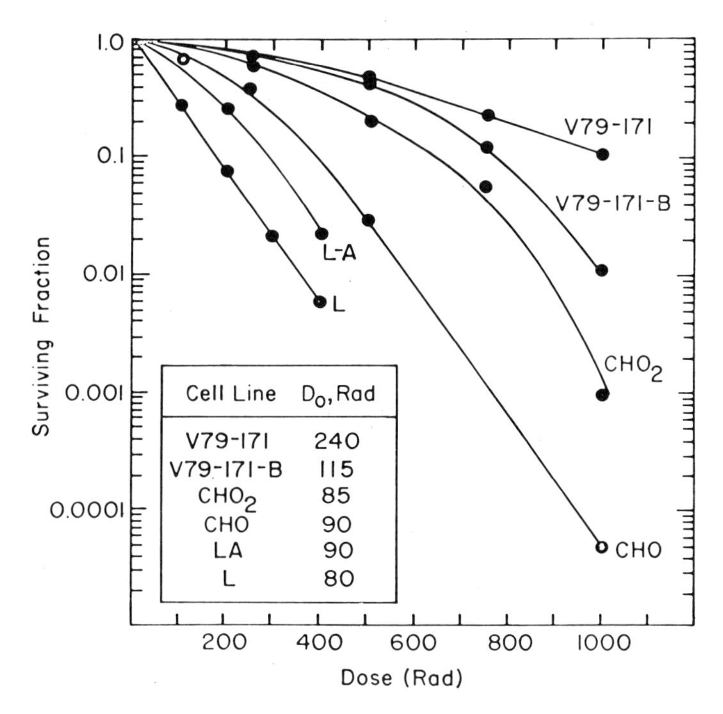


**Figure 18-3.** Dose survival curves for V79-171 and V79-171-B lung cells, CHO, $CHO_2$, and the lymphocytic leukemia lines L5178Y (L) and L5178Y (L-A) grown in McCoy's 5a medium. The cells were irradiated with 220-kV x-rays, 37°C, in McCoy's 5a medium, and colony formation was determined 10 days later.

the small shoulder obtained with the attached cell line. The CHO line is more sensitive than its morphological variant, and both have shoulders. The most resistant cell line was found to be the V79-171. Its cloned variant was more sensitive to radiation. The $D_0$ values are seen in the lower left corner of the figure. These cell lines demonstrate the variation in radiation response obtainable with different cell lines; the variation in $D_0$ from 80 to 240 is comparable with values found in the literature. Of importance is the difference in the shoulder of the survival curve between the most resistant line (V79-171-B) and the most sensitive shoulderless L5178Y lines, suggesting a different capacity to accumulate sublethal damage.

The large variation in the $D_0$-values as well as the shape of the survival curve makes comparisons between laboratories nearly impossible unless rigorous conditions for growth, irradiation, and plating of cells have been maintained.

## INDIRECT AND DIRECT EFFECTS OF RADIATION

Cells are usually irradiated in an aqueous environment containing salt or nutrients such as proteins, vitamins, and amino acids, and approaching conditions thought to exist in vivo.[8] In addition, cells contain considerable amounts of water. Therefore, the radiation chemistry of water is of primary importance in explaining the biological effects of ionizing radiations.[5,6,9] One of the reactions observed in aqueous solutions is the formation of the free radicals ·OH and ·H from water, as described in Chapter 10. These radicals carry no charge but have a strong affinity for electrons (or hydrogen bonds). They can remove hydrogen atoms from other encountered molecules—for example, from organic molecules in the cell:

$$\cdot OH + RH \rightarrow R\cdot + HOH$$

Most of the damage to "critical molecules" such as DNA, RNA, or protein in the living cell occurs by this "indirect" mechanism. Notedly, ·OH and ·H radicals will, on encounter, recombine to form water. Therefore, they rapidly disappear through recombination or other radical–radical interactions. In the presence of $O_2$, recombination of ·H and ·OH is minimized by formation of the hydroperoxyl radical, which is a more stable free radical but is also capable of extracting H atoms from certain organic molecules. This accounts, in part, for the observed enhancement of damage by irradiation in the presence of oxygen (ie, the "oxygen effect").[2,3,5,6]

Although the formation of free radicals in critical molecules of the cells occurs largely by means of the "indirect" mechanism, it can result also from an ionizing event produced by a *direct hit* in the molecule. A free-radical site is produced by both direct and indirect mechanisms. A variety of reactions can occur with a molecule containing a free radical R·.

***Immediate Chemical Restitution***

R· + H· → RH is minimized in the presence of oxygen but increased in the presence of H-donors (eg, cysteine, glutathione, etc). Free radicals can combine with $O_2$ to form $RO_2\cdot$, which prevents repair. In this reaction stable oxidative products formed are no longer susceptible to immediate chemical restitution. This accounts for another major part of the oxygen effect (see later).

***Reactions with other Free Radicals***

$$R\cdot + \cdot OH \rightarrow ROH \quad \text{or} \quad R\cdot + R_1\cdot \rightarrow R{:}R_1$$

Several examples of alterations in biologically important molecules can be cited, such as the splitting of peptide bonds in proteins by opening of heterocyclic ring structures (eg, tryptophan) and deamination as well as cross-linking. With respect to nucleic acids there can be splitting of the polynucleotide chain, deaminations, removal of purine or pyrimidine bases, cross-linking, and alterations in purines and pyrimidines.

## CELLULAR EFFECTS

Many attempts have been made to explain the effects of radiation on living cells through interference with specific parts of the metabolic machinery.

Studies have been performed with isolated enzymes and enzyme systems, such as those involved in protein synthesis in vivo and in vitro as well as lipid and carbohydrate metabolism. In all cases the measured effects are too small to account for a specific mechanism for the overall observed damage to the intact cell.[2,3,5]

It is conceivable that breakdown of internal structures (membrane, endoplasmic reticulum, etc) may account for damage that cannot be readily measured by biochemical tests. Loss of ions from the cell nucleus, reduced uptake of metabolites from the cytoplasm by nucleus, and the inhibition of "nuclear oxidative phosphorylation" have been reported.[2,3] However, their exact importance for altered cell metabolism resulting in cell death has not yet been determined.

At present we are unable to explain adequately why some nondividing cells (such as lymphocytes) disintegrate after exposure to relatively limited radiation doses, whereas others (eg, liver cells) appear relatively unaltered and continue to function. One clue to this appears to be the large difference in the cytoplasm/nuclear ratio in these cases: cells with large cytoplasmic volumes are generally more radioresistant. The most conspicuous cellular effects of radiation are mitotic delay, chromosome aberrations, mutations (including carcinogenic transformation), and cell killing or loss of reproductive ability.[2,3,5,6] The most important and certainly the most studied is cell killing. However, some cells survive irradiation and are not killed. This survival is due to their repair capacity.

### *Repair of Radiation Damages*

The radiation dose response curve for many of the mammalian cells is not a straight line on semilog paper but has a "shoulder" (Figure 18-2). This shoulder has been explained in at least three ways:

1. The cell may require hits in more than one sensitive area (or volume) to produce lethality. If $n$ is the number of targets, it may be shown that the number of survivors is represented by

$$\frac{N_s}{N} = 1 - (1 - e^{-D/D_0})^n$$

   if one plots log $N_s/N$ vs. $D$. The quantity $D_0$ is a measure of the slope of the high-dose straight-line portion, and $n$ is the extrapolation number, as shown in Figure 18-2. This is called the multitarget single-hit model. The log $N_s/N$ term was previously defined as the surviving fraction or the number of colonies formed divided by the number of single cells plated multiplied by the plating efficiency.
2. The cell may require $n$ hits in the same area. This concept is called single-target multihit theory; although it leads to survival curves similar to those in Figure 18-2, it is a much less popular explanation of the shape of the survival curve for single cells. It has been applied most commonly to colony survival, where all of the cells in a group must be killed to suppress colony formation.
3. The cell can repair some of the damage to the radiation.[2,3,5,6]

Cellular recovery from exposure to ionizing radiation is time dependent, and numerous experiments have shown that it occurs in less than two hours following the end of the irradiation. That "sublethal" recovery is associated with the low-dose shoulder of the survival curve seen in Figure 18-4, which illustrates the effect of interrupting the exposure at dose A, waiting an interval longer than, say, three hours, and then delivering the rest of the dose.[10] A new curve with shoulder appears at point A if radiation repair has occurred. For two equal doses, as shown, it is obvious that the number of survivors ($F_1$ at $D$) is greater than if all the dose had been delivered at once, resulting in $F_2$ at $D$. If the dose-response curve was simply log-linear, there would be no difference in the number of survivors. The quantity $D_s$ is a measure of the shoulder and has been considered as the "lost" or "wasted" dose from fractionation.

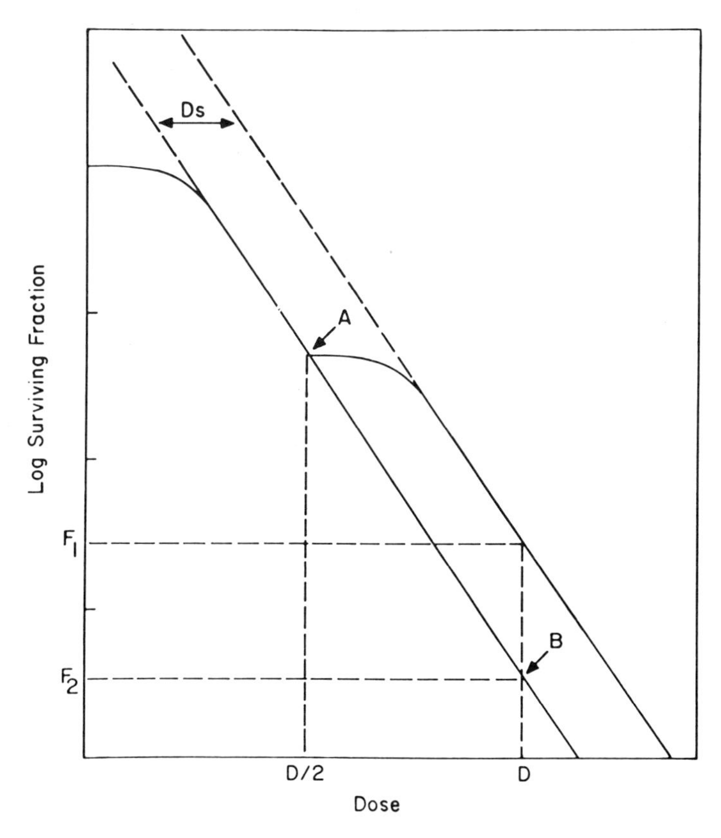


**Figure 18-4.** Illustration of the effect of adding two doses of radiation when the survival curve has a shoulder due to time-dependent repair. For $D$ given acutely, survival is $F_2$; for $D$ given in two equal doses separated by several hours, survival is $F_1$. $D_s$ is sometimes known as the "lost" dose due to fractionation.

Another means for demonstrating repair of sublethal radiation damage is by the use of the "split-dose" experiment originated by Elkind and Sutton.[11] If a culture of cells is given an initial dose of irradiation (a conditioning dose) that will bring the survival past the shoulder of the dose-survival curve and then a second dose is given at various times thereafter, the surviving fraction will vary with the interval between the two doses (Figure 18-5). The early increase in survival (for short intervals between the doses) has been interpreted to reflect repair of sublethal damage; the further variations probably result from progression of cells through the cell cycle (cells not killed by the first dose move from a less sensitive state to a more sensitive state in the cell cycle) (Figures 18-6 and 18-7).

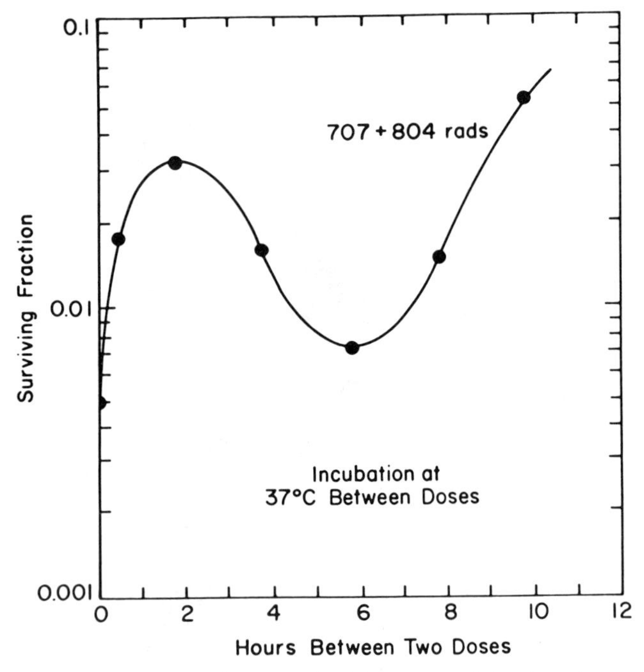


**Figure 18-5.** The survival of cells treated with two conditioning doses of radiation followed in time by a second dose of radiation. The increased survival following the first doses is suggested as the repair of sublethal radiation damage. The dip between 3 and 8 hr indicates movement of cell-cycle progression into a more sensitive phase.

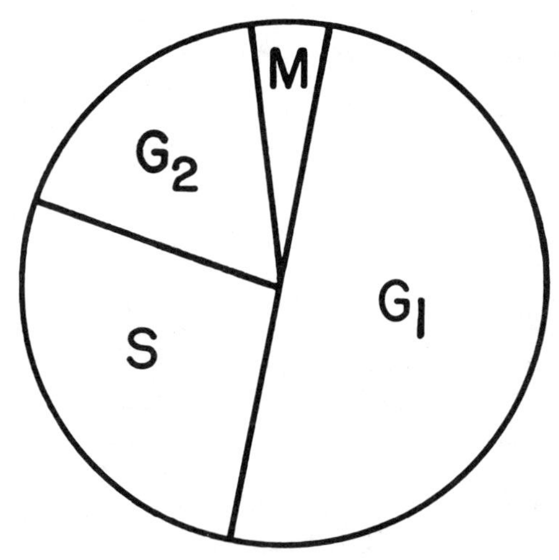


**Figure 18-6.** Cell-cycle dependent variation in sensitivity to radiation. The response of cells to radiation varies throughout the cell cycle. The major, readily identifiable stages of the division cycle are shown graphically. In most mammalian cells the duration of the different stages are $M = 0.5$–$1$ hr; $G_2 = 2$–$4$ hrs; $S = 6$–$8$ hrs; $G_1 =$ variable (0 hr—very long). Cell-cycle dependent variations in radiosensitivity have been studied with synchronized cell populations.

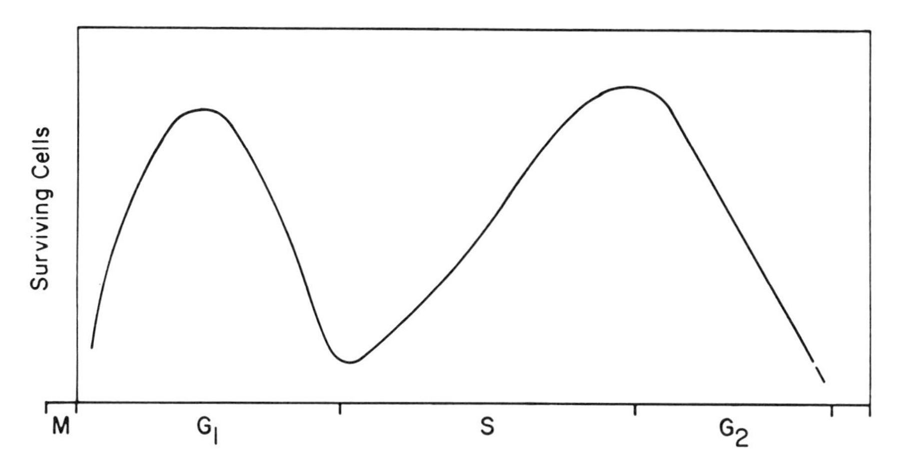


**Figure 18-7.** Survival of cells irradiated at various stages of the cell cycle.

### *Other Types of Repair*

The post-irradiation treatment of cells irradiated with a single dose of x-rays has been found to affect the cell survival. Various ways of slowing down cell metabolism, such as lowered temperature or absence of nutrients in the growth medium lead to increased survival, indicating that some "potentially lethal" damage has been repaired. It is not presently clear whether sublethal (see previous discussion) and potentially lethal damage are related. Other post-irradiation treatments may increase the killing; these are often treatments with drugs that inhibit repair. The effect of elevated temperature (hyperthermia) during or immediately after irradiation has been explained as an inhibition of post-irradiation repair. Radiation-induced cellular damage seems to be maximized at two stages of the cell cycle: during *S* and during the mitosis period. Any treatment that tends to affect the temporal relationship between repair and fixation of damage (Figures 18-6 and 18-7) appears to affect lethality. The cell-cycle dependent variation in radiosensitivity discussed in the next section may be explained on the basis of such interplay between repair and fixation of damage. The term "repair" has also been used at the tissue level in cases where *repopulation* or *restoration* of cells could be more accurate terms.[2,3,5,6]

## TARGET FOR RADIATION DAMAGE

The primary target for radiation-induced cell killing is believed to be the DNA molecule.[2–6] The evidence for DNA occupying this central position is manifold and has been obtained both directly and indirectly; for example:

1. The size of the target is important, and because the nucleus, containing cellular genetic material, constitutes a large part of the cell, the likelihood for a "hit" on the DNA molecule is statistically quite possible.
2. The DNA molecule is unique and has low redundancy, so that each molecule is more vulnerable than the highly redundant compounds such as lipids; also the central position of DNA in the control of cellular activity makes it critical, and the damage gets "amplified" through transcription and translation.
3. Direct irradiation of the cell nucleus (eg, with microbeams) is many times more efficient than irradiation of the cytoplasm of the cell.
4. Chromosomal damage parallels lethality.
5. Sensitivity of many related cells and organisms parallels the DNA content and to some extent the base composition of DNA.
6. Incorporation of radioactive molecules, such as $^{3}$H-thymidine, into DNA leads to very efficient cell killing.
7. Incorporation of base analogues (eg, 5-bromodeoxyuridine) into DNA increases the sensitivity to ionizing radiation.
8. Cell killing has been shown to be associated with structural damage to DNA (eg, double-strand breaks; chromosomal damage).

The identification of DNA as the main target for cell killing by radiation makes it possible to visualize repair of sublethal and potentially lethal damage as being related to repair synthesis of DNA. Also, the "fixation" of radiation damage during DNA replication might explain the decrease in sensitivity of cells as they progress through the *S* period (see earlier).

## THE OXYGEN EFFECT WITH CELLS

No other chemical has been as widely studied as oxygen, and yet it continues to be such an intensive subject of study as a sensitizer of mammalian cells.[2–6] The early studies of Gray[12] (Figure 18-8) indicate that as the oxygen concentration is lowered there is a corresponding decrease in the radiation response of cells. The relative radiosensitivity of cells increases rapidly between 0 and 0.3% oxygen. Further increases occur until approximately 30 mm oxygen, after which additional increases are very small. The oxygen enhancement ratio (OER), or relative radiosensitivity, varies between 2 and 3.5 for the majority of cells.[2,3,5,6] Oxygen has been studied as a sensitizer because of the problem thought to occur in vitro with human tumors. Human tumor cells outgrow their blood supply, resulting in a decreased availability of oxygen, which produces hypoxic and anoxic tumor areas that may exhibit a decreased radiation response. The chief cause of the oxygen

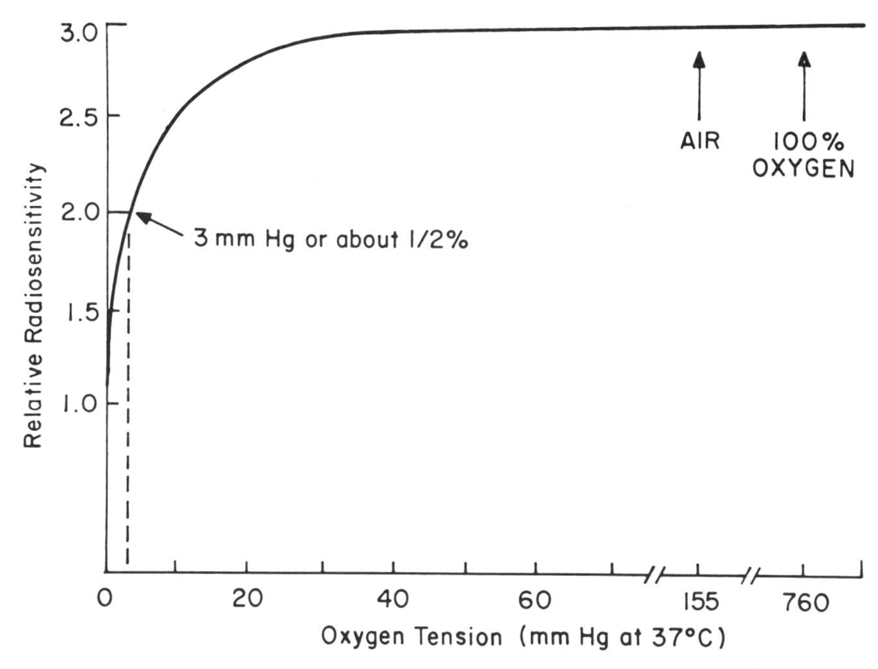


**Figure 18-8.** Illustration of the dependence of radiosensitivity on oxygen concentration. This diagram is idealized and does not represent any specific experimental data. Comparable results have been obtained with yeast, bacteria, and mammalian cells in culture.

effect in vivo is due to the consumption of oxygen by the tumor cells. The metabolic utilization of oxygen decreases the distance to which it may penetrate in the cells more distant from the capillaries. Unlike the physiological situation, most experiments are performed in vitro with cells that have been equilibrated with a nitrogen–carbon dioxide gas mixture for a period of time that insures the depletion of the dissolved oxygen. These procedures are laborious and require special equipment. A simpler way of demonstrating the oxygen effect is to concentrate cells into a dense suspension approaching in vivo cell densities. It was found that cells under these conditions exhaust their supply of oxygen within minutes (Figure 18-9) and may be immediately irradiated, diluted, and plated for survival assays.

Figure 18-10 shows the radiation response of this type of dense cell suspension compared to the radiation response obtained for cells at the same density but irradiated at zero degree to inhibit cellular oxygen utilization. The ratio of $D_0$ values for each curve results in a dose-modifying factor of 3.3. The OER will vary between cell lines and depends on the previous history and growth conditions of the cells.[2,3,5,6]

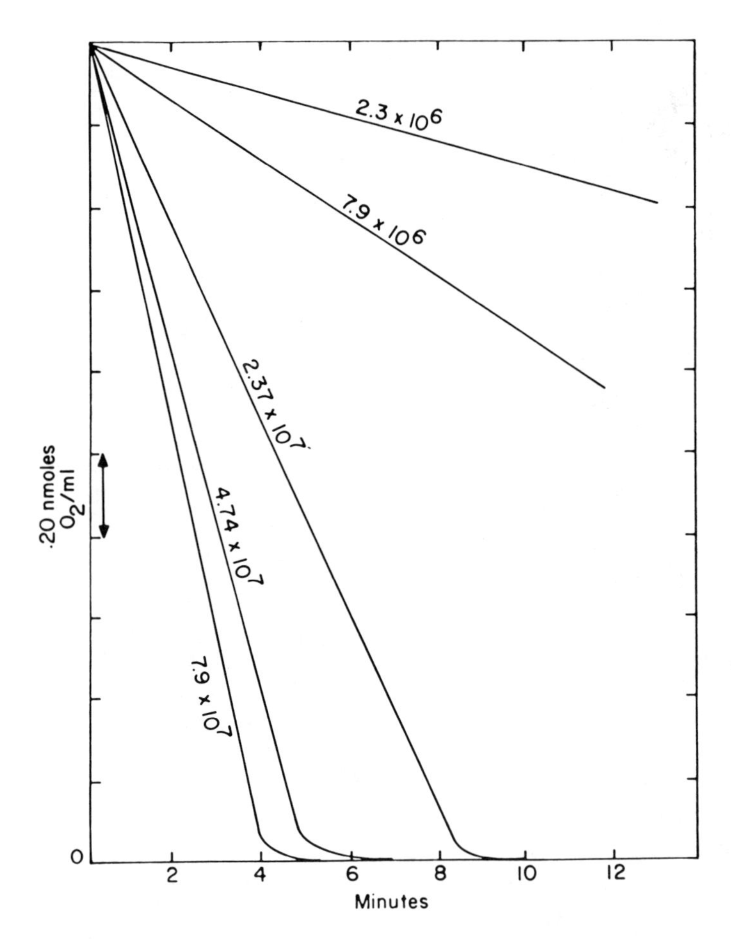


**Figure 18-9.** The effect of different densities of cells on the rate of depletion of oxygen in a sealed container. The disappearance of oxygen was monitored with a Clark oxygen electrode. The cells were incubated at 37°C in 20 m*M* HEPES buffered physiological saline, pH 7.4.

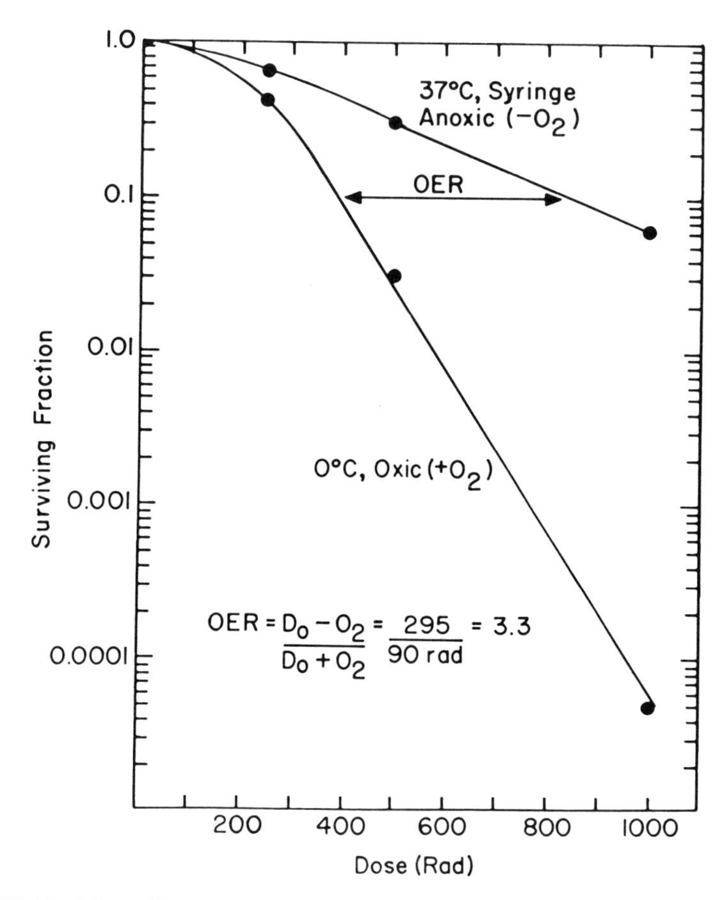


**Figure 18-10.** The effect of metabolically produced hypoxia on the radiation response in dense suspensions of CHO cells. In the top curve the $2 \times 10^8$ CHO cells/mL in 0.02 *M* HEPES buffered media were drawn up into a glass syringe, immediately radiated, at 37°C diluted and assayed for clonogenic survival. In the lower curve the same density of cells was spread as a layer of cells on the surface of a T30 flask, gassed with humidified $CO_2$, and irradiated at 0°C, diluted and assayed for colony-forming ability. (Reproduced with permission from the author and the American Chemical Society.)

## *Mechanisms of Oxygen Effect*

Flanders and Moore[13] proposed that two types of radiation damage are produced in cells: oxygen dependent, and oxygen independent. For example, DNA radicals produced by radiation would be subject to a competitive chemical challenge, with oxygen on the one hand and intracellular donors on the other competing for the radicals (Figure 18-11). In the absence

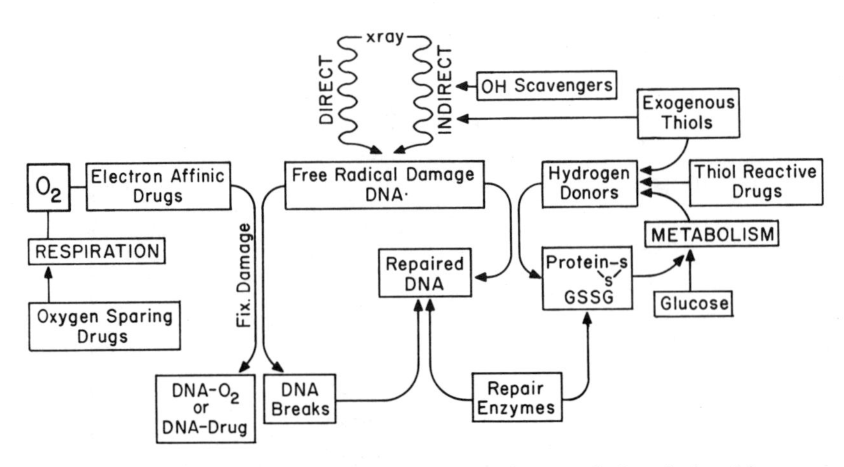


**Figure 18-11.** The relationship between metabolism and chemicals with respect to DNA damage and repair. The left side shows the effects of electron-affinic and oxygen-sparing drugs on cellular respiration. Electron-affinic drugs, oxygen, and oxygen-sparing drugs (increased availability of oxygen) increase x-ray-produced damage. DNA adducts may or may not be enzymatically repaired. The extent of the repair of DNA is dependent upon enzymes. These enzymes are all influenced by the metabolic state of the cell (ie, redox). The redox state of the cell is controlled by glucose and is altered by thiol reactive drugs. Radical damage is also influenced by exongenous and endogenous radical scavengers such as glutathione. Radical damage may be "chemically repaired" by hydrogen donors in the absence of oxygen.

of oxygen the first type of oxygen-dependent damage may revert chemically to a harmless state, possibly chemically repaired by hydrogen donation from SH-containing compounds or other hydrogen-donating molecules:

$$R\cdot + XSH \rightarrow RH + XS\cdot$$

If oxygen reacts, the so-called damage may be fixed. The cell is unable to repair itself, thus resulting in lethality:

$$R\cdot + O_2 \rightarrow RO_2\cdot$$

As yet there is no direct evidence for this competition in irradiated cellular systems, although there is no doubt that it applies in simple model systems. The competition between chemical repair and oxygen fixation leads to a working hypothesis for the oxygen effect. This approach was further developed by experiments with the reactions of suspected DNA radicals with oxygen, electron-affinic sensitizers, and sulfhydryl compounds in irradiated bacteriophage and with purified DNA.[2,14] A reaction scheme based on the competition between oxygen and sulfhydryl compounds for the oxygen-dependent damage was proposed. With the development of alkaline sucrose gradient techniques, other workers carried out experiments to study the oxygen dependence on radiation-induced single-strand breaks. Strand breakage was enhanced in the presence of oxygen by a factor of 2 to 3.[2,6,14] Others[15,16] studied DNA strand-breakage efficiency as a function of oxygen concentration and concluded that the oxygen effects on strand breakage are similar to those of the radiation killing. These results were interpreted as supporting the viewpoint that the radiation target associated with reproductive death by the oxygen effect is DNA. Agreement on this subject is by no means universal. Other investigators suggest that different target sites, such as the membrane, may be important in radiation-induced cellular inactivation, particularly in vivo with lung and heart tissue.[17]

Despite considerable research in past years, the basic mechanism by which oxygen sensitizes cells to the action of radiation is still not completely understood. For example, there are reported deviations from the predicted dependency of OER on oxygen concentration.[18,19] Part of the difficulty in attempting to understand the mechanism of radiosensitization by oxygen is due to the very fast time scale ($10^{-3}$ s) involved in these processes. Oxygen must be present at least $1–2 \times 10^{-3}$ s prior to irradiation. This has been demonstrated using rapid mixing systems, a gas explosion method, pulsed irradiation, and mechanical mixing techniques.[2,9,18–20]

The fast time scale for the oxygen effect in biological systems, combined with knowledge of the chemical reactions induced by radiation of aqueous systems, strongly implies that the mechanism for the oxygen effect involves free-radical reactions. The oxygen fixation hypothesis[2–6,9] states that $O_2$ reacts with radiation-induced free-radical sites on the target molecules, presumably DNA (see Figure 18-11), to form peroxides, which are believed to be nonreparable forms of damage.[14] Rapid mixing techniques[20] demonstrated that irradiation at the shortest time possible after mixing, about $4 \times 10^{-3}$ s, resulted in an oxygen enhancement ratio of 1.7, regardless of the oxygen concentration (from 1 to 50%) in the mixed solution. The OER increased to its full value of 2.8 as the time between mixing and irradiation was increased to $4 \times 10^{-2}$ s. The profile of the increase of OER as a function of time after the initial $(4 \times 10^{-3})$-s level was dependent on the oxygen concentration. These data support the conclusion that there are two components to the oxygen effect, with the slower one demonstrating a dependence on oxygen concentration. The mammalian cell two-component oxygen or sensitizer effect has not yet been universally accepted, and there is much current work in this area. Chapter 17 should be consulted for information on oxygen effects in micro-organisms.

## *Hypoxic Cell Radiosensitizing Drugs (Oxygen Mimicking)*

In the last decade there has been rapid development in the field of specific radiosensitizers for hypoxic cells, due largely to the work of numerous chemists and radiobiologists[2,21] who defined the desirable criteria for these agents and then set out to select and design appropriate compounds. Most of the initial work was performed with cultured cells.[21,22] Basically there was an attempt to develop drugs that would not be metabolized as rapidly as oxygen yet would be oxygen mimicking and possibly of use in vivo in sensitizing hypoxic tumor cells. The electron-affinic, hypoxic cell radiosensitizing drugs all contain an aromatic ring and a nitro ($NO_2$) group, which appears to be the critical structural feature. Sensitizer studies to date include nitrobenzenes, nitrofurans, nitroimidazoles, nitropyrroles, and nitropyrazoles.[21–24] The most important nitro compounds (Figure 18-12)

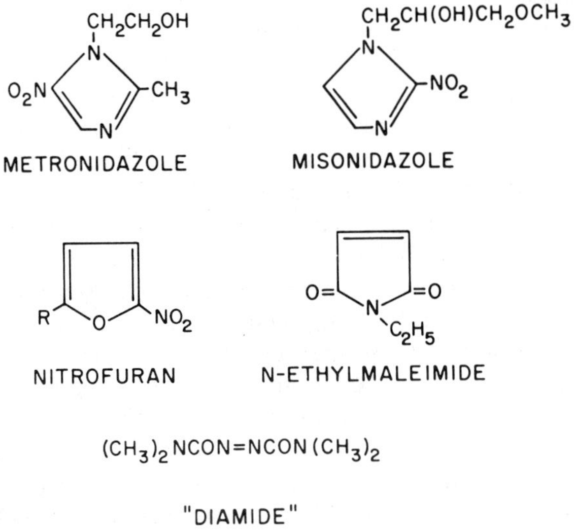

$CH_2CH_2OH$ $O_2N$ $CH_3$ N N

METRONIDAZOLE

$CH_2CH(OH)CH_2OCH_3$ N $NO_2$ N

MISONIDAZOLE

R O $NO_2$

NITROFURAN

O= N =O $C_2H_5$

N-ETHYLMALEIMIDE

$(CH_3)_2NCON{=}NCON(CH_3)_2$

"DIAMIDE"

**Figure 18-12.** Structure of some nitro aromatic hypoxic cell radiosensitizing drugs.

so far studied are the nitroimidazoles, metronidazole (Flagyl) and misonidazole. In Figure 18-13 are seen the effects of misonidazole on hypoxic cells. There is a small effect with 1 m*M*, however, the response is greatly improved but not as good as $O_2$ when the concentration is increased to 10 m*M*. There is no effect of sensitizer in the presence of oxygen. The effect of medium containing dissolved oxygen (0.2 m*M*) is also seen. Various nitro aromatic compounds sensitize hypoxic cells to x-rays by increasing the slope of the survival curve; they do not affect the shoulder except under conditions of prolonged contact prior to irradiation.[25] Demonstration of the radiosensitizing ability of a compound in vitro is no guarantee of favorable results

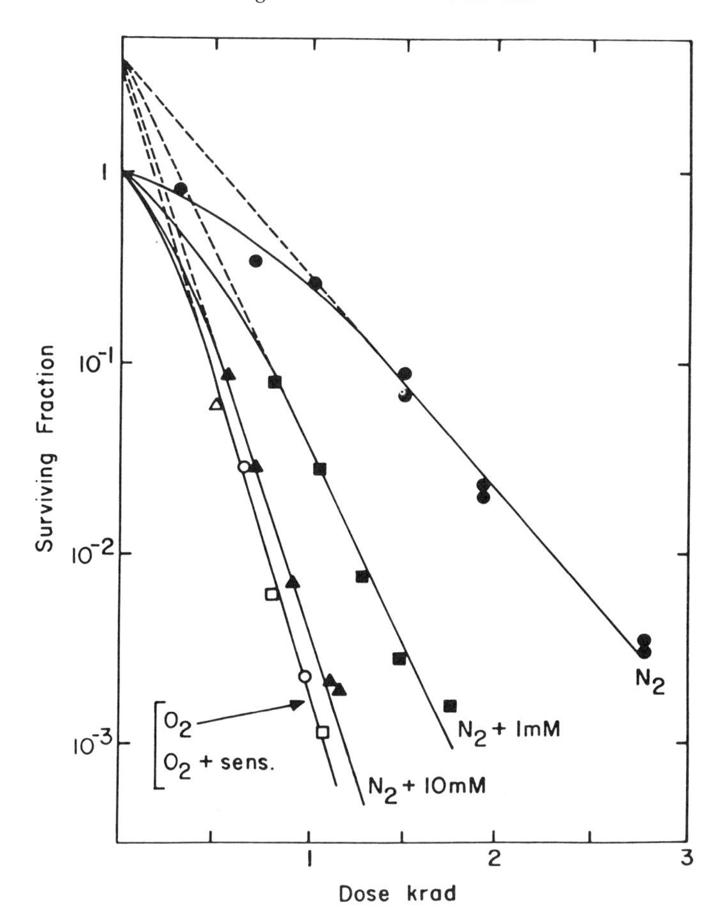


**Figure 18-13.** The effect of misonidazole on the radiation response of hypoxic CHO cells. (Reproduced with permission.)[20]

in vivo, because there may be biochemical, physiological, or pharmacological complexities involved in vivo that might interfere with sensitizer action. For example, alterations in cellular biochemicals[26] as well as effects on cellular oxygen utilization[27] may alter the effectiveness of the agents in vivo.

Unfortunately, except for a few isolated cases, the basic studies on the mechanism of action of many of these compounds has been superseded in an attempt to find drugs that will be clinically valuable. With practicality as a guideline the nitroimidazoles are being used for studies with humans. Metronidazole and misonidazole (Ro-07-0582) are undergoing clinical trials in several cancer centers. At this time, studies are being carried out to determine appropriate dose and fractionation schemes for both drugs and radiation. This would result in therapeutic gain due to radiosensitization in selected tumor sites (eg, brain) without the complication of neurotoxicity observed during the early stages of testing these compounds. Another limitation of these drugs is the effective concentration that must be reached within the tumor to achieve an observable radiosensitization. These concentrations lie in the range of 0.5 to 1 m*M*, which produces a dose-modifying factor of 1.2 to 1.7. Another difficulty is that clinicians use low-radiation doses in order to avoid mistakes; consequently, the improvement in the actual radiation response in humans is at present marginal.[24,28]

## *Mechanism of Action of Hypoxic Cell Radiosensitizers*

There is general agreement that the electron-affinic sensitizers mimic the action of oxygen at the level of DNA.[20–24,28] These contentions are supported by the facts that these sensitizers have no radiation effect on aerated cells, and the ability to sensitize is closely correlated with the one-electron reduction potential, a measure of electron affinity determined by pulse radiolysis.[20] Oxygen has the highest one-electron reduction potential and is the most effective radiosensitizer. The electron affinity parameter is obviously important, but other variables have been investigated. Because penetration of the sensitizer to hypoxic cells of a tumor is of fundamental importance for

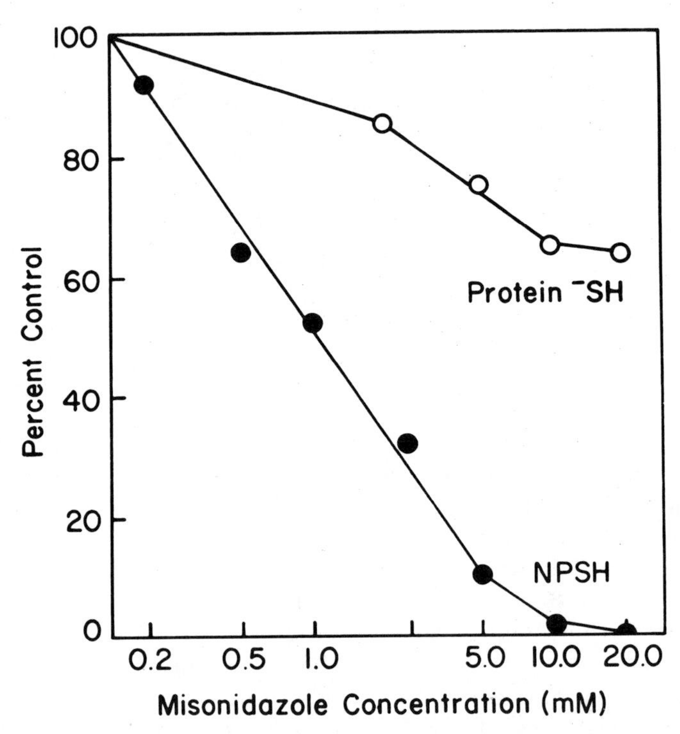


**Figure 18-14.** The effe~t of misonidazole on the nonprotein and the protein thiol content of Ehrlich ascites tumor cells. Cells ($10^7$/mL) were incubated anaerobically in 0.05 *M* PBS (phosphate buffered saline) and 10 m*M* glucose at 37°C, pH 7.3. Thiol determinations were performed 15 min after anaerobic conditions had been achieved. (Reproduced with permission.)[31]

effectiveness in vivo, it is reasoned that high solubility in lipids might be a desirable feature in enhancing the diffusion of the drug. Studies of the octanol:water partition coefficient of sensitizers were therefore conducted.[28] It was found that there is no correlation between sensitizer effectiveness and lipophilicity. Side effects such as neurotoxicity may be enhanced if the drug partitions itself in high-lipid-containing tissue such as brain.[28] Additional studies have demonstrated that nitro compounds are far from inert and that upon addition to cells they profoundly alter cellular electron transfer processes involving cellular respiration.[27,29,30] Intracellular levels of reduced species such as NAD(P)H and glutathione are also affected.[26,29–32] The effect of hypoxic cell radiosensitizing drugs on cellular glutathione (Figure 18-14) and protein thiols has been demonstrated and may be the reason that this drug shows an enhanced radiation response when cells are preincubated with the drug for prolonged periods (Figure 18-15).

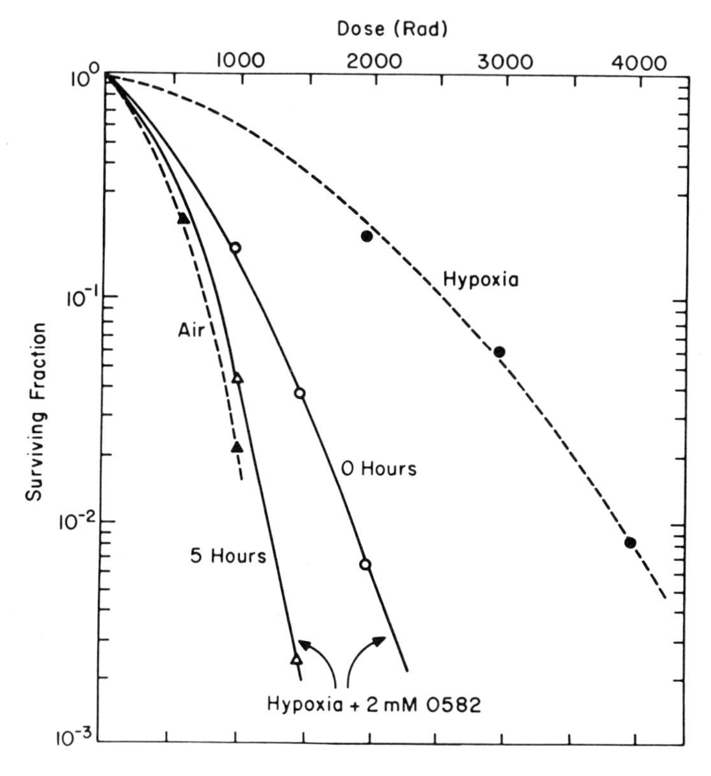


**Figure 18-15.** The closed triangles and circles present the survival of Chinese hamster cells irradiated with $Co^{60}$ $\gamma$-rays under aerated and hypoxic conditions. The survival of hypoxic cells irradiated in the presence of 2 m*M* Ro-07-0582 is represented by open and closed triangles. Open triangles refer to irradiations carried out immediately after the drug was added and the cells made hypoxic. Closed triangles refer to irradiations carried out after the cells were stored at room temperature for 5 hr following the addition of the drug, a treatment that killed about 80% of the cells. (Reproduced with permission.)[25]

Cells also show an enhanced radiation response when pretreated with misonidazole alone, washed, and then irradiated under hypoxic conditions. However, cells are more sensitive in the presence of misonidazole.[29,30]

Other studies have shown that if the thiols are removed by the sulfhydryl oxidizing agent diamide, there is a synergistic effect on the radiation response that is greater than that obtainable by either agent alone.[31,32] Harris and Power[32] reported a lower survival at both 800 and 1200 rad for anoxic cells irradiated in the presence of diamide and nifuroxime (a nitrofuran) than was observed for either sensitizer alone. Similarly, Chapman[33] showed that diamide plus NF-269, another nitrofuran, sensitized to a greater degree than that of either agent alone, and more than oxygen. Watts et al[34] found

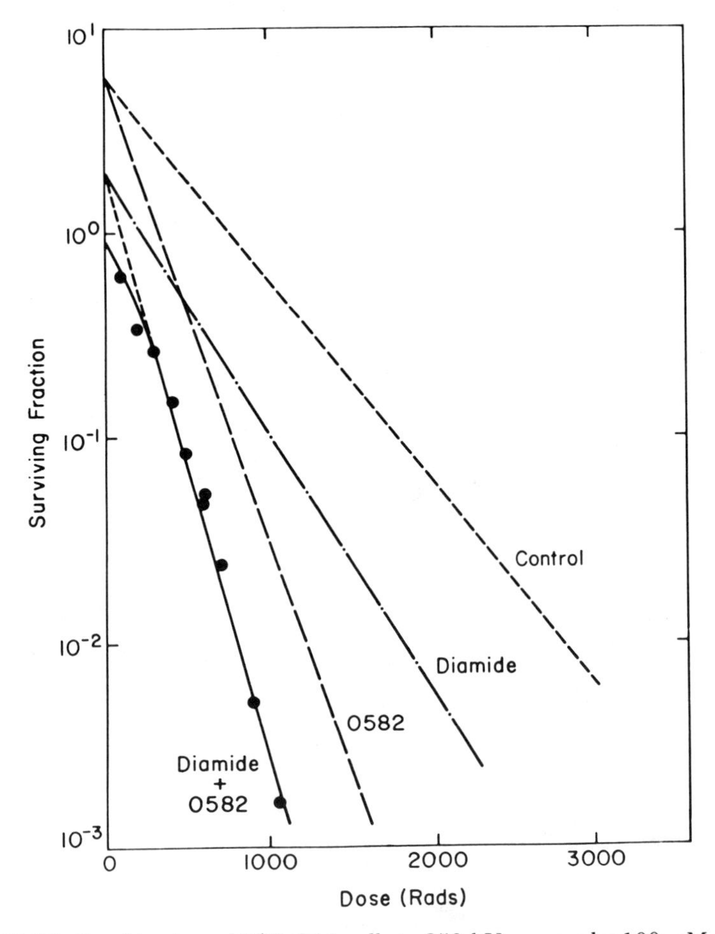


**Figure 18-16.** Sensitization of V79-GL1 cells to 250-kVp x-rays by 100-$\mu$M diamide plus 5 m$M$ Ro-07-0582 (misonidazole) in deoxygenated MEM (minimum essential medium) + 15% serum. The broken lines show the survival observed for the same concentrations of diamide or misonidazole independently as well as the no-drug control. (Reproduced with permission.)[34]

that the combination of diamide and misonidazole was more effective than each individual compound, with a survival curve showing a reduction in $n$ and a change in slope (Figure 18-16). Diamide and misonidazole together sensitized to a greater extent than oxygen and gave results similar to the effect found for diamide in the presence of oxygen.

# HYDROGEN DONORS AND CHEMICAL REPAIR OF RADIATION DAMAGE

The pretreatment effects with misonidazole[25,35] and the combination studies with the thiol oxidant diamide[31–34] suggest that endogenous reducing species are important in the overall mechanism of radiation damage and repair. As mentioned previously, the reaction of the DNA radicals with hydrogen donor or reducing substrates will result in chemical repair of the radiation damage. Some of the intracellular hydrogen donors capable of reacting with radicals are reduced flavins, reduced pyridine nucleotides, ascorbate, and thiols. The largest concentration of intracellular reducing materials is the protein thiols, the most important low-molecular nonprotein thiol (NPSH) being glutathione (GSH). Intracellular glutathione is believed to be an important hydrogen donor under anaerobic conditions for repair of radiation-induced radicals in DNA. However, it cannot compete effectively with oxygen in air at low cell densities because the rate of reaction of oxygen with DNA radicals is at least an order of magnitude greater than the reaction rate with thiols.[36] In addition, under the experimental conditions usually used for studies on the radiation response of cells, the cell density is usually quite low: 10,000 or less. If we assume that there are 5 nmoles NPSH per $10^9$ cells, then 10,000 cells would contribute approximately $5 \times 10^{-11}$ moles of NPSH. This compares to an oxygen tension near $2 \times 10^{-4}$ $M$ or $2 \times 10^{-7}$ moles. Even if we allow for a factor of 10 for the contribution of the protein thiols as radioprotectors, the thiol concentration is still only $5 \times 10^{-10}$ $M$. Oxygen would be in great excess compared to either protein or nonprotein thiols.

## *Cellular Thiols and Radiation Response*

There have been many attempts to remove the endogenous radioprotecting species with agents such as $N$-ethylmaleimide (NEM)[37] and diamide.[2,32,38] In the case of diamide[38] there is a spontaneous chemical reaction with the cellular nonprotein thiols as well as with cellular reduced pyridine nucleotide (Figure 18-17). This reaction can be used to advantage during hypoxic conditions and has been found to increase the radiation response.[32] Diamide (Figure 18-18) was found to radiosensitize hypoxic Chinese hamster cells by decreasing the shoulder of the survival curve at low concentrations and increasing the slope at high concentrations. The effect on the shoulder appears to be due to oxidation of endogenous nonprotein sulfhydryls and reduced pyridine nucleotides (NADPH + NADH) and a partial oxidation of the protein thiols,[39] biochemicals that would normally effect rapid chemical repair of certain single-hit-type lesions. The slope effect, on the other hand, may have been due to reactions with DNA similar to those that have been described for the electron-affinic compounds. In addition, cells pretreated with diamide and maintained at 0°C remained sensitized after the removal of diamide. At zero degrees the metabolic regeneration of the thiols and pyridine nucleotides was inhibited.[32] Diamide also sensitizes in air (Figure 18-19; refs. 33, 34).

Another thiol-binding reactive agent that has been tested as a radiosensitizer of cells is NEM.[37] At low concentrations NEM is quite effective as a sensitizer of cells under oxygenated conditions.[37] Unlike diamide, NEM is less specific for cellular nonprotein thiols. It reacts with both protein and nonprotein thiols and DNA. The reaction with protein thiols alters enzyme activities believed to be involved in the repair of radiation damage. It has been suggested that the primary effect of NEM on the radiation response

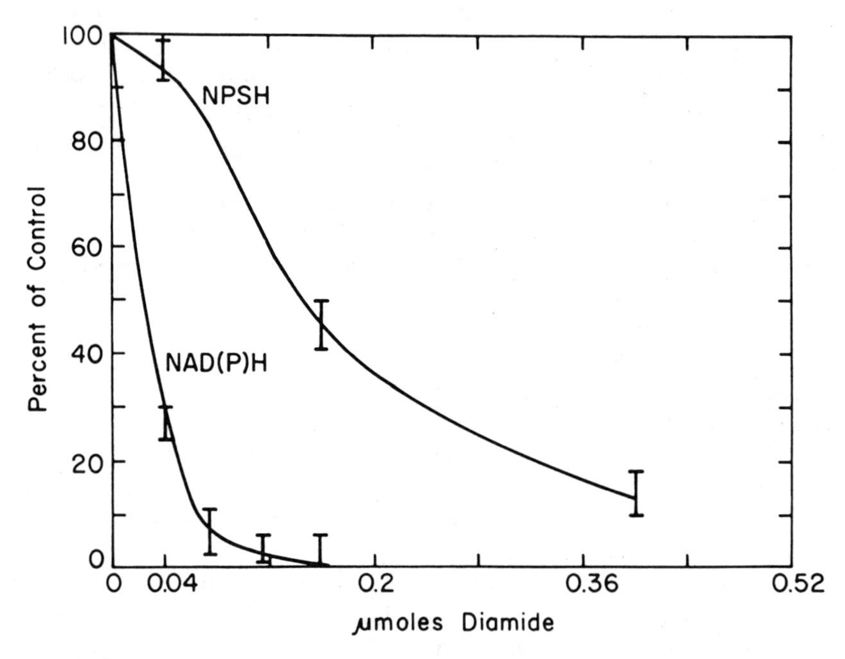


**Figure 18-17.** Titration of NPSH and NAD(P)H in Ehrlich ascites tumor cells. Single doses of diamide were added to samples of $4 \times 10^7$ cells/3 mL. The recorded values for NAD(P)H represent the points of minimal fluorescence, before regeneration of reduced pyridine nucleotide occurred. Non-protein thiol was measured in duplicate samples, also at the time of minimal fluorescence. The control value for NPSH was $2.5 \times 10^{-6}$ nmoles/cell. (Reproduced with permission.)[38]

of cells is due to the removal of a "Q" factor involved in the repair of radiation damage. Presumably, this Q factor is a sulfhydryl-containing enzyme.[37] It is interesting to note that other workers claim to have a glutathione-deficient, mutant, human cell line that does not show the oxygen enhancement ratio[4] for DNA "breaks." This cell line has the same nonprotein-thiol content relative to others cells. The deficiency in glutathione appears to be in the final step in biosynthesis of the tripeptide. The reported effects on DNA breakage with these cells may be due to a glutathione deficiency for an enzyme requiring glutathione involved in the repair of the radiation damage.

More recently,[29] glutathione depletion has been accomplished by diethylmaleate, a substrate for glutathione-*S*-transferase and by buthionine sulfoximine, an inhibitor of glutathione biosynthesis. Some cell lines show a

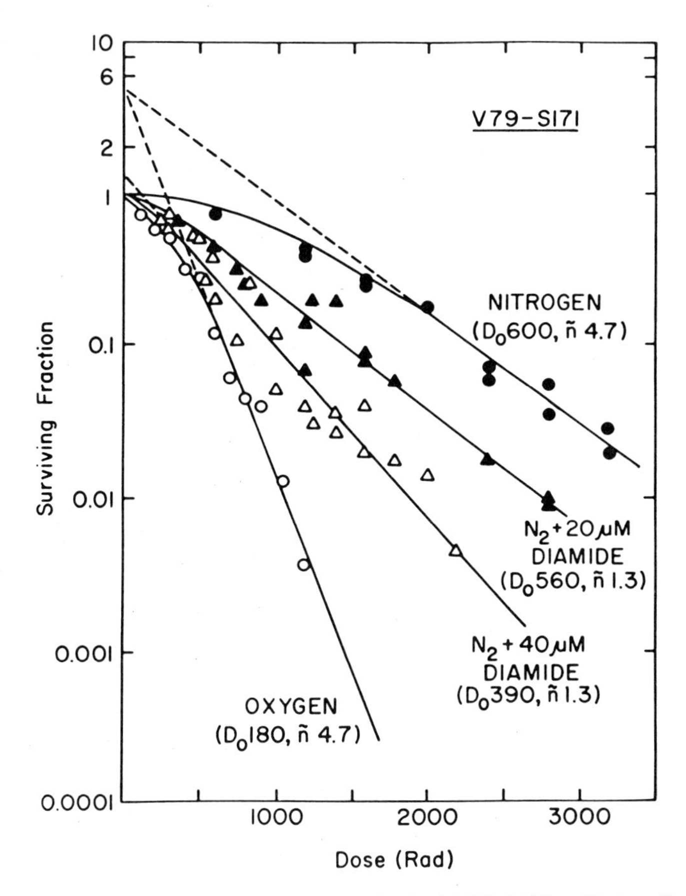


**Figure 18-18.** Effect of diamide on the survival of V79-S171 cells irradiated in air or nitrogen in the presence of 20 $\mu M$ and 40 $\mu M$ diamide. The apparent OER is 3.3. Survival curve parameters were estimated by eye; each point is the mean of at least two experiments. (Reproduced with permission.)[32]

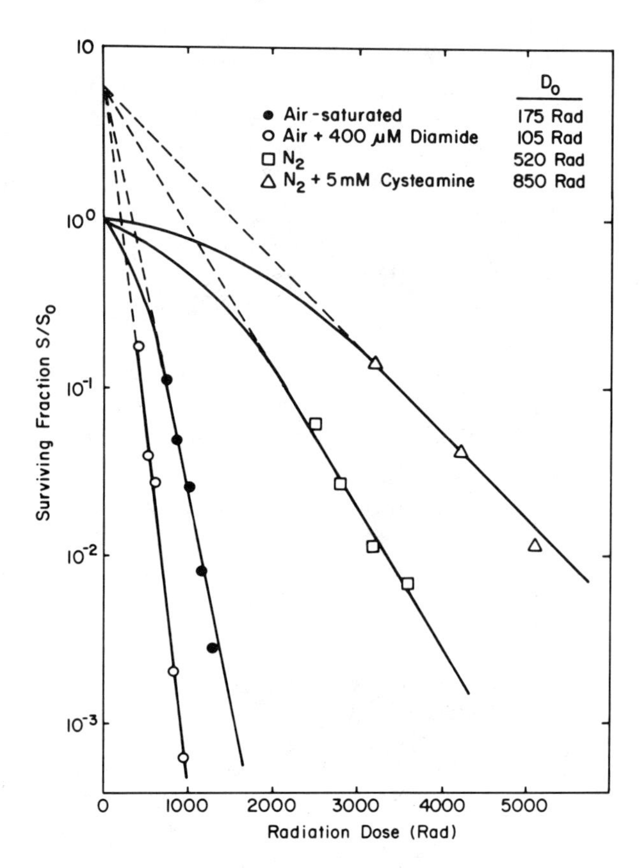


**Figure 18-19.** Whole survival curves for Chinese hamster cells irradiated under conditions exhibiting near maximum radiosensitivity and near maximum radioprotection. (Reproduced with permission.)[33]

partial increase to hypoxic irradiation after incubation with either of these agents. However, many workers[28,29] have also found increased aerobic sensitization as well. Biaglow[28,29] have suggested a role for glutathione in the aerobic detoxification mechanisms based on the supposition that glutathione will chemically reduce peroxy radicals to hydroperoxides. In addition, glutathione is a substrate for glutathione peroxidase and would reduce potentially harmful hydroperoxides to alcohols. Both of these glutathione reactions would decrease potentially harmful chain reactions involving cellular macromolecules.

## RADIOPROTECTORS AND RADIATION RESPONSE

Addition of chemicals to the culture medium prior to irradiation of the cells has resulted in either protection or sensitization of the cells. In a unique publication, based in part on a long series of observations, Chapman and his colleagues demonstrated such effects.[33] Figure 18-19 shows survival curves for Chinese hamster cells irradiated under conditions exhibiting near maximum and near minimum radiosensitivities, obtained with the radiosensitizer diamide under oxic conditions and the radioprotector cysteamine under hypoxic conditions. Survival curves for cells irradiated under air-saturated and acutely hypoxic conditions have been included for comparison. These results show that the radiosensitivity of air-saturated cells is not a maximum and that radiosensitivity of cells made acutely hypoxic is not a minimum. The interpretation of these results according to a free-radical model led Chapman to suggest that the cellular target environment is an important component in the expression of potentially lethal free-radical damage in the cellular target(s). Chapman[39] found that dimethylsulfoxide (DMSO) protected against radiation damage by competing with cellular targets for ·OH and had no effect on the radical-repairing or radical-fixing species within the cell. *t*-Butanol, while not as effective a radioprotector as DMSO, also had no effect on the environment near the target molecules.

Radioprotection by cysteamine (Figure 18-19) as well as by other thiols[33,41] such as dithiothreitol, does not appear to be the result of ·OH scavenging, although thiols are quite reactive with this radical (Figure 18-11).[20–21] It is believed that cysteamine protection reflects the total amount of radical damage in cellular targets that can be chemically repaired by hydrogen-donating species (resulting in enhanced cell viability). The thiol protectors increase the hydrogen-donating species in the target area (Figure 18-11).

In his careful studies, Chapman suggests that approximately 82% of the radiation inactivation (cell death) measured for air-saturated cells is due to the fixation of target radicals (62% of the target radicals are produced from ·OH and 20% from the direct effect) by oxygen (approximately 65%) and other endogenous electron-affinic substances (approximately 17%). The remaining 18% of cellular inactivation or cell death may result from irreparable or lethal damage to cellular target(s) by direct action. Moreover, the extent to which the indirect action of $e_{aq}$ and ·H contribute to the remaining cell inactivation is not known. There is at the present time no comprehensive mechanism that will account for all of the effects on ionizing radiation on mammalian cells.[2]

## INHIBITION OF CELLULAR OXYGEN UTILIZATION AND RADIOSENSITIZATION OF AN IN VITRO TUMOR MODEL

It was realized early in the development of radiosensitizing drugs that the ability to sensitize hypoxic cells to ionizing radiation could not be completely related to their electron affinity or one-electron acceptance capacity.[27] The more-electron-affinic compounds are also more active metabolically,[26,27,29,30] and one of the more important metabolic effects is alteration of oxygen utilization.[27] Such an effect is obviously unimportant under totally anoxic conditions, as have been previously used to determine radiosensitizing ability.[20–23] However, the effects of drugs on oxygen utilization becomes important in tumor-like situations where hypoxic conditions result because the oxygen utilization of the cells closest to the blood supply exceeds the rate of oxygen diffusion to more distant cells.[22] One system that can be utilized as a model to demonstrate such respiratory effects in vitro is the multicellular spheroid system as developed by Sutherland and Durand.[42] This system is based on the fact that a number of animal as well as human tumor cells can be grown in suspension culture in vitro in clusters consisting of many cells. In these clusters the consumption of oxygen by the outermost cells decreases the availability of oxygen to the innermost cells, thereby limiting the depth to which oxygen can penetrate in the spheroid. This results in an inner core of hypoxic cells that are radioresistant because of lack of oxygen. These clusters of cells can be trypsinized, and radiation survival curves obtained. With this system Biaglow and Durand[43] were able to demonstrate that relatively poor sensitizers, such as the nitrobenzene derivative nitrotoluene, were more effective sensitizers of the spheroids than the more electron-affinic dinitrobenzonitrile, known to have a potent effect on the radiation response of hypoxic cells irradiated as monolayer or attached cells.[43] The reason nitrotoluene was a more effective sensitizer is

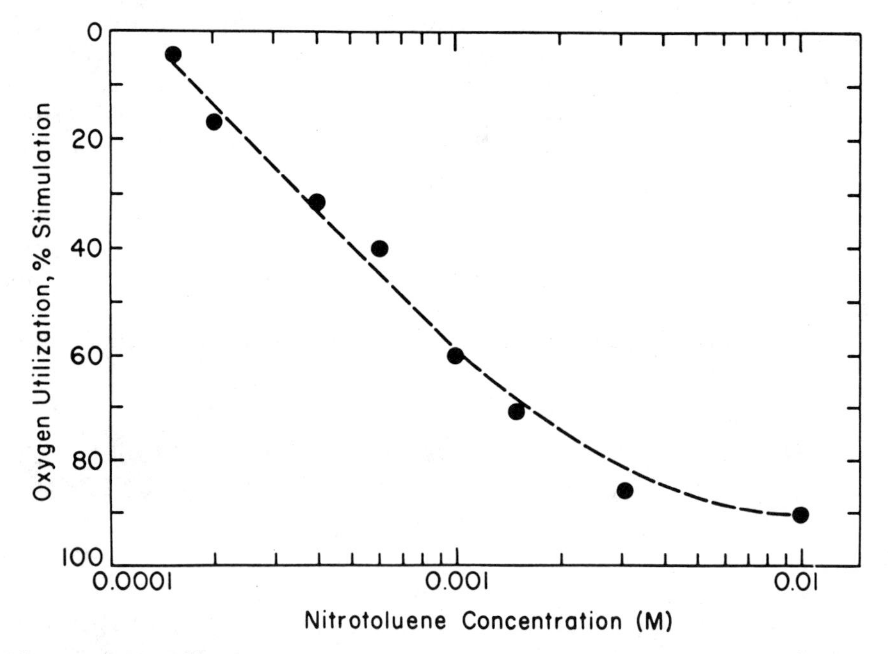


**Figure 18-20.** Effect of nitrotoluene on the oxygen consumption of V79 cells. (Reproduced with permission.)[43]

that it is a potent inhibitor of cellular oxygen utilization (Figure 18-20). A nearly log-linear response was found between nitrotoluene concentration, starting as low as $1.5 \times 10^{-4}$ *M*, and inhibition of oxygen utilization.[43] Inhibition was 85% with 2 m*M* and 90% with 10 m*M* nitrotoluene. No recovery from this inhibition was observed for periods up to 1 hr in the presence of drug. Concentrations of 2 m*M* and 10 m*M* nitrotoluene proved to be slightly toxic when incubated for periods greater than 90 min; 0.4 m*M* was not toxic. The same concentrations of nitrotoluene were tested for effects on the radiation response of the multicell spheroids (Figure 18-21A). At all concentrations tested, the survival after 2100 rad was drastically reduced immediately after addition of the drug. To determine whether this was a true radiosensitizing effect or metabolic phenomena caused by the effects of nitrotoluene on oxygen utilization, complete survival curves for the spheroids in the presence and absence of drug were generated (Fig 18-21B). Spheroids exposed to the nontoxic concentration of 2 m*M* nitrotoluene for 1 hr prior to radiation had survival curves that were essentially indistinguishable from those of fully reoxygenated spheroids, obtained by preirradiation trypsinization of the spheroids into single cells. Thus, in the partially hypoxic spheroid system, the inhibition of cellular oxygen utilization by nitrotoluene

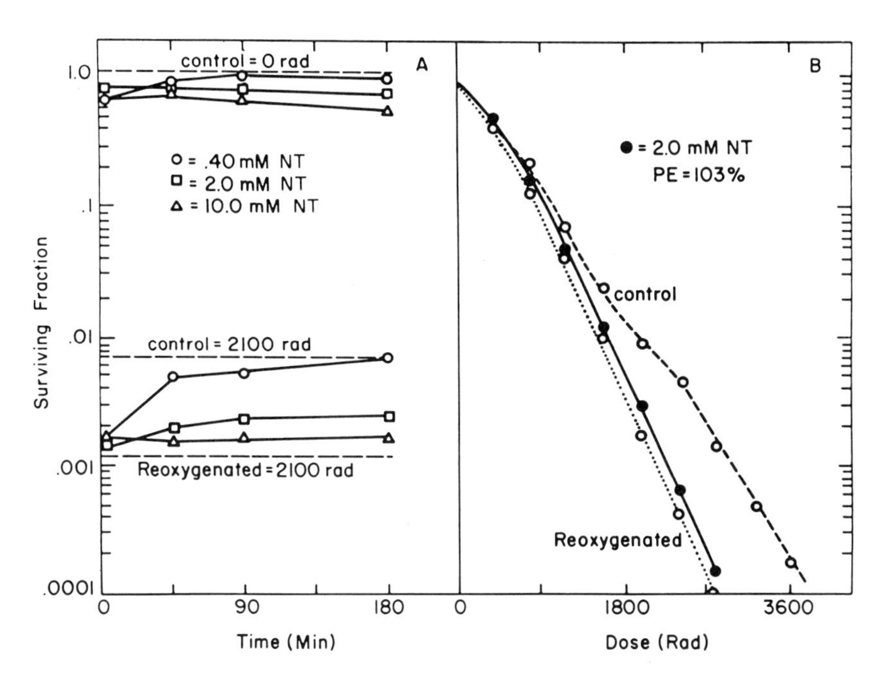


**Figure 18-21.** (A) Effects of nitrotoluene (NT) on plating efficiency (top) and survival (bottom) (after 2100 rad) of cells recovered from treated spheroids. Broken lines show fully reoxygenated cells. (B) Survival of cells from irradiated 582-$\mu M$ spheroids pretreated 1 hr with 2.0 m*M* NT. Survival is expressed relative to untreated spheroid cells at time zero. (Reproduced with permission.[43])

(Figure 18-20) sensitized the spheroid to almost the same degree as did molecular oxygen (Figure 18-21B), demonstrating that under simulated in vivo conditions alterations in the radiation response can be produced even by a drug known to be a relatively poor anoxic sensitizer[43] if it is an inhibitor of cellular oxygen utilization.[27]

Dinitrobenzonitrile, a moderately good anoxic radiosensitizer stimulated oxygen utilization[43] at concentrations as low as 30 $\mu M$ (Figure 18-22). An increase in the rate by a factor of 3, or 200%, occurred with DNBN concentrations in excess of 1 to 2 m*M*. The pronounced stimulation occurred immediately after addition of the drug and declined with time. However, at 45 min all the cells still consumed oxygen at rates greater than did the controls (Figure 18-22).

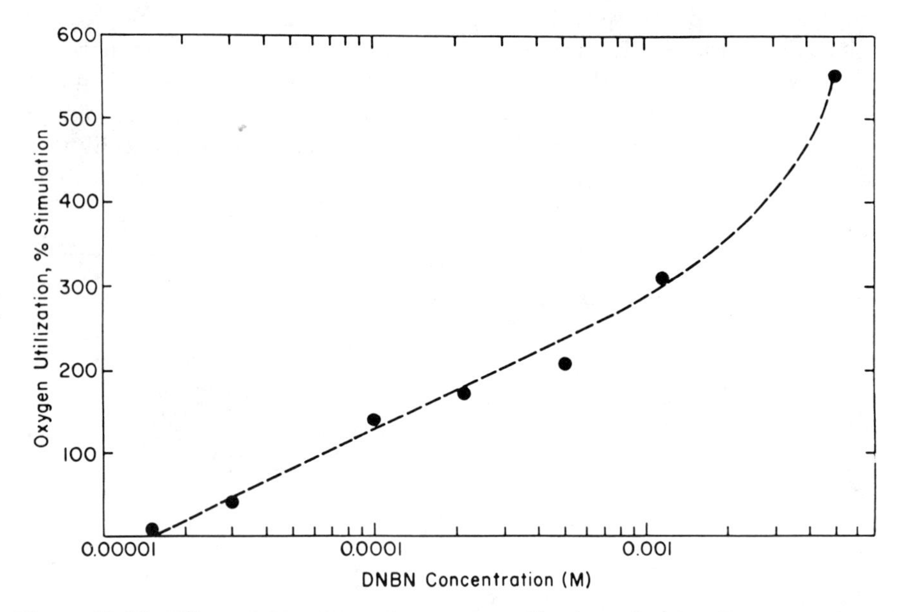


**Figure 18-22.** Effect of DNBN on the oxygen utilization of V79 cells. (Reproduced with permission.[43])

All concentrations of DNBN applied prior to radiation initially protected against 2100 rad, as seen by the increasing surviving fraction (Figure 18-23A). The greatest protection as well as the largest effect on oxygen utilization occurred after addition of 5 m*M* DNBN. The radiation protection with 5 m*M* drug diminished with time, becoming zero at 45 min; at 90 and 180 min, measurable radiosensitization occurred.[43] Sensitization at longer incubation time may be due to either cytotoxic effects or increased metabolic consumption of thiols known to protect against radiation damage. Complete radiation survival curves were obtained after exposure to nontoxic concentrations of 0.5 m*M* DNBN for 1 hr (Figure 18-23B). The radiation

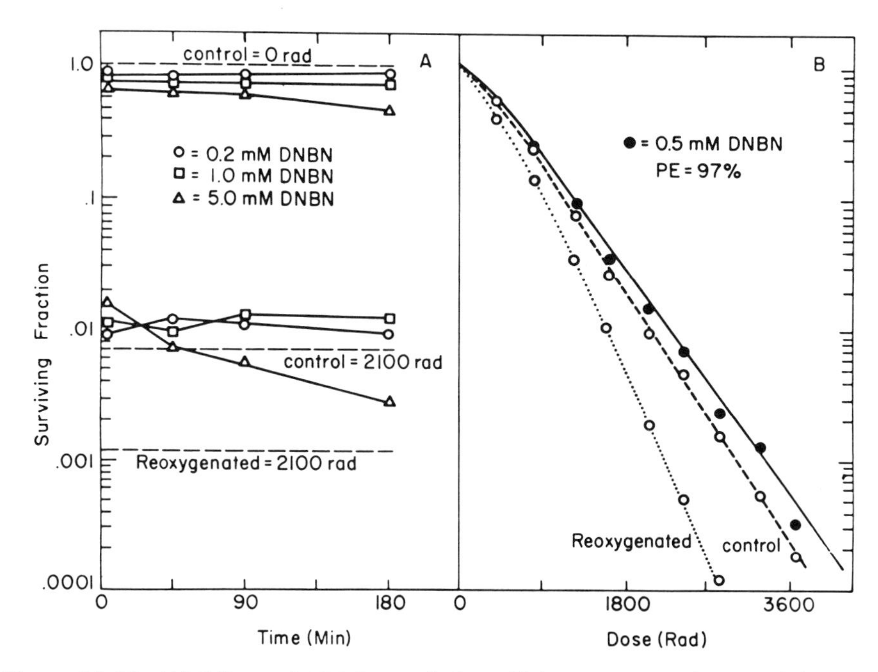


**Figure 18-23.** (A) Effects of DNBN on plating efficiency (top) and survival (bottom) (after 2100 rad) of cells recovered from treated spheroids. Broken lines represent control survival when hypoxic cells were reoxygenated. (B) Survival of cells from 582 $\mu M$ spheroids pretreated with 0.5 m$M$ DNBN for 1 hr prior to irradiation. (Reproduced with permission.[43])

response measured demonstrated a slight degree of radioprotection when compared to the response observed with untreated spheroids. Even greater protection was observed initially with higher DNBN concentrations.

To compare the consequences of stimulation or inhibition of oxygen utilization within a single population of spheroids, we constructed a composite curve by plotting the survival observed when addition of nitrotoluene or DNBN immediately preceded a radiation dose of 2100 rad against the measured respiratory rate (Figure 18-24). A fairly straightforward calculation leads to a prediction of response as a function of cellular respiration rate (dotted line) and enables a comparison to be made between the observed and theoretical responses.[22] The consistent overestimate of survival at the various DNBN concentrations may indicate a degree of hypoxic cell sensitization by this compound or some other metabolic effect, as already described. For nitrotoluene a similar tendency was only apparent at the lowest concentration tested.[44]

As a first-order approximation, the depth to which oxygen will penetrate the spheroid is proportional to the inverse square root of the oxygen consumption rate. Histological preparations allow an estimate of the size of the necrotic region. The relative volume of hypoxic cells can easily be

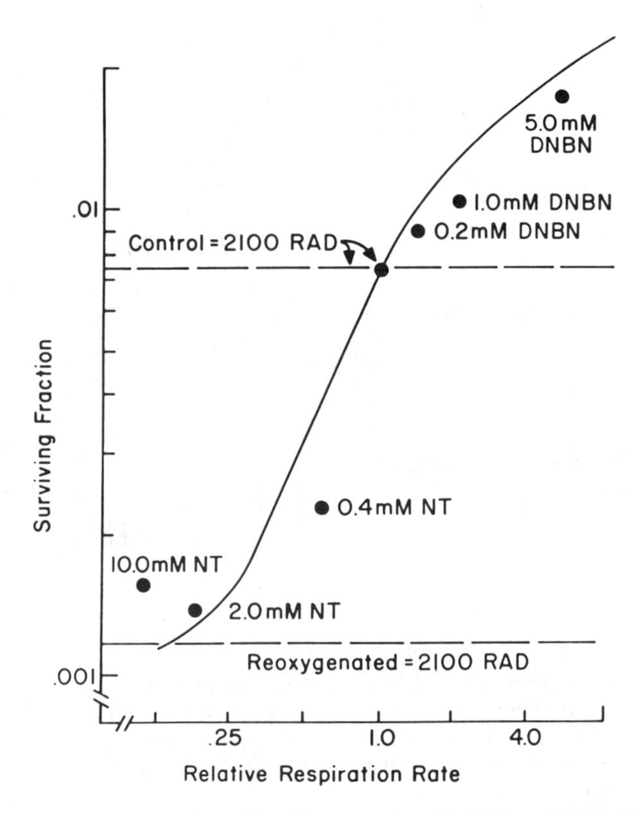


**Figure 18-24.** Survival of spheroid cells treated with NT or DNBN immediately prior to receiving 2100 rad, as a function of the relative oxygen utilization rate for each drug treatment. The solid line represents the predicted survival for spheroids consuming $O_2$ at the indicated rates. (Reproduced with permission.[43])

estimated if it is assumed that they are located predominantly near the necrotic region of the spheroid. Therefore, the spheroids with radius 291 $\mu$m, necrotic core radius $b$ (72 $\mu$m), and "critical" oxygen penetration depth $r$, the hypoxic fraction $F$ is given by $F = [(a - r)^3 - b^3]/(a^3 - b^3)$ for $a - b \geqslant r \geqslant 0$. The spheroids, when totally oxygenated, had a surviving fraction $S_{ox} = 0.0012$ after 2100 rad; if totally anoxic, the survival was $S_{AN} = 0.129$. Hence, the net survival is a function of the fraction of cells that are anoxic; ie, $S = (1 - F)S_{ox} + F - S_{AN}$. Since the oxygen penetration ($r$) varies with rate of consumption $K$ so that $r = r_0(K)^{1/2}$, where $r_0$ represents the penetration depth under normal conditions (observed survival = 0.0076), substituting into the equation for $S$ defines $r_0$ and, hence, allows calculation of $S$ as a function of $K$.

The results with spheroids indicate that radiosensitizing chemicals that alter oxygen utilization cannot be adequately tested for potential usefulness in radiotherapy when studied in hypoxic single-cell systems. Oxygen is not available to cells distant from the vascular supply (metabolic barrier) and the degree and extent of cellular and tissue hypoxia are critically dependent upon the rate of oxygen utilization. Many electron-affinic drugs, including some of the nitrobenzenes and nitrofurans, have been shown to stimulate cellular oxygen utilization[27] or to stimulate oxygen consumption by reaction with ascorbate.[45] This property may alter the effectiveness of these drugs as sensitizers in tumors, because they may produce an increased hypoxic region, as they do in the case of spheroids, and thereby reduce any effect they may have as radiation sensitizers of the hypoxic and anoxic cells.[20,23,27,43]

In an extension of these studies Durand and Biaglow investigated the radiosensitizing effects of more potent inhibitors of respiration. The effects of the classical inhibitors of mitochondrial respiration on the oxygen consumption of V79 lung cells in suspension are shown in Figure 18-25. The

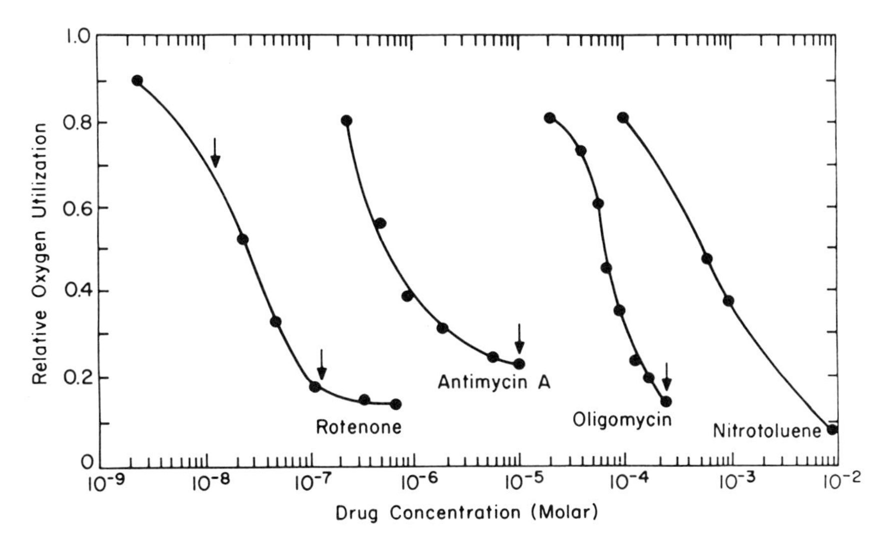


**Figure 18-25.** The effect of different concentrations of rotenone, antimycin A, oligomycin, and nitrotoluene on the oxygen consumption of V79 lung cells, $3.0 \times 10^7$ cells/mL, in complete growth medium buffered with 0.02 *M* HEPES. Each point represents a duplicate measurement (from the same preparation) at pH 7.4 and 37°C. DMSO, at the concentrations used, had no effect on respiration. (Reproduced with permission.[46])

concentrations required for 50% inhibition were rotenone ($3 \times 10^{-8}$ *M*), antimycin A ($6 \times 10^{-7}$ *M*), oligomycin ($7 \times 10^{-6}$ *M*), and nitrotoluene ($5 \times 10^{-4}$ *M*). Clearly, rotenone is the most effective of these classical inhibitors of cellular oxidation on a molar basis, and all were considerably more effective than nitrotoluene (Figure 18-20). All of the drugs that inhibited cellular oxygen consumption were found to be sensitizers for the spheroid system.[45] Figure 18-26 shows the results obtained with rotenone. By varying the time of incubation with the drug prior to irradiation (Figure 18-26A),

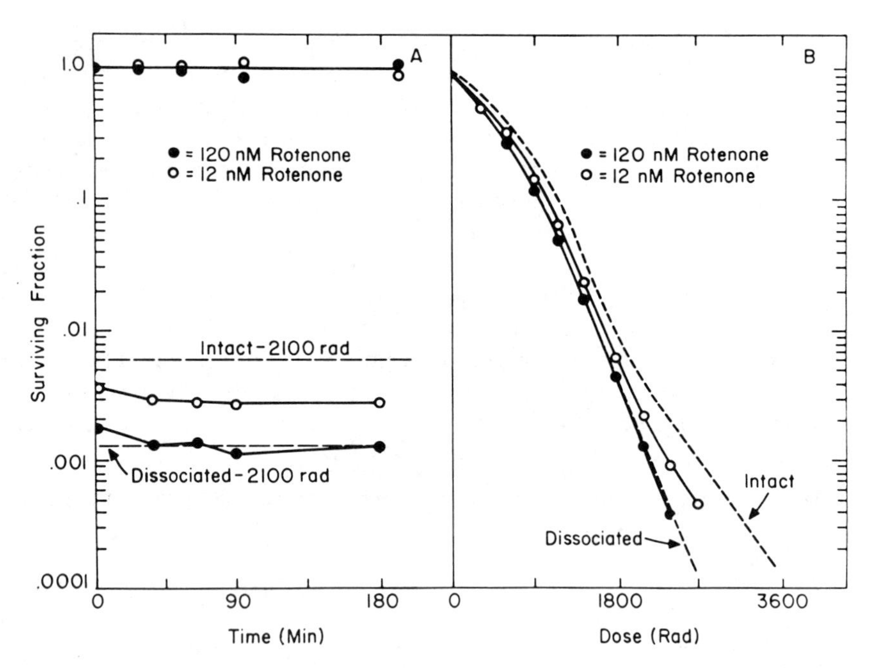


**Figure 18-26.** The response of spheroids to the rotenone and irradiation. Panel (A) shows the response to rotenone alone (top) or to 2100 rad of $\gamma$-rays after exposure to the drug for different times. The dotted lines indicate the response of intact or dissociated spheroids to 2100 rad alone. Complete dose-survival curves for the same drug concentrations present 30 min before and during irradiation are shown in panel (B). (Reproduced with permission.[46])

the kinetics of reoxygenation of the hypoxic cells in the spheroid can be estimated.[46] With the higher concentration of rotenone a large amount of reoxygenation occurred almost immediately, whereas less sensitization was observed immediately after the lower concentration. Less sensitization was observed with $1.2 \times 10^{-8}$ *M* rotenone, as would be expected on the basis of the respiration data. Suspensions of aerobic and anoxic V79 cells from single-cell cultures showed no modification of the radiation response by the same rotenone concentrations.[46] These data indicate that oxygen-sparing mechanisms are very efficient in sensitizing hypoxic cells within a spheroid to radiation. However, such potent inhibitors of respiration cannot be used in vivo because they would inhibit the oxygen utilization of every cell in the body, producing death. A more attractive approach is to utilize drugs that concentrate in tumor tissue and are also inhibitors of cellular oxygen utilization.[46–49] Another approach that we have been concerned with over the last few years is to utilize physiological controls or tumor enzymes that would release respiratory inhibitors such as cyanide[47] in the tumor in order to produce the oxygen-sparing effects. We have found that exploitation of the Crabtree effect, namely, the inhibition of respiration by glucose, is a potential physiological means for producing oxygen-sparing effects. The Crabtree effect is nearly a universal phenomena occurring with most tumor tissue in vitro and has been recently shown to occur in vivo. The glucose effects are innocuous to the surrounding tissue, and the metabolic transient alteration is transient with rapid return to normal. The metabolic use of glucose in vivo has shown some promise alone and in combination with breathing oxygen.[47–49]

## SUMMARY

Depicted in Figure 18-11 are the various places where drugs may influence the radiation effects on the target molecule DNA as well as its subsequent chemical and enzymatic repair. We have also indicated the sites where metabolism may be important with respect to the cellular content of endogenous hydrogen-donating species and to the availability of oxygen consumed by respiration. The effects of radiation on DNA can be modified by hydroxyl radical scavengers in air and by hydrogen-donating thiols under aerobic and anaerobic conditions. The ·OH radical scavengers alter the indirect effect of radiation by reacting with ·OH before it reacts with DNA. Thiols can react with either the ·OH radical or add to the pool of hydrogen-donating species within the cell that can chemically repair the damaged DNA. The pool of hydrogen-donating species can be removed by preincubation of cells with nitro compounds or by the oxidizing agent diamide. *N*-ethylmaleimide is also quite effective in reacting with thiols. The major hydrogen-donating thiol species within the cell are the nonprotein and protein thiols. Radiation or drug-oxidized thiols are regenerated back to the reduced thiol by means of glucose-linked metabolism. Oxidation of protein thiols can result in enzyme inhibition and a decreased capacity to repair DNA damage. On the left side of Figure 18-11 is seen the ability of oxygen- or electron-affinic drugs to react with the target radical to produce drug or oxygen adducts as well as DNA breaks. Some of this damage can be enzymatically repaired, resulting in an increase in cell survival. Oxygen is the chief radiosensitizer, and it is rapidly depleted in vivo and in vitro in multicellular systems by cellular respiration. Many drugs can stimulate or inhibit the oxygen consumption, resulting in radiation protection or sensitization.

## CONCLUSION

The effects of ionizing radiation on mammalian cells remains a viable area for investigation because of the influence of radiation on our lives. Radiation is used to treat cancer and to diagnose disease and is a waste byproduct from medical and other industries as well as from nuclear power plants. Solar radiation penetrates our bodies every day, and we have natural radioisotopes in the environment. In addition, space flights may be limited because of the continuous solar bombardment by $\gamma$-rays that astronauts are exposed to. The chief problem with human exposure to radiation is the production of mutational events that may lead to malignancies. Therefore, understanding the effects of radiation on living things is necessary in order to define and limit the unnecessary risks due to high-level exposure that may inadvertently occur through misuse of radiation. The understanding of the basic mechanisms for radiation damage and protection may lead to the design of drugs that will greatly reduce the risk from high- or low-level exposure and improve the response of human tumors to ionizing radiation with respect to this later possibility; the protection of normal tissue during the irradiation of tumor tissue is being actively pursued in order to deliver more radiation to the tumor.

## ACKNOWLEDGMENTS

I would like to thank Dr. Marie E. Varnes and Mrs. Birgit Jacobson for reading and correcting this manuscript prior to its completion. The work on this review was supported by Grant No. CA-13747, awarded by the National Cancer Institute, DHEW.

## REFERENCES

1. Fogh, J. "Human Tumor Cells In Vitro"; Plenum Press: New York, 1975.
2. Alper, T. "Cellular Radiobiology"; Cambridge University Press: Cambridge, 1979
3. Elkind, M. M.; Sinclair, W. K. "Current Topics in Radiation Research"; North-Holland: Amsterdam, 1965, Vol. 1, p. 165.
4. Puck, T. T.; Marcus, P. I. J. *Exp. Med.* **1956**, *103*, 653.
5. Hall, E. J. "Radiobiology for the Radiologist"; Harper: New York, 1978.
6. Grosch, D. S.; Hopwood, L. E. "Biological Effects of Radiations", 2nd ed.; Academic Press: New York, 1979.
7. Salmon, S. E.; Hamburger, A. W.; Soehnlen, B.; Brian, B. S.; Durie, G. M.; Aberts, D. S.; Moon, T. E. *New England J. Med.* **1978**, *298*, 1321.
8. Jakoby, W. B.; Postan, J. H. (ed.) "Methods in Enzymology" Vol. 18. Cell Culture"; Academic Press: New York, 1979.
9. Pryor, W. A. (ed.) "Free Radicals in Biology"; Academic Press: New York, 1977, Vols. 1–6.
10. Gregg, E. C. "Effects of Ionizing Radiation in Humans", Chemical Rubber Co. Reviews, 1984.
11. Elkind, M. M.; Sutton, H. *Nature* **1959**, *184*, 1293.
12. Gray, L. H.; Conger, A. D.; Ebert, M.; Hornsey, S.; Scott, O. C. A. *Brit. J. Radiol.* **1958**, *26*, 638.
13. Flanders, H. P.; Moore, D. *Radiat. Res.* **1958**, *9*, 422.
14. Ward, J. F. *Adv. Radiat. Biol.* **1977**, *5*, 181.
15. Painter, R. B. In "Radiation Biology in Cancer Research"; Meyer, R. E.; and Withers, H. R. Eds.; Raven Press: New York **1980**, p. 59.

16. Roots, R.; Okada, S. *Radiat. Res.* **1975**, *64*, 306.
17. Epp, E. R.; Weiss, H.; Ling, C. C. *Current Topics Radiat. Res. Quart.* **1976**, *11*, 201.
18. Millar, B. C.; Feilden, E. M.; Steele, L. *Int. J. Radiat. Biol.* **1979**, *36*, 177.
19. Okada, S. In "Radiation Biochemistry"; Academic Press: New York, 1969, Vols. 1 and 2.
20. Adams, G. E.; Jameson, D. *Radiat. Environ. Biophys.* **1980**, *17*, 95 or Adams, G. E.; Wardman, P. In "Free Radicals and Cancer"; Pryor, W. A.; Ed.; Academic Press: New York, 1977, Vol. 3, p. 53.
21. Adams, G. E. In "Radiation Sensitizers for Hypoxic Cells: Problems and Prospects in Treatment of Radioresistant Tumors"; Abe, M.; Sakamoto, K.; and Philips, T. L. Eds., Elsevier Biomedical Press, 1979.
22. Fowler, J. F.; Adams, G. E.; Denekamp, J. *Cancer Treatment Rev.* **1976**, *3*, 2277.
23. Becker, F. F. (ed.) "Cancer, A Comprehensive Treatise, 6, Radiotherapy, Surgery and Immunotherapy"; Plenum Press: New York, 1977.
24. Gray Conference on Hypoxic Cell Sensitizers in Radiobiology and Radiotherapy, *Brit. J. Cancer*, Suppl. III, **1978**. "Chemical Modifiers of Cancer Treatment"; Chapman, J. D.; and Whitmore, G. F. Eds.; Pergamon Press: 1984, "Chemical Modification: Radiation and Cytotoxic Drugs"; Sutherland, R. M., Ed.; Pergamon Press, **1982**.
25. Hall, E. J.; Biaglow, J. E. *Int. J. Radiat. Oncol. Biol. Phys.* **1977**, *2*, 521.
26. Biaglow, J. E. In "Oxygen and Oxy-Radicals in Chemistry and Biology"; Rodgers, M. J. and Power, E. L.; Eds.; Academic Press: New York, 1981.
27. Biaglow, J. E. *J. Pharm. Ther.* **1980**, *10*, 283.
28. Brady, L. "Proceedings of the Conference on Combined Modality Cancer Treatment: Radiation Sensitizers and Protectors"; Masson Press: New York, 1980.
29. Biaglow, J. E.; Varnes, M. E.; Epp, E.; Clark, E. *Radiat. Res.* **1983**, *195*, 437.
30. Biaglow, J. E.; Varnes, M. E.; Koch, C. J.; Sridhar, R. In "Free Radicals and Cancer"; Floyd, R., Ed.; Marcel Dekker: New York, 1980.
31. Varnes, M. E.; Biaglow, J. E.; Koch, C. J.; Hall, E. J. In "Conference on Combined Modality Cancer Treatment: Radiation Sensitizers and Protectors"; Brady, L., Ed.; Masson Press: New York, 1980.
32. Harris, J. W.; Power, J. A. *Radiat. Res.* **1973**, *56*, 97.
33. Chapman, J. D. *Radiat. Res.* **1973**, *56*, 291.
34. Watts, M. E.; Whillans, D. W.; Adams, G. E. *Int. J. Radiat. Biol.* **1975**, *27*, 259.
35. Whitmore, G. F.; Gulyas, S.; Varghese, A. *Brit. J. Cancer* **1978**, *37*, Suppl. III, 115.
36. Greenstock, C. L.; Dunlop, I. In "Fast Processes in Radiation Chemistry and Biology"; Adams, G. E.; Fielden, E. M.; Michael, B. D., Eds.; John Wiley: London, 1975, p. 247.
37. Sinclair, W. In "Radiation Research, Biomedical, Chemical and Physical Perspectives"; Nygaard, O. F.; Adler, H. I.; and Sinclair, W. K., Eds.; Academic Press: New York, 1975.
38. Biaglow, J. E.; Nygaard, O. F. *Biochem. Biophys. Res. Commun.* **1973**, *54*, 874.
39. Harris, J. W.; Biaglow, J. E. *Biochem. Biophys. Res. Commun.* **1972**, *46*, 1743.
40. Revesz, L.; Edgren, M.; Larsson, A.; In "Proceedings of the Sixth International Congress on Radiation Research, Tokyo"; Okada, S.; Imamura, M.; Terashima, T.; Yamaguchi, H.; Eds.; Toppan Printing: Tokyo, Japan, 1979, p. 862.
41. Chapman, J. D.; Dugle, D. L.; Reuvers, A. P.; Gillispie, C. J.; Borsam, J. In "Radiation Research: Biomedical, Chemical and Physical Perspectives"; Nygaard, O. F.; Adler, H. L.; Sinclair, W. K., Eds.; Academic Press: New York, 1975, p. 478.
42. Sutherland, R. M.; Durand, R. E. *Current Topics Radiat. Res. Quart.* **1976**, *11*, 87.
43. Biaglow, J. E.; Durand, R. E. *Radiat. Res.* **1976**, *65*, 529.
44. Biaglow, J. E.; Jacobson, B.; Greenstock, C. L.; Raleigh, J. *Mol. Pharm.* **1977**, *13*, 269.
45. Biaglow, J. E.; Jacobson, B. *Brit. J. Radiol.* **1977**, *50*, 844.
46. Durand, R. E.; Biaglow, J. E. *Radiat. Res.* **1977**, *69*, 359.
47. Biaglow, J. E.; Durand, R. E. *Int. J. Radiat. Biol.* **1978**, *33*, 377.
48. Biaglow, J. E.; Lavick, P.; Ferencz, N. *Radiat. Res.* **1969**, *39*, 623.
49. Biaglow, J. E.; Ferencz, N.; Friedell, H. *Am. J. Roent. Radium. Ther. Nucl. Med.* **1970**, *108*, 405.

# 19

# Some Applications of Radiation Chemistry to Biochemistry and Radiobiology

**P. WARDMAN**

*Gray Laboratory of the Cancer Research Campaign, Mount Vernon Hospital*

## INTRODUCTION

Prefacing the proceedings of the Symposium on Free Radicals in Biological Systems in 1960, Blois commented: "As short a time as seven years ago a question frequently heard was '... do free radicals *really occur* in biological systems?' This has been answered affirmatively ...".[1] The considerable progress in our knowledge of the role of free radicals in biology in the last two decades is attested to by Pryor's note of "... the remarkable explosion of research in free radical biology over the past few years" in prefacing, in 1982, one of an important series of volumes of reviews in this area.[2] The radical reactions of oxygen and oxyradicals in biology are of particular interest,[3,4] especially the involvement of oxygen radicals in tissue damage.[5,6] Studies of free-radical mechanisms in lipid peroxidation,[7] inflammatory disease,[8] and many aspects of cancer,[7,9] including carcinogenesis[10,11] and anticarcinogens,[12] illustrate the diversity of current activity. Free-radical intermediates in the metabolism or activation of xenobiotic compounds, including medically important drugs, are being positively identified in many


*Radiation Chemistry: Principles and Applications*

biological systems.[13] The Society for Free Radical Research was founded in 1982 specifically to further discussion of free-radical processes of industrial and medical importance.[14]

The likely involvement of radicals in radiation-induced cell killing has been accepted for many years. The study of radiation-chemical reactions of potential importance to radiobiology has, therefore, been an important area in radiation chemistry since its emergence. The most important cellular "target" is usually considered to be DNA and/or DNA/membrane complexes,[15,16] and reviews of the physical, chemical, and biological effects of ionizing radiation on DNA have been collated.[17] The potential modification of cellular radiosensitivity (radiosensitization or radioprotection) by drugs that could be given to a patient before radiotherapy has its basis in radiation-chemical reactions. Although there is widespread radiobiological and clinical interest in the potential use of such sensitizers and protectors in the radiotherapy of solid tumors,[12,18–21] there are still many unanswered questions concerning the radiation-chemical mechanisms involved.[22,23]

We have, then, two major aspects of the role of free radicals in biology. On the one hand, the interest is in the chemical reactions of free radicals of known or potential importance in biochemistry or biology, where the radicals are generated in nature by biochemical processes without the aid of radiation. On the other hand, in radiobiology radicals ultimately leading to cell death are generated radiation chemically. The potential contribution of radiation-chemical techniques to both these areas is not, however, limited or defined by this artificial division. Radiation chemistry has a great deal to contribute to both fields because of the ease with which selected radicals can be generated and their reactions monitored, using techniques originally established to help understand the primary processes following radiation absorption. Thus, an identifiable solute radical or stable product may originally have been used to measure, for example, the yield of $e_{aq}^-$ or $OH\cdot$ produced in irradiated water, and relative or absolute rate constants for reaction of hundreds of solutes with these primary radicals have been established. We can now utilize these concepts and techniques together with the wealth of rate data to design experiments in which radicals of biological interest can be selectively generated with a high degree of confidence in the initial processes likely to occur and in the concentrations of radicals produced. Many of the potential reactions of these selected radicals with biologically important molecules can now be studied in these model systems, often ending years of speculation about reactions that conventional techniques have suggested but cannot hope to characterize.

Radiation chemistry in this context, then, is often used, not so much to investigate radiation effects per se, but as a tool. In many radiation-chemical experiments of this type, the chemical details (eg, of the primary processes of water radiolysis) are often taken for granted and "glossed over." Although such simplification for the sake of brevity is a flattering reflection of the efforts of half a century's research, careful attention to experimental design is still necessary. A most common error, for example, is the assignment of $G$-values for $e_{aq}^-$, OH·, and so on, as universal constants, whereas the yields of scavengeable primary radicals vary with solute concentration in most circumstances. This point is discussed in Chapter 10, and other recent reviews[24] or experimental studies[25] demonstrate these spur scavenging effects with clarity and precision.

In this chapter we illustrate the use of radiation chemistry as a tool in investigating biologically important radical reactions, and we also outline some studies of models for radiobiological damage. Because aqueous solutions usually offer the most important matrix, an appreciation of the main features of water radiolysis (Chapter 10) will be essential. Most of the illustrations involve pulse radiolysis (Chapter 3), and some familiarity with chemical kinetics (Chapter 4) is assumed. In addition to these and other chapters in this book, readers will find the proceedings of a recent NATO Advanced Study Institute[26] most useful. We shall not try to review here all the applications of radiation chemistry to biochemistry and biology, but we will illustrate, using selected examples, the main principles and practical advantages and problems. Another recent volume[27] covers the main contributions of flash photolysis and pulse radiolysis to the chemistry of biology and medicine, complementing earlier reviews.[28–34] Papers from symposia on radical processes in radiobiology and carcinogenesis,[35–38] and on superoxide dismutases,[39–40] and proceedings of recent international congresses of radiation research,[41–43] together with the other publications referred to above will enable the reader to gain a comprehensive overview of the role of radicals in biological processes and the contributions of radiation chemistry.

## REDOX PROCESSES IN BIOLOGY: PARALLELS IN RADIATION CHEMISTRY

### *Oxidation and Reduction*

Reduction of an organic compound Q to a compound $QH_2$ formally involves the addition of 2 electrons ($e^-$) and 2 protons; reduction of 1,4-benzoquinone to 1,4-hydroquinone would be a simple example, but Eq. (19-1) could well be used to formally describe the relationship between the reduced and oxidized forms of many other compounds important in biology—oxygen, quinones, flavins, ascorbic acid, nicotinamide derivatives, and so on:

$$Q + 2\,e^- + 2\,H^+ \rightleftharpoons QH_2 \qquad (19\text{-}1)$$

with the proviso that in aqueous solutions, prototropic equilibria may define the precise form of $QH_2$ and Q:

$$QH_2 \rightleftharpoons QH^- + H^+ \quad (19\text{-}2)$$

$$QH^- \rightleftharpoons Q^{2-} + H^+ \quad (19\text{-}3)$$

$$QH^+ \rightleftharpoons Q + H^+, \quad \text{and so on} \quad (19\text{-}4)$$

In experiments involving potentiometric titrations of a pyocyanine (5,10-*H*-1-phenazinol), Michaelis found, in 1931, that oxidation of the reduced form $QH_2$ in acid solution by only one-electron equivalent gave a green intermediate, whereas two-electron oxidation to the fully oxidized form produced the colorless Q. The intermediate QH· was thought to be a free radical and was termed a "semiquinone":[44]

$$QH_2 \rightleftharpoons QH\cdot + e^- + H^+ \quad (19\text{-}5)$$

$$QH\cdot \rightleftharpoons Q + e^- + H^+ \quad (19\text{-}6)$$

which was stable only in acid solutions:

$$QH\cdot \rightleftharpoons Q\cdot^- + H^+ \quad (19\text{-}7)$$

As Chance has described,[45] this discovery had major impact on our understanding of biological oxidation-reduction reactions. Although Michaelis' conviction "... that all oxidations of organic molecules, although they are bivalent, proceed in two successive univalent steps, the intermediate being a free radical ..."[45] is now seen to be too broad a generalization, a wide range of biologically important redox couples in which the semiquinone-type intermediate exist have now been characterized. One of the most important of these is the stepwise reduction of oxygen to hydrogen peroxide:

$$O_2 + e^- + H^+ \rightleftharpoons HO_2\cdot \quad (19\text{-}8)$$

$$HO_2\cdot + e^- + H^+ \rightleftharpoons H_2O_2 \quad (19\text{-}9)$$

$$HO_2\cdot \rightleftharpoons O_2^-\cdot + H^+ \quad (\text{p}K_a = 4.69)^{46} \quad (19\text{-}10)$$

$$H_2O_2 \rightleftharpoons HO_2^- + H^+ \quad (\text{p}K_a = 11.62)^{47} \quad (19\text{-}11)$$

with $O_2^-\cdot$ and $H_2O_2$ the predominant forms of the one- and two-electron reduced species, respectively, at pH 7 in aqueous solution. (We shall frequently use abbreviations such as $Q^{2-}$, $Q^-\cdot$ as appropriate, without regard to prototropic state.)

## *The Stability of Radical Intermediates in Biologically Important Redox Reactions*

The stability of the semiquinone intermediate $Q^-\cdot$ ($QH\cdot$, $QH_2^+\cdot$, etc) can be expressed as a formation constant:

$$Q + Q^{2-} \rightleftharpoons 2\,Q^-\cdot \quad (19\text{-}12)$$

$$K_f = \frac{[Q^-\cdot]^2}{[Q][Q^{2-}]} \tag{19-13}$$

although at pH 7, $Q^{2-}$, $Q^-\cdot$ will generally be at least singly protonated, $QH^-$, $QH_2$, $QH\cdot$, and so on, and an experimental formation constant at any $pH_i$ may be defined, where the subscript "tot" denotes the sum of all prototropic forms:

$$K_{fi} = \frac{[Q^-\cdot]_{tot}{}^2}{[Q]_{tot}[Q^{2-}]_{tot}} \tag{19-14}$$

with pH-dependence given by

$$K_{fi} = K_f\left(1 + \frac{[H^+]_i}{K_7}\right)^2\left(1 + \frac{[H^+]_i}{K_3} + \frac{[H^+]_i^2}{K_2K_3}\right)^{-1} \tag{19-15}$$

where $K_7$, $K_3$, and $K_2$ are the equilibrium constants for dissociations (19-7), (19-3), and (19-2), respectively.

It is useful to relate $K_f$ to the reduction potentials of the individual one-electron couples $E^1(Q/Q^-\cdot)$ and $E^2(Q^-\cdot/Q^{2-})$, abbreviated to $E_i^1$, $E_i^2$ or the reduction potentials vs. NHE for addition of the first or second electrons to the oxidant Q (potentials in mV):

$$E_i^1 - E_i^2 = \left(\frac{\mathbf{RT}}{F}\right)\ln K_{fi} \tag{19-16}$$

$$E_i^1 - E_i^2 = 59.15 \log K_{fi}. \tag{19-17}$$

Since the reduction potential of the two-electron couple, $Q/Q^{2-}$, is frequently available from conventional electrochemical measurements (often symbolized as $E°$ or $E°'$ at pH 0 or 7, respectively, but here given as $E_m$, following Clark[48]), and

$$E_{mi} = \frac{E_i^1 + E_i^2}{2} \tag{19-18}$$

we see that $K_{fi}$ can be calculated if $E_{mi}$ and *either* $E_i^1$ *or* $E_i^2$ are known. Conversely, $E_i^1$, $E_i^2$ can be calculated if $K_{fi}$ and $E_{mi}$ are known.

The likelihood of detecting the intermediate $Q^-\cdot$ ($QH\cdot$, etc) in any redox system can be ascertained from the concentrations $[Q]_{tot}$, $[QH_2]_{tot}$ permissible and $K_{fi}$. The latter can be extrapolated to, for example, pH 7 from values at $pH_i$ if $K_2$, $K_3$, and $K_7$ are known. Consider, for example, the equilibrium (19-12) with $Q^{2-}$ = ascorbic acid and Q = dehydroascorbic acid. Measurements of $[Q^-\cdot]_{tot}$ (the ascorbyl radical) by electron spin resonance (ESR) gave values of $K_{fi}$ at 298 K decreasing from $5.1 \times 10^{-9}$ at pH 6.4 to $5.6 \times 10^{-12}$ at pH 4.0.[49] Using Eq (19-15) with $pK_2$, $pK_3$, and $pK_7$ = 4.21, 11.52[50], and −0.45,[51] respectively, we find that the mean value of $K_f$ from six measurements is $1.2 \times 10^{-3}$. In Figure 19-1 we have plotted $K_{fi}$ from pH 0–14 and also the reduction potential $E_i^2$ ($Q^{-}\cdot/Q^{2}$) (ascorbyl radical/ascorbate) calculated from this value of $K_f$ and the value[48] $E_{m7}(A/AH_2) = 58$ mV, using Eqs. (19-17) and (19-18). At pH 13.5 we calculate $E^2$ ($Q^{-}\cdot/Q^{2-}$) = 11 mV, in excellent agreement with the value of 15 mV found by Steenken and Neta[52], using a quite independent pulse radiolysis method (see below).

Figure 19-1 also illustrates the pH dependence of a few other typical redox couples. Baxendale and Hardy's values[55,56] of $pK_2 = 11.24$, $pK_3 = 12.83$, and $K_f = 1.28$ for duroquinone (2,3,5,6-tetramethyl-1,4-benzoquinone) at 298 K, together with Patel and Willson's measurement[57] of $pK_7 = 5.1$ and $E_{m7} = 55$ mV (based on two studies[58,59]), give $E_7^1$ ($Q/Q^{-}\cdot$) = −240 mV, in excellent agreement with Wardman and Clarke's independent estimate[60] of (−244 ± 7) mV, using pulse radiolysis (see below). The formation constant for flavin mononucleotide (FMN) semiquinone was calculated from measurements[48,61] of $E_m$(FMN/FMNH$_2$) and $E_7^1$(FMN/FMNH·) and the appropriate $pK_a$'s; some early independent estimates[62–64] of $K_{fi}$ are in fair agreement, considering the uncertainties in extinction coefficient, and so on, at that time. Data for $O_2/O_2^{-}\cdot/O_2^{2-}$ are based on $E_7^1(O_2/O_2^{-}\cdot) = -155$ mV, $E_{m7} = 355$ mV (1 $M$ $O_2$),[65] and the $pK_a$'s given above. Formation constants $K_{fi}$ at pH 7 for the radical intermediates in reduction of nicotinamide adenine dinucleotide ($NAD^+$)[66,67] and *N,N′*-dimethyl-4,4′-bipyridylium dichloride (methyl viologen, Paraquat)[68] have also been calculated to illustrate the wide range of this parameter.

These illustrations of simple thermodynamic parameters of free radicals, typical of many important in biology, provide the basis for understanding why, on the one hand, steady-state concentrations of semiquinone-type radicals from flavins, quinones, and viologens have been detected in biological systems,[13,69] while on the other hand, *steady-state* concentrations of $O_2^{-}\cdot$ would be impossibly low to detect at physiological pH. In the examples given, radiation-chemical techniques (see below) have provided refinements of, and support for, previous measurements using classical methods. A recent volume entitled "Biological Electrochemistry"[70] was notably lacking in reliable values of thermodynamically reversible reduction potentials for couples involving free radicals in aqueous solutions: Independent measurements of redox properties of radicals using radiation-chemical techniques will do much to avoid the mistakes that are often apparent in the literature.

The ease with which $Q^{-}\cdot$ can be generated radiation chemically is illustrated below; rate constants for the reverse reaction (19-12), $k_{12b}$ (disproportionation of semiquinones) are typically[71–74] of the order of $10^8$–$10^9$ $M^{-1}$ $s^{-1}$ for radicals derived from quinones, flavins, and so on, and somewhat lower ($10^5$–$10^6$ $M^{-1}$ $s^{-1}$) for $O_2^{-}\cdot$ and ascorbate around pH 7.[46,75] Radicals generated by, for example, a 1-$\mu$s electron pulse producing an initial concentration of 10 $\mu M$ would thus have a natural lifetime with respect to Reaction (19-12) of the order of 100 $\mu$s to 1 s, depending on $k_{12b}$ and usually, therefore, on pH and $pK_7$.

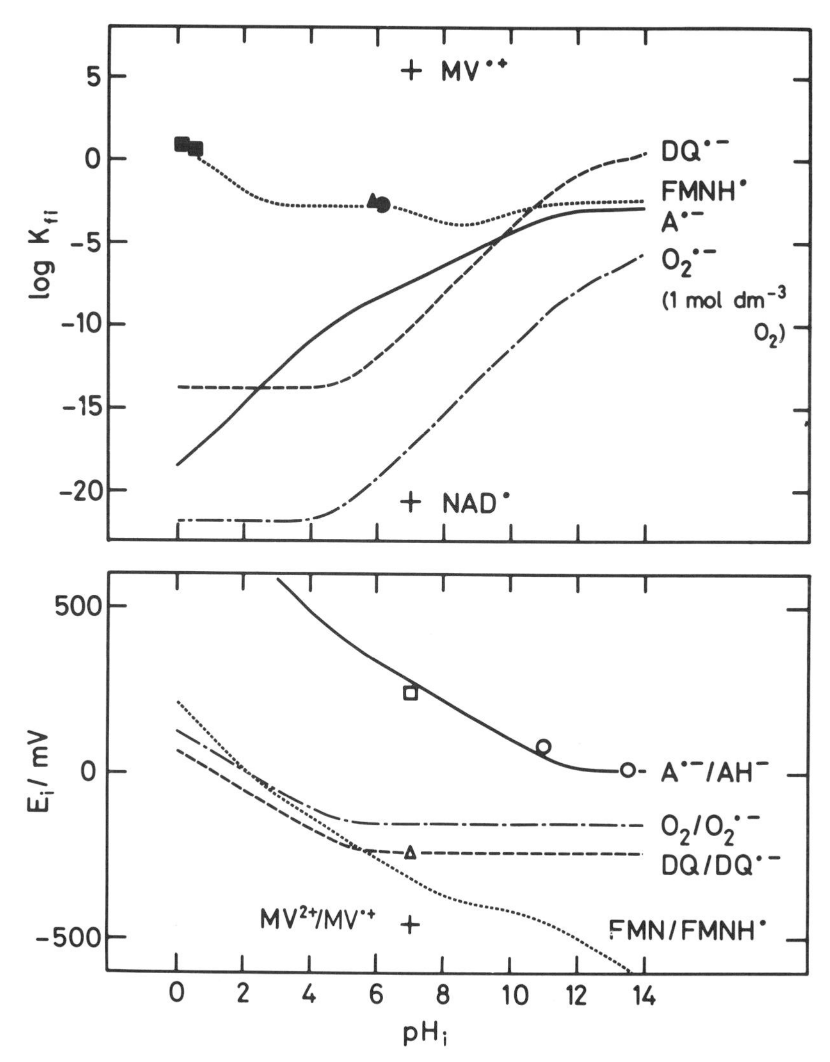


**Figure 19-1.** Semiquinone formation constants [$K_{fi}$, Eq. (19-14)] and one-electron reduction potentials, $E$ (vs. normal hydrogen electrode), of some typical redox systems. The lines for ascorbic acid ($AH_2$) and duroquinone (DQ) were calculated from literature values of $K_{fi}$ and p$K_a$'s (see text).[49–51,54–57] Independent pulse radiolysis measurements of $E$ are from refs. 52 (○); 53 (□); 60 (△). The lines for FMN and $O_2$ are from pulse radiolysis measurements of $E$ and literature p$K_a$'s (see text).[61,65] Independent estimates of $K_{fi}$ for FMN are from refs. 62 (●); 63 (■), 64 (▲). Formation constants for NAD· and $MV^{\cdot +}$ ($MV^{2+}$ = methyl viologen) were calculated from literature data.[66–68] Labels correspond to prototropic forms predominant at pH 7. Note nonstandard reference state for $O_2$ of 1 $M$ used to facilitate comparison (see ref. 65).

## *One-Electron Reduction and Oxidation of Biologically Important Molecules by Using Radiation-Chemical Methods*

Radiolysis of water (Chapter 10) generates the one-electron oxidant OH· and two one-electron reductants, $e_{aq}^-$ and H· (the latter about 10% of the total radical yield). Chapter 10 has discussed the methods of interconverting oxidizing and reducing species to give essentially one-radical systems, and only the briefest outline is presented here before we proceed with illustrations of interest in biology.

Formation of the semiquinone $Q^-\cdot$ can be achieved *either* by reduction of a stable or otherwise accessible oxidant Q *or* by oxidation of the reductant $QH_2$. In some instances, both routes can be taken to confirm the assignment of the radical, but usually one or the other is practically more convenient. Thus, $Q^-\cdot$ may be formed by reduction of Q by $e_{aq}^-$:

$$Q + e_{aq}^- \rightarrow Q^-\cdot \qquad (19\text{-}19)$$

or other one-electron reductants derived from the primary radicals OH· and H· by reaction with formate or 2-propanol:

$$Q + CO_2^-\cdot \rightarrow Q^-\cdot + CO_2 \qquad (19\text{-}20)$$

$$Q + (CH_3)_2\dot{C}OH \rightarrow Q^-\cdot + (CH_3)_2CO + H^+ \qquad (19\text{-}21)$$

In some instances these secondary reductants ($CO_2^-\cdot$, etc) may be too weak to achieve the desired Reactions (19-20), (19-21), and so on, and removal of the unwanted OH· may be achieved by 2-methyl-2-propanol, the radical product being *usually* relatively unreactive:

$$OH\cdot + (CH_3)_3COH \rightarrow H_2O + \cdot CH_2(CH_3)_2COH \qquad (19\text{-}22)$$

Figure 19-2 compares the absorption spectra of durosemiquinone produced *either* by reduction of duroquinone by using Reactions (19-19) and (19-21) and recording the transient absorption 10 $\mu$s after pulse radiolysis[57] *or* by mixing duroquinone with its hydroquinone in 0.05–0.2 *M* NaOH and recording the absorption of the steady-state concentration of semiquinone spectrophotometrically.[56]

Oxidation of $QH_2$ to yield $Q^-\cdot$ may be accompanied by converting $e_{aq}^-$ to OH· by saturation with $N_2O$:

$$e_{aq}^- + N_2O + H^+ \rightarrow N_2 + OH\cdot \qquad (19\text{-}23)$$

followed by using OH· as a one-electron oxidant directly:

$$OH\cdot + QH_2 \rightarrow OH^- + QH\cdot + H^+ \qquad (19\text{-}24)$$

This route is not particularly useful because OH· is frequently not so discriminating in its reactions (although examples are known[76] and the benzosemiquinones can be produced by means of water elimination of the OH· adduct of hydroquinones[71,76]). More commonly, OH· is converted to

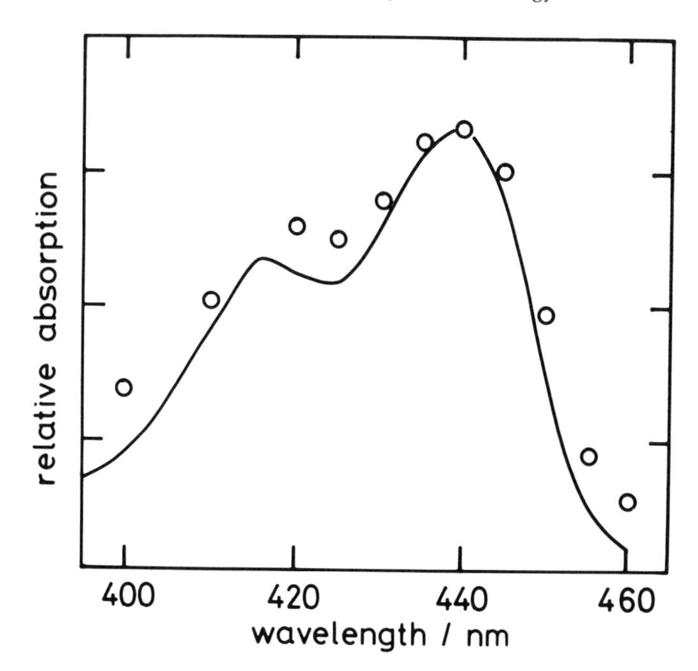


**Figure 19-2.** Comparison of the absorption spectrum of durosemiquinone produced without radiation at high pH by using Eq. (19-12) (solid line[56]) with that produced at pH 7 by pulse radiolysis[57] (○). The absorptions have been normalized at the maximum.

a secondary, more selective one-electron oxidant:

$$OH\cdot + X^- \rightarrow OH^- + X\cdot \qquad (19\text{-}25)$$

where, for example, $X^- = N_3^-$ or $CO_3^{2-}$.[77–79] When (as in most studies of this type) $X^- = Br^-$, $SCN^-$, $I^-$, or $Cl^-$, then (19-25) is followed by[80]

$$X\cdot + X^- \rightleftharpoons X_2^-\cdot \qquad (19\text{-}26)$$

Either $X\cdot$ or $X_2^-\cdot$ may then react with $QH_2$ (aromatic amino acids, hydroquinones, phenols, etc):

$$X\cdot + QH_2 \rightarrow X^- + QH\cdot + H^+ \qquad (19\text{-}27)$$

$$X_2^-\cdot + QH_2 \rightarrow 2\,X^- + QH\cdot + H^+ \qquad (19\text{-}28)$$

Another useful[52,81] one-electron oxidant is $CH_2CHO\cdot$, formed by base-catalyzed elimination of $H_2O$ from the radical produced upon scavenging $OH\cdot$ with 1,2-ethanediol:[81]

$$CH_2CHO\cdot + QH_2 \rightarrow CH_3CHO + QH\cdot \qquad (19\text{-}29)$$

It is important to recognize that the rate constants for reactions such as (19-27)–(19-29) may vary markedly with pH because ionized hydroquinones, phenols, and so on, are usually much more reactive than the undissociated forms.[79,81]

Figure 19-3 illustrates the absorption spectra of the radicals from one-electron oxidation of tryptophan (TrpH) or its *N*-methyl derivative, the radicals being produced *either* by photoionization[82]

$$\mathrm{TrpH} + h\nu \rightarrow \mathrm{TrpH}^{+}\cdot + e_{aq}^{-} \quad (19\text{-}30)$$

*or* by oxidation using $Br_2^{-}\cdot$ produced radiolytically:[78,79,83,84]

$$\mathrm{TrpH} + \mathrm{Br_2}^{-}\cdot \rightarrow \mathrm{TrpH}^{+}\cdot + 2\,\mathrm{Br}^{-} \quad (19\text{-}31)$$

$$\mathrm{TrpH}^{+}\cdot \rightleftharpoons \mathrm{Trp}\cdot + \mathrm{H}^{+} \qquad (\mathrm{p}K_a = 4.3)^{83} \quad (19\text{-}32)$$

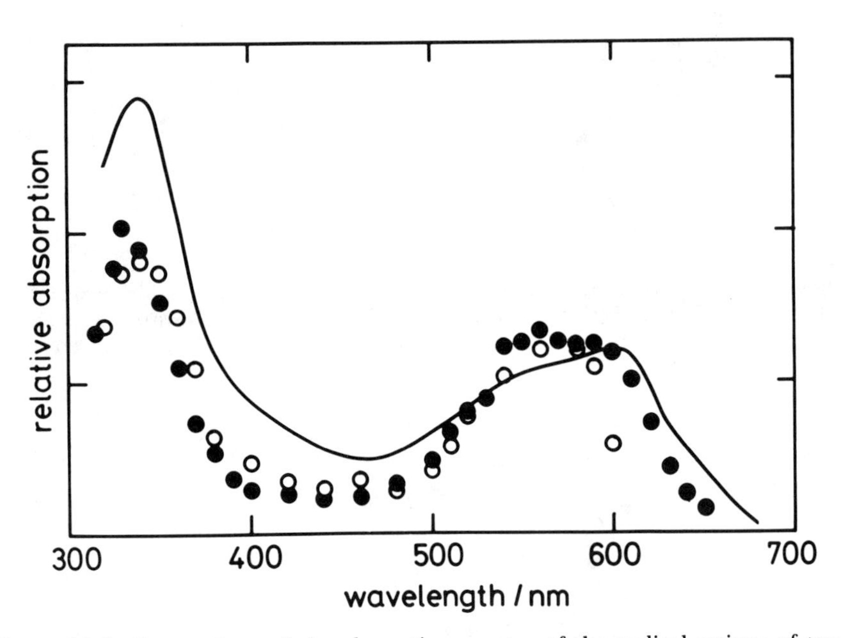


**Figure 19-3.** Comparison of the absorption spectra of the radical cations of tryptophan and *N*-methyl tryptophan. Solid line: recorded on a spectrographic plate after flash photoionization of *N*-methyl tryptophan in HCl (2 *M*).[82] Spectral points recorded by pulse radiolysis by reaction of $Br_2^{-}\cdot$ with (○) *N*-methyl tryptophan, pH 7;[84] (●) tryptophan, pH 3.[83] The absorptions have been normalized at the long-wavelength maximum.

Figure 19-4 shows that the radical NAD· may be generated *either* by one-electron reduction of $NAD^{+}$[85,86]

$$\mathrm{NAD}^{+} + e_{aq}^{-} \rightarrow \mathrm{NAD}\cdot \quad (19\text{-}33)$$

*or* by one-electron oxidation of NADH[86]

$$\mathrm{NADH} + \mathrm{Br_2}^{-}\cdot \rightarrow \mathrm{NAD}\cdot + \mathrm{H}^{+} + 2\,\mathrm{Br}^{-}. \quad (19\text{-}34)$$

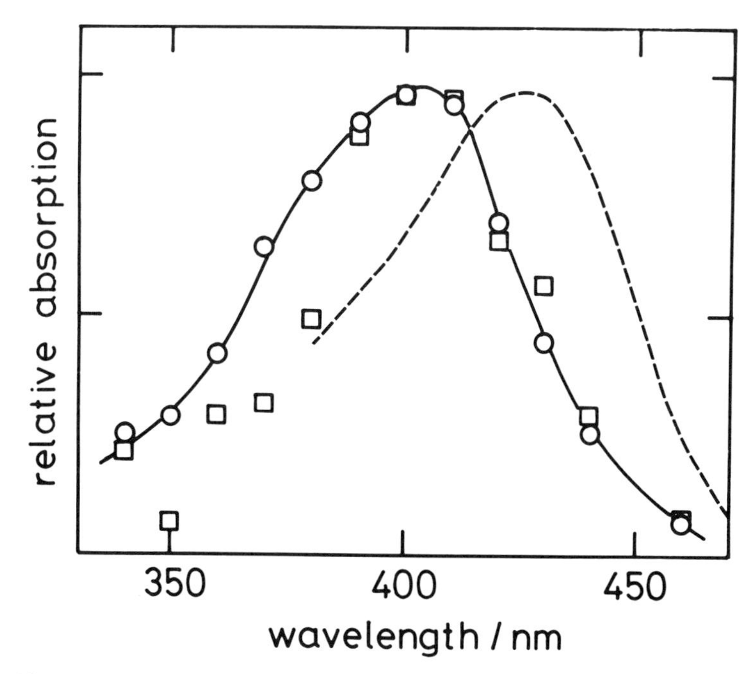


**Figure 19-4.** Comparison of the absorption spectrum of NAD· produced by reduction of $NAD^+$ by $e_{aq}^-/CO_2^-\cdot$ (○), or by oxidation of NADH by $Br_2^-\cdot$ (□). Uncertainties in the correction for the loss of absorbing NADH probably account for the discrepancies in the latter spectrum <380 nm.[85,86] The broken line is the spectrum of NAD· bound to an enzyme; see below.[151] The spectra have been normalized at the maximum.

These few illustrations are typical of many we shall outline later; the manipulation of solute concentrations to achieve the desired reaction pathways has been described in Chapter 10 and illustrated in several reviews.[26–34] Experimental design is greatly facilitated by the comprehensive collections of kinetic, spectral, thermodynamic, and bibliographic data prepared by the Radiation Chemistry Data Center of the University of Notre Dame.[87]

## PROPERTIES OF SOME INTERMEDIATES IN BIOCHEMICAL ELECTRON TRANSPORT STUDIED BY PULSE RADIOLYSIS

### *Electron Transport and Radical Intermediates*

The basis of oxidative or respiratory-chain phosphorylation (occurring in respiratory assemblies in the inner membrane of mitochondria, subcellular

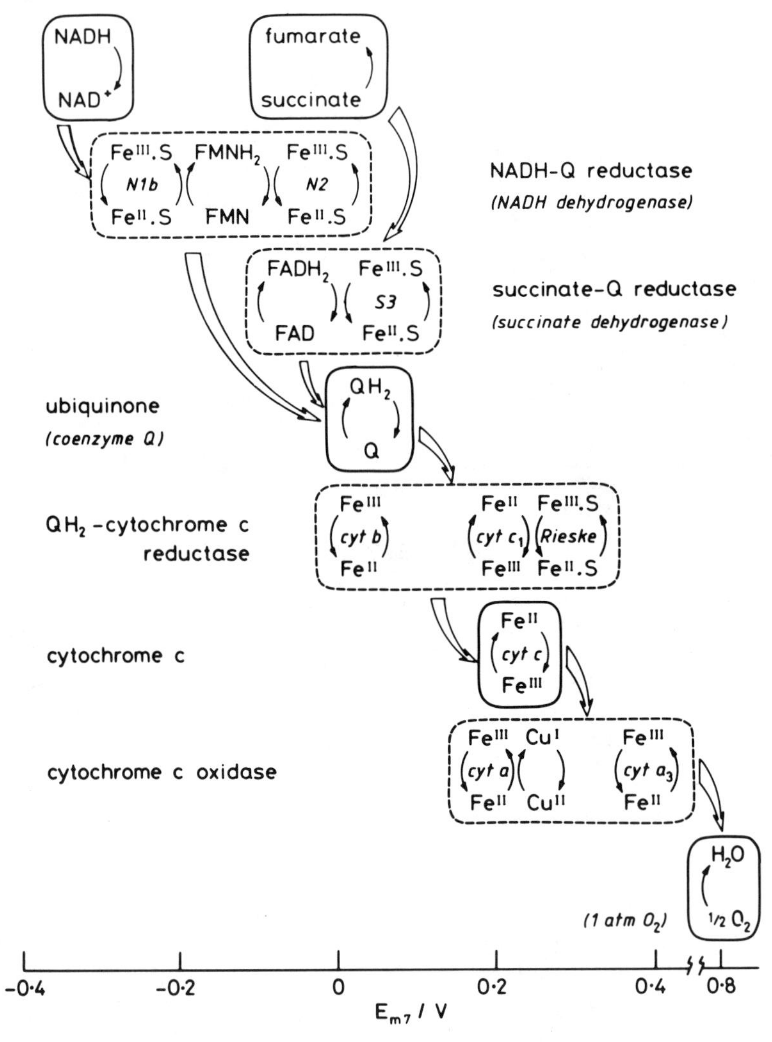


**Figure 19-5.** Much-simplified representation of some of the electron transfer sequences important in the multienzyme complexes of respiratory-chain phosphorylation.[88,89] Couples are located at or near their appropriate positions on the reduction potential scale where known ($E_{m7}$ for two-electron couples): The potentials of flavoprotein centers are not well characterized. Note iron–sulfur (Fe–S) centers are the largest class of compounds in the respiratory chain, and only a few representative Fe–S centers are included here; thus NADH–dehydrogenase contains at least seven recognized Fe–S centers.[107]

organelles, ca. $2 \times 0.5$–$1\ \mu m$) is the oxidation of NADH to $NAD^+$:

$$NADH + H^+ + \tfrac{1}{2}O_2 \rightarrow NAD^+ + H_2O \qquad (19\text{-}35)$$

which is highly exergonic ($\Delta G^{\circ\prime} = -219\ kJ\ mol^{-1}$ at pH 7, 1 atm $O_2$) and coupled to the endergonic synthesis of adenosine triphosphate (ATP) from the diphosphate (ADP):

$$ADP + HPO_4^{2-} + H^+ \rightarrow ATP + H_2O \qquad (19\text{-}36)$$

($\Delta G^{\circ\prime} = 32\ kJ\ mol^{-1}$ at ionic strength 0.25, 1 m$M$ $Mg^{2+}$).[88,89] Although Reaction (19-35) is thermodynamically favorable, it is kinetically slow and is catalyzed in the respiratory chain as outlined in Figure 19-5. Initially, electrons flow from NADH to the FMN prosthetic group in the enzyme NADH-Q reductase (NADH dehydrogenase) and then via iron–sulfur complexes (Fe–S) to coenzyme Q (ubiquinone). Unlike the other electron carriers, the latter is not protein bound and is mobile. It acts to carry electrons between the flavoproteins and the cytochromes. The heme prosthetic group in cytochromes serves to complete the electron transfer to $O_2$, cytochrome *c* being a water-soluble membrane protein also acting, like ubiquinone, as a mobile electron carrier.[89]

The Fe—S and heme proteins can only act as one-electron couples by means of the Fe(II)/Fe(III) redox pairs; it is now recognized[90,91] that semiquinone radical intermediates are important in the coupled oxidation/reductions with both flavin and ubiquinone centers. The energetics of the individual one-electron steps in the reduction of flavins, ubiquinones, and so on, the spectral properties of the radical intermediates, and the kinetics of the electron transfer reactions involving these intermediates can all be investigated by radiation-chemical methods: A few illustrative examples are described below.

## *Radicals Derived from Flavins (Isoalloxazines): Flavosemiquinones*

The importance of redox reactions involving flavins and flavoproteins in many, diverse biological processes (in addition to the function in the respiratory electron transfer chain noted above) can be seen from the proceedings of major symposia.[92] Radical intermediates, FH· in the reduction of flavins, F, to dihydroflavins, $FH_2$, were recognized in 1936 by Michaelis[93]

$$F + e^- + H^+ \rightleftharpoons FH\cdot \qquad (19\text{-}37)$$

$$FH\cdot + e^- + H^+ \rightleftharpoons FH_2 \qquad (19\text{-}38)$$

and pulse radiolysis was shown to be a most useful method of characterizing the absorption spectra of FH· over a wide pH range by Land and Swallow.[94] In the case of riboflavin, prototropic equilibria involving both oxidized flavin and semiquinone are important:

$$FH_2^{2+} \rightleftharpoons FH^+ + H^+ \rightleftharpoons F + 2H^+ \rightleftharpoons [F(-H^+)]^- + 3H^+ \quad (19\text{-}39)$$

(successive p$K_a$'s for these ground-state dissociations are −8, 0.25, and 10.05);

$$FH_2^{+\cdot} \rightleftharpoons FH\cdot + H^+ \rightleftharpoons F^{-}\cdot + 2H^+ \quad (19\text{-}40)$$

(successive p$K_a$'s for these radical dissociations are 2.3 and 8.3).[94] We shall illustrate later (Figure 19-7; see below) an example of the semiquinone spectra obtained.[94] One-electron reduction of F ($FH^+$, etc) was achieved by the reductant $CO_2^{-}\cdot$ (from radiolysis of aqueous formate/$N_2O$ solutions), a particularly convenient reductant for biochemical studies because it is a somewhat weaker (and therefore more selective) reductant than $e_{aq}^{-}$. In addition, the low p$K_a$ for dissociation of $CO_2H\cdot$,

$$CO_2H\cdot \rightleftharpoons CO_2^{-}\cdot + H^+ \qquad (pK_a = 1.4)^{95} \quad (19\text{-}41)$$

facilitates measurements over a wide pH range without the common problem of the protonated reductant reacting either much slower than, or differently to, its dissociated form.

It should be stressed that the primary data in pulse radiolysis measurements of this type are of the relative differences in extinction coefficients between ground state and radical at various wavelengths, $\Delta\varepsilon$. True absorptions of radicals require corrections for conversion of ground state to radical, that is, an accurate knowledge of the radical concentration or radiation-chemical yield. We have previously drawn attention to this problem;[33] in studies of this type it is preferable to eliminate dosimetry errors completely by calibrating the system with, for instance a similar formate/$N_2O$ solution but replacing (in this case) flavin, F, with ferricyanide at a concentration such that

$$k(CO_2^{-}\cdot + F)[F] \simeq k(CO_2^{-}\cdot + Fe(CN)_6^{3-})[Fe(CN)_6^{3-}] \quad (19\text{-}42)$$

and using the difference in extinction coefficient between $Fe(CN)_6^{3-}$ and $Fe(CN)_6^{4-}$, 1027 $M^{-1}$ $cm^{-1}$ at 420 nm[25] (note that this will vary with bandwidth). There is a clear case for reporting primary measurements of $\Delta\varepsilon$ between ground state and radical at selected wavelengths, not only the corrected radical spectra.

The energetics of the one-electron reduction steps (19-37) and (19-38) are of obvious importance in assessing likely reactions of $FH\cdot$, $FH_2$, and so on, and although early estimates[93] were available, there was some disagreement in the literature that was unequivocally settled by measurements using pulse radiolysis of one-electron transfer equilibria involving flavosemiquinones and quinones[61,96] or bipyridylium compounds.[97] The potential contribution of pulse radiolysis in establishing the energetics of biologically importar.. one-electron redox couples is so important that we digress briefly here to illustrate the *general* applicability of the method before returning to review the measurements of flavin/flavosemiquinone potentials.

## *Measuring the Position of One-Electron Transfer Equilibria by Pulse Radiolysis*

Radiolysis of a solution containing two solutes A and B together with cosolutes designed to secure purely reducing or oxidizing conditions (formate, alcohols, $Br^-$, and so on; see above and Chapter 10) will initially yield radicals such as $A^-\cdot$, $B^-\cdot$ (restricting ourselves for simplicity to reducing conditions) in concentrations proportional to products of [A] or [B] and the respective rate constants for scavenging the appropriate primary or secondary radicals. Thus there is, initially, *kinetic* control of the relative radical concentrations (see Chapter 4). However, if the rate of approach to the one-electron transfer equilibrium

$$A^-\cdot + B \rightleftharpoons A + B^-\cdot \tag{19-43}$$

is sufficiently rapid (commonly, deprotonation of some types of radicals may be required), then we can estimate the difference, $\Delta E_i$, in the reduction potentials between the two couples if the equilibrium constant $K_{43}$ can be measured:

$$\Delta E_i = E_i^1(B/B^-\cdot) - E_i^1(A/A^-\cdot) = \left(\frac{\mathbf{R}T}{F}\right) \ln K_{43} \tag{19-44}$$

Measurement of the concentrations of radicals after *thermodynamic* control of relative concentrations has been achieved by using Reaction (19-43) but *before* radical loss, (eg, by disproportionation or dimerization occurs) can thus yield valuable information about energetics—and therefore feasibility—of radical reactions.

The course of typical experiments[60] is illustrated in Figure 19-6. The absorption spectrum immediately after pulse radiolysis (0.2 $\mu$s, 2 Gy) of a solution containing 2-propanol (0.2 $M$), 1,1′-dibenzyl-4,4′-bipyridylium dichloride ($BV^{2+}$, 2 m$M$), and duroquinone (DQ, 44 $\mu M$) was initially that of the radical $BV^+\cdot$, as expected from kinetic competition by $BV^{2+}$ and DQ for $e_{aq}^-$ and $(CH_3)_2\dot{C}OH$. However, by 50 $\mu$s the spectrum shown was obtained and consisted of contributions from $BV^+\cdot$ (but only about one-third the initial intensity) together with a peak at 440 nm, which had grown in over 25 $\mu$s and attributable to the semiquinone $DQ^-\cdot$ (see Figure 19-2). By 50 $\mu$s, then, the equilibrium

$$BV^+\cdot + DQ \rightleftharpoons BV^{2+} + DQ^-\cdot \tag{19-45}$$

had been achieved, long before the unwanted disproportionation reaction (19-12) could occur. Measurement of the equilibrium value of $[BV^+\cdot]$ gave $K_{45} = 115 \pm 11$ at ionic strength 0.014, in good agreement with less precise but independent analysis of the exponential rate of approach to equilibrium:

$$k_{45,obs} = k_{45f}[DQ] + k_{45b}[BV^{2+}] \tag{19-46}$$

which gave $K_{45} = k_{45f}/k_{44b} = 75 \pm 32$ (see Chapter 4).[60] Using published[68]

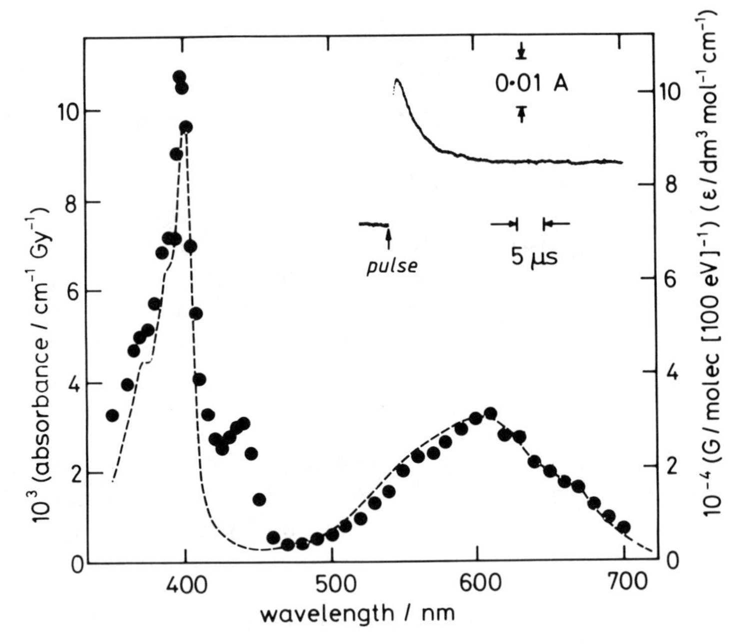


**Figure 19-6.** Absorption spectrum (●) measured 50 $\mu$s after pulse radiolysis of a solution containing $BV^{2+}$ and DQ (see text).[60] Broken line: spectrum of $BV^{+}\cdot$ obtained by $\gamma$-radiolysis and recorded by conventional spectrophotometry, normalized at 600 nm. Inset: rate of the approach to equilibrium 19–45, recorded at 600 nm.

estimates of $E_7^1$ $(BV^{2+}/BV^{+}\cdot) = (-354 \pm 7)$ mV and Eq. (19-44) gave an estimate $E_7^1(DQ/DQ^{-}\cdot) = (-244 \pm 7)$ mV.

These experiments[60] were preceded by a number of other investigations using the same approach,[57,65,98–100] some of which included one-electron transfer equilibrium involving $O_2/O_2^{-}\cdot$ and led to the definitive value of the potential of this important couple.[57,65,99–101] More details of the principles involved have been outlined, and values of many reduction potentials involving radicals have been collated.[102]

## *Reduction Potentials of Couples Involving Flavosemiquinones*

Using the pulse radiolysis technique described above, Meisel and Neta[96] measured the equilibrium constants for equilibria of the form of Eq. (19-43), involving riboflavin and either duroquinone or 9,10-anthraquinone-2-sulfonate between pH 6 and pH 12. The pH dependence of $E_i^1(F/F^{-}\cdot)$ was found, as expected,[48] to be complex because of the several prototropic equilibria [(19-39), (19-40)]; their data fitted the appropriate function[48] using the p$K_a$ values given above.[94] In particular, for

$$FH\cdot \rightleftharpoons F^{-}\cdot + H^{+} \qquad (19\text{-}47)$$

a value of $pK_{47} = 8.3$, as estimated by pulse radiolysis observations[94] and supporting other measurements,[103–105] was required by the reduction potential/pH profile. Thus, the value of $pK_{47} = 6.5$ derived from the classical experiments[93] could be confidently discounted.

Two-electron potentials, $E_{mi}(F/FH_2)$ for FMN and FAD (flavin adenine dinucleotide) are indistinguishable between pH 2 to 12;[103] data for riboflavin are very similar.[48] As expected, then, $E_7{}^1$ $(F/F^-\cdot)$ for riboflavin, FMN and FAD measured by pulse radiolysis were identical within experimental error, and the pH dependence of $E_i{}^1$ for FMN was satisfactorily fitted by using the $pK_a$'s for riboflavin,[61] as shown in Figure 19-1. (An error in calculation of an earlier estimate[97] for FAD was later corrected.[61]) It may not be justified to use these values as estimates of the corresponding couple for flavoproteins, but let us use them to provide an insight into the energetics, for example, of the flavin/ubiquinone system.

Swallow[106] has suggested a value of $E_7{}^1(Q/Q^-\cdot) \approx -230$ mV for ubiquinone, and, using $E_{m7}(Q/Q^{2-}) = 50$ mV,[107] we have, using Eq. (19-18), an estimate $E_7{}^2(Q^-\cdot/Q^{2-}) \approx 330$ mV. Then for

$$FH_2 + Q \rightleftharpoons FH\cdot + QH\cdot \qquad (19\text{-}48)$$

we estimate $\Delta G^{\circ\prime} \approx 10$ kJ mol$^{-1}$ at pH 7 ($\Delta E \approx -110$ mV), and for

$$FH\cdot + Q \rightleftharpoons F + QH\cdot \qquad (19\text{-}49)$$

we estimate $\Delta G^{\circ\prime} \approx -8$ kJ mol$^{-1}$ ($\Delta E \approx 80$ mV). These calculations are, of course, merely illustrative and certainly simplistic because the electron transfer equilibria do not take place in dilute aqueous solution, and semiquinone formation constants are markedly dependent upon environment.[106]

## *Radiolytic Generation of Dihydroflavins*

Steady-state radiolysis methods offer a particularly convenient, zero-order source of one-electron reducing or oxidizing equivalents and deserve much more widespread attention. Thus, although F can be reduced to $FH_2$ by reductants, such as dithionite, or photochemically in the presence of a nitrogenous base as electron donor, $\gamma$-radiolysis of formate/$N_2O$ solutions of F offers a particularly convenient and controllable method of producing solutions containing F and $FH_2$ in any desired ratio, the formate cosolute usually being much less reactive than, for example, the bisulfite coproduct of dithionite reduction, bisulfite itself being a reductant. Clarke[108] used this technique to generate $FMNH_2$, showing the product of FMN reduction by $CO_2{}^-\cdot$ had identical reduction stoichiometry and reactivity towards nitroaryl oxidants when compared with the products using conventional reductants.

Ahmad and Armstrong[109] (showed $CO_2{}^-\cdot$ reduced riboflavin to the dihydroflavin, the latter being unreactive towards $CO_2{}^-\cdot$, so that it is difficult to "overreduce" flavins with this reductant. They also irradiated solutions of $FH_2$ (F = FAD, lumichrome) containing $Br^-$ or cysteine. In this case the zero-order, radiolytic generation of $Br_2^{-}\cdot$ or RS· permitted "titration" experiments, demonstrating reactions of overall stoichiometry given by

$$FH_2 + 2\,Br_2^{-}\cdot \rightarrow F + 4\,Br^- + 2\,H^+ \tag{19-50}$$

$$FH_2 + 2\,RS\cdot \rightarrow F + 2\,RSH \tag{19-51}$$

However, since $Br_2^{-}\cdot$ and RS· are known one-electron oxidants (see Eq. (19-28), and below), the reactions doubtless proceed by means of one-electron steps of the form [cf Eq. (19-34)]

$$FH_2 + X\cdot \rightarrow FH\cdot + X^- + H^+. \tag{19-52}$$

The high formation constants [Eq. (19-12)] for flavosemiquinones at low pH should result in relatively high concentrations of radicals coexisting with oxidized and fully reduced forms. This is precisely what Swallow[110] observed almost 30 years ago; in Figure 19-7 the spectrum[110] of radiolytically reduced

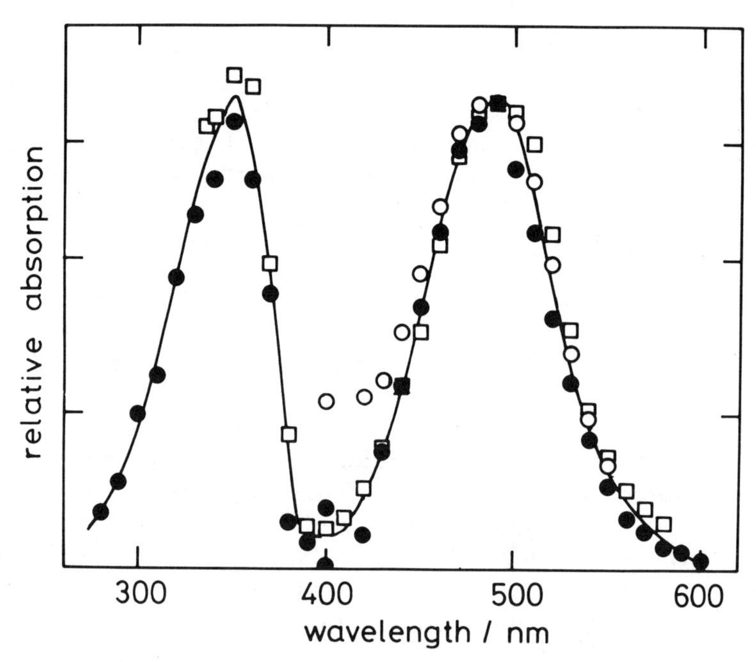


**Figure 19-7.** Spectra of protonated flavin semiquinones: (○), steady-state absorption produced upon x-irradiation of a solution of riboflavin in aqueous ethanol (0.5 *M*) containing $H_2SO_4$ (1 *M*),[110] points below 400 nm omitted because of an additional, interfering absorption—perhaps a radical addition byproduct;[109] (□) steady-state absorption upon mixing FMN and $FMNH_2$, pH −0.4;[63] (●) absorption recorded by pulse radiolysis of riboflavin in formate, $H_0 = -1.1$.[88] The spectra have been normalized at 490 nm.

riboflavin in 1 $M$ $H_2SO_4$ is compared with that obtained[63] on mixing FMN with $FMNH_2$ at pH −0.4.

### *Reactions of Reduced Flavins with Oxygen*

Although $FH_2$ react rapidly with some quinones, ferricyanide, and so on, reaction with molecular oxygen is complex; reactions of $FH_2$, FH·, etc with $O_2$, $O_2^-\cdot$, etc are of major interest in flavoprotein catalysis.[105,111,112] Most radiation-chemical studies characterize reactions of unstable radicals with stable ground states of other solutes or dimerization or dismutation of single-radical types: for example, few studies have been made of reactions of the form

$$A^-\cdot + B^-\cdot \rightarrow \text{products} \qquad (19\text{-}53)$$

There is, however, much scope for using pulse radiolysis to generate $A^-\cdot$, $B^-\cdot$ essentially simultaneously in any desired *ratio* (set, of course, by kinetic competition for primary or secondary radicals, Chapter 4), as Anderson[113,114] illustrated in studies of the reaction of flavin semiquinone radicals with $O_2^-\cdot$:

$$FH\cdot + O_2^-\cdot \rightarrow HFO_2^-(HFO_2H) \qquad (19\text{-}54)$$

In aerated formate solutions containing flavins (80 $\mu M$), kinetic competition results in $[O_2^-\cdot]/[FH\cdot] \simeq 11$ shortly after the radiation pulse, and the *absolute* concentration of $[O_2^-\cdot]$ can be varied simply by varying the radiation dose. The ratio $[O_2^-\cdot]/[FH\cdot]$ is sufficiently high to result in exponential decay of FH· within experimental error, and the first-order decay constant, $k_{obs}$, was found to vary with radiation dose, and hence with $[O_2^-\cdot]$, in a first-order manner, as shown in Figure 19-8.[113]

The product of Reaction (19-54) absorbed at 385 nm, and it was possible to characterize the base-catalyzed dissociation of this complex to oxidized flavin, F. The absorption was similar to that of species previously observed in a number of biochemical studies, but radiolytic generation enabled much better characterization than was possible earlier.[113,114]

Although the reaction of FH· with $O_2$,

$$FH\cdot + O_2 \rightarrow F + O_2^-\cdot + H^+ \qquad (19\text{-}55)$$

is exergonic ($\Delta G^{\circ\prime} = -15.3$ kJ mol$^{-1}$ at pH 7, 1 $M$ $O_2$), the rate constant is much lower than that for the dissociated flavosemiquinone,

$$F^-\cdot + O_2 \rightarrow F + O_2^-\cdot \qquad (19\text{-}56)$$

($k_{55} < 3.8 \times 10^4$, $k_{56} \simeq 3 \times 10^8$ $M^{-1}$ s$^{-1}$),[105,111] and the p$K_a$ for the dissociation of FH· is about 8.3.[94,103,104,105] Hence, in aerated solutions at pH $\lesssim 7$, Reactions (19-55), (19-56) are relatively slow, and the reaction (19-54) of interest can be studied (the reactions with $O_2$ lead to the positive intercept seen in Figure 19-8).

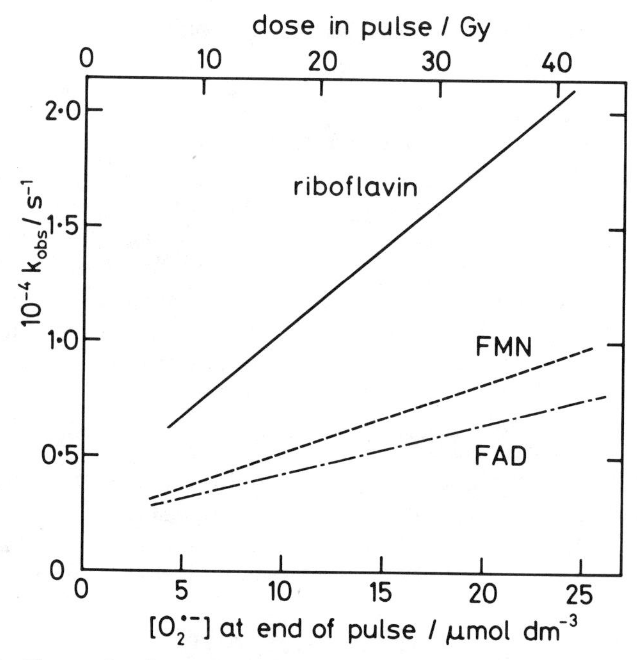


**Figure 19-8.** First-order dependence upon radiation dose of the decay rate constant, $k_{obs}$, for decay of the absorption at 540 nm of flavosemiquinones observed after pulse radiolysis of aerated, flavin solutions containing formate (see text).[106]

The reaction of dihydroflavins with oxygen,

$$FH_2 + O_2 \rightarrow F + H_2O_2 \tag{19-57}$$

proceeds via semiquinone and superoxide only at higher pH[115]

$$FH^- + O_2 \rightarrow F^-\cdot + O_2^-\cdot + H^+ \tag{19-58}$$

the latter reaction being exergonic only at pH $\geqslant$ 8.5 (1 $M$ $O_2$). Anderson[61] suggested that Reaction (19-57) involves the same complex $HFO_2H$ as an intermediate, as was characterized by pulse radiolysis using Reaction (19-54), and the absence of semiquinone formation at lower pH[115] is a consequence of the endergonicity. Recall, though, that the contrasting *rates* of Reactions (19-55) and (19-56) are typical of many where protonation of radicals is accompanied by a marked decrease in the *rate* of electron transfer reactions, even those highly exergonic (Reaction (19-55) is not endergonic until pH $\leqslant$ 2, 1 $M$ $O_2$). We have much to learn about the relative importance of kinetic and thermodynamic control in such reactions. The $pK_a$ for dissociation of the lumiflavin semiquinone radical, FH·, is indistinguishable from that of riboflavin; both FH· and $F^-\cdot$ forms of the lumiflavin radical react with other, more powerful oxidants (such as benzoquinone and ferricyanide) with rate constants approximately greater than $10^8$ $M^{-1}$ $s^{-1}$, varying relatively little with pH.[105]

## *Some Radical Reactions Involving Flavoproteins*

Elliot et al[116] used pulse radiolysis to record the absorption spectrum of the radical produced upon reaction of the flavoprotein, lipoamide dehydrogenase with $CO_2^{-}\cdot$. In the 410–500-nm region the decrease in absorbance corresponds to that previously reported for glucose oxidase semiquinone anion,[117] as shown in Figure 19-9(a). It is interesting that the absorption between 360 to 410 nm resembled an anionic semiquinone at pH 7.0,[117] whereas the corresponding radicals from free flavins are protonated at this pH ($pK_a \simeq 8.3$).[94,103–105]

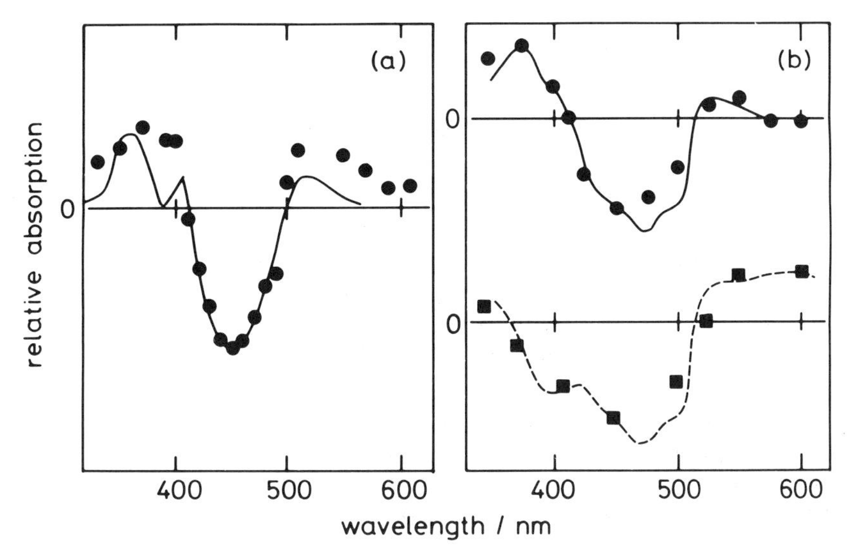


**Figure 19-9.** Flavosemiquinone centers in proteins. (a) After reduction of lipoamide dehydrogenase with $CO_2^{-}\cdot$, points are the absorbance differences at 0.7 ms,[116] and the solid line is the difference spectrum for glucose oxidase semiquinone anion[117] normalized at 450 nm. (b) After radiolytic reduction of the D-amino acid oxidase–benzoate complex (see text):[118] upper spectrum (●) at 0.5 ms, solid line is the spectrum of the red semiquinone anion prepared photolytically (normalized at 450 nm); lower spectrum (■) at 30 ms, broken line is the blue, neutral semiquinone spectrum (normalized at 450 nm).

Kobayashi et al[118] discuss several examples in the literature where flavoprotein semiquinones exist in either the "blue" (FH·) or "red" ($F^{-}\cdot$) forms, often independently of the external pH. These authors found $e_{aq}^{-}$ was too powerful a reductant to react selectively with the flavin center in the flavoprotein, D-amino acid oxidase (DAAO), and instead utilized the benzoate anion radical as a weaker reductant (benzoate ground-state complexes with DAAO at the substrate site). The spectrum[118] resulting from electron transfer from the benzoate radical-anion to the DAAO–benzoate complex is shown in Figure 19-9(b), the pulse radiolysis observations at

0.5 ms superimposing upon the difference spectrum of the red semiquinone produced photolytically in the presence of EDTA as electron donor. By 30 ms, however, the spectrum had changed to the blue semiquinone, which absorbed at 600 nm. The rate constants of the formation of the blue semiquinone obtained by the use of pulse radiolysis were identical to those of binding of benzoate with the red semiquinone obtained by stopped-flow methods.[118]

Although Kobayashi et al[118] had found $e_{aq}^-$ reacted (probably) with amino acid residues rather than with the flavin center in DAAO, Faraggi and Klapper[119] showed that the flavoprotein, flavodoxin, was reduced by $e_{aq}^-$ at the flavin site. These differences can be understood in terms of the contrasting secondary structures of the proteins and the accessibility of the flavin center to the aqueous phase.[118,119] Elliot et al[116] found that reduction of lipamide dehydrogenase by $CO_2^-\cdot$ did not inactivate the enzyme; in the case of DAAO the amino acids proposed by Kobayashi et al[118] as "electron sinks" (tyrosyl, cysteinyl, and histidyl residues) were known to be located at the active site, and an earlier radiolysis study[120] of DAAO was consistent with the importance of these residues.

Anderson et al[120] compared the spectra of oxidized tryptophan and (at higher pH) tyrosine residues obtained on one-electron oxidation of DAAO by $(SCN)_2^-\cdot$ and $Br_2^-\cdot$ [Reaction (19-28)] with the effect such reactions had on the enzyme activity. They concluded that another oxidizable residue, probably cysteinyl center(s), was present but much more exposed in the apo (FAD-free) enzyme and unimportant with respect to enzyme activity.

The semiquinone from flavodoxin (in which FMN is bound in a noncovalent manner) is produced by means of a complex sequence, including conformational changes requiring about 1 ms, when flavodoxin extracted from *Clostridium* is reduced by $e_{aq}^-$.[119]

## *Properties of Ubisemiquinone*

Ubiquinone (coenzyme Q) can be reduced to the semiquinone by $e_{aq}^-$, $(CH_3)_2\dot{C}OH$, $\cdot CH_2OH$, and so on, but its relative insolubility in water has restricted most experiments to mixed or nonaqueous solvent systems. The absorption spectra of the semiquinone anion, $Q^-\cdot$, and its conjugate acid, $QH\cdot$, produced by pulse radiolysis differ little in methanol[121] and in water/2-propanol (7 *M*, 0.54 volume fraction).[57] The $pK_a$ for the dissociation of $QH\cdot$ [Eq. (19-7)] was measured as 5.9 in water/2-propanol (7 *M*)/acetone (1 *M*); Swallow[106] has suggested the value in water might be about 4.9, and an estimate of 6.45 for the $pK_a$ for the dissociation in methanol has been made, assuming the reaction between $Q^-\cdot$ and $H^+$ was diffusion controlled.[122]

In view of Swallow's estimate[106] of $E_7^1(Q/Q^-\cdot) \approx -230$ mV for ubiquinone, and rapid electron exchange between $Q^-\cdot/Q$ and $O_2/O_2^-\cdot$,[57,65,99,123]

$$Q^-\cdot + O_2 \rightleftharpoons Q + O_2^-\cdot \qquad (19\text{-}59)$$

($K_{59}$ would be $\approx 19$ [1 $M$ $O_2$]), the reported[124] reaction of ubisemiquinone in methanol with $k_{59f}$ as slow as about $10^3\ M^{-1}\ s^{-1}$ might appearing surprising. Recall, however, the susceptibility of semiquinone formation constants to media noted above,[106] and also the marked variation in the rate constant for another, nominally simple, electron transfer reaction involving $O_2/O_2^{-}\cdot$

$$V^{+}\cdot + O_2 \rightarrow V^{2+} + O_2^{-}\cdot \qquad (19\text{-}60)$$

($V^{2+}$ = $N, N'$-dimethyl-4,4′-bipyridylium[dichloride]) in different solvents[125–129] and at micellar surfaces.[128] Because the mitochondrial reactions of ubiquinone may not be closely modeled by dilute aqueous solutions, there is obvious scope for more experimental studies in this area, (eg, based upon sequences of the form shown in Figure 19-10). This type of reaction sequence occurring in heterogeneous media can be established electrochemically[13] and is a feature of many radiation-chemical studies of micellar systems.

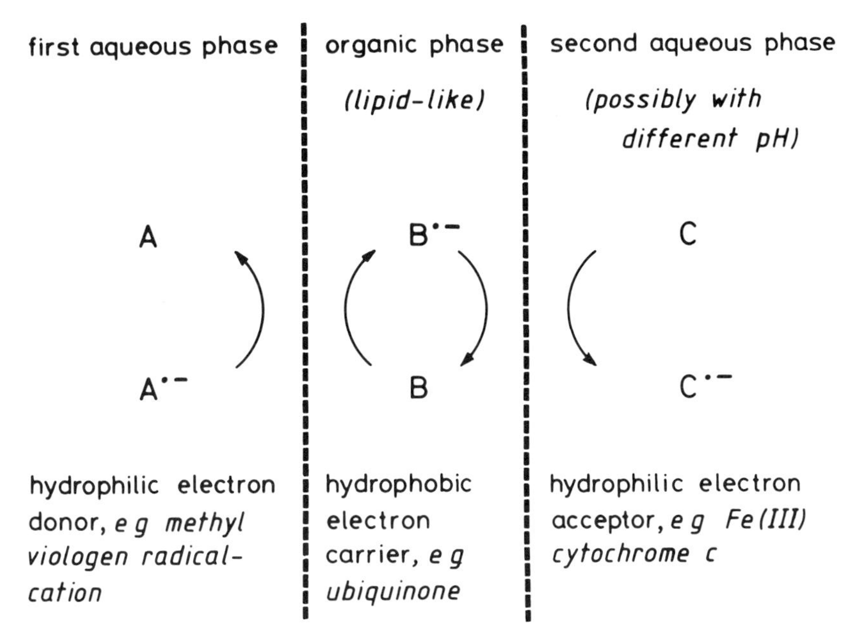


**Figure 19-10.** Schematic representation of the type of electron transfer sequence that can be observed in heterogeneous model systems[130] (see Chapter 12).

## *Redox Reactions of Cytochromes*

The reduction of Fe(III) forms of these iron porphyrins by $e_{aq}^{-}$ and $CO_2^{-}\cdot$ have been studied extensively by pulse radiolysis; the reactions are pH dependent, and characteristic differences in the spectra of Fe(II)/Fe(III) states are useful features facilitating comparison of the results with those using conventional methods. These studies have been recently reviewed by Bensasson et al[27] to which the reader is referred for a full account.

The interaction between $O_2^{-}\cdot$ and cytochrome *c* is of particular interest because both reactions

$$O_2^{-}\cdot + \text{Fe(III)-cyt } c \rightarrow O_2 + \text{Fe(II)-cyt } c \tag{19-61}$$

$$O_2^{-}\cdot + \text{Fe(II)-cyt } c \rightarrow O_2^{2-} + \text{Fe(III)-cyt } c \tag{19-62}$$

are exergonic ($\Delta G^{\circ\prime} = -37$ and $-62$ kJ mol$^{-1}$, respectively, at pH 7, 1 *M* $O_2$, using $E_7^1$ [Fe(III)/Fe(II)] = 225 mV[107]). Pulse radiolysis studies[131–135] showed Reaction (19-61) proceeded with 100% efficiency, but the rate constant $k_{61}$ varied considerably with pH, illustrated in Figure 19-11(a).

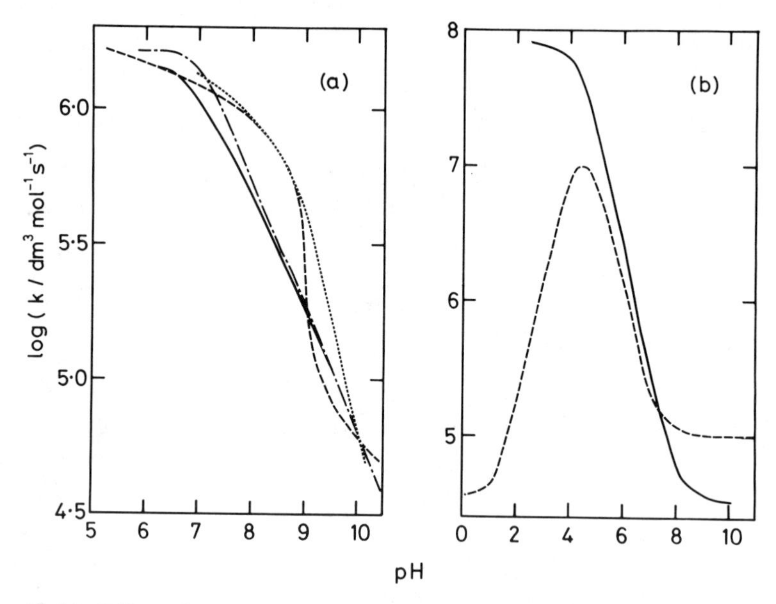


**Figure 19-11.** Effect of pH on rate constants for some radical reactions. (a) Reduction of Fe(III)-cytochrome *c* by $O_2^{-}\cdot$: (——);[131] (– – –);[133] (— - —);[134] (- - - - - - - - -).[135] Ionic strength ca. 0.1, <0.01, 0.15, and <0.01, respectively. (b) (——) disproportionation of ascorbyl radical ($-d[\text{R}]/dt = 2k[\text{R}]^2$); (– – – –), reaction of $HO_2\cdot/O_2^{-}\cdot$ with ascorbic acid/ascorbate.

Results from four independent studies[131,133–135] are selected to illustrate the effects of other experimental variables, such as ionic strength, on the reaction. However, Koppenol et al[133] pointed out that ionic strength effects were insufficient to account for the differences at pH $\leqslant$ 8 and suggested that Cu(II) impurities could lead to an overestimate of $k_{61}$ by catalyzing the disproportionation of $O_2^{-}\cdot$. This suggestion was refuted by Ilan et al[136] following measurements of the rate constant for reaction of Cu(I) with

Fe(III)-cyt *c*

$$Cu(I) + Fe(III)\text{-cyt } c \rightarrow Cu(II) + Fe(II)\text{-cyt } c \quad (19\text{-}63)$$

If significant catalytic reaction of $O_2^{-}\cdot$ with Cu(II) were occurring,

$$Cu(II) + O_2^{-}\cdot \rightarrow Cu(I) + O_2 \quad (19\text{-}64)$$

$$Cu(I) + O_2^{-}\cdot \rightarrow Cu(II) + O_2^{2-} \quad (19\text{-}65)$$

then Reaction (19-63) would need to be faster than (19-65) to be consistent with the approximately 100% reduction efficiency of Fe(III)-cyt *c*; measurements showed $k_{63}$ to be about six orders of magnitude slower than $k_{65}$. Subsequent studies[135] showed that only a few nanomolar added Cu(II) significantly increased the apparent value of $k_{61}$, but that the reduction yield was too high to be accounted for by Reactions (19-64) and (19-65) acting in simple competition.

The rate of Reaction (19-61) changes dramatically at pH 8.5 to 9.5, consistent with the effect arising from the major conformational change with a transition point at pH 9.1 to 9.3, accompanied by a decrease in the reduction potential of the Fe(III)/Fe(II) couple. It seems unlikely that the latter consequence is in itself sufficient to explain the results illustrated in Figure 19-11(a). Incidentally, high concentrations of alcohols change the conformation of cytochrome *c* in neutral solution,[137] so that the presence of alcohols in radiolysis systems could give rise to artifacts.

A number of other radicals reduce Fe(III)-cyt *c* (see, eg, Simic et al[134]); an earlier derivation[68] of rate constants for reactions of bipyridylium radical cations and cytochrome *c*, from electrochemical measurements, was about three to four orders of magnitude in error.[134]

# RADICALS DERIVED FROM NICOTINAMIDES

## *Production of the Radical NAD· by Radiolysis and Its Stability*

We have already seen [Figure 19-4, Reactions (19-33), (19-34)] how the radical NAD· can be generated *either* by one-electron reduction of $NAD^+$ *or* by one-electron oxidation of NADH. In addition to $e_{aq}^-$, reduction of $NAD^+$ may be readily accomplished [eg, by $CO_2^{-}\cdot$ ($k = 1.6 \times 10^9\ M^{-1}\ s^{-1}$ at ionic strength 0.1[85]) or $(CH_3)_2\dot{C}OH$ ($k = 1.0 \times 10^9\ M^{-1}\ s^{-1}$ [138])]. Several alternative one-electron oxidants [in addition to $Br_2^{-}\cdot$, Eq. (19-34)] readily produce NAD· by means of NADH. As well as halide or pseudohalide analogues of $Br_2^{-}\cdot$,[86] oxidation of NADH by $CCl_3O_2\cdot$, and some phenoxyl and phenothiazine radicals, has been reported;[139,140] hydrogen or electron transfer to glutathionyl radicals, GS·, from NADH also generates NAD·[141] (see below).

Early steady-state radiolysis experiments[142] had shown that the final product of one-electron reduction of $NAD^+$ using ethanol radicals was not enzymically active; that is, Reaction (19-66) rather than (19-67) occurred;

$$2\ NAD\cdot \rightarrow (NAD)_2 \qquad (19\text{-}66)$$

$$2\ NAD\cdot \rightarrow NAD^+ + NADH \qquad (19\text{-}67)$$

Subsequent direct observations of the reaction by pulse radiolysis showed[85] $k_{66} = 5.6 \times 10^7\ M^{-1}\ s^{-1}$ $(-d[NAD\cdot]/dt = 2k_{66}[NAD\cdot]^2)$.

This rate constant, considerably slower than the diffusion-controlled limit[65] (Chapter 4) facilitates studies of reactions of NAD· because the "natural" first half-life of decay of NAD· in typical pulse radiolysis experiments is a few milliseconds in the absence of reactive solutes. (A somewhat contradictory study, claiming formation of NADH via Reaction (19-67) [to an unspecified extent], requires further substantiation and product identification.[143])

## *Electron Transfer Reactions of Free NAD· and NADH*

Facile measurement of one-electron transfer equilibria using pulse radiolysis, as illustrated in Figure 19-6, led to the suggestion[60] that it should be feasible to measure the reduction potential of the couple $NAD^+/NAD\cdot$ by using a low-potential bipyridylium compound as redox indicator [cf Eq. (19-45)]. This was found to be possible: Anderson[67] obtained a value of $E_7^1(NAD^+/NAD\cdot) = (-922 \pm 8)$mV using 1,1′-butano-4,4′-dimethyl-2,2′-bipyridylium dibromide, and Farrington et al[66] simultaneously reported $E_7^1 = -940$ mV based on equilibria with 1-methylisonicotinamide (the potential of the latter determined vs. 1,1′-propano-2,2′-bipyridylium dibromide). Such good agreement for a potential obtained by way of independent redox indicators is typical of the method.

With NAD· being such a powerful reductant, electron transfer from NAD· to numerous electron acceptors is *thermodynamically* favorable. Thus, electron transfer to oxygen,

$$NAD\cdot + O_2 \rightarrow NAD^+ + O_2^-\cdot \qquad (19\text{-}68)$$

$(\Delta G^{\circ\prime} = -75\ kJ\ mol^{-1}$ at pH 7, $1\ M\ O_2)$ is diffusion controlled, $k_{68} = 2.0 \times 10^9\ M^{-1}\ s^{-1}$.[86,138] (Enzyme tests showed $NAD^+$ was produced.[86]) Other typical acceptors of a single electron from NAD· are 1,4-benzoquinone,[138] a nitroxyl radiosensitizer,[144] a haem,[145] 2-methyl-1,4-naphthoquinone,[146] riboflavin,[146] FAD,[97] and cytochrome *c*,[147] with rate constants for electron transfer all of the order of $10^8$–$10^9\ M^{-1}\ s^{-1}$.

Using $E_7^2(NAD\cdot/NADH) = 290$ mV,[66,67] we have for

$$NADH + O_2^-\cdot + H^+ \rightarrow NAD\cdot + H_2O_2 \qquad (19\text{-}69)$$

$\Delta G^{\circ\prime} = -56\ kJ\ mol^{-1}$ at pH 7. However, attempts to measure the *kinetics* of this thermodynamically favorable reaction failed, an upper limit of much less than 27 $M^{-1}$ $s^{-1}$ being reported.[86] In contrast, oxidation of NADH with other radicals having approximately the same potential as the $O_2^{-}\cdot/O_2^{2-}$ couple was observed (eg, oxidation by 1,3-benzosemiquinone and some phenothiazine cation radicals).[140] On the other hand, the same study[140] showed that one-electron reduction of NAD· (eg, by ascorbate, $AH^-$),

$$AH^- + NAD\cdot + H^+ \rightarrow AH\cdot + NADH \qquad (19\text{-}70)$$

($\Delta G^{\circ\prime} = -1.2$ kJ $mol^{-1}$ at pH 7, using $E_7{}^2(AH\cdot/AH^-) = 278$ mV; see Figure 19-1), and reduction of NAD· by more powerful reductants were *kinetically* too slow to be detected. The authors suggested that there was no true meaning to $E_7{}^2$(NAD·/NADH) because of this apparent lack of reversible reduction.[140] We have noted earlier other illustrations of kinetically slow but thermodynamically favorable reactions.

## *Radical Reactions of Enzyme-Bound NAD(P)· and NAD(P)H*

Bielski and Chan[148] confirmed the lack of reactivity[86] of free NADH towards $O_2^{-}\cdot$ [(Reaction (19-69)] by recording an efficiency of only about 6% of the radical yield in the radiolytic loss of NADH in aerated formate solutions (a convenient source of $O_2^{-}\cdot$). In contrast, when the enzyme lactate dehydrogenase was present, each $O_2^{-}\cdot$ radical produced resulted in up to about 10 molecules of NADH oxidized.[148] Lactate dehydrogenase (LDH) complexes with both NADH and $NAD^+$, but the dissociation constants[149] indicate approximately a 100-fold higher stability of the NADH complex. The chain oxidation mechanism proposed involved reaction of $O_2^{-}\cdot$ with *enzyme-bound* NADH, with a rate constant several orders of magnitude higher than that of Reaction (19-69):

$$LDH + NADH \rightleftharpoons LDH\cdot NADH \qquad (19\text{-}71)$$

$$LDH\cdot NADH + O_2^{-}\cdot + H^+ \rightarrow LADH\cdot NAD\cdot + H_2O_2 \qquad (19\text{-}72)$$

$$LDH\cdot NAD\cdot + O_2 \rightarrow LDH\cdot NAD^+ + O_2^{-}\cdot \qquad (19\text{-}73)$$

$$LDH\cdot NAD^+ \rightleftharpoons LDH + NAD^+ \qquad (19\text{-}74)$$

Particular attention was paid to the removal of impurities that catalyze the disproportionation of $O_2^{-}\cdot$, when an optimal chain length of about 6 and an estimate of $k_{72} = 1.0 \times 10^5$ $M^{-1}$ $s^{-1}$ was obtained.[150] An increase in $k_{72}$ with decreasing pH was ascribed to the protonation of $O_2^{-}\cdot$ [Eq. (19-10)]; it was suggested[150] that oxidation of LDH-bound NADH by $HO_2\cdot$ was about 20-fold faster than the corresponding reaction with $O_2^{-}\cdot$, although a precise value could not be derived because the pH-dependence of $K_{71}$ was not established.

Reaction (19-66) might be expected to interfere in such reactions, but Bielski and Chan[151] also studied the rate of NAD· disappearance in the presence of LDH (and other enzymes), finding a 1000-fold increase in radical stability compared to free NAD·. Figure 19-4 includes the absorption spectrum of the complex between malate dehydrogenase and NAD·, which was reported[151] to be similar to that of the LDH·NAD· complex. The spectrum of NAD· undergoes a red shift of about 30–35 nm when enzyme bound. The decay kinetics of the enzyme–NAD· complexes implied rapid achievement of equilibrium (19-74).

In contrast to this catalytic effect of the enzyme on a radical reaction, electron transfer between the phosphate analogue NADP· and the free flavin, FAD, is diffusion controlled[152] (cf NAD·[97])

$$\mathrm{NADP\cdot + FAD + H^+ \rightarrow NADP^+ + FADH\cdot} \qquad (19\text{-}75)$$

whereas reduction of the flavin center in the enzyme ferredoxin–NADP reductase (FNR–FAD) by NADP· requires complexing of NADP· to the enzyme or electron exchange with bound $NADP^+$:

$$\mathrm{NADP\cdot + (FNR\text{–}FAD)\cdot NADP^+ \rightarrow (FNR\text{–}FAD)\cdot NADP\cdot + NADP^+} \qquad (19\text{-}76)$$

Subsequent electron migration to the ferredoxin center could be monitored by pulse radiolysis.[152] Bielski[153] has reviewed these and other contrasts in the chemistry of free and enzyme-bound nicotinamide radicals, of great relevance to the role of the pyridinyl radicals in biology.[154,155]

## SOME RADICAL REACTIONS OF ASCORBATE AND THIOLS

Thiols and ascorbate are of great importance as biological reductants and protective agents; many xenobiotic compounds are detoxified by hepatic metabolism involving glutathione conjugation,[156] and there is much interest in the reactions of antidioxidants such as vitamins E ($\alpha$-tocopherol) and ascorbate.[12,157] Typical tissue thiol levels[156] (principally glutathione, GSH) are of the order of 1 to 2 m$M$—higher in the liver—and a normal adult has plasma ascorbate levels of the order of 50 $\mu M$.[158] The crucial role of thiols in tissue radioprotection by radical reactions has been appreciated for many years[159] and there is still much current interest in manipulating endogenous or exogenous thiols for potential therapeutic gain in tumor radiotherapy.[12,21,160–162] The chemistry of free-radical repair mechanisms,[159] of sulfur-centered free radicals,[163] and of ascorbyl radicals[164] has been recently reviewed, and we illustrate here only briefly a few typical reactions of interest in biochemistry and radiobiology.

## *Radiolytic Production and Stability of Ascorbyl and Thiyl Radicals*

The ascorbate anion radical, $A^{-}\cdot$, is most usefully generated radiolytically by oxidation of ascorbate ($AH_2$) (eg, by $Br_2^{-}\cdot$ [165,166]),

$$AH_2 + Br_2^{-}\cdot \rightarrow A^{-}\cdot + 2\,Br^{-} + 2H^{+} \qquad (19\text{-}77)$$

or at higher pH by phenoxyl radicals,[166] when reversible electron transfer reactions can be established, and, therefore, reduction potentials calculated.[52,167] The semiquinone disproportionation equilibrium [Eq. (19-12)] has $K_{fi} = 4 \times 10^{-8}$ at pH 7 (Figure 19-1), but the ascorbyl radical has been detected in tumor[168] and other tissues.[164] The disproportionation reaction is thought to involve a dimer, and the rate is markedly pH dependent in the physiological range, as shown in Figure 19-11(b).[169]

Thiyl radicals, RS·, may be produced by reaction of thiols, RSH, with OH·,[170] but in the case of GSH more than one radical is formed,[171] and a preferable route is reaction of a secondary radical (eg, $\cdot CH_2OH$ or $(CH_3)_2\dot{C}OH$) with the thiol in a radical "repair" reaction typical of those important in biology:[159,172]

$$\cdot CH_2OH + RSH \rightarrow CH_3OH + RS\cdot \qquad (19\text{-}78)$$

In some experiments oxidation of thiols by $Br_2^{-}\cdot$, and so on, may be convenient.[78] Thiyl radicals decay by combination to form the disulfide RSSR, but the equilibrium[170]

$$RS\cdot + RS^{-} \rightleftharpoons RSSR^{-}\cdot \qquad (19\text{-}79)$$

results in an influence of thiol concentration and pH on the lifetime of RS·. For GSH, estimates of $k_{79f}$ and $k_{79b}$ of about 4–6 $\times 10^8\ M^{-1}\ s^{-1}$ and $2 \times 10^5\ s^{-1}$ have been made; that is, $K_{79} \approx 3 \times 10^3\ M^{-1}$.[171,173,174]

Thiyl radicals add rapidly to oxygen to form $RSO_2\cdot$ radicals;[175] whereas ascorbate radicals are unreactive towards $O_2$ ($k < 5 \times 10^2\ M^{-1}\ s^{-1}$).[176] The oxidation of ascorbate by $HO_2\cdot/O_2^{-}\cdot$ follows the remarkable pH profile illustrated in Fig. 19-11(b), the result of a detailed study.[177]

## *Electron Transfer or Hydrogen Atom Transfer?*

The reduction of $NAD^{+}$ to NADH has long been the focus of arguments about the stoichiometry and order of $e^{-}$ and $H^{+}$ addition;[30,153–155] a recent study[141] showed how GSH could act as a link between hydrogen and electron transfer. Simple alcohol radicals are likely to be repaired by *hydrogen* transfer by, for example, thiols[172] or ascorbate,[178] while NADH is unreactive. However, thiyl radicals, GS·, and so on, were found to be reduced rapidly by both ascorbate and NADH ($k > 10^8\ M^{-1}\ s^{-1}$ in both cases at pH 7) probably by *electron* transfer.[141] Thus, the overall reaction

$$R^1R^2\dot{C}OH + NADH \rightarrow R^1R^2CHOH + NAD\cdot \qquad (19\text{-}80)$$

is facilitated by

$$R^1R^2\dot{C}OH + GSH \rightarrow R^1R^2CHOH + GS\cdot \quad (19\text{-}81)$$

$$GS\cdot + NADH \rightarrow GSH + NAD\cdot \quad (19\text{-}82)$$

The reaction between GS· and ascorbate is of particular interest in respect to the importance of GSH in the reversibility of the ascorbate redox system.[179] The catalytic role of vitamin E ($\alpha$-tocopherol) in the repair of organic radicals, with ascorbate reducing the tocopherol phenoxy radical, is another important example of sequential hydrogen and electron transfer established by pulse radiolysis.[180]

## CONCLUSIONS

Two decades ago, most reviewers of the radiation-chemical literature tended to adopt a style: "The yield of product $X$ after $\gamma$-irradiation of a solution $Y$ was $G(X) = x$, suggesting ..." or (only just beginning) "Pulse radiolysis of a solution containing $A$ and $B$ gave an absorption with maxima at ...". Such phraseology is quite absent from the present chapter, underlining the earlier comment that in many studies of the type described, the radiation-chemical details are often taken for granted. Indeed, the reader might be reminded here that virtually all the properties of, and reactions involving, free radicals discussed in this chapter have either been positively characterized for the first time or have been greatly supported by radiation-chemical techniques, principally pulse radiolysis.

The author is acutely aware that only a small fraction of the applications of radiation chemistry to biochemistry and biology have been mentioned. The choice of topics reflects not only space limitations but also the recent publication[27] of a much more comprehensive survey of many further examples, many excellent reviews referred to above, and, not least, the fact that the author has just completed an outline of some radiation-chemical aspects of an important area of radiobiological research[23] and wished to avoid undue duplication. Intentionally, emphasis has been placed on the energetics of radical reactions, because reliable appreciation of such quantities as reduction potentials of couples involving radicals unstable in aqueous solution has only been possible in the last decade, and this will continue to be a most important area to which pulse radiolysis can make major contributions.

In the immediate future there seems little doubt that increasing use will be made of radiation-chemical techniques in characterizing radicals derived from compounds important in medicine—radicals which are known[13,69,181] to be produced in biology. Many current issues of journals such as *Biochemical Pharmacology*, *Journal of Medicinal Chemistry*, and *Chemico-Biological Interactions* contain at least one article suggesting radical reactions involving medically important drugs that could be readily characterized by radiation-chemical methods.

Radiation chemistry is alive and well and living in all areas of molecular science where free radicals reside. Pulse radiolysis is now of age: Let us put it to work.

## ACKNOWLEDGMENT

This work is supported by the Cancer Research Campaign.

## REFERENCES

1. Blois, M. S., Jr.; Brown, H. W.; Lemmon, R. M.; Lindblom, R. O.; Weissbluth, M. "Free Radicals in Biological Systems"; Academic Press: New York, 1961.
2. Pryor, W. A. (ed.) "Free Radicals in Biology"; Academic Press; New York, 1982, Vol. 5.
3. Singh, A.; Petkau, A. (eds.) *Photochem. Photobiol.* **1978**, *28*, 429.
4. Rodgers, M. A. J.; Powers, E. L. (eds.) "Oxygen and Oxy-Radicals in Chemistry and Biology"; Academic Press: New York, 1981.
5. "Oxygen Free Radicals and Tissue Damage"; Ciba Foundation Symposium 65 (new series), Excerpta Medica: Amsterdam, 1979.
6. Petkau, A.; Dhalla, N. S. (eds.) *Can. J. Physiol.* **1982**, *60*, 1327.
7. McBrien, D. C. H.; Slater, T. F. (eds.) "Free Radicals, Lipid Peroxidation and Cancer"; Academic Press: London, 1982.
8. Hiller, K.-O.; Willson, R. L. *Biochem. Pharm.* **1983**, *32*, 2109.
9. Floyd, R. A. (ed.) "Free Radicals and Cancer"; Marcel Dekker; New York, 1982.
10. Demopoulos, H. B.; Pietronigro, D. D.; Flamm, E. S.; Seligman, M. L. *J. Environ. Pathol. Toxicol.* **1980**, *3*, 273.
11. Greenstock, C. L. *Prog. React. Kinet.* **1982**, *11*, 73.
12. Nygaard, O. F.; Simic, M. G. (eds.) "Radioprotectors and Anticarcinogens"; Academic Press; New York, 1983.
13. Mason, R. P. In "Reviews in Biochemical Toxicology"; Hodgson, E.; Bend, J. R.; Philpot, R. M., Eds.; Elsevier: New York, 1979, Vol. 1, p. 151.
14. Enquiries to: Prof. R. L. Willson, Dept. of Biochemistry, Brunel University, Uxbridge, Middlesex UB8 3PH, England.
15. Elkind, M. M.; Redpath, J. L. In "Cancer: a Comprehensive Treatise"; Becker, F. F., Ed.; Plenum Press: New York, 1977, Vol. 6, p. 51.
16. Cole, A.; Meyn, R. E.; Chen, R.; Corry, P. M.; Hittleman, W. In "Radiation Biology in Cancer Research"; Meyn, R. E.; Withers, H. R., Eds.; Raven Press: New York, 1980, p. 33.
17. Hütterman, J.; Köhnlein, W.; Téoule, R.; Bertinchamps, A. J. (eds.) "Effects of Ionizing Radiation on DNA. Physical, Chemical and Biological Aspects"; Springer-Verlag: Berlin, 1978.
18. Adams, G. E.; Fowler, J. F.; Wardman, P. (eds.) *Brit. J. Cancer.* **1978**, *37*, Suppl. III.
19. Brady, L. W. (ed.) "Radiation Sensitizers, Their Use in the Clinical Management of Cancer"; Masson: New York, 1980.
20. Adams, G. E.; Breccia, A.; Rimondi, C. (eds.) "Advanced Topics in Hypoxic Cell Radiosensitization" Plenum Presss: New York, 1982.
21. Sutherland, R. M. (ed.) *Int. J. Radiat. Oncol. Biol. Phys.* **1982**, *8*, 323.
22. Greenstock, C. L. *J. Chem. Ed.* **1981**, *58*, 156.
23. Wardman, P. *Radiat. Phys. Chem.* **1984**, *24*, 293.
24. Buxton, G. V. In "The Study of Fast Processes and Transient Species by Electron Pulse Radiolysis"; Baxendale, J. H.; and Busi, F., Eds.; Reidel: Dordrecht, 1982, pp. 241–266.

25. Schuler, R. H.; Hartzell, A. L.; Behar, B. *J. Phys. Chem.* **1981**, *85,* 192.
26. Baxendale, J. H.; Busi, F. (eds.) "The Study of Fast Processes and Transient Species by Electron Pulse Radiolysis"; Reidel: Dordrecht, 1982.
27. Bensasson, R. V.; Land, E. J.; Truscott, T. G. "Flash Photolysis and Pulse Radiolysis. Contributions to the Chemistry of Biology and Medicine"; Pergamon Press: Oxford, 1983.
28. Shafferman, A.; Stein, G. *Biochim. Biophys. Acta* **1975**, *416,* 287.
29. Land, E. J. *Current Topics Radiat. Res. Quart.* **1970–1972**, *7,* 105.
30. Bielski, B. H. J.; Gebicki, J. M. In "Free Radicals in Biology"; Pryor, W. A. Ed.; Academic Press: New York, 1977, Vol. 3, Chapter 1, pp. 1–51.
31. Adams, G. E.; Wardman, P. In "Free Radicals in Biology"; Pryor, W. A., Ed.; Academic Press: New York, 1977, Vol. 3, Chapter 2, pp. 53–95.
32. Willson, R. L. *Chem. Ind.* **1977**, 183.
33. Wardman, P. *Rep. Prog. Phys.* **1978**, *41,* 259.
34. Willson, R. L. In "Biochemical Mechanisms of Liver Injury"; Slater, T. F., Ed.; Academic Press: London, 1978, Chapter 4, pp. 123–224.
35. Ward, J. F. *Radiat. Res.* **1981**, *86,* 185.
36. Greenstock, C. L. *Radiat. Res.* **1981**, *86,* 196.
37. Biaglow, J. E. *Radiat. Res.* **1981**, *86,* 212.
38. Floyd, R. A. *Radiat. Res.* **1981**, *86,* 243.
39. Bannister, J. V.; Hill, H. A. O. (eds.) "Chemical and Biochemical Aspects of Superoxide and Superoxide Dismutase"; Elsevier: Amsterdam, 1980.
40. Bannister, W. H.; Bannister, J. V. (eds.) "Biological and Clinical Aspects of Superoxide and Superoxide Dismutase"; Elsevier: Amsterdam, 1980.
41. Nygaard, O. F.; Adler, H. C.; Sinclair, W. K. (eds.) "Radiation Research. Biomedical, Chemical, and Physical Perspectives"; Academic Press: New York, 1975.
42. Okada, S.; Imamura, M.; Terashima, T.; Yamaguchi, H. (eds.) "Radiation Research"; Japanese Association for Radiation Research: Tokyo, 1979.
43. Broerse, J. J.; Barendsen, G. W.; Kal, H. B.; van der Kogel, A. J. (eds.) "Radiation Research"; Martinus Nijhoff; Amsterdam, 1983.
44. Michaelis, L. *J. Biol. Chem.* **1931**, *92,* 211.
45. Chance, B. In "Free Radicals in Biological Systems"; Blois, M. S., Jr.; Brown, H. W.; Lemmon, R. M.; Lindblom, R. O.; Weissbluth, M., Eds.; Academic Press; New York, 1961, p. 1.
46. Bielski, B. H. J. *Photochem. Photobiol.* **1978**, *28,* 645.
47. Weast, R. C. (ed.) "Handbook of Chemistry and Physics" 56th ed.; CRC Press: Cleveland, 1975, p. D-151.
48. Clark, W. M. "Oxidation-Reduction Potentials of Organic Systems"; Bailliere, Tindall and Cox: London, 1960.
49. Von Foerster, G.; Weiss, W.; Staudinger, Hj. *Ann. Chem.* **1965**, *690,* 166.
50. Kortüm, G.; Vogel, W.; Andrussow, K. "Dissociation Constants of Organic Acids in Aqueous Solution"; Butterworths: London, 1961.
51. Laroff, G. P.; Fessenden, R. W.; Schuler, R. H. *J. Amer. Chem. Soc.* **1972**, *94,* 9062.
52. Steenken, S.; Neta, P. *J. Phys. Chem.* **1979**, *83,* 1134.
53. Everling, B.; Weis, W.; Staudinger, Hj. *Z. Physiol. Chem.* **1969**, *350,* 886.
54. Pelizzetti, E.; Mentasti, E.; Pramauro, E. *Inorg. Chem.* **1978**, *17,* 1181.
55. Baxendale, J. H.; Hardy, H. R. *Trans. Faraday Soc.* **1953**, *49,* 1140.
56. Baxendale, J. H.; Hardy, H. R. *Trans. Faraday Soc.* **1953**, *49,* 1433.
57. Patel, K. B.; Willson, R. L. *J. Chem. Soc., Faraday Trans 1,* **1973**, *69,* 814.
58. Conant, J. B.; Fieser, L. F. *J. Amer. Chem. Soc.* **1923**, *45,* 2194.
59. Michaelis, L.; Schubert, M. P.; Reber, R. K.; Kuck, J. A.; Granick, S. *J. Amer. Chem. Soc.* **1938**, *60,* 1678.
60. Wardman, P.; Clarke, E. D. *J. Chem. Soc., Faraday Trans 1,* **1976**, *72,* 1377.
61. Anderson, R. F. *Biochim. Biophys. Acta,* **1983**, *722,* 158.
62. Gibson, Q. H.; Massey, V.; Atherton, N. *Biochem. J.* **1962**, *85,* 369.
63. Nakamura, S.; Nakamura, T.; Ogura, Y. *J. Biochem. (Tokyo)* **1963**, *53,* 143.
64. Holmström, B. *Photochem. Photobiol.* **1964**, *3,* 97.
65. Meisel, D.; Czapski, G. *J. Phys. Chem.* **1975**, *79,* 1503.
66. Farrington, J. A.; Land, E. J.; Swallow, A. J. *Biochim. Biophys. Acta,* **1980**, *590,* 273.

67. Anderson, R. F. *Biochim. Biophys. Acta,* **1980**, *590,* 277.
68. Steckham, E.; Kuwana, T. *Ber. Bunsenges. Phys. Chem.* **1974**, *78,* 253.
69. Mason, R. In "Free Radicals in Biology"; Pryor, W. A., Ed., Academic Press: New York, 1982, Vol. 5, Chapter 6, pp. 161–222.
70. Dryhurst, G.; Kadish, K. M.; Scheller, F.; Renneberg, R. "Biological Electrochemistry"; Academic Press; New York, 1982, Vol. 1.
71. Adams, G. E.; Michael, B. D. *Trans. Faraday Soc.* **1967**, *63,* 1171.
72. Fendler, J. H. In "The Chemistry of the Quinonoid Compounds"; Patai, S., Ed.; John Wiley: London, 1974, Part 1, Chapter 10, pp. 539–578.
73. Khudyakov, I. V.; Kuzmin, V. A.; Emanuel, N. M. *Int. J. Chem. Kinet.* **1978**, *10,* 1005.
74. Anderson, R. F. *Biochim. Biophys. Acta,* **1983**, *723,* 78.
75. Bielski, B. H. J.; Allen, A. O.; Schwarz, H. A. *J. Am. Chem. Soc.* **1981**, *103,* 3516.
76. Richter, H. W.; Waddell, W. H. *J. Am. Chem. Soc.* **1983**, *105,* 5434.
77. Prütz, W. A.; Land, E. J. *Int. J. Radiat. Biol.* **1979**, *36,* 513.
78. Adams, G. E.; Aldrich, J. E.; Bisby, R. H.; Cundall, R. B.; Redpath, J. L.; Willson, R. L. *Radiat. Res.* **1972**, *49,* 278.
79. Bisby, R. H.; Cundall, R. B.; Davies, A. K. *Photochem. Photobiol.* **1978**, *28,* 825.
80. Baxendale, J. H.; Bevan, P. L. T.; Stott, D. A. *Trans. Faraday Soc.* **1968**, *64,* 2389.
81. Steenken, S. *J. Phys. Chem.* **1979**, *83,* 595.
82. Santus, R.; Grossweiner, L. I. *Photochem. Photobiol.* **1973**, *15,* 101.
83. Posener, M. L.; Adams, G. E.; Wardman, P.; Cundall, R. B. *J. Chem. Soc., Faraday Trans. 1* **1976**, *72,* 2231.
84. Redpath, J. L.; Santus, R.; Ovadia, J.; Grossweiner, L. I. *Int. J. Radiat. Biol.* **1975**, *27,* 201.
85. Land, E. J.; Swallow, A. J. *Biochim. Biophys. Acta* **1968**, *162,* 327.
86. Land, E. J.; Swallow, A. J. *Biochim. Biophys. Acta* **1971**, *234,* 34.
87. Enquiries to Dr. Alberta B. Ross, Radiation Chemistry Data Center, University of Notre Dame, Indiana 46556, U.S.A.
88. Stryer, L. "Biochemistry", 2nd ed.; W. H. Freeman: San Francisco, 1981, Chapter 14, pp. 307–332.
89. Lehninger, A. L. "Biochemistry", 2nd ed.; Worth: New York, 1975, Chapters 18, 19, pp. 477–541.
90. Chance, B.; Hagihara, B. In "Proceedings of the Fifth International Congress on Biochemistry, Moscow, 1961"; Pergamon Press: Oxford, p. 3.
91. Ohnishi, T.; Blum, H.; Salerno, J. C. In "Functions of Quinones in Energy-Conserving Systems"; Trumpower, B. L., Ed.; Academic Press: New York, 1982, pp. 247–261.
92. Massey, V.; Williams, C. H., Jr. (Eds.) Proceedings of the Seventh International Symposium on Flavins and Flavoproteins"; Elsevier: New York, 1982.
93. Michaelis, L.; Schubert, M. P.; Smythe, C. V. *J. Biol. Chem.* **1936**, *116,* 587.
94. Land, E. J.; Swallow, A. J. *Biochemistry* **1969**, *8,* 2117.
95. Buxton, G. V.; Sellers, R. M. *J. Chem. Soc., Faraday Trans. 1* **1973**, *69,* 555.
96. Meisel, D.; Neta, P. *J. Phys. Chem.* **1975**, *79,* 2459.
97. Anderson, R. F. *Ber. Bunsenges. Phys. Chem.* **1976**, *80,* 969.
98. Arai, S.; Dorfman, L. M. *Adv. Chem. Ser.* **1968**, *82,* 378.
99. Ilan, Y. A.; Meisel, D.; Czapski, G. *Isr. J. Chem.* **1974**, *12,* 891.
100. Meisel, D.; Neta, P. *J. Amer. Chem. Soc.* **1975**, *97,* 5198.
101. Wood, P. M. *F.E.B.S. Lett.* **1974**, *44,* 22.
102. Wardman, P. "Reduction Potentials of One-Electron Couples Involving Free Radicals in Aqueous Solution"; *J. Phys. Chem. Ref. Data,* in the press.
103. Lowe, H. J.; Clark, W. M. *J. Biol. Chem.* **1956**, *221,* 983.
104. Draper, R. D.; Ingraham, L. L. *Arch. Biochem. Biophys.* **1968**, *125,* 802.
105. Vaish, S. P. *Bioenergetics* **1971**, *2,* 61.
106. Swallow, A. J. In "Functions of Quinones in Energy Conserving Systems"; Trumpower, B. L., Ed.; Academic Press; New York, 1982, pp. 59–72.
107. Wilson, D. F.; Ericinska, M.; Dutton, P. L. *Ann. Rev. Biophys. Bioeng.* **1974**, *3,* 203.
108. Clarke, E. D. "Reduction of Nitroaromatic and Nitroheterocyclic Compounds in Model Biochemical Systems"; M.Phil. Thesis, The Hatfield Polytechnic, 1980.
109. Ahmad, R.; Armstrong, D. A. *Biochemistry* **1982**, *22,* 5445.
110. Swallow, A. J. *Nature* **1955**, *176,* 793.
111. Faraggi, M.; Hemmerich, P.; Pecht, I. *F.E.B.S. Lett.* **1975**, *51,* 47.

112. Massey, V.; Palmer, G.; Ballou, D. In Proceedings of the Third International Symposium on Flavins and Flavoproteins; Kamin, H., Ed.; University Park Press; Baltimore; and Butterworths; London, 1971, pp. 349–361.
113. Anderson, R. F. In "Oxygen and Oxy-Radicals in Chemistry and Biology"; Rodgers, M. A. J. and Powers, E. L., Eds.; Academic Press: New York, 1981, pp. 597–600.
114. Anderson, R. F. In Proceedings of the Seventh International Symposium on Flavins and Flavoproteins, Massey, V.; and Williams, C. H., Eds.; Elsevier: New York, 1982, Chapter 45, pp. 278–283.
115. Hemmerich, P.; Wessiak, A. In "Biochemical and Clinical Aspects of Oxygen"; Caughey, W. S., Ed.; Academic Press; New York, 1979, pp. 491–511.
116. Elliot, A. J.; Ahmed, R.; McIntosh, L.; Burke, D.; Stevenson, K. J.; Armstrong, D. A. In Proceedings of the Seventh International Symposium on Flavins and Flavoproteins; Massey, V.; and Williams, C. H., Eds.; Elsevier: New York, 1982, Chapter 87, pp. 533–536.
117. Massey, V.; Ghisla, S. *Ann. N.Y. Acad. Sci.* **1974**, *277*, 446.
118. Kobayashi, K.; Hirota, K.; Ohara, H.; Hayashi, K.; Miura, R.; Yamano, T. *Biochemistry* **1983**, *22*, 2239.
119. Faraggi, M.; Klapper, M. H. *J. Biol. Chem.* **1979**, *254*, 8139.
120. Anderson, R. F.; Patel, K. B.; Adams, G. E. *Int. J. Radiat. Biol.* **1977**, *32*, 523.
121. Land, E. J.; Simic, M.; Swallow, A. J. *Biochim. Biophys. Acta* **1971**, *226*, 239.
122. Land, E. J.; Swallow, A. J. *J. Biol. Chem.* **1970**, *245*, 1890.
123. Meisel, D. *Chem. Phys. Lett.* **1975**, *34*, 263.
124. Bensasson, R. V.; Land, E. J. *Biochim. Biophys. Acta*, **1973**, *325*, 175.
125. Farrington, J. A.; Ebert, M.; Land, E. J.; Fletcher, K. *Biochim. Biophys. Acta* **1973**, *314*, 372.
126. Patterson, L. K.; Small, R. D.; Scaiano, J. C. *Radiat. Res.* **1977**, *72*, 218.
127. Farrington, J. A.; Ebert, M.; Land, E. J. *J. Chem. Soc., Faraday Trans 1* **1978**, *74*, 665.
128. Rodgers, M. A. J. In "Oxygen and Oxy-Radicals in Chemistry and Biology"; Rodgers, M. A. J.; and Powers, E. L., Eds.; Academic Press: New York, 1981, pp. 714–715.
129. Rodgers, M. A. J. *Radiat. Phys. Chem.* **1984**, *23*, 245.
130. Anderson, S. S.; Lyle, I. G.; Paterson, R. *Nature* **1976**, *259*, 147.
131. Butler, J.; Jayson, G. G.; Swallow, A. J. *Biochim. Biophys. Acta* **1975**, *408*, 215.
132. Seki, H.; Ilan, Y. A.; Ilan, Y.; Stein, G. *Biochem. Biophys. Acta* **1976**, *440*, 573.
133. Koppenol, W. H.; van Buren, K. J. H.; Butler, J.; Braams, R. *Biochim. Biophys. Acta* **1976**, *449*, 157.
134. Simic, M. G.; Taub, I. A.; Tocci, J.; Hurwitz, P. A. *Biochem. Biophys. Res. Commun.* **1975**, *62*, 161.
135. Butler, J.; Koppenol, W. H.; Margoliash, E. *J. Biol. Chem.* **1982**, *257*, 10747.
136. Ilan, Y.; Ilan, A. I.; Czapski, G. *Biochim. Biophys. Acta*, **1978**, *503*, 399.
137. Land, E. J.; Swallow, A. J. *Biochem. J.* **1976**, *157*, 781.
138. Willson, R. L. *Chem. Commun.* **1970**, 1005.
139. Packer, J. E.; Willson, R. L.; Bahnemann, D.; Asmus, K.-D. *J. Chem. Soc., Perkin Trans. 2* **1980**, 296.
140. Grodkowski, J.; Neta, P.; Carlson, B. W.; Miller, L. *J. Phys. Chem.* **1983**, *87*, 3135.
141. Forni, L. G.; Mönig, J.; Mora-Arellano, V. O.; Willson, R. L. *J. Chem. Soc., Perkin Trans. 2* **1983**, 961.
142. Swallow, A. J. *Biochem. J.* **1953**, *54*, 253.
143. Aikawa, M.; Honna, T.; Sumiyoshi, T.; Katayama, M. *Chem. Lett.* **1980**, 247.
144. Willson, R. L. *Trans. Faraday Soc.* **1971**, *67*, 3008.
145. Goff, H.; Simic, M. G. *Biochim. Biophys. Acta* **1975**, *392*, 201; *ibid.* **1980**, *590*, 273.
146. Rao, P. S.; Mayon, E. *Nature* **1973**, *243*, 344.
147. Land, E. J.; Swallow, A. J. *Ber. Bunsenges. Phys. Chem.* **1975**, *79*, 436.
148. Bielski, B. H. J.; Chan, P. C. *Arch. Biochem. Biophys.* **1973**, *159*, 873.
149. Schwert, G. W.; Miller, B. R.; Peanasky, R. J. *J. Biol. Chem.* **1976**, *242*, 3245.
150. Bielski, B. H. J.; Chan, P. C. *J. Biol. Chem.* **1976**, *251*, 3841.
151. Bielski, B. H. J.; Chan, P. C. *J. Amer. Chem. Soc.* **1980**, *102*, 1713.
152. Maskiewicz, R.; Bielski, B. H. J. *Biochim. Biophys. Acta* **1982**, *680*, 297.
153. Bielski, B. H. J. In "Radioprotectors and Anticarcinogens"; Nygaard, O. F.; Simic, M. G., Eds.; Academic Press: New York, 1983, pp. 43–52.
154. Kosower, E. M. In "Free Radicals in Biology"; Pryor, W. A., Ed.; Academic Press: New York, 1977, Vol. 2, Chapter 1, pp. 1–53.

155. Kosower, E. M.; Teuerstein, A.; Burrows, H. D.; Swallow, A. J. *J. Am. Chem. Soc.* **1978**, *100*, 5185.
156. Chasseaud, L. F. *Adv. Cancer Res.* **1979**, *29*, 175.
157. Seib, P. A.; Tolbert, B. M. (eds.) "Ascorbic Acid: Chemistry, Metabolism and Uses"; *Adv. Chem. Ser.* 200; American Chemical Society: Washington, D.C., 1982.
158. Koh, E. T.; Chi, M. S.; Lowenstein, F. W. *Am. J. Clin. Nutr.* **1980**, *33*, 1828.
159. Willson, R. L. In "Radioprotectors and Anticarcinogens"; Nygaard, O. F.; and Simic, M. G., Eds.; Academic Press: New York, 1983, pp. 1–22.
160. Bump, E. A.; Yu, N. Y.; Taylor, Y. C.; Brown, J. M.; Travis, E. L.; Boyd, M. R. In "Radioprotectors and Anticarcinogens"; Nygaard, O. F.; and Simic, M. G., Eds.; Academic Press: New York, 1983, pp. 297–323.
161. Hodgkiss, R. J.; Middleton, R. W.; *Int. J. Radiat. Biol.* **1983**, *43*, 179.
162. Jensen, G. L.; Meister, A. *Proc. Natl. Acad. Sci. U.S.A.* **1983**, *80*, 4714.
163. Asmus, K.-D. In "Radioprotectors and Anticarcinogens" Nygaard, O. F.; and Simic, M. G., Eds.; Academic Press: New York, 1983, pp. 23–42.
164. Bielski, B. H. J. In "Ascorbic Acid: Chemistry, Metabolism and Uses" *Adv. Chem. Ser.* 200; Seib, P. A.; and Tolbert, B. M., Eds.; American Chemical Society: Washington, D.C., 1982, Chapter 4, pp. 81–100.
165. Schoeneshoefer, M. *Z. Naturforsch.* **1972**, *B27*, 649.
166. Schuler, R. H. *Radiat. Res.* **1977**, *69*, 417.
167. Steenken, S.; Neta, P. *J. Phys. Chem.* **1982**, *86*, 3661.
168. Dodd, N. F. J. *Brit. J. Cancer* **1973**, *28*, 257.
169. Bielski, B. H. J.; Allen, A. O.; Schwarz, H. A. *J. Am. Chem. Soc.* **1981**, *103*, 3517.
170. Adams, G. E.; McNaughton, G. S.; Michael, B. D. In "The Chemistry of Ionization and Excitation"; Johnson, G. R. A.; and Scholes, G., Eds.; Taylor and Francis: London, 1967, pp. 281–293.
171. Sjöberg, L.; Eriksen, T. E.; Révész, L. *Radiat. Res.* **1982**, *89*, 255.
172. Adams, G. E.; McNaughton, G. S.; Michael, B. D. *Trans. Faraday Soc.* **1968**, *64*, 902.
173. Baker, M. Z.; Badiello, R.; Tamba, M.; Quintiliani, M.; Gorin, G. *Int. J. Radiat. Biol.* **1982**, *41*, 595.
174. Hoffman, M. Z.; Hayon, E. *J. Phys. Chem.* **1973**, *77*, 990.
175. Schafer, K.; Bonifacic, M.; Bahnemann, D.; Asmus, K.-D. *J. Phys. Chem.* **1978**, *82*, 2777.
176. Bielski, B. H. J.; Richter, H. W.; Chan, P. C. *Ann. N.Y. Acad. Sci.* **1975**, *258*, 231.
177. Cabelli, D. E.; Bielski, B. H. J. *J. Phys. Chem.* **1983**, *87*, 1809.
178. Redpath, J. L.; Willson, R. L. *Int. J. Radiat. Biol.* **1973**, *23*, 51.
179. Sapper, H.; Kang, S.-O.; Pau, H.-H.; Lohmann, W. *Z. Naturforsch.* **1982**, *37c*, 942.
180. Packer, J. E.; Slater, T. F.; Willson, R. L. *Nature*, **1979**, *278*, 737.
181. Mason, R. P. *Pharmacological Rev.* **1982**, *33*, 189.

# 20

# Radiation Processing and Sterilization

M. TAKEHISA AND S. MACHI

*Takasaki Radiation Chemistry Research Establishment, Japan Atomic Energy Research Institute*

## INTRODUCTION

During the past three decades extensive applied research and development of radiation chemistry has been carried out in a number of countries, including the United States, the United Kingdom, the Soviet Union, the Federal Republic of Germany, France, and Japan. These research and development activities have resulted in various industrial applications of radiation processing in polymer modifications, sterilization of medical supplies, and food irradiation.[1]

Radiation cross-linking of polyethylene insulators was the first industrial application and was achieved by the Raychem Corporation in the United States in 1958. Since then, radiation processing has penetrated a variety of industries.

Today, approximately 300 electron beam accelerators (EBAs) and more than 100 Co-60 irradiation facilities are operating for commercial purpose. World total sales of irradiated products is estimated at more than two billion U.S. dollars per year, and it continues to grow at a rate of 15 to 20% per year.[2]

This growth of commercial radiation processing has been largely dependent on the achievement in production of reliable and less expensive radiation facilities as well as the research and development effort for new applications. Although world statistics of the growth are not available, Figure 20-1 shows steady growth in the number of EBAs installed in Japan for


*Radiation Chemistry: Principles and Applications*

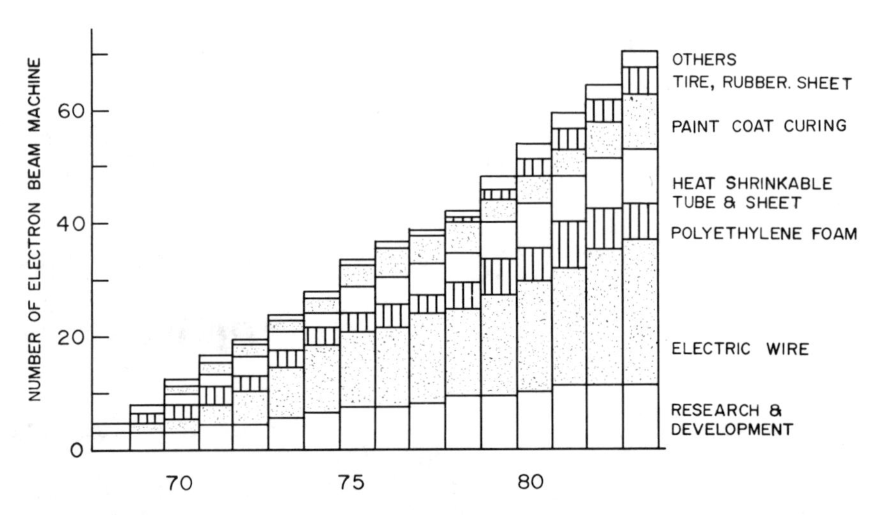


**Figure 20-1.** Growth of the number of electron beam accelerators in Japan.

various purposes.[3] Growth rate of Co-60 sources supplied by AECL (Atomic Energy of Canada Limited), which supplies approximately 80% of the world market, is approximately 10% per year, including future growth estimates.[4]

Potential applications of radiation processing under development are in environmental conservation (eg, treatment of sewage sludge, waste water, and exhaust gases) and bioengineering (eg, immobilization of bioactive materials). We plan to introduce here the characteristics of radiation processing, examples of its industrial applications, the status of its research and development activities, and an economic analysis.

## CHARACTERISTICS OF RADIATION PROCESSING

A simplified scheme of the primary processes induced by radiation are as follows:

$$AB \rightsquigarrow AB^+ + e^- \text{ and } AB^* \quad (20\text{-}1)$$

$$AB^+ + e^- = AB^* \quad (20\text{-}2)$$

$$AB^* = A\cdot + B\cdot \quad (20\text{-}3)$$

Active species formed by Reactions (20-1) and (20-3) are precursors of secondary reactions, such as cross-linking, polymerization, oxidation, and so on. Ions formed by Reaction (20-1) have short lifetimes and can continue further reaction with other molecules only in special circumstances (eg, at low temperature, at high dose rate, in a high-dielectric-constant medium, and in a superdry system). Excited molecules formed directly (20-1) and by recombination reactions (20-2) easily decompose to active species with nearly zero activation energy by Reaction (20-3), and the produced radicals live long enough to react with surrounding molecules. In a solid matrix the radicals last for days or more.

The reactions producing the active species are scarcely dependent on temperature, so radiation-induced reactions can be initiated at any temperature. An example of the temperature dependence of radiation-induced polymerization is that the overall activation energy is 5 kcal/mol, which is smaller than that of polymerization initiated by chemical means. The overall activation energy may be expressed as

$$E = E_i + E_p - E_t \tag{20-4}$$

where $E$, $E_i$, $E_p$, $E_t$ are the activation energies for the overall reaction, and the initiation, propagation, and termination steps, respectively. The activation energy for the initiation reaction induced by radiation was found to be nearly zero. The temperature independence of reaction rate is one of the most important characteristics of radiation processing.

Gamma-rays have a large penetration range; 90% of Co-60 $\gamma$-rays are absorbed by about 0.3 m of water. Accordingly, homogeneous introduction of active species is possible in solid or liquid substances. This penetration capability is advantageously used for the initiation of reactions in solid and liquid phases, food irradiation, and sterilization of medical products.

Ease of control in reaction rate and product quality is another advantage of radiation processing. By changing the dose rate, we can simply control the reaction rate of radiation-induced reactions.[5–7] For EBAs, dose rate and penetration range are well controlled by varying the beam current and voltage.

Radiation-induced reactions proceed without adding a catalyst. Therefore, processes to remove catalyst residue are not required, and the product is not contaminated by such residue. The pure products are advantageously used in bioapplications and optical applications. In sterilization of medical products with ethylene oxide gas, products may be contaminated by toxic residual gas. Such contamination is avoided by replacing ethylene oxide with radiation.

Radiation provides an excellent means to initiate solid-phase reactions. Figure 20-2 shows a trioxane single crystal and polyoxymethylene formed by radiation polymerization.[8] Even after polymerization the shape of the crystal remains as it was.

Electron beam (EB) radiation provides very high-intensity irradiation, although the penetration range is low compared with $\gamma$-radiation. Extremely high-speed reactions can be achieved by using EBs, which are not possible by using chemical initiators.[9] Curing of surface coatings by EBs (which is essentially polymerization) can be completed in a few seconds.

Characteristics of radiation processing can be summarized as follows:

Temperature independence

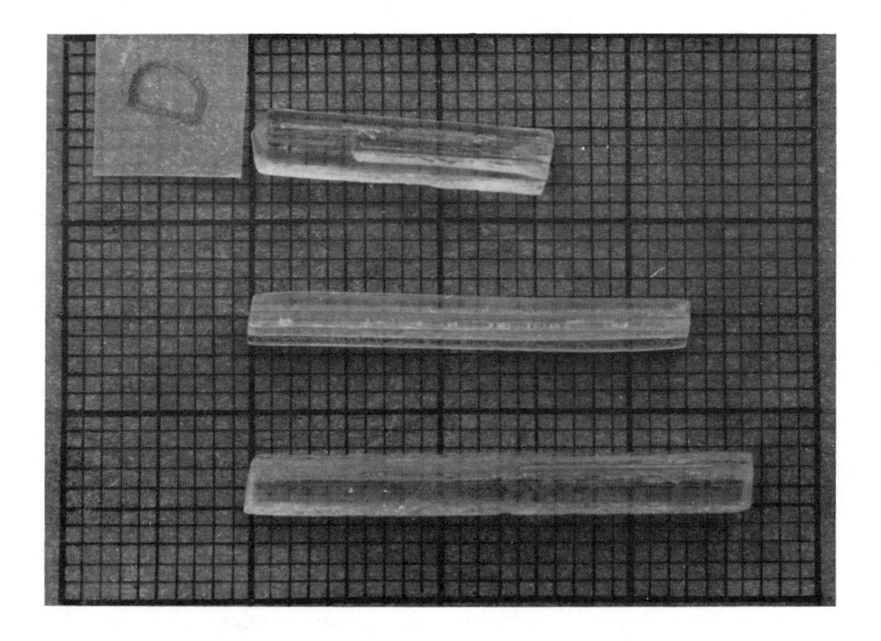


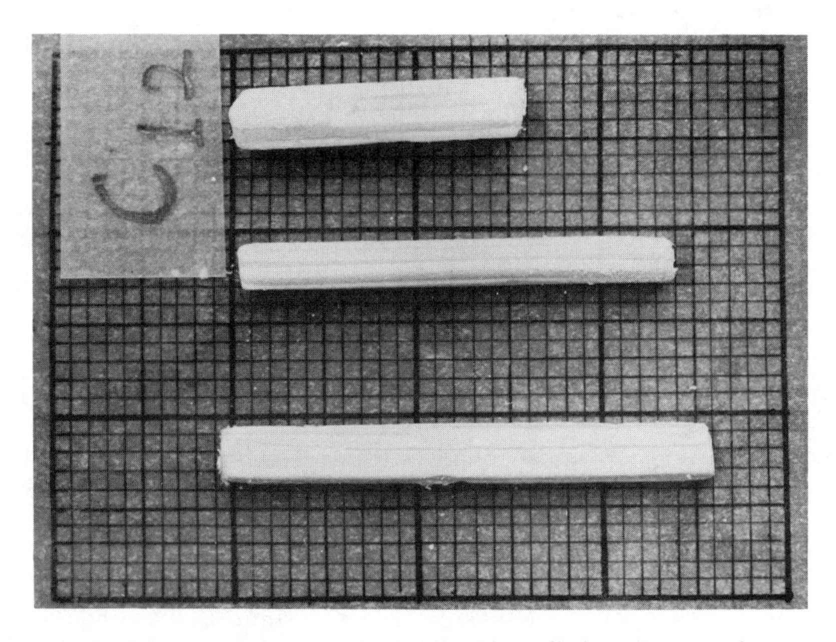


**Figure 20-2.** Solid-state polymerization initiated by radiation: Upper—trioxane crystal; lower—polyoxymethylene as polymerized ($10^2$ Gy/hr for 10 hr at 55°C).

Low-temperature capability
Ease of control
Freedom from catalyst and catalyst residue
Capability in solid substances
High-speed treatment capability

# INDUSTRIAL RADIATION SOURCES AND FACILITIES

For industrial purposes, Co-60 $\gamma$-rays and EBAs are usually used. Gamma-rays from Co-60 have a large penetration range and comparatively low intensity ($10^2$–$10^4$ Gy/hr), whereas EBs from an accelerator have short penetration range, depending on beam energy, and high radiation intensity ($10^2$–$10^4$ Gy/s).

In practice, most radiation sterilizers use Co-60 as a source because of this large penetration range. On the other hand, radiation cross-linking applications such as production of cross-linked wire insulation, heat-shrinkable film, and curing of surface coatings are usually carried out in industry by using EBAs.

## *Gamma-Irradiation Facilities*

Major components of $\gamma$-irradiation facilities are

Co-60 source
Radiation shield
Source-handling mechanism
Source containment
Product-handling equipment

### Co-60 Source

Co-60 is produced from Co-59 by neutron irradiation in a nuclear reactor for one year.[10] The Co-60 pellets or slugs are double encapsulated with stainless steel in long thin pencils; for example, the AECL standard size is 450 mm in length and 12.5 mm in diameter. Specific activity of Co-60 is in the range of 50 to 100 Ci/g.[11] The Co-60 source is installed in a source rack to form a suitable geometry, usually a slab.[12,13]

### Radiation Shield

Since $\gamma$-rays from Co-60 have strong penetration capability, the thickness of the concrete shield required is usually between 1.5 to 1.8 m.[11] Irradiation cells should be designed in accordance with the user's request.

### Source Storage

When not in use, large radiation sources are usually stored in water pools of more than 6 m in depth located in the irradiation cell. Water in the pool is deionized and filtered to avoid corrosion of Co-60 source capsules and to maintain transparency.[14] The pools should be designed to withstand a severe earthquake.

### Source-Handling Equipment

To transport Co-60 in the source rack between the bottom of the pool and the irradiation cell, we must have a most reliable mechanism. A metal chain or wire with special driving devices is often used to raise the source. The detailed design is classified manufacturer's information.

### Product-Handling Equipment

A conveyor-bed system and overhead monorail are often used in industrial irradiation applications to carry products in or out of the irradiation cell. The conveyor design should be made considering dose requirement, dose homogeneity, product size, and irradiation efficiency.[11,15–17]

### Irradiation Efficiency

Irradiation efficiency is defined as the ratio between the amount of radiation absorbed in the product and the amount of radiation emitted from the source. Some fraction of radiation is absorbed in the Co-60 source itself, depending on the source geometry. The irradiation efficiency also depends on the geometry of the product arrangement or the product conveyor design around the source, and the product packing density. For medical product sterilization the efficiency is between 0.2 and 0.4. Recent progress in the design uses incremental dose irradiators, which allow various dose levels of irradiation.[17,18]

## *Electron Beam Irradiation (EBI) Facility*

Great improvements in the quality of EBAs have been made in the last decade. Today, a reliable machine with a capacity as large as 200 kW is available.[19] Compared to $\gamma$-irradiation, EBI is characterized by its ability to process thin-layer materials at high speed.

Electron beam irradiation systems consist of an electron accelerator, shielding, and product-handling equipment. When selecting the electron accelerator, one should determine the beam current and energy in accordance with production rate and required depth of treatment, respectively. The effective penetration range of an electron beam is almost proportional to its energy level (0.4 g/cm$^2$ per MeV).

### Low-Energy (<300-kV) Electron Beam Accelerator

Low-energy EBAs are used to treat thin-layer materials, such as curing surface coatings, curing adhesives, and film-converting applications. A relatively compact, self-shielded, low-energy EBA characterized with linear filament type and plane filament type electron gun and needless of beam scanning has been developed.[20,21] A typical one has an energy range of 150 to 250 keV, and current of 10 to 100 mA. The width of the irradiation zone is in the range of 15 to 155 cm.

### Medium Energy (300-keV–1-MeV) Electron Beam Accelerator

A cable-connected transformer-type accelerator with scanner has been manufactured in several countries to generate intense EBs up to 1 MeV.[22] The dc power is fed from a dc power supply to an accelerator tube by means of a high-voltage cable. Energy and current ranges available with this accelerator are 300 keV to 1 MeV and 20 to 100 mA. Accelerators with energy between 300 and 550 keV can be self-shielded. Even for the higher energy ones, the dc power supply can be positioned outside the building.

The main advantages of the accelerator are

High-power conversion efficiency (ratio of dc output to ac input) of 90% and higher
Space saving; power supply unit can be installed outside a building.

### High-Energy (1-MeV–4-MeV) Electron Beam Accelerator

The dc power supply and accelerator tube are installed in a high-pressure vessel. The dc power supply consists of an ac capacitor column, dc capacitor column, and rectifier for a Cockcroft–Walton-type accelerator. This accelerator, with 100-kW output capacity and up to 5 MeV, has been manufactured. The Dynamitron produced by Radiation Dynamics consists of a high-frequency voltage generator coupled to a parallel-fed cascaded arrangement of rectifiers. An accelerator with 200-kW output capacity and beam energy of 4 MeV has been manufactured.[23] A Dynamitron can produce both horizontal and vertical beams and is a comparatively compact machine.

## INDUSTRIAL RADIATION PROCESSING

### *Sterilization*

There have been steadily increasing demands for disposable medical supplies that are sterilized and ready for use. The use of disposable supplies is an effective means for prevention of in-hospital infections. Radiation sterilization using $\gamma$-rays is the largest radiation processing application. Presently, there are more than 120 irradiation facilities in 40 countries, and the total Co-60 activity exceeds 60 MCi,[18] whereas fewer than 10 radiation facilities use electron beams.

Items commercially irradiated are disposable injection syringes, injection needles, surgical gloves, infusion sets, catheters, sutures, hollow-type dialyzers, transfusion filters, and others. Most of the devices are made of plastics that usually melt at 110° to 120°C or deform at a temperature lower than the melting point. Therefore, thermal sterilization is not applicable for such plastic products, and radiation sterilization, which is a room-temperature method, has expanded along with the use of plastic products.

The common room-temperature sterilization process for plastics is currently the ethylene oxide method.[24] But this is carried out by batch operation and has a disadvantage in that any residual gas is of toxic character.[25] The radiation sterilization process provides reliable disinfection, owing to the high penetrating power of $\gamma$-rays and preclusion of the possibility of secondary contamination because completely sealed products can be sterilized.[26]

The sterilization is based on a statistical kill of microorganisms following first-order kinetics with the radiation dose.[27,28] That is, the number of microorganisms (on a logarithmic scale) decreases linearly with dose. The relationship, called a survival curve, is generally a straight line, but in some cases it may appear as a sigmoidal curve with a shoulder (for a pure culture of microorganisms). For a mixture of several kinds of microorganisms having different radiation sensitivities, the survival curve tails. If we extrapolate the survival curve to a high dose under restricted conditions,[29] the number of surviving microorganisms decreases below unity in the high-dose range. The number of survivors of $10^{-3}$, for example, means one survivor per 1000 samples on average. In this sense there is no chance to reach complete sterilization, but it is generally accepted that sterilization is accomplished at $10^{-6}$ sterility.

The tangent to the survival curve depends on the kind of microorganism and the environment and represents the radiation sensitivity of the microorganism. The radiation sensitivity is quantitatively defined as the *D*-value (decimal reduction value), which is the dose needed to kill 90% of the initial population of microorganisms or to reduce the numbers to one-tenth. The *D*-value for usual vegetative cells is less than 1 kGy and is higher for spores and several radiation-resistant vegetative cells.[30] The standard microorganism used as the biological indicator for radiation sterilization is spores of *Bacillus Pumilus* ATCC 27142 (E601), and the standard *D*-value is 1.5 to 2 kGy, depending on preparations and substrates.[27]

The standard sterilization dose of 25 kGy has been adopted in many countries; some nations regulate the dose by law, and others use the dose as the result of experiments. The strictest standard has been applied in Scandinavian countries, because of the presence of radiation-resistant *Streptococcus faecium*; the dose is 32 to 45 kGy by $\gamma$-irradiation and 35 to 50 kGy by EB irradiation, depending on the initial counts.[31] Currently, the sterilization doses are 25 and 32 kGy according to the bioburden.[32]

Recent requirements for the sterilization dose is to ensure the $10^{-6}$ sterility. Investigation of the bioburden—the number of microorganisms and microflora of the product to be irradiated—and the decrease in the initial microbial contamination of the products through good manufacturing practice (GMP) assist in decreasing the sterility assurance dose to less than 25 kGy. Recent methods for determining the sterilization dose[33] have been described in the guidelines of the Association for the Advancement of Medical Instrumentation (AAMI).[34]

Once the sterilization dose has been determined, assurance of sterility can be obtained by confirming the dose. That is, a dose measurement for the irradiated products can be a routine alternative to a microbiological test,[35] and a "go–no go" dosimeter (by color change) can be used for this purpose. Irradiation is the only continuous sterilization process and does not require sterility tests for batch-by-batch products, such as is necessary in heat and ethylene oxide gas sterilization.

Owing to the high power output of EB machines, proposals have been made for EB sterilization[36] and for using bremsstrahlung x-rays.[37] Introduction of the EB machine to an irradiation service company has begun.

An important contribution from radiation chemists in sterilization is research and development of materials suitable for medical devices and of the technology to prevent degradation during their storage. The various plastics used in this field include polystyrene, polyethylene, silicone, cellulose, polyvinyl chloride, and polypropylene.[38,39] Among these, polymers, such as polystyrene and polyethylene, have no problems because they are stable under radiation up to several ten thousand grays. Polypropylene and polyvinylchloride are degradable by irradiation, so that considerable amounts of antiradiation agents are added. Antioxidant, plasticizer, and so on, are also added, and simpler chemical structure and higher molecular weight are desirable. Acceleration of radical decay by increasing polymer chain mobility is one of the guiding principles for reduction of polymer degradation.[40]

Another area of investigation is the methodology to predict long-term radiation aftereffects on mechanical properties by using an acceleration test based on sound scientific data.[41] Radiolytic products from the supplies, including their additives, should be carefully studied because some may have long-term toxicity.[31]

Concerning the potential use of EBs for sterilization, high-dose-rate irradiation can reduce degradation of polymeric materials, owing to anoxic condition in the polymer, but data accumulation on the microorganism kill at a high dose rate is required.[31]

Pharmaceuticals are also sterilized by radiation, but this is not yet widely accepted. Rare examples are eye ointment and antibiotics.[42,43] Many reports exist on the radiolysis of drugs, pharmaceuticals, and their base materials of natural and synthetic origins.[44] These are stable under radiolysis at the sterilization dose in both the dry state and when suspended in an oily base.[45] In dilute aqueous solutions at room temperature, the drugs decompose markedly and are not suitable for sterilization.[44] But radiolytic decomposition of drugs decreases in frozen solutions at low temperatures.[46] A major target of drug sterilization is the application to raw materials from natural origins and microbiological products such as antibiotics, because these products contain essentially microbial loads and the effective components are often heat sensitive. The main obstacle to practical application is a shortage of toxicological and pharmacological data on irradiated drugs.[47] However, radiation-chemical study of food irradiation indicates that radiolytic products are similar to those caused by thermal decomposition, and concentrations are very low.[48]

Radiation sterilization of human tissues, skin, bone, and so on, for surgical grafting as well as artificial implants is applied practically.[49,50] Pasteurization of containers for drugs and foods is now carried out by radiation rather than chemical and heat processes.

## *Cross-linking and Degradation*

### Wire and Cable

Thermoplastics and elastomers are widely used as electrical insulating materials because of their electrical and mechanical properties and their processability. However, good processability of common thermoplastics does not agree with the requirement of thermal resistance of insulators of wire and cable. Since cross-linking of the insulators is an effective means for improving thermal resistance, abrasion resistance, tensile strength, and so on, cross-linking of polyethylene insulators has been adopted in industry by chemical means, mostly by using peroxides. Chemical cross-linking is applied to the majority of thick, insulated power cables, but heating of the insulator is always necessary for cross-linking by steam, dry air, molten metal, and so on.[51] On the other hand, radiation provides rapid, room-temperature, flexible operations for cross-linking and produces a better product than does the chemical method concerning stripping and electrical properties.

Early on, insulators of thin wire for high-temperature ratings were of fluorinated polymer and silicone rubber, but cross-linked polyethylene and polyvinylchloride are replacing these expensive insulators.[52] High-density polyethylene and polyvinylchloride insulators are not easily cross-linked by chemical methods.

Presently, major manufacturers of wire and cable in the United States and Japan use electron accelerators for cross-linking. The dose for polyethylene cross-linking is 200 to 400 kGy, and 30 to 100 kGy for polyvinylchloride containing considerable amounts of cross-linking additives, such as tetraethyleneglycol dimethacrylate (TEGDM).[53] The key point is to select the minimum dose for improving the required thermal resistance and to irradiate the insulator homogeneously with a high efficiency of EB utilization.

The common irradiation method for thin wire is figure-eight-type irradiation by electron beams.[54] The efficiency of radiation utilization is high enough to achieve the homogeneous dose distribution in the insulator. With the use of a permanent magnet to deflect the electron beams passed through the bank of wires being irradiated, the dose uniformity in the insulator and radiation utilization efficiency are improved.[55]

Thick, insulated power cable, 70% of which is cross-linked, is not yet produced by radiation process. Problems to be solved are uniform irradiation[56] and prevention of void formation in the insulator by Lichtenberg's discharge.[57] The latter problem can be solved by using additives to reduce charge buildup in the insulator during irradiation and by dose reduction.[58]

### Heat-Shrinkable Film and Tubing

A major application of radiation is in the improvement of the thermal resistance of plastics, as seen in the wire insulator. A radiation–cross-linked, thermal-resistant, low-density polyethylene bag, which does not shrink in boiling water, was a major product in Japan. Currently, heat-shrinkable plastic film and tubing are major products of radiation. When semicrystalline thermoplastics are cross-linked, they show rubberlike properties at temperatures above the melting point. If the cross-linked polymer is deformed at a higher temperature and then cooled, the deformed shape is maintained; it returns to its original shape upon reheating. The phenomenon is called a "memory effect" and is applied in the production of heat-shrinkable film and tubing.

The initial commercial products were small-diameter, shrinkable, insulation tubing for electrical connectors and other circuits, supplied by the Raychem Corp., made of various kinds of plastic materials, such as polyethylene, polyvinylchloride, fluoro polymers, silicone, and so on. Several other producers then began commercial production. Later, large-diameter shrinkable tubing became popular, and such products are used as wrapping materials for protection of pipe lines, especially weld lines. The main unit processes of these products are extrusion of film and tubing, EBI at 200 to 300 kGy, extension or expansion at high temperature, and cooling.[59] A major supplier of shrinkable film for food packaging produces it by using 10 to 200 kGy irradiation.[60] Tight packaging is obtained by wrapping a foodstuff with the film and then heating. The film has high durability during refrigeration.[61]

### Polyolefin Foams

Foamed plastics are produced from several kinds of polymers, such as polystyrene, polyurethane, and polyolefins. Among them, polyethylene foam is usually produced with a cross-linking process; polyethylene is a semicrystalline polymer having a sharp melting point where the viscosity changes markedly. The production process involves sheeting the polymer with a foaming agent at a temperature below its decomposition temperature, cross-linking the sheet by EBs and foaming the sheet by heating. To obtain a highly expanded product, one must carry out the decomposition of the foaming agent at the proper viscoelastic condition of the polymer. Figure 20-3 shows reasons to cross-link for better foaming. The viscosity of the polymer in the temperature range of the decomposition depends on the degree of cross-linking. The degree of cross-linking is 20 to 40 in gel %, and the required dose is 50 to 100 kGy.[62]

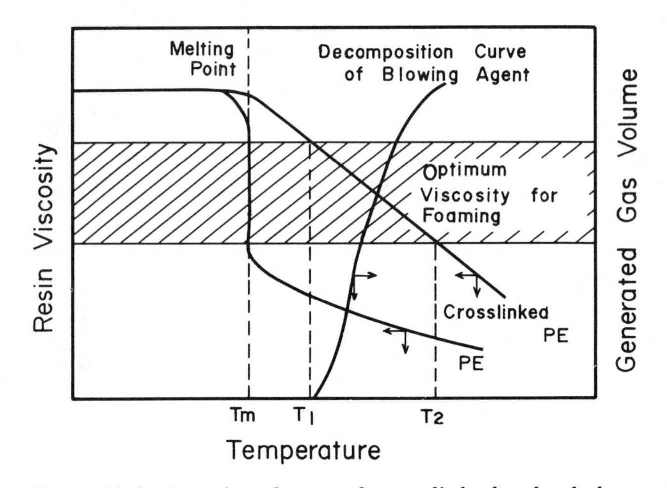


**Figure 20-3.** Foaming theory of cross-linked polyethylene.

Radiation cross-linking is not a requisite for processing, but the radiation–cross-linked polyethylene foams have better properties than those chemically cross-linked, due to a lower degree of cross-linking. A typical degree of cross-linking by a chemical agent is 70 in gel %.[63]

**Degradation of Polymers**

Applications of radiation degradation are minor compared to those of cross-linking, however, polytetrafluoroethylene (PTFE) wax is conveniently produced by radiation processing. There are presently many producers throughout the world because this useful product is easily obtained by irradiating low-value, off-grade, or scrap, PTFE. The only point to consider is corrosive HF evolution upon irradiation.

The low-molecular-weight PTFE produced is used as a lubricant, after blending it with other plastic materials, and in coating materials. Considerable amounts of off-grade, virgin PTFE powder are formed during the production and handling of PTFE before the molding process, because virgin PTFE has a fragile higher-order structure. The off-grade PTFE is irradiated to several hundred thousand grays; it is sold as is or after pulverization. The dose can be decreased by irradiating at high temperature or in an atmosphere containing carbon tetrachloride.[64,65] Industrial irradiation conditions are not disclosed and must differ according to the original PTFE grade.

Polyethylene oxide formed by polymerization has very high molecular weights. Lower-molecular-weight polyethylene oxide is not easily produced by changing the polymerization conditions.[66] Radiation degradation can be applied to control the molecular weight.[67] The process is simple in that the high-molecular-weight polymer is placed in a bag and irradiated for the required dose.

### Rubber

Radiation processing has not yet been applied commercially for the cross-linking of natural rubber because of the necessary high dose and the imperfect physical properties of the products compared to those of a traditional sulfur-cured rubber. The advantage of radiation cross-linking is that it is useful for cross-linking many kinds of rubber, including synthetic, which cannot be cured by chemical means.

An important application is seen for the tire industry by improving the handling process of raw rubber sheets is by slight cross-linking with EBs. Several tire companies use EBI for improving the green strength of a rubber sheet in radial-tire processing. The dose for precuring is 1 to 200 kGy.[68]

## *Polymerization*

### Curing

Radiation curing is the cross-linking of polymer molecules through radiation polymerization of the mixture of unsaturated polymer and vinyl monomer. During the late 1960s, industrial applications of radiation curing evolved with the development of low-energy, industrial, EB machines.

The characteristics of radiation curing are these:

1. No heating is necessary; it is thus applicable to heat-sensitive materials and avoids heating substrates having large thermal capacities.
2. Rapid reaction of the order of seconds; therefore small, space-saving processing machines can be used.
3. Solvent free; therefore it is safe and will not harm the environment.
4. Catalyst free; this leads to a long pot life for paint and better coating properties.

5. Ease of operation; therefore it has rapid start-up and shut-down.

Because radiation curing is not a drying process requiring solvent evaporation but a chemical reaction induced by radiation, the formulation of radiation-curable paints or resins is important. The prepolymer compositions commonly used are unsaturated polyester (with internal double bonds), unsaturated acrylics, urethane acrylics, epoxyacrylics, polybutadiene (with double bonds in the side chain), polyurethane acrylics, and polyester acrylics (with terminal double bonds). These prepolymers themselves can be cured by radiation, but polyfunctional and monofunctional monomers are usually added to the prepolymer to accelerate the curing or to adjust the viscosity of the paint. In practical applications the paint formulation contains pigments and other additives and is varied to meet individual needs, such as high solid content, physical properties of cured film, and less smell.

Industrial radiation-curing lines are operating throughout the world in the motor industry,[69,70] wood panel manufacturing, and production of steel sheets[71] for many uses.[72]

Another example is the coating of roof tiles. A roof tile made of cement has better size stability than that of porcelain, but it is more acceptable to consumers when it is colored like porcelain roof tiles. Acrylic or urethane painting is used for coloration. For good appearance and durability a thick coating is needed. The thickness of the coating cured by conventional methods is in the range of 70 to 80 $\mu$m, whereas that cured by radiation can be 150 $\mu$m in one-path processing.[73]

The curing of coatings on heat-sensitive substrates such as paper and film is being carried out by several industries.[74] One characteristic of such curing is its rapidity (of the order of seconds), which gives a very smooth surface, because paint laid on the substrate was not allowed to flow and follow the roughness of the substrate.[75]

Industrial application of radiation curing to manufacturing of magnetic tapes and floppy disks will begin shortly.[76,77] Present technology consists of thermal curing of the binder containing a large quantity of solvent. The radiation technique will provide high-density magnetic media, solidifying the magnetic particles just after orientation, thereby providing an increase in the S/N ratio of the product.[78]

The targets for radiation chemists are the development of radiation-curable paint in air, a less viscous composition without organic, low-molecular-weight compounds, and stress-free composition during rapid curing. The first problem may possibly be solved by obtaining polymerization by an ionic mechanism. In curing by UV light there is a composition containing azo or onium compounds, producing a cationic initiator upon decomposition.[79] A water emulsion of radiation-curable acrylourethane oligomer with low viscosity has been studied,[80] rather than oligomer mixed with a monomer, as the viscosity modifier, but the water vaporization problem must be solved.

**Grafting**

The first industrial application of grafting was the modification of cotton–polyester fabrics by radiation-induced grafting of *N*-methylol acrylamide. This grafting gives permanent set and soil-release properties.[81] It was studied in the late 1960s, fully industrialized in 1972, and lasted for five years. Presently, a similar modification is carried out by chemical means.

In spite of extensive research and development on grafting, industrial processes are mostly limited to acrylic acid grafting onto polyolefin. One application is to produce a separator for tiny silver oxide batteries.[82]

In one process, a polyethylene film roll with a spacer is $\gamma$-irradiated in an aqueous solution of acrylic acid for several days, and the film is grafted.[83] Elsewhere, the polyethylene film is irradiated by an electron beam in air and soaked in aqueous acrylic acid for grafting.[84] Figure 20-4 shows a pilot plant of the grafting process. The advantage of the latter process is that irradiation can be carried out separately from the grafting process.

**Figure 20-4.** Pilot plant for acrylic acid grafting to polyethylene.

The adhesive property of films used to line aluminum foil caps of milk containers can be improved in this way. The polyolefin powder is irradiated by EBs in air and peroxidized. The powder is blended with acrylic acid in an extruder, and the grafted polyolefin is pelletized and used as the laminating material for aluminum. The grafting ratio required is very small.[85]

**Polymerization**

The industrial applications of radiation-induced polymerization in its strict sense are limited. They are based on the simplicity, low-temperature capability, and catalyst-free polymer formation of radiation processing. Some examples are given in refs. 86–91.

### Composite Materials

There are two ways of applying radiation to composite materials: using radiation cross-linking of composites produced by conventional processes, and polymerization or graft polymerization of a monomer in a matrix. Composite materials of plastic and porous materials from natural origin that improve the properties are in commercial production by a $\gamma$-irradiation process. The typical manufacturing process consists of impregnation of monomers into the porous materials, followed by irradiation polymerization of the monomer in the porous matrix. The key technologies are the selection of a suitable monomer, removal of polymerization heat, and temperature control in the thick matrix. Radiation processing has an advantage in the latter point over conventional processes using chemical initiators.

The composite of wood and polymer, mostly acrylics, is exemplified here. The unit processes consist of loading wood pieces, mostly red oak, into an irradiation container, evacuating air from the container to remove oxygen, adding methyl methacrylate, pressurizing to accelerate monomer penetration into the wood, draining the excess monomer from the exterior of the wood, $\gamma$-irradiating to slightly higher than 10 kGy, and finishing the products.[92] The improvement of the compression and abrasion resistance properties is marked, which is important for use in flooring. The production of such materials, and a composite of brick and acrylics, is very high.[93,94]

## *Others*

### Organic Synthesis

It is well known that the Dow Chemical Company achieved the first radiation synthesis of ethyl bromide by the $\gamma$-induced addition reaction of hydrogen bromide to ethylene in 1962.[95] It is a chain reaction with a high $G$-value.[96] Industrial radiation syntheses reported are alkylsulfonate by sulfochlorination of paraffin–several thousands tons per year (t/y); tetrachloroalkane by telomerization of carbon tetrachloride with ethylene–250 t/y; and tin organic compounds used as a polymer stabilizer by the addition reaction of a tin halide to olefin–20 t/y.[97]

Radiation synthesis in the gas phase by EBs is a promising process. Electron beams will generate the initiating species at a lower price than chemical initiators. A process for the synthesis of silicone compounds by an EB-initiated addition reaction was claimed.[98] If industrial chemical syntheses are reviewed from this point of view, new applications of radiation may be found in processes where a large amount of initiator is currently necessary.

**Semiconductor Devices**

Radiation damage of semiconductor devices in spacecraft is a recent problem, and simulation tests are being carried out in many radiation laboratories.[99] Electron beams and $\gamma$-rays displace silicon atoms from the lattice, and the resultant defects distribute homogeneously. The defects are mobile and stabilized by combination with impurities or similar defects.

Heavy particles and fast neutrons cause defects heterogeneously, such as spurs. The defects act as recombination centers and reduce the carrier lifetime. This effect is used to reduce the reverse recovery time of silicon diodes and thyristors for high-frequency applications.[100] The advantage of radiation processing is its ability to control the switching properties of silicon devices after completion.[101,102] The optimum doses for thyristor modification are several hundred thousand grays for $\gamma$-rays, more than $10^{15}$ electrons/cm$^2$ for 0.4–12-MeV EBs,[103] and $10^{11}$–$10^{15}$ fast neutrons/cm$^2$. Similar effects are claimed with particulate radiation, such as protons or $\alpha$-particles.[104]

The commercial availability of irradiated, high-switching-speed diodes and thyristors is not yet disclosed, but articles and catalogues from service irradiation companies indicate positive results.[105]

# POTENTIAL APPLICATIONS OF RADIATION PROCESSING CURRENTLY UNDER RESEARCH AND DEVELOPMENT

## *Immobilization of Biofunctional Materials by Radiation*

Trapping of bioactive materials, such as enzymes, antigens, antibodies, microbial cells, tissue cells, and drugs, in polymer matrices by radiation processing has received much attention. There are two typical trapping methods: chemical bonding and physical occlusion or adhesion.

In the chemical bonding method, monomers containing functional groups (usually—OH or —COOH) are radiation-grafted onto backbone polymer. Biomaterials can then form primary bonds with the above functional groups, usually by means of their $—NH_2$ group. This technique has been used to immobilize enzymes, antibodies, and heparine.[106–109]

The physical trapping method is the simplest because it only involves a polymerization of monomer solutions or cross-linking of polymer solutions containing bioactive materials.[110–113] In this technique provision of a porous structure or of finely divided form is important in order to increase the accessibility for other biomolecules. Enzymes are dissolved in an aqueous solution of polyvinyl alcohol and irradiated at ambient temperature by EBs or $\gamma$-rays to form cross-linked hydrogen-trapping enzymes.

### *Preparation of Biomaterials*

Radiation cross-linking, polymerization, and grafting are used to prepare biomaterials used in implants. Radiation cross-linking of high-density polyethylene has been studied to increase wear resistance. The cross-linked polyethylene may be used for orthopedic joints.

Silicone rubber, widely used for implants, is normally cross-linked by chemical methods. Radiation cross-linking can avoid the use of catalyst, heat, and perhaps silica filler. This may provide better biocompatibility.

Hydrogel has been prepared by radiation cross-linking of polyvinyl alcohol, which shows good biocompatibility and potentially can be used for artificial eyes.[114,115]

Surface grating by radiation can improve the biocompatibility of materials. Polyethylene grafted with HEMA/EMA (2-hydroxylethylmethacrylate and ethyl methacrylate) has superior biocompatibility.[116] Radiation techniques for the above application have the following advantages over chemical methods: process simplicity and additive-free product, room-temperature or even lower-temperature capability, and simultaneous sterilization.

### *Preparation of Functional Polymers by Radiation Grafting*

Radiation grafting of a variety of monomers onto polymers is more advantageous than chemical methods due to its simplicity, room-temperature capability, and controllability of the grafting distribution. Polyethylene, polytetrafluoroethylene, and some synthetic hydrophobic polymers have excellent mechanical properties. Hydrophilicity can be given to these polymers by grafting with a hydrophilic monomer. Active species (eg radicals) on backbone carbon are formed by irradiation prior to grafting. Lifetimes of the radicals are fairly long, so that grafting can take place by immersion of irradiated polymer film in a monomer solution.

Various kinds of functional polymers can be prepared by this technique. Examples are ion exchange membranes with special properties, chelate polymers to absorb uranium from sea water,[117,118] and a reactive microgel used for water-borne paint.[119]

### *Radiation Treatment of Sewage Sludge*

It is well known that microorganisms are efficiently destroyed by radiation. Radiation is used to treat sewage sludge from waste-water treatment plants, which is heavily contaminated by microorganisms, parasites, and viruses.[120–122]

The first semicommercial plant was installed in 1972 to treat 120 $m^3$/day of sewage sludge at 300 krad using Co-60 radiation of 450 kCi. The Co-60 source was positioned at the center of an irradiation vessel of 5.6 $m^3$, and the sludge was agitated by external circulation in order to receive homogeneous irradiation. Treated sludge was used as a soil conditioner or fertilizer in nearby farmland. The 300-krad irradiation was sufficient to reduce the population of pathogens by a factor of $10^{-4}$–$10^{-5}$.[123]

EBA radiation may also be used for disinfection. A pilot scale plant with a capacity of 100,000 gal/day was put into operation near Boston. Because the EB penetration is shallow, the sludge should be irradiated in a layer of thickness less than several millimeters.[124]

To increase irradiation efficiency, Sandia National Laboratories has irradiated dried sludge in a pilot scale plant with a bucket-type conveyor and Cs-137 radiation source, which is a byproduct from spent fuel reprocessing. The irradiated sludge is used as a supplement to animal feeds and soil conditioner.[125] The Japan Atomic Energy Research Institute (JAERI) is studying the composting of irradiated dewatered sludge to improve the accessibility of fertilizer to farmers. Sanitary compost can be produced at a high fermentation rate.[126]

### *Radiation Treatment of Waste Water*

Some of the organic pollutants in waste water, such as aromatic detergents and phenols, cannot be decomposed by conventional activated sludge treatments. It has been found that such nonbiodegradable pollutants in water can be degraded by radiation by the oxidative species formed in water radiolysis (eg, OH, $HO_2$ radicals).

Because the *G*-value of degradation of the pollutant is low, (in the range of 1 to 5) and hence the required dose is high, the pure radiation treatment technique is not economically feasible. The combination of radiation treatment with a conventional treatment process such as ozonation is under investigation at JAERI.[127] Irradiation is effective for sanitation and the removal of metallic components from water in addition to the oxidation of organics in water.[128]

## *Radiation Treatment of Exhaust Gases for Removing $SO_2$ and $NO_x$*

JAERI and Ebara Manufacturing Companies jointly found that $SO_2$ and $NO_x$ can be removed from heavy-oil combustion gas by EBI. $SO_2$ and $NO_x$ are converted to $H_2SO_4$ and $HNO_3$ by irradiation and separated from the exhaust gases. Bench-scale flow-type experiments were conducted at JAERI using a 15-kW electron accelerator to treat exhaust gas from heavy-oil burning. Approximately 80% of the $SO_2$ ($SO_2$ concentration = 600–900 ppm, $NO_x$ concentration = 80 ppm) and almost 100% of the $NO_x$ were removed simultaneously.[129,130] A pilot plant with a capacity of 1000 $m^3$/hr was installed at Ebara Manufacturing Company to study the technical feasibility of this process. The addition of ammonia equimolar to

$SO_2$ and $NO_x$ was found to give stable dry-powder byproducts—a mixture of ammonium sulfate and ammonium nitrate sulfate.[131,132]

For removing $SO_2$ and $NO_x$ from iron ore sintering flue gas, a large pilot plant with a capacity of 10,000 $m^3$/hr was installed at one of the largest steel works in Japan. Two EBAs, positioned on both sides of the irradiation vessel, were used as the radiation source. After irradiation, solid particles were captured by an electrostatic precipitator. Continuous operation of the plant for one month gave stable removal of the pollutants.[133] Modification of this process is under study in the United States for application to coal combustion gases.[134]

### *Biomass Conversion by Radiation*

Biomass, such as chaff, rice straw, bagasse, and sawdust, can be converted into glucose by enzymatic saccharification. However, mechanical crushing is needed for enzymes to access the cellulose, which is protected by lignin. It was found that the combined treatment of mechanical crushing with radiation degradation increased the glucose yield,[135] and the radiation-immobilized cellulase or a cellulase-productive microbial cell can convert the cellulosic wastes into glucose with high efficiency.[111] Degradation of cellulosic materials by radiation also increases their digestibility as animal feed[136] and increases the sugar yield by acid hydrolysis.[137]

## ECONOMIC CONSIDERATIONS OF RADIATION PROCESSING

Economic analysis of radiation processing is an important consideration in commercial production. The purpose of this section is to present general factors to be considered in an economic analysis.

### *Estimation of Required Amount of Radiation Sources*

Designing a radiation facility and estimating the radiation cost involves estimating the amount of the radiation source required for production. The amount of source required is expressed as a function of the production rate in the equation.

$$W = \frac{xD}{3600f} \tag{20-5}$$

where $W$ is the amount of source in kW, $x$ is the production rate in kg/hr, $D$ is the dose given to the product in kGy, and $f$ is the irradiation efficiency defined as the ratio of the amount of absorbed radiation in the product to the total amount of radiation emitted from the source.

For example, if 25 kGy is required for the sterilization of a medical product and the efficiency is 0.3, then 23.2 kW of $\gamma$-power is required to treat 1 ton of product per hour. One kilowatt is emitted by 67,438 Ci of Co-60, so that 1562 kCi of Co-60 is required.

The efficiency factor, $f$, is influenced by the design of the irradiator and the type of radiation source. Generally, for EBI, the $f$-value is comparatively high.

## *Case Study of Irradiation Costs for a Co-60 Irradiator*

An example of the economic analysis made by AECL for a $\gamma$-sterilizer plant in 1977 showed the irradiation cost to be $7 to $66 per cubic meter for 25 kGy, depending on the Co-60 capacity and the utilization efficiency and assuming that the Co-60 source must be refreshed periodically at the rate of 12.3% per year to maintain the same sterilization capacity. Processing costs decrease sharply with increase in volume processed.[138] Prices for Co-60 were about $0.5/Ci in 1976, but today prices are higher than $1.00/Ci on average, depending on quantity and quality. All costs need to be current prices of equipment, shield, and sources.

## *Case Study of Electron Beam Processing*

### A Cost Comparison of Radiation and Chemical Process for Cross-linking of Wire Insulation

As shown in Figure 20-5, EB processing does not need the mixing of a cross-linking agent in polymer insulating material and thermal treatment for cross-linking, which are the major cost components of the chemical

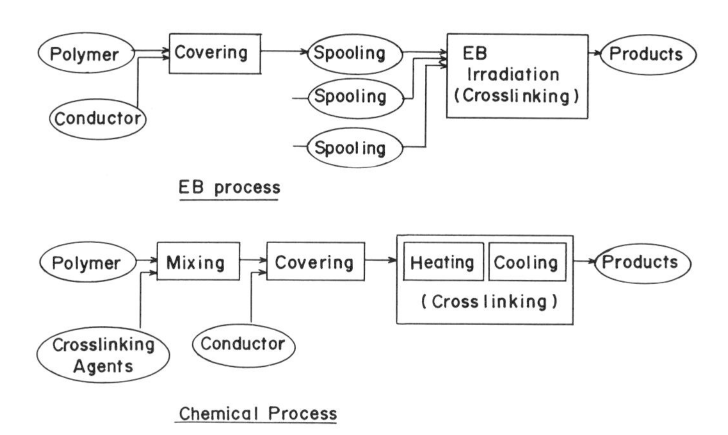


**Figure 20-5.** Manufacturing processes for polyethylene insulated cables.

process. However, the installation cost of an EBA is much higher than that of a thermal oven, so capital investment is larger in the EB process.

Energy costs of EB processes are much smaller than those for chemical processes because of their room-temperature capability and the high efficiency of energy utilization. Material cost components, including insulator material, are high in chemical processing, due to expensive cross-linking agents.

A case study indicates, as shown in Figure 20-6, that a large-capacity EB process has economical advantages because the costs for amortization and interest for the capital investment decrease.[139]

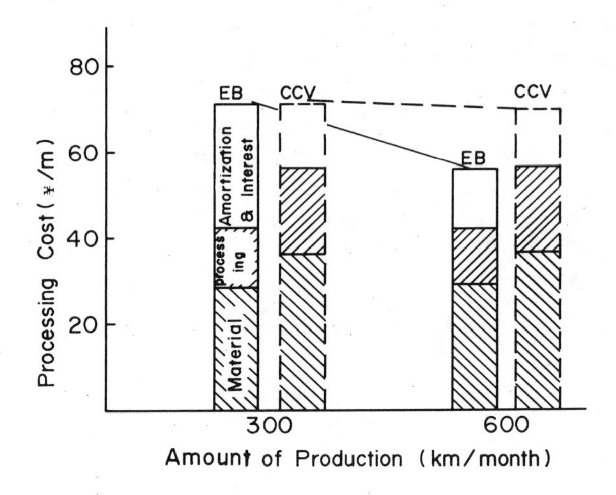


**Figure 20-6.** Cost comparison between electron beam (EB) and chemical (CCV)

### Cost Comparison for Polyethylene Foam Production

A similar case study was made for the production of polyethylene foam by using electron beams and chemical curing. For the cross-linking wire-insulation study, cost comparisons between the two methods showed similar trends for the same reasons.[139]

### Electron Beam Curing

One of the advantages of EB curing processes over the conventional peroxide and UV processes is the lower probability of producing defective products. Among the three processes considered, the curing process was shown to be the cheapest.[140]

# REFERENCES


1. Charlesby, A. In "Ind. Appl. Radioisotopes and Radiat. Tech." IAEA, Vienna, **1981**, pp. 105–115.
2. Silverman, J. *Radiat. Phys. Chem.* **1979**, *14*, 17.
3. Sakamoto, I., private communication.
4. Ratz, R. G. In "Proceedings of the 15th Japan Conference on Radioisotopes"; JAIF: Tokyo, 1981, pp. 7–14.
5. Hashimoto, S.; Kawakami, W.; Akehata, T. *Ind. Eng. Chem., Process Design Develop.* **1976**, *15*, 244.
6. Hashimoto, S.; Kawakami, W.; Akehata, T. *ibid.* **1976**, *15*, 549.
7. Hashimoto, S.; Kawakami, W. *ibid.* **1979**, *13*, 107.
8. Photograph; courtesy of Y. Nakase.
9. Sakurada, I. *Radiat. Phys. Chem.* **1979**, *14*, 23.
10. Warland, H. M. F. *AECL Tech. Report* No. 18, **1981**.
11. Harrod, R. A. *Radiat. Phys. Chem.* **1977**, *9*, 91.
12. Williams, J. L.; Dann, T. S. *Radiat. Phys. Chem.* **1979**, *14*, 185.
13. Frohnsdorff, R. S. M. *Radiat. Phys. Chem.* **1983**, *22*, 197.
14. Tsai, C. M.; Fu, Y. K.; Yang, Y. H. *Radiat. Phys. Chem.* **1981**, *18*, 1247.
15. Takehisa, M. *Radiat. Phys. Chem.* **1981**, *18*, 159.
16. Varaklis, I. *AECL Tech. Paper GPS 351*; June 1983.
17. Welt, M. A. *Radiat. Phys. Chem.* **1983**, *22*, 215.
18. Masefield, J. In "Sterilization of Medical Products"; Gaughran, E. R. L.; and Morrissey, R. F. Eds.; Multiscience: Montreal, 1981, pp. 202–209.
19. Clealand, M. R. *Radiat. Phys. Chem.* **1981**, *18*, 301.
20. Nablo, S. V.; Tripp, E. P. *Radiat. Phys. Chem.* **1977**, *9*, 325.
21. Ramler, W. J. *ibid.* **1977**, *9*, 69.
22. Emanuelson, R. M. *Radiat. Phys. Chem.* **1981**, *18*, 313.
23. Clealand, M. R. *Radiat. Phys. Chem.* **1981**, *18*, 301.
24. Parisi, A. N. In "Sterilization of Medical Products"; Gaughran, E. R. L.; and Morrissey, R. F., Eds.; Multiscience: Montreal, 1981, pp. 187–201.
25. Willson, J. E. *ibid.* pp. 129–149.
26. Artandi, C. *Radiat. Phys. Chem.* **1977**, *9*, 183.
27. Lay, F. J. In "Manual on Radiation Sterilization of Medical and Biological Materials"; IAEA: Vienna, 1973, pp. 37–64.
28. Tallentire, A. *Radiat. Phys. Chem.* **1980**, *15*, 83.
29. Tallentire, A. *ibid.* **1979**, *14*, 225.
30. Whitby, J. L. *ibid.* **1979**, *14*, 285.
31. Handlos, V. *Radiat. Phys. Chem.* **1981**, *18*, 175.
32. Sjoberg, L. In "Sterilization of Medical Products"; Gaughran, E. R. L.; and Morrisey, R. F., Eds.; Multiscience: Montreal, 1981, pp. 294–299.
33. David, K. W.; Strawderman, W. E. Masefield, J.; Whitby, J. L. *ibid.* pp. 34–102.
34. "Process Control Guideline for Radiation Sterilization of Medical Devices (RS-P 1/81)"; Association for the Advancement of Medical Instrumentation: Arlington, Va, 1981.
35. Dierksheide, W. C. *Radiat. Phys. Chem.* **1977**, *9*, 221.
36. Bly, J. H. *ibid.* **1979**, *14*, 403.
37. Farrel, J. P. *ibid.* **1979**, *14*, 377.
38. Skiens, W. E. *ibid.* **1980**, *15*, 47.
39. Landfield, H. *ibid.* **1980**, *15*, 34.
40. Williams, J. L.; Dunn, T. S. *ibid.* **1983**, *22*, 209.
41. Williams, J. L.; Dunn, T. S. *ibid.* **1980**, *15*, 59.
42. Clouston, J. G. In "Proceedings of the 14th Japan Conference on Radioisotopes"; Japan Atomic Ind. Forum: Tokyo, 1979, pp. 17–32.
43. Deshpande, R. G.; Iya, V. K. *ibid.* pp. 33–43.
44. Trutnau, H.; Bogl, W.; Stockhausen, K. STH Berichte. "Der Einfluss der Strahlenbehandlung auf Arzeneimittel und Helfsstoff, Eine Literaturstudie"; Dietrich Reimer Verlag: Berlin, Teil 1, 1978, Teil 2 und 3, 1979, Teil 4, 1982.
45. Tsuji, K.; Kane, M. P.; Rahn, P. D.; Sleindler, K. A. *Radiat. Phys. Chem.* **1981**, *18*, 583.

46. Pikaev, A. K. *ibid.* **1983**, *22*, 241.
47. Bogl, W. In 3rd. International Conference on Pharmaceutical Technology; Paris, June 1983.
48. "Wholesomeness of Irradiated Food"; Tech. Report Ser. 659, WHO: Geneva, 1981.
49. "Sterilization and Preservation of Biological Tissues by Ionizing Radiation"; IAEA: Vienna, 1970.
50. Edwards, H. E.; Phillips, G. O. *Radiat. Phys. Chem.* **1983**, *22*, 889.
51. Oda, E. *Radiat. Phys. Chem.* **1981**, *18*, 241.
52. Saito, E. *ibid.* **1977**, *9*, 675.
53. Loan, L. D. *ibid.* **1977**, *9*, 253.
54. Bly, J. H.; Brandt, E. S.; Burgess, R. G. *ibid.* **1979**, *14*, 931.
55. Uehara, K. *ibid.* **1977**, *9*, 593.
56. Studer, H. R. *ibid.* **1979**, *14*, 809.
57. Sasaki, T.; Hosoi, F.; Hagiwara, M.; Araki, K. *ibid.* **1979**, *14*, 821.
58. Patel, G. N. *ibid.* **1979**, *14*, 729.
59. Ota, S. *ibid.* **1981**, *18*, 81.
60. Baird, W. G., Jr. *ibid.* **1979**, *9*, 225.
61. Lovin, J. R. *ibid.* **1979**, *14*, 213.
62. Sagane, N.; Harayama, H. *ibid.* **1981**, *18*, 99.
63. Trageser, D. A. *ibid.* **1977**, *9*, 261.
64. Hagiwara, M.; Tagawa, T.; Tsuchida, E.; Shinohara, E.; Kagiya, T. *Kobunshi Ronbunshu* **1974**, *31*, 336.
65. Tagawa, T.; Tsuchida, E.; Shinohara, I.; Hagiwara, M.; Kagiya, T. *J. Polym. Sci., Polym. Lett. Ed.* **1975**, *13*, 287.
66. Ballantine, D. S. IAEA/PL-411, 1970.
67. Silverman, J. *Radiat. Phys. Chem.* **1977**, *9*, 1.
68. Hunt, J. D.; Alliger, G. *ibid.* **1979**, *14*, 39.
69. Anonymous *Ind. Finish. Surface Coat.* **1971**, *3*, 13.
70. Haering, E. In "Ind. Appl. Radioisotopes and Radiat. Tech."; IAEA: Vienna, 1982, pp. 235–253.
71. Fujioka, S.; Fujikawa, J. *Radiat. Phys. Chem.* **1981**, *18*, 865.
72. Ueno, N. 1982 Annual Meeting Nat. Coil Coaters Assoc. Scottsdale, Arizona, May 1982, p. 6.
73. Yoshizaki, K. In "Proceedings of the 15th. Japan Conference on Radioisotopes"; Japan Atomic Ind. Forum: Tokyo, 1981, pp. 234–235.
74. Nablo, S. V.; Tripp, E. P. *Radiat. Phys. Chem.* **1979**, *14*, 481.
75. Nablo, S. V. In Panel on Progress in Industrial Radiation Processing. "Ind. Appl. Radioisotopes and Radiation Tech."; IAEA: Vienna, 1982, pp. 117–120.
76. Tsuzuki, Y. In "Proceedings of the 15th Japan Conference on Radioisotopes"; Japan Atomic Ind. Forum: Tokyo, 1981, pp. 238–239.
77. Klein, A. F. "Developments in Electron Beam Curing of Magnetic Media" Radcure '83: Lausanne, May, 1983.
78. Akashi, G.; Fujiyama, M.; Yamada, Y. (Fuji Photofilm Co.) Jap. P. Sho 56-130835 (Oct. 14, 1981).

80. Gerst, D. D. *J. Radiat. Curing* **1982**, *9*, (No. 4) 27.
81. Hoffman, A. S. "Radiation Modification of Textiles"; IAEA/SM-123/21, 1969.
82. Okamoto, J.; Ishigaki, I.; Murata, K. *New Materials & New Processes* **1981**, *1*, 139.
83. D'Agostino, V.; Lee, J. AFML-TR-72-13, 1972.
84. Ishigaki, I.; Sugo, T.; Senoo, T.; Takayama, T.; Machi, S.; Okamoto, J.; Okada, T. *Radiat. Phys. Chem.* **1981**, *18*, 899.
85. Guimon, C. *ibid.* **1979**, *14*, 841.
86. Ransohoff, J. A. *ibid.* **1981**, *18*, 239.
87. Machi, S.; Wada, T.; Sekiya, H. USP 4,066,522 (Jan. 3, 1978).
88. Fukuzaki, H.; Ishigaki, I.; Washino, S.; Okamoto, J. *Radiat. Phys. Chem.* **1981**, *18*, 1161.
89. Block, D. In "Ind. Appl. Radioisotopes & Radiat. Tech."; IAEA: Vienna, 1982, p. 119.
90. Kaetsu, I.; Yoshida, K.; Okubo, H.; Yoshi, F.; Hayashi, K. *Radiat. Phys. Chem.* **1979**, *14*, 737.
91. *Nihon Kogyo Shinbun* (Daily Ind. Newspaper) 1983, July 27.

92. Witt, A. E.; Henise, P. D.; Griest, L. W. *Radiat. Phys. Chem.* **1981**, *18*, 67.
93. Witt. A. E. *ibid.* **1977**, *9*, 271.
94. Witt. A. E. In Panel on Progress in Ind. Radiat. Processing. "Ind. Appl. Radioisotopes and Radiat."; IAEA: Vienna, 1982, pp. 117–120.
95. Harmer, D. E.; Beale, J. S.; Pumpelly, C. T.; Wilkinson, B. W. "Ind. Uses of Large Radiation Sources"; IAEA: Vienna, 1963, Vol. 2, p. 205.
96. Armstrong, D. A.; Spinks, J. W. T. *Can. J. Chem.* **1959**, *37*, 1002.
97. Babkin, I. Yu.; Solovev, S. P. *Radiat. Phys. Chem.* **1981**, *18*, 371.
98. Vainstein, B. I. Jap. P. 856450, 1977 (Sho 51-27662).
99. Adams, L. *Radiat. Phys. Chem.* **1980**, *15*, 525.
100. Oku, T.; Horie, K. *Mitsubishi Denki Giho* **1967**, *41*, 1417.
101. Carlson, R. O.; Sun, Y. S.; Assalit, H. B. *IEEE Trans. Electron. Devices* **1977**, *ED-24*, 1103.
102. Tarneja, K. S.; Johnson, J. E.; Bartko, J. USP 4,075,037 (21 Feb. 1978).
103. Sheng, W. W.; Sun, Y. S. E.; Tefft, E. G. German P. 2837762 (15 Mar. 1979).
104. Bartko, J.; Chu, C. K.; Schlegel, E. S.; Tarneja, K. S. USP 4,3211,534 (19 Jan. 1982).
105. Williams, C. B.; Vroom, D. *Radiat. Phys. Chem.* **1979**, *14*, 509.
106. Hoffman, A. S. *Radiat. Phys. Chem.* **1977**, *9*, 207.
107. Hoffman, A. S. *ibid.* **1981**, *18*, 323.
108. Hoffman, A. S. In "Ind. Appl. Radioisotopes Radiat. Tech."; IAEA: Vienna, 1981, pp. 279–321.
109. Gaussens, G. *ibid.* pp. 343–352.
110. Kaetsu, I. *Radiat. Phys. Chem.* **1981**, *18*, 343.
111. Kobayashi, M.; Kaetsu, I. In "Ind. Appl. Radioisotopes Radiat. Tech."; IAEA: Vienna, **1981**, pp. 353–370.
112. Kawashima, K.; Umeda, K. *Biotech. Bioeng.* **1974**, *16*, 609.
113. Kawashima, K.; Umeda, K. *ibid.* **1975**, *17*, 599.
114. Yamauchi, A; Matsuzawa, Y.; Nishioka, K.; Hara, Y.; Kamiya, S. *Kobunshi Ronbunshu* **1977**, *34*, 261.
115. Hara, Y.; Kamiya, S.; Nishioka, K.; Saishin, M.; Nakao, S.; Yamauchi, A. *Acta Soc. Ophthalm. Jap.* **1979**, *83*, 1478.
116. Hoffman, A. S.; Cohn, D.; Hanson, S. R.; Harker, L. A.; Horbett, T. A.; Ranter, B. D.; Raynolds, L. O. *Radiat. Phys. Chem.* **1983**, *22*, 267.
117. Okamoto, J.; Katakai, A.; Sugo, T.; Ohmichi, H. Preprint, Int. Meeting on Recovery of Uranium from Sea Water; Atomic Energy Soc. Jap. & IAEA: Tokyo, 1983, Oct 17–19, pp. 125–134.
118. Ohmichi, H.; Katakai, A., Sugo, T.; Okamoto, J. *ibid.*, pp. 135–144.
119. Makuuchi, K.; Nakayama, N. *Prog. Org. Coatings* **1983**, *11*, 241.
120. Watanabe, H.; Ito, H.; Iizuka, H.; Takehisa, M. Hakko. *Kogyo Gakai-shi* **1981**, *59*, 449.
121. "High Energy Electron Radiation of Wastewater Liquid Residuals"; MIT, Feb. 28, 1979, NSF Grant ENV77 10199.
122. Yeager, J. G. SAND 80-2744, Dec. 1980, pp. 69–90.
123. Suess, A.; Lessel, T. *Radiat. Phys. Chem.* **1977**, *9*, 353.
124. Trump, J. G.; Wright, K. A.; Sinsky, A. J.; Shah, D. N.; Fernald, R. In "Proceedings of the Japan 14th Radioisotope Conference"; JAIF: Tokyo, **1981**, pp. 283–290.
125. Sivinski, J. S. *Radiat. Phys. Chem.* **1983**, *22*, 99.
126. Kawakami, W.; Hashimoto, S.; Watanabe, H.; Nishimura, K.; Watanabe, H.; Ito, H.; Takehisa, M. *Radiat. Phys. Chem.* **1981**, *18*, 771.
127. Takehisa, M.; Sakumoto, A. In "Ind. Appl. Radioisotopes Radiat. Tech."; IAEA: Vienna, 1981, pp. 217–233.
128. Levaillant, C.; Gallien, C. L. *Radiat. Phys. Chem.* **1979**, *14*, 309.
129. Kawamura, K.; Aoki, S.; Kawakami, W.; Hashimoto, S.; Machi, S. "Radiation for Clean Environment"; IAEA: Vienna, 1975, pp. 621–631.
130. Washino, M.; Tokunaga, O.; Hashimoto, S.; Kawakami, W.; Machi, S.; Kawamura, K.; Aoki, S. *ibid.*, pp. 633–642.
131. Tokunaga, O.; Nishimura, K.; Suzuki, N.; Machi, S.; Washino, M. *Radiat. Phys. Chem.* **1978**, *11*, 299.
132. Suzuki, N.; Nishimura, K.; Tokunaga, O. *J. Nucl. Sci. Tech.* **1980**, *17*, 822.
133. Kawamura, K.; Katayama, T. *Radiat. Phys. Chem.* **1981**, *18*, 389–398.

134. Kawamura, K.; Shui, V. H. In "Ind. Appl. Radioisotopes & Radiat. Tech."; IAEA: Vienna, 1982, pp. 197–215.
135. Kumakura, M.; Kojima, K.; Kaetsu, I. *Biomass* **1982**, *2*, 299.
136. Baer, M.; Huebner, G.; Leonhardt, J. W. *ibid.*, p. 96.
137. Brenner, W.; Rugg, B.; Arnon, J.; *Radiat. Phys. Chem.* **1979**, *14*, 299.
138. Harrod, R. A. *Radiat. Phys. Chem.* **1977**, *9*, 91.
139. Takahashi, M. In "Proceedings of the 14th Japan Radioisotopes Conf."; JAIF: Tokyo, 1977, pp. 149–155.
140. Grosmaire, P. R. *Radiat. Phys. Chem.* **1977**, *9*, 857.

# Compound Index

This listing contains stable species only. Radicals, etc. are found in Subject Index.

# Subject Index